RELIABILITY ASSESSMENT USING STOCHASTIC FINITE ELEMENT ANALYSIS

RELIABILITY ASSESSMENT USING STOCHASTIC FINITE ELEMENT ANALYSIS

Achintya Haldar

Department of Civil Engineering and Engineering Mechanics
University of Arizona

Sankaran Mahadevan

Department of Civil and Environmental Engineering
Vanderbilt University

JOHN WILEY & SONS, INC.

New York / Chichester / Weinheim / Brisbane / Singapore / Toronto

This book is printed on acid-free paper. ∞

Copyright © 2000 by John Wiley & Sons, Inc. All rights reserved.

Published simultaneously in Canada.

This publication is designed to provide accurate and authoritative information in regard to the subject matter covered. It is sold with the understanding that the publisher is not engaged in rendering professional services. If professional advice or other expert assistance is required, the services of a competent professional person should be sought.

Library of Congress Cataloging-in-Publication Data:

Haldar, Achintya
    Reliability assessment using stochastic finite element analysis / Achintya Haldar,
Sankaran Mahadevan
      p. cm.
    Includes bibliographical references.
    ISBN 0-471-36961-6 (alk. paper)
    1. Reliability (Engineering) 2. Finite element method. I. Mahadevan, Sankaran. II.
Title.
TA169.H35 2000
620′.00452—dc21   99-042170

Printed in the United States of America

10 9 8 7 6 5 4 3 2 1

To my wife Carolyn for her help in all phases of the development
of this book, our son Justin, and my mother Angur Haldar.

Achintya Haldar

To my parents, Ganesan Sankaran and Janaki Sankaran,
and my wife Monica.

Sankaran Mahadevan

CONTENTS

PREFACE

The applicability of risk and reliability in engineering analysis, design, and planning has been accepted throughout the world. Design codes and guidelines incorporating the concept of risk already exist, and more are being developed, as a result of three decades of work in different engineering disciplines. Risk-based design has matured to the point that the information on risk alone is no longer sufficient; it is also necessary to know how accurately the information on uncertainty in the load and resistance-related parameters has been incorporated in the mathematical model. The first-generation design guidelines and codes are being modified to reflect this concern. The subject of this book is reliability evaluation considering the uncertainty in the load and resistance behavior as accurately as possible.

The authors have written another book, *Probability, Reliability and Statistical Methods in Engineering Design* (published by John Wiley & Sons, Inc., 2000), which will be denoted hereafter as Book 1. The basic concepts of risk and reliability and simple design issues are emphasized in Book 1, which was written primarily for undergraduate engineering students who must satisfy the ABET requirements. Book 1 can be used for undergraduate and graduate engineering students in all disciplines and by practicing engineers with no formal instruction on the subject. An understanding of the contents of Book 1 will provide a basic working knowledge of risk and reliability, which is essential for maximizing the usefulness of the current book (Book 2).

Book 1 does not address the major complaint that basic engineering problems need to be simplified considerably before a risk-based design can be implemented. Consider a very simple example. Suppose that it is necessary to know the risk of collapse of a building subjected to seismic excitation, and assume that this can be learned by tracking the lateral deformation at the top of the building. In their simplest form, most of the commonly used risk- or reliability-based methods require that a functional relationship among the load and resistance-related variables, commonly known as the limit state or performance function, be available in explicit form. But it can be quite difficult to explicitly define the relationship of the lateral displacement at the top of the building as a function of the geometry and size of the building, material properties and the associated constitutive relationships, the loads acting on the building, the various sources of nonlinearity expected in the structural behavior just before failure, and so on. For a large structural system, it may be impossible. Thus, simple, commonly used risk-based analysis and design procedures cannot be applied when the performance function is implicit. Even a deterministic evaluation of the problem can be complicated. However, accurate reliability evaluation of such structures is demanded by society. This book attempts to address the issue by introducing advanced concepts, such as stochastic finite element analysis, which can be used for both implicit and explicit performance functions.

Finite element analysis is a powerful tool commonly used by many engineering disciplines to analyze simple or complicated structures. The word "structure" is used in a broad sense here to represent any system that can be represented by finite elements. The finite element methods have matured significantly and it has become routine to use finite element analysis for all types of structures, considering their behavior as realistically as possible. With this approach it is very easy and straightforward to consider complicated geometric arrangements, realistic connection and support conditions, various sources of nonlinearity, and the load path to failure.

In the early 1980s we realized that the available reliability methods failed to represent structures as realistically as possible. On the other hand, the deterministic finite element method fails to consider the uncertainty in the variables, and thus cannot be used for reliability analysis. To capture the desirable features of these two approaches, they needed to be combined, leading to the concept of the stochastic finite element method (SFEM) or probabilistic finite element method (PFEM). In this book, we use the acronym SFEM.

SFEM is also used to consider random fields. As presented in this book, SFEM is a broad and sophisticated reliability analysis method that can be used for both explicit and implicit performance functions. It is parallel to the deterministic finite element method, but is able to consider the uncertainty in the variables involved in the problem. It can be used to estimate the risk associated with any structure that can be represented by finite elements, making it very powerful and robust.

In this book we strive to deal comprehensively with issues relevant to students, professionals, and society. There are several target audiences for this book. The book addresses issues that will be of interest to advanced undergraduate and graduate students and practicing engineers in all engineering disciplines who use the finite element method to analyze structures. As stated earlier, the state of the art in deterministic finite element analysis is very advanced. Considering uncertainty in the problem is a natural extension of the concept, and will be of interest to researchers working to advance the deterministic finite element concept. It will also be of interest to the general risk and reliability research community, since it is another powerful and robust reliability method that can be used for both implicit and explicit performance functions. The material is presented in a simple manner with many examples, and the presentation is based on long experience in dealing with students and practicing engineers on safety-related issues.

We assume that the book will be used by advanced undergraduate and graduate students and professionals who have a basic understanding of the concept of risk and reliability and a working knowledge of the finite element method. However, to make the book self-sufficient, the essential concepts of risk, the distributions commonly used to represent uncertainties, and commonly used risk estimation procedures for explicit limit state functions are presented. Since simulation is another independent tool that can be used to estimate risk or to verify the risk estimated by other methods, it is also discussed. The availability of personal computers is another factor behind the chapter on simulation. If the readers need a more extensive review on the fundamentals of risk, reliability, and statistical methods in engineering design, they are encouraged to refer to Book 1.

We then introduce reliability analysis methods that can be used when the performance functions are implicit. The necessity for using the finite element method in that context is explained, ultimately leading to the concept of the stochastic finite element method. For broader application, both displacement and stress-based finite element methods are discussed. To demonstrate the robustness of the procedures, separate consideration is given to linear, nonlinear, static, and dynamic problems. The risk or reliability estimation procedure for each case is presented in different chapters. The theory is presented first, followed by a series of examples to demonstrate its application potential.

With the advances in computer technology, it is quite appropriate to develop a finite element-based reliability analysis technique, parallel to the deterministic analysis procedure. With this technique the reliability-based design concept can be applied to real problems. The simple reliability methods currently used in design codes need to be upgraded for this purpose. This book attempts to fill this need. Both authors have been involved extensively in the development of SFEM-based methodologies since the early 1980s. In the last five chapters, we develop the concept and implementation of SFEM-based reliability analysis in the context of practical structural design. It is expected that anyone with a working knowledge of deterministic finite element analysis should be able to estimate the risk or reliability of a wide variety of structures by following the steps in these chapters.

The chapters are written so as to be understandable by members of any engineering discipline. Chapter 1 presents the basic concepts of risk and reliability and justification for the use of the finite element method in the estimation of risk. Risk evaluation procedures using some of the common distributions, including extreme value distributions, are discussed in Chapter 2. The currently available risk evaluation procedures, primarily for explicit performance functions, are discussed in Chapter 3. The use of simulation in estimating risk is introduced in Chapter 4. These chapters establish the need for reliability methods when the performance function is implicit.

The need for SFEM-based evaluation procedures is discussed in Chapter 5. The concept of the stochastic finite element method is introduced. The algorithm can be used for both the displacement-based and the assumed stress-based deterministic finite element approaches. The displacement-based SFEM for linear structures is introduced in Chapter 6. Issues related to the spatial variability (random field) problems in the context of SFEM are discussed in Chapter 7. Nonlinear static SFEM analysis is presented in Chapter 8. If the failure of a structure is imminent, several sources of nonlinearity and their associated uncertainties need to be considered. In addition to the geometric and material nonlinearities, the modeling uncertainties of the connections and support conditions must be incorporated explicitly. These issues are discussed in detail in Chapter 8. Due to its numerical efficiency, the stress-based SFEM is considered for the risk estimation of structures with nonlinear behavior. The use of SFEM for solving dynamic problems is still a topic of research. However, the development of such an algorithm is explored in Chapter 9 for both linear and nonlinear problems. The last two chapters will be of significant interest to the research community, since the use of SFEM is an emerging area.

To improve the readability of the book, citations in the middle of a discussion are avoided. In this book we tried to integrate advanced concepts in risk-based design, finite element, and mechanics. Many people contributed to the development of all these areas. It is not possible to cite all their contributions, but an extensive list of the references required for the development of the SFEM concept presented here is given at the end of the book. We have tried to make this list as complete as possible.

Jungwon Huh's help in developing figures, tables, and numerical solutions for many of the problems given in the book is very much appreciated. Numerous former and present students and colleagues of ours directly or indirectly contributed to the development of the material. We thank B. M. Ayyub at the University of Maryland, T. A. Cruse at Vanderbilt University, Hari B. Kanegaonkar, Yiguang Zhou, Liwei Gao, Zhengwei Zhao, Duan Wang,

Alfredo Reyes Salazar, Rajasekhar K. Reddy, Nilesh Shome, Ali Mehrabin, Seung Yeol Lee, Xiolin Ling, Peter H. Vo, Sandeep Mehta, Robert Tryon, Qiang Xiao, Animesh Dey, Xiaoping Liu, Zhisong Guo, Pan Shi, Christopher W. Gantt, Yue Wang, Hongyin Mao, and Tong Zou.

We appreciate all the help provided by the editorial, production, and marketing staff of John Wiley & Sons, Inc. We would especially like to thank Neil Levine for his support and encouragement during the development phase of this book.

Some of the methods presented in this book were developed with financial support received from several different sources over a decade, including the National Science Foundation, the American Institute of Steel Construction, the U.S. Army Corps of Engineers, Illinois Institute of Technology, Georgia Institute of Technology, the University of Arizona, Vanderbilt University, and other industrial sources that provided matching funds for the Presidential Young Investigator Award, which the first author received in 1984. The second author's research in reliability and probabilistic finite element analysis methods has been supported for over a decade by several agencies, including NASA, the U.S. Army Corps of Engineers, the U.S. Army Research Office, the National Science Foundation, and the Oak Ridge and Sandia National Laboratories.

NOTES FOR INSTRUCTORS

This book is suitable for advanced undergraduate and graduate students and practicing engineers in all engineering disciplines who use the finite element method to analyze structures. Readers are assumed to have a working knowledge of the risk-based design concept and to be familiar with the finite element analysis method.

The material presented in the book can be taught in one semester. We would like to suggest a tentative course outline: Assuming that an instructor has about 15 weeks to cover the material, the first four chapters could be covered in about 7 weeks. More emphasis should be given to the information in Chapter 3 on available methods in reliability analysis and less emphasis to the information on simulation presented in Chapter 4. Depending on the background of the students, Chapter 5 may take 1 to 2 weeks. Chapter 6 may take about 3 weeks. The first part of Chapter 7 on random processes needs a brief introduction. However, the second part on random fields may take 1 1/2 to 2 weeks. Chapters 8 and 9 cover advanced topics. Coverage of these chapters should be based on the availability of time and the interests of the students.

1

BASIC CONCEPT
OF RELIABILITY

1.1 INTRODUCTORY COMMENTS

Risk-based analysis and design methods have been under considerable research, development, and verification for the last three decades. There has been much ferment in this field, resulting in some ready-to-use risk-based design codes or guidelines. The concept of reliability, probability, or risk is not new and has been a thought-provoking subject for a long time. Pierre Simon, Marquis de Laplace (1749–1827), published a volume of work on pure and applied mathematics. He also wrote a commentary on general intelligence, published as "A Philosophical Essay on Probabilities." In it, Laplace wrote, "I present here without the aid of analysis the principles and general results of this theory, applying them to the most important questions of life, which are indeed for the most part only problems of probability. Strictly speaking it may even be said that nearly all our knowledge is problematical; and in the small number of things which we are able to know with certainty, even in the mathematical sciences themselves, the principal means of ascertaining truth—induction and analogy—are based on probabilities; so that the entire system of human knowledge is connected with the theory set forth in this essay." He concluded, "It is seen in this essay that the theory of probabilities is at bottom only common sense reduced to calculus; it makes us appreciate with exactitude that which exact minds feel by a sort of instinct without being able ofttimes to give a reason for it. It leaves no arbitrariness in the choice of opinions and sides to be taken; and by its use can always be determined the most advantageous choice. Thereby it supplements most happily the ignorance and weakness of the human mind" (translation, 1951).

These timeless remarks sum up the importance of probability or reliability concepts in human endeavor. However, as in any other area, success raises the question of the quality of risk information being generated using currently available methods. The first generation design guidelines are now being updated to reflect this concern. The aim of this book is to present practical yet advanced methods to estimate risk or reliability considering load and resistance behavior as realistically as possible.

1.2 WHAT IS RELIABILITY?

Most observable phenomena in the world contain a certain amount of uncertainty; that is, they cannot be predicted with certainty. In general, repeated measurements of physical phenomena generate multiple outcomes. Among these multiple outcomes, some are more frequent than others. The occurrence of multiple outcomes without any pattern is described by terms such as uncertainty, randomness, and stochasticity. The word *stochasticity* comes from the Greek word *stochos*, meaning uncertain. For example, if several "identical" specimens of a steel bar were loaded until failure in a laboratory, each specimen would fail at different values of the load. The load capacity of the bar is therefore a random quantity, formally known as a random variable. In general, all the parameters of interest in engineering analysis and design have some degree of uncertainty and thus may be considered to be random variables. Although other methods exist for treating uncertainties, as discussed in Section 1.5.2, only probabilistic methods are included in this book.

The planning and design of most engineering systems utilize the basic concept that the capacity, resistance, or supply should at least satisfy the demand. Different terminology is used to describe this concept, depending on the problem under consideration. In structural, geotechnical, and mechanical engineering, the supply can be expressed in terms of resistance, capacity, or strength of a member or a collection of members, and demand can be expressed in terms of applied loads, load combinations, or their effect. For a construction and management project, the completion time is an important parameter in defining success. The estimated completion time during the bidding process and the actual time spent to complete the project will give the essential components of supply and demand. In environmental engineering, the actual air or water quality of a given city or site is always measured with respect to allowable or recommended values suggested by a responsible regulatory agency, such as the Environmental Protection Agency (EPA), giving the essential components of supply and demand. In transportation engineering, an airport or highway is designed considering future traffic needs; the capacity of the airport or highway must meet the traffic demand. In hydraulics and hydrology engineering, the height and location of a dam to be built on a river may represent the capacity. The annual rainfall, catchment areas and the vegetation in them, other rivers or streams that flow into the river under consideration, usage upstream and downstream, and location of population centers may represent demand. Errors can never be avoided in surveying projects. The measurement error could be positive or negative, and the quality or sophistication of the equipment being used and the experience of the surveyor may represent capacity. In this case, the acceptable tolerance may indicate demand.

The point is that no matter how supply and demand are modeled or described, most engineering problems must satisfy the concept. However, it has already been established that most of the parameters related to supply and demand are random quantities. The primary task of planning and design is to ensure satisfactory performance, that is, to ensure that the capacity or resistance is greater than demand during the system's useful life.

In view of the uncertainties in the problem, satisfactory performance cannot be absolutely assured. Instead, assurance can only be made in terms of the probability of success in satisfying some performance criterion. In engineering terminology, this probabilistic assurance of performance is referred to as *reliability*.

An alternative way to look at the problem is to consider unsatisfactory performance of the system. In that case, one might measure the probability of failure to satisfy some performance criterion, and the corresponding term would be *risk*. Thus, *risk* and *reliability* are complementary terms. (In some references, the term *risk* is not just the probability of failure but includes the consequence of failure. For example, if the cost of failure is to be included in risk assessment, then risk is defined as the product of the probability of failure and the cost of failure.)

Reliability or risk assessment of engineering systems uses the methods of probability and statistics. A distinction needs to be drawn here between probability and statistics. Statistics is the mathematical quantification of uncertainty (a variable's mean, standard deviation, etc.; these terms will be defined later), whereas probability theory uses the information from statistics to compute the likelihood of specific events.

We discussed the basic concept of risk, reliability, and statistics in another book published by John Wiley & Sons, Inc., titled *Probability, Reliability, and Statistical Methods in Engineering Design*. However, as the area has matured, the discussion has now shifted from risk estimation to the accuracy of the information on risk. It is now accepted that identical specimens of a steel bar loaded to failure will fail at different load levels. If a mathematical model is used to predict the failure, is it sophisticated enough to capture the behavior as realistically as possible? This book is not expected to end the discussion, but it does propose a new technique to account for some of the uncertainties. We expect that the proposed method is powerful and robust enough to provide some of the improvements necessary in the currently available reliability estimation procedures.

1.3 NEED FOR RELIABILITY EVALUATION

The presence of uncertainty in the analysis and design of engineering systems has always been recognized. However, traditional approaches simplified the problem by considering the uncertain parameters to be deterministic, and accounted for the uncertainties through the use of empirical safety factors. Safety factors are derived based on past experience but do not absolutely guarantee safety or satisfactory performance. They do not provide any information on the influence the different parameters of the system have on safety. Therefore, it is difficult to design a system with a uniform distribution of safety levels among the different components using empirical safety factors.

Engineering design is usually a trade-off between maximizing safety levels and minimizing cost. A design methodology that accomplishes both of these goals is highly desirable. Deterministic safety factors do not provide adequate information to achieve optimal use of the available resources to maximize safety. On the other hand, probabilistic analysis does provide the required information for optimum design. While probabilistic analysis brings rationality to the consideration of uncertainty in design, it does not discount the experience or expertise gathered from a particular system. In fact, the probabilistic methodology includes a "professional factor," which incorporates the expert opinions of experienced designers on different uncertain quantities in the system.

For this reason, design guidelines or codes have recently been revised to incorporate probabilistic analysis. Examples of such revisions include the American Institute of Steel Construction (AISC) Load and Resistance Factor Design (LRFD) (1986, 1994) specifications and the European and Canadian structural design specifications. Several other design specifications incorporating probabilistic design concepts are now in different stages of development. However, the emphasis in developing these guidelines is now on efficient and accurate consideration of the uncertainty in the design variables. The second edition of AISC's LRFD guidelines reflects this concern. The literature on probabilistic analysis and design and the number of engineering applications have greatly increased in recent years, expanding engineers' familiarity with and acceptance of this methodology. The use of probabilistic analysis in these codes is expected to provide more information about system behavior, the influence of different uncertain variables on system performance, and the interaction between different system components.

1.4   MEASURES OF RELIABILITY

Many different terms are used to describe the reliability of an engineering system. Some of the terms are self-explanatory, but others are not. The commonly used term *probability of failure* is always associated with a particular performance criterion. An engineering system will usually have several performance criteria, and a probability of failure is associated with each criterion. In addition, an overall system probability of failure may be computed. The probability of failure may be expressed as a fraction, such as 1 in 100, or as a decimal, such as 0.01. Reliability is the probability of successful performance; thus, it is the converse of the term *probability of failure*. It is common in the aerospace industry to express reliability in terms of decimals, such as 0.999 or 0.9999, and refer to these numbers as "three 9s reliability" or "four 9s reliability." (The corresponding probability of failure values for these reliability estimates are 0.001 and 0.0001, respectively.)

A measure of reliability in the context of design specifications is the safety factor, whose value provides a qualitative measure of safety. The safety factor may be used in the context of the load (or demand) on the system, or the resistance (or capacity) of the system, or both. In the context of the load, the nominally observed value of the load (referred to as the service load) is multiplied by a safety factor greater than 1.0 (referred to as the load factor) to obtain the design load. In the context of the resistance, the nominal value of the resistance of the system is multiplied by a safety factor usually less than 1.0 (referred to as the resistance factor or capacity reduction factor) to obtain the allowable resistance. Both load and resistance are uncertain quantities, with a mean, standard deviation, and so forth. The word *nominal* means that a deterministic value is specified by the designer or manufacturer for the load and/or the resistance for design purposes. In the case of loads, the nominal value is usually greater than the mean value. In the case of resistances, the nominal value is usually less than the mean value.

When both load and resistance factors are used, the overall safety is measured by the ratio of values of the load and the resistance. The central safety factor is the ratio of the mean values of the resistance and the load. The nominal safety factor is the ratio of the nominal values of the resistance and the load. Again, these concepts are briefly elaborated upon in Chapter 3.

For practical structures and performance criteria, it is difficult to compute the probability of failure precisely. Therefore, a first-order estimate is frequently used in probabilistic design specifications. This first-order estimate employs a measure known as the reliability index or safety index (denoted by the Greek symbol beta, β). The concepts of reliability index and first-order approximation of the probability of failure are described in Chapter 3.

1.5   FACTORS AFFECTING RELIABILITY EVALUATION

Reliability analysis requires information about uncertainties in the system. Before collecting such uncertainty information and proceeding with the reliability analysis, the engineer needs to understand that there are different types of uncertainty in engineering systems, and each type of uncertainty requires a different approach for data collection and use in reliability evaluation. In a broad sense, uncertainties in a system may come from cognitive (qualitative) and noncognitive (quantitative) sources.

1.5.1 Noncognitive Sources of Uncertainty

Noncognitive or quantitative sources of uncertainty or randomness can be classified into three types for discussion purposes. The first source is the inherent randomness in all physical observation. That is, repeated measurements of the same physical quantity do not yield the same value, due to numerous fluctuations in the environment, test procedure, instruments, observer, and so on. This may be referred to as inherent uncertainty. The engineer tries to address this type of uncertainty by collecting a large number of observations. This provides good information about the variability of the measured quantity, and leads to high confidence in the value used in the design. However, the number of observations that can be collected is limited by the availability of resources such as money and time.

This leads to the second source of uncertainty, known as statistical uncertainty. In this case, one does not have precise information about the variability of the physical quantity of interest due to limited data. The information on variability will vary, depending on the number of samples used. Therefore, quantitative measures of confidence based on the number of data are added to the reliability evaluation.

A third type of uncertainty is referred to as modeling uncertainty. System analysis models are only approximate representations of system behavior. Computational models strive to capture the essential characteristics of system behavior through idealized mathematical relationships or numerical procedures, such as finite element methods for structural analysis. In the process, some of the minor determinants of system behavior are ignored, leading to differences between computational prediction and actual behavior. Probabilistic methodology is able to include modeling uncertainty. Past experience on the difference between a computational model and actual behavior can be used to develop a statistical description of modeling error, to be included as an additional variable in the reliability analysis.

These three sources of uncertainty can be illustrated with a simple example. Suppose the wind load or pressure acting on a building needs to be estimated (in units of pounds per square foot). Recorded wind speed data, in miles per hour, can be collected for the site. Wind speed cannot be predicted with certainty; thus, it is inherently random. Its statistical uncertainty can be estimated by considering past observations, and more data lead to a better estimate. However, the statistical information on wind speed needs to be converted to wind pressure. Bernoulli's theorem is commonly used for this purpose. This introduces another source of uncertainty, known as modeling uncertainty.

1.5.2 Cognitive Sources of Uncertainty

Cognitive or qualitative sources of uncertainty relate to the vagueness of the problem arising from intellectual abstractions of reality. They may come from (1) the definitions of certain parameters, such as structural performance (failure or survival), quality, deterioration, skill and experience of construction workers and engineers, environmental impact of projects, and conditions of existing structures; (2) other human factors; and (3) definitions of the interrelationships among the parameters of the problems, especially for complex systems (Ayyub, 1994). These uncertainties are usually dealt with using fuzzy set theory, which is beyond the scope of this book.

1.6 STEPS IN THE MODELING OF UNCERTAINTY

We discussed the essential steps to quantify the uncertainty elsewhere (Haldar and Mahadevan, 2000) for situations where the sources of uncertainty are noncognitive and the
necessary information is available. Suppose the uncertainty in the annual rainfall or annual
maximum wind speed for a particular city needs to be quantified. Obviously, the information can be generated by collecting all the available recorded data on rainfall or wind
speed. There could be records of data for the past 50, 75, or 100 years, giving 50, 75, or 100
samples. The necessary statistical information can be extracted from these samples following the steps shown in Figure 1.1. The information collected constitutes the sample space.
The randomness characteristics can be described graphically in the form of a histogram
or frequency diagram, as elaborated upon in Chapter 2. For a more general representation of randomness, the frequency diagram can be approximated by a known theoretical
probability density function, such as the normal density function. To describe the probability density function uniquely, certain parameters of the distribution need to be estimated.
The estimation of these parameters, called statistics, is itself a major component of the
uncertainty analysis. The randomness in each of the load and resistance parameters can be
quantified using these statistics. Then, the risk involved in the design can be estimated for
a specific performance criterion.

It is undesirable and uneconomical, if not impossible, to design a risk-free structure.
In most cases of practical importance, risk can be minimized but cannot be eliminated

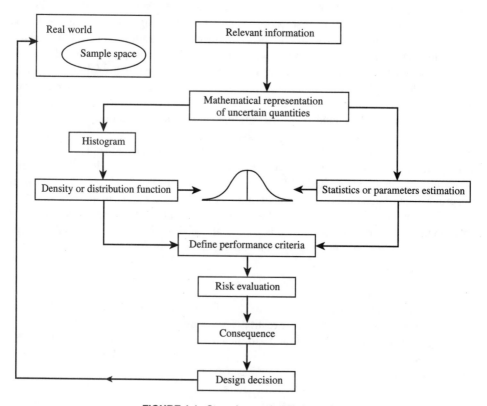

FIGURE 1.1. Steps in a probabilistic study.

completely. Nuclear power plants are relatively safer than ordinary buildings and bridges, with a corresponding high cost, but they are not absolutely safe. Making a structure safer costs more money in most cases. For a given structure, the corresponding risks for different design alternatives can be estimated. The information on risk and the corresponding consequences of failure, including the replacement cost of the structure, can be combined using a decision analysis framework to obtain the best alternative. The probability concept provides a unified framework for quantitative analysis of uncertainty and assessment of risk as well as the formulation of trade-off studies for decision making, planning, and design considering the economic aspects of the problem.

A large number of data are important to accurately implement the risk-based design concept. It is always preferable to estimate uncertainty using an adequate number of reliable observations. However, in many engineering problems, there are very few available data, sometimes only one or two observations. The probability concept can still be used in this case by combining experience, judgment, and observational data. The Bayesian approach can be used for this purpose. In the classical statistical approach, the parameters are assumed to be constant but unknown, and sample statistics are used as the estimators of these parameters. This requires a relatively large number of data. In the Bayesian approach, the parameters are considered to be random variables themselves, enabling an engineer to systematically combine subjective judgment based on intuition, experience, or indirect information with observed data to obtain a balanced estimate, and to update the estimate as more information becomes available.

In almost all cases, the risk-based design concept can be used successfully regardless of the number of available data.

1.7 ROLE OF THE FINITE ELEMENT METHOD IN RELIABILITY EVALUATION

Finite element analysis is a very powerful tool commonly used by many engineering disciplines to analyze simple or complicated structures. The word "structure" is used here in a broad sense to include all systems that can be discretized using finite elements. With this approach, it is easy and straightforward to consider complicated geometric arrangements and constitutive relationships of the material, realistic connection or support conditions, various sources of nonlinearity, and the load path to failure. It gives good results for a set of assumed values of the variables while ignoring the uncertainty in them. On the other hand, many of the available reliability methods are able to account for the uncertainties, but fail to represent the structural behavior as realistically as possible and cannot be used when the performance function is not available in explicit form. In the early 1980s, we realized that the desirable features of these two approaches need to be combined, leading to the concept of the stochastic finite element method (SFEM). Advances in computer technology make this approach more attractive today than ever before.

If the basic variables are uncertain, every quantity computed during the deterministic analysis is also uncertain, being a function of the basic variables. The currently available reliability methods can still be used if the uncertainty in the response can be tracked in terms of the variation of the basic variables at every step of the deterministic analysis. The finite element method (FEM) provides such an opportunity, and this concept forms the basis of the stochastic finite element method.

The deterministic finite element area is very advanced and the researchers in this area are looking for new ideas and concepts. The risk and reliability research communities are looking for robust reliability analysis methods that can be used for both implicit and explicit performance functions and that represent structures as realistically as possible. Students are looking for new and computationally challenging concepts. This topic will be able to satisfy these needs.

With the advances in computer technology, it is quite appropriate to develop a finite element-based reliability analysis technique, parallel to the deterministic analysis procedure. This will help to implement the reliability-based design concept to real problems. Most engineers will also have a tool to estimate the risk or reliability of simple or complicated systems considering all major sources of uncertainty.

The reliability evaluation of engineering systems occurs at two levels: individual performance criteria and overall system performance. For a simple beam, the performance criterion could be strength-related, e.g., bending moment or shear, or serviceability-related, e.g., deflection or vibration. A structure such as a truss or frame consists of multiple structural elements or components, and failure may occur in one or more components. The concept used to consider multiple failure modes or multiple component failures is known as *system reliability* evaluation. A complete reliability analysis includes both component-level and system-level reliability estimates. The stochastic finite element method presented indirectly accounts for system effects because it considers the presence of all structural elements in component-level reliability evaluation. System reliability techniques need to be used to estimate the reliability of the overall system.

1.8 CONCLUDING REMARKS

Stochastic finite element-based reliability analysis is necessary to evaluate risk or reliability of simple or complicated systems considering their behavior as realistically as possible. It can be used when the limit state function is implicit or explicit. It is parallel to the deterministic finite element method but incorporates uncertainties in the variables. It is very powerful, robust, and easy to use, and will enable engineers to estimate the risk of engineering systems of practical importance.

2

COMMONLY USED
PROBABILITY
DISTRIBUTIONS

2.1 INTRODUCTORY COMMENTS

A basic understanding of modeling and quantifying uncertainty in a random variable and how to use the information to evaluate the risk or reliability for simple cases is the first essential step in any reliability evaluation. Before the advanced concept of stochastic finite element is presented, we first discuss the estimation of reliability for an assumed distribution.

As stated in the Preface, it is assumed that the readers are familiar with the fundamental concepts of risk and reliability evaluation, including set theory, the mathematics of probability, the modeling of uncertainty in continuous and discrete random variables, the determination of distributions and parameters from observed data, and the randomness in a response variable when it is a function of other random variables. These concepts are discussed in detail by Haldar and Mahadevan (2000) elsewhere. The readers are strongly encouraged to review these materials before proceeding with the advanced topics discussed in this book.

To provide familiarity with the basic concepts of risk and reliability and the notations used in subsequent chapters, consider Table 2.1, which contains 41 values of the Young's modulus E for structural steel members used in the Golden Gate Bridge in San Francisco (Beard, 1937). In standard deterministic design, the Young's modulus for steel is usually assumed to be 29,000 ksi. Table 2.1 indicates that this value rarely occurs. Obviously, E should be treated as a random variable, and this randomness must be modeled appropriately. From now on, a random variable will be represented in this book by an uppercase letter, and a particular realization of a random variable will be represented by a lowercase letter.

To make this discussion meaningful, two additional terms, *population* and *sample*, need to be introduced. A population represents all conceivable observations of a random variable. Data collected on the Young's modulus for a particular grade of steel from all over the world would represent the population. Since it is impractical to collect the information from all available sources, a representative sample is collected. The data on Young's mod-

Table 2.1. Young's modulus E for the Golden Gate Bridge

Test No.	Young's Modulus, E (ksi)
1	28,900
2	29,200
3	27,400
4	28,700
5	28,400
6	29,900
7	30,200
8	29,500
9	29,600
10	28,400
11	28,300
12	29,300
13	29,300
14	28,100
15	30,200
16	30,200
17	30,300
18	31,200
19	28,800
20	27,600
21	29,600
22	25,900
23	32,000
24	33,400
25	30,600
26	32,700
27	31,300
28	30,500
29	31,300
30	29,000
31	29,400
32	28,300
33	30,500
34	31,100
35	29,300
36	27,400
37	29,300
38	29,300
39	31,300
40	27,500
41	29,400

ulus given in Table 2.1 represent such a sample. Representative samples are generally used to gather information on population. A relatively large sample size is always preferable.

In Table 2.1, the *maximum value* for E is 33,400 ksi and the *minimum value* is 25,900 ksi. The information on minimum and maximum values has very limited use in actual design. Furthermore, these observed values may not be the absolute minimum or maximum values. If more samples are collected, these minimum and maximum values may change.

To overcome the deficiency of the minimum–maximum approach, one common-sense approach is to calculate the *average* or *mean* or *expected value* of the Young's modulus. Suppose X is a random variable and n observations of it are available. The mean or expected value of X is a measure of central tendency in the data. It is also known as the first central moment and is denoted as $E(X)$ or μ_X, and we can calculate it for the n observations as

$$\text{Mean} = E(X) = \mu_X = \frac{1}{n} \sum_{i=1}^{n} x_i \tag{2.1}$$

The mean value alone does not provide complete information. For example, the mean value of 0 and 100 is 50. The same mean value will be obtained if the numbers are 40 and 60 or 45 and 55. Information is needed on the dispersion of the values with respect to the mean. The dispersion can be expressed in terms of the *variance, standard deviation, or coefficient of variation (COV)*.

The variance of X is a measure of the spread in the data about the mean. It is also known as the second central moment and is denoted hereafter as $\text{Var}(X)$, and we can calculate it as

$$\text{Variance} = \text{Var}(X) = \frac{1}{n-1} \sum_{i=1}^{n} (x_i - \mu_X)^2 \tag{2.2}$$

Equations 2.1 and 2.2 implicitly assume that the sample size is relatively large.

If the random variable X is expressed in ksi, then the variance will be expressed in $(\text{ksi})^2$. This dimensional problem can be avoided by taking the square root of the variance. This is the standard deviation, denoted as σ_X hereafter, and can be calculated as $\sigma_X = \sqrt{\text{Var}(X)}$. Although the standard deviation value is expressed in the same units as the mean value, its absolute value does not clearly indicate the degree of dispersion in the random variable without referring to the mean value. For example, the value of the standard deviation could be 10 or 100 without indicating the degree of dispersion. Since the mean and the standard deviation values are expressed in the same units, a nondimensional term can be introduced by taking the ratio of the standard deviation and the mean. This is called the coefficient of variation (COV), and will be denoted as $\text{COV}(X)$ or δ_X. Thus, $\text{COV}(X) = \delta_X = \sigma_X/\mu_X$. For a deterministic variable, $\text{COV}(X)$ is zero. A smaller value of the COV indicates a smaller amount of uncertainty or randomness in the variable, and a larger amount indicates a larger amount of uncertainty. In many engineering problems, a COV of 0.1 to 0.3 is common for a random variable.

We can use the 41 data on the Young's modulus E to calculate

$$\mu_E = \tfrac{1}{41}(1{,}212{,}600) = 29{,}575.6 \text{ ksi}$$

$$\text{Var}(E) = \tfrac{1}{41-1}(90{,}835{,}609.8) = 2{,}270{,}890.2 \ (\text{ksi})^2$$

$$\sigma_E = \sqrt{2{,}270{,}890.2} = 1507 \text{ ksi}$$

$$\text{COV}(E) = \delta_E = \tfrac{1507}{29{,}575.6} = 0.051$$

For the given data, the uncertainty in the Young's modulus is relatively small.

A preliminary description of the randomness in a variable can be obtained from the numerical values of these parameters. A more complete description can be obtained by plotting the information graphically in the form of a *histogram*. Figure 2.1 shows a histogram for the data in Table 2.1. The area under a histogram depends on the width of the intervals and the number of data points. For Figure 2.1, the area under the histogram is $1500 \times 41 = 61,500$. Since the probability of an event is between 0.0 and 1.0, it is mathematically advantageous to have the area under a histogram equal to unity. A histogram with a unit area is known as a *frequency diagram*. The frequency diagram is obtained by dividing the ordinates of a histogram by its area. This does not change the shape of the diagram, as shown in Figure 2.1. The histogram or frequency diagram gives the relative frequencies of various intervals.

One of the primary objectives of a frequency diagram is to fit a curve to model the pattern of the randomness. A curve can be easily fitted to the frequency diagram as shown in Figure 2.1. As more data are added, the fitted curve approaches the frequency diagram more closely. Attempts can be made to verify whether the fitted curve represents one of many commonly used *distributions*, such as normal or lognormal.

The curve shown in Figure 2.1 is called the *probability density function* (PDF) or *density function*, and is represented by $f_X(x)$. It does not directly provide information on probability; it only indicates the nature of the randomness. To calculate the probability of X having a value between x_1 and x_2, the area under the PDF between these two limits needs to be calculated. We can express this as

$$P(x_1 < X \le x_2) = \int_{x_1}^{x_2} f_X(x)\, dx \tag{2.3}$$

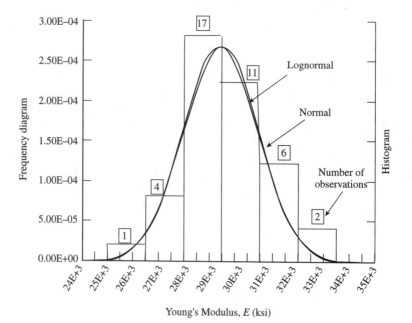

FIGURE 2.1. Histogram and frequency diagram of Young's modulus.

To calculate $P(X \leq x)$, which is specifically denoted as $F_X(x)$ and is known as the *cumulative distribution function* (CDF) or *distribution function,* the area under the PDF needs to be integrated for all possible values of X less than or equal to x.

Considering the physical aspects of some variables used in engineering design, such as the number of fires in a subdivision, the number of strong winds or earthquakes or severe snowstorms, the number of cars crossing an intersection, or the duration in days of a construction activity, it is not logical to model them as continuous random variables. The number of these events can be measured only in integers, and they must be treated as discrete random variables. The mathematical treatment of continuous random variables is also applicable to discrete random variables with some modifications. Since a discrete random variable occurs only at certain discrete points, its relative frequency of occurrence can be evaluated only at these discrete points. This is known as the *probability mass function* (PMF), and is denoted as $p_X(x)$. PMF is similar to PDF, but it is not a continuous function and consists of a series of spikes. Summations of the PMFs are necessary to calculate the CDF for a discrete random variable. Mathematically, we can express these observations as

$$F_X(x) = P(X \leq x) = \sum_{x_i \leq x} p_X(x_i) \tag{2.4}$$

Figure 2.2 shows a typical PMF and CDF of a discrete random variable. The same figure will be considered in Example 2.7 in Section 2.3.1 to explain the procedures necessary to develop the PMF and CDF of a discrete random variable.

The introductory comments on the PDF, PMF, and CDF are not complete without considering two additional parameters: the *mode* and *median* of a random variable X. The mode or modal value of X is the value with the largest PDF or PMF. The median of a random variable, x_m, is the value at which the CDF is 0.50, that is, the value of X for which it is equally probable that X will be above or below it. This condition is expressed as

$$F_X(x_m) = 0.5 \tag{2.5}$$

The selection of the design value of a random variable considering its uncertainty is an important engineering task. This can be done using the concept of *percentile value*. For example, the 90th percentile value indicates that the design value will not be exceeded 90% of the time. In general, for resistance-related random variables, the design value is considered to be less than the 50th percentile. For load-related random variables, the design value is over the 50th percentile value.

In summary, the uncertainty in a random variable can be modeled by its underlying *distribution,* generally expressed in terms of PDF or PMF and CDF. To uniquely define a PDF or PMF, its parameters need to be estimated. Generally, the parameters are estimated using the information on mean, variance, and COV obtained from available data.

There are many commonly used standard distributions. They are routinely used to evaluate the risk or reliability of events. Since they are very important in any risk evaluation procedure, this chapter emphasizes the basic principles of calculating risk or reliability using these distributions. Many computer spreadsheets and other programs such as EXCEL, QUATTRO PRO, and MATLAB are routinely used to calculate risk for many assumed distributions. These programs are used as "black boxes" without requiring an understanding of the fundamentals of risk assessment. This chapter emphasizes the conceptual and computational aspects of risk evaluation for an assumed distribution without using a computer program.

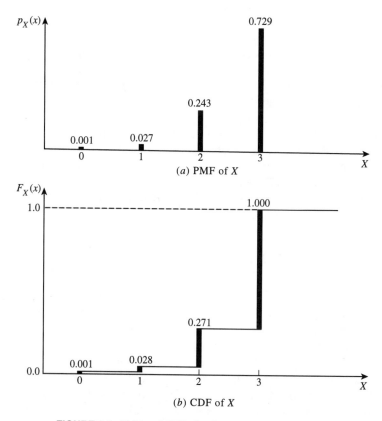

FIGURE 2.2. PMF and CDF of a discrete random variable.

Beginners in the area of risk and reliability analyses need a thorough understanding of the concepts briefly described in this section before proceeding further. Haldar and Mahadevan (2000) wrote a book on the more fundamental aspects of risk and reliability evaluation, and it can be referred to for this purpose.

2.2 CONTINUOUS RANDOM VARIABLES

Commonly used continuous and discrete random variables are discussed separately in the following sections.

2.2.1 Normal or Gaussian Distribution

One of the most commonly used distributions in engineering problems is the *normal or Gaussian distribution*. The PDF of the distribution can be expressed as

$$f_X(x) = \frac{1}{\sigma_X \sqrt{2\pi}} \exp\left[-\frac{1}{2}\left(\frac{x - \mu_X}{\sigma_X}\right)^2\right] \qquad -\infty < x < +\infty \qquad (2.6)$$

where the mean μ_X and the standard deviation σ_X are the two parameters of the distribution, usually estimated from the available data. The corresponding CDF can be expressed as

$$F_X(x) = \int_{-\infty}^{x} \frac{1}{\sigma_X \sqrt{2\pi}} \exp\left[-\frac{1}{2}\left(\frac{x - \mu_X}{\sigma_X}\right)^2\right] dx \qquad (2.7)$$

The normal distribution is widely used and is denoted as $N(\mu, \sigma)$, indicating that it is a normal random variable with a mean and standard deviation of μ and σ, respectively. The PDF and CDF of a normal distribution with a mean of 100 and a standard deviation of 10 are shown in Figure 2.3. This distribution has many desirable features. It is applicable for any value of a random variable from $-\infty$ to $+\infty$. The distribution is symmetric about the mean, and the mean, median, and modal values are identical and can be estimated directly from the data. However, it is not simple to estimate the probability by integrating Equation 2.6. The problem can be addressed by transforming the original random variable X into a standard normal variable with zero mean and unit standard deviation, as

$$S = \frac{X - \mu_X}{\sigma_X} \qquad (2.8)$$

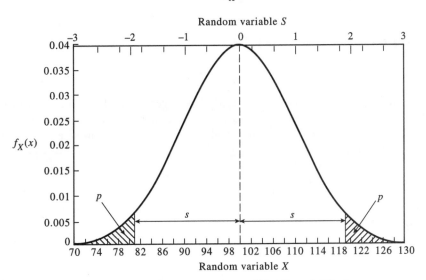

(a) Normal PDF of X, $N(100, 10)$ or S, $N(0, 1)$

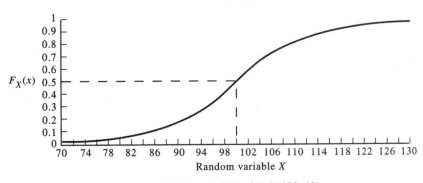

(b) Normal CDF of X, $N(100, 10)$

FIGURE 2.3. PDF and CDF of normal random variable.

Using Equation 2.6 and the variable transformation technique, we can express the PDF of S as

$$f_S(s) = \frac{1}{\sqrt{2\pi}} \exp\left[-\frac{1}{2}s^2\right] \qquad -\infty < s < +\infty \qquad (2.9)$$

The corresponding CDF of S is

$$F_S(s) = \int_{-\infty}^{s} \frac{1}{\sqrt{2\pi}} \exp\left[-\frac{1}{2}s^2\right] ds \qquad (2.10)$$

The standard normal distribution is denoted as $N(0, 1)$, and its CDF is denoted as $\Phi(s)$, that is, $\Phi(s) = F_S(s)$, given by Equation 2.10. The CDF of the standard normal distribution is widely available in tabulated form, as shown in Appendix 1, or can be calculated using a standard subroutine available in many computer programs. Since the normal distribution is perfectly symmetrical, referring to Figure 2.3, we can show that

$$\Phi(-s) = 1.0 - \Phi(s) = p \qquad (2.11a)$$

or, when $p < 0.5$, we can show that

$$-s = \Phi^{-1}(p) = -\Phi^{-1}(1-p) \qquad (2.11b)$$

The table in Appendix 1 is valid only for positive values of the standard normal random variable. However, as discussed earlier, negative values are possible for a normal or standard random variable. The CDF of the standard normal variable can be evaluated for a negative value using Equation 2.11.

Example 2.1

Suppose $\Phi(-0.1)$ needs to be evaluated using the table in Appendix 1. Using the table, we find that

$$\Phi(0.1) = 0.53983$$

Using Equation 2.11, we can show that

$$\Phi(-0.1) = 1.0 - \Phi(0.1) = 1.0 - 0.53983 = 0.46017$$

From Appendix 1, it is found that $\Phi(0.0) = 0.5$. Thus, it can also be concluded that the CDF of the standard normal distribution evaluated at a negative value of S will be less than 0.5.

If a random variable X is $N(\mu_X, \sigma_X)$, then its probability of having a value between two limits a and b is

$$P(a < X \le b) = \frac{1}{\sigma_X \sqrt{2\pi}} \int_a^b \exp\left[-\frac{1}{2}\left(\frac{x - \mu_X}{\sigma_X}\right)^2\right] dx \qquad (2.12)$$

Again, by variable transformation, Equation 2.12 can be rewritten in terms of the standard normal variable S as

$$P(a < X \leq b) = \frac{1}{\sqrt{2\pi}} \int\limits_{\frac{a-\mu_X}{\sigma_X}}^{\frac{b-\mu_X}{\sigma_X}} \exp\left[-\frac{1}{2}s^2\right] ds$$

or

$$P(a < X \leq b) = \Phi\left(\frac{b-\mu_X}{\sigma_X}\right) - \Phi\left(\frac{a-\mu_X}{\sigma_X}\right) \qquad (2.13)$$

Equation 2.13 indicates that the probability of a normal random variable between two limits can be calculated easily using the table in Appendix 1.

Example 2.2

To demonstrate the steps involved, consider the values of the Young's modulus given in Table 2.1. Assume that the randomness in E can be described by a normal random variable. Its mean and standard deviation were estimated in Section 2.1 as 29,576 ksi and 1507 ksi, respectively. Using Equation 2.13, we can calculate the probability of E having a value between 28,000 and 29,500 ksi as

$$P(28,000 < E \leq 29,500)$$

$$= \Phi\left(\frac{29,500 - 29,576}{1507}\right) - \Phi\left(\frac{28,000 - 29,576}{1507}\right)$$

$$= \Phi(-0.05) - \Phi(-1.05) = [1 - \Phi(0.05)] - [1 - \Phi(1.05)]$$

$$= (1 - 0.51994) - (1.0 - 0.85314) = 0.33320$$

As stated earlier, the commonly used Young's modulus for steel is 29,000 ksi. The probability of the Young's modulus being less than the design value, that is, $-\infty < E \leq 29,000$, can be calculated as

$$P(E \leq 29,000) = \Phi\left(\frac{29,000 - 29,576}{1507}\right) - \Phi\left(\frac{-\infty - 29,576}{1507}\right)$$

$$= \Phi(-0.38) - \Phi(-\infty)$$

$$= (1.0 - 0.64803) - 0.0 = 0.35197$$

This means that the design value of E is approximately the 35th percentile value for the data given in Table 2.1.

Similarly, the probability that Young's modulus will be at least 29,000 ksi can be calculated as

$$P(E \geq 29,000) = P(29,000 < E \leq +\infty)$$

$$= \Phi\left(\frac{\infty - 29,576}{1507}\right) - \Phi\left(\frac{29,000 - 29,576}{1507}\right)$$

$$= \Phi(+\infty) - \Phi(-0.38) = 1 - 1 + \Phi(0.38) = 0.64803$$

Common sense suggests that the Young's modulus cannot be negative; thus, modeling it as a normal random variable may distort the physical aspects of the problem. This type of argument is often used against probabilistic analysis. The probability of E being less than or equal to zero can be calculated as

$$P(E \leq 0.0) = \Phi \left(\frac{0.0 - 29{,}576}{1507} \right) - \Phi(-\infty) = \Phi(-19.63) \approx 0.0$$

Obviously, there is virtually no impact on the uncertainty analysis of the Young's modulus. If the underlying distribution is normal and negative values are not possible, the available data should reflect the physical aspect of the random variable. This is discussed further later.

The Young's modulus is a resistance-related random variable, whose design value is usually selected to be less than the mean value. Suppose its design value is selected to be the 10th percentile value. Denoting it as $e_{0.10}$, we can calculate the design value of the Young's modulus as

$$P(E \leq e_{0.10}) = 0.10$$

or

$$\Phi \left(\frac{e_{0.10} - 29{,}576}{1507} \right) = 0.10$$

or

$$\frac{e_{0.10} - 29{,}576}{1507} = \Phi^{-1}(0.10) = -\Phi^{-1}(0.90) = -1.28$$

Thus,

$$e_{0.10} = 29{,}576 - 1.28 \times 1507 = 27{,}647 \text{ ksi}$$

If the 90th percentile value of the Young's modulus is desired, it can be calculated as

$$e_{0.90} = 29{,}576 + 1.28 \times 1507 = 31{,}505 \text{ ksi}$$

For the data under consideration, the design value of 29,000 ksi of the Young's modulus is approximately the 35th percentile value, as shown earlier. For a normal random variable, the probability associated with an event or the design value can be easily calculated using the simple procedure discussed above.

When a distribution like the normal distribution, which is valid from $-\infty$ to $+\infty$, lacks a physical interpretation to consider the practical aspects of the problem, it is common in the literature to consider the values of the random variable as belonging to a range bounded by the mean plus and minus some standard deviation values. For a normal distribution, mean $\pm 3\sigma$ bounds are very common. If the data are limited to these lower and upper bounds, they will give a probability of 0.997 instead of 1.0, indicating that the error associated with the probability calculation is marginal. For the Young's modulus parameter under consideration, the lower and upper bounds will be $29{,}576 - 3 \times 1507 = 25{,}055$ ksi and $29{,}576 + 3 \times 1507 = 34{,}097$ ksi, respectively. Thus, considering the physical aspects of the parameter, values of the Young's modulus between 25,055 and 34,097 ksi will effectively give practical limits to the bounds and will include about 99.7% of the data. Table 2.1 indicates that there are no observations outside this range, which validates this statement.

For a normal distribution, if mean $\pm 1\sigma$ bounds are used, about 68.3% of the data are included. If the bounds are increased to mean $\pm 2\sigma$, about 95.4% of the data are included.

Example 2.3

Suppose a steel cable has to carry a weight of 10 kips. Information on the strength of similar cables indicates that the strength of the cable, R, can be modeled by a normal random variable with a mean of 25 kips and a standard deviation of 5 kips. Calculate the probability that the cable will be unable to carry the weight, or the probability that the cable will break.

Solution

$$P(\text{the cable will break}) = P(\text{failure}) = P(R \leq 10)$$
$$= \Phi\left(\frac{10-25}{5}\right) - \Phi\left(\frac{-\infty-25}{5}\right)$$
$$= \Phi(-3) - \Phi(-\infty)$$
$$= 1 - \Phi(3) = 1 - 0.99865 = 0.00135$$

2.2.2 Lognormal Distribution

In many engineering problems, a random variable cannot have negative values due to the physical aspects of the problem. In this situation, modeling the variable as lognormal (i.e., considering the natural logarithm of the variable X) is more appropriate, automatically eliminating the possibility of negative values. If a random variable has a lognormal distribution, then its natural logarithm has a normal distribution. This is the meaning of the term *lognormal*. The PDF of a lognormal variable is given by

$$f_X(x) = \frac{1}{\sqrt{2\pi}\zeta_X x} \exp\left[-\frac{1}{2}\left(\frac{\ln x - \lambda_X}{\zeta_X}\right)^2\right] \qquad 0 \leq x < \infty \qquad (2.14)$$

where λ_X and ζ_X are the two parameters of the lognormal distribution. The PDF of a typical lognormal distribution with a mean of 100 and a standard deviation of 10 is shown in Figure 2.4. The PDF values of both the normal and lognormal random variables with the same mean and standard deviation are plotted in Figure 2.5. The lognormal variable has values between zero and $+\infty$. Its PDF is unsymmetrical, and thus its mean, median, and modal values are expected to be different. Comparing Equations 2.6 and 2.14, we can observe some similarities between the normal and lognormal distributions. In fact, the two parameters of the lognormal distribution can be calculated from the information on the two parameters of the normal distribution: the mean (μ) and standard deviation (σ) of the sample population. It can be shown that

$$\lambda_X = E(\ln X) = \ln \mu_X - \frac{1}{2}\zeta_X^2 \qquad (2.15)$$

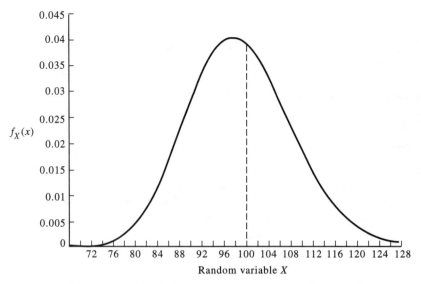

(a) Lognormall PDF of X with mean = 100 and standard deviation = 10

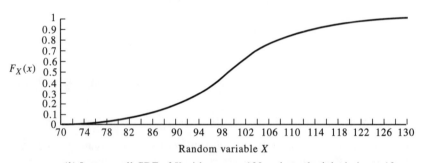

(b) Lognormall CDF of X with mean = 100 and standard deviation = 10

FIGURE 2.4. PDF and CDF of lognormal random variable.

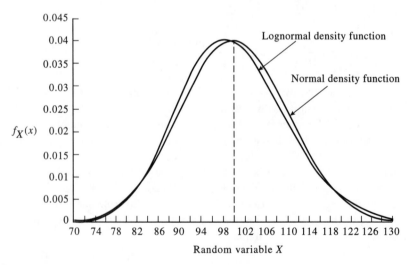

FIGURE 2.5. PDF of normal and lognormal distributions with mean = 100 and standard deviation = 10.

and

$$\zeta_X^2 = \text{Var}(\ln X) = \ln\left[1 + \left(\frac{\sigma_X}{\mu_X}\right)^2\right] = \ln(1 + \delta_X^2) \qquad (2.16)$$

If the COV (i.e., $\delta_X = \sigma_X/\mu_X$) is not very large—for example, less than 0.30—then $\zeta_X \approx \delta_X$, the COV of the random variable X.

To calculate the probability of an event where the underlying distribution of a random variable is lognormal, the procedures used for the normal variable are still applicable, except that for the lognormal case, the standard variable S will take the following form instead of Equation 2.8:

$$S = \frac{\ln X - \lambda_X}{\zeta_X} \qquad (2.17)$$

The probability of a lognormal random variable having a value between two limits a and b can be calculated as

$$P(a < X \le b) = \frac{1}{\sqrt{2\pi}} \int_{\left(\frac{\ln a - \lambda_X}{\zeta_X}\right)}^{\left(\frac{\ln b - \lambda_X}{\zeta_X}\right)} \exp\left(-\frac{1}{2}s^2\right) ds$$

$$= \Phi\left(\frac{\ln b - \lambda_X}{\zeta_X}\right) - \Phi\left(\frac{\ln a - \lambda_X}{\zeta_X}\right) \qquad (2.18)$$

Thus, all required probabilities for the lognormal variable can be calculated from the CDF table developed for the standard normal variable, given in Appendix 1.

Example 2.4

To demonstrate the calculation of probability for a lognormal variable, the same Young's modulus example with a mean of 29,576 ksi and a standard deviation of 1507 ksi can be considered, except that now it is assumed that the Young's modulus is lognormally distributed. In this case,

$$\delta = \frac{\sigma}{\mu} = \frac{1507}{29,576} = 0.051 \le 0.3$$

Thus,

$$\zeta \approx \delta = 0.051$$

$$\lambda = \ln 29,576 - 0.5 \times 0.051^2 = 10.293$$

The probability of E having a value between 28,000 and 29,500 ksi can be calculated as

$$P(28,000 < E \le 29,500)$$

$$= \Phi \left(\frac{\ln 29{,}500 - 10.293}{0.051} \right) - \Phi \left(\frac{\ln 28{,}000 - 10.293}{0.051} \right)$$

$$= \Phi(-0.017) - \Phi(-1.04) = (1 - 0.50678) - (1.0 - 0.85083) = 0.34405$$

Again, the probability of E being less than the design value of 29,000 ksi can be calculated as

$$P(E \leq 29{,}000) = \Phi \left(\frac{\ln 29{,}000 - 10.293}{0.051} \right) - \Phi(-\infty) = \Phi(-0.35)$$

$$= 1.0 - 0.63683 = 0.36317$$

If E is modeled as a lognormal variable, the design value is about the 36th percentile value for the data given in Table 2.1.

If the design value for the Young's modulus is still the 10th percentile value, then it can be estimated as

$$\Phi \left(\frac{\ln e_{0.10} - 10.293}{0.051} \right) = 0.10 = \Phi(-1.28)$$

or

$$\ln e_{0.10} = 10.293 - 1.28 \times 0.051$$

or

$$e_{0.10} = 27{,}659 \text{ ksi}$$

For the Young's modulus example under consideration, the results are similar for the normal and lognormal cases. This is discussed further later.

Some of the important features of a lognormal variable can be summarized as follows:

1. If X is a lognormal variable with parameters λ_X and ζ_X, then $\ln X$ is normal with a mean of λ_X and a standard deviation of ζ_X.
2. $\lambda_X = \ln \mu_X - \frac{1}{2}\zeta_X^2$
3. $\zeta_X^2 = \ln(1 + \delta_X^2)$. When $\delta_X \leq 0.3$, $\zeta_X \approx \delta_X$, the COV of X.
4. Denoting x_m as the median of a lognormal variable X, we can show that $\lambda_X = \ln x_m$.
5. $x_m = \mu_X / \sqrt{1 + \delta_X^2}$; that is, the median value of a lognormal variable is always less than the mean value.

2.2.3 Beta Distribution

The *beta distribution* is a very flexible and useful distribution, and can be used when a random variable is known to be bounded by two limits, a and b. The normal distribution is valid between $-\infty$ and $+\infty$, and the lognormal distribution is valid between 0 and $+\infty$. Many random variables of engineering significance may be bounded by two limits, and the beta distribution could be quite appropriate. The PDF of a beta distribution is represented as

$$f_X(x) = \frac{1}{B(q,r)} \frac{(x-a)^{q-1}(b-x)^{r-1}}{(b-a)^{q+r-1}} \qquad a \le x \le b$$
$$= 0 \qquad \text{elsewhere} \qquad\qquad (2.19)$$

where q and r are the parameters of the distribution and $B(q,r)$ is the *beta function*. The parameters q and r can be estimated from the mean and standard deviation of the available data using the following relationships:

$$E(X) = a + \frac{q}{q+r}(b-a) \qquad\qquad (2.20)$$

and

$$\text{Var}(X) = \frac{qr}{(q+r)^2(q+r+1)}(b-a)^2 \qquad\qquad (2.21)$$

If the upper and lower limits and the mean and variance of a random variable are known, the corresponding q and r parameters of the beta distribution can be estimated using Equations 2.20 and 2.21.

The beta function in Equation 2.19 can be shown to be

$$B(q,r) = \int_0^1 x^{q-1}(1-x)^{r-1}\, dx \qquad\qquad (2.22)$$

The beta function can also be calculated as

$$B(q,r) = \frac{\Gamma(q)\Gamma(r)}{\Gamma(q+r)} \qquad\qquad (2.23)$$

where $\Gamma(\ )$ is the gamma function. Procedures to calculate the gamma function are given in Appendix 2.

If the lower limit a is 0 and the upper limit b is 1, Equation 2.19 takes the following form:

$$f_X(x) = \frac{1}{B(q,r)} x^{q-1}(1-x)^{r-1} \qquad 0 \le x \le 1$$
$$= 0 \qquad \text{elsewhere} \qquad\qquad (2.24)$$

Equation 2.24 is known as the *standard beta distribution*. Its PDFs for several values of q and r are shown in Figure 2.6. When q and r are both equal to one, the beta distribution becomes a *uniform distribution*.

Once the PDF of a beta distribution is defined, the probability of any event can be estimated by numerically integrating the area under the PDF corresponding to the upper and lower limits. The probability can also be found using tables for the standard beta distribution similar to the standard normal table given in Appendix 1.

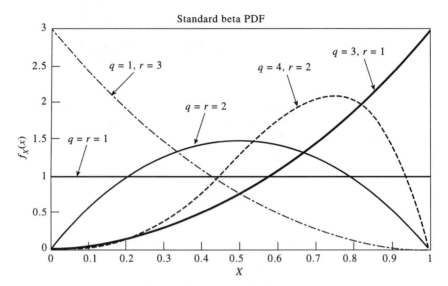

FIGURE 2.6. Standard beta PDF.

Example 2.5

The daily maximum temperature in June in Tucson, Arizona, varies between 80 and 110°F. Using 100 years of data, we can estimate that the average daily maximum temperature in June is 95°F and the corresponding standard deviation is 10°F. Assuming that the daily maximum temperature can be modeled by a beta distribution, what is the probability that on any given day in June, the daily maximum temperature will exceed 100°F?

It is clear that modeling the temperature by a normal or lognormal distribution may not be appropriate. The beta distribution may be a reasonable alternative. The following information can be extracted from the available data. For the beta distribution under consideration, $a = 80$, $b = 110$. Also, using Equations 2.20 and 2.21, we can show that

$$80 + \frac{q}{q+r}(110 - 80) = 95$$

and

$$\frac{qr}{(q+r)^2(q+r+1)}(110 - 80)^2 = 10^2$$

In this particular case, $q = r = \frac{5}{8}$, and using Equation 2.23

$$B\left(\frac{5}{8}, \frac{5}{8}\right) = \frac{\Gamma\left(\frac{5}{8}\right)\Gamma\left(\frac{5}{8}\right)}{\Gamma\left(\frac{5}{8} + \frac{5}{8}\right)}$$

Gamma functions can be evaluated using Appendix 2, where $\Gamma\left(\frac{5}{8}\right)$ is shown to be 1.434519178. Using the first equation in Section 2a of Appendix 2, we can calculate

$\Gamma\left(\frac{5}{8}+\frac{5}{8}\right)$ or $\Gamma\left(\frac{5}{4}\right)$ as

$$\Gamma\left(\frac{5}{4}\right) = \Gamma\left(1+\frac{1}{4}\right) = 1 - 0.5748646 \times \left(\frac{1}{4}\right) + 0.9512363 \times \left(\frac{1}{4}\right)^2$$

$$- 0.6998588 \times \left(\frac{1}{4}\right)^3 + 0.4245549 \times \left(\frac{1}{4}\right)^4 - 0.1010678 \times \left(\frac{1}{4}\right)^5$$

$$= 0.906360543$$

Thus,

$$B\left(\frac{5}{8},\frac{5}{8}\right) = \frac{(1.434519178)^2}{0.906360453} = 2.270$$

Many computer spreadsheets and other programs like EXCEL, QUATTRO PRO, and MATLAB can also be used to calculate the gamma functions. To demonstrate how to use Appendix 2, we have evaluated the gamma functions up to nine decimal place accuracy. However, considering the practical aspect of the problem, the beta function in Equation 2.19 or 2.24 need not be calculated with similar accuracy.

Once the beta function is evaluated, the corresponding PDF of the beta distribution of the daily maximum temperature, T, is given by Equation 2.19 and can be shown to be

$$f_T(t) = \frac{1}{2.270} \frac{(t-80)^{5/8-1}(110-t)^{5/8-1}}{(110-80)^{5/8+5/8-1}}$$

$$= \frac{1}{5.313}(t-80)^{-3/8}(110-t)^{-3/8}, \quad 80 \le t \le 110$$

Thus, if we numerically integrate the above PDF from 100 to 110°F, the probability that on any given day in June, the daily maximum temperature in Tucson will exceed 100°F can be calculated as

$$P(T > 100) = \int_{100}^{110} \frac{1}{5.313}(t-80)^{-3/8}(110-t)^{-3/8}\, dt = 0.3716$$

This example demonstrates that depending on the mathematical form of the PDF, it may be necessary to numerically integrate the area under the PDF between the upper and lower limits to estimate the corresponding probability.

2.3   DISCRETE RANDOM VARIABLES

2.3.1   Binomial Distribution

In many engineering applications, events consisting of repeated *trials* can be formulated in terms of occurrence or nonoccurrence, success or failure, good or bad, and so forth. Only two outcomes are possible, representing the behavior of a discrete random variable. In addition, if the events satisfy the additional requirements of a *Bernoulli sequence,* that is, if they are statistically independent and the probability of occurrence or nonoccurrence of events remains constant, they can be mathematically represented by the *binomial distribution*. If the probability of occurrence of an event in each trial is p and the probability of nonoccurrence is $(1 - p)$, then the probability of x occurrences out of a total of n trials can be described by the PMF of a binomial distribution as

$$P(X = x, n \mid p) = \binom{n}{x} p^x (1 - p)^{n-x} \qquad x = 0, 1, 2, \ldots, n \qquad (2.25)$$

where p, the probability of occurrence in each trial, is the parameter of the distribution, and $\binom{n}{x} = n!/[x!(n-x)!]$ is the *binomial coefficient*, indicating the number of ways that x occurrences out of a total of n trials are possible. Note that X, the number of occurrences, is a discrete random variable, since it can take only integer values.

Example 2.6

Suppose the probability of failure of a structure due to earthquakes is estimated as 10^{-5} per year. Assuming that the design life of the structure is 50 years and the probability of failure in each year remains constant and independent during its lifetime, then the probability of no failure can be estimated using the binomial distribution as

$$P(\text{no failure in 50 years}) = P(X = 0, 50 \mid 10^{-5})$$

$$= \binom{50}{0}(10^{-5})^0(1 - 10^{-5})^{50-0} = \frac{50!}{0!(50 - 0)!}(1 - 10^{-5})^{50}$$

$$\approx 1 - 50 \times 10^{-5} = 0.99950$$

$$P(\text{failure in 50 years}) = 1 - P(\text{no failure in 50 years})$$

$$= 1 - 0.99950 = 0.00050$$

Example 2.7

Suppose 3 cars are available to go from from point A to point B at any time in an emergency. Assume that a car will be in good condition 90% of the time and in bad condition 10% of the time. The problem can be described by the binomial distribution. Since three cars are involved, $n = 3$. Also, the probability of each car being good is 0.9, or $p = 0.9$. The binomial coefficients when $X = 0, 1, 2$, and 3 can be shown to be 1, 3, 3, and 1, respectively, indicating the total number of sample points in each event. Thus, using Equation 2.25, we can calculate the PMF of X when it is 0, 1, 2, and 3 by taking the product of the probability of occurrence of one sample point in any event and the corresponding binomial coefficients, as shown below.

$$p_X(0) = P(X = 0) = 1 \times 0.1 \times 0.1 \times 0.1 \qquad = 0.001$$

$$p_X(1) = P(X = 1) = 3 \times 0.9 \times 0.1 \times 0.1 \qquad = 0.027$$

$$p_X(2) = P(X = 2) = 3 \times 0.9 \times 0.9 \times 0.1 \qquad = 0.243$$

$$p_X(3) = P(X = 3) = 1 \times 0.9 \times 0.9 \times 0.9 \qquad = \underline{0.729}$$

$$\Sigma\ 1.000$$

The CDF of $X = 2$, that is, $F_X(2)$, can be calculated as

$$P(X \leq 2) = F_X(2) = p_X(0) + p_X(1) + p_X(2) = 0.001 + 0.027 + 0.243$$
$$= 0.271$$

The PMFs of X when it is 0, 1, 2, and 3 are shown in Figure 2.2. Once the PMFs are available, the corresponding CDFs of X can be easily calculated, as shown in Figure 2.2.

Example 2.8

The drainage system of a city has been designed for a rainfall intensity that will be exceeded on an average once in 50 years. What is the probability that the city will be flooded only 2 out of 10 years?

Solution   Since the possible outcomes in each year consist of flooding or nonflooding, the problem can be modeled as a binomial distribution. In this case, the parameter p (the probability of flooding in one year) is $1/50 = 0.02$. Thus,

$$P(\text{flooding in 2 out of 10 years}) = P(X = 2, 10 \mid 0.02)$$

$$= \binom{10}{2}(0.02)^2(1 - 0.02)^{10-2}$$

$$= \frac{10!}{2!(10-2)!}(0.02)^2(0.98)^8 = 0.015$$

The probability of flooding in at most 2 years out of 10 years can be calculated as

$$P(X = 0, 10 \mid 0.02) + P(X = 1, 10 \mid 0.02) + P(X = 2, 10 \mid 0.02)$$

$$= \binom{10}{0}(0.02)^0(0.98)^{10} + \binom{10}{1}(0.02)^1(0.98)^9 + \binom{10}{2}(0.02)^2(0.98)^8$$

$$= 0.817 + 0.167 + 0.015 = 0.999$$

It is interesting that there is a probability of 0.817 that the city will not be flooded in any of 10 years. Similarly, the probability of no flood in 50 years can be shown to be $0.98^{50} = 0.364$, although on average the city is expected to be flooded once in 50 years. This will be elaborated upon in Section 2.3.3.

2.3.2 Geometric Distribution

The first occurrence time of an event is of great interest in engineering. Information on the first time the design wind speed will be exceeded in an area or the first time a structure will be damaged by earthquakes is important. If the events occur in a Bernoulli sequence and p is the probability of occurrence in each trial, then the probability that the event will occur for the first time at the ith trial, which implies that there was no occurrence in the previous $(t-1)$ trials, is given by the *geometric distribution* as

$$P(T = t) = p(1 - p)^{t-1} \qquad t = 1, 2, \ldots \qquad (2.26)$$

Considering Example 2.6, the probability of the failure of the structure in the 10th year can be calculated as

$$P(T = 10) = 10^{-5}(1 - 10^{-5})^{10-1} = 9.999 \times 10^{-6}$$

2.3.3 Return Period

The design wind speed, rainfall, or flood level at a particular location is usually expressed in terms of *return period*. Suppose an event occurs in a Bernoulli sequence, and it occurs for the first time after T_1 years. It occurs again T_2 years after the first occurrence, and again T_3 years after the second occurrence, and so on. The *recurrence time*, the time between two consecutive occurrences of the same event, must follow the probabilistic characteristics of the first occurrence, that is, the geometric distribution whose PMF is given by Equation 2.26. Thus, we can calculate the *mean* recurrence time, also known as the return period, as

$$\text{Return period } T = E(T) = \sum_{t=1}^{\infty} t p_T(t) = \sum_{t=1}^{\infty} t p (1 - p)^{t-1}$$

$$= p[1 + 2(1 - p) + 3(1 - p)^2 + 4(1 - p)^3 + \cdots] \qquad (2.27)$$

The terms in square brackets in Equation 2.27 represent an infinite series and can be shown to be $1/p^2$. Thus, Equation 2.27 can be simplified to

$$\text{Return period } T = p \times \frac{1}{p^2} = \frac{1}{p} \qquad (2.28)$$

Equation 2.28 states that if the design wind speed corresponds to a 50-year return period, then the probability in each year that the design wind speed will be exceeded is $1/50 = 0.02$; on average, the design wind speed will be exceeded once every 50 years. It must be noted that the design wind speed can be exceeded several times or not at all within the return period but on average will be exceeded once in 50 years.

In Example 2.8, the design flood level is considered to have a return period of 50 years, indicating that on average there will be a flood once every 50 years. However, there is a probability of 0.364 that no flood will occur in the next 50 years.

2.3.4 Poisson Distribution

Another important distribution used frequently in engineering to evaluate the risk of damage is the *Poisson distribution*. Defects can occur at any location along the length of welds. A tornado can strike a structure at any time during its lifetime. An accident can occur at any location along a highway. These events can occur at any point in time or space. If they need to be modeled in a Bernoulli sequence, that is, occurrence or nonoccurrence at a given time or space, the total space or time needs to be subdivided into very small intervals so that *only one* occurrence is possible in an interval. Suppose that the *mean occurrence rate of* tornadoes at a location is v times a year. Thus, over a period of t years, tornadoes will occur an *average of* vt times. If the time period t is divided into n intervals, then the probability of tornado occurrence in each interval will be vt/n. Modeling x occurrences in time t in a Bernoulli sequence as n approaches infinity will lead to the *Poisson distribution*, which can be expressed as

$$P(x \text{ occurrences in time } t) = \lim_{n \to \infty} \binom{n}{x} \left(\frac{vt}{n}\right)^x \left(1 - \frac{vt}{n}\right)^{n-x}$$

$$= \lim_{n \to \infty} \left[\frac{n!}{x!(n-x)!} \left(\frac{vt}{n}\right)^x \left(1 - \frac{vt}{n}\right)^{n-x} \right]$$

$$= \lim_{n \to \infty} \left[\frac{n}{n} \frac{n-1}{n} \cdots \frac{n-x+1}{n} \frac{(vt)^x}{x!} \left(1 - \frac{vt}{n}\right)^n \left(1 - \frac{vt}{n}\right)^{-x} \right]$$

$$= \lim_{n \to \infty} \left[\frac{(vt)^x}{x!} \left(1 - \frac{vt}{n}\right)^n \right]$$

Taking the limit of this equation, and knowing that

$$\lim_{n \to \infty} \left(1 - \frac{vt}{n}\right)^n = 1 - vt + \frac{(vt)^2}{2!} - \frac{(vt)^3}{3!} + \cdots = e^{-vt}$$

we can show that

$$P(x \text{ occurrences in time } t) = \frac{(vt)^x}{x!} e^{-vt} \tag{2.29}$$

Equation 2.29 represents the PMF of the Poisson distribution.

Example 2.9

From records of the past 50 years, it is observed that tornadoes occur in a particular area an average of two times a year. In this case, $v = 2/\text{year}$. The probability of no tornadoes in the next year (i.e., $x = 0$, and $t = 1$ year), can be calculated as

$$P(\text{no tornado next year}) = \frac{(2 \times 1)^0 e^{-2 \times 1}}{0!} = 0.135$$

$$P(\text{exactly 2 tornadoes next year}) = \frac{(2 \times 1)^2 e^{-2 \times 1}}{2!} = 0.271$$

$$P(\text{no tornado in next 50 years}) = \frac{(2 \times 50)^0 e^{-2 \times 50}}{0!} = 3.72 \times 10^{-44}$$

The results indicate that, for a tornado-prone area where an average of two tornadoes per year are expected, the probability of no tornadoes in the 50-year design life is very close to zero, essentially an impossible event. In other words, the probability of at least one tornado in the next 50 years will be very close to 1 ($= 1 - 3.72 \times 10^{-44}$), indicating that it is almost a certainty.

Example 2.10

For a large construction project, the contractor estimates that the average rate of on-the-job accidents is three times per year. From past experience, the contractor also estimates that the cost incurred for each accident may be modeled as a lognormal random variable with a median of $6000 and COV of 20%. The cost of each accident can be assumed to be statistically independent.

(a) What is the probability that there will be no accident in the first month of construction?

(b) What is the probability that only 1 out of the first 3 months of construction is free of accidents?

(c) What is the probability that an accident will incur a loss exceeding $4000?

(d) What is the probability that none of the accidents in a month will cost more than $4000?

Solution

(a) For the Poisson distribution, $v = 3$ times per year $= \frac{3}{12} = \frac{1}{4}$ time per month, $t = 1$ month, $vt = \frac{1}{4} \cdot 1 = \frac{1}{4}$, and no accident means $x = 0$. Thus, using Equation 2.29,

$$P(\text{no accident in the first month}) = \frac{e^{-1/4}\left(\frac{1}{4}\right)^0}{0!} = e^{-1/4} = 0.7788$$

(b) The binomial distribution needs to be used in this case. Thus, using Equation 2.25,

$$P(X = 1, 3 \mid 0.7788) = \binom{3}{1}(0.7788)^1(1 - 0.7788)^{3-1} = 0.1143$$

(c) The cost incurred for each accident is modeled as a lognormal distribution. In this case, using Equation 2.18, $\delta \approx \zeta = 0.2$, and $\lambda = \ln(\text{median}) = \ln 6000 = 8.70$. Thus,

$$P(\text{cost of an accident} > \$4000) = \Phi\left(\frac{\ln \infty - 8.7}{0.20}\right) - \Phi\left(\frac{\ln 4000 - 8.7}{0.20}\right)$$
$$= 1 - \Phi(-2.027) = 1 - 0.0213 = 0.9787$$

(d) This can be solved by considering that there could be n number of accidents in a month; n could be any number, and no accident should exceed a cost of $4000. From part (c),

$$P(\text{an accident will cost less than } \$4000) = 1 - 0.9787 = 0.0213$$

Thus,

$P(\text{none of the accidents in a month will cost more than } \$4000)$

$$= \sum_{n=0}^{\infty} P(\text{cost of an accident} \leq \$4000 \mid X = n)P(X = n)$$

$$= \sum_{n=0}^{\infty} (0.0213)^n \cdot \frac{e^{-1/4} \left(\frac{1}{4}\right)^n}{n!}$$

$$= e^{-1/4} \left(\sum_{n=0}^{\infty} \frac{\left[(0.0213)\left(\frac{1}{4}\right)\right]^n}{n!} \right) = e^{-1/4} \cdot e^{[(0.0213)(1/4)]} = e^{-0.2447} = 0.78296$$

Note that the infinite series in the first bracket in the above equation is an exponential series: that is

$$e^x = 1 + \frac{x}{1!} + \frac{x^2}{2!} + \frac{x^3}{3!} + \cdots .$$

If this series is not obvious, then the first few terms of the series (maybe 3 or 4) can be considered to calculate the probability.

Example 2.11

The safety of a building in an earthquake-prone area is under consideration. The past 100 years of data indicate that there were four strong earthquakes in the area. Also, a detailed evaluation indicates that during a strong earthquake, the probability that the building will suffer damage is 0.10. Assume that the probability of damage for each earthquake is statistically independent.

(a) What is the probability that there will be no strong earthquake in the area in 50 years, the service life of the building?

(b) What is the probability that there will be only two strong earthquakes in 50 years?

(c) What is the probability that the building will suffer damage due to strong earthquakes in 50 years?

Solution

(a) In this case, the average rate of strong earthquake occurrences, v, is $4/100 = 0.04$ per year. Thus, $vt = 0.04 \times 50 = 2$.

$$P(\text{no strong earthquake in 50 years}) = P(X = 0)$$

$$= \frac{e^{-2} \times (2)^0}{0!} = 0.13534$$

(b) $P(\text{two strong earthquakes in 50 years}) = P(X = 2)$

$$= \frac{e^{-2} \times (2)^2}{2!} = 0.27067$$

(c) Let D denote the event that the building will suffer earthquake damage in 50 years. Then,

$$P(D) = 1.0 - P(\bar{D})$$

$$= 1.0 - \sum_{n=0}^{\infty} P(\bar{D} \mid X = n) P(X = n)$$

$$= 1.0 - \sum_{n=0}^{\infty} (1.0 - 0.1)^n \frac{e^{-2} \times (2)^n}{n!}$$

$$= 1.0 - e^{-2} \sum_{n=0}^{\infty} \frac{(0.9 \times 2)^n}{n!} = 1.0 - e^{-2} e^{1.8} = 1.0 - e^{-0.2} = 0.18127$$

Again, the infinite series in the above equation is an exponential series, as discussed in the previous example.

2.3.5 Exponential Distribution

If events occur according to a Poisson process, then the time T before the first occurrence of the event implying no occurrence ($x = 0$) in time t can be represented by the *exponential distribution*. It can be shown that

$$P(T > t) = \frac{e^{-\nu t}(\nu t)^0}{0!} = e^{-\nu t} \tag{2.30}$$

Thus, the CDF of T can be shown to be

$$F_T(t) = P(T \le t) = 1 - e^{-\nu t} \tag{2.31}$$

and the corresponding PDF of the exponential distribution is

$$f_T(t) = \frac{dF_T(t)}{dt} = \nu e^{-\nu t} \quad t \ge 0 \tag{2.32}$$

The mean value of T can be shown to be $1/\nu$. Simply stated, the mean of the first occurrence time or the recurrence time, or simply the return period for the Poisson model, is $1/\nu$. Note that when events occur in a Bernoulli sequence, the return period is $1/p$ (Equation 2.28 in Section 2.3.3). It can be shown that for events with small occurrence rate ν, the return periods according to the Bernoulli sequence and the Poisson model are approximately the same.

Example 2.12

In the previous example (Example 2.11), strong earthquakes in an area are assumed to occur according to the Poisson distribution with the average rate of occurrences $\nu = 0.04$ per year. Then, the recurrence time T, the time between two consecutive occurrences of strong earthquakes, can be modeled by an exponential distribution as

$$f_T(t) = 0.04e^{-0.04t} \quad t \ge 0$$

The mean recurrence time or the return period of strong earthquakes can be shown to be

$$\text{Return period } T = \int_0^\infty t \times 0.04 \times e^{-0.04t}\, dt = \frac{1}{0.04} = 25 \text{ years}$$

The probability of no strong earthquakes in 50 years can be calculated as

$$P(T > 50) = \int_{50}^\infty 0.04 \times e^{-0.04t}\, dt = e^{0.04 \times 50} = 0.13534$$

The same result was obtained when the occurrences of strong earthquakes were assumed to follow a Poisson distribution, as shown in the previous section.

Example 2.13

In earthquake engineering, the PDF for earthquake intensities, for example, in Modified Mercalli (MM) scale, is sometimes modeled by an exponential distribution. The parameter v is determined from local seismicity records.

In earthquake-resistant design of nuclear power plants, unserviceability and collapse due to earthquakes are the two most important concerns for engineers. The corresponding earthquake intensities are known in the profession as the operating basis earthquake (OBE) and the safe shutdown earthquake (SSE), respectively. One way to design for these incidents is to choose a design intensity x_i such that the probability that this intensity level is exceeded, that is, $P(X > x_i) = p$, is small. Since the collapse of a nuclear power plant presents a great hazard to the public, the chance of its occurrence should be extremely small. Suppose a design intensity x_1 corresponding to a risk level of 10^{-3} is chosen for the OBE, and x_2 corresponding to a risk level of 10^{-6} is chosen for the SSE.

(a) Determine x_2 (SSE intensity) in terms of x_1 (OBE intensity).
(b) If power plant service is interrupted during an earthquake, what is the probability that the plant will collapse?

Solution

(a) Using Equation 2.30, we can sumarize the information in the problem as

$$P(X > x_1) = 10^{-3} \quad \text{and} \quad P(X > x_2) = 10^{-6}$$

Thus,

$$e^{-vx_1} = 10^{-3} \quad \text{and} \quad e^{-vx_2} = 10^{-6}$$

By simplifying, we get

$$x_1 = \frac{6.908}{v} \quad \text{and} \quad x_2 = \frac{13.816}{v}$$

Or, $x_2 = 2.0x_1$. In this particular example, the SSE intensity is twice the OBE intensity.

(b) P(plant will collapse | service has been interrupted)

$$= P(X > x_2 \mid X > x_1) = \frac{P[(X > x_2)(X > x_1)]}{P(X > x_1)}$$

$$= \frac{P(X > x_2)}{P(X > x_1)} = \frac{10^{-6}}{10^{-3}} = 10^{-3}$$

Example 2.14

The rate of oxygen consumption, D, caused by wastes discharged into a river, expressed in terms of biological oxygen demand (BOD), depends on the remaining BOD concentration. Suppose D can be described by an exponential distribution.

(a) If the mean value of D is found to be 6mg/m^3 d, define its PDF.

(b) What is the probability that D will be less than or equal to 4mg/m^3 d?

Solution

(a) One attractive property of an exponential distribution is that its parameter ν, in Equation 2.32, is the reciprocal of the mean or standard deviation. Thus, for the problem under consideration, $\nu = \frac{1}{6}$. The PDF of D can be shown to be

$$f_D(d) = \tfrac{1}{6} e^{-d/6}$$

(b) $P(D \le 4) = \displaystyle\int_0^4 \tfrac{1}{6} e^{-d/6} dd = \left[-e^{-d/6} \right]_0^4 = -0.51342 + 1.0 = 0.48658$

2.4 EXTREME VALUE DISTRIBUTIONS

In many engineering applications, the extreme values of random variables are of special importance. The largest or smallest values of random variables may dictate a particular design. Wind speeds are recorded continuously at airports and weather stations. Obviously, the voluminous information collected cannot be used directly in engineering designs. The maximum wind speeds per hour, day, month, year, or other period can be used for this purpose. Usually, the information on yearly maximum wind speed is used in the engineering profession. Thus, for every year of recorded data, the maximum wind speed is noted. If data are collected for several years, the design wind speed can be established statistically to ensure that it will not be exceeded within the design life of the structure with a specific probability level. If the design wind speed has a 50-year return period, then the probability that the wind speed will exceed the design value in a year is $\frac{1}{50} = 0.02$. Design earthquake loads, flood levels, and so forth are also determined in this way. In all these cases, the peak or maximum value of a random variable during certain intervals is of interest. In some cases, the minimum value of a random variable is also of interest for design applications. For example, when a large number of identical devices are manufactured, such as calculators or cars, their minimum service lives are of great interest to consumers. Some of them could be subjected to accelerated testing to determine their life, and the probability distribution of life could be constructed. Then the minimum service life could be established so that it does not fall below an acceptable number with a predetermined probability level. Therefore, extreme value statistics have received a lot of attention for engineering design applications.

In constructing an extreme value distribution, an underlying random variable with a particular distribution is necessary. If different sets of samples are obtained (through physical or numerical experimentation), one can select the extreme values from each sample set— either the maximum or the minimum values—and then construct a different distribution for the extreme values. Therefore, the underlying distribution of a variable governs the form of the corresponding extreme value distribution.

The detailed mathematical aspects of extreme value distributions can be found elsewhere (Gumbel, 1958; Castillo, 1988). Only the essential concept, emphasizing engineering applications, is presented very briefly here.

2.4.1   Concept of Extreme Value Distributions

Let X be a random variable with some known distribution function. If there are n samples from the population X, then the extreme values of the sample, such as the minimum value Y_1 or the maximum value Y_n, may be of interest. Y_1 and Y_n can be defined as

$$Y_n = \max(X_1, X_2, \ldots, X_n) \tag{2.33}$$

$$Y_1 = \min(X_1, X_2, \ldots, X_n) \tag{2.34}$$

If different sets of samples of the same size n are obtained for X, each set will have different minimum and maximum values. Using all these sets, distribution functions for the minimum and maximum values can be constructed. The cumulative distribution function (CDF) of the largest value Y_n can be derived as

$$F_{Y_n}(y) = P(Y_n \leq y) = P(X_1 \leq y, X_2 \leq y, \ldots, X_n \leq y) \tag{2.35}$$

For identically distributed and statistically independent X_i's, Equation 2.35 becomes

$$F_{Y_n}(y) = [F_X(y)]^n \tag{2.36}$$

Similarly, the CDF of the smallest value Y_1 can be derived as

$$P(Y_1 > y) = P(X_1 > y, X_2 > y, \ldots, X_n > y) = 1 - F_{Y_1}(y) \tag{2.37}$$

Again, for identically distributed and statistically independent X_i's, Equation 2.37 becomes

$$F_{Y_1}(y) = 1 - [1 - F_X(y)]^n \tag{2.38}$$

The basic idea of these derivations is that if the largest value Y_n is less than some quantity y, then all the sample values—X_1, X_2, etc.—should also be less than y; similarly, for the smallest value Y_1, if Y_1 is greater than some quantity y, then all of the samples X_1, X_2 up to X_n should be greater than y.

Example 2.15

Suppose 10 cracks are detected in a beam in a bridge deck segment. For the purpose of illustration, assume the crack sizes are normally distributed with a mean value of 0.5

inch and a COV of 0.1. What is the probability that the maximum crack size is less than 0.6 inch?

This is an extreme (maximum) value problem. The distribution of the largest value and therefore the probability of the largest crack being less than 0.6 inch can be calculated as discussed below.

Let X be the random variable denoting the size of a crack. From the data, X is a normal random variable with $\mu_X = 0.5$ in and $\delta_X = 0.1$. Using Equation 2.33, we can express the CDF of $Y_{10} =$ largest value among 10 samples of X as

$$F_{Y_{10}}(y) = P(Y_{10} \leq y) = [P(X \leq y)]^{10}$$

The probability that the size of any crack is less than 0.6 inch is

$$P(X \leq 0.6) = \Phi\left(\frac{0.6 - 0.5}{0.05}\right) = \Phi(2.0) = 0.9772$$

Therefore, the probability that the maximum crack size is less than 0.6 inch is

$$P(Y_{10} \leq 0.6) = [P(X \leq 0.6]^{10} = 0.794$$

2.4.2 Asymptotic Distributions

In Equations 2.35 to 2.38, as the sample size n grows larger and approaches infinity, the distribution of the largest or the smallest values may asymptotically approach a mathematical distribution function in some cases if the samples are identically distributed and statistically independent. Some of these asymptotic distributions have a wide range of applications in engineering problems. Gumbel (1958) classified three types of asymptotic extreme value distributions for both minima and maxima, labeling them as Type I, Type II, and Type III extreme value distributions. The Type I extreme value distribution of the largest value is also referred to as EVD (extreme value distribution) in mechanical reliability engineering applications. The distribution of maxima in sample sets from a population with a normal distribution will asymptotically converge to this distribution. This distribution is used to model environmental phenomena such as wind loads and flood levels. The Type II extreme value distribution of the largest value is also used to model extreme environmental phenomena such as earthquake loads and may result from sample sets from a lognormal distribution. The Type III extreme value distribution, which is referred to as the Weibull distribution in the case of the smallest value, may be obtained by the convergence of most of the commonly known distributions that have a lower bound. It is commonly used to describe material strengths and time to failure of electronic and mechanical devices and components. These distributions are discussed in the following sections.

Extreme value distributions are treated no differently than any other distributions discussed earlier. An extreme value distribution can be uniquely defined in terms of its PDF or CDF and the parameters of the distributions. In most cases, the parameters can be estimated from the information on the mean, variance, or coefficient of variation of the random variable. Once an extreme value distribution is uniquely defined, probabilistic information can be extracted from it using the procedure discussed in the previous sections. In the following sections, some of the commonly used extreme value distributions are discussed, emphasizing their basic definition.

2.4.3 The Type I Extreme Value Distribution

The CDF of the Type I asymptotic form of the distribution of the largest value, also referred to as the Gumbel distribution or simply the EVD, can be expressed as

$$F_{Y_n}(y_n) = \exp[-e^{-\alpha_n(y_n - u_n)}] \tag{2.39}$$

where u_n is the characteristic largest value of the initial variable X, and α_n is an inverse measure of dispersion of the largest value of X. The corresponding probability density function PDF can be shown to be

$$f_{Y_n}(y_n) = \alpha_n e^{-\alpha_n(y_n - u_n)} \exp\left[-e^{-\alpha_n(y_n - u_n)}\right] \qquad -\infty < y_n < +\infty \tag{2.40}$$

The parameters u_n and α_n are related to the mean and standard deviation of the extreme value variable Y_n as

$$\alpha_n = \frac{1}{\sqrt{6}}\left(\frac{\pi}{\sigma_{Y_n}}\right) \tag{2.41a}$$

and

$$u_n = \mu_{Y_n} - \frac{0.5772}{\alpha_n} \tag{2.41b}$$

As mentioned earlier, the Type I extreme value distribution for the largest value is commonly used for modeling environmental loads such as winds and floods. Another use of the Type I extreme value distribution for maxima is in aircraft design, where the peak gust velocity experienced by an aircraft during every 1000 hours of operation is considered to be a Type I largest variable. Another common example is the maximum water level in a year at a specified location in a river. This is an important variable in the design of flood control, water supply, and irrigation systems.

For the smallest value of an initial variable X, the corresponding Type I asymptotic form for the CDF is

$$F_{Y_1}(y_1) = 1 - \exp\left[-e^{\alpha_1(y_1 - u_1)}\right] \tag{2.42a}$$

The corresponding PDF is

$$f_{Y_1}(y_1) = \alpha_1 e^{\alpha_1(y_1 - u_1)} \exp\left[-e^{\alpha_1(y_1 - u_1)}\right] \qquad -\infty < y_1 < +\infty \tag{2.42b}$$

In these equations, the parameters are defined as u_1, which is the characteristic smallest value of the initial variable X, and α_1, which is an inverse measure of dispersion of the smallest value of X.

u_1 and α_1 are related to the mean and standard deviation of Y_1 as

$$\alpha_1 = \frac{1}{\sqrt{6}}\left(\frac{\pi}{\sigma_{Y_1}}\right), \text{ and } u_1 = \mu_{Y_1} + \frac{0.5772}{\alpha_1} \tag{2.43}$$

In general, the Type I asymptotic form is obtained by the convergence of distributions with an exponential tail. For example, the PDF of the Gaussian distribution has an exponential decaying term and therefore an exponential tail in the extreme directions. The extreme values of a variable with a Gaussian distribution will have a Type I distribution.

Example 2.16

The data on maximum annual wind velocity V at a site have been compiled for n years, and its mean and standard deviation are estimated to be 61.3 and 7.52 mph, respectively. Assuming that V_n has a Type I extreme value distribution, what is the probability that the maximum wind velocity will exceed 100 mph in any given year?

Solution   Equations 2.41a and 2.41b can be used to calculate the two parameters u_n and α_n of the Type I extreme value distribution as

$$\alpha_n = \frac{1}{\sqrt{6}}\left(\frac{\pi}{\sigma_{Y_n}}\right) = 0.17055 \quad \text{and} \quad u_n = \mu_{Y_n} - \frac{0.5772}{\alpha_n} = 57.9157$$

Therefore, from Equation 2.39, the probability that the maximum wind velocity is greater than 100 mph is

$$P(Y_n > 100) = 1 - F_{Y_n}(100) = 1 - \exp\left[-e^{-0.17055(100-57.9157)}\right] = 0.000763$$

This type of analysis may also be used to calculate design values for various types of engineering applications. Suppose the design wind speed with a return period of 100 years needs to be estimated for a particular site. With V_d denoted as the design wind speed to be estimated, the probability that it will be exceeded in a given year is $\frac{1}{100} = 0.01$. Thus,

$$P(Y_n > V_d) = 1 - F_{Y_n}(V_d) = 0.01$$

or

$$1 - \exp\left[-e^{-0.17055(V_d-57.9157)}\right] = 0.01$$

or $V_d = 84.89$ mph.

2.4.4   The Type II Extreme Value Distribution

The CDF of the Type II asymptotic form for the largest value, also referred to as the Fréchet distribution, can be shown to be

$$F_{Y_n}(y_n) = \exp\left[-\left(\frac{v_n}{y_n}\right)^k\right] \tag{2.44}$$

The corresponding PDF is

$$f_{Y_n}(y_n) = \frac{k}{v_n}\left(\frac{v_n}{y_n}\right)^{k+1}\exp\left[-\left(\frac{v_n}{y_n}\right)^k\right] \qquad y_n \geq 0,\ k > 2 \tag{2.45}$$

where v_n and k are the parameters of the distribution; v_n is the characteristic largest value of the underlying variable X; and k, the shape parameter, is a measure of dispersion.

The Type II asymptotic form is obtained as n goes to infinity from an initial distribution that has a polynomial tail in the direction of the extreme value. Note the difference between this and Type I, which converges from an exponential tail. The Type II distribution requires a polynomial tail, and therefore a lognormal distribution converges to a Type II asymptotic form for the largest value.

The relationship between the Type I and Type II forms is also interesting. In Sections 2.2.1 and 2.2.2 we saw that the normal and lognormal distributions are related to each other; that is, if a variable has a lognormal distribution, the natural log of that variable has a normal distribution. The Type I and Type II extreme value distributions may be obtained through the asymptotic convergence of these two initial distributions. Therefore, if Y_n has a Type II asymptotic distribution with parameters v_n and k, then $\ln Y_n$ will have a Type I asymptotic form with parameters $u_n = \ln v_n$ and $\alpha_n = k$.

For the Type II distribution of maxima, the mean, standard deviation, and COV of Y_n are related to the distribution parameters v_n and k as follows:

$$\mu_{Y_n} = v_n \Gamma \left(1 - \frac{1}{k} \right) \qquad k > 1 \tag{2.46a}$$

$$\sigma_{Y_n}^2 = v_n^2 \left[\Gamma \left(1 - \frac{2}{k} \right) - \Gamma^2 \left(1 - \frac{1}{k} \right) \right] \qquad k > 2 \tag{2.46b}$$

and

$$1 + \delta_{Y_n}^2 = \frac{\Gamma \left(1 - \frac{2}{k} \right)}{\Gamma^2 \left(1 - \frac{1}{k} \right)} \qquad k > 2 \tag{2.46c}$$

In these equations, Γ is the gamma function and can be estimated by Appendix 2. A plot of $(1 + \delta_{Y_n}^2)$ versus $1/k$ is given in Figure 2.7. This figure can be helpful in solving problems, as will be shown later with examples.

The CDF and PDF of the Type II asymptotic form of the smallest value can be shown to be

$$F_{Y_1}(y_1) = 1 - \exp \left[- \left(\frac{v_1}{y_1} \right)^k \right] \tag{2.47}$$

and

$$f_{Y_1}(y_1) = -\frac{k}{v_1} \left(\frac{v_1}{y_1} \right)^{k+1} \exp \left[- \left(\frac{v_1}{y_1} \right)^k \right] \qquad y_1 < 0 \tag{2.48}$$

where the parameter v_1 is the characteristic smallest value of the initial variable X and k is the shape parameter, an inverse measure of dispersion. The comment regarding the relationship between Type I and Type II asymptotic forms of the largest value also holds for the asymptotic forms of the smallest value. Therefore, if Y_1 has a Type II asymptotic distribution with positive parameters v_1 and k, then $\ln Y_1$ has a Type I asymptotic distribution with parameters $u_1 = \ln v_1$ and $\alpha_1 = k$.

For the Type II distribution of minima, the mean, standard deviation, and COV of Y_1 are related to the distribution parameters v_1 and k as follows:

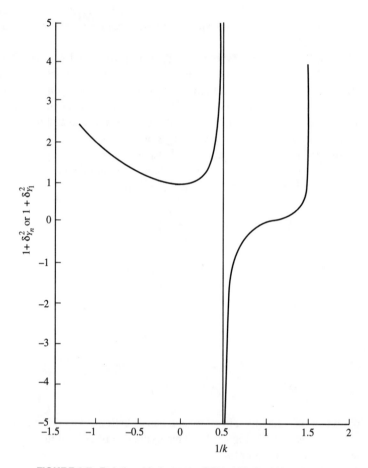

FIGURE 2.7. Relationship between COV of Y_n (or Y_1) and $1/k$.

$$\mu_{Y_1} = v_1 \Gamma \left(1 - \frac{1}{k} \right) \qquad k > 1 \tag{2.49a}$$

$$\sigma_{Y_1}^2 = v_1^2 \left[\Gamma \left(1 - \frac{2}{k} \right) - \Gamma^2 \left(1 - \frac{1}{k} \right) \right] \qquad k > 2 \tag{2.49b}$$

and

$$1 + \delta_{Y_1}^2 = \frac{\Gamma \left(1 - \frac{2}{k} \right)}{\Gamma^2 \left(1 - \frac{1}{k} \right)} \qquad k > 2 \tag{2.49c}$$

By comparing Equations 2.46c and 2.49c, we can observe that the relationship between the COV of Y_n and k is identical to the relationship between the COV of Y_1 and k. Thus, Figure 2.7 is also applicable for the Type II distribution of minima.

Example 2.17

Suppose the example on the annual maximum wind velocity considered in Example 2.16 is to be modeled using a Type II extreme value distribution of maxima. What is the probability that the maximum wind velocity will exceed 100 mph in any given year?

Solution   The parameters of the Type II distribution have to be determined first, using Equations 2.46a and 2.46b. From the previous example, the mean and the standard deviation of maximum wind velocity are 61.3 and 7.52 mph, respectively. Therefore, $\delta_{Y_n} = 7.52/61.3 = 0.123$. From Equation 2.46c,

$$1 + (0.123)^2 = 1.015 = \frac{\Gamma\left(1 - \frac{2}{k}\right)}{\Gamma^2\left(1 - \frac{1}{k}\right)}$$

Referring to Figure 2.7, and considering k is greater than 2, we can estimate k to be 10. From Equation 4.46a,

$$61.3 = v_n \Gamma\left(1 - \frac{1}{k}\right) = v_n \Gamma(0.9)$$

Using Appendix 2, we find $\Gamma(0.9)$ to be 1.0686. Thus,

$$v_n = \frac{61.3}{1.0686} = 57.36$$

Therefore, using Equation 4.44, we find the probability that the maximum wind velocity will exceed 100 mph in any given year to be

$$P(Y_n > 100) = 1 - \exp\left[-\left(\frac{57.36}{100}\right)^{10}\right] = 1 - 0.99615 = 0.00385$$

2.4.5   The Type III Extreme Value Distribution

Both the Type I and Type II asymptotic distributions are limiting forms of the distribution of extreme values from initial distributions that are unbounded in the direction of the extreme value. In contrast, the Type III asymptotic form represents a limiting distribution of the extreme values from initial distributions that have a finite upper or lower bound value.

For the largest value, the Type III asymptotic CDF can be written as

$$F_{Y_n}(y_n) = \exp\left[-\left(\frac{\omega - y_n}{\omega - w_n}\right)^k\right] \tag{2.50}$$

The corresponding PDF is

$$f_{Y_n}(y_n) = \frac{k}{\omega - w_n}\left(\frac{\omega - y_n}{\omega - w_n}\right)^{k-1} \exp\left[-\left(\frac{\omega - y_n}{\omega - w_n}\right)^k\right] \qquad y_n \leq \omega \tag{2.51}$$

where ω is the upper bound of the initial distribution, that is, $F_X(\omega) = 1.0$, and w_n and k are the parameters of the distribution. w_n is the characteristic largest value of X, and is defined by:

$$F_X(w_n) = 1 - \frac{1}{n} \tag{2.52}$$

and k is a shape parameter.

The mean and variance of Y_n are related to the parameters w_n and k as follows:

$$\mu_{Y_n} = \omega - (\omega - w_n)\Gamma\left(1 + \frac{1}{k}\right) \tag{2.53a}$$

and

$$\sigma_{Y_n}^2 = \text{Var}(\omega - Y_n) = (\omega - w_n)^2\left[\Gamma\left(1 + \frac{2}{k}\right) - \Gamma^2\left(1 + \frac{1}{k}\right)\right] \tag{2.53b}$$

Equations 2.53a and 2.53b can be used to show that

$$1 + \left(\frac{\sigma_{Y_n}}{\omega - \mu_{Y_n}}\right)^2 = \frac{\Gamma\left(1 + \frac{2}{k}\right)}{\Gamma^2\left(1 + \frac{1}{k}\right)} \tag{2.53c}$$

The relationship is shown graphically in Figure 2.8.

The CDF of the Type III asymptotic distribution of the smallest value from an initial distribution with a lower limit ϵ, that is, $F_X(\epsilon) = 0$, is

$$F_{Y_1}(y_1) = 1 - \exp\left[-\left(\frac{y_1 - \epsilon}{w_1 - \epsilon}\right)^k\right] \tag{2.54}$$

The corresponding PDF is

$$f_{Y_1}(y_1) = \frac{k}{w_1 - \epsilon}\left(\frac{y_1 - \epsilon}{w_1 - \epsilon}\right)^{k-1}\exp\left[-\left(\frac{y_1 - \epsilon}{w_1 - \epsilon}\right)^k\right] \qquad y_1 \geq \epsilon \tag{2.55}$$

where w_1 and k are the parameters of the distribution. w_1 is the characteristic smallest value, defined as

$$F_X(w_1) = \frac{1}{n} \tag{2.56}$$

and k is the shape parameter.

The mean and variance of Y_1 are related to the parameters w_1 and k as follows:

$$\mu_{Y_1} = \epsilon + (w_1 - \epsilon)\Gamma\left(1 + \frac{1}{k}\right) \tag{2.57a}$$

and

$$\sigma_{Y_1}^2 = \text{Var}(Y_1 - \epsilon) = (w_1 - \epsilon)^2\left[\Gamma\left(1 + \frac{2}{k}\right) - \Gamma^2\left(1 + \frac{1}{k}\right)\right] \tag{2.57b}$$

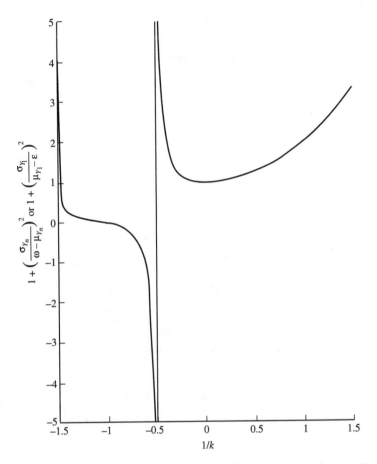

FIGURE 2.8. Relationship beteen the mean and variance of Y_n (or Y_1) and $1/k$.

Equations 2.57a and 2.57b can be used to show that

$$1 + \left(\frac{\sigma_{Y_1}}{\mu_{Y_1} - \epsilon} \right)^2 = \frac{\Gamma \left(1 + \frac{2}{k} \right)}{\Gamma^2 \left(1 + \frac{1}{k} \right)} \tag{2.57c}$$

Comparing Equations 2.53c and 2.57c shows that Figure 2.8 is also appropriate to represent Equation 2.57c.

Example 2.18

A number of L-shaped structural steel sections are rolled in a steel mill to be used as members in bridge trusses. The distribution of their minimum axial load capacity is of interest. Suppose the minimum load capacity can be modeled with a Type III distribution of the smallest value, with a mean value of 300 kips and a COV of 0.15. Also, assume that the load capacity has a lower bound of 100 kips. Determine the probability that the minimum load capacity will be less than 200 kips.

Solution   Using $\epsilon = 100$, we can calculate the left-hand side of Equation 2.57c as

$$1 + \left(\frac{\sigma_{Y_1}}{\mu_{Y_1} - \epsilon}\right)^2 = 1 + \left(\frac{0.15 \times 300}{300 - 100}\right)^2 = 1.0506$$

Using Figure 2.7, $1/k = 0.20$, or $k = 5.0$. Using Equation 2.57a,

$$300 = 100 + (w_1 - 100)\Gamma\left(1 + \frac{1}{5.0}\right)$$

or

$$200 = (w_1 - 100)\Gamma(1.20)$$

or

$$w_1 = 100 + \frac{200}{\Gamma(1.20)} = 100 + \frac{200}{0.9182} = 317.82$$

Note that $\Gamma(1.20) = 0.9182$ is calculated using Appendix 2. Therefore, Equation 2.54 can be used to calculate the probability that the minimum load capacity is less than 200 kips:

$$P(Y_1 \leq 200) = F_{Y_1}(200) = 1 - \exp\left[-\left(\frac{200 - 100}{317.82 - 100}\right)^{5.0}\right] = 0.0202$$

The Type III asymptotic distribution of the smallest value, developed by Weibull in connection with the study of fatigue and fracture of materials, is known as the Weibull distribution. Equation 2.54 is a three-parameter representation of the Weibull distribution. A two-parameter form of this distribution is commonly used in mechanical and electronic component life estimation. It is obtained by setting ϵ to be zero in Equation 2.54, reflecting the physical nature of the problem.

The CDF of the two-parameter Weibull distribution is

$$F_{Y_1}(y_1) = 1 - \exp\left[-\left(\frac{y_1}{w_1}\right)^k\right] \tag{2.58}$$

The corresponding PDF is

$$f_{Y_1}(y_1) = \frac{k}{w_1}\left(\frac{y_1}{w_1}\right)^{k-1}\exp\left[-\left(\frac{y_1}{w_1}\right)^k\right] \qquad y_1 \geq 0 \tag{2.59}$$

In these two equations, k and w_1 have to be positive values.

For the two-parameter Weibull distribution, the mean value and the coefficient of variation are related to the parameters k and w_1 as follows:

$$\mu_{Y_1} = w_1 \Gamma \left(1 + \frac{1}{k} \right) \tag{2.60a}$$

and

$$\delta_{Y_1} = \left[\frac{\Gamma \left(1 + \frac{2}{k} \right)}{\Gamma^2 \left(1 + \frac{1}{k} \right)} - 1 \right]^{1/2} \tag{2.60b}$$

where $\Gamma(\ )$ is the gamma function. If the mean and coefficient of variation are known, the following approximation can be used to compute the parameters k and w_1 for practical applications:

$$k = \delta_{Y_1}^{-1.08} \tag{2.61a}$$

and

$$w_1 = \frac{\mu_{Y_1}}{\Gamma \left(1 + \frac{1}{k} \right)} \tag{2.61b}$$

2.4.6   Special Cases of Two-Parameter Weibull Distribution

Two special cases of the two-parameter Weibull distribution used widely in engineering are the exponential distribution and the Rayleigh distribution. In Equation 2.58, if the random variable Y_1 is denoted by another random variable X, and $k = 1$ and $1/w_1 = \nu$, it results in an exponential distribution. The CDF and the PDF of an exponential distribution are given by Equations 2.31 and 2.32, respectively. Note that ν is the parameter of the exponential distribution and is the reciprocal of the mean or the standard deviation of the random variable X. The exponential distribution is commonly used in the reliability analysis of electronic and mechanical devices. In Equations 2.31 and 2.32, if time to failure T of a unit has an exponential distribution, then ν is the failure occurrence rate.

Similarly, a two-parameter Weibull distribution with $k = 2$ and $w_1 = \sqrt{2}\alpha$ results in a Rayleigh distribution. For example, wave heights are modeled with a Rayleigh distribution. The PDF of the Rayleigh distribution is

$$f_X(x) = \frac{x}{\alpha^2} \exp \left[-\frac{1}{2} \left(\frac{x}{\alpha} \right)^2 \right] \qquad x \geq 0 \tag{2.62}$$

Its parameter α can be estimated from the information on the mean and variance of X, that is, $E(X) = \sqrt{\frac{\pi}{2}}\alpha$ and $\mathrm{Var}(X) = \left(2 - \frac{\pi}{2} \right) \alpha^2$ (Haldar and Mahadevan, 2000). An example of the use of this distribution is in the description of the peaks in a narrow-band stationary Gaussian random process.

2.5   OTHER USEFUL DISTRIBUTIONS

In the previous sections, some of the commonly used continuous and discrete random variables were identified and the calculation of probability for each was described. Several other standard distributions are available for engineering applications. The exact forms of their PDF can be determined by referring to the literature. Once their PDFs have been defined uniquely, their probability estimation procedures are similar to the distributions discussed here.

The selection of one distribution over the others and the selection of parameters to describe a distribution uniquely is discussed elsewhere by Haldar and Mahadevan (2000). However, in many engineering problems, there is not enough information available to justify the use of a particular standard distribution. Based on limited experience, an engineer may have some idea of the lower and upper limits of a random variable, but there may not be enough data available between these two limits to justify a specific distribution. In this situation, any distribution can be used, such as a uniform distribution or one of the many different forms of triangular or trapezoidal distributions shown in Table 2.2. The parameters for these nonstandard distributions cannot be calculated in terms of mean and standard deviation based on sample information, because it is not available. However, they can be calculated from the assumed shape of the distribution. The mean of a random variable rep-

Table 2.2. Mean and coefficient of variation of X corresponding to different distributions assumed over its range

Case No.	Distribution	Mean $\overline{X}$	COV of X
1	x_l x_u	$\frac{1}{3}(2x_\ell + x_u)$	$\frac{1}{\sqrt{2}}\dfrac{x_u - x_\ell}{2x_\ell + x_u}$
2	x_l x_u	$\frac{1}{3}(x_\ell + 2x_u)$	$\frac{1}{\sqrt{2}}\dfrac{x_u - x_\ell}{x_\ell + 2x_u}$
3	x_l x_u	$\frac{1}{2}(x_\ell + x_u)$	$\frac{1}{\sqrt{6}}\dfrac{x_u - x_\ell}{x_u + x_\ell}$
4	x_l x_u	$\frac{1}{2}(x_\ell + x_u)$	$\frac{2}{\sqrt{12}}\dfrac{x_u - x_\ell}{x_u + x_\ell}$
5	x_l x_n x_u	$\frac{1}{3}(x_\ell + x_u + x_n)$	$\frac{1}{\sqrt{2}}\dfrac{\sqrt{x_\ell^2+x_u^2+x_n^2-x_\ell x_u-x_\ell x_n-x_u x_n}}{(x_\ell+x_u+x_n)}$

resents the centroidal distance, and the variance is the first moment of inertia of the area about the centroidal axis. For an assumed shape, these values can be easily calculated and are shown in Table 2.2.

2.6 CONCLUDING REMARKS

Modeling and quantifying uncertainty in a random variable is an important step in risk or reliability evaluation. We briefly introduced the concept in this chapter assuming that the readers have some background in this area. Otherwise, the readers are strongly encouraged to review the materials. Another book authored by Haldar and Mahadevan (2000) can be used for this purpose.

Some commonly used distributions, both continuous and discrete, and procedures to calculate the probability of events using them are discussed. The use of these distributions to solve some practical problems is presented to show their implementation potential.

In many engineering applications, the largest or smallest values of random variables may dictate a particular design. To model them, some of the commonly used extreme value distributions are discussed.

The information presented here is expected to provide sufficient background to calculate the probability of events using commonly used distributions. Available computer programs can also be used for this purpose.

PROBLEMS

2.1. The breaking strength, R, of a cable can be assumed to be a normal random variable with a mean value of 80 kips and a standard deviation of 20 kips.

(a) If a load, P, of magnitude 60 kip is hung from the cable, calculate the probability of failure of the cable.

(b) The magnitude of P cannot be determined with certainty. Suppose it could be either 40 or 60 kip, and the corresponding PMFs are shown in Figure P2.1. Calculate the probability of failure of the cable.

(c) If the cable breaks, what is the probability that the load was 40 kip?

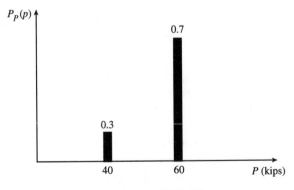

FIGURE P2.1. PMF of P.

2.2. The magnitude of a load acting on a structure can be modeled by a normal distribution with a mean of 100 kip and a standard deviation of 20 kip.

(a) If the design load is considered to be the 90th percentile value, determine the design load.

(b) If the design load is considered to be the mean + 2 standard deviation value, what is the probability that it will be exceeded?

(c) A load of magnitude less than zero is physically illogical; calculate its probability. Is a normal distribution appropriate to model the load?

2.3. The capacity of an isolated spread footing foundation under a column is modeled by a normal distribution with a mean of 300 kip and a COV of 20%. Suppose the column is subjected to a dead load of 100 kip and a live load of 150 kip.

(a) Calculate the probability of failure of the foundation under dead load only.

(b) Calculate the probability of failure of the foundation under the combined action of dead and live loads.

(c) If the probability of failure of the foundation needs to be limited to 0.001, and the dead load of 100 kip cannot be changed, what is the maximum amount of live load that can be applied to the foundation?

2.4. The average annual rainfall for a city is estimated to be 100 cm, and its mean ± 3 standard deviation values are estimated to be 160 and 40 cm, respectively.

(a) Calculate the standard deviation of the annual rainfall.

(b) What is the probability that the rainfall will be less than 0?

(c) What is the probability that the annual rainfall will be within the ± 3 standard deviation values?

(d) Is the normal distribution appropriate in this case?

2.5. Solve parts (a) and (b) of Problem 2.1, assuming that the breaking strength of the cable is a lognormal variable with the same mean and standard deviation.

2.6. Solve all three parts of Problem 2.2, assuming the load to be a lognormal random variable with the same mean and standard deviation.

2.7. Solve all three parts of Problem 2.3, if the capacity of the foundation is modeled by a lognormal random variable with the same mean and COV.

2.8. The compressive strength of concrete delivered by a supplier can be modeled by a lognormal random variable. Its mean and the coefficient of variation are estimated to be 4.7 ksi, and 0.21, respectively.

(a) If the 10th percentile value is considered to be the design value, calculate the value of the compressive strength to be used in a design.

(b) Suppose the COV of the compressive strength is reduced to 0.10 without affecting its mean value by introducing quality control procedures. Calculate the design value of the compressive strength if it is assumed to be the 10th percentile value.

(c) By comparing the results obtained in parts (a) and (b), discuss whether quality control measures are preferable.

2.9. The northbound train traffic in a subway station between 7:00 and 8:00 A.M. on a typical workday is studied. Trains are supposed to arrive every 5 minutes. Based on the collected data, it is observed that trains generally arrive at the station with an average delay of 1 minute and a variance of 2.0 min². Assume that the delay of each train is statistically independent and lognormally distributed, and if a train arrives within 30 seconds of the scheduled time it is not considered to be late.

(a) What is the probability that a train will arrive late at this station?

(b) What is the probability that the first train to arrive on time will be the third train?

(c) What is the probability that no train will arrive at the station on time during the peak traffic one-hour period?

(d) If a train has not arrived at the scheduled time, what is the probability that it will arrive within 1 minute of the scheduled time?

2.10. A contractor purchases a large number of bolts in one batch for future use. Based on past experience, the contractor estimates that the tensile strength of each bolt can be modeled by a lognormal variable with a mean of 80 kips and a COV of 0.15. A bolt must carry at least 60 kip to be acceptable. It is impractical to test all the bolts for strength. For quality control purposes, the contractor proposes the following three inspection schemes: (1) all 5 bolts selected at random from the batch must pass the test, (2) at least 6 of the first 7 bolts tested must pass the test, and (3) at least 8 of the first 10 bolts tested must pass the test. Which inspection scheme is the most severe from the supplier's point of view?

2.11. The relative density of a homogeneous soil deposit is measured and found to have a mean value of 0.80 and a COV of 0.20. From a theoretical point of view, the relative density can be between 0 and 1. Suppose the relative density has a beta distribution.

(a) Define the PDF of the relative density.

(b) What is the probability that the relative density of the soil deposit is greater than 0.90?

2.12. The travel time T between home and office is expected to be between 20 and 40 minutes depending on traffic conditions. Based on experience, the average travel time is 30 minutes, the corresponding variance is 20 min².

(a) Determine the PDF of T. (Hint: Assume it is a beta distribution.)

(b) What is the probability that T will exceed 30 minutes on a particular day?

2.13. The probability that the maximum temperature on a typical summer day will exceed 100°F in Tucson, Arizona is estimated to be 0.30.

(a) What is the probability that the maximum temperature will not exceed 100°F in the next 7 days?

(b) What is the probability that the maximum temperature will exceed 100°F in only 2 of the next 7 days?

(c) What is the probability that the maximum temperature will exceed 100°F at least 3 days in the next 7 days?

(d) What is the probability that the maximum temperature will exceed 100°F in at most 2 days in the next 7 days?

2.14. The probability of damage to a structure due to fire, p, is estimated to be 0.05 per year. Assume the design life of the structure is 50 years.

   (a) What is the probability that the structure will not be damaged by fire during its design life?

   (b) What is the probability that the structure will be damaged due to fire in the 10th year?

   (c) If the insurance company requires that the maximum risk of damage to the structure be limited to 0.10 during its lifetime, calculate the maximum permissible value of p.

2.15. The mean compressive strength of a batch of concrete is found to have a lognormal distribution with a mean value of 5000 psi and a standard deviation of 500 psi. The minimum required strength is 4000 psi. Five cylinders from this batch are tested.

   (a) What is the probability that at least one cylinder will fail?

   (b) What is the probability that two of the cylinders will fail?

   (c) What is the probability that the fifth cylinder failed and the others passed?

2.16. The annual precipitation in inches per year in Tucson, Arizona for the past 30 years are given below.

11.60, 7.19, 12.69, 11.86, 14.81, 8.07, 11.15, 8.00, 9.55, 11.02, 19.54, 8.63, 12.33, 8.53, 16.55, 19.74, 18.40, 11.37, 10.55, 8.68, 9.62, 6.93, 14.80, 10.64, 14.76, 15.19, 14.56, 9.68, 11.13, and 4.35. Assume the annual precipitation follows a Poisson process.

   (a) On average, how often will the annual precipitation exceed 12 inches?

   (b) What is the probability that in the next 5 years, the annual precipitation will exceed 12 inches exactly twice?

   (c) What is the probability that at least once in the next 5 years, the annual precipitation will exceed 12 inches?

   (d) How will the probability in part (b) change if the annual precipitation increases to 15 inches?

2.17. The PMF of the number of fires in a subdivision in a year is shown in Figure P2.17. More than 2 fires a year did not occur in the subdivision. During a fire, the mean damage to the property is expected to be $50,000 with a COV of 0.2. Assume that the

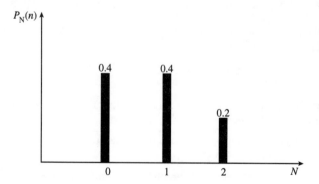

FIGURE P2.17. PMF of number of fires.

property damage can be modeled by a lognormal random variable and the property damage between fires is statistically independent.

(a) What is the probability of property damage exceeding $100,000 in a fire?

(b) What is the probability that none of the fires in a year will cause property damage exceeding $100,000?

2.18. On the average, 1 damaging earthquake occurs in a county every 10 years. Assume that the occurrence of earthquakes is a Poisson process in time.

(a) What is the probability of having at most 2 earthquakes in 1 year?

(b) What is the probability of having at least 1 earthquake in 5 years?

2.19. Based on the available records, it was observed that there were 2 fires in the past 10 years in a subdivision.

(a) What is the probability that there will be no fire in the subdivision in the next 10 years?

(b) What is the probability that there will be at least 1 fire in the subdivision in the next 10 years?

(c) What is the probability that there will be at least 2 fires in the subdivision in the next 10 years?

2.20. The occurrence of floods in a county follows a Poisson process at an average rate of once in 20 years. The damage in each flood is lognormally distributed with a mean of 2 million dollars and a COV of 25%. Assume damage in any one flood is statistically independent of the damage in any other flood.

(a) Determine the probability of having more than 2 floods in the county during the next 10 years.

(b) What is the probability that the damage in the next flood will exceed 3 million dollars?

(c) What is the probability that the damage in each of the next 2 floods will exceed 3 million dollars?

(d) What is the probability that none of the floods will cause damage exceeding 3 million dollars in the next 10 years?

2.21. The average car accident rate in a particular intersection is one per month. During an accident, the probability of personal injury is 0.30. Assume that the accidents occur according to a Poisson process.

(a) What is the probability that there will be no accident in that intersection next year?

(b) What is the probability that there will be no personal injury in the next n accidents?

(c) What is the probability that there will be no personal injury in that intersection next year?

2.22. A bank employee notices that there are no customers waiting for service and decides to take a 5-minute break. On an average, 1 customer arrives every 2 minutes, in a Poisson process. It takes 1 minute to serve each customer.

(a) When she returns from the break, what is the probability that 5 customers are waiting?

(b) Suppose she returns and finds 5 customers waiting. What is the probability that she will be able to take another break in exactly 10 minutes?

2.23. A driver arrives at a merging intersection, as shown in Figure P2.23, and notices 10 cars ahead of him at the stop sign. It takes 3 seconds for each car to clear the intersection. Cars in the other direction arrive at an average rate of 1 car every 10 seconds, in a Poisson process. Cars in the two directions alternate, one at a time, in crossing the intersection. What is the probability that the driver will clear the intersection in 45 seconds or less?

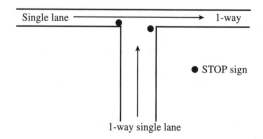

FIGURE P2.23. A T intersection.

2.24. Suppose that on an average 2 tornadoes occur in 10 years in a county in Oklahoma. Further assume that the tornado-generated wind speed can be modeled by a lognormal random variable with a mean of 120 mph and a standard deviation of 12 mph.

(a) What is the probability that there will be at least 1 tornado next year?

(b) If a structure in the county is designed for wind speed of 150 mph, what is the probability that the structure will be damaged during such a tornado?

(c) What is the probability that the structure will be damaged by tornadoes next year? (*Hint:* Consider that any number of tornadoes can occur next year).

2.25. Suppose the life of a light bulb can be modeled by an exponential distribution with an average life of 12 months. Suppose a maintenance worker checks the bulb every 6 months.

(a) What is the probability that the light bulb needs to be replaced at the first scheduled inspection?

(b) If the bulb is in good condition during the first scheduled inspection, what is the probability that it will be in good condition during the next scheduled inspection?

(c) If there are 10 light bulbs in a room, what is the probability that at least one of them needs replacement at the first scheduled inspection?

2.26. Structures in a county need to be designed for earthquake loading. After a detailed seismic risk analysis of the county, it is observed that the peak ground acceleration A can be modeled by an exponential distribution with a mean of 0.2g. If A exceeds 0.4g, structures in the county will suffer significant damage. Assume that earthquakes with A exceeding 0.4g occur once every 15 years in the county, and the damage from different earthquakes is statistically independent.

(a) What is the probability that there will be exactly two earthquakes where A exceeds 0.4g in the next 50 years?

(b) What is the probability of significant structural damage in an earthquake?

(c) What is the probability of no significant structural damage due to earthquakes in a year?

(d) What is the probability of no significant structural damage due to earthquakes in the next 50 years?

2.27. The annual maximum stage height in a river channel is modeled using a Type I extreme value distribution of the largest value, with a mean value of 30 ft and a COV of 10%. The stage height at which flooding will occur is 40 ft. What is the probability that the annual maximum stage height will exceed this level?

2.28. A steel cable consists of 8 high-strength steel strands. The strength of each strand can be modeled by a lognormal random variable with a mean of 50 kips and a COV of 10%. What is the probability that the weakest strand will have a strength less than 40 kips?

2.29. In a seismic hazard analysis, the magnitude of the earthquake (Richter's scale) is modeled with a Type II extreme value distribution of the largest value. In a certain geographical region, data have been collected for 50 years, and the annual maximum values of the earthquake magnitude are found to have a mean value of 4.0 and a standard deviation of 2.0. It is estimated that an earthquake with a magnitude of 9.0 or more will devastate the region. What is the probability that the annual maximum magnitude will be greater than or equal to 9.0?

2.30. The safety of a statically determinate truss structure is governed by the weakest member in the truss. (This is referred to as a weakest link system, or a series system.) The safety margin, defined as the ratio of the resistance to the applied load, of the weakest member in a particular truss is judged to have a mean value of 1.5 and a COV of 10%, and is assumed to follow a Type II extreme value distribution of the smallest value. The member, and therefore the whole truss, will fail if the safety margin of the weakest member drops below 1.0. What is the probability of failure of the truss?

2.31. The minimum life of an automobile brake pad is modeled with a Type III extreme value distribution of the smallest value. From the available data, the parameters of the distribution are estimated to be $k = 2.5$, $w_1 = 24$ months, and $\epsilon = 0$. What are the mean value and standard deviation of the brake pad's life? What is the probability that the brake pad will last longer than 36 months?

2.32. The minimum fatigue life of rivets in a compressor airseal inlet of a gas turbine engine is modeled with a two-parameter Weibull distribution. During accelerated testing for the purpose of certification and approval, the mean value of minimum life is found to be 90 minutes, and the COV is 15%. The rivet design is unacceptable if the minimum life during accelerated testing is less than 50 minutes. What is the probability of nonacceptance of the rivet design?

2.33. For a very unusual project, an engineer estimates that construction time may vary between 30 and 50 days. However, the engineer has no prior knowledge and believes that the completion time is equally likely to range between 30 and 50 days. Using Table 2.2, calculate the following:

    (a) The mean completion time.

    (b) The standard deviation, variance, and coefficient of variation of the completion time.

    (c) The probability that the completion time will be greater than 40 days.

2.34. In Problem 2.33, if the completion time has a triangular distribution with a modal value of 45 days, calculate the following:

    (a) The mean completion time.

    (b) The standard deviation, variance, and coefficient of variance of the completion time.

    (c) The probability that the completion time will be greater than 40 days.

2.35. A cofferdam needs to be built to facilitate the construction of a bridge pier. The maximum monthly flood level X at the site is considered to have a triangular distribution, as shown in Case 5 in Table 2.2, with x_l, x_n, and x_u values are 2, 8, and 10 ft, respectively. The pier will take 8 months to construct.

    (a) What would be the height of the cofferdam with reliability 0.90, that is, the cofferdam will not be flooded during the construction period with a probability of 0.9?

    (b) What is the height of the cofferdam corresponding to the flood return period of 8 months?

3

FUNDAMENTALS OF
RELIABILITY ANALYSIS

3.1 INTRODUCTORY COMMENTS

Before looking at the advanced concept of risk evaluation using the stochastic finite element method, you should be familiar with the other available reliability methods. A detailed review of these methods is presented in this chapter.

The final outcome of any engineering analysis and design consists of proportioning the elements of a system so that it satisfies various criteria for performance, safety, serviceability, and durability under various demands. For example, a structure should be designed so that its strength or resistance is greater than the effects of the applied loads. However, there are numerous sources of uncertainty in the load and resistance-related parameters. In general, the uncertainties in most of these parameters have been identified and quantified and are widely available in the literature (Haldar and Mahadevan, 2000). The need to incorporate these uncertainties into engineering design has given rise to a variety of reliability methods.

3.2 DETERMINISTIC AND PROBABILISTIC APPROACHES

It is not simple to satisfy the basic design requirements in the presence of uncertainty. Figure 3.1 shows a simple case considering two variables (one relating to the demand on the system, e.g., load on the structure, S, and the other relating to the capacity of the system, e.g., resistance of the structure, R). Both S and R are random in nature; their randomness is characterized by their means, μ_S and μ_R; standard deviations, σ_S and σ_R; and corresponding probability density functions, $f_S(s)$ and $f_R(r)$, as shown in Figure 3.1. Figure 3.1 also identifies the deterministic (nominal) values of these parameters, S_N and R_N, used in a conventional safety factor-based approach. It is necessary to develop a rational procedure to incorporate the information in Figure 3.1 into actual designs.

The concept of risk-based design was introduced by Freudenthal (1956) and was summarized by Freudenthal, Garrelts, and Shinozuka (1966). The concept has matured since then and is presented in the following sections.

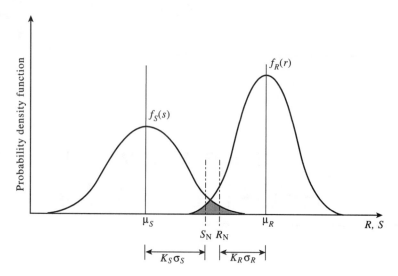

FIGURE 3.1. Fundamentals of risk evaluation.

3.3 RISK AND SAFETY FACTORS CONCEPT

The application risk-based design format in various engineering disciplines can be described, at best, as nonuniform. Most of the progress has been made in structural engineering. Design guidelines using the load and resistance factor design (LRFD) concept are essentially based on the risk-based design format. In the following discussion, the risk-based design format in structural engineering is emphasized. However, the same concept can also be applied to other engineering disciplines by replacing resistance and load by supply and demand as discussed in Section 1.1 in Chapter 1. We hope the following discussion will help to accelerate the implementation of the risk-based design concept to other enginering disciplines, where necessary.

From Figure 3.1, design safety is ensured in a deterministic approach by requiring that R_N be greater than S_N with a specified margin of safety as

$$\text{Nominal SF} = \frac{R_N}{S_N} \tag{3.1}$$

where SF is the safety factor.

The nominal resistance (or capacity) R_N is usually a conservative value, perhaps 1, 2, or 3 standard deviations below the mean value. The nominal load (or demand) S_N is also a conservative value, but it is several standard deviations above the mean value. Thus, the intended conservatism introduced in designs in the form of the nominal safety factor depends on many other factors, namely the uncertainty in the load and resistance and how conservatively the nominal load and resistance values are selected. The nominal safety factor may fail to convey the actual margin of safety in a design.

Conceptually, then, in a deterministic design the nominal safety factor can be applied to the resistance, to the load, or to both. The allowable stress design methods use a safety factor to compute the allowable stresses in members from the ultimate stress, and a successful design ensures that the stresses caused by the nominal values of the loads do not exceed the allowable stresses. In other words, referring to Figure 3.1 and Equation 3.1, R_N is divided by a safety factor to compute the allowable resistance R_a, and safe design requires that the condition $S_N < R_a$ be satisfied. In this case, the safety factor is used for the resistance only. In the ultimate strength design method, the loads are multiplied by certain load factors to determine the ultimate loads, and the members are required to resist various design combinations of the ultimate loads. That is, in Figure 3.1, S_N is multiplied by a load factor to obtain the ultimate load S_u, and safe design requires the satisfaction of the condition $S_u < R_N$. In this case, the safety factors are used in the loads and in load combinations. In some designs—for example, in concrete design or in steel design using the load and resistance factor design (LRFD) concept—the capacity reduction factor (generally less than one) and load factors (generally more than one) are used to achieve the same objective. Essentially, the safety factors are used to estimate both the resistance and the loads.

The intent of these conventional approaches can be explained by considering the area of overlap between the two curves (the shaded region in Figure 3.1), which provides a *qualitative measure* of the probability of failure. This area of overlap depends on three factors:

1. *The relative positions of the two curves.* As the distance between the two curves increases, reducing the overlapped area, the probability of failure decreases. The positions of the curves may be represented by the means (μ_R and μ_S) of the two variables.

2. *The dispersion of the two curves.* If the two curves are narrow, then the area of overlap and the probability of failure are small. The dispersion may be characterized by the standard deviations (σ_R and σ_S) of the two variables.

3. *The shapes of the two curves.* The shapes are represented by the probability density functions $f_R(r)$ and $f_S(s)$.

The objective of safe design in deterministic design procedures can also be achieved, perhaps more comprehensively, by selecting the design variables in such a way that the area of overlap between the two curves is as small as possible, so that the underlying risk is not compromised, within the constraints of economy. Conventional design approaches achieve this objective by shifting the positions of the curves through the use of safety factors. A more rational approach would be to compute the risk of failure by accounting for all three overlap factors and selecting the design variables so that an acceptable risk of failure is achieved. This is the foundation of the risk-based design concept. With this approach, however, the information on the probability density functions of the resistance and loads (as in Figure 3.1) is usually difficult to obtain, and engineers must formulate an acceptable design methodology using only the information on means and standard deviations.

Instead of using the safety factor for the resistance alone, as in the working stress method, or for the loads alone, as in the ultimate strength method, it is more rational to apply safety factors to both resistance and loads, as is done in concrete or steel (LRFD) structural design. Referring to Figure 3.1, where the uncertainties in the load and resistance variables are expressed in the form of the probability density functions, we can express the measure of risk in terms of the probability of failure as the failure event or $P(R < S)$ as

$$p_f = P(\text{failure}) = P(R < S)$$

$$= \int_0^\infty \left[\int_0^s f_R(r)\, dr \right] f_S(s)\, ds \tag{3.2}$$

$$= \int_0^\infty F_R(s) f_S(s)\, ds$$

where $F_R(s)$ is the CDF of R evaluated at s. Equation 3.2 states that when the load is $S = s$, the probability of failure is $F_R(s)$, and since the load is a random variable, the integration needs to be carried out for all the possible values of S, with their respective likelihoods represented by the PDF of S. Equation 3.2 is the basic equation of the risk-based design concept. The CDF of R or the PDF of S may not always be available in explicit form, and thus the integration of Equation 3.2 may not be practical. However, Equation 3.2 can be evaluated easily, without performing the integration, for some special cases (Haldar and Mahadevan, 2000).

3.4 FUNDAMENTAL CONCEPT OF RELIABILITY ANALYSIS

The basic concept of the classical theory of structural reliability and risk-based design can now be presented more formally. We have shown elsewhere (2000) that it is not difficult to calculate the underlying resistance and load factors for a given design, assuming an acceptable level of risk. However, it is more relevant to calculate the underlying risk of a given design, as is discussed in the following sections.

The first step in evaluating the reliability or probability of failure of a structure is to decide on specific performance criteria and the relevant load and resistance parameters, called the basic variables X_i, and the functional relationships among them corresponding to each performance criterion. Mathematically, this relationship or *performance function* can be described as

$$Z = g(X_1, X_2, \ldots, X_n) \tag{3.3}$$

The *failure surface* or the *limit state* of interest can then be defined as $Z = 0$. This is the boundary between the safe and unsafe regions in the design parameter space, and it also represents a state beyond which a structure can no longer fulfill the function for which it was designed. The failure surface and the safe and unsafe regions are shown in Figure 3.2, where R and S are the two basic random variables. The *limit state equation* plays an important role in the development of structural reliability analysis methods. A limit state can be an explicit or implicit function of the basic random variables, and it can be in simple or complicated form. Reliability analysis methods have been developed corresponding to limit states of different types and complexity, as discussed in the following sections.

Using Equation 3.3, we find that failure occurs when $Z < 0$. Therefore, the probability of failure, p_f, is given by the integral

$$p_f = \int \cdots \int_{g(\ ) < 0} f_X(x_1, x_2, \ldots, x_n)\, dx_1\, dx_2 \cdots dx_n \tag{3.4}$$

in which $f_X(x_1, x_2, \ldots, x_n)$ is the joint probability density function for the basic random variables $X_1, X_2, \ldots, X_n$ and the integration is performed over the failure region, that is, $g(\ ) < 0$. If the random variables are statistically independent, then the joint probabil-

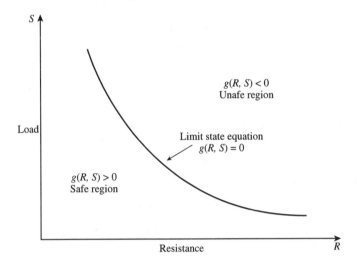

FIGURE 3.2. Limit state concept.

ity density function may be replaced by the product of the individual probability density functions in the integral.

Equation 3.4 is a more general representation of Equation 3.2. The computation of p_f by Equation 3.4 is called the *full distributional approach* and is the fundamental equation of reliability analysis. In general, the joint probability density function of random variables is almost impossible to obtain. Even if this information is available, evaluating the multiple integral is extremely complicated. Therefore, one approach is to use analytical approximations of this integral that are simpler to compute. To clarify the presentation, these methods can be grouped into two types: *first-order reliability methods (FORM)* and *second-order reliability methods (SORM)*.

The limit state of interest can be linear or nonlinear functions of the basic variables. FORM can be used to evaluate Equation 3.4 when the limit state function is a linear function of uncorrelated normal variables or when the nonlinear limit state function is represented by a first-order (linear) approximation with equivalent normal variables, as is elaborated further in Section 3.5. SORM estimates the probability of failure by approximating the nonlinear limit state function, including a linear limit state function with correlated nonnormal variables, by a second-order representation. SORM is discussed in Section 3.6.

3.5 FIRST-ORDER RELIABILITY METHODS (FORM)

The development of FORM can be traced historically to second-moment methods, which used the information on first and second moments of the random variables. These are *first-order second-moment (FOSM)* and *advanced first-order second-moment (AFOSM)* methods. In FOSM methods, information on the distribution of random variables is ignored; however, in AFOSM methods, the distributional information is appropriately used.

3.5.1   First-Order Second-Moment Method (FOSM) or MVFOSM Method

The FOSM method is also referred to as the *mean value first-order second-moment (MVFOSM)* method in the literature. The MVFOSM method derives its name from the fact that it is based on a first-order Taylor series approximation of the performance function linearized at the mean values of the random variables, and becaues it uses only second-moment statistics (means and covariances) of the random variables. The original formulation by Cornell (1969) uses the simple two-variable approach of the previous sections. A performance function in this case can be defined as

$$Z = R - S \tag{3.5}$$

Assuming that R and S are statistically independent normally distributed random variables, the variable Z is also a normal random variable, that is, $N\left(\mu_R - \mu_S, \sqrt{\sigma_R^2 + \sigma_S^2}\right)$. The event of failure is $R < S$, or $Z < 0$. Then, using Equation 3.2, we can evaluate the probability of failure as

$$p_f = P(Z < 0)$$

or

$$p_f = \Phi\left(\frac{0 - (\mu_R - \mu_S)}{\sqrt{\sigma_R^2 + \sigma_S^2}}\right)$$

or

$$p_f = 1 - \Phi\left(\frac{\mu_R - \mu_S}{\sqrt{\sigma_R^2 + \sigma_S^2}}\right) \tag{3.6}$$

where Φ is the CDF of the standard normal variate.

The probability of failure depends on the ratio of the mean value of Z to its standard deviation. This ratio is commonly known as the *safety index* or *reliability index* and is denoted as β:

$$\beta = \frac{\mu_Z}{\sigma_Z} = \frac{\mu_R - \mu_S}{\sqrt{\sigma_R^2 + \sigma_S^2}} \tag{3.7}$$

The probability of failure in terms of the safety index can be obtained by rewriting Equation 3.6 as

$$p_f = 1 - \Phi(\beta) \tag{3.8}$$

An alternative formulation proposed by Rosenbleuth and Esteva (1972) may also be used. In this case, considering the physical aspects of a design problem, R and S can be more appropriately considered to be statistically independent lognormal variables, that is, $LN(\lambda_R, \zeta_R)$ and $LN(\lambda_S, \zeta_S)$, since they cannot take negative values. In this case, another random variable Y can be introduced as

$$Y = \frac{R}{S} \qquad (3.9a)$$

or

$$\ln Y = Z = \ln R - \ln S \qquad (3.9b)$$

Equation 3.9a or 3.9b represents the performance function. The failure event can be defined as when $Y < 1.0$ or $Z < 0.0$. Since R and S are lognormal, $\ln R$ and $\ln S$ are normal (see Section 2.2.2); therefore, $\ln Y$ or Z is a normal random variable, that is, $Z \sim N\left(\lambda_R - \lambda_S, \sqrt{\zeta_R^2 + \zeta_S^2}\right)$. The probability of failure, similar to Equation 3.6, can be defined as

$$p_f = 1 - \Phi\left(\frac{\lambda_R - \lambda_S}{\sqrt{\zeta_R^2 + \zeta_S^2}}\right) \qquad (3.10)$$

Using the relationships between the mean, standard deviation, and coefficient of variation and the parameters of the lognormal distribution (Equations 2.15 and 2.16), we can write Equation 3.10 as

$$p_f = 1 - \Phi\left[\frac{\ln\left(\mu_R/\mu_S\right)\sqrt{(1 + \delta_S^2)/(1 + \delta_R^2)}}{\sqrt{\ln(1 + \delta_R^2)(1 + \delta_S^2)}}\right] \qquad (3.11)$$

If δ_R and δ_S are not large, say ≤ 0.30, Equation 3.11 can be simplified as

$$p_f \simeq 1 - \Phi\left[\frac{\ln\left(\mu_R/\mu_S\right)}{\sqrt{\delta_R^2 + \delta_S^2}}\right] \qquad (3.12)$$

In this formulation, β as in Equation 3.7, can be shown to be

$$\beta = \Phi^{-1}(1 - p_f) = \frac{\ln\left\{\left(\frac{\mu_R}{\mu_S}\right)\sqrt{\frac{1+\delta_S^2}{1+\delta_R^2}}\right\}}{\sqrt{\ln(1 + \delta_R^2)(1 + \delta_S^2)}} \approx \frac{\ln\left(\frac{\mu_R}{\mu_S}\right)}{\sqrt{\delta_R^2 + \delta_S^2}} \qquad (3.13)$$

These formulations may be generalized for many random variables, denoted by a vector **X**. Let the performance function be written as

$$Z = g(\mathbf{X}) = g(X_1, X_2, \ldots, X_n) \qquad (3.14)$$

A Taylor series expansion of the performance function about the mean value gives

$$Z = g(\mu_\mathbf{X}) + \sum_{i=1}^{n} \frac{\partial g}{\partial X_i}(X_i - \mu_{X_i}) + \frac{1}{2}\sum_{i=1}^{n}\sum_{j=1}^{n} \frac{\partial^2 g}{\partial X_i \partial X_j}(X_i - \mu_{X_i})(X_j - \mu_{X_j}) + \cdots \qquad (3.15)$$

where the derivatives are evaluated at the mean values of the random variables ($X_1, X_2, \ldots, X_n$), and μ_{X_i} is the mean value of X_i. Truncating the series at the linear terms, we obtain the first-order approximate mean and variance of Z as

$$\mu_Z \approx g\left(\mu_{X_1}, \mu_{X_2}, \ldots, \mu_{X_n}\right) \tag{3.16}$$

and

$$\sigma_Z^2 \approx \sum_{i=1}^{n} \sum_{j=1}^{n} \frac{\partial g}{\partial X_i} \frac{\partial g}{\partial X_j} \text{Cov}(X_i, X_j) \tag{3.17}$$

where $\text{Cov}(X_i, X_j)$ is the covariance of X_i and X_j.

If the variables are uncorrelated, then the variance is simply

$$\sigma_Z^2 \approx \sum_{i=1}^{n} \left(\frac{\partial g}{\partial X_i}\right)^2 \text{Var}(X_i) \tag{3.18}$$

The safety index can be calculated by taking the ratio of the mean and standard deviation of Z as in Equation 3.7. Remember that the performance function is linearized at the mean values of the random variables, reflecting the concept behind the MVFOSM method.

Using the safety index, β, we can find the exact probability of failure in only a few cases. For example, if all the X_i's are statistically independent normal variables and if Z is a linear function of the X_i values, then Z is normal and the probability of failure is given by Equation 3.8. Similarly, if all the X_i's are statistically independent lognormal variables and if $g(\mathbf{X})$ is a multiplicative function of the X_i's, then $Z = \ln g(\mathbf{X})$ is normal and the probability of failure is given by Equation 3.8.

However, in most cases it is not likely that all the variables are statistically independent normals or lognormals. Nor is it likely that the performance function is a simple additive or multiplicative function of these variables. In such cases, the safety index cannot be directly related to the probability of failure; nevertheless, it does provide a rough idea of the level of risk or reliability in the design. The MVFOSM method was used to derive earlier versions of the reliability-based design formats, such as the American Institute of Steel Construction, Inc. (AISC, 1986), Canadian Standard Associations (CSA, 1974), and Comité European du Béton (CEB, 1976), to cite just a few examples.

However, the MVFOSM approach has some deficiencies. The method does not use the distribution information about the variables when it is available. The function $g(\ )$ in Equation 3.3 is linearized at the mean values of the X_i variables. When $g(\ )$ is nonlinear, significant error may be introduced by neglecting higher order terms. More important, the safety index defined by Equation 3.7 fails to be constant under different but mechanically equivalent formulations of the same performance function. For example, the safety margins defined as $R - S < 0$ in Equation 3.5 and $R/S < 1$ in Equation 3.9a are mechanically equivalent. Yet the probabilities of failure given by Equations 3.6 and 3.10 or 3.11 are different for the two formulations. Furthermore, an engineering problem can be formulated in terms of stress or strength, as elaborated with the help of the following examples, and should produce identical results in either case. But the simplified method just discussed will give two different safety indexes. These observations can be explained with the help of simple examples.

Example 3.1

A W16 × 31 steel section made of A36 steel is suggested to carry an applied deterministic bending moment of 1140 kip-in. The nominal yield stress F_y of the steel is 36 ksi, and the nominal plastic modulus of the section Z is 54 in^3. Consider that the distributions of these random variables are unknown; only the means, standard deviations, and COVs are known:

$$\mu_{F_y} = 38 \text{ ksi}, \ \sigma_{F_y} = 3.8 \text{ ksi, and } \delta_{F_y} = 0.1$$

$$\mu_Z = 54 \text{ in}^3, \ \sigma_Z = 2.7 \text{ in}^3, \text{ and } \delta_Z = 0.05$$

It is quite logical to assume that F_y and Z are statistically independent.

Strength Formulation

Considering the strength formulation first, the resistance $R = F_y Z$ and the load $S = 1140$. In this example, the load is a constant, thus $\mu_S = 1140, \sigma_S = 0$, and $\delta_S = 0$. Using Equations 3.16 and 3.18, we can calculate the first-order mean and standard deviation of R as

$$\mu_R \approx \mu_{F_y} \mu_Z = 38 \times 54 = 2{,}052 \text{ kip-in}$$

and

$$\sigma_R \approx \left[\text{Var}(F_y) \left(\frac{\partial R}{\partial F_y} \right)^2 + \text{Var}(Z) \left(\frac{\partial R}{\partial Z} \right)^2 \right]^{1/2} = \left(3.8^2 \times \mu_Z^2 + 2.7^2 \times \mu_{F_y}^2 \right)^{1/2}$$

$$= [(3.8 \times 54)^2 + (2.7 \times 38)^2]^{1/2}$$

$$= 229.42 \text{ kip-in}$$

Thus, $\delta_R = 229.42/2052 = 0.112$.

Assuming the performance function of the form represented by Equation 3.5, we can write the limit state equation as

$$g(\ ) = F_y Z - 1140 = 0 \tag{3.19}$$

The corresponding safety index, as in Equation 3.7, is

$$\beta = \frac{2052 - 1140}{\sqrt{(229.42)^2 + 0^2}} = 3.975$$

If the performance function is assumed to be of the form represented by Equation 3.9, the corresponding safety index, according to Equation 3.13, becomes

$$\beta = \frac{\ln(2052/1140)}{\sqrt{(0.112)^2 + 0^2}} = 5.248$$

Obviously, these two safety indexes and the corresponding probabilities of failure are quite different.

Stress Formulation

The same problem can also be formulated in term of stresses. The limit state equation in this case can be expressed as

$$g() = F_y - \frac{1140}{Z} = 0 \tag{3.20}$$

In this case, the resistance R is represented by the random variable F_y, and the load $S = 1140/Z$. Thus,

$$\mu_R = \mu_{F_y} = 38 \text{ ksi}$$

and

$$\sigma_R = \sigma_{F_y} = 3.8 \text{ ksi}$$

and $\delta_R = 0.1$.

The first-order mean and standard deviation of S can be calculated as

$$\mu_S \approx \frac{1140}{\mu_Z} = \frac{1140}{54} = 21.11 \text{ ksi}$$

$$\sigma_S \approx \left[\text{Var}(Z) \left(-\frac{1140}{\mu_Z^2} \right)^2 \right]^{1/2} = \sigma_Z \frac{1140}{\mu_Z^2} = 2.7 \left(\frac{1140}{54^2} \right) = 1.056 \text{ ksi}$$

Thus, $\delta_S = 1.056/21.11 = 0.05$.

The safety index according to Equation 3.7 for the stress formulation is found to be

$$\beta = \frac{38 - 21.11}{\sqrt{(3.8)^2 + (1.056)^2}} = 4.282$$

The corresponding safety index according to Equation 3.13 is

$$\beta = \frac{\ln(38/21.11)}{\sqrt{(0.1)^2 + (0.05)^2}} = 5.258$$

The observations are summarized in Table 3.1.

The results clearly indicate that the safety indexes depend on the formulation of the limit state equation as well as the underlying assumption about the distribution of the limit state.

In the early 1970s, this lack of invariance problem was observed by many researchers. It was overcome by the advanced first-order second moment (AFOSM) method proposed by Hasofer and Lind (1974) for normal variables.

Table 3.1. Variance problem in the MVFOSM

	Normal	Lognormal
Strength formulation	3.975	5.248
Stress formulation	4.282	5.258

3.5.2   AFOSM Method for Normal Variables (Hasofer–Lind Method)

The *Hasofer–Lind (H–L) method* is applicable for normal random variables. It first defines the reduced variables as

$$X_i' = \frac{X_i - \mu_{X_i}}{\sigma_{X_i}} \qquad (i = 1, 2, \dots, n) \tag{3.21}$$

where X_i' is a random variable with zero mean and unit standard deviation. Equation 3.21 is used to transform the original limit state $g(\mathbf{X}) = 0$ to the reduced limit state, $g(\mathbf{X}') = 0$. The $\mathbf{X}$ coordinate system is referred to as the *original coordinate system*. The $\mathbf{X}'$ coordinate system is referred to as the *transformed or reduced coordinate system*. Note that if X_i is normal, X_i' is standard normal. These notations will be used throughout this chapter to denote different coordinate systems. The safety index $\beta_{\text{H-L}}$ is defined as the minimum distance from the origin of the axes in the reduced coordinate system to the limit state surface (failure surface). It can be expressed as

$$\beta_{\text{H-L}} = \sqrt{(\mathbf{x}'^*)^{\text{t}}(\mathbf{x}'^*)} \tag{3.22}$$

The minimum distance point on the limit state surface is called the *design point* or *checking point*. It is denoted by vector $\mathbf{x}^*$ in the original coordinate system and by vector $\mathbf{x}'^*$ in the reduced coordinate system. These vectors represent the values of all the random variables, that is, $X_1, X_2, \dots, X_n$ at the design point corresponding to the coordinate system being used.

This method can be explained with the help of Figure 3.3. Consider the linear limit state equation in two variables,

$$Z = R - S = 0 \tag{3.23}$$

This equation is similar to Equation 3.5. Note that R and S need not be normal variables. A set of reduced variables is introduced as

$$R' = \frac{R - \mu_R}{\sigma_R} \tag{3.24}$$

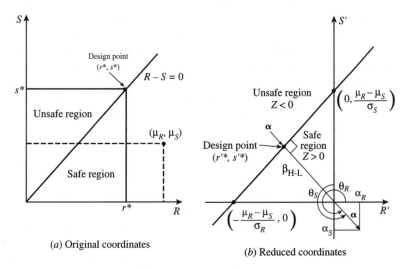

(a) Original coordinates

(b) Reduced coordinates

FIGURE 3.3. Hasofer–Lind reliability index: linear performance function.

and

$$S' = \frac{S - \mu_S}{\sigma_S} \tag{3.25}$$

If we substitute these into Equation 3.23, the limit state equation in the reduced coordinate system becomes

$$g() = \sigma_R R' - \sigma_S S' + \mu_R - \mu_S = 0 \tag{3.26}$$

The transformation of the limit state equation from the original to the reduced coordinate system is shown in Figure 3.3b. The safe and failure regions are also shown. From Figure 3.3b it is apparent that if the failure line (limit state line) is closer to the origin in the reduced coordinate system, the failure region is larger, and if it is farther away from the origin, the failure region is smaller. Thus, the position of the limit state surface relative to the origin in the reduced coordinate system is a measure of the reliability of the system. The coordinates of the intercepts of Equation 3.26 on the R' and S' axes can be shown to be $[-(\mu_R - \mu_S)/\sigma_R, 0]$ and $[0, (\mu_R - \mu_S)/\sigma_S]$, respectively. Using simple trigonometry, we can calculate the distance of the limit state line (Equation 3.26) from the origin as

$$\beta_{\text{H-L}} = \frac{\mu_R - \mu_S}{\sqrt{\sigma_R^2 + \sigma_S^2}} \tag{3.27}$$

This distance is referred to as the *reliability index or safety index*. It is the same as the reliability index defined by the MVFOSM method in Equation 3.7 if both R and S are normal variables. However, it is obtained in a completely different way based on geometry. It indicates that if the limit state is linear and if the random variables R and S are normal, both methods will give an identical reliability or safety index. This may not be true for other cases, as is discussed further later.

In general, for many random variables represented by the vector $\mathbf{X} = (X_1, X_2, \ldots, X_n)$ in the original coordinate system and $\mathbf{X'} = (X'_1, X'_2, \ldots, X'_n)$ in the reduced coordinate system, the limit state $g(\mathbf{X'}) = 0$ is a nonlinear function as shown in the reduced coordinates for two variables in Figure 3.4. At this stage, X'_i's are assumed to be uncorrelated. Consideration of correlated random variables is discussed in Section 3.7. Here, $g(\mathbf{X'}) > 0$ denotes the safe state and $g(\mathbf{X'}) < 0$ denotes the failure state. Again, the Hasofer–Lind reliability index $\beta_{\text{H-L}}$ is defined as the minimum distance from the origin to the design point on the limit state in the reduced coordinates and can be expressed by Equation 3.22, where $\mathbf{x'^*}$ represents the coordinates of the design point or the point of minimum distance from the origin to the limit state. In this definition the reliability index is invariant, because regardless of the form in which the limit state equation is written, its geometric shape and the distance from the origin remain constant. For the limit state surface where the failure region is away from the origin, it is easy to see from Figure 3.4 that $\mathbf{x'^*}$ is the most probable failure point. As will be elaborated with the help of an example later, the Hasofer–Lind reliability index can be used to calculate a first-order approximation of the failure probability as $p_f = \Phi(-\beta_{\text{H-L}})$. This is the integral of the standard normal density function along the ray joining the origin and $\mathbf{x'^*}$. It is obvious that the nearer $\mathbf{x'^*}$ is to the origin, the larger is the failure probability. Thus, the minimum distance point on the limit state surface is also the most probable failure point. The point of minimum distance from the origin to the limit state surface, $\mathbf{x'^*}$, represents the worst combination of the stochastic variables and is appropriately named the *design point* or the *most probable point (MPP)* of failure.

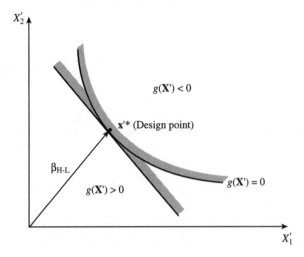

FIGURE 3.4. Hasofer–Lind reliability index: nonlinear performance function.

For nonlinear limit states, the computation of the minimum distance becomes an optimization problem:

$$\text{Minimize } D = \sqrt{\mathbf{x}'^t \mathbf{x}'}$$

$$\text{Subject to the constraint } g(\mathbf{X}) = g(\mathbf{X}') = 0 \tag{3.28}$$

where $\mathbf{x}'$ represents the coordinates of the checking point on the limit state equation in the reduced coordinates to be estimated. Using the method of Lagrange multipliers, we can obtain the minimum distance as

$$\beta_{\text{H-L}} = -\frac{\sum_{i=1}^{n} x'^{*}_i \left(\frac{\partial g}{\partial X'_i}\right)^*}{\sqrt{\sum_{i=1}^{n} \left(\frac{\partial g}{\partial X'_i}\right)^{2*}}} \tag{3.29}$$

where $\left(\partial g/\partial X'_i\right)^*$ is the ith partial derivative evaluated at the design point with coordinates $(x'^{*}_1, x'^{*}_2, \ldots, x'^{*}_n)$. The asterisk after the derivative indicates that it is evaluated at $(x'^{*}_1, x'^{*}_2, \ldots, x'^{*}_n)$. The design point in the reduced coordinates is given by

$$x'^{*}_i = -\alpha_i \beta_{\text{H-L}} \qquad (i = 1, 2, \ldots, n) \tag{3.30}$$

where

$$\alpha_i = \frac{\left(\frac{\partial g}{\partial X'_i}\right)^*}{\sqrt{\sum_{i=1}^{n} \left(\frac{\partial g}{\partial X'_i}\right)^{2*}}} \tag{3.31}$$

are the direction cosines along the coordinate axes X_i'. In the space of the original coordinates and using Equation 3.21, we find the design point to be

$$x_i^* = \mu_{X_i} - \alpha_i \sigma_{x_i} \beta_{\text{H-L}} \qquad (3.32)$$

An algorithm was formulated by Rackwitz (1976) to compute $\beta_{\text{H-L}}$ and $x_i'^*$ as follows:

Step 1   Define the appropriate limit state equation.
Step 2   Assume initial values of the design point x_i^*, $i = 1, 2, \ldots, n$. Typically, the initial design point may be assumed to be at the mean values of the random variables. Obtain the reduced variates $x_i'^* = (x_i^* - \mu_{X_i})/\sigma_{X_i}$.
Step 3   Evaluate $(\partial g/\partial X_i)'^*$ and α_i at $x_i'^*$.
Step 4   Obtain the new design point $x_i'^*$, in terms of $\beta_{\text{H-L}}$, as in Equation 3.30.
Step 5   Substitute the new $x_i'^*$ in the limit state equation $g(\mathbf{x}'^*) = 0$ and solve for $\beta_{\text{H-L}}$.
Step 6   Using the $\beta_{\text{H-L}}$ value obtained in step 5, reevaluate $x_i'^* = -\alpha_i \beta_{\text{H-L}}$.
Step 7   Repeat Steps 3 through 6 until $\beta_{\text{H-L}}$ converges.

This algorithm is shown geometrically in Figure 3.5. The algorithm constructs a linear approximation to the limit state at every search point and finds the distance from the origin to the limit state. In Figure 3.5, point B represents the initial design point, usually assumed to be at the mean values of the random variables, as noted in step 2. Note that B is not on the limit state equation $g(\mathbf{X}') = 0$. The tangent to the limit state at B is represented by the line BC. Then AD will give an estimate of $\beta_{\text{H-L}}$ in the first iteration, as noted in step 5. As the iteration continues the $\beta_{\text{H-L}}$ value converges. This is a first-order approach, similar to the MVFOSM method, with the important difference that the limit state is linearized at the most probable failure point rather than at the mean values of the random variables.

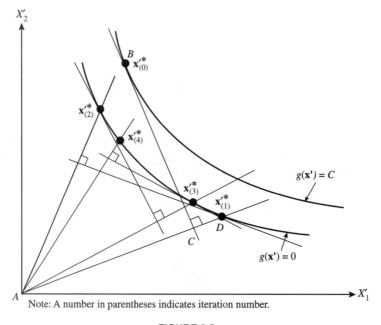

Note: A number in parentheses indicates iteration number.

FIGURE 3.5.

Ditlevsen (1979a) showed that for a nonlinear limit state surface, $\beta_{\text{H-L}}$ lacks comparability; the ordering of $\beta_{\text{H-L}}$ values may not be consistent with the ordering of actual reliabilities. An example of this is shown in Figure 3.4 with two limit state surfaces: one flat and the other curved. The shaded region to the right of each limit state represents the corresponding failure region. Clearly, the structure with the flat limit state surface has a different reliability than the one with the curved limit state surface, but the $\beta_{\text{H-L}}$ values are identical for both surfaces and suggest equal reliability if Equation 3.3 is used. To overcome this inconsistency, Ditlevsen (1979a) introduced the generalized reliability index, β_{g}, defined as

$$\beta_{\text{g}} = \Phi^{-1} \left[\int \cdots \int \phi(x_1')\phi(x_2') \cdots \phi(x_n')dx_1'dx_2' \cdots dx_n' \right]$$

$$g(\mathbf{x}' > 0)$$

(3.33)

where Φ and ϕ are the cumulative distribution function and the probability density function of a standard normal variable, respectively. Because the reliability index in this definition includes the entire safe region, it provides a consistent ordering of second-moment reliability. The integral in the equation looks similar to that in Equation 3.4, and is difficult to compute directly. Hence, Ditlevsen (1979a) proposed approximating the nonlinear limit state by a polyhedral surface consisting of tangent hyperplanes at selected points on the surface.

Example 3.2

Denoting R and S as the random variables representing the resistance and the applied load on a structure, assume that the limit state equation is represented by Equation 3.23 in the original coordinate system and by Equation 3.26 in the reduced coordinate system. These are shown in Figures 3.3a and 3.3b, respectively.

Using Equation 3.31, we can evaluate the direction cosines α_R and α_S as

$$\alpha_R = \frac{(\partial g/\partial R')}{\sqrt{(\partial g/\partial R')^2 + (\partial g/\partial S')^2}} = \frac{\sigma_R}{\sqrt{\sigma_R^2 + \sigma_S^2}}$$

and

$$\alpha_S = -\frac{\sigma_S}{\sqrt{\sigma_R^2 + \sigma_S^2}}$$

Notice that in Figure 3.3b, $\alpha_R = \cos\theta_R$ and $\alpha_S = \cos\theta_S$. The angles θ_R and θ_S are defined as counterclockwise angles of rotation from the positive directions of the R' and S' axes to the positive direction of the $\boldsymbol{\alpha}$ vector (i.e., vector of direction cosines). Using Equation 3.30, we can show the coordinates of the checking point in the reduced coordinate to be

$$r'^* = -\alpha_R\beta_{\text{H-L}} = -\frac{\sigma_R}{\sqrt{\sigma_R^2 + \sigma_S^2}}\beta_{\text{H-L}}$$

$$s'^* = -\alpha_S\beta_{\text{H-L}} = \frac{\sigma_S}{\sqrt{\sigma_R^2 + \sigma_S^2}}\beta_{\text{H-L}}$$

Substituting these new checking points in Equation 3.26, we can calculate the reliability index $\beta_{\text{H-L}}$ as

$$\beta_{\text{H-L}} = \frac{\mu_R - \mu_S}{\sqrt{\sigma_R^2 + \sigma_S^2}}$$

Since this is a linear limit state equation, iteration is not required. This is the same result as obtained by Equation 3.7, indicating that the algorithm works correctly. Using Equation 3.32, we find the new checking point in the original coordinates to be

$$r^* = \mu_R - \left(\frac{\sigma_R}{\sqrt{\sigma_R^2 + \sigma_S^2}}\right)\sigma_R\left(\frac{\mu_R - \mu_S}{\sqrt{\sigma_R^2 + \sigma_S^2}}\right) = \frac{\mu_R \sigma_S^2 + \mu_S \sigma_R^2}{\sigma_R^2 + \sigma_S^2}$$

$$s^* = \mu_S - \left(-\frac{\sigma_S}{\sqrt{\sigma_R^2 + \sigma_S^2}}\right)\sigma_S\left(\frac{\mu_R - \mu_S}{\sqrt{\sigma_R^2 + \sigma_S^2}}\right) = \frac{\mu_R \sigma_S^2 + \mu_S \sigma_R^2}{\sigma_R^2 + \sigma_S^2}$$

In this case, $r^* = s^*$, indicating that the checking point is on the limit state line, since it is at a 45-degree angle to both coordinate axes.

Several important observations can be made by comparing the safety indexes calculated by the MVFOSM (Equation 3.7) and the AFOSM proposed by Hasofer and Lind (Equation 3.29). As long as the limit state equation of resistance and load is linear and all the variables are normal, the safety indexes calculated by the two methods will be the same. However, strictly speaking, the MVFOSM does not use any information on the distribution of the resistance and load, whereas the AFOSM proposed by Hasofar and Lind is applicable when they are normal. The most important difference is that in the MVFOSM method, the design point is at the mean values of R and S, indicating they are not on the limit state line. The AFOSM (Hasofer–Lind) method indicates that the design point is on the limit state line. This can be elaborated further with the help of an example.

Example 3.3

Suppose a cable of resistance R needs to carry a weight S. Assume that both R and S are normal random variables with means of 120 and 50 kip, respectively, and corresponding standard deviations of 18 and 12 kip, respectively. The limit state equation can be represented by Equation 3.23. Then, the safety index according to the MVFOSM and the Hasofer–Lind methods will be the same:

$$\beta = \beta_{\text{H-L}} = \frac{120 - 50}{\sqrt{18^2 + 12^2}} = 3.236$$

The design point according to the MVFOSM is (120, 50), as shown in Figure 3.3a. The coordinates of the design point according to the Hasofer–Lind method can be estimated as

$$r^* = s^* = \frac{\mu_R \sigma_S^2 + \mu_S \sigma_R^2}{\sigma_R^2 + \sigma_S^2} = \frac{120 \times 12^2 + 50 \times 18^2}{18^2 + 12^2} = 71.54$$

The coordinates for the checking point are (71.54, 71.54), indicating that it is on the limit state, as shown in Figure 3.3b.

3.5.3 AFOSM Methods for Nonnormal Variables

The Hasofer–Lind reliability index can be exactly related to the failure probability using Equation 3.8 if all the variables are statistically independent and normally distributed and the limit state surface is linear. For any other situation, it will not give correct information on the probability of failure. Rackwitz and Fiessler (1978), Chen and Lind (1983), and others corrected this shortcoming and included information on the distributions of the random variables in the algorithm for both the linear and nonlinear limit state equations. In the context of AFOSM, the probability of failure has been estimated using two types of approximations to the limit state at the design point: *first order* (leading to the name FORM) and *second order* (leading to the name SORM). The MVFOSM discussed in Section 3.5.1 is an earlier version of FORM. At the present time, AFOSM is known as FORM. In Section 3.5.2, the Hasofer–Lind method is discussed in the context of AFOSM. Other FORM methods are discussed next, and SORM is discussed in Section 3.6.

3.5.3.1 Equivalent Normal Variables
The deficiency in the Hasofer–Lind method, that it is applicable only for normal variables, needs to be addressed at this stage. If not all the variables are normally distributed, as is common in engineering problems, it is necessary to transform the nonnormal variables into equivalent normal variables. The *Rosenblatt transformation* (Rosenblatt, 1952) can be used to obtain a set of statistically independent standard normal variables, if the joint CDF of all the random variables is available. Conceptually, statistically independent nonnormal variables can be transformed to equivalent normal variables in several ways. Procedures to transform correlated nonnormal variables are discussed in Section 3.7. Because a normal random variable can be described uniquely by two parameters (mean and standard deviation), any two appropriate conditions can be used for this purpose. Paloheimo (1973) suggested approximating a nonnormal distribution by a normal distribution having the same mean value and the same P percentile (the value of the variate at which the cumulative probability is $P\%$). He set P equal either to the target failure probability p_f if the variable was a loading variable, or to $(1.0 - p_f)$ if the variable was a resistance variable. The *Rackwitz–Fiessler method* (two-parameter equivalent normal) and the *Chen–Lind method* (1983) and the *Wu–Wirsching method* (1987) (three-parameter equivalent normal) can also be used for this purpose. The Rackwitz–Fiessler method is discussed next.

3.5.3.2 Two-Parameter Equivalent Normal Transformation
Rackwitz and Fiessler (1976) estimated the parameters of the equivalent normal distribution, $\mu_{X_i}^N$ and $\sigma_{X_i}^N$, by imposing two conditions. The cumulative distribution functions and the probability density functions of the actual variables and the equivalent normal variables should be equal at the checking point $(x_1^*, x_2^*, \ldots, x_n^*)$ on the failure surface. Considering each statistically independent nonnormal variable individually and equating its CDF with equivalent normal variable at the checking point results in

$$\Phi\left(\frac{x_i^* - \mu_{X_i}^N}{\sigma_{X_i}^N}\right) = F_{X_i}(x_i^*) \tag{3.34}$$

in which $\Phi(\ )$ is the CDF of the standard normal variate, $\mu_{X_i}^N$ and $\sigma_{X_i}^N$ are the mean and standard deviation of the equivalent normal variable at the checking point, and $F_{X_i}(x_i^*)$ is the CDF of the original nonnormal variables. Equation 3.34 yields

$$\mu_{X_i}^N = x_i^* - \Phi^{-1}[F_{X_i}(x_i^*)]\sigma_{X_i}^N \tag{3.35}$$

Equating the PDFs of the original variable and the equivalent normal variable at the checking point results in

$$\frac{1}{\sigma_{X_i}^N} \phi \left(\frac{x_i^* - \mu_{X_i}^N}{\sigma_{X_i}^N} \right) = f_{X_i}(x_i^*) \tag{3.36}$$

in which $\phi(\ )$ and $f_{X_i}(x_i^*)$ are the PDFs of the equivalent standard normal and the original nonnormal random variable. Equation 3.36 yields

$$\sigma_{X_i}^N = \frac{\phi\{\Phi^{-1}[F_{X_i}(x_i^*)]\}}{f_{X_i}(x_i^*)} \tag{3.37}$$

Having determined $\mu_{X_i}^N$ and $\sigma_{X_i}^N$ and proceeding similarly to the case in which all random variables are normal, we can obtain $\beta_{\text{H-L}}$ using the 7 steps described earlier. Then Equation 3.8 can be used to calculate the failure probability. This approach became well known as the *Rackwitz–Fiessler method* and has been used extensively in the literature.

This approximation of nonnormal distributions can become more and more inaccurate if the original distribution becomes increasingly skewed. For highly skewed distributions, such as the Fréchet (Type II distribution of maxima, see Section 2.4.5), the conditions represented in Equations 3.35 and 3.37 need to be modified. In this case, the mean value and the probability of exceedence of the equivalent normal variable are made equal to the median value and the probability of exceedence of the original random variable, respectively, at the checking point (Rackwitz and Fiessler, 1978). $\mu_{X_i}^N$ and $\sigma_{X_i}^N$ can be estimated as

$$\mu_{X_i}^N = F_{X_i}^{-1}(0.5) = \text{median of } X_i \tag{3.38}$$

and

$$\sigma_{X_i}^N = \frac{x_i^* - \mu_{X_i}^N}{\Phi^{-1}[F_{X_i}(x_i^*)]} \tag{3.39}$$

in which $F_{X_i}^{-1}(\ )$ is the inverse of the nonnormal CDF of X_i.

For highly skewed random variables, usually load-related variables, and relatively large values of x_i^*, the cumulative distribution function at x_i^* will be close to one, and the value of the density function at x_i^* will be very small. Rackwitz and Fiessler (1978) observed, as we did, that if Equations 3.35 and 3.37 are used to calculate $\mu_{X_i}^N$ and $\sigma_{X_i}^N$, then $\mu_{X_i}^N$ will be forced to be small. The larger x_i^* is, the smaller $\mu_{X_i}^N$ will tend to be. But this may destroy the validity of the distribution of X_i; for example, for the Fréchet distribution, it is valid only for the positive values of the random variable. As shown by Ayyub and Haldar (1984), this problem might occur in many designs. A lower limit on $\mu_{X_i}^N$ of zero is suggested and has been proven to give accurate estimates of β and p_f using the optimization algorithm of FORM and SORM. If this lower value is imposed on $\mu_{X_i}^N$, then if $\mu_{X_i}^N < 0$,

$$\sigma_{X_i}^N = \frac{x_i^*}{\Phi^{-1}[F_{X_i}(x_i^*)]}$$

(3.40)

and

$$\mu_{X_i}^N = 0;$$

(3.41)

otherwise use Equations 3.35 and 3.37.

The 7 steps described in Section 3.5.2 to calculate $\beta_{\text{H-L}}$ are still applicable for the Rackwitz–Fiessler method if all the random variables in the limit state equation are normal. If all or some of them are not normal random variables, then another step is necessary. In this step the equivalent normal mean and standard deviation of all the nonnormal random variables at the design point need to be estimated.

Two optimization algorithms are commonly used to obtain the design point and the corresponding reliability or safety index. The first method (Rackwitz, 1976) requires solution of the limit state equation during the iterations and will be referred to as FORM Method 1 in the following discussion. The second method (Rackwitz and Fiessler, 1978) does not require the solution of the limit state equation. Instead, it uses a Newton–Raphson type recursive formula to find the design point. This method will be referred to as FORM Method 2 in the subsequent discussion.

3.5.3.3   FORM Method 1   The steps in this method to estimate the reliability or safety index are explained next, including the computation of parameters for equivalent normal variables. Some improvements in the algorithm suggested by Ayyub and Haldar (1984) are included in these steps. The original coordinate system is used in describing these steps.

Step 1   Define the appropriate limit state equation.

Step 2   Assume an initial value of the safety index β. Any value of β can be assumed; if it is chosen intelligently, the algorithm will converge in a very few steps. An initial β value of 3.0 is reasonable.

Step 3   Assume the initial values of the design point x_i^*, $i = 1, 2, \ldots, n$. In the absence of any other information, the initial design point can be assumed to be at the mean values of the random variables.

Step 4   Compute the mean and standard deviation at the design point of the equivalent normal distribution for those variables that are nonnormal.

Step 5   Compute partial derivatives $(\partial g/\partial X_i)^*$ evaluated at the design point x_i^*.

Step 6   Compute the direction cosines α_i at the design point as

$$\alpha_{X_i} = \frac{\left(\frac{\partial g}{\partial X_i}\right)^* \sigma_{X_i}^N}{\sqrt{\sum_{i=1}^n \left(\frac{\partial g}{\partial X_i} \sigma_{X_i}^N\right)^{2*}}}$$

(3.42)

Note that Equations 3.31 and 3.42 are identical. In Equation 3.31, the direction cosines are evaluated in the reduced coordinates where the standard deviations of the reduced variables are unity. In Equation 3.42, if the random variables are normal, then their standard deviations can be used directly; otherwise, for nonnormal random variables, the equivalent standard deviations at the checking point need to be used.

Step 7   Compute the new values for checking point x_i^* as

$$x_i^* = \mu_{X_i}^N - \alpha_i \beta \sigma_{X_i}^N \qquad (3.43)$$

If necessary, repeat steps 4 through 7 until the estimates of α_i converge with a predetermined tolerance. A tolerance level of 0.005 is common. Once the direction cosines converge, the new checking point can be estimated, keeping β as the unknown parameter. This additional computation may improve the robustness of the algorithm. Note that the assumption of an initial value for β in step 2 is necessary only for the sake of this additional computation. Otherwise, step 2 can be omitted.

Step 8   Compute an updated value for β using the condition that the limit state equation must be satisfied at the new checking point.

Step 9   Repeat steps 3 through 8 until β converges to a predetermined tolerance level. A tolerance level of 0.001 can be used, particularly if the algorithm is developed in a computer environment.

The algorithm converges very rapidly, most of the time within 5 to 10 cycles, depending on the nonlinearity in the limit state equation. A small computer program can be written to carry out the necessary calculations.

Example 3.4

To help implement the preceeding algorithm, an example of a detailed step-by-step solution is given next. Example 3.1 considered in Section 3.5.1 and summarized in Table 3.1, which demonstrated the deficiencies in MVFOSM by giving different safety indexes for different formulations, is considered again. To estimate the safety index using FORM, the distributions of F_y and Z need to be considered. For illustration purposes, assume that F_y is a lognormal variable with a mean of 38 ksi and standard deviation of 3.8 ksi, and Z is a normal random variable with a mean of 54 in^3 and standard deviation of 2.7 in^3. The strength limit state is considered in this example.

The 9 steps necessary to estimate the safety index using FORM are summarized in Table 3.2. For ease of comprehension, these steps are explained in detail as follows.

Step 1   Using the strength formulation, we can express the limit state equation for the problem as $g(\ ) = F_y Z - 1140 = 0$.

Step 2   Assume that $\beta = 3.0$.

Step 3   The initial design point is assumed to be 38 and 54, the mean values of F_y and Z, respectively.

Step 4   Since Z is a normal random variable, no additional transformations are needed. However, since F_y is a lognormal variable, its equivalent normal mean and standard deviation at the design point can be estimated in two ways as discussed next.

Alternative 1 Using Equations 3.35 and 3.37

In this case, $\delta_{F_y} = 0.1$. Thus,

Table 3.2. Steps in FORM Method 1

Step 1				$g() = F_y Z - 1140$		
Step 2	β	3.0			5.002		5.150
Step 3	f_y^*	38.	27.64	29.02	23.96	24.50	24.21
	z^*	54.	50.37	50.27	47.59	47.32	47.10
Step 4	$\mu_{F_y}^N$	37.81	36.30	36.70	34.89	35.13	35.00
	$\sigma_{F_y}^N$	3.79	2.76	2.89	2.39	2.44	2.42
	μ_Z^N	54.0	54.0	54.0	54.0	54.0	54.0
	σ_Z^N	2.7	2.7	2.7	2.7	2.7	2.7
Step 5	$\left(\frac{\partial g}{\partial F}\right)^*$	54.0	50.37	50.17	47.59	47.32	47.10
	$\left(\frac{\partial g}{\partial Z}\right)^*$	38.0	27.64	29.02	23.96	24.50	24.21
Step 6	α_{F_y}	0.894	0.881	0.880	0.869	0.868	0.867
	α_Z	0.448	0.473	0.475	0.494	0.496	0.498
Step 7				Go to step 3. Compute the new checking point using information from step 6		
Step 8	β			5.002		5.150	5.151
Step 9				Repeat steps 3 through 8 until β converges		

Note. The final checking piont is (24.22, 47.07).

$$\zeta_{F_y} = \sqrt{\ln(1 + \delta_{F_y}^2)} = \sqrt{\ln(1 + 0.2^2)} = 0.0997513$$

and

$$\lambda_{F_y} = \ln \mu_{F_y} - \tfrac{1}{2}\zeta_{F_y}^2 = \ln 38 - \tfrac{1}{2}(0.0997513)^2 = 3.632611$$

$$f_{F_y}(f_y^*) = f_{F_y}(38) = \frac{1}{\sqrt{2\pi}\,\zeta_{F_y} f_y^*} \exp\left[-\frac{1}{2}\left(\frac{\ln f_y^* - \lambda_{F_y}}{\zeta_{F_y}}\right)^2\right]$$

$$= \frac{1}{\sqrt{2\pi}\,(0.0997513)(38)} \exp\left[-\tfrac{1}{2}\left(\frac{\ln 38 - 3.632611}{0.0997513}\right)^2\right]$$

$$= 0.1051157$$

$$F_{F_y}(f_y^*) = P(0 < F_y \leq 38) = \Phi\left(\frac{\ln 38 - 3.632611}{0.0997513}\right) = \Phi(0.0498756)$$

Thus,

$$\Phi^{-1}\left[F_{F_y}(f_y^*)\right] = 0.0498756$$

$$\phi\left\{\Phi^{-1}\left[F_{F_y}(f_y^*)\right]\right\} = \frac{1}{\sqrt{2\pi}}\exp\left[-\frac{1}{2}(0.0498756)^2\right] = 0.3984464$$

Using Equation 3.37, we can show that

$$\sigma_{F_y}^N = \frac{0.3984464}{0.1051157} = 3.7905487$$

Using Equation 3.35, we can show that

$$\mu_{F_y}^N = 38 - (0.0498756)(3.7905487) = 37.810944$$

Alternative 2 Simplified Approach

For a lognormal random variable X with parameters λ_X and ζ_X, the equivalent normal mean and standard deviation at the design point x^* can be shown to be

$$\sigma_X^N = \zeta_X x^* \tag{3.44}$$

and

$$\mu_X^N = x^*(1 - \ln x^* + \lambda_X) \tag{3.45}$$

Thus,

$$\sigma_{F_y}^N = \zeta_{F_y}^* f_y^* = (0.0997513)(38) = 3.7905494$$

and

$$\mu_{F_y}^N = 38(1 - \ln 38 + 3.632611) = 37.810944$$

These are the same values estimated using alternative 1. The information is summarized in Table 3.2.

Step 5   For the example under consideration, the partial derivatives $(\partial g / \partial X_i)^*$ evaluated at the design point can be shown to be

$$\left(\frac{\partial g}{\partial F_y}\right)^* = z^* = 54 \quad \text{and} \quad \left(\frac{\partial g}{\partial Z}\right)^* = f_y^* = 38$$

Step 6   Equation 3.42 can be used to calculate the direction cosines for F_y and Z:

$$\alpha_{F_y} = \frac{54 \times 3.7905487}{\sqrt{54 \times 3.7905487)^2 + (38 \times 2.7)^2}} = 0.8939809$$

$$\alpha_Z = \frac{38 \times 2.7}{\sqrt{54 \times 3.7905487^2 + (38 \times 2.7)^2}} = 0.4481049$$

Step 7   Equation 3.43 is used to find the coordinates of the new design point:

$$f_y^* = 37.810944 - 0.8939809 \times 3.0 \times 3.7905487 = 27.644908$$

and

$$z^* = 54 - 0.4481049 \times 3.0 \times 2.7 = 50.37035$$

The second iteration will start with the coordinates of the new design point just calculated in step 7, as shown in Table 3.2. Steps 3 through 7 are repeated until the direction cosines converge at a tolerance level of 0.005. The detailed calculations are not shown here; however, they are similar to the calculations just discussed. At the third iteration, α_{F_y} and α_Z converge to 0.8800674 and 0.4748486, respectively.

Step 8   The coordinates of the new design point, keeping β as the unknown parameter, are

$$f_y^* = 36.698103 - (0.8800674)(2.8942793)\beta = 36.698103 - 2.54716\beta$$

and

$$z^* = 54 - (0.4748486)(2.7)\beta = 54 - 1.28209\beta$$

A new β value can be estimated by satisfying the limit state equation as

$$(36.698103 - 2.54716\beta)(54 - 1.28209\beta) - 1140 = 0$$

When this equation is solved, β is found to be 5.002. This updated β is considerably different than the initial assumed value of 3.0. With the updated β value, the coordinates of the new design point become

$$f_y^* = 36.70 - 0.880 \times 5.002 \times 2.89 = 23.96$$

and

$$z^* = 54 - 0.475 \times 5.002 \times 2.7 = 47.59$$

Thus, the fourth iteration will start with the updated information on the coordinates of the design point, as shown in Table 3.2. Again, the direction cosines converge after the fifth iteration, and the updated β becomes 5.150.

Step 9   Steps 3 through 8 are repeated until β converges to a tolerance level of 0.005. As shown in Table 3.2, at the sixth iteration, β converges to 5.151 with a tolerance level of 0.005, and the corresponding checking point is (24.22, 47.07).

3.5.3.4   FORM Method 2   Notice that in step 8 of FORM Method 1, the limit state equation needs to be solved to find the new design point. This may be difficult in the case of complicated nonlinear g functions. Also, in many practical problems, the g function may not even be available in a closed form. In that case, it is impossible to perform step 8, thus limiting the usefulness of FORM Method 1. Therefore, an alternative Newton–Raphson type recursive algorithm, referred to as FORM Method 2 in this section, is presented here to find the design point.

This algorithm, suggested by Rackwitz and Fiessler (1978), is similar to FORM Method 1 in that it linearizes the performance function at each iteration point; however, instead of solving the limit state equation explicitly for β, it uses the derivatives to find the next iteration point. The algorithm can best be explained with the help of Figures 3.6 and 3.7. Consider first the linear performance function shown in Figure 3.6. Since the limit state is not available in closed form, the starting point $x_0'^*$ (usually the vector of mean values of

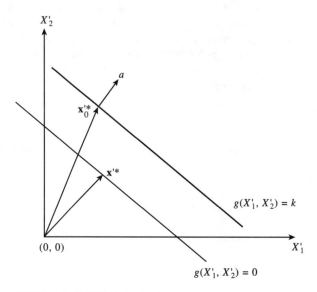

FIGURE 3.6. FORM Method 2 for a linear performance function.

the random variables) may not be on the limit state $g(X_1', X_2') = 0$, but on a parallel line $g(X_1', X_2') = k$. Hence the optimization algorithm has to start from point $\mathbf{x}_0'^*$ which may not be on the limit state, and converge to the minimum distance point $\mathbf{x}'^*$ on the limit state. The linear performance function $g(\mathbf{x}')$ may be expressed as

$$g(\mathbf{x}') = b + \mathbf{a}^t\mathbf{x}'$$
$$= b + a_1 x_1' + a_2 x_2' \tag{3.46}$$

Here $\mathbf{a}^t = (a_1, a_2)$ is the transpose of the gradient vector (i.e., vector of first derivatives) of the performance function. The magnitudes of the vectors $\mathbf{x}_0'^*$ and $\mathbf{x}'^*$ denote the distance from the origin to starting point and to the limit state $g(\mathbf{X}') = 0$, respectively. From geometry, $\mathbf{x}'^*$ can be expressed in terms of $\mathbf{x}_0'^*$ as

$$\mathbf{x}'^* = \frac{1}{|\mathbf{a}|^2} \left[\mathbf{a}^t\mathbf{x}_0'^* - g\left(\mathbf{x}_0'^*\right) \right] \{\mathbf{a}\} \tag{3.47a}$$

Rewriting Equation 3.47a in terms of the components of all the vectors results in

$$\left\{ \begin{matrix} x_1'^* \\ x_2'^* \end{matrix} \right\} = \frac{1}{a_1^2 + a_2^2} \left[a_1 x_{01}'^* + a_2 x_{02}'^* - g(x_{01}'^*, x_{02}'^*) \right] \left\{ \begin{matrix} a_1 \\ a_2 \end{matrix} \right\} \tag{3.47b}$$

Since the performance function is linear in this case, its gradient is constant; hence the distance to the limit state from the origin is obtained in one step.

Equation 3.47a can be generalized for a nonlinear performance function as shown in Figure 3.7 as

$$\mathbf{x}_{k+1}'^* = \frac{1}{|\nabla g(\mathbf{x}_k'^*)|^2} \left[\nabla g(\mathbf{x}_k'^*)^t \mathbf{x}_k'^* - g(\mathbf{x}_k'^*) \right] \nabla g(\mathbf{x}_k'^*) \tag{3.48}$$

where $\nabla g(\mathbf{x}_k'^*)$ is the gradient vector of the performance function at $\mathbf{x}_k'*$, the kth iteration point. Note that k refers to the iteration number. Therefore $\mathbf{x}_k'^*$ is a vector with components $\{x_{1k}'^*, x_{2k}'^*, \ldots, x_{nk}'^*\}^t$, where n is the number of random variables. The meaning of $\mathbf{x}_{k+1}'^*$ is similar.

Since the performance function is nonlinear, the gradient is not constant but varies from point to point. Therefore, instead of a one-step solution in the case of the linear performance function, the point of minimum distance has to be searched through the recursive formula given in Equation 3.48. This formula can be geometrically interpreted using Figure 3.7. At each iteration point, the performance function is approximated by the tangent at the point; that is, the performance function is linearized with $g(\mathbf{x}_k'^*)$ and $\nabla g(\mathbf{x}_k'^*)$ corresponding to $\mathbf{x}_0'^*$ and $\mathbf{a}$, respectively, in Equation 3.47a. The next iteration point $\mathbf{x}_{k+1}'^*$ is computed the same way as in the case of the linear performance function. If the performance function were linear, $\mathbf{x}_{k+1}'^*$ would exactly correspond to $\mathbf{x}_k'^*$, for $k > 0$. However, since the performance function is nonlinear, the gradient at $\mathbf{x}_{k+1}'^*$ is different from those at $\mathbf{x}_k'^*$. Therefore, it is again linearized at $\mathbf{x}_{k+1}'^*$ and another iteration point $\mathbf{x}_{k+2}'^*$ is computed. The algorithm is repeated until convergence, satisfying the following two criteria:

1. If $|\mathbf{x}_k'^* - \mathbf{x}_{k-1}'^*| \leq \delta$, stop.
2. If $|g(\mathbf{x}_k'^*)| \leq \epsilon$, stop.

Both δ and ϵ are small quantities, usually 0.001.

From this discussion, it is obvious that the recursive formula in Equation 3.48 results from the linearization of the performance function. Equation 3.48 may also be derived

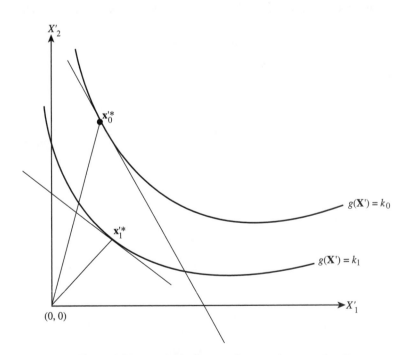

FIGURE 3.7. FORM Method 2 for a nonlinear performance function.

directly from this idea, considering a first-order Taylor series approximation of the performance function as

$$g(\mathbf{x}_{k+1}^{\prime*}) = g(\mathbf{x}_k^{\prime*}) + \nabla g(\mathbf{x}_k^{\prime*})(\mathbf{x}_{k+1}^{\prime*} - \mathbf{x}_k^{\prime*}) \tag{3.49}$$

Thus, the limit state $g(\mathbf{x}_{k+1}^{\prime*}) = 0$ becomes

$$g(\mathbf{x}_k^{\prime*}) + \nabla g(\mathbf{x}_k^{\prime*})(\mathbf{x}_{k+1}^{\prime*} - \mathbf{x}_k^{\prime*}) = 0 \tag{3.50}$$

Rearrangement of the terms in this equation gives Equation 3.48.

Compared to other nonlinear optimization algorithms available in the literature, the algorithm just described requires the least computation at each step. The next iteration point is computed using a single recursive formula that requires information only about the value and the gradient of the performance function. The storage requirement is therefore minimal. The algorithm is also found to converge fast in many cases. For these reasons, this algorithm has been widely used in the literature.

Convergence Problems   This algorithm may fail to converge in some situations. It may converge very slowly, or oscillate about the solution without convergence, or diverge away from the solution. Two such examples are shown in Figures 3.8 and 3.9. For the case of a single variable x', the formula in Equation 3.48 reduces to the Newton–Raphson method to find the root of $g(x') = 0$. It is well known that the Newton–Raphson method may fail to find the roots of a function in certain circumstances. Figure 3.8 illustrates one such situation where the Newton–Raphson method diverges further and further away from the solution. In the example with two variables (Figure 3.9), the performance function is

$$g(\mathbf{x}') = x_1' x_2' - d \tag{3.51}$$

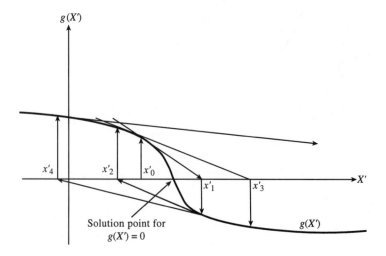

FIGURE 3.8. Example of failure of the Newton–Raphson method.

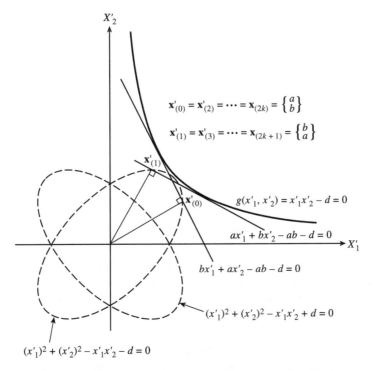

FIGURE 3.9. Nonconvergence of the Rackwitz–Fiessler algorithm: bivariate case.

If the starting point (a, b) falls on one of the two ellipses $(x_1')^2 + (x_2')^2 + x_1'x_2' + d = 0$ and $(x_1')^2 + (x_2')^2 - x_1'x_2' - d = 0$, then the algorithm generates points that oscillate between (a, b) and (b, a), as shown by Liu and Der Kiureghian (1986).

Thus, it is possible that the Rackwitz–Fiessler algorithm may not converge to the MPP (minimum distance point, or most probable point of failure) in some cases. Other optimization algorithms such as sequential quadratic programming or the BFGS (Broyden–Fletcher–Goldfarb–Shanno) method (Vanderplaats, 1984) may be used in that case.

Similar to FORM Method 1, FORM Method 2 can be described as follows. Both the original and equivalent standard normal or reduced coordinate systems are used in this method.

Step 1   Define the appropriate performance function.

Step 2   Assume initial values of the design point x_i^*, $i = 1, 2, \ldots, n$, and compute the corresponding value of the performance function $g(\ )$. In the absence of any other information, the initial design point can be the mean values of the random variables.

Step 3   Compute the mean and standard deviation at the design point of the equivalent normal distribution for those variables that are nonnormal. The coordinates of the design point in the equivalent standard normal space are

$$x_i'^* = \frac{x_i^* - \mu_{X_i}^N}{\sigma_{X_i}^N} \tag{3.52}$$

Step 4   Compute the partial derivative $\partial g/\partial X_i$ evaluated at the design point $\mathbf{x}_i^*$.

Step 5   Compute the partial derivatives $\partial g/\partial X_i'$ in the equivalent standard normal space by the chain rule of differentiation as

$$\frac{\partial g}{\partial X_i'} = \frac{\partial g}{\partial X_i}\frac{\partial X_i}{\partial X_i'} = \frac{\partial g}{\partial X_i}\sigma_{X_i}^N \tag{3.53}$$

The partial derivatives $\partial g/\partial X_i'$ are the components of the gradient vector of the performance function in the equivalent standard normal space. The components of the corresponding unit vector are the direction cosines of the performance function, computed as

$$\alpha_i = \frac{\left(\frac{\partial g}{\partial X_i'}\right)^*}{\sqrt{\sum_{i=1}^{n}\left(\frac{\partial g}{\partial X_i'}\right)^{2*}}} = \frac{\left(\frac{\partial g}{\partial X_i}\right)^*\sigma_{X_i}^N}{\sqrt{\sum_{i=1}^{n}\left(\frac{\partial g}{\partial X_i'}\sigma_{X_i}^N\right)^{2*}}} \tag{3.54}$$

Note that this is exactly the same formula as in Equation 3.42. Although the direction cosines are not directly used in the current algorithm, they are used later in the implementation of SORM, the second-order reliability method, discussed in Section 3.6.

Step 6   Compute the new values for the design point in the equivalent standard normal space $(\mathbf{x}_i'^*)$ using the recursive formula of Equation 3.48.

Step 7   Compute the distance to this new design point from the origin as

$$\beta = \sqrt{\sum_{i=1}^{n}(x_i'^*)^2} \tag{3.55}$$

Check the convergence criterion for β (i.e., change in the value of β between two consecutive iterations is less than a predetermined tolerance level, say 0.001).

Step 8   Compute the new values for the design point in the original space (x_i^*) as

$$x_i^* = \mu_{X_i}^N + \sigma_{X_i}^N x_i'^* \tag{3.56}$$

Compute the value of the performance function $g(\ )$ for this new design point, and check the convergence criterion for $g(\ )$; that is, check that the value of $g(\ )$ is very close to zero, say within 0.001. If both convergence criteria are satisfied, stop. Otherwise, repeat steps 3 through 8 until convergence.

Example 3.5

Consider again the performance function $g(\ ) = F_y Z - 1140$ used in clarifying FORM Method 1 in Example 3.4. Note that FORM Method 2 is particularly useful when the performance function is implicit, that is, when it cannot be written as a closed-form expression in terms of the random variables. However, this simple closed-form performance function is chosen for the sake of illustration and comparison with FORM Method 1. Issues related to implicit functions are discussed in detail in Chapter 5.

F_y is assumed to have a lognormal distribution with a mean value of 38.0 ksi and a standard deviation of 3.8 ksi. Z is assumed to have a normal distribution with a mean value of 54.0 in^3 and a standard deviation of 2.7 in^3. The 8 steps of FORM Method 2 are summarized in Table 3.3. For ease of comprehension, the first iteration is discussed next.

Step 1   The performance function is $g(\ ) = F_y Z - 1140$.

Step 2   The initial values of the design point are chosen to be the same as the mean values of the two random variables, that is, $f_y^* = 38$ and $z^* = 54$. For this initial design point, the value of $g(\ )$ is computed as

$$g(\ ) = (38)(54) - 1140 = 912$$

Table 3.3. Steps in FORM method 2

Step 1		$g(\ ) = F_y Z - 1140$			
Step 2		Initial Values: $f_y^* = 38$, $z^* = 54$, $g(\ ) = 912.0$			
Step 3	$\mu_{F_y}^N$	37.81	35.116	34.960	35.003
	$\sigma_{F_y}^N$	3.79	2.44	2.405	2.415
	μ_Z^N	54.00	54.00	54.00	54.00
	σ_Z^N	2.70	2.70	2.70	2.70
	$f_y'^*$	0.05	−4.365	−4.510	−4.471
	z'^*	0.00	−1.765	−2.479	−2.558
Step 4	$\left(\frac{\partial g}{\partial F_y}\right)^*$	54.00	49.235	47.307	47.093
	$\left(\frac{\partial g}{\partial Z}\right)^*$	38.00	24.464	24.112	24.207
Step 5	$\left(\frac{\partial g}{\partial F_y'}\right)^*$	204.69	120.15	113.78	113.71
	$\left(\frac{\partial g}{\partial Z'}\right)^*$	102.60	66.05	65.10	65.36
Step 6	New $f_y'^*$	−3.521	−4.509	−4.471	−4.466
	New z'^*	−1.765	−2.479	−2.558	−2.567
Step 7	New β	3.939	5.145	5.151	5.151
	$\Delta\beta$		1.206	0.006	0.0001
Step 8	New f_y^*	24.464	24.112	24.207	24.22
	New z^*	49.235	47.307	47.093	47.07
	New $g(\ )$	64.500	0.679	−0.020	−0.0002

Note. Convergence criteria in steps 7 and 8: (1) $|\Delta\beta| \leq 0.001$, (2) $|g(\ )| \leq 0.001$. The final checking point is (24.22, 47.07).

Step 3   The equivalent normal mean and standard deviation for the lognormal variable F_y are computed in the same way as in FORM Method 1 as $\mu_{F_y}^N = 37.81$ and $\sigma_{F_y}^N = 3.79$. Since Z is a normal random variable, its equivalent mean and standard deviation are the same as the original mean and standard deviation. Using Equation 3.52, the coordinates of the design point in the equivalent standard normal space are

$$f_y'^* = \frac{38 - 38.81}{3.79} = 0.05 \qquad z'^* = \frac{27 - 27}{2.7} = 0$$

Step 4   The partial derivatives evaluated at the design point are

$$\left(\frac{\partial g}{\partial F_y}\right)^* = z^* = 54 \quad \text{and} \quad \left(\frac{\partial g}{\partial Z}\right)^* = f_y^* = 38$$

Step 5   Equation 3.53 is used to find the partial derivatives in the equivalent normal space:

$$\left(\frac{\partial g}{\partial F_y'}\right)^* = \left(\frac{\partial g}{\partial F_y}\right)^* \sigma_{F_y}^N = 54 \times 3.79 = 204.69$$

$$\left(\frac{\partial g}{\partial Z'}\right)^* = \left(\frac{\partial g}{\partial Z}\right)^* \sigma_Z^N = 38 \times 2.7 = 102.60$$

Step 6   The coordinates of the new design point in the equivalent standard normal space are computed using the recursive formula of Equation 3.48 as

New $\begin{Bmatrix} f_y'^* \\ z'^* \end{Bmatrix}$

$$= \frac{1}{\left[\left\{\left(\frac{\partial g}{\partial F_y'}\right)^*\right\}^2 + \left\{\left(\frac{\partial g}{\partial Z'}\right)^*\right\}^2\right]} \left[\left(\frac{\partial g}{\partial F_y'}\right)^* f_y'^* + \left(\frac{\partial g}{\partial Z'}\right)^* z'^* - g()\right] \begin{Bmatrix} \left(\frac{\partial g}{\partial F_y'}\right)^* \\ \left(\frac{\partial g}{\partial Z'}\right)^* \end{Bmatrix}$$

$$= \frac{1}{[204.69^2 + 102.60^2]}(204.69 \times 0.05 + 102.60 \times 0.0 - 912) \begin{Bmatrix} 204.69 \\ 102.60 \end{Bmatrix}$$

$$= \begin{Bmatrix} -3.521 \\ -1.765 \end{Bmatrix}.$$

Step 7   Using Equation 3.55, we find the value of β to be

$$\beta = \sqrt{(-3.521)^2 + (-1.765)^2} = 3.939$$

The check for convergence at this step will start during the second iteration. In the second iteration, β is calculated as 5.145. Therefore, the change in the value of β between the first and second iterations is $1.206 > 0.001$.

Step 8   Equation 3.56 is used to find the coordinates of the new iteration point in the original space:

$$f_y^* = \mu_{F_y}^N + \sigma_{F_y}^N f_y'^* = 37.81 + 3.79 \times (-3.521) = 24.464$$

$$z^* = \mu_Z^N + \sigma_Z^N z'^* = 54.0 + 2.7 \times (-1.765) = 49.235$$

At these values, the performance function is evaluated as

$$g(\ ) = 24.464 \times 49.235 - 1140 = 64.5$$

The convergence criterion for $g(\ )$ is checked. The current $g(\ )$ value is greater than the tolerance level of 0.001. Therefore, proceed to the next iteration at step 3.

The search is stopped after four iterations since the value of β has converged to 5.151 and the value of $g(\ )$ has become less than 0.001. As expected, both FORM methods gave identical results. However, the advantage of FORM Method 2 is clear; it does not require solution of the limit state equation and simply uses a recursive formula to converge to the design point. Comparing Tables 3.2 and 3.3, we can observe that the same quantities are computed in both methods. The only difference is in how the new iteration point is computed. Also, during the first iteration, since both the algorithms were started from the mean values of the variables, many of the quantities have the same values.

Observations   Several important observations can be made at this time. The safety index obtained using any one of the FORM methods is different than the safety indices shown in Table 3.1. Also note that if the stress formulation of the limit state was considered, the safety index would have the same value (i.e., 5.151) as the strength formulation. The Hasofer–Lind algorithm ignores the information on the distributions of the random variables, essentially assuming both random variables are normal; if it is used, β_{H-L} is found to be 4.261.

The FORM methods clearly demonstrate that information on the distribution of random variables is important in calculating the safety index and the corresponding probability of failure. To amplify the point, the safety indices of the same beam problem just considered are calculated assuming F_y and Z have different distributions, with the results summarized in Table 3.4.

In this example, only two random variables are present in the limit state equation, but there could be any number of random variables in the limit state equation. As long as they are uncorrelated, either of the two algorithms discussed here can be used without modification to calculate the safety index or the corresponding probability of failure. If

Table 3.4. Safety index under various probability distributions

Limit state equation = $g(\ ) = F_y Z - 1140 = 0$

Random Variables	F_y	Z	β
Probability	Normal	Normal	4.261
Distribution	Normal	Lognormal	4.266
	Lognormal	Normal	5.151
	Lognormal	Lognormal	5.213

Table 3.5. Uncertainty in the design parameters of a reinforced concrete beam

Random Variables	Mean	Coefficient of Variation
A_s (in^2)	1.56	0.036
f_y (ksi)	47.7	0.15
f_c' (ksi)	3.5	0.21
b (in)	8.0	0.045
d (in)	13.2	0.086
η	0.59	0.05
M (kip-in)	326.25	0.17

the random variables are correlated, some modifications in the algorithm are necessary, as discussed in Section 3.7.

To demonstrate the application of FORM to a much more complicated problem involving several random variables, the moment capacity of a singly reinforced rectangular prismatic concrete beam is considered here. The moment capacity or resistance M_R of such a beam can be calculated using the following expression:

$$M_R = A_s f_y d \left(1 - \eta \frac{A_s}{b\,d} \frac{f_y}{f_c'}\right) \tag{3.57}$$

where A_s is the area of the tension reinforcing bars, f_y is the yield stress of the reinforcing bars, d is the distance from the extreme compression fiber to the centroid of the tension reinforcing bars, η is the concrete stress block parameter, f_c' is the compressive strength of concrete, and b is the width of the compression face of the member. It is extensively reported in the literature that all these variables are random. Their mean values and coefficients of variation are tabulated in Table 3.5. Assume further that the beam is subjected to a moment, M, which is also a random variable. Its mean value and coefficient of variation are shown in Table 3.5.

The limit state equation for the problem can be expressed as

$$g(\,) = A_s f_y d \left(1 - \eta \frac{A_s}{b\,d} \frac{f_y}{f_c'}\right) - M = 0 \tag{3.58}$$

For various distributions of the random variables in Equation 3.58, the safety indices are calculated using the FORM method. It is not possible to show the detailed calculations in tabular form as in the previous example. A computer program is used for this purpose. The results, summarized in Table 3.6, clearly indicate that the distributions of random variables play a very important role in safety index or probability of failure estimation.

3.6   SECOND-ORDER RELIABILITY METHODS (SORM)

As mentioned in Section 3.5, limit states, either explicit or implicit, linear or nonlinear, are essential in risk and reliability analysis. The computations required for reliability analysis of problems with linear limit state equations are relatively simple. However, the limit state could be nonlinear either due to the nonlinear relationship between the random variables in

Table 3.6. Safety index of a reinforced concrete beam under various probability distributions

Random Variables	Probability Distribution			
A_s	Normal	Normal	Lognormal	Lognormal
f_y	Normal	Normal	Lognormal	Lognormal
f_c'	Normal	Normal	Lognormal	Lognormal
b	Normal	Normal	Lognormal	Lognormal
d	Normal	Normal	Lognormal	Lognormal
η	Normal	Normal	Lognormal	Lognormal
M	Normal	Lognormal	Normal	Lognormal
β	3.833	3.761	4.338	4.091

the limit state equation or due to some variables being nonnormal. A linear limit state in the original space becomes nonlinear when transformed to the standard normal space (which is where the search for the minimum distance point is conducted) if any of the variables is nonnormal. Also, the transformation from correlated to uncorrelated variables might induce nonlinearity; this transformation is discussed in detail in Section 3.7. If the joint probability density function, PDF, of the random variables decays rapidly as one moves away from the minimum distance point, then the first-order estimate of failure probability is quite accurate. If the decay of the joint PDF is slow and the limit state is highly nonlinear, then one has to use a higher-order approximation for the failure probability computation.

Consider the two limit states shown in Figure 3.10, one linear and one nonlinear. Both limit states have the same minimum distance point, but the failure domains, shown by the shaded regions, are different for the two cases. The FORM approach will give the same reliability estimate for both cases. But it is apparent that the failure probability of the non-linear limit state should be less than that of the linear limit state, due to the difference in the failure domains. The curvature of the nonlinear limit state is ignored in the FORM ap-

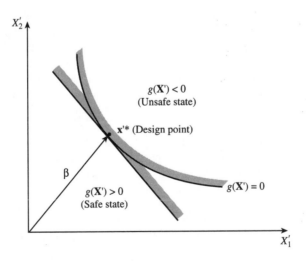

FIGURE 3.10. Linear and nonlinear limit states.

proach, which uses only a first-order approximation at the minimum distance point. Thus, the curvature of the limit state around the minimum distance point determines the accuracy of the first-order approximation in FORM. The curvature of any equation is related to the second-order derivatives with respect to the basic variables. Thus, the second-order reliability method (SORM) improves the FORM result by including additional information about the curvature of the limit state.

The second-order Taylor series approximation to a general nonlinear function $g(X_1, X_2, \ldots, X_n)$ at the value $(x_1^*, x_2^*, \ldots, x_n^*)$ is

$$
\begin{aligned}
g(X_1, X_2, \ldots, X_n) = {} & g(x_1^*, x_2^*, \ldots, x_n^*) + \sum_{i=1}^{n} (x_i - x_i^*) \frac{\partial g}{\partial X_i} \\
& + \frac{1}{2} \sum_{i=1}^{n} \sum_{i=1}^{n} (x_i - x_i^*)(x_j - x_j^*) \frac{\partial^2 g}{\partial X_i \partial X_j} + \cdots
\end{aligned}
\tag{3.59}
$$

where the derivatives are evaluated at the design point of the X_i's.

The variables $(X_1, X_2, \ldots, X_n)$ are used in Equation 3.59 in a generic sense. One should use the appropriate set of variables and notation depending on the space being considered. In the case of reliability analysis, the second-order approximation to $g(\ )$ is being constructed in the space of standard normal variables, at the minimum distance point. The following notation is used in this section: X_i refers to a random variable in the original space, and Y_i refers to the random variable in the equivalent uncorrelated standard normal space. If all the variables are uncorrelated, $Y_i = (X_i - \mu_{X_i}^N)/\sigma_{X_i}^N$, where $\mu_{X_i}^N$ and $\sigma_{X_i}^N$ are the equivalent normal mean and standard deviation of X_i at the design point x_i^*. The transformation from X_i to Y_i for correlated variables is discussed in Section 3.7.

In the Taylor series approximation given in Equation 3.59, FORM ignores the terms beyond the first-order term (involving first-order derivatives), and SORM ignores the terms beyond the second-order term (involving second-order derivatives).

The SORM approach was first explored by Fiessler et al. (1979) using various quadratic approximations. A simple closed-form solution for the probability computation using a second-order approximation, p_{f_2}, was given by Breitung (1984) using the theory of asymptotic approximations as

$$
p_{f_2} \approx \Phi(-\beta) \prod_{i=1}^{n-1} (1 + \beta \kappa_i)^{-1/2}
\tag{3.60}
$$

where κ_i denotes the principal curvatures of the limit state at the minimum distance point, and β is the reliability index using FORM. Breitung showed that this second-order probability estimate asymptotically approaches the first-order estimate as β approaches infinity, if $\beta \kappa_i$ remains constant. Refer to Hohenbichler et al. (1987) for a theoretical explanation of FORM and SORM using the concept of asymptotic approximations.

In Equation 3.60, it is necessary to compute the principal curvatures κ_i. To do this, first the Y_i variables (in the $\mathbf{Y}$ space) are rotated to another set of variables, denoted as Y_i', such that the last Y_i' variable coincides with the vector $\boldsymbol{\alpha}$, the unit gradient vector of the limit state at the minimum distance point. This is shown in Figure 3.11 for a problem with two random variables. It is apparent that this is simply a rotation of coordinates.

The transformation from the $\mathbf{Y}$ space to $\mathbf{Y}'$ space is an orthogonal transformation:

$$
\mathbf{Y}' = \mathbf{RY}
\tag{3.61}
$$

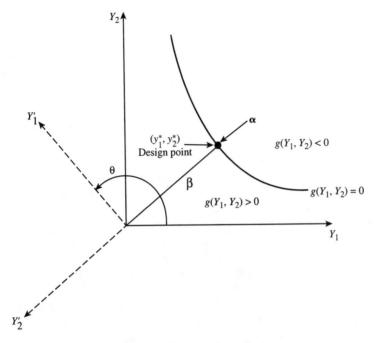

FIGURE 3.11. Rotation of coordinates.

where $\mathbf{R}$ is the rotation matrix. For the simple case of two random variables, it is

$$\mathbf{R} = \begin{bmatrix} \cos\theta & \sin\theta \\ -\sin\theta & \cos\theta \end{bmatrix} \tag{3.62}$$

where θ is the angle of rotation as shown in Figure 3.11 (counterclockwise rotation of the axes gives positive θ). When the number of variables is more than two, the $\mathbf{R}$ matrix is computed in two steps. In step 1, first a matrix, $\mathbf{R}_0$, is constructed as follows:

$$\mathbf{R}_0 = \begin{bmatrix} 1 & 0 & . & . & . & 0 \\ 0 & 1 & 0 & . & . & 0 \\ . & . & . & . & . & . \\ . & . & . & . & . & . \\ \alpha_1 & \alpha_2 & . & . & . & \alpha_n \end{bmatrix} \tag{3.63}$$

where $\alpha_1, \alpha_2, \ldots, \alpha_n$ are the direction cosines, that is, components of the unit gradient vector $\boldsymbol{\alpha}$ shown in Figure 3.11. In step 2, a Gram–Schmidt orthogonalization procedure (refer to Appendix 3) is applied to this matrix, and the resulting matrix is $\mathbf{R}$.

Once the $\mathbf{R}$ matrix is obtained, a matrix $\mathbf{A}$, whose elements are denoted as a_{ij}, is computed as

$$a_{ij} = \frac{(\mathbf{RDR^t})_{ij}}{|\nabla G(\mathbf{y}^*)|} \qquad i, j = 1, 2, \ldots, n-1 \tag{3.64}$$

where $\mathbf{D}$ is the $n \times n$ second-derivative matrix of the limit-state surface in the standard normal space evaluated at the design point, $\mathbf{R}$ is the rotation matrix, and $|\nabla G(\mathbf{y}^*)|$ is the length of the gradient vector in the standard normal space.

In the rotated space, the last variable, Y_n, coincides with the β vector computed in FORM. In the next step, the last row and last column in the $\mathbf{A}$ matrix and the last row in the $\mathbf{Y}'$ vector are dropped to take this factor into account. The limit state can be rewritten in terms of a second-order approximation in this rotated standard normal space $\mathbf{Y}'$ as

$$y'_n = \beta + \tfrac{1}{2}\mathbf{y}'^t \mathbf{A} \mathbf{y}' \qquad (3.65)$$

where the matrix $\mathbf{A}$ is now of the size $(n-1) \times (n-1)$.

Finally, the main curvatures κ_i, used in Breitung's formula (Equation 3.60), are computed as the eigenvalues of the matrix $\mathbf{A}$. Once the κ_i's are computed, Breitung's formula can be used to compute the second-order estimate of the probability of failure.

Breitung's SORM method uses a parabolic approximation; that is, it does not use a general second-order approximation. (It ignores the mixed terms and their derivatives in the Taylor series approximation in Equation 3.59.) Also, as mentioned earlier, it uses the theory of asymptotic approximation to derive the probability estimate. The asymptotic formula is accurate only for large values of β, which is the case for practical high-reliability problems. However, if the value of β is low, the SORM estimate could be inaccurate. Tvedt (1990) developed two alternative SORM formulations to take care of these problems. Tvedt's method uses a parabolic and a general second-order approximation to the limit state, and it does not use asymptotic approximations. Refer to Tvedt (1990) for a detailed presentation of this method.

Der Kiureghian et al. (1987) approximated the limit state by two semiparabolas using curve fitting at several discrete points around the design point, and used both sets of curvatures in Breitung's formula (Equation 3.60). This strategy helps to avoid the computation of a full second-derivative matrix using the original limit state and is efficient for problems with a large number of random variables.

Example 3.6

Example 3.4 in Section 3.5.3, discussed in detail in Table 3.2, is considered again. Using FORM and the strength formulation, assuming F_y to be a lognormal variable with a mean of 38 ksi and standard deviation of 3.8 ksi, and assuming Z to be a normal variable with a mean of 54 in^3 and standard deviation of 2.7 in^3, we find the safety index is 5.151. Using SORM, estimate the safety index.

Solution   The estimation of a safety index using FORM and to a greater extent using SORM is rarely undertaken using hand calculations; computer programs are used for this purpose. However, as in the FORM example where the detailed calculations are summarized in Table 3.2, in this section the detailed steps in SORM are explained for better understanding of the concept.

Table 3.2 reveals that the final design point in the original variable space is (24.22 ksi, 47.07 in^3), and the corresponding direction cosines for F_y and Z are 0.867 and 0.498, respectively. The equivalent normal mean and standard deviation of F_y at the design point are 35.008 and 2.416, respectively. Transforming F_y and Z from the original to standard normal space results in

$$Y_{F_y} = \frac{F_y - F_y^N}{\sigma_{F_y^N}}$$

and

$$Y_Z = \frac{Z - \mu_Z}{\sigma_Z}$$

The coordinates of the design point in the standard normal space become

$$y_{F_y}^* = \frac{24.22 - 35.008}{2.416} = -4.466$$

and

$$y_Z^* = \frac{47.07 - 54.00}{2.7} = -2.567$$

The design point is graphically shown in the standard normal space in Figure 3.12.

The first step in SORM is to construct the rotation matrix $\mathbf{R}$ in Equation 3.62 or 3.63. For the two-variable problem under consideration, Equation 3.62 is sufficient. In this example, Y_{F_y} is the first coordinate and Y_Z is the second coordinate. In the rotated coordinates, the second coordinate Y_Z' needs to coincide with the unit gradient vector $\boldsymbol{\alpha}$. The corresponding rotation angle θ is shown in Figure 3.12. Therefore, the $\mathbf{R}$ matrix is

$$\mathbf{R} = \begin{bmatrix} 0.498 & -0.867 \\ 0.867 & 0.498 \end{bmatrix}$$

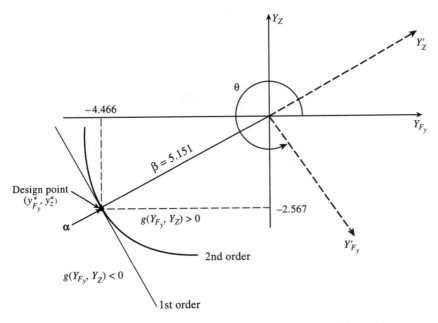

FIGURE 3.12. Design point and rotation of coordinates in the standard normal space.

Notice that the elements of **R** are easily available from the direction cosines, that is, the components of the unit gradient vector $\boldsymbol{\alpha}$.

The next step is to construct the **D** matrix, containing the second derivatives of the performance function, in the standard normal space. For the performance function of $g(\ ) = F_y Z - 1140$ in the original space, using the chain rule of differentiation, the elements of **D** are

$$\frac{\partial^2(\ )}{\partial F_y'^2} = \frac{\partial}{\partial F_y}\left\{\left[\frac{\partial g(\ )}{\partial F_y}\frac{\partial F_y}{\partial F_y'}\right]\right\}\frac{\partial F_y}{\partial F_y'} = \frac{\partial}{\partial F_y}\left[Z\sigma_{F_y}^N\right]\sigma_{F_y}^N = 0$$

$$\frac{\partial^2 g(\ )}{\partial Z'^2} = \frac{\partial}{\partial Z}\left\{\left[\frac{\partial g(\ )}{\partial Z}\frac{\partial Z}{\partial Z'}\right]\right\}\frac{\partial Z}{\partial Z'} = \frac{\partial}{\partial Z}\left[F_y\sigma_Z\right]\sigma_Z = 0$$

$$\frac{\partial^2 g(\ )}{\partial F_y'\partial Z'} = \frac{\partial}{\partial F_y}\left\{\left[\frac{\partial g(\ )}{\partial Z}\frac{\partial Z}{\partial Z'}\right]\right\}\frac{\partial F_y}{\partial F_y'} = \frac{\partial}{\partial F_y}\left[F_y\sigma_Z\right]\sigma_{F_y}^N = \sigma_Z\sigma_{F_y}^N$$

Therefore, matrix **D** is assembled as

$$\mathbf{D} = \begin{bmatrix} 0 & \sigma_Z\sigma_{F_y}^N \\ \sigma_Z\sigma_{F_y}^N & 0 \end{bmatrix} = \begin{bmatrix} 0 & 2.7 \times 2.416 \\ 2.7 \times 2.416 & 0 \end{bmatrix} = \begin{bmatrix} 0 & 6.523 \\ 6.523 & 0 \end{bmatrix}$$

Next matrix **A** from Equation 3.64 needs to be computed. To do this, the length of the gradient vector in the standard normal space at the design point $|\nabla G(\mathbf{y}*)|$ is needed. Normally, this would be readily available from the FORM analysis. In fact, the direction cosines given previously are simply the components of the unit gradient vector. However, the computation is shown in detail as follows, for the sake of clarity.

To evaluate $\nabla G(\mathbf{y}^*)$ in Equation 3.64, the following two partial derivatives need to be evaluated:

$$\frac{\partial g(\ )}{\partial F_y'} = \frac{\partial g(\ )}{\partial F_y}\frac{\partial F_y}{\partial F_y'} = Z\sigma_{F_y}^N$$

and

$$\frac{\partial g(\ )}{\partial Z'} = \frac{\partial g(\ )}{\partial Z}\frac{\partial Z}{\partial Z'} = F_y\sigma_Z$$

At the design point, the two partial derivatives are

$$\nabla G(\mathbf{y}^*) = \begin{Bmatrix} z^* \times \sigma_{F_y} \\ f_y \times \sigma_Z \end{Bmatrix} = \begin{Bmatrix} 47.07 \times 2.416 \\ 24.22 \times 2.7 \end{Bmatrix} = \begin{Bmatrix} 113.721 \\ 65.394 \end{Bmatrix}$$

The length of the vector is

$$|\nabla G(\mathbf{y}^*)| = \sqrt{(113.721)^2 + (65.394)^2} = 131.182$$

Equation 3.64 is used to compute matrix **A**:

$$[\mathbf{A}] = \frac{1}{131.182}\begin{bmatrix} 0.498 & -0.867 \\ 0.867 & 0.498 \end{bmatrix}\begin{bmatrix} 0 & 6.523 \\ 6.523 & 0 \end{bmatrix}\begin{bmatrix} 0.498 & 0.867 \\ -0.867 & 0.498 \end{bmatrix}$$

$$= \begin{bmatrix} -0.043 & -0.025 \\ -0.025 & 0.043 \end{bmatrix}$$

As explained in the text, the rotation of coordinates makes the last variable coincide with the β vector. Therefore, the last row and the last column of **A** are dropped from future consideration. For this two-variable problem, that leaves the matrix **A** with just one element, $a_{11} = -0.043$. Therefore, the eigenvalue of this one-element matrix is simply $\kappa_1 = a_{11} = -0.043$.

Equation 3.60 is used to compute the probability of failure using the second-order approximation:

$$p_{f_2} \approx \Phi(-5.151)[1 + 5.151 \times (-0.043)]^{-1/2} = 1.4708 \times 10^{-7}$$

For the sake of comparison with FORM, a new safety index is computed as the inverse of this failure probability estimate as

$$\beta_{SORM} = -\Phi^{-1}(1.4708 \times 10^{-7}) = 5.139$$

Note that for this example, the second-order approximation should give a larger failure probability estimate, as shown in Figure 3.12. Correspondingly, the safety index for SORM is less than that for FORM.

Considering the same example and using various distributions of F_y and Z, we can calculate the safety indexes according to the FORM and SORM methods. The results are summarized in Table 3.7. The underlying distributions of random variables have a considerable amount of influence on the safety index calculations; however, their differences are not significant for the FORM and SORM methods in this problem, since the limit state is barely nonlinear.

3.7 RELIABILITY ANALYSIS WITH CORRELATED VARIABLES

The FORM and SORM methods described in the previous sections implicitly assume that the basic variables $X_1, X_2, \ldots, X_n$ are uncorrelated. However, usually some variables are correlated. Consider the X_i's in Equation 3.3 to be correlated variables with means μ_{X_i},

Table 3.7. Comparison of FORM and SORM results

$$g(\ ) = F_y Z - 1140 = 0$$

$$\mu_{F_y} = 38 \text{ ksi}, \quad \delta_{F_y} = 0.1, \quad \mu_Z = 54 \text{ in}^3, \quad \text{and} \quad \sigma_Z = 0.05$$

Probability Distributions		Safety Indices	
F_y	Z	FORM	SORM
Normal	Normal	4.261	4.246
Lognormal	Normal	5.151	5.139
Normal	Lognormal	4.266	4.259
Lognormal	Lognormal	5.213	5.211

standard deviations σ_{X_i}, and the covariance matrix represented as

$$[\mathbf{C}] = \begin{bmatrix} \sigma_{X_i}^2 & \text{Cov}(X_1, X_2) & \cdots & \text{Cov}(X_1, X_n) \\ \text{Cov}(X_2, X_1) & \sigma_{X_2}^2 & \cdots & \text{Cov}(X_2, X_n) \\ \vdots & \vdots & \vdots & \\ \text{Cov}(X_n, X_1) & \text{Cov}(X_n, X_2) & \cdots & \sigma_{X_n}^2 \end{bmatrix} \tag{3.66}$$

If the reduced variables X_i' are defined as

$$X_i' = \frac{X_i - \mu_{X_i}}{\sigma_{X_i}} \qquad (i = 1, 2, \ldots, n) \tag{3.67}$$

then it can be shown that the covariance matrix $[\mathbf{C}']$ of the reduced variables X_i' is

$$[\mathbf{C}'] = \begin{bmatrix} 1 & \rho_{X_1, X_2} & \cdots & \rho_{X_1, X_n} \\ \rho_{X_2, X_1} & 1 & \cdots & \rho_{X_2, X_n} \\ \vdots & \vdots & \vdots & \vdots \\ \rho_{X_n, X_1} & \rho_{X_n, X_2} & \cdots & 1 \end{bmatrix} \tag{3.68}$$

where ρ_{X_i, X_j} is the correlation coefficient of the X_i and X_j variables.

The FORM and SORM methods can be used if the X_i's are transformed into uncorrelated reduced normal $\mathbf{Y}$ variables and Equation 3.3 is expressed in terms of the $\mathbf{Y}$ variables. This can be done using the following equation:

$$\{\mathbf{X}\} = \left[\sigma_{\mathbf{X}}^N\right] [\mathbf{T}]\{\mathbf{Y}\} + \{\mu_{\mathbf{X}}^N\} \tag{3.69}$$

in which $\mu_{X_i}^N$ and $\sigma_{X_i}^N$ are the equivalent normal mean and standard deviation, respectively, of the X_i variables evaluated at the design point on the failure surface using Equations 3.35 and 3.37, and $\mathbf{T}$ is a transformation matrix to convert the correlated reduced $\mathbf{X}'$ variables to uncorrelated reduced normal $\mathbf{Y}$ variables. Note that the matrix containing the equivalent normal standard deviations in Equation 3.69 is a diagonal matrix. The $\mathbf{T}$ matrix can be shown to be

$$[\mathbf{T}] = \begin{bmatrix} \theta_1^{(1)} & \theta_1^{(2)} & \cdots & \theta_1^{(n)} \\ \theta_2^{(1)} & \theta_2^{(2)} & \cdots & \theta_2^{(n)} \\ \vdots & \vdots & \vdots & \vdots \\ \theta_n^{(1)} & \theta_n^{(2)} & \cdots & \theta_n^{(n)} \end{bmatrix} \tag{3.70}$$

$[\mathbf{T}]$ is basically an orthogonal transformation matrix consisting of the eigenvectors of the correlation matrix $[\mathbf{C}']$ (Eq. 3.68). $\{\theta^{(i)}\}$ is the eigenvector of the ith mode. $\theta_1^{(i)}, \theta_2^{(i)}, \ldots, \theta_n^{(i)}$ are the components of the ith eigenvector.

Using Equation 3.69, we can write Equation 3.3 in terms of reduced uncorrelated normal $\mathbf{Y}$ variables. For this case, estimating the probability of structural failure is simple, as outlined in this section.

For practical large problems, the correlated variables may also be transformed into uncorrelated variables through an orthogonal transformation of the form

$$Y = L^{-1}(X')^t \tag{3.71}$$

where L is the lower triangular matrix obtained by Cholesky factorization of the correlation matrix $[C']$. If the original variables are nonnormal, their correlation coefficients change on transformation to equivalent normal variables. Der Kiureghian and Liu (1985) developed semiempirical formulas for fast and reasonably accurate computation of $[C']$.

The procedure discussed here can be applied when the marginal distributions of all the variables as well as the covariance matrix are known. When the joint distributions of all the correlated nonnormal variables are available, an equivalent set of independent normal variables can be obtained using the Rosenblatt transformation. From a practical point of view, this situation would be rare unless all the variables are either normal or lognormal. Furthermore, it is not possible to define the joint probability density function uniquely using the information on marginal distributions and the covariance matrix (Bickel and Doksum, 1977).

When the random variables are correlated, two types of problems can be envisioned. In the first type, all the random variables are normal, but they are correlated to each other. In the second type, some or all of the correlated random variables are nonnormal. These cases are explained with the help of examples in the following sections.

3.7.1 Correlated Normal Variables

Example 3.7

Example 3.1, considered in Section 3.5.1 and represented by the limit given by Equation 3.19, can be considered again. To illustrate the procedures for safety index or probability of failure evaluation for the correlated normal variables case, both random variables F_y and Z are considered to be normal with the mean and the coefficient of variation given in Table 3.7. The correlation coefficient between them is assumed to be 0.3.

Because the random variables are all normal, it is necessary only to rewrite the limit state Equation 3.19 in terms of the Y variables, that is, uncorrelated normal variables using Equation 3.69. The transformation matrix T needs to be evaluated at this stage using Equation 3.70. The correlation matrix $[C']$ given by Equation 3.68 for the problem under consideration is

$$[C'] = \begin{bmatrix} 1 & 0.3 \\ 0.3 & 1 \end{bmatrix}$$

The two eigenvalues of $[C']$ can be calculated by solving the following equation:

$$\det \begin{bmatrix} (1-\lambda) & 0.3 \\ 0.3 & (1-\lambda) \end{bmatrix} = 0$$

or

$$(1-\lambda)^2 - 0.3^2 = 0$$

or

$$\lambda_1 = 0.7 \quad \text{and} \quad \lambda_2 = 1.3$$

The λ_i's are the variance of the Y_i's. The corresponding eigenvectors can be obtained by solving the following equation:

$$\begin{bmatrix} (1-\lambda_i) & 0.3 \\ 0.3 & (1-\lambda_i) \end{bmatrix} \begin{Bmatrix} \theta_1^{(i)} \\ \theta_2^{(i)} \end{Bmatrix} = 0$$

For each eigenvalue, the corresponding eigenvector can be calculated. For the problem under consideration, the eigenvectors are $\{\theta^{(1)}\} = \{1 - 1\}$ and $\{\theta^{(2)}\} = \{1\ 1\}$, respectively. The [**T**] matrix in Equation 3.70 represents the normalized eigenvectors and can be expressed as

$$[\mathbf{T}] = \begin{bmatrix} 0.707 & 0707 \\ -0.707 & 0.707 \end{bmatrix}$$

Equation 3.69 can now be expressed as

$$\begin{Bmatrix} F_y \\ Z \end{Bmatrix} = \begin{bmatrix} 3.8 & 0 \\ 0 & 2.7 \end{bmatrix} \begin{bmatrix} 0.707 & 0.707 \\ -0.707 & 0.707 \end{bmatrix} \begin{Bmatrix} Y_1 \\ Y_2 \end{Bmatrix} + \begin{Bmatrix} 38 \\ 54 \end{Bmatrix}$$

or

$$F_y = 2.687Y_1 + 2.687Y_2 + 38$$

$$Z = -1.909Y_1 + 1.909Y_2 + 54$$

Equation 3.19 can now be rewritten in terms of the uncorrelated normal **Y** variables as

$$(2.687Y_1 + 2.687Y_2 + 38)(-1.909Y_1 + 1.909Y_2 + 54) - 1140 = 0$$

or

$$g() = -5.129483Y_1^2 + 5.129483Y_2^2 + 72.556Y_1 + 217.640Y_2 + 912 = 0 \quad (3.72)$$

Considering Equation 3.72 to be the limit state equation and using the nine steps of FORM Method 1, we can estimate the reliability index for the problem under consideration. All the necessary steps are summarized in Table 3.8. FORM Method 2 can also be used for this purpose, as discussed in Section 3.5.3.4.

3.7.2 Correlated Nonnormal Variables

Example 3.8

Example 3.7 for correlated normal random variables can be considered again, except that the random variable F_y is a lognormal random variable with a mean of 38 ksi and a standard deviation of 3.8 ksi. The random variable Z is again normal, with mean and standard deviation of 54 in^3 and 2.7 in^3, respectively. Assume further that F_y and Z are correlated with a correlation coefficient of 0.3.

In general, the limit state equation for correlated nonnormal variables is quite involved. Since FORM is an iterative procedure, at each iteration the checking point and the corresponding equivalent mean and standard deviation of nonnormal variables are expected to be different, indicating that the limit state equation needs to be redefined

Table 3.8. Steps in FORM method 1 for correlated normal variables

Step 1		$g(\ ) = -5.129483Y_1^2 + 5.129483Y_2^2 + 72.556Y_1 + 217.64Y_2 + 912 = 0$					
Step 2	β	5.0				3.927	
Step 3	y_1^*	0.0	-0.996	-1.477	-1.535	-1.212	-1.104
	y_2^*	0.0	-5.536	-5.336	-5.302	-4.160	-4.218
Step 4	$\mu_{Y_1}^N$	0.0	0.0	0.0	0.0	0.0	0.0
	$\sigma_{Y_1}^N$	$\sqrt{0.7}$ $= 0.837$	0.837	0.837	0.837	0.837	0.837
	$\mu_{Y_2}^N$	0.0	0.0	0.0	0.0	0.0	0.0
	$\sigma_{Y_2}^N$	$\sqrt{1.3}$ $= 1.140$	1.140	1.140	1.140	1.140	1.140
Step 5	$\dfrac{\partial g}{\partial Y_1}$	72.556	82.774	87.709	88.304	84.990	83.882
	$\dfrac{\partial g}{\partial Y_2}$	217.64	160.846	162.898	163.247	174.963	174.368
Step 6	α_{Y_1}	0.238	0.353	0.367	0.369	0.336	0.333
	α_{Y_2}	0.971	0.936	0.930	0.929	0.942	0..943
Step 7		Go to step 3. Compute the new checking point using information from step 6					
Step 8	β					3.927	3.922
Step 9		Repeat steps 3 through 8 until β converges					

Note. The final checking point is (23.732, 48.036).

in each iteration. It is not necessary to give complete hand calculations to estimate the safety index here. However, all the necessary steps required to solve the problem are outlined as follows. The results are summarized in Table 3.9.

As an approximation, it is assumed that the covariance matrix [C], the correlation matrix [C'], and the corresponding eigenvalues, eigenvectors, and transformation matrix [T] do not change with the distribution of the random variables. Thus, the eigenvalues of 0.7 and 1.3 and the corresponding [T] matrix obtained earlier when both F_y and Z are normal variables can be used in this case also. The steps of FORM Method 1 are illustrated next.

Iteration 1

Based on past experience with similar problems, it is expected that the safety index β will be closer to 5.0 than 3.0. Thus, β is assumed to be 5.0 in the first iteration. However, β can still be assumed to be 3.0, but it will take more iterations to reach convergence.

As before, the initial checking points of F_y and Z can be assumed to be their mean values: 38 and 54, respectively. Since F_y is lognormal with $\lambda_{F_y} = 3.632611$ and $\zeta_{F_y} =$

Table 3.9. Steps in FORM method 1 for correlated nonnormal random variables

Step 1				$g(\ )F_y Z - 1140 = 0$			
Step 2	β	5.0			4.576		4.585
Step 3	f_y^*	38	20.305	23.855	25.058	25.058	25.037
	z^*	54	45.320	44.695	45.496	45.550	45.535
Step 4	$\mu_{F_y}^N$	37.811	32.930	34.843	35.367	35.367	35.359
	$\sigma_{F_y}^N$	3.791	20.025	2.380	2.500	2.500	2.498
	μ_Z^N	54.0	54.0	54.0	54.0	54.0	54.0
	σ_Z^N	2.7	2.7	2.7	2.7	2.7	2.7
Step 5	$\dfrac{\partial g}{\partial Y_1}$	72.231	26.139	29.688	32.605	32.689	32.622
	$\dfrac{\partial g}{\partial Y_2}$	217.317	103.654	120.762	128.269.	128.360	128.212
Step 6	α_{Y_1}	0.2370	0.1820	0.1775	0.1834	0.1837	0.1835
	α_{Y_2}	0.9715	0.9833	0.9841	0.9830	0.9830	0.9830
Step 7		Go to step 3. Compute the new checking point using information from step 6					
Step 8	β				4.576	4.585	4.586
Step 9		Repeat steps 3 through 8 until β converges					

Note. The final checking point is (25.038, 45.531).

0.0997513, its equivalent mean and standard deviation at the checking point can be estimated as (see Alternative 2 in Example 3.4)

$$\sigma_{F_y}^N = 0.0997513 \times 38 = 3.791$$

and

$$\mu_{F_y}^N = 38(1 - \ln 38 + 3.632611) = 37.811$$

Then, using Equation 3.69, we can show that

$$\left\{ \begin{array}{c} F_y \\ Z \end{array} \right\} = \begin{bmatrix} 3.791 & 0 \\ 0 & 2.7 \end{bmatrix} \begin{bmatrix} 0.707 & 0.707 \\ -0.707 & 0.707 \end{bmatrix} \left\{ \begin{array}{c} Y_1 \\ Y_2 \end{array} \right\} + \left\{ \begin{array}{c} 37.811 \\ 54 \end{array} \right\}$$

or

$$F_y = 2.681Y_1 + 2.681Y_2 + 37.811$$

$$Z = -1.909Y_1 + 1.909Y_2 + 54$$

Thus, the limit state equation becomes

$$(2.681Y_1 + 2.681Y_2 + 37.811)(-1.909Y_1 + 1.909Y_2 + 54) - 1140 = 0$$

or

$$g() = -5.118Y_1^2 + 5.118Y_2^2 + 72.593Y_1 + 216.955Y_2 + 901.794 = 0$$

Thus,

$$\frac{\partial g}{\partial Y_1} = -10.236Y_1 + 72.593$$

$$\frac{\partial g}{\partial Y_2} = 10.236Y_2 + 216.955$$

The random variables F_y and Z in the reduced coordinates at the checking point become

$$F_y'^* = \frac{38 - 37.811}{3.791} = 0.050 \quad \text{and} \quad Z'^* = \frac{54 - 54}{2.7} = 0$$

Since the transformation matrix $\mathbf{T}$ is orthogonal, $\mathbf{T}^{-1} = \mathbf{T}'$. Using Equation 3.69, we can show that $\mathbf{Y} = \mathbf{T}'\mathbf{X}'$. Thus, the coordinate of the checking point in the $\mathbf{Y}$ coordinates become

$$\begin{Bmatrix} y_1^* \\ y_2^* \end{Bmatrix} = \begin{bmatrix} 0.707 & -0.707 \\ 0.707 & 0.707 \end{bmatrix} \begin{Bmatrix} 0.050 \\ 0 \end{Bmatrix} = \begin{Bmatrix} 0.0354 \\ 0.0354 \end{Bmatrix}$$

Thus,

$$\left(\frac{\partial g}{\partial Y_1} \right)^* = -10.236 \times 0.0354 + 72.593 = 72.231$$

$$\left(\frac{\partial g}{\partial Y_2} \right)^* = 10.236 \times 0.0354 + 216.955 = 217.317$$

The direction cosines of Y_1 and Y_2 can be shown to be

$$\alpha_{Y_1} = \frac{72.231 \times \sqrt{0.7}}{\sqrt{72.231^2 \times 0.7 + 217.317^2 \times 1.3}} = 0.2370$$

and

$$\alpha_{Y_2} = \frac{217.317 \times \sqrt{1.3}}{\sqrt{72.231^2 \times 0.7 + 217.317^2 \times 1.3}} = 0.9715$$

Then,

$$y_1^* = -0.2370 \times \sqrt{0.7} \times 5.0 = -0.9914$$

$$y_2^* = -0.9715 \times \sqrt{1.3} \times 5.0 = -5.5384$$

The new checking point in the original coordinate becomes

$$f_y^* = 2.681(-0.9914 - 5.5384) + 37.811 = 20.305$$

$$z^* = -1.909(-0.9914) + 1.909(-5.5384) + 54 = 45.320$$

This completes the first iteration.

Second Iteration

With this new checking point, the equivalent normal mean and standard deviation of F_y are calculated again. Proceeding in a manner similar to the first iteration and using Equation 3.69, we can show the relationships between F_y and Z, and Y_1 and Y_2, to be

$$F_y = 1.432(Y_1 + Y_2) + 32.930$$

$$Z = 1.909(-Y_1 + Y_2) + 54$$

The direction cosines of Y_1 and Y_2 are found to be 0.1820 and 0.9833, respectively. They do not converge to the values from the first iteration with a tolerance level of 0.005. However, they are found to be 0.1775 and 0.9841, respectively, after the third iteration satisfying the tolerance criterion, as shown in Table 3.9. With β as an unknown parameter, the checking point in the Y coordinates becomes

$$y_1^* = 0.1775 \times \sqrt{0.7}\beta = -0.1485\beta$$

and

$$y_2^* = -0.9841 \times \sqrt{1.3}\beta = -1.1220\beta$$

The checking point in the original coordinates becomes

$$f_y^* = 1.683(-0.1485\beta - 1.220\beta) + 34.843 = -2.1383\beta + 34.843$$

$$x^* = 1.909(0.1485\beta - 1.1220\beta) + 54 = -1.8584\beta + 54$$

Substituting these in the limit state equation, we find β to be 4.576. This β value is not acceptable with a tolerance level of 0.005. With this new β value, a new checking point can again be defined. Proceeding as in the previous steps, we find the direction cosines of Y_1 and Y_2 to be 0.1834 and 0.9830, respectively. These values are not acceptable with a tolerance level of 0.005. However, in the next iteration, they converge to 0.1837 and 0.9830, respectively. The corresponding β value is found to be 4.585. Again, this is not acceptable with a tolerance level of 0.005.

When the new β value of 4.585 is used, the direction cosines of Y_1 and Y_2 become 0.1835 and 0.9830, satisfying the tolerance criterion. The corresponding safety index is found to be 4.586, which satisfies the tolerance criterion. Thus, for this problem with correlated nonnormal variables, the safety index is found to be 4.586 and the corresponding checking point is (25.038, 45.531). The results are summarized in Table 3.9.

This example clearly indicates that hand calculations for this type of problem can be very cumbersome, but a computer program can be easily written to implement the algorithm.

3.8 PROBABILISTIC SENSITIVITY INDICES

Because all input random variables do not have equal influence on the statistics of the output, a measure called the *sensitivity index* can be used to quantify the influence of each basic random variable. The quantity $\nabla g(\mathbf{Y})$, which is the gradient vector of the performance function in the space of standard normal variables, is used for this purpose. Let $\boldsymbol{\alpha}$

be a unit vector in the direction of this gradient vector. Then, because the design point can be expressed as $\mathbf{y}^* = -\beta\boldsymbol{\alpha}$, it is easily seen that

$$\alpha_i = -\frac{\partial\beta}{\partial y_i^*} \tag{3.73}$$

Thus, the elements of the vector $\boldsymbol{\alpha}$ are directly related to the derivatives of β with respect to the standard normal variables. If these are related to the original variables and their statistical variation, a unit sensitivity vector can be derived as (Der Kiureghian and Ke, 1985)

$$\boldsymbol{\gamma} = \frac{\mathbf{SB}^t\boldsymbol{\alpha}}{|\mathbf{SB}^t\boldsymbol{\alpha}|} \tag{3.74}$$

where $\mathbf{S}$ is the diagonal matrix of standard deviations of the input variables (equivalent normal standard deviations for the nonnormal random variables) and $\mathbf{B}$ is also a diagonal matrix required to transform the original variables $\mathbf{X}$ to equivalent uncorrelated standard normal variables $\mathbf{Y}$, that is, $\mathbf{Y} = \mathbf{A} + \mathbf{BX}$. For the ith random variable, this transformation is $Y_i = (X_i - \mu_{X_i})/\sigma_{X_i}$. Thus, the matrix $\mathbf{B}$ contains the reciprocals of the standard deviations or the equivalent normal standard deviations. If the variables are statistically independent, then in Equation 3.74, the product of $\mathbf{SB}^t$ will be a unit diagonal matrix. Thus, the sensitivity vector will be identical to the direction cosines vector of the random variables. However, if the variables are correlated, another transformation matrix $\mathbf{T}$, as in Equation 3.69, will come into the picture. Then, the sensitivity vector and the direction cosines vector will be different.

The elements of the vector $\boldsymbol{\gamma}$ may be referred to as sensitivity indices of individual variables. The sensitivity indices can be used to improve computational efficiency. Variables with very low sensitivity indices at the end of the first few iterations can be treated as deterministic at their mean values for subsequent iterations in the search for the minimum distance. This significantly reduces the amount of computation because, as a practical matter, only a few variables have a significant effect on the probability of failure. These sensitivity indices are also useful in reducing the size of problems with random fields, in which the random fields are discretized into sets of correlated random variables (Mahadevan and Haldar, 1991) and in reliability-based optimization (Mahadevan, 1992).

3.9 SYSTEM RELIABILITY EVALUATION

In the previous sections of this chapter, reliability was estimated for a single performance criterion or limit state for a structural element using FORM or SORM. In general, any engineering system has to satisfy more than one performance criterion. Even for a simple beam, the performance criterion could be strength-related, e.g., bending moment or shear, or serviceability-related, e.g., deflection or vibration. Thus, the beam can fail in more than one performance mode. A structure such as a truss or a frame consists of multiple structural elements or components, and failure may occur in one or more components. The long-distance telephone communication system between the east and west coasts consists of several networks. At any given time, one or more such networks may not be in operating condition. The ordinary services may or may not suffer any disruption depending on the availability of alternative systems. The water supply to a community may come from different sources through a network of piping. Again, the water supply from different sources and/or networks may be disrupted, but water may still be available to the community. The concept used to consider multiple failure modes and/or multiple component failures is known as *system reliability* evaluation. A complete reliability analysis includes both component-level and system-level estimates.

In general, system reliability evaluation is quite complicated and depends on many factors. Some of the important factors are (1) the contribution of the component failure events to the system's failure, (2) the redundancy in the system, (3) the postfailure behavior of a component and the rest of the system, (4) the statistical correlation between failure events, and (5) the progressive failure of components. In the beam example, reliability can be estimated by calculating the probability of satisfying all the performance criteria. An engineering system usually consists of multiple components, and system failure may occur when one or more components fail.

The stochastic finite element concept, presented in Chapters 5 to 9, indirectly considers system effects because it considers the presence of all structural elements in the reliability estimation. However, the reliability of the system under multiple performance criteria still needs to be considered. This is an active area of research. Haldar and Mahadevan (2000) discuss system reliability evaluation in detail. Interested readers are encouraged to refer to that publication.

3.10 CONCLUDING REMARKS

The risk-based analyis and design concept is presented in this chapter. It is shown that the conventional safety factor-based deterministic designs in terms of capacity reduction factor and load factors, and the risk or probability-based load and resistance factor designs, are essentially parallel to each other. However, risk-based design explicitly incorporates more information in developing these factors. It is perhaps a better and more comprehensive approach, and it empowers engineers to make better design decisions on a case-by-case basis.

Using the fundamental concept of risk-based design, various reliability analysis methods with different degrees of complexity and completeness are available. They are presented systematically in this chapter. The steps to extract the necessary information are identified and clarified with the help of examples. The concept of the second-order reliability method is introduced. Correlated random variables are often present in reliability analysis. The additional steps necessary for correlated variables are described in this chapter. Not all the variables in a problem may need to be considered random. To evaluate their relative importance in the overall reliability evaluation, the concept of probabilistic sensitivity indices is introduced. Variables whose sensitivity indices are relatively low at the end of the first few iterations can be treated as deterministic, essentially reducing the size of the problem. The concept of system reliability is briefly introduced. A complete reliability analysis includes both component-level and system-level estimates.

Reliability evaluation procedures for implicit performance functions are briefly introduced. They are presented more formally in Chapter 5.

The information presented in the first three chapters of this book provides readers with the necessary background to calculate the failure probability or reliability for simple systems. We believe that the readers are now prepared to study reliability evaluation using the stochastic finite element method. However, reliability evaluation using simulation is presented next, because simulation plays a very important role in verifing the reliability results obtained using any method.

PROBLEMS

3.1. The bearing capacity, C, of soil under a square foundation of size 9 ft^2 is determined to be a random variable with a mean of 3 ksf and a standard deviation of 0.5 ksf. The applied axial load, P, acting on the foundation is also a random variable with a mean of 15 kip and a standard deviation of 2 kip. Assume that C and P are statistically independent and that no information on their distribution is available. Using a limit state function of the form $g(\ ) = 9C - P$ and the MVFOSM method, calculate the reliability index for the foundation.

3.2. Consider Problem 3.1.

    (a) If C and P are statistically independent normal random variables with the same means and standard deviations, calculate the reliability index and the corresponding probability of failure of the foundation.

    (b) If C and P are statistically independent lognormal random variables with the same means and standard deviations, will the reliability index be different? Can the probability of failure of the foundation be calculated exactly?

3.3. A simply supported beam of span $L = 360$ inches is loaded by a uniformly distributed load w in kip/in and a concentrated load P in kip applied at the midspan. The maximum deflection of the beam at the midspan can be calculated as

$$\delta_{max} = \frac{5}{384} \frac{wL^4}{EI} + \frac{1}{48} \frac{PL^3}{EI}$$

A beam with $EI = 63.51 \times 10^6$ kip-in^2 is selected to carry the load. Both w and P are statistically independent random variables with mean values estimated to be 0.2 kip/in and 25 kip, respectively. The corresponding standard deviations are 0.03 kip/in and 2.5 kip, respectively. Assume that the distributions of w and P are unavailable. The allowable deflection, δ_a for the beam is a constant of value 1.5 inch. Considering the limit state equation of the form $g(\ ) = \delta_a - \delta_{max} = 0$, calculate the reliability index for the beam in deflection using the MVFOSM method.

3.4. In Problem 3.3, suppose both w and P are statistically independent normal random variables with the same means and standard deviations. Using FORM Method 1, calculate the reliability index for the beam. Discuss why the reliability indices obtained in Problems 3.3 and 3.4 are the same.

3.5. Consider Problem 3.4. Calculate the reliability index of the beam using FORM Method 2. Is it identical to the value obtained in Problem 3.4?

3.6. In Problem 3.3, suppose w is a normal random variable and P is a lognormal random variable with the same means and standard deviations. If w and P are statistically independent, calculate the reliability index for the beam using the FORM Method 1. Discuss why the reliability indices obtained in Problems 3.4 and 3.6 are different.

3.7. Consider Example 3.4. The limit state equation according to the strength formulation to be $g(\ ) = F_y Z - 1140 = 0$, in which F_y has a lognormal distribution with mean and COV of 38 ksi and 0.1, respectively, and Z has a normal distribution with mean and COV of 54 in^3 and 0.05, respectively. It can be seen from Table 3.2 that for this problem $\beta = 5.151$ and the checking point is (24.22, 47.07). The limit state equation is rewritten in the stress formulation as

$$g() = F_y - \frac{1140}{Z} = 0$$

Using either Method 1 or Method 2, show that the reliability index and the checking point will be unchanged according to FORM.

3.8. The fully plastic flexural capacity of a steel beam section can be estimated as YZ, where $Y =$ the yield strength of steel and $Z =$ the plastic section modulus of the section. If the applied bending moment at a location of interest is M, the performance function may be defined as

$$g() = YZ - M$$

Assume that Z is a constant of value 50 in^3 and that Y and M are independent normal random variables with mean values of 40 ksi and 1000 kip-in, respectively; the corresponding COVs are 0.125, and 0.20, respectively. Estimate the reliability of the beam using FORM (Method 1 or Method 2).

3.9. In Problem 3.8, if Y and M are independent lognormal random variables with the same means and COVs, and Z is a constant of value 50 in^3, estimate the reliability of the beam using FORM (Method 1 or Method 2).

3.10. In Problem 3.8, consider Y to be a normal and M to be a lognormal random variable with the same means and COVs, and Z is a constant with a value of 50 in^3. Estimate the reliability of the beam using FORM (Method 1 or Method 2).

3.11. For Problem 3.4, calculate the probability of failure using SORM.

3.12. For Problem 3.6, calculate the probability of failure using SORM.

3.13. For Problem 3.9, calculate the probability of failure using SORM.

3.14. Consider Problem 3.3. If w and P are correlated normal variables with $\rho_{w,P} = 0.7$, with all other information remaining the same, calculate the safety index using FORM.

3.15. In Problem 3.14, if w is a normal random variable and P is a lognormal random variable with the same means and standard deviations, and the correlation coefficient between them is still 0.7, calculate the safety index using FORM.

3.16. In Problem 3.14, if w and P are correlated lognormal variables with $\rho_{w,P} = 0.7$, with all other information remaining the same, calculate the safety index using FORM.

4

SIMULATION TECHNIQUES

4.1 INTRODUCTORY COMMENTS

Several methods with various degrees of complexity that can be used to estimate the reliability or safety index or the probability of failure are discussed in Chapter 3. Most of these methods are applicable when the limit state equations are explicit functions of the random variables involved in a problem. In FORM Method 2 presented in Chapter 3, we also introduce the concept of reliability evaluation when the limit state equations are implicit functions of the random variables involved in a problem. This concept is introduced more formally in Chapter 5. In any case, estimating the probability of failure using these techniques requires a background in probability and statistics, as discussed in the previous chapters of this book. But with a simple simulation technique, it is possible to calculate the probability of failure for both the explicit and implicit limit state functions without knowing these analytical techniques and with only a little background in probability and statistics. The availability of personal computers and software makes the process very simple. In fact, to evaluate the accuracy of these sophisticated techniques or to verify a new technique, simulation is routinely used to independently evaluate the underlying probability of failure.

In the simplest form of the basic simulation, each random variable in a problem is sampled several times to represent its real distribution according to its probabilistic characteristics. Considering each realization of all the random variables in the problem produces a set of numbers that indicates one realization of the problem itself. Solving the problem deterministically for each realization is known as a *simulation cycle*, *trial*, or *run*. Using many simulation cycles gives the overall probabilistic characteristics of the problem, particularly when the number of cycles, N, tends to infinity. The simulation technique using a computer is an inexpensive way (compared to laboratory testing) to study the uncertainty in the problem.

4.2   MONTE CARLO SIMULATION TECHNIQUE

The method commonly used for this purpose is called the *Monte Carlo simulation technique*. The name itself has no significance, except that it was used first by von Neumann during World War II as a code word for secret work on nuclear weapons at Los Alamos Laboratory in New Mexico. Most commonly the name Monte Carlo is associated with a place where gamblers take risks. This technique has evolved as a very powerful tool for engineers with only a basic working knowledge of probability and statistics for evaluating the risk or reliability of complicated engineering systems.

The Monte Carlo simulation technique has six essential elements: (1) defining the problem in terms of all the random variables; (2) quantifying the probabilistic characteristics of all the random variables in terms of their probability density functions and the corresponding parameters; (3) generating values of these random variables; (4) evaluating the problem deterministically for each set of realizations of all the random variables, or simply numerical experimentation; (5) extracting probabilistic information from N such realizations; and (6) determining the accuracy and efficiency of the simulation. All these elements are discussed in the following section. Initially, all the random variables are considered to be uncorrelated. The use of the Monte Carlo simulation technique for correlated random variables is discussed in Section 4.3.

4.2.1   Formulation of the Problem

Consider a simply supported beam, shown in Figure 4.1, subjected to a uniformly distributed load W and a concentrated load P at the midspan. Assume that both W and P are random variables and thus the design bending moment, M, at the midspan is also a random variable. The task is to evaluate the probabilistic characteristics of the design bending moment using the Monte Carlo simulation technique. If the span of the beam is 30 feet, the expression for the design bending moment can be written as

$$M = \frac{WL^2}{8} + \frac{PL}{4}$$

or

$$M = 112.5W + 7.5P \tag{4.1}$$

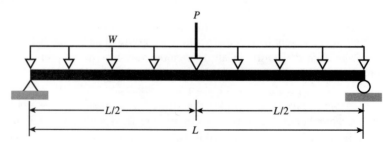

FIGURE 4.1. Simply supported beam.

4.2.2   Quantifying the Probabilistic Characteristics of Random Variables

We assume that the readers have a basic understanding of how to determine the underlying distribution of a random variable in terms of its PDF or CDF and the corresponding parameters to define it uniquely. No additional discussion is made here. However, for a quick review, the readers are referred to another book by the authors (Haldar and Mahadevan, 2000).

4.2.3   Generation of Random Numbers

The random variable to be generated could be continuous or discrete. Although the same concept underlies the generation of random numbers for continuous and discrete random variables, they need to be discussed separately.

4.2.3.1   Generation of Random Numbers for Continuous Random Variables
Assume that W is a normal random variable with $\mu_W = 2$ kip/ft and $\sigma_W = 0.2$ kip/ft, and P is a uniformly distributed random variable between 10 and 20 kip. Further assume that they are statistically independent random variables. Of course, they could have any distribution, and one of them could be a known deterministic constant. If both of them are constants, then the bending moment is a constant and probabilistic study is not necessary. The task now is to generate N random numbers for W according to its probabilistic characteristics (i.e., in this case, a normal distribution with specified mean and standard deviation) and another N random numbers for P, which is uniformly distributed.

The generation of random numbers according to a specific distribution is the heart of Monte Carlo simulation. In general, all modern computers have the capability to generate uniformly distributed random numbers between 0 and 1. Sometimes the random number generators use bits and binary digits, and in most cases they are linear congruential generators. Corresponding to an arbitrary *seed value*, the generators will produced the required number of uniform random numbers between 0 and 1. By changing the seed value, different sets of random numbers can be generated. Depending on the size of the computer, the random numbers may be repeated. However, this repetition will usually start only after generating a very large quantity of random numbers, such as 10^9. Random numbers generated this way are called *pseudo random numbers*. From a practical point of view, random numbers are rarely needed in this quantity; thus, the repetition of random numbers is of academic interest only. One hundred random numbers for a uniform distribution between 0 and 1 are given in Table 4.1. These random numbers will be used in the subsequent discussion.

The next task is to transform the uniform random numbers u_i between 0 and 1, either generated by a computer or obtained from a table, to random numbers with the appropriate characteristics. The process is shown graphically in Figure 4.2. This is commonly known as the *inverse transformation technique* or *inverse CDF method*. In this method, the CDF of the random variable is equated to the generated random number u_i; that is, $F_{X_i}(x_i) = u_i$, and the equation is solved for x_i as

$$x_i = F_X^{-1}(u_i) \tag{4.2}$$

A simple example to describe the technique is the transformation of a uniform random number U between 0 and 1, such as $u_1 = 0.86061$ (the first number in Table 4.1), to another uniform random number x_1 between two limits, a and b. The CDF of U is u_i.

Table 4.1. Uniform random numbers between 0 and 1

0.86061	0.15017	0.42172	0.48932
0.92546	0.74098	0.95349	0.54707
0.41806	0.58515	0.16119	0.64271
0.28964	0.70074	0.58394	0.66930
0.14225	0.09666	0.95626	0.27681
0.44961	0.97948	0.20661	0.90451
0.24653	0.65400	0.24566	0.79163
0.21687	0.67980	0.94934	0.42397
0.56503	0.46872	0.16118	0.68086
0.40015	0.12846	0.01988	0.82174
0.83771	0.12237	0.27493	0.94600
0.73006	0.17468	0.03348	0.26457
0.56341	0.21305	0.38943	0.31697
0.82178	0.82744	0.36283	0.05336
0.32715	0.20220	0.41536	0.82238
0.68853	0.98479	0.30607	0.97673
0.74358	0.53164	0.14563	0.72927
0.24672	0.58442	0.44542	0.68251
0.90324	0.11799	0.53053	0.23987
0.79263	0.29124	0.58757	0.02894
0.44281	0.73958	0.17326	0.87885
0.70826	0.51527	0.10593	0.80716
0.22664	0.63765	0.72448	0.14197
0.62557	0.52224	0.44245	0.74708
0.48342	0.46079	0.37091	0.80193

Since X is uniform, its CDF will be $F_X(x) = (x - a)/(b - a)$. The transformation to obtain the corresponding x_i value can be accomplished by equating the two CDFs as

$$u_i = \frac{x_i - a}{b - a}$$

or

$$x_i = a + (b - a)u_i \qquad (4.3)$$

When $a = 0$ and $b = 1$, $x_i = u_i$, which is obvious. If X is uniform between 10 and 20, the corresponding first random number is

$$x_1 = 10 + (20 - 10) \times 0.86061 = 18.6061$$

If X is normally distributed, that is, $N(\mu_X, \sigma_X)$, then $S = (X - \mu_X)/\sigma_X$ is a standard normal variate, that is, $N(0, 1)$, as discussed in Section 2.2.1. It can be shown that

$$u_i = F_X(x_i) = \Phi(s_i) = \Phi\left(\frac{x_i - \mu_X}{\sigma_X}\right) \qquad (4.4)$$

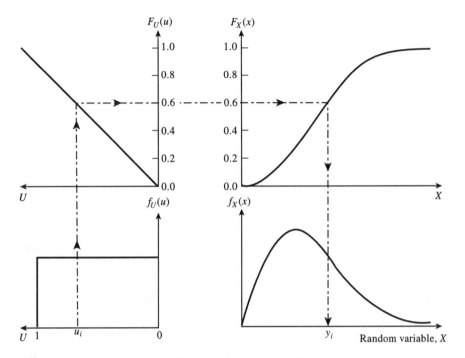

FIGURE 4.2. Mapping for simulation

or

$$s_i = \frac{x_i - \mu_X}{\sigma_X}$$

Thus,

$$x_i = \mu_X + \sigma_X s_i = \mu_X + \sigma_X \Phi^{-1}(u_i) \tag{4.5}$$

Equation 4.5 suggests that in this case, the u_i's first need to be transformed to s_i's, that is, $s_i = \Phi^{-1}(u_i)$, and Φ^{-1} is the inverse of the CDF of a standard normal variable. Table 4.2 shows the set of 100 standard normal random numbers corresponding to the uniform random numbers between 0 and 1 given in Table 4.1. The x_i's can be calculated from the information on the s_i's. For $u_1 = 0.86061$, $s_1 = \Phi^{-1}(0.86061) = 1.08306$; with the information on μ_X and σ_X, the corresponding x_1 can be calculated. For the uniform load under consideration, the first random number according to the normal distribution is

$$x_1 = 2 + 0.2 \times 1.08306 = 2.21661$$

If the random variable X is lognormally distributed with parameters λ_X and ζ_X, then the ith random number x_i according to the lognormal distribution can be generated as

$$u_i = \Phi\left(\frac{\ln x_i - \lambda_X}{\zeta_X}\right)$$

Table 4.2. Standard normal random numbers corresponding to the uniform numbers in Table 4.1

1.08306	−1.03571	−0.19750	−0.02677
1.44279	0.64637	1.67968	0.11826
−0.20686	0.21509	−0.98958	0.36571
−0.55444	0.52653	0.21198	0.43798
−1.07027	−1.30081	1.70884	−0.59234
−0.12665	2.04313	−0.81824	1.30769
−0.68545	0.39614	−0.68821	0.81209
−0.78281	0.46714	1.63849	−0.19175
0.16373	−0.07849	−0.98962	0.47011
−0.25296	−1.13370	−2.05621	0.92202
0.98509	−1.16322	−0.59797	1.60725
0.61299	−0.93583	−1.83193	−0.62932
0.15962	−0.79588	−0.28081	−0.47619
0.92217	0.94410	−0.35090	−1.61313
−0.44780	−0.83379	−0.21378	0.92447
0.49169	2.16458	−0.50702	1.99047
0.65442	0.07939	−1.05536	0.61061
−0.68485	0.21321	−0.13724	0.47473
1.30024	−1.18509	0.07660	−0.70672
0.81558	−0.54977	0.22130	−1.89656
−0.14385	0.64205	−0.94136	1.16926
0.54831	0.03829	−1.24847	0.86748
−0.74996	0.35218	0.59620	−1.07151
0.32014	0.05578	−0.14476	0.66533
−0.04157	−0.09844	−0.32944	0.84854

or

$$\ln x_i = \lambda_X + \zeta_X \Phi^{-1}(u_i)$$

or

$$x_i = \exp[\lambda_X + \zeta_X \Phi^{-1}(u_i)] \tag{4.6}$$

A computer program can be written to generate random numbers according to any distribution. In fact, many available computer programs can generate random numbers for commonly used distributions. If the computer cannot generate a specific distribution, Equation 4.2 can be used to obtain it.

Example 4.1

Generate random numbers for the Type II extreme value distribution discussed in Section 2.4.5, whose CDF is given by Equation 2.44:

$$F_X(x) = \exp\left[-\left(\frac{u}{x}\right)^k\right]$$

where u and k are the parameters of the distribution.

Solution If an available computer program cannot generate random numbers for the Type II extreme value distribution, they can be easily generated using Equation 4.2, as shown next. Denoting u_i as a uniform random number between 0 and 1, then

$$u_i = F_X(x_i) = \exp\left[-\left(\frac{u}{x_i}\right)^k\right]$$

or

$$x_i = \frac{u}{[\ln(1/u_i)]^{1/k}} \tag{4.7}$$

Thus, for any u_i, the corresponding x_i according to the Type II extreme value distribution can be calculated using Equation 4.7.

4.2.3.2 Generation of Random Numbers for Discrete Random Variables If

X is a discrete random variable, its CDF $F_X(x_i)$ needs to be calculated by taking the summation of the individual PMFs, as in Equation 2.4. The inverse transformation technique can be used to generate discrete random numbers. It is necessary to equate u_i to the corresponding $F_X(x)$ value. Thus, it is necessary to evaluate the CDF for all possible values of X. Then, a numerical search procedure is needed to obtain the discrete random number satisfying the condition as

$$F_X(x_{j-1}) < u_i \le F_X(x_j) \tag{4.8}$$

Example 4.2

The CDF of a binomial distribution is shown in Figure 2.2. Suppose $u_i = 0.86061$, the first uniform number between 0 and 1 in Table 4.1. Figure 2.2 indicates that

$$F_X(2) < 0.86061 < F_X(3)$$

In this case $x_i = 3$.

Generalizing the procedure, we can state that if X is a discrete random variable with CDF of $F_X(x_i)$, then

$$x_i \text{ is such that } i \text{ is the smallest integer with } u \le F_X(x_i) \tag{4.9}$$

Generating discrete random numbers can be cumbersome in many cases. Several other procedures are available to generate discrete random numbers in a computer environment (Abramowitz and Stegun, 1964). They should be used whenever possible. Any quantity of random numbers, discrete or continuous, according to specific CDFs can be generated from the information on uniform random numbers between 0 and 1.

4.2.4 Numerical Experimentation

N random numbers for each of the random variables in the problem will give N sets of random numbers, each set representing a realization of the problem. Thus, solving the problem N times deterministically will give N sample points, essentially generating information on the randomness in the output or response of the system to each set of input variables. The N sample points generated for the output or response can then be used to calculate all the required sample statistics, the histogram, the frequency diagram, the PDF or PMF and the corresponding CDF, and the probability of failure considering various performance criteria. The accuracy of the evaluation will increase as the number of simulations N increases. This is illustrated next.

Example 4.3

The probabilistic characteristics of the bending moment for the beam shown in Figure 4.1 and represented by Equation 4.1 can now be generated using the Monte Carlo simulation technique. For the sake of brevity, only 10 simulation cycles are considered here. Again, W is $N(2 \text{ kip/ft}, 0.2 \text{ kip/ft})$ and P is uniform between 10 and 20 kip. Since P is uniformly distributed, its mean value and COV can be calculated from Table 2.2. In this case, the mean value is

$$\mu_P = \frac{(10 + 20)}{2} = 15 \text{ kip}$$

and the COV is

$$\delta_P = \frac{2}{\sqrt{12}} \frac{(20 - 10)}{(20 + 10)} = 0.1925,$$

and the corresponding standard deviation is $\sigma_P = 0.1925 \times 15 = 2.89$ kip.

Suppose 10 uniform random numbers between 0 and 1 (the first 10 numbers in Table 4.1) are generated for W and another 10 (the next 10 numbers, i.e, number 11 through 20 in Table 4.1) are generated for P. The steps involved in generating a set of 10 random numbers for w_i and p_i according to their statistical characteristics and the corresponding m_i are summarized in Table 4.3.

Using the 10 sample points for the bending moment thus generated, we can calculate its mean and standard deviation to be 347.68 and 30.38 kip-ft, respectively. The statistical distribution of the bending moment can also be obtained by generating enough data to draw a histogram or by using other statistical techniques discussed elsewhere (Haldar and Mahadevan, 2000). The optimal numbers of simulation cycles required are discussed in Section 4.2.6.

If the mean value first-order second moment (MVFOSM) method discussed in Section 3.5.1 is used, the mean value and standard deviation of the bending moment are estimated as

$$\mu_M \approx 112.5 \times 2 + 7.5 \times 15 = 337.5 \text{ kip-ft}$$

and

$$\sigma_M \approx \sqrt{112.5^2 \times 0.2^2 + 7.5^2 \times 2.89^2} = 31.24 \text{ kip-ft}.$$

Table 4.3. Monte Carlo simulations

$W \sim N(2, 0.2)$			P Uniform Between 10 and 20		$m_i = 112.5 w_i + 7.5 p_i$
u_i	s_i	w_i	u_i	p_i	m_i
0.86061	1.08306	2.21661	0.83771	18.3771	387.1972
0.92546	1.44279	2.28856	0.73006	17.3006	387.2173
0.41806	−0.20686	1.95863	0.56341	15.6341	337.6014
0.28964	−0.55444	1.88911	0.82178	18.2178	349.1587
0.14225	−1.07027	1.78595	0.32715	13.2715	300.4553
0.44961	−0.12665	1.97467	0.68853	16.8853	348.7902
0.24653	−0.68545	1.86291	0.74358	17.4358	340.3459
0.21687	−0.78281	1.84344	0.24672	12.4672	300.8909
0.56503	0.16373	2.03275	0.90324	19.0324	371.4270
0.40015	−0.25296	1.94941	0.79263	17.9263	353.7557

Note. Mean of moment = 347.68 kip-ft; standard deviation of moment = 30.38 kip-ft.

The differences between the parameters estimated by the two methods are expected to become narrower as the number of simulation cycles increases. The first-order approximation does not use the information on distributions of random variables and thus may not match the simulation results even when the number of simulation cycles is very large.

This simple example indicates the power and simplicity of the simulation technique. The most significant point is that detailed knowledge of the analytical methods in Chapter 3 is not required to generate the necessary probabilistic information. The task is much simpler if a computer program is used for this purpose.

4.2.5 Extracting Probabilistic Information Using Simulation

The method described in the previous section can also be used to evaluate the risk or reliability of an engineering system. Consider the limit state represented by Equation 3.3 corresponding to a failure mode for a structure. With all the random variables in Equation 3.3 assumed to be statistically independent, the Monte Carlo simulation approach consists of drawing samples of the variables according to their probability density functions and then feeding them into the mathematical model $g(\ )$. The sample statistics thus obtained and the corresponding PDF would give the probabilistic characteristics of the response random variable Z. The extraction of such information could be cumbersome. However, if the objective is only to estimate the failure probability, that can be done quite simply, as follows.

It is known that a value of $g(\ )$ less than zero indicates failure. Let N_f be the number of simulation cycles when $g(\ )$ is less than zero and let N be the total number of simulation cycles. Therefore, an estimate of the probability of failure can be expressed as

$$p_f = \frac{N_f}{N} \tag{4.10}$$

4.2.6 Accuracy and Efficiency of Simulation

The ability of Equation 4.10 to accurately estimate the probability of failure is a matter of concern. Obviously, the accuracy of the estimate will depend on the number of simulation cycles. For a small failure probability and/or small N, the estimate of p_f given by Equation 4.10 may be subject to considerable error. The estimate of the probability of failure would approach the true value as N approaches infinity. The accuracy of Equation 4.10 can be studied in several ways. One way would be to evaluate the variance or COV of the estimated probability of failure (Ayyub and Haldar, 1985). The variance or COV can be estimated by assuming each simulation cycle to constitute a Bernoulli trial, and the number of failures in N trials can be considered to follow a binomial distribution. Then, the COV of p_f can be expressed as

$$\text{COV}(p_f) = \delta_{p_f} \approx \frac{\sqrt{(1 - p_f)p_f/N}}{p_f} \tag{4.11}$$

A smaller value of δ_{p_f} is desirable. Equation 4.11 indicates that δ_{p_f} approaches zero as N approaches infinity.

Another way to study the error associated with the number of simulation cycles is by approximating the binomial distribution with a normal distribution and estimating the 95% confidence interval of the estimated probability of failure (Shooman, 1968). It can be shown that

$$P\left[-2\sqrt{\frac{(1 - p_f^T)p_f^T}{N}} < \frac{N_f}{N} - p_f^T < 2\sqrt{\frac{(1 - p_f^T)p_f^T}{N}}\right] = 0.95 \tag{4.12}$$

where, p_f^T is the true probability of failure. The percentage error can be defined as

$$\epsilon\% = \frac{N_f/N - p_f^T}{p_f^T} \times 100\% \tag{4.13}$$

Combining Equations 4.12 and 4.13, we obtain

$$\epsilon\% = \sqrt{\frac{1 - p_f^T}{N \times p_f^T}} \times 200\% \tag{4.14}$$

Equation 4.14 indicates that there will be about 20% error if p_f^T is 0.01 and if 10,000 trials were used in the simulation. It can also be stated that there is 95% probability that the probability of failure will be in the range of 0.01 ± 0.002 with 10,000 simulations. Conversely, if the desired error is 10% and p_f^T is 0.01, then from Equation 4.14, the required number of simulations $N = 39,600$.

Both Equations 4.11 and 4.14 indicate that the number of simulation cycles to achieve a certain level of accuracy depends on the unknown probability of failure. In many engineering problems, the probability of failure could be smaller than 10^{-5}. Therefore, on average, only 1 out of 100,000 trials would show a failure. Thus, at least 100,000 simulation cycles are required to predict this behavior. For a reliable estimate, at least 10 times this minimum (i.e., 1 million simulation cycles) is usually recommended. If the problem has n random variables, then n million random numbers are necessary if the Monte Carlo simulation is to successfully estimate the probability of failure.

Example 4.4

The probability of failure of a beam is under consideration. The following performance function for the beam can be used:

$$g() = F_y Z - M \tag{4.15}$$

where F_y is the yield stress, Z is the section modulus, and M is the applied bending moment. They are considered to be random variables. Their statistical characteristics are given in Table 4.4.

First, using the MVFOSM method and Equation 4.15, we estimate the safety index to be 3.251 as shown in Table 4.5. When the FORM method is used, the safety index is found to be 2.340. As expected, they are quite different since the limit state equation is nonlinear and the distributions of two out of three random variables are nonnormal. The difference tends to be larger when the difference between the mean values (required for the MVFOSM method) and the design point on the failure surface (required for the FORM method) gets larger. Thus, MVFOSM may or may not be a realistic measure of probability of failure, depending on the nature of the problem under consideration.

The result obtained by FORM is more reliable and the corresponding probability of failure is found to be 0.009650, a comparatively large number considering actual design practice. This large probability of failure is intentionally considered in this example so that the results can be compared with a simulation study with a reasonable number of simulations. For this example, the probability of failure is of the order of 10^{-2}; thus, about 1000 simulation cycles are necessary for a reasonable estimate of the probability of failure.

Using the limit state given by Equation 4.15 and the Monte Carlo simulation technique, we can calculate the probability of failure of the beam in two different ways: by considering the statistics of the limit state equation, and by counting the failures in different cycles of simulations. In the first method, using the mean and the standard

Table 4.4. Probabilistic characteristics of basic parameters

Parameter	Mean Value	COV	Probability Distribution
F_y	38.0 ksi	0.10	Normal
Z	60.0 in^3	0.05	Lognormal
M	1000.0 kip-in	0.30	Type II

Table 4.5. Results of MVFOSM and FORM methods

Reliability Methods	β	$p_f = 1 - \Phi(\beta)$
MVFOSM	3.251	0.000577
FORM	2.340	0.009650

Table 4.6. Summary of simulation results

Number of Cycles, N	Direct Simulation		Counting
	Using Samples Statistics		
	$\beta = \frac{\mu_g}{\sigma_g}$	$p_f = 1 - \Phi(\beta)$	$p_f = N_f/N$
1	2	3	4
10	3.345	0.000412	0.000000[a]
50	3.224	0.000632	0.000000[a]
100	3.427	0.000305	0.000000[a]
250	3.177	0.000744	0.008000
500	2.964	0.001518	0.016000
1000	3.081	0.001032	0.011000

[a] $N_f = 0$, for these cases.

deviation of $g(\ )$, we calculate the safety indices as shown in column 2 in Table 4.6. Assuming that $g(\ )$ is a normal random variable, the corresponding probability of failure is shown in column 3. As the number of simulation cycles increases, the probability of failure does not show any convergence behavior.

Using the same simulation data with the counting technique given by Equation 4.10, we again calculate the probability of failure of the beam, and the results are shown in column 4 of Table 4.6. In this case, when the number of simulation cycles is relatively small, such as less than 100, none of the trials resulted in failure of the beam [$g(\ ) < 0$], giving the corresponding probability of failure to be zero. However, as the N values increase, the probability of failure converges to about 0.011. The probabilities of failure obtained by the two methods, that is, the sample statistics and the counting methods, are quite different. If the performance function $g(\ )$ is not normally distributed, this type of difference is expected. In general, since $g(\ )$ may not be normal in most cases, direct simulation with failure counting is superior to the first method.

The probabilities of failure obtained using FORM and the Monte Carlo simulation method are almost the same when the number of simulation cycles is relatively large. This indicates that the Monte Carlo method can also be used to evaluate the probability of failure if the number of simulation cycles is relatively large. This also points out the weakness of the direct Monte Carlo simulation technique. The probability of failure of a complicated system is not known in advance; thus, it will be difficult to estimate a reasonable number of simulation cycles in advance. A trial-and-error approach may need to be employed. Also, if the probability of failure is relatively small, as is expected in many engineering designs, the number of cycles necessary to estimate the probability of failure with reasonable accuracy will be very large, making the simulation method time-consuming. With advancements in computer technology, the time required to complete such a large number of simulations may not be a problem, but it could still be prohibitive if the deterministic system analysis for each simulation is computationally intensive.

The concept behind simulation appears to be simple, but its application in engineering reliability analysis and its acceptance as an alternative reliability evaluation method depend mainly on the efficiency of the simulation. To achieve efficiency, the number of simulation cycles needs to be greatly reduced. It is the simulator's task to increase the efficiency of the simulation by expediting the execution and minimizing computer storage requirements. Alternatively, efficiency can be increased by reducing the variance or the error of the estimated output variable without disturbing the expected or mean value and without increasing the sample size. This need led to the development of several *variance-reduction techniques (VRTs)*. The type of VRT that can be used depends on the particular model under consideration. It is usually impossible to know beforehand how much variance reduction might be achieved using a given technique.

The VRTs can be grouped in several ways. One method is to consider whether the variance reduction method alters the experiment by altering the input scheme, by altering the model, or by special analysis of the output. The VRTs can also be grouped according to description or purpose (i.e., sampling methods, correlation methods, and special methods). These groupings are somewhat arbitrary, but they produce a better understanding of the concept involved.

The sampling methods either constrain the sample to be representative or distort the sample to emphasize the important aspects of the function being estimated. Some of the commonly used sampling methods are systematic sampling, importance sampling, stratified sampling, Latin hypercube sampling, adaptive sampling, randomization sampling, and conditional expectation. The correlation methods employ strategies to achieve correlation (both positive and negative) between functions, random observations, or different simulations to improve the accuracy of the estimators. Some of the commonly used correlation methods are common random numbers, antithetic variates, and control variates. Other special VRTs available are partition of the region, random quadratic method, biased estimator, and indirect estimator. The VRTs can also be combined to further increase the efficiency of the simulation.

VRTs increase the efficiency and accuracy of the risk or reliability estimation using a relatively small number of simulation cycles; however, they increase the computational difficulty for each simulation, and a considerable amount of expertise may be necessary to implement them. The most desirable feature of simulation, its basic simplicity, is thus lost. Also, in the age of high-speed computers, the number of simulation cycles or time required to analyze a problem may be less important than in the past. In any case, it is important to understand the logic and concepts behind some of the VRTs commonly used in engineering. These VRTs include sampling methods and correlation methods. The two types of methods can also be combined to increase the efficiency further. A detailed discussion on VRTs is provided by the authors elsewhere (Haldar and Mahadevan, 2000).

4.3   SIMULATION OF CORRELATED RANDOM VARIABLES

Our discussion thus far has assumed that all the random variables are uncorrelated. In some cases, it may be necessary to estimate the probability of failure of a structure when some or all the random variables are correlated.

The fundamental concepts that need consideration are how to convert the correlated random variables to uncorrelated or statistically independent random variables, and how to modify the original function expressed in terms of correlated variables into a function of uncorrelated random variables.

The methods proposed by Morgenstern (1956) and Nataf (1962) can be used to convert correlated variables to uncorrelated random variables. Nataf's model is discussed very briefly here, since it is more flexible than Morgenstern's model. Suppose $X_1, X_2, \ldots, X_n$

are correlated random variables with the covariance matrix $[\mathbf{C}]$ given by Equation 3.66. The correlation coefficient between X_i and X_j is denoted as ρ_{X_i,X_j}. The X_i's can be transformed into standard normal variates U_i's as:

$$U_i = \Phi^{-1}\left[F_{X_i}(X_i)\right] \qquad i = 1, 2, \ldots, n \qquad (4.16)$$

The U_i's have a zero mean and unit standard deviation. However, the transformation may change the correlation coefficient between any two correlated random variables to $\rho'_{U_1 U_2}$, or simply ρ' in subsequent discussions. We can show that

$$\rho_{X_1,X_2}\sigma_{X_1}\sigma_{X_2} = E(X_1 X_2) - E(X_1)E(X_2) \qquad (4.17)$$

With the transformation of X_i's to U_i's, the expectation operation in Equation 4.17 becomes difficult, and, in some cases, approximate solutions such as the use of Taylor's series may be necessary. Nataf's model can be used, but the calculations become very tedious. Der Kiureghian and Liu (1985) suggested an empirical relationship between the two correlation coefficients as

$$\rho' = F\rho_{X_1,X_2} \qquad (4.18)$$

in which $F \geq 1.0$. Liu and Der Kiureghian (1986) estimated the values of F for several two-parameter distributions of X_i and X_j, as shown in Table 4.7. If one of the two variables is normal, F may be a constant or a function of the COV of the other random variable. However, in all cases, the maximum error in the estimation of F is very small. For combinations of distributions not shown in Table 4.7, refer to Liu and Der Kiureghian (1986).

The symbol ρ' indicates the correlation coefficient between two standard normal variables with zero mean and unit standard deviation. The correlation matrix for this case can be written as

$$[\mathbf{C}] = \begin{bmatrix} 1 & \rho' \\ \rho' & 1 \end{bmatrix} \qquad (4.19)$$

Table 4.7. Evaluation of F parameter

X_i	X_j	F	Maximum Error (%)
Normal	Normal	1.0	
Normal	Uniform	1.023	0.0
Normal	Shifted exponential	1.107	0.0
Normal	Shifted Rayleigh	1.014	0.0
Normal	Type I largest value	1.031	0.0
Normal	Type I smallest value	1.031	0.0
Normal	Lognormal	$\delta_{X_j}\big/\sqrt{\ln(1 + \delta_{X_j}^2)}$	Exact
Normal	Gamma	$1.001 - 0.007\delta_{X_j} + 0.118\delta_{X_j}^2$	0.0
Normal	Type II largest value	$1.030 + 0.238\delta_{X_j} + 0.364\delta_{X_j}^2$	0.1
Normal	Type III smallest value	$1.031 - 0.195\delta_{X_j} + 0.328\delta_{X_j}^2$	0.1

Denoting $\mathbf{V}$ as the uncorrelated standard normal variables, we can show that

$$\{\mathbf{U}\} = [\mathbf{T}]\{\mathbf{V}\} \tag{4.20}$$

As shown in Section 3.7, the transformation matrix $\mathbf{T}$ is composed of the eigenvectors of $[\mathbf{C}_U]$ and can be shown to be

$$[\mathbf{T}] = \begin{bmatrix} \dfrac{1}{\sqrt{2}} & \dfrac{1}{\sqrt{2}} \\ -\dfrac{1}{\sqrt{2}} & \dfrac{1}{\sqrt{2}} \end{bmatrix} \tag{4.21}$$

The corresponding eigenvectors $(1 - \rho)$ and $(1 + \rho)$ are the variances of V_1 and V_2. Thus, V_1 and V_2 are two independent normal variables with zero mean, and the corresponding variances are $(1 - \rho)$ and $(1 + \rho)$, respectively.

The modifications of the original function to be simulated in terms of V_i's are discussed in the following sections with the help of examples. For ease of presentation, the discussion is divided into two parts: simulation of correlated normal random variables and simulation of correlated nonnormal random variables.

4.3.1 Simulation of Correlated Normal Variables

To demonstrate the simulation procedure for correlated random variables, the example discussed in Section 4.2.1 can again be considered. For this example, the expression for the design bending moment at the midspan of the beam is given by Equation 4.1. Previously, the two random variables W and P in Equation 4.1 were considered to be statistically independent. The task now is to simulate them and calculate the bending moments if they are correlated.

For the purpose of illustration, first consider both W and P to be normal random variables with means of 2 kip/ft and 15 kip, respectively and corresponding standard deviations of 0.2 kip/ft and 2.5 kip, respectively. The correlation coefficient of the two variables is assumed to be $\rho_{W,P} = 0.3$.

To demonstrate the difference in simulation between uncorrelated and correlated random variables, W and P are first considered to be independent normal random variables. Using 10 simulation cycles, we can calculate the bending moments at the midspan of the beam, with the results summarized in Table 4.8. In this example, the first 10 uniform numbers between 0 and 1 are assigned to W, and the next 10 uniform numbers between 0 and 1 are assigned to P. The mean and the standard deviation of the bending moment are calculated to be 344.29 and 26.34 kip-ft, respectively.

To consider the effect of correlation between W and P, the following steps can be performed.

Step 1   Transform the correlated normal variables to correlated standard normal variables, U_i's. The U_i's have a zero mean and unit standard deviation. For this particular example, it can be shown that

$$U_1 = \frac{W - \mu_W}{\sigma_W}$$

Table 4.8. Monte Carlo simulations for uncorrelated normal variables

$$W \sim N(2, 0.2), \ P \sim N(15, 2.5), \ \rho_{W,P} = 0$$

	$W \sim N(2, 0.2)$			$P \sim N(15, 2.5)$		$m_i =$ $112.5w_i + 7.5p_i$
U_i	s_i	w_i	u_i	s_i	p_i	m_i
0.86061	1.08306	2.21661	0.83771	0.98509	17.46273	380.33910
0.92546	1.44279	2.28856	0.73006	0.61299	16.53248	381.45660
0.41806	−0.20686	1.95863	0.56341	0.15962	15.39905	335.83875
0.28964	−0.55444	1.88911	0.82178	0.92217	17.30543	342.31560
0.14225	−1.07027	1.78595	0.32715	−0.44780	13.88050	305.02313
0.44961	−0.12665	1.97467	0.68853	0.49169	16.22923	343.86960
0.24653	−0.68545	1.86291	0.74358	0.65442	16.63605	334.34775
0.21687	−0.78281	1.84344	0.24672	−0.68485	13.28788	307.04610
0.56503	0.16373	2.03275	0.90324	1.30024	18.25060	365.56388
0.40015	−0.25296	1.94941	0.79263	0.81558	17.03895	347.10075

Note. Mean of moment = 344.29 kip-ft; standard deviation of moment = 26.34 kip-ft.

or,

$$W = \mu_W + \sigma_W U_1$$

Similarly,

$$P = \mu_P + \sigma_P U_2$$

Step 2   Evaluate the correlation coefficient ρ' between U_1 and U_2 in terms of the correlation coefficient of the original correlated normal variables. As shown in Table 4.7, when both variables are normal, the parameter F is 1.0. Thus,

$$\rho' = \rho_{W,P} = 0.3$$

Step 3   Transform the correlated standard normal variables U_i's to uncorrelated standard normal variables V_i's. Using Equation 4.20, we can show that:

$$U_1 = \frac{1}{\sqrt{2}}(V_1 + V_2)$$

and

$$U_2 = \frac{1}{\sqrt{2}}(-V_1 + V_2)$$

Step 4   Express the function to be simulated in terms of V_i's. For the problem under consideration, it can be shown that

$$W = \mu_W + \frac{\sigma_W}{\sqrt{2}}(V_1 + V_2)$$

and

$$P = \mu_P + \frac{\sigma_P}{\sqrt{2}}(-V_1 + V_2)$$

Thus,

$$M = 112.5W + 7.5P = 112.5\left[\mu_W + \frac{\sigma_W}{\sqrt{2}}(V_1 + V_2)\right]$$

$$+ 7.5\left[\mu_P + \frac{\sigma_P}{\sqrt{2}}(-V_1 + V_2)\right]$$

By substituting the mean and standard deviation values in the equation and simplifying, we can show that

$$M = 337.5 + 2.65165V_1 + 29.16815V_2$$

As mentioned earlier, V_1 and V_2 are normal random variables with zero mean and corresponding standard deviations of $\sqrt{1 - \rho'} = \sqrt{1 - 0.3} = 0.837$ and $\sqrt{1 + \rho'} = \sqrt{1 + 0.3} = 1.140$, respectively.

Step 5   Carry out standard Monte Carlo simulation using the modified function expressed in terms of V_i's. Considering the first 10 uniform random numbers between 0 and 1 in Table 4.1 for V_1 and the next 10 for V_2, we can calculate the bending moments, as summarized in Table 4.9. The mean and standard deviation of the bending moment for the correlated case are 353.29 and 22.03 kip-ft, respectively.

Table 4.9. Monte Carlo simulations for correlated normal variables

$$W \sim N(2, 0.2), \ P \sim N(15, 2.5), \ \rho_{W,P} = 0.3$$

$V_1 \sim N(0, 0.837)$			$V_2 \sim N(0, 1.140)$			$m_i = 337.5 + 2.65165v_{1_i}$ $+ 29.16815v_{2_i}$
μ_i	s_i	v_{1_i}	u_i	s_i	v_{2_i}	m_i
0.86061	1.08306	0.90652	0.83771	0.98509	1.12300	372.65961
0.92546	1.44279	1.20762	0.73006	0.61299	0.69881	361.08518
0.41806	−0.20686	−0.17314	0.56341	0.15962	0.18197	342.34862
0.28964	−0.55444	−0.46407	0.82178	0.92217	1.05127	366.93305
0.14225	−1.07027	−0.89582	0.32715	−0.44780	−0.51049	320.23455
0.44961	−0.12665	−0.10601	0.68853	0.49169	0.56053	353.56852
0.24653	−0.68545	−0.57372	0.74358	0.65442	0.74604	357.73930
0.21687	−0.78281	−0.65521	0.24672	−0.68485	−0.78073	312.99016
0.56503	0.16373	0.13704	0.90324	1.30024	1.48227	381.09846
0.40015	−0.25296	−0.21173	0.79263	0.81558	0.92976	364.05795

Note: Mean of moment = 353.29 kip-ft; standard deviation of moment = 22.03 kip-ft.

4.3.2 Simulation of Correlated Nonnormal Variables

In the previous example, both W and P could have nonnormal distributions. The general steps discussed next can be used for the simulation of correlated nonnormal variables. For simplicity of discussion, consider W to be a normal random variable with a mean of 2.0 kip/ft and a standard deviation of 0.2 kip/ft. However, P is a lognormal random variable with a mean of 15 kip, and a standard deviation of 2.5 kip. Equations 2.15 and 2.16 are used to find the two parameters of the lognormal distribution: $\lambda_P = 2.694$ and $\zeta_P = 0.166$. The coefficient of variation of P is $\delta_P = 0.167$. Again, the five steps discussed in the previous section can be carried out in the following way.

Step 1   Since W is a normal random variable, it can be shown that

$$W = \mu_W + \sigma_W U_1$$

Since P is a lognormal random variable, the following additional calculations are necessary:

$$\Phi(U_2) = \Phi\left(\frac{\ln P - \lambda_P}{\zeta_P}\right)$$

or

$$U_2 = \frac{\ln P - \lambda_P}{\zeta_P} \quad \text{or} \quad \ln P = \lambda_P + U_2\zeta_P$$

or

$$P = \exp(\lambda_P + U_2\zeta_P)$$

Step 2   From Table 4.7, the parameter F can be estimated as

$$F = \frac{\delta_P}{\sqrt{\ln(1 + \delta_P^2)}} = \frac{0.167}{\sqrt{\ln(1 + 0.167^2)}} = 1.00692$$

Using Equation 4.18,

$$\rho' = F\rho_{W,P} = 1.00692 \times 0.3 = 0.302$$

Step 3   The relationships between the U_i's and V_i's discussed in the previous example will remain the same.

Step 4   For the problem under consideration, it can be shown that

$$W = \mu_W + \frac{\sigma_W}{\sqrt{2}}(V_1 + V_2) = 2.0 + \frac{0.2}{\sqrt{2}}(V_1 + V_2)$$

and

$$P = \exp\left[\lambda_P + \frac{\zeta_P}{\sqrt{2}}(-V_1 + V_2)\right] = \exp\left[2.694 + \frac{0.166}{\sqrt{2}}(-V_1 + V_2)\right]$$

Thus, the function to estimate the bending moment can be expressed in terms of V_i's. As discussed before, V_1 and V_2 are normal random variables with zero mean and corresponding standard deviations of $\sqrt{1 - \rho'} = \sqrt{1 - 0.302} = 0.835$ and $\sqrt{1 + \rho'} = \sqrt{1 + 0.302} = 1.141$.

Step 5   Considering the first 10 uniform numbers between 0 and 1 in Table 4.1 for V_1 and the next 10 for V_2, we can calculate the bending moments, as summarized in Table 4.10. The mean value of the bending moment is found to be 352.22 kip-ft, and the corresponding standard deviation is 22.38 kip-ft.

4.4   CONCLUDING REMARKS

The use of simulation techniques to estimate the probability of failure for both explicit and implicit performance functions is discussed in this chapter. The simulation method can be carried out without knowing the more complicated analytical techniques and with a working knowledge of probability and statistics. This method is also robust. The simulation method can provide estimates for any problem, whereas analytical methods may not always converge in their iterations. With the advancement in computer technology, simulation is becoming an attractive alternative to classical analytical methods.

The accuracy of simulation is always a major concern. Simulation is expected to give accurate results as the number of simulation trials approaches infinity. In the age of high-speed computing, a large number of simulation cycles may not be an important hurdle. However, it is important to understand the logic and concepts behind some of the commonly used variance reduction techniques to increase the efficiency and accuracy of simulations.

The basic simulation method assumes that the random variables to be simulated are essentially statistically independent. In practice, however, some or all the random variables may be correlated. Simulation of correlated normal and nonnormal random variables is therefore discussed in this chapter.

A working knowledge of simulation methods is helpful because they are not only simple and robust but also necessary for verifying more sophisticated analytical methods. The stochastic finite element concept that will be presented in the subsequent chapters is verified using simulation.

PROBLEMS

(*Note:* The following problems are intended for homework assignments using the uniform random numbers given in Table 4.1. Answers for homework assignments are requested with few simulations, since the number of uniform random numbers in Table 4.1 is limited and a considerable amount of time would be needed for a large simulation cycle using hand calculation. This way students will get unique results and it will be easier for the teacher to check their accuracy. However, if a teacher prefers to use a computer, then the number of simulations could be large. In that case, the results may vary based on the computer being used and the seed value used to generate uniform random numbers. However, the results are expected to be similar, particularly when relatively large numbers of simulation cycles are used.)

Table 4.10. Monte Carlo simulations for correlated nonnormal variables

$$W \sim N(2, 0.2),\ P \sim \ln(15, 2.5),\ \rho_{W,P} = 0.3$$

$$m_i = 112.5 w_i + 7.5 p_i$$
$$w_i = 2 + \frac{0.2}{\sqrt{2}}(v_{1_i} + v_{2_i})$$
$$p_i = \exp\left[2.694 + \frac{0.166}{\sqrt{2}}(-v_{1_i} + v_{2_i})\right]$$

$V_1 \sim N(0, 0.835)$			$V_2 \sim N(0, 1.141)$			
u_i	s_i	v_{1_i}	u_i	s_i	v_{2_i}	m_i
0.86061	1.08306	0.90436	0.83771	0.98509	1.12399	371.09798
0.92546	1.44279	1.20473	0.73006	0.61299	0.69942	359.83725
0.41806	-0.20686	-0.17273	0.56341	0.15962	0.18213	340.79820
0.28964	-0.55444	-0.46296	0.82178	0.82217	1.05220	366.89708
0.14225	-0.07027	-0.89368	0.32715	-0.44780	-0.51094	318.68070
0.44961	-0.12665	-0.10575	0.68853	0.49169	0.56102	352.20398
0.24653	-0.68545	-0.57235	0.74358	0.65442	0.74669	357.28080
0.21687	-0.78281	-0.65365	0.24672	-0.68485	-0.78141	311.44740
0.56503	0.16373	0.13671	0.90324	1.30024	0.48357	380.70840
0.40015	-0.25296	-0.21122	0.79263	0.81558	0.93058	363.28470

Note. Mean of moment = 352.22 kip-ft; standard deviation of moment = 22.38 kip-ft.

4.1. The total shear resistance of soil between B and C, as shown in Figure P4.1, against slope failure is given by

$$F = (C + P \tan \phi)L$$

where C is the cohesion, P is the pressure normal to arc BC, ϕ is the friction angle of soil, and L is the length of arc $BC = 10$ ft.

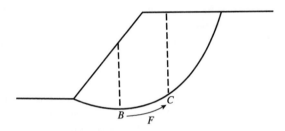

FIGURE P4.1. Shear strength evaluation.

Assume that C, P, and ϕ are statistically independent random variables and L is a constant. Further assume that C is a uniformly distributed random variable between 0 and 1, P is uniformly distributed between 1 and 3, and ϕ is uniformly distributed between 20 and 30 degrees.

(a) Calculate the first-order mean and variance of F.

(b) Using 10 cycles of simulation, calculate the mean and variance of F.

4.2. The drag force, F_D, acting on an immersed body by moving fluid can be calculated as

$$F_D = C_D A \frac{\rho U^2}{2}$$

where C_D is the drag coefficient, A is the projected area of the body on a plane normal to the flow, ρ is the mass density of the fluid, and U is the undisturbed velocity of the fluid. Suppose A and ρ are constants with values 10 ft^2 and 1.94 slug/ft^3, respectively. C_D and U are both assumed to be statistically independent normal random variables with means of 0.5 and 10 ft/sec, respectively, and the corresponding COVs are 0.1 and 0.2, respectively.

(a) Calculate the first-order mean and variance of F_D.

(b) Using 15 cycles of simulation, calculate the mean and variance of F_D.

4.3. A simply supported beam of span L and stiffness EI is loaded with a concentrated load P at the midspan and a uniformly distributed load w along the length of the beam. The maximum deflection at the midspan can be calculated as

$$\delta_{\max} = \frac{PL^3}{48EI} + \frac{5}{385} \frac{wL^4}{EI}$$

Suppose L and EI are constants of values 30 ft and 4.495×10^7 kip-in^2, but P is a normal random variable with a mean of 50 kip and a standard deviation of 10 kip, and w is a lognormal variable with a mean of 1 kip/ft and a standard deviation of 0.1 kip/ft. Using 20 cycles of simulation, calculate the mean and standard deviation of the maximum deflection of the beam.

4.4. The fully plastic flexural capacity of a steel beam section can be given by YZ, where Y is the yield strength of steel and Z is the plastic section modulus of the section. If the applied bending moment at a section of interest is M, the performance function can be defined as

$$g(\ ) = YZ - M$$

Assume that Z is a constant with a value of 40 in^3, Y is uniformly distributed between 25 and 55 ksi, and M is uniformly distributed between 500 and 1500 kip-in. Assume that Y and M are statistically independent. Using 20 cycles of simulation, calculate the probability of failure of the beam. Briefly discuss the accuracy of the result and how it could be improved.

4.5. In Problem 4.4, suppose Y and M are statistically independent normal random variables with mean values of 40 ksi and 1000 kip-in, respectively, corresponding COVs of 0.125 and 0.20, respectively, and Z is a constant with a value of 40 in^3.

(a) What is the distribution of $g(\ )$ and its parameters?

(b) What is the probability of failure of the beam?

(c) Using 15 cycles of simulation, calculate the probability of failure of the beam.

4.6. In Problem 4.4, assume that Y and M are statistically independent lognormal random variables; all other information remains the same. Calculate the probability of failure of the beam using 15 cycles of simulation.

4.7. In Problem 4.5, if Y and M are correlated normal variables with $\rho_{Y,M} = 0.70$ and all other information remains the same, calculate the probability of failure of the beam using 15 simulation cycles.

4.8. Repeat Problem 4.7, where Y is a normal variable with a mean of 40 ksi and a COV of 0.125, M is a uniform random variable between 500 and 1500 kip-in, and the correlation coefficient between them is still 0.70.

4.9. Repeat Problem 4.7, where Y is a normal variable with a mean of 40 ksi and a COV of 0.125, M is a lognormal variable with a mean of 1000 kip-in and a COV of 0.20, and the correlation coefficient between them is still 0.70.

5

IMPLICIT PERFORMANCE FUNCTIONS: INTRODUCTION TO SFEM

5.1 INTRODUCTORY COMMENTS

The reliability analysis methods discussed in Chapters 3 and 4 are easy to implement if the performance function $g(\mathbf{X})$ is an explicit function of the load and resistance-related input random variables $\mathbf{X}$. In the case of FORM, when an explicit function is available, it is easy to compute the derivatives of $g(\mathbf{X})$ with respect to the random variables $\mathbf{X}$ in order to proceed with the search for the minimum distance point on the limit state. In the case of Monte Carlo simulation, an explicit function can be evaluated quickly and therefore a large number of simulations can be performed without difficulty. However, in many cases of practical importance, particularly for complicated structures, the performance function $g(\mathbf{X})$ is generally not available as an explicit, closed-form function of the input variables. For most realistic structures, the response has to be computed through a numerical procedure such as finite element analysis. In such cases, the derivatives are not readily available, and each evaluation of the performance function $g(\mathbf{X})$ could be time-consuming. This chapter presents methods to perform reliability analysis under such situations.

Several computational approaches could be pursued for the reliability analysis of structures with implicit performance functions. These can be broadly divided into three categories, based on their essential philosophy, as (1) Monte Carlo simulation (including efficient sampling methods and variance reduction techniques), (2) response surface approach, and (3) sensitivity-based analysis. The last method leads to probabilistic or stochastic finite element analysis, which is introduced toward the end of this chapter.

5.2 MONTE CARLO SIMULATION

Monte Carlo simulation is discussed in Chapter 4. Briefly, it uses randomly generated samples of the input variables for each deterministic analysis, and estimates response statistics and reliability after numerous repetitions of the deterministic analysis. The efficiency of the simulation can be improved by variance reduction techniques. The examples in Chapter 4 feature closed-form performance functions, but the Monte Carlo method can easily handle problems with implicit performance functions as well. As long as an algorithm (a black box, such as a commercial finite element program) is available to compute the structural response, given the values of the input variables, the Monte Carlo method can easily evaluate $g(\mathbf{X})$ for each deterministic analysis and therefore compute the failure probability after performing several deterministic analyses. However, if the deterministic structural analysis is time-consuming, as in the case of structures with numerous finite elements, then Monte Carlo simulation may be impractical.

Example 5.1

Consider a steel portal frame as shown in Figure 5.1. An estimate is needed for the probability that the horizontal displacement at node 2 (u_2) exceeds a limiting value of 0.9 in.

Four random variables are considered: the dead load, D; the live load, L; the wind load, W; and the Young's modulus, E. For the sake of simplicity, the cross-sectional properties of the members are assumed to be deterministic. The beam has an area of cross section of 10.6 in^2 and a moment of inertia of 448 in^4. For the columns, the area of cross section is 20.1 in^2 and the moment of inertia is 1830 in^4. The assumed statistics of the random variables are shown in Table 5.1.

The performance function may be written as

$$g(\mathbf{X}) = 0.9 - u_2 \tag{5.1}$$

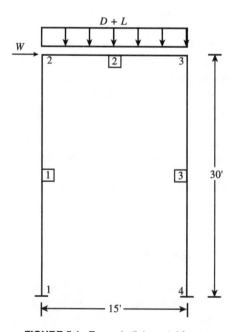

FIGURE 5.1. Example 5.1: portal frame.

Table 5.1. Portal frame: random variable statistics

Variable	Mean Value	Coefficient Of Variation	Type of Distribution
D	3.15 k/ft	0.1	Normal
L	0.91 k/ft	0.32	Type I
W	7.425 k	0.5	Type I
E	29,000 ksi	0.06	Normal

where $\mathbf{X} = (D, L, W, E)^t$ is the vector of the random variables. In this equation, $g(\mathbf{X}) < 0$ indicates failure. Considering this single performance function, an estimate is required for the probability $P[g(\mathbf{X}) < 0]$.

Note that the performance function does not explicitly contain any of the four random variables. The quantity u_2 is dependent on the basic random variables, but it cannot be easily expressed as a closed-form function of the random variables. Instead, it has to be evaluated using a matrix-based equation-solving procedure, explained in detail in Chapter 6. Thus, this is a problem with an implicit performance function.

In the Monte Carlo method, many simulations are performed. In each simulation, the values of the variables $\mathbf{X}$ are randomly generated according to their probability density functions and the corresponding value of u_2 is computed. Then Equation 5.1 is used to evaluate the performance function in each simulation. Finally, the probability $P[g(\mathbf{X}) < 0]$ is estimated as

$$P[g(\mathbf{X}) < 0] = \frac{N_f}{N} \tag{5.2}$$

where N_f is the number of simulations with $g(\mathbf{X}) < 0$ (i.e., number of failures), and N is the total number of simulations. Using the performance function given by Equation 5.1 and 10,000 simulations, the probability $P[g(\mathbf{X}) < 0]$ is estimated to be 0.0646.

5.3 RESPONSE SURFACE APPROACH

The response surface approach constructs a polynomial closed-form approximation (e.g., first-order or second-order) for $g(\mathbf{X})$ through (1) a few selected deterministic analyses, and (2) regression analysis of these results. The approximate closed-form expression thus obtained is then used to search for the design point, and the failure probability is computed using first-order (FORM) or second-order (SORM) reliability methods, as described in Chapter 3. Monte Carlo simulation may also be used with the closed-form approximation to estimate the failure probability, as described in Chapter 4.

The implementation of the response surface concept may proceed along the following steps:

Step 1   Select sets of values of the random variables to evaluate the performance function $g(\mathbf{X})$. This step is referred to as design of experiments. The literature on design of experiments is quite vast, and many methods are available in various textbooks (e.g., Myers and Montgomery, 1995). A simple method is to use a full factorial design. Consider two or three values of each variable, and evaluate the performance function for all possible combinations. In that case, the number of combinations is 2^n or 3^n, respectively, where n is the number of random variables. If two values are used, these may be selected to be a low value and a high value (e.g., $\mu \pm k\sigma$, where k is an integer), usually leading to a first-order model in step 3. If three values are used, these may be selected to be a low value, a medium value, and a high value (e.g., $\mu - k\sigma$, μ, and $\mu + k\sigma$), usually leading to a second-order model in step 3.

Step 2   Evaluate the performance function $g(\mathbf{X})$ using the deterministic analysis for all the sets of values of the random variables selected in step 1.

Step 3   Construct a first-order or second-order (or higher order) model using regression analysis with the data collected in step 2. This is a least-squares fit, which gives an approximate closed-form expression of the performance function in terms of the random variables $\mathbf{X}$. (Note that if the number of coefficients in the model is equal to the number of data points, then the coefficients can be obtained by solving a number of simultaneous equations, without using regression analysis.)

Step 4   Use either FORM/SORM (Chapter 3) or Monte Carlo simulation (Chapter 4), with the closed-form expression developed in step 3, to estimate the probability $P[g(\mathbf{X}) < 0]$.

The response surface method gives an approximate closed-form expression, based on the values selected in step 1. The approximation could be inadequate for highly nonlinear performance functions. Also, as a general rule, a closed-form expression developed using regression analysis is valid only within the range of the values considered for the random variables; extrapolation beyond the range may not be accurate.

Example 5.2

Consider the portal frame of Example 5.1. Once again, an estimate is required for the probability that the horizontal displacement at node 2 (u_2) exceeds a limiting value of 0.9 in. The performance function and the random variable definitions are the same as in Example 5.1.

To implement the response surface approach for this problem, consider two values of each random variable in step 1. Since there are four random variables, the number of combinations for a full factorial design are $2^4 = 16$ for this problem. The 16 sets of values for analysis (plus one more set at the mean values) and the corresponding values of the displacement u_2 obtained in step 2 are listed in Table 5.2.

Next, the regression analysis of step 3 is implemented to construct a first-order closed-form approximation to $g(\mathbf{X})$ of the form

$$g(\mathbf{X}) = b_0 + b_1 D + b_2 L + b_3 W + b_4 E \tag{5.3}$$

Many computer programs and spreadsheets are available to construct this least-squares regression model. The numerical values of the coefficients b_0 to b_4 are estimated using any one of these, and the first-order approximation of the performance function of Equation 5.1 is

$$g(\mathbf{X}) = 0.416 + 0.00022D - 0.00024L - 0.06531W - 1.67 \times 10^{-5}E \tag{5.4}$$

Table 5.2. Response surface method: deterministic analyses

Combination	D	L	W	E	u_2
1	3.465	1.2012	11.1375	30740	0.6828
2	2.835	1.2012	11.1375	30740	0.6829
3	3.465	0.6188	11.1375	30740	0.6829
4	3.465	1.2012	3.7125	30740	0.2269
5	3.465	1.2012	11.1375	27260	0.7699
6	2.835	0.6188	11.1375	30740	0.683
7	2.835	1.2012	3.7125	30740	0.2271
8	2.835	1.2012	11.1375	27260	0.7701
9	3.465	0.6188	3.7125	30740	0.2271
10	3.465	0.6188	11.1375	27260	0.7701
11	3.465	1.2012	3.7125	27260	0.2559
12	2.835	0.6188	3.7125	30740	0.2272
13	2.835	0.6188	11.1375	27260	0.7702
14	2.835	1.2012	3.7125	27260	0.2561
15	3.465	0.6188	3.7125	27260	0.2561
16	2.835	0.6188	3.7125	27260	0.2562
17	3.15	0.91	7.425	29000	0.4823

FORM Method 2 of Chapter 3 is applied to this closed-form function, and the probability $P[g(\mathbf{X}) < 0]$ is estimated as 0.061.

Compare this result with that of Example 5.1, where the failure probability for the same event was estimated to 0.0646 with 10,000 simulations. This indicates that a linear response surface approximation may be adequate for this particular problem. Note that in this example, three of the random variables are load variables. Since the structural analysis used is linear, it is obvious that the displacement is a linear function of the load variables. Although displacement is a nonlinear (in fact, inverse) function of the fourth random variable E, the load variables are much more dominant. Therefore, a linear response surface approximation appears to be adequate. In general, a good physical knowledge of the system is very useful in deciding the appropriate order of the response surface approximation.

The response surface approach is presented in more detail in Section 9.6 in Chapter 9.

5.4 SENSITIVITY-BASED APPROACH

The third approach, described in detail in the following sections, is based on sensitivity analysis. In this method, the sensitivity of the structural response to the input variables is computed and used in the FORM and SORM methods of Chapter 3. The fundamental concept of the FORM and SORM methods, the search for the design point or checking point, requires only the value and gradient of the performance function at each iteration. The value of the performance function is available from deterministic structural analysis. The gradient is computed using sensitivity analysis. In the case of explicit closed-form functions, the gradient is computed simply by analytical or numerical differentiation of the performance function with respect to each random variable. In the case of problems that do not have explicit closed-form solutions, several approximate methods are available to compute the gradient of the performance function. These methods are described in the following sections.

The sensitivity-based reliability analysis approach is more elegant and, in general, more efficient than the simulation or response surface methods. It is apparent from Section 5.2 that Monte Carlo simulation could be time-consuming for high-reliability (low failure probability) problems. In the response surface approach of Section 5.3, the number of deterministic analyses required to construct an approximate closed-form expression may be quite large for problems with a large number of random variables, thus making this method time-consuming. Even if a problem has a small number of random variables, it should be kept in mind that the reliability estimate using the response surface approach is only as accurate as the closed-form approximation to the performance function. If the underlying implicit performance function is highly nonlinear and the closed-form approximation is too approximate, the reliability estimate may also be too approximate.

The combination of sensitivity analysis and FORM/SORM for the reliability analysis with implicit performance functions does not suffer from the drawbacks of Monte Carlo simulation or the response surface approach. It uses the information about the actual value and the actual gradient of the performance function at each iteration of the search for the checking point (also referred to as the design point, the minimum distance point, or the most probable point (MPP) of failure), and uses an optimization scheme to converge to the minimum distance point. For many practical problems, this method has usually been observed to converge to the minimum distance point within 10 or 20 iterations. Therefore, it is computationally inexpensive compared to Monte Carlo simulation and the response surface approach, and also more accurate compared to the response surface approach.

Apart from an immediate application to FORM/SORM, it is also important to know the sensitivity of the structural response to the basic random variables from a design point of view. If the uncertainty in a certain basic variable is found to have a large effect on structural failure, then it is important to reduce its uncertainty by collecting additional information including testing or quality control or both. This will improve the reliability of the design by redistributing the available resources appropriately. Also, it is possible to derive different design safety factors for different random variables based on their uncertainty and on their influence on structural behavior. Sensitivity analysis can also be used to ignore the uncertainty in those variables that do not show a significant influence on structural reliability, which saves a great amount of computational effort, without compromising the accuracy in the reliability estimate. Thus, sensitivity information is very useful in probabilistic analysis and design.

The response sensitivities can be computed in three different ways: (1) through a finite difference approach, by perturbing each variable and computing the corresponding change in response through multiple deterministic analyses; (2) through classical perturbation methods that apply the chain rule of differentiation to finite element analysis; and (3) through iterative perturbation analysis techniques. The following sections describe these three methods in detail.

5.5 FINITE DIFFERENCE METHOD

Consider two variables related as $Z = g(X)$. The derivative of Z with respect to X is defined as

$$\frac{dZ}{dX} = \lim_{\Delta X \to 0} \frac{\Delta Z}{\Delta X} \qquad (5.5)$$

If the analytical differentiation of $g(X)$ is difficult, the simplest numerical (approximate) approach to compute the derivative is to change X by a small amount (close to zero) and measure the corresponding change in the value of Z. This is the basis of the finite difference approach. When $g(X)$ is not explicit, if it is represented by a computational algorithm, then the finite difference approach may be used to approximately calculate the derivative.

Now consider a problem where Z is a function of n variables $X_1, X_2, \ldots, X_n$, i.e., $Z = g(X_1, X_2, \ldots, X_n)$. The variable Z may be referred to as an output or response variable, and the variables $X_1, X_2, \ldots, X_n$ may be referred to as input or basic variables. If the forward difference approach is used to compute the derivatives $\partial Z / \partial X_1, \partial Z / \partial X_2, \ldots,$ $\partial Z / \partial X_n$, at the point $(X_1^0, X_2^0, \ldots, X_n^0)$, the implementation will be as follows:

Step 1   First compute $Z_0 = g(X_1^0, X_2^0, \ldots, X_n^0)$.

Step 2   Change the value of X_1 to $X_1^0 + \Delta X_1$, where ΔX_1 is a small number, and may be referred to as the perturbation in the value of X_1. All other variables X_2 to X_n stay at the same value as in step 1. Compute the new value of Z as $Z_1 = g(X_1^0 + \Delta X_1, X_2^0, \ldots, X_n^0)$.

Step 3   The change in the value of Z due to the small change in the value of X_1 is $\Delta Z = Z_1 - Z_0$. Compute the approximate derivative of Z with respect to X_1 as $\Delta Z / \Delta X_1$.

Step 4   Repeat steps 2 and 3 for each variable X_2 to X_n. Each time, the value of only one input variable is changed, while the other input variables are maintained at the initial value (indicated by the superscript 0).

Thus, the derivatives of Z with respect to all the variables $X_1, X_2, \ldots, X_n$ are approximately computed using the finite difference approach. Note that the total number of times the function $g(X_1, X_2, \ldots, X_n)$ is evaluated in the forward difference approach is $n + 1$, i.e., one evaluation at step 1 and n repetitions of steps 2 and 3 for the n variables. (The backward difference and central difference methods may also be used to compute the derivatives. The backward difference method uses negative perturbation in step 2 above. The central difference method uses both positive and negative perturbations, and uses $2n + 1$ evaluations of the g function. This section only considers the forward difference method.)

The finite difference approach can be used for the numerical and approximate computation of the derivatives of any function with respect to the input variables. In the context of reliability analysis, the function of interest is the performance function $g(X_1, X_2, \ldots, X_n)$, and the equation $g(X_1, X_2, \ldots, X_n) = 0$ refers to the limit state. FORM Method 2 of Chapter 3 searches for the minimum distance point from the origin on the limit state in the standard normal space, using the value and the derivatives of $g(X_1, X_2, \ldots, X_n)$ at each iteration point. Therefore, in the context of FORM, the derivatives of $g(X_1, X_2, \ldots, X_n)$ have to be computed at each iteration, until convergence. If the finite difference approach (forward or backward difference) is used to compute these derivatives, then the performance function will be evaluated a total of $(n + 1)m$ times, where n is the number of random variables, and m is the number of iterations of the FORM algorithm.

FIGURE 5.2. Axially loaded bar.

Example 5.3

Consider the axially loaded bar shown in Figure 5.2, with an axial (tensile) load P, area of cross section A, and yield strength F_y. Assume that all three variables, P, A, F_y, are random variables. Consider the problem of estimating the reliability of this axially loaded bar. The bar is assumed to fail when the tensile stress in the bar exceeds the yield strength. The corresponding performance function is of the form $g(R, S) = R - S$, where $R = F_y$, and $S = P/A$ is the tensile stress, and is the structural response quantity of interest. Failure occurs when $g(R, S) < 0$.

The statistics of the random variables are shown in Table 5.3. It is necessary to find the derivatives of the performance function with respect to the three random variables using the finite difference approach, at the mean values of the random variables.

The performance function can be written in terms of the random variables as

$$g = F_y - \frac{P}{A} \tag{5.6}$$

At the mean values of the random variables, the value of g is

$$g_0 = 36 - \frac{30}{1} = 6 \tag{5.7}$$

Now, keeping F_y and A at their mean values, change the value of P to 30.1 (i.e., $\Delta P = 0.1$). The corresponding value of g is

$$g_1 = 36 - \frac{30.1}{1} = 5.9 \tag{5.8}$$

The change in the value of g due to the change in the value of P is

$$\Delta g = g_1 - g_0 = -0.1 \tag{5.9}$$

Table 5.3. Axially loaded bar: random variable statistics

Variable	Mean Value	Coefficient Of Variation	Type of Distribution
P	30 kips	0.1	Normal
A	1 in^2	0.1	Normal
F_y	36 ksi	0.1	Normal

Therefore, the derivative of g with respect to P is approximated as

$$\frac{\partial g}{\partial P} \simeq \frac{\Delta g}{\Delta P} = -1 \qquad (5.10)$$

To compute the derivative of g with respect to A, maintain F_y and P at their mean values, and change the value of A to 1.01 (i.e., $\Delta A = 0.01$). Notice that ΔA is much smaller than ΔP. It is common practice to use perturbation values in proportion to the standard deviations of the variables. For example, one may use a perturbation size of one-tenth of the standard deviation for each variable. The corresponding value of g is

$$g_1 = 36 - \frac{30}{1.01} = 6.297 \qquad (5.11)$$

The change in the value of g due to the change in the value of A is

$$\Delta g = g_1 - g_0 = 0.297 \qquad (5.12)$$

Therefore, the derivative of g with respect to A is approximated as

$$\frac{\partial g}{\partial A} \simeq \frac{\Delta g}{\Delta A} = 29.7 \qquad (5.13)$$

Similarly, to compute the derivative of g with respect to F_y, maintain P and A at their mean values and change the value of F_y to 36.1. Repeating the above steps, the derivative of g with respect to F_y is computed as 1.0.

In this example, the numerical values of the derivatives computed above are valid only at the mean values of the random variables. During the iterations of the FORM reliability analysis, the derivatives need to be recalculated at each iteration, since the values of the variables change from iteration to iteration. At each iteration, if the finite difference approach is applied to compute the derivatives, the value of the g function is computed first at that point. Then each variable is perturbed a little (while maintaining the other variables at the iteration value), and the change in g is computed. This gives the approximate estimate of the derivatives at that iteration point.

It is apparent that the finite difference approach is a brute force method and requires many evaluations of $g(X_1, X_2, \ldots, X_n)$ to compute its derivatives. As mentioned earlier, the number of deterministic analyses using the finite difference approach is $(n+1)m$, where n is the number of random variables and m is the number of FORM iterations. Thus, for a problem with $n = 10$ and $m = 5$, the number of deterministic analyses required is 55. Note that this is for computing the failure probability corresponding to one limit state. In a practical structure, a number of limit states may need to be checked for reliability. Thus, the finite difference approach could become increasingly time-consuming as the number of limit states increase.

5.6   CLASSICAL PERTURBATION

Since the basic variables are stochastic, every quantity computed during the deterministic analysis, being a function of the basic variables, is also stochastic. Hence, one efficient way to estimate the variation of response is to keep account of the variation of the quantities at every step of the deterministic analysis, in terms of the variation of the basic random variables. In practical terms, this is simply the application of the chain rule of differentiation to compute the derivatives of structural response or the performance function with respect to the basic random variables. Such an approach is referred to as classical perturbation.

Example 5.4

Consider again the axially loaded bar of Example 5.3, with three random variables: the axial load P, area of cross section A, and yield strength F_y. Once again, the bar is assumed to fail when the tensile stress in the bar exceeds the yield strength. The corresponding performance function is of the form $g(R, S) = R - S$, where $R = F_y$ and $S = P/A$ is the tensile stress, and is the structural response quantity of interest. Failure occurs when $g(R, S) < 0$. Use the chain rule of differentiation to compute the derivatives of g with respect to the input random variables P, A, and F_y.

Even in this simple problem, it is worth noting that the derivatives are computed using the chain rule of differentiation, according to the following steps:

Step 1   The partial derivative of g with respect to P:

$$\frac{\partial g}{\partial P} = \frac{\partial g}{\partial S}\frac{\partial S}{\partial P} \tag{5.14}$$

$$= (-1.0)\frac{1}{A} \tag{5.15}$$

Step 2   The partial derivative of g with respect to A:

$$\frac{\partial g}{\partial A} = \frac{\partial g}{\partial S}\frac{\partial S}{\partial A} \tag{5.16}$$

$$= (-1.0)\left(-\frac{P}{A^2}\right) \tag{5.17}$$

Step 3   The partial derivative of g with respect to F_y:

$$\frac{\partial g}{\partial F_y} = \frac{\partial g}{\partial R}\frac{\partial R}{\partial F_y} \tag{5.18}$$

$$= (1.0)(1.0) \tag{5.19}$$

Notice that the chain rule of differentiation has been applied in each of the above three steps to compute the derivatives of g with respect to the basic random variables P, A, and F_y. The computation of derivatives with respect to a variable such as F_y is usually easy, since it occurs explicitly in the performance function in many cases. But the computation of derivatives with respect to P and A is of special interest here. The chain rule of differentiation in each case made use of the structural response computational algorithm, which in this problem is simply $S = P/A$. In other problems, the computation of the structural response S may involve many steps (e.g., finite element analysis). In that case, the chain rule of differentiation may also go through several steps in order to compute the derivatives of g with respect to the basic random variables that affect S.

Example 5.5

Assume that the axially loaded bar of Example 5.4 has a rectangular cross section with width b and thickness t. The area of cross section is $A = bt$. Assume that b and t are the basic random variables, instead of A. Now there are 4 basic random variables: P, b, t, F_y. The derivatives with respect to P and F_y are computed the same way as in the Example 5.4. The derivative of g with respect to b is computed, using the chain rule of differentiation, as

$$\frac{\partial g}{\partial b} = \frac{\partial g}{\partial S}\frac{\partial S}{\partial A}\frac{\partial A}{\partial b} \tag{5.20}$$

$$= (-1.0)\frac{P}{A^2}t \tag{5.21}$$

Similarly, the derivative of g with respect to t is computed as

$$\frac{\partial g}{\partial t} = \frac{\partial g}{\partial S}\frac{\partial S}{\partial A}\frac{\partial A}{\partial t} \tag{5.22}$$

$$= (-1.0)\frac{P}{A^2}b \tag{5.23}$$

In this case, the response computation algorithm has two steps: (1) $A = bt$, and (2) $S = P/A$. The chain rule of differentiation, therefore, used both these steps in the computation of derivatives of g with respect to the variables b and t. Thus, it is clear that the chain rule of differentiation can be extended to include many steps in the computational algorithm.

In the case of the axially loaded bar considered here, the computational algorithm is simple enough that the performance function g can be directly written as an explicit function in terms of the basic variables P, b, t, and F_y, by substituting the formulas $R = F_y$, $A = bt$, and $S = P/A$ in the performance function $g = R - S$. In the case of structures where the structural response such as displacement or stress at a particular location is computed using a matrix or finite element analysis, it may not be possible to simply obtain a closed-form expression by substitution. The chain rule of differentiation is then a valuable tool to compute the derivatives in such cases. This chain rule of differentiation is referred to as classical perturbation.

5.7 ITERATIVE PERTURBATION

This method is suitable in the context of nonlinear structural analysis, where the solution for structural response is obtained through an iterative process. The purpose, once again, is to compute the sensitivity of a response quantity or performance function to changes in the random variable values. There are several methods of nonlinear analysis, and iterative perturbation can be tailored to the method of choice. Nonlinear structural analysis and iterative perturbation are presented in detail in Chapter 8. In this section, the basic concepts are introduced through a simple example. Two methods of iterative perturbation are discussed.

5.7.1 Method 1

The first method of iterative perturbation uses the simplest possible iterative process to account for nonlinear structural behavior. It is based on the observation that the geometry of a structure changes gradually as the loading is increased. Therefore, an iterative procedure is used to obtain the correct solution, as shown by the following example.

Example 5.6

Consider the axially loaded bar of Example 5.3. It is required to estimate the sensitivity of the elongation U of this bar to the applied axial load P. For a linear elastic problem, the elongation is simply evaluated from the formula

$$U = \frac{PL}{AE} \tag{5.24}$$

where L is the original length of the bar, A is the area of cross section, and E is the Young's modulus. In this case, the derivative of U with respect to P is simply

$$\frac{\partial U}{\partial P} = \frac{L}{AE} \tag{5.25}$$

Equation 5.24 may also be written in familiar finite element notation as

$$KU = F \tag{5.26}$$

where $K = AE/L$ may be referred to as the axial stiffness of the bar, and $F = P$. Equations 5.24 and 5.25 are valid only for the case of small deformation, where the change in length of the bar is very small. Therefore, the stiffness, K, may be assumed to be constant as the length of the bar changes.

In the case of large deformation, the stiffness, K, cannot be assumed as constant; it changes with the change in length, L. As the load is gradually increased from 0 to F, the length of the bar, L, is continuously changing, and therefore the stiffness $K = AE/L$ is continuously changing. This is the situation with nonlinear analysis. It is required to compute the derivative of the structural response U with respect to the random variable F under this condition. This problem is solved using iterative perturbation as follows.

Assume that at the combination (F_0, U_0), the following static equilibrium relationship is satisfied:

$$K_0 U_0 = F_0 \tag{5.27}$$

where $K_0 = AE/L_0$ is the initial stiffness of the bar. Now, let the load be perturbed by a *small* value, ΔF. To start the computation, assume that the F vs. U (force vs. deformation) relationship is linear (only in the first step). Then the stiffness, K_0, may be assumed to be unchanged for the small change in the force, ΔF, in the initial step. The only change will be in the elongation (denoted as ΔU), and is simply computed using

$$K_0 \Delta U = \Delta F \tag{5.28}$$

Now, the new length of the bar is $L_0 + \Delta U$. The stiffness of the bar, corresponding to this new length, is

$$K = \frac{AE}{L_0 + \Delta U} \quad (5.29)$$

Now, solve again for ΔU using the new stiffness in Equation 5.28. This will again change the length of the bar, which will change the stiffness and therefore the deformation, and so on. Thus, the correct solution has to be obtained at the end of an iterative process. The iterations may be stopped when the change in values of U and K between two successive iterations is very small.

Example 5.7

Consider an axially loaded steel bar with a cross-sectional area $A = 1.0 \text{ in}^2$, initial length $L_0 = 100$ in, and Young's modulus $E = 30 \times 10^6$ psi. An axial load $\Delta F = 10,000$ lb is applied to this bar. Find the final length of the bar, the corresponding deformation ΔU, and the derivative of U with respect to the load F corresponding to this perturbation.

The initial stiffness of the bar is

$$K_0 = \frac{AE}{L_0} = 300,000 \quad (5.30)$$

Using this initial stiffness, the deformation is calculated in the first step as

$$\Delta U = \frac{\Delta F}{K_0} = 0.03333333 \quad (5.31)$$

The new length of the bar is

$$L = L_0 + \Delta U = 100.03333333 \quad (5.32)$$

Corresponding to this new length, the new stiffness is calculated, and the above steps are repeated. The iterations are shown in Table 5.4. The converged solution is obtained in four iterations as $L = 100.0333444$ in. The derivative of the deformation, U, with respect to the load, F, is approximately calculated as the change in length divided by the change in load:

$$\frac{\partial U}{\partial F} \simeq \frac{\Delta U}{\Delta F} = \frac{0.0333444}{10,000} = 3.33444 \times 10^{-6} \quad (5.33)$$

This is an elementary example of nonlinear structural analysis, and the corresponding sensitivity analysis using iterative perturbation. In practice, the load, F, is not applied all at once; instead, it is applied in small increments, and an iterative solution is

Table 5.4. Axially loaded bar: simple iterative perturbation

Variable	Iteration			
	1	2	3	4
K	300,000	299,900.0333	299,900	299,900
ΔU	0.033333333	0.033344444	0.033344448	0.033344448
L	100.0333333	100.0333444	100.0333444	100.0333444

found for each incremental load step. Also, the solution algorithm is quite elementary. Practical nonlinear analysis uses more sophisticated solution algorithms, such as the Newton method or the secant method. However, the discussion clearly explains the essential concept of iterative perturbation in the context of geometric nonlinearity. That is, for a small perturbation in the load, an iterative solution is computed for the structural response. This is used to compute the gradient or sensitivity of the response.

In general, the correct equilibrium relationship may be written as

$$(K_0 + \Delta K)(U_0 + \Delta U) = F_0 + \Delta F \tag{5.34}$$

The task is to compute the change in the structural response, ΔU, so that its derivative with respect to the load variable can be approximately calculated as $\Delta U/\Delta F$. Note that both K_0 and ΔK could be stochastic quantities, but in this section we are concerned with only deterministic sensitivity analysis. As a first step, split Equation 5.34 into two equations, as

$$K_0 U_0 = F_0 \tag{5.35}$$

and

$$K_0 \Delta U + \Delta K U_0 + \Delta K \, \Delta U = \Delta F \tag{5.36}$$

The first equation is the same as Equation 5.27 and represents the original equilibrium state. The second equation relates the perturbed quantities in the new equilibrium state.

Simplify Equation 5.36 by neglecting the second-order term $\Delta K \, \Delta U$ to obtain

$$K_0 \Delta U + \Delta K U_0 = \Delta F \tag{5.37}$$

Use this equation to construct an iterative solution for ΔU according to the following steps:

Step 1   Since at the beginning, the change in K (i.e., ΔK) is not known, first solve for ΔU using Equation 5.28. (This is equivalent to assuming that the F vs. U relationship is linear within the load increment. This assumption is erroneous; subsequent iterations will correct this error.)

Step 2   Use the current solution of ΔU to obtain the corrected stiffness $K_{i+1} = K_i + \Delta K$, where ΔK is the change in stiffness corresponding to the deformation ΔU.

Step 3   Solve for ΔU using Equation 5.37.

Step 4   Repeat steps 2 and 3 until convergence.

5.7.2   Method 2

For large structures such as multistory, multibay frames, the preceding simple algorithm is not practical. The quantities in Equation 5.27 are matrices, with the stiffness matrix **K** usually quite large. In such a case, it is computationally too expensive to solve Equation 5.37 by using the updated stiffness matrix at each iteration. In typical finite element analysis, it is the solution of the system of equations as in Equation 5.37 that consumes the most computational effort. Therefore, a better strategy is to keep the stiffness matrix the same in all the iterations and to manipulate other quantities to obtain the correct solution. Such an iterative perturbation technique was proposed by Dias and Nagtegaal (1985). In this method, a residual force vector $\mathbf{r}_1$ is defined as

$$\mathbf{r}_1 = \mathbf{f}_1 - \mathbf{K}_1\mathbf{u} \tag{5.38}$$

where $\mathbf{f}_1$ is the perturbed force vector, $\mathbf{K}_1$ is the perturbed stiffness matrix, and $\mathbf{u}$ is the displacement vector of the original unperturbed structure. The perturbed displacement is then obtained by solving

$$\mathbf{K}\ \Delta\mathbf{u}_1 = \mathbf{r}_1 \tag{5.39}$$

and

$$\mathbf{u}_1 = \mathbf{u} + \Delta\mathbf{u}_1 \tag{5.40}$$

Note that the original $\mathbf{K}$ matrix is used to obtain the perturbed solution. A new residual vector is defined, and the above analysis is repeated in a predictor–corrector sequence until convergence to the perturbed solution.

Example 5.8

Consider again the axially loaded bar with the same data as in the previous example. Find the final length of the bar and the derivative of the deformation to the load, using the iterative perturbation method without changing the stiffness.

The initial stiffness of the bar is

$$K_0 = \frac{AE}{L_0} = 300{,}000 \tag{5.41}$$

The initial displacement of the unperturbed structure is $u = 0$. The perturbed force vector $\mathbf{f}_1 = 10{,}000$. From Equation 5.38, the residual force vector is

$$\mathbf{r}_1 = 10{,}000 \tag{5.42}$$

The deformation $\Delta\mathbf{u}_1$ is then computed from Equation 5.39 as

$$\Delta\mathbf{u}_1 = \frac{10{,}000}{300{,}000} = 0.03333333 \tag{5.43}$$

The perturbed displacement $\mathbf{u}_1$ is then

$$\mathbf{u}_1 = 0.03333333 \tag{5.44}$$

Note that the original unperturbed displacement is zero. The new length of the bar is

$$L = L_0 + \Delta U = 100.03333333 \tag{5.45}$$

Corresponding to this new length, the perturbed stiffness, $\mathbf{K}_1$, is calculated as

$$K_1 = \frac{(1)(30 \times 10^6)}{100.03333333} \tag{5.46}$$

Table 5.5. Iterative perturbation with the same stiffness matrix K

Variable	Iteration			
	1	2	3	4
f	10,000	10,000	10,000	10,000
K	300,000	299,900.0333	299,900	299,900
r	10,000	3.332222592	0.0022220741	1.48025E-06
Δu	0.033333333	1.11074E-05	7.40247E-09	4.93416E-12
u	0.033333333	0.033344441	0.033344448	0.033344448
L	100.0333333	100.0333444	100.0333444	100.0333444

A new residual force vector is calculated, and the steps are repeated. The perturbed stiffness is only used to calculate the residual force vector; the original stiffness is used in solving Eq. 5.39. The iterations are shown in Table 5.5. The converged solution is obtained in four iterations as $L = 100.0333444$ in.

The derivative of the deformation U with respect to the load F is once again approximately calculated as the change in length divided by the change in load:

$$\frac{\partial U}{\partial F} \simeq \frac{\Delta U}{\Delta F} = \frac{0.0333444}{10,000} = 3.33444 \times 10^{-6} \tag{5.47}$$

Compare this result with that of Example 5.7. For this simple problem of an axially loaded bar, both the methods of iterative perturbation converged in four iterations to give identical results. However, the advantage of the second method is clear. The original stiffness is used in every iteration to calculate the perturbed displacement, $\mathbf{u}_1$.

5.8   USE OF SENSITIVITY INFORMATION: BASIC CONCEPT OF SFEM

The previous sections presented three methods for obtaining the derivatives of the structural response to the basic random variables, in the context of structural analysis. All three methods have their own domains of usefulness. The finite difference approach can be used with any type of analysis, and becomes particularly useful when the structural analysis is done using commercial software. The classical perturbation method uses the chain rule of differentiation, and can be used for simpler problems where the analysis computer program can be modified by the user. The iterative perturbation method is useful with nonlinear structural analysis.

Once the derivatives of the performance function with respect to the basic random variables are computed, this information can be used in two ways: (1) to construct an approximate closed-form performance function (similar to the response surface approach) or (2) to use the derivatives directly in the reliability analysis. In the former approach, one can start by perturbing each random variable about its mean value and computing the corresponding variation in the response, and use the approximate derivatives to construct a first-order Taylor series function. (One may also compute second-order derivatives and construct a second-order Taylor series approximation.) This closed-form expression is then combined with the analytical reliability methods (FORM or SORM) discussed earlier. The derivatives can be refined in subsequent iterations of FORM to construct updated closed-form approximations to the performance function.

It is, however, not necessary to construct a closed-form expression for the performance function to determine the probability of failure. Notice that FORM Method 2 of Chapter 3 needs only the value and gradient of the performance function at each iteration to search for the most probable point. The value of $g(\mathbf{X})$ is simply obtained from deterministic analysis of the structure. The gradient vector $\nabla g(\mathbf{X})$ is evaluated through sensitivity analysis using one of the three methods described above. Thus, at each iteration of the FORM algorithm, the value and the derivatives of $g(\mathbf{X})$ are computed using structural analysis supplemented with sensitivity analysis, leading ultimately to the estimation of reliability. This elegant approach is demonstrated in greater detail in subsequent chapters in this book, and is referred to as the Stochastic Finite Element Method (SFEM), to represent the combination of finite element analysis and probabilistic analysis.

5.9 FINITE ELEMENT FORMULATION

In earlier sections, the three methods of sensitivity analysis were illustrated using simple examples of structural elements for which closed-form solutions are available. The sensitivity-based approach is obviously more useful for problems where a closed-form solution for the structural response is not available. Therefore, the implementation of the three methods of sensitivity analysis in the context of finite element analysis is discussed below.

5.9.1 Finite Difference Approach

The formulation and implementation of the finite difference approach in the context of finite element analysis is exactly the same as in Section 5.3. The finite element code is treated as a black box. If the derivatives of $g(X_1, X_2, \ldots, X_n)$ are to be computed at the values $(X_1^0, X_2^0, \ldots, X_n^0)$, each variable, X_i, is perturbed a little while keeping the other variables at the values X_j^0 ($j \neq i$). The finite element analysis is run for each perturbation, and the change in $g(\mathbf{X})$ is computed for each run, to give approximate estimates of the derivatives. Once again, note that the total number of times the function $g(X_1, X_2, \ldots, X_n)$ is evaluated in this procedure is $n + 1$, i.e., one evaluation at $(X_1^0, X_2^0, \ldots, X_n^0)$, and n repetitions for the perturbations of the n variables.

It is apparent that the finite difference approach can be quite time-consuming if the number of variables is large. However, in many practical cases, the engineer may have no choice but to use the finite difference approach. For example, if one is using a commercially available computer program to evaluate the function $g(X_1, X_2, \ldots, X_n)$, it may be difficult or impossible to change the program source code. Even if one had access to the source code and decided to use more mathematically sophisticated methods to compute the derivatives, the programming effort could be quite prohibitive. Therefore, practical engineers may decide to simply use a commercial software program as a black box and apply the finite difference approach to compute the derivatives. Such finite difference results would be accurate only when the input variables have small variability. However, carefully chosen small perturbation sizes might be able to provide satisfactory results, even when the variabilities are large.

5.9.2   Classical Perturbation

In a typical linear deterministic finite element analysis *under static loading*, the steps leading up to the computation of any response quantity, S, are as follows:

Step 1   Using the parameters of the structure, form the global stiffness matrix, $\mathbf{K}$, and the global nodal load vector, $\mathbf{F}$.

Step 2   Solve for the nodal displacements, $\mathbf{U}$, using the finite element equation

$$\mathbf{KU} = \mathbf{F} \tag{5.48}$$

Step 3   Compute the response vector, $\mathbf{S}$, from the computed displacements, $\mathbf{U}$, using a transformation of the form

$$\mathbf{S} = \mathbf{Q}^t \mathbf{U} + \mathbf{S}_0 \tag{5.49}$$

where $\mathbf{Q}^t$ is a matrix that relates the desired response, S, to the nodal displacements, $\mathbf{U}$, and $\mathbf{S}_0$ is the response for $\mathbf{U} = 0$.

Step 4   For reliability analysis, the performance function is constructed as

$$g(\mathbf{X}) = g\{\mathbf{R}(\mathbf{X}), \mathbf{S}(\mathbf{X})\} \tag{5.50}$$

where $\mathbf{X} = (X_1, X_2, \ldots, X_n)$ is the vector of the basic random variables, $\mathbf{R}$ is vector of the resistance variables, and $\mathbf{S}$ is the vector of response quantities occuring in the performance function.

In classical perturbation, the principle of chain rule of differentiation is applied to each of the steps above to compute the derivatives of g with respect to the basic random variables, $\mathbf{X}$. That is, the partial derivatives of each quantity are computed with respect to the quantities computed in the previous steps, all the way down to the input random variables, $\mathbf{X}$. Chapter 6 develops this idea in detail.

5.9.3   Iterative Perturbation

The concept of iterative perturbation in the context of finite element analysis is the same as in Section 5.7. The implementation of this method is explained in more detail in Chapter 8 in the context of reliability analysis of structures with nonlinear behavior.

5.10   CONCLUDING REMARKS

This chapter presented various methods for reliability analysis in the case of limit states involving implicit performance functions. The use of sensitivity-based analysis was observed to be elegant and efficient in the context of FORM and SORM methods. Three methods of sensitivity analysis were presented: finite difference, classical perturbation, and iterative perturbation. All three ideas have been implemented in the context of finite element analysis in various studies. The basic concepts of these methods are general enough to be combined with other analysis methods for structural and other engineering systems. The combination of finite element analysis with sensitivity analysis that leads to probabilistic analysis has been referred to as probabilistic finite element method (PFEM) or stochastic finite element method (SFEM) in the literature. The following chapters describe the implementation of this method in detail for various types of structural behavior and loading conditions.

6

SFEM FOR LINEAR STATIC PROBLEMS

6.1 INTRODUCTORY COMMENTS

Chapters 3 and 4 of this book introduced the basic concepts and methods of reliability analysis. In Chapter 5, additional methods were developed to implement reliability analysis for problems with implicit performance functions. One of these methods, the sensitivity-based approach, led to the formulation of the Stochastic Finite Element Method (SFEM), where the structural response is computed using finite element analysis. The concepts of SFEM-based reliability analysis were introduced in a general, but simple manner. The present chapter explains the implementation of these concepts in detail, with practical examples, in the context of linear elastic structural behavior. The purpose of this chapter is not to discuss the details of deterministic finite element analysis, but to emphasize the implementation of stochastic finite element analysis. Information on deterministic FEM is available in many textbooks, e.g., Cook (1995).

6.2 THE DISPLACEMENT APPROACH OF FEM

The simplest type of structural analysis uses the assumption of linear elastic behavior under static loading. For this type of analysis, the displacement approach or the stiffness approach of finite element analysis is adequate. This method involves the assembly of the structural stiffness matrix $\mathbf{K}$ and the load vector $\mathbf{F}$, and the solution for the vector of nodal displacements $\mathbf{U}$ from the equation

$$\mathbf{KU} = \mathbf{F} \tag{6.1}$$

FIGURE 6.1. Axially loaded bar.

Example 6.1

Consider the axially loaded bar shown in Figure 6.1. The bar has two ends, or two nodes, 1 and 2. Node 2 is clamped (fixed), and node 1 is free. The bar has a cross-sectional area, A; Young's modulus, E; and length, L. An axial force, P, is applied at node 1. The load vector, $\mathbf{F}$, is written as

$$\mathbf{F} = \left\{ \begin{matrix} P \\ 0 \end{matrix} \right\} \tag{6.2}$$

Here, the load vector has two entries, corresponding to two nodes. The vector of nodal displacements is written as

$$\mathbf{U} = \left\{ \begin{matrix} U_1 \\ U_2 \end{matrix} \right\} \tag{6.3}$$

For this simple problem, the nodal displacements are easily known as

$$U_1 = \frac{PL}{AE} \tag{6.4}$$

$$U_2 = 0 \tag{6.5}$$

This solution can be expressed in matrix form similar to Equation 6.1 as

$$\begin{bmatrix} \frac{AE}{L} & 0 \\ 0 & \frac{AE}{L} \end{bmatrix} \left\{ \begin{matrix} U_1 \\ U_2 \end{matrix} \right\} = \left\{ \begin{matrix} P \\ 0 \end{matrix} \right\} \tag{6.6}$$

where the first matrix (size 2×2) on the left is referred to as the stiffness matrix $\mathbf{K}$. The terms of the stiffness matrix relate the nodal displacements to the nodal forces.

In this example, the elements of the stiffness matrix were easily obtained as explicit formulas in terms of A, E, and L. In other problems with complicated geometries (e.g., plate with a crack), the elements of the stiffness matrix may need to be computed through numerical integration (see any text on finite element methods, e.g., Cook (1995), for more details). However, the SFEM procedure being formulated here is equally applicable for both situations.

The above example considered a structure with a single element. In the case of structures with multiple elements, the (local) stiffness terms corresponding to each element are first computed, and then entered into appropriate locations in an overall (global) stiffness matrix $\mathbf{K}$ for the entire structure. Refer to any text on matrix structural analysis or finite element analysis for more details on this procedure. The vector of nodal forces, $\mathbf{F}$, is assembled quite easily. Once $\mathbf{K}$ and $\mathbf{F}$ are constructed, Equation 6.1 is used to solve for the nodal displacement vector $\mathbf{U}$.

6.3   SFEM-BASED RELIABILITY ANALYSIS

The steps of SFEM-based reliability analysis are developed in this section. FORM Method 2 described in Chapter 3 requires the value of the performance function $G(\mathbf{Y})$, and its gradient $\nabla G(\mathbf{Y})$ at each iteration point, in the *standard normal space*, $\mathbf{Y}$, to search for the minimum distance point on the limit state. In the displacement method of finite element analysis, the steps leading up to the computation of $G(\mathbf{Y})$ are as follows:

Step 1   Using the parameters of the structure, assemble the global stiffness matrix $\mathbf{K}$ and the global nodal load vector $\mathbf{F}$.

Step 2   Solve for the displacements, $\mathbf{U}$, using the finite element equation

$$\mathbf{K}\mathbf{U} = \mathbf{F} \tag{6.7}$$

Step 3   Compute the vector of desired response quantities $\mathbf{S}$ (e.g., stress) from the computed displacements using a transformation of the form

$$\mathbf{S} = \mathbf{Q}^t U + \mathbf{S}_0 \tag{6.8}$$

where $\mathbf{Q}^t$ is a transformation matrix relating $\mathbf{U}$ and $\mathbf{S}$, and $\mathbf{S}_0$ is the response vector for $\mathbf{U} = \mathbf{0}$. For example, if the desired response quantity, $\mathbf{S}$, refers to the vector of stresses in the members of the structure, then $\mathbf{S}_0$ may refer to the initial stresses.

Step 4   Compute the performance function

$$g(\mathbf{X}) = g\left\{\mathbf{R}(\mathbf{X}), \mathbf{S}(\mathbf{X})\right\} \tag{6.9}$$

where $\mathbf{R}$ is the vector of resistance variables, $\mathbf{S}$ is the vector of response quantities occurring in the performance function, and $\mathbf{X}$ is the vector of the original random variables.

Step 5   An additional step is to transform the original random variables $\mathbf{X}$ to equivalent uncorrelated reduced normal variables $\mathbf{Y}$. This transformation has been discussed in detail in Chapter 3, and may be symbolically written as

$$\mathbf{Y} = T(\mathbf{X}) \tag{6.10}$$

For example, if there are two uncorrelated random variables, X_1 and X_2, the transformations $T(X_1)$ and $T(X_2)$ may be written as

$$\begin{Bmatrix} Y_1 \\ Y_2 \end{Bmatrix} = \begin{Bmatrix} T(X_1) \\ T(X_2) \end{Bmatrix} = \begin{Bmatrix} \frac{X_1 - \mu_1^N}{\sigma_1^N} \\ \frac{X_2 - \mu_2^N}{\sigma_2^N} \end{Bmatrix} \tag{6.11}$$

where μ_1^N and μ_2^N are the equivalent normal means, and σ_1^N and σ_2^N are the equivalent normal standard deviations. In the $\mathbf{Y}$ space, the performance function is denoted as $G(\mathbf{Y})$. However, the numerical value of the performance function is the same, whether in the $\mathbf{X}$ space or in the $\mathbf{Y}$ space.

The only remaining step is to compute $\nabla G(\mathbf{Y})$ in order to implement the FORM algorithm. This is quite easy if the gradient $\nabla g(\mathbf{X})$ in the $\mathbf{X}$ space is known. Any of the three sensitivity analysis methods in Chapter 5 can be used to compute $\nabla g(\mathbf{X})$. In the SFEM formulation being developed in this chapter, classical perturbation (i.e., the chain rule of differentiation) is used. That is, the partial derivatives of every quantity in the aforementioned steps are computed with respect to the original random variables. This step-by-step computation of the partial derivates results in the computation of the gradient vector of the performance function in the $\mathbf{X}$ space as well as in the $\mathbf{Y}$ space.

Consider the relationship between the gradients in the $\mathbf{X}$ space and the $\mathbf{Y}$ space. For a single random variable X_1 in the $\mathbf{X}$ space, its $\mathbf{Y}$ space counterpart is Y_1. Once $\partial g/\partial X_1$ is computed, then the gradient in the $\mathbf{Y}$ space, $\partial G/\partial Y_1$, is computed as

$$\frac{\partial G}{\partial Y_1} = \frac{\partial g}{\partial X_1}\frac{\partial X_1}{\partial Y_1} \tag{6.12}$$

using the chain rule of differentiation. Here, the quantity $\partial X_1/\partial Y_1$ is easily computed using Equation 6.11 as

$$\frac{\partial X_1}{\partial Y_1} = \sigma_1^{N} \tag{6.13}$$

If the performance function is in terms of $\mathbf{R}$ and $\mathbf{S}$ as in Equation 6.9, then Equation 6.12 may be expanded as

$$\frac{\partial G}{\partial Y_1} = \frac{\partial g}{\partial \mathbf{R}}\frac{\partial \mathbf{R}}{\partial X_1}\frac{\partial X_1}{\partial Y_1} + \frac{\partial g}{\partial \mathbf{S}}\frac{\partial \mathbf{S}}{\partial X_1}\frac{\partial X_1}{\partial Y_1} \tag{6.14}$$

using the chain rule of differentiation once again.

Example 6.2

Consider the axially loaded bar of Example 6.1. It is necessary to find the probability that the axial strain of the bar, ϵ_x, exceeds a limiting value, ϵ_0. Three uncorrelated random variables are considered: the limiting strain, ϵ_0; the nodal force, P; and the Young's modulus, E. Formulate the computation of $\nabla G(\mathbf{Y})$ using the steps of SFEM-based reliability analysis developed above.

The performance function for this problem may be written as

$$g(R,\, S) = \epsilon_0 - \epsilon_x \tag{6.15}$$

Thus, $R = \epsilon_0$ and $S = \epsilon_x$. In this formulation, failure occurs when $g < 0$. The response quantity, S, is dependent on two random variables, P and E.

The vector of random variables is $\mathbf{X} = \{X_1, X_2, X_3\}^{t} = \{\epsilon_0, P, E\}^{t}$. The corresponding $\mathbf{Y}$-space variables are denoted as $\mathbf{Y} = \{Y_1, Y_2, Y_3\}^{t}$. These are related to the $\mathbf{X}$-space variables through equivalent normal transformation as in Equation 6.11. It is necessary to compute $\nabla G(\mathbf{Y}) = \{\partial G/\partial Y_1, \partial G/\partial Y_2, \partial G/\partial Y_3\}^{t}$.

The quantity $\partial G/\partial Y_1$ is easily computed as

$$\frac{\partial G}{\partial Y_1} = \frac{\partial g}{\partial R}\frac{\partial R}{\partial X_1}\frac{\partial X_1}{\partial Y_1} \tag{6.16}$$

$$= \frac{\partial g}{\partial \epsilon_0}(1.0)\,\sigma_1^N$$

$$= (1.0)(1.0)\,\sigma_1^N$$

The quantity $\partial G/\partial Y_2$ is computed as

$$\frac{\partial G}{\partial Y_2} = \frac{\partial g}{\partial S}\frac{\partial S}{\partial X_2}\frac{\partial X_2}{\partial Y_2} \tag{6.17}$$

$$= \frac{\partial g}{\partial \epsilon_x}\frac{\partial \epsilon_x}{\partial P}\sigma_2^N$$

$$= (-1.0)\frac{\partial \epsilon_x}{\partial P}\sigma_2^N$$

The quantity $\partial G/\partial Y_3$ is computed as

$$\frac{\partial G}{\partial Y_3} = \frac{\partial g}{\partial S}\frac{\partial S}{\partial X_3}\frac{\partial X_3}{\partial Y_3} \tag{6.18}$$

$$= \frac{\partial g}{\partial \epsilon_x}\frac{\partial \epsilon_x}{\partial E}\sigma_3^N$$

$$= (-1.0)\frac{\partial \epsilon_x}{\partial E}\sigma_3^N$$

To complete the computation of $\partial G/\partial Y_2$ and $\partial G/\partial Y_3$, consider Equation 6.6 once again:

$$\begin{Bmatrix} \frac{AE}{L} & 0 \\ 0 & \frac{AE}{L} \end{Bmatrix}\begin{Bmatrix} U_1 \\ U_2 \end{Bmatrix} = \begin{Bmatrix} P \\ 0 \end{Bmatrix} \tag{6.19}$$

which is similar to Equation 6.7. Solving this matrix equation, the nodal displacements are obtained as

$$\begin{Bmatrix} U_1 \\ U_2 \end{Bmatrix} = \begin{Bmatrix} \frac{PL}{AE} \\ 0 \end{Bmatrix} \tag{6.20}$$

The desired response quantity S in this case is ϵ_x. This is obtained simply as $\epsilon_x = (U_1 - U_2)/L$. In the context of Equation 6.8, this is expressed as

$$\epsilon_x = \begin{Bmatrix} \frac{1}{L} & -\frac{1}{L} \end{Bmatrix}\begin{Bmatrix} U_1 \\ U_2 \end{Bmatrix} \tag{6.21}$$

Here, the matrix $\mathbf{Q}$ of Equation 6.8 is

$$Q = \left\{ \begin{matrix} \frac{1}{L} \\ -\frac{1}{L} \end{matrix} \right\}$$ (6.22)

and the initial strain $S_0 = 0$. From Equations 6.20 and 6.21,

$$\epsilon_x = \frac{P}{AE}$$ (6.23)

Substituting Equation 6.23 in Equations 6.17 and 6.18, the computation of the required partial derivatives is completed as

$$\frac{\partial G}{\partial Y_2} = (-1.0)\frac{\partial \epsilon_x}{\partial P}\sigma_2^N$$ (6.24)

$$= \frac{-1}{AE}\sigma_2^N$$

and

$$\frac{\partial G}{\partial Y_3} = (-1.0)\frac{\partial \epsilon_x}{\partial E}\sigma_3^N$$ (6.25)

$$= \frac{P}{AE^2}\sigma_3^N$$ (6.26)

This example, although simple, mirrors the matrix formulations and operations involved in SFEM-based reliability analysis. The problem had three random variables, and the computation of partial derivative for each variable was done separately. A more compact formulation is presented below, for the case of multiple variables.

6.3.1 Matrix Formulation for Multiple Random Variables

In the context of multiple random variables, a matrix representation of the partial derivatives computation is compact and convenient. As a first step, compare the transformation to the standard normal space in Equations 6.10 and 6.11. Equation 6.10 may be linearized as

$$Y = A + BX$$ (6.27)

where, for the example of two variables in Equation 6.11,

$$A = \left\{ \begin{matrix} -\frac{\mu_1^N}{\sigma_1^N} \\ -\frac{\mu_2^N}{\sigma_2^N} \end{matrix} \right\}$$ (6.28)

$$B = \left\{ \begin{matrix} \frac{1}{\sigma_1^N} & 0 \\ 0 & \frac{1}{\sigma_2^N} \end{matrix} \right\}$$ (6.29)

$$X = \left\{ \begin{matrix} X_1 \\ X_2 \end{matrix} \right\}$$ (6.30)

In the FORM algorithm, new equivalent normal means and standard deviations are calculated at each iteration. Therefore, the vector $\mathbf{A}$ and the matrix $\mathbf{B}$ change from iteration to iteration of the FORM algorithm.

Refer to Equations 6.12 to 6.14 where the chain rule of differentiation is applied to a single random variable. If we apply the same procedure to matrix notation of multiple variables in Equations 6.27 to 6.30, the gradient vector $\nabla G(\mathbf{Y})$ is obtained as

$$\nabla G(\mathbf{Y}) = (\mathbf{B}^{-1})^t \, \nabla g_x(\mathbf{R}, \mathbf{S})$$
$$= (\mathbf{B}^{-1})^t \, [\mathbf{J}_r \nabla g_r(\mathbf{R}, \mathbf{S}) + \mathbf{J}_s \nabla g_s(\mathbf{R}, \mathbf{S})] \tag{6.31}$$

where $\nabla g_x(\mathbf{R}, \mathbf{S})$, $\nabla g_r(\mathbf{R}, \mathbf{S})$, and $\nabla g_s(\mathbf{R}, \mathbf{S})$ are the gradient vectors of the performance function with respect to $\mathbf{X}$, $\mathbf{R}$, and $\mathbf{S}$, respectively, and $\mathbf{J}_r = \frac{\partial \mathbf{R}}{\partial \mathbf{X}}$ and $\mathbf{J}_s = \frac{\partial \mathbf{S}}{\partial \mathbf{X}}$. ($\mathbf{J}_r$ and $\mathbf{J}_s$ are also referred to as the Jacobian matrices of the transformations $\mathbf{R} = \mathbf{R}(\mathbf{X})$ and $\mathbf{S} = \mathbf{S}(\mathbf{X})$ respectively.) The (i, j) elements of $\mathbf{J}_r$ and $\mathbf{J}_s$ are given by $\partial R_i / \partial X_j$ and $\partial S_i / \partial X_j$, respectively. The computation of $\nabla g_r(\mathbf{R}, \mathbf{S})$, $\nabla g_s(\mathbf{R}, \mathbf{S})$, and $\mathbf{J}_r$ can be easily carried out either numerically or by simple differentiation, since the performance function $g(\mathbf{R}, \mathbf{S})$ is expressed in terms of the $\mathbf{R}$ and $\mathbf{S}$ variables and the $\mathbf{R}$ variables are easily related to some of the basic $\mathbf{X}$ variables. The computation of $\mathbf{J}_s = \partial \mathbf{S} / \partial \mathbf{X}$ is done using Equation 6.8.

Differentiation of Equation 6.8 gives

$$\frac{\partial \mathbf{S}}{\partial X_j} = \frac{\partial \mathbf{Q}^t}{\partial X_j} \mathbf{U} + \mathbf{Q}^t \frac{\partial \mathbf{U}}{\partial X_j} + \frac{\partial \mathbf{S}_0}{\partial X_j} \tag{6.32}$$

where X_j refers to the jth random variable. In this equation, the partial derivatives $\partial \mathbf{Q}^t / \partial X_j$ and $\partial \mathbf{S}_0 / \partial X_j$ are computed based on how the quantities $\mathbf{Q}^t$ and $\mathbf{S}_0$ are defined in terms of the basic random variables $\mathbf{X}$. This is dependent on the type of the problem (see Example 6.2).

The next step is to evaluate $\partial \mathbf{U} / \partial X_j$ in Equation 6.32, using Equation 6.7. To facilitate this, Equation 6.7 is written in the form

$$\mathbf{U} = \mathbf{K}^{-1} \mathbf{F} \tag{6.33}$$

Therefore,

$$\frac{\partial \mathbf{U}}{\partial X_j} = \mathbf{K}^{-1} \frac{\partial \mathbf{F}}{\partial X_j} + \frac{\partial \mathbf{K}^{-1}}{\partial X_j} \mathbf{F} \tag{6.34}$$

The next step is to compute $\partial \mathbf{K}^{-1} / \partial X_j$. The following result is useful. Since

$$\mathbf{K}\mathbf{K}^{-1} = 1, \tag{6.35}$$

differentiating this equation with respect X_j gives

$$\frac{\partial \mathbf{K}^{-1}}{\partial X_j} = -\mathbf{K}^{-1} \frac{\partial \mathbf{K}}{\partial X_j} \mathbf{K}^{-1} \tag{6.36}$$

Substitute this in Equation 6.34 to obtain

$$\frac{\partial \mathbf{U}}{\partial X_j} = \mathbf{K}^{-1} \frac{\partial \mathbf{F}}{\partial X_j} - \mathbf{K}^{-1} \frac{\partial \mathbf{K}}{\partial X_j} \mathbf{U} \tag{6.37}$$

since $\mathbf{U} = \mathbf{K}^{-1}\mathbf{F}$.

When Equations 6.32 and 6.37 are combined, the derivative of the structural response with respect to the basic random variable X_j, $\partial \mathbf{S}/\partial X_j$, is obtained as (Der Kiureghian and Ke, 1985)

$$\frac{\partial \mathbf{S}}{\partial X_j} = \frac{\partial \mathbf{Q}^t}{\partial X_j} \mathbf{U} + \mathbf{Q}^t \mathbf{K}^{-1} \left\{ \frac{\partial \mathbf{F}}{\partial X_j} - \frac{\partial \mathbf{K}}{\partial X_j} \mathbf{U} \right\} + \frac{\partial \mathbf{S}_0}{\partial X_j} \tag{6.38}$$

Note that as the quantities $\mathbf{F}$, $\mathbf{K}$, etc. are computed, their partial derivatives with respect to the basic variables, such as $\partial \mathbf{F}/\partial X_j$ and $\partial \mathbf{K}/\partial X_j$ are also computed in parallel. Therefore, all the quantities required for the computation of the derivatives of the response are available and the computation of $\mathbf{J}_s = \partial \mathbf{S}/\partial \mathbf{X}$ is complete.

Thus, the computation of $\nabla G(\mathbf{Y})$ is achieved in two steps:

Step 1   Computation of $\partial \mathbf{S}/\partial \mathbf{X}$ using Equation 6.38.

Step 2   Computation of $\nabla G(\mathbf{Y})$ using Equation 6.31.

Recall that this was the purpose of the SFEM analysis for structures without closed-form limit states. With the procedure formulated in this and the previous section, the values of $G(\mathbf{Y})$ and $\nabla G(\mathbf{Y})$ are computed in each iteration of FORM Method 2 of Chapter 3, and the search for the minimum distance point on the limit state is implemented. Other steps in FORM or SORM reliability computation are the same as in Chapter 3.

6.4   IMPLEMENTATION OF SFEM

In the finite element approach, the matrices $\mathbf{K}$, $\mathbf{F}$, $\mathbf{Q}$, and $\mathbf{S}_0$ are constructed by assembling the stiffness matrices or the load vectors. Since the differential operator is linear, it follows that the matrices $\partial \mathbf{K}/\partial X_j$, $\partial \mathbf{F}/\partial X_j$, $\partial \mathbf{Q}/\partial X_j$, and $\partial \mathbf{S}_0/\partial X_j$ can be similarly constructed in terms of element-level matrices that contain the partial derivatives of the element stiffness matrices and element load vectors. Thus, a first-order Stochastic Finite Element Method for linear structures includes, in addition to element stiffness matrices and load vectors, routines for computing the *partial derivative stiffness matrix* and the *partial derivative load vector* for each element. The existing routines for assembling the global matrices in the deterministic finite element approach can be directly used to compute the global partial derivative matrices described above.

The practical implementation of the SFEM-based reliability analysis formulated here needs to be done in a computer environment. The success of such implementation depends on two factors: (1) programming effort to compute the partial derivative matrices at the element level, requiring the development of appropriate data structures and control logic; and, (2) memory available in the computer for the storage and manipulation of the large partial derivative matrices. Available memory can be a serious problem. The amount of memory required grows rapidly with the number of basic random variables. As an example, for a structure with 100 degrees of freedom and 20 basic random variables, the matrix of first-order partial derivatives of **K** has 200,000 elements. (This problem may become worse when some of the stochastic parameters are represented as random fields, as discussed in Chapter 7. In that case, each random field is discretized into a set of several correlated random variables, thus increasing the number of random variables and corresponding partial derivatives.) This results in a limitation regarding the complexity of the structures that can be analyzed, due to both the available computer memory and the computational time. For large, complicated structures supercomputers may be needed, resulting in considerable changes in the structure of the stochastic finite element program.

6.5   RELIABILITY ANALYSIS OF FRAMED STRUCTURES

In this section, the implementation of the Stochastic Finite Element Method to the reliability analysis of frames is illustrated within the context of linear static structural analysis, as developed in the previous sections. The stochastic finite element analysis requires the computation of the partial derivatives of various matrices and vectors in terms of the basic random variables at the element level, and subsequent assembly at the global level. It is simple to perform such computation and assembly in the case of a frame element, since its element matrices can be written explicitly in terms of the basic parameters, without requiring any numerical integration. Note that the SFEM technique is general and applicable to any structure that is analyzed by FEM; the use of frames in this section is only for the sake of simplicity in illustration and for comparison with known results and practical design procedures.

6.5.1   Stiffness Matrix and Its Derivatives

The stiffness matrix k_e and its partial derivatives for a plane frame element with six degrees of freedom, shown in Figure 6.2, are of the form

$$k_e, \frac{\partial k_e}{\partial E}, \frac{\partial k_e}{\partial A}, \quad \text{or} \quad \frac{\partial k_e}{\partial I} = \begin{bmatrix} a & b & c & -a & -b & c \\ b & d & e & -b & -d & e \\ c & e & f & -c & -e & g \\ -a & -b & -c & a & b & -c \\ -b & -d & -e & b & d & -e \\ c & e & g & -c & -e & f \end{bmatrix} \quad (6.39)$$

In the above equation, each of the quantities on the left-hand side are given by the general form of the matrix on the right-hand side. The constants a to g in this matrix are different for the different quantities. For k_e,

$$a = \frac{12EI}{L^3} \sin^2 \theta + \frac{EA}{L} \cos^2 \theta$$

$$b = \left(\frac{EA}{L} - \frac{12EI}{L^3} \right) \sin \theta \cos \theta$$

$$c = -\frac{6EI}{L^2} \sin \theta \quad\quad\quad\quad\quad\quad\quad (6.40)$$

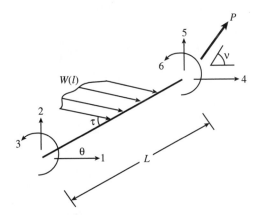

FIGURE 6.2. Plane frame element.

$$d = \frac{12EI}{L^3} \cos^2 \theta + \frac{EA}{L} \sin^2 \theta$$

$$e = \frac{6EI}{L^2} \cos \theta$$

$$f = \frac{4EI}{L}$$

$$g = \frac{2EI}{L}$$

where

E = modulus of elasticity

A = area of cross section

I = moment of inertia

L = length of the frame element

θ = angle of inclination of the element with respect to the global coordinates.

Here E, A, and I are assumed to be random variables, while L and θ are assumed to be deterministic.

Since the element stiffness matrix is explicitly available in terms of the random variables E, A, and I, the partial derivatives of the elements in this matrix can be easily computed through direct differentiation of the formulae in Equation 6.40 with respect to E, A, and I. Thus, in Equation 6.39, the constants in the matrix for $\partial k_e / \partial E$ are

$$a = \frac{12I}{L^3} \sin^2 \theta + \frac{A}{L} \cos^2 \theta$$

$$b = \left(\frac{A}{L} - \frac{12I}{L^3} \right) \sin \theta \cos \theta$$

$$c = -\frac{6I}{L^2} \sin \theta \tag{6.41}$$

$$d = \frac{12I}{L^3} \cos^2 \theta + \frac{A}{L} \sin^2 \theta$$

$$e = -\frac{6I}{L^2} \cos \theta$$

$$f = \frac{4I}{L}$$

$$g = \frac{2I}{L}$$

Similarly, for $\partial k_e / \partial A$,

$$a = \frac{E}{L} \cos^2 \theta$$

$$b = \frac{E}{L} \sin \theta \cos \theta$$

$$c = 0 \tag{6.42}$$

$$d = \frac{E}{L} \sin^2 \theta$$

$$e = 0$$

$$f = 0$$

$$g = 0$$

For $\partial k_e / \partial I$,

$$a = \frac{12E}{L^3} \sin^2 \theta$$

$$b = -\frac{12E}{L^3} \sin \theta \cos \theta$$

$$c = -\frac{6E}{L^2} \sin \theta \tag{6.43}$$

$$d = \frac{12E}{L^3} \cos^2 \theta$$

$$e = \frac{6E}{L^2} \cos \theta$$

$$f = \frac{4E}{L}$$

$$g = \frac{2E}{L}$$

Thus, the element partial derivative stiffness matrices are computed. Since the stiffness matrix is formulated here in the context of linear analysis of frames, and since the only random variables in this formulation are E, A, and I, its partial derivatives with respect to all other random variables are zero.

Two kinds of loads are considered on the plane frame element: (1) concentrated loads acting at the nodes, and (2) distributed loads acting along the element. Nodal loads are directly assembled into the global load vector according to the global degrees of freedom of the points of application. The partial derivatives of the global load vector with respect to a particular nodal load that is acting at an angle ν in relation to the global coordinate system (see Figure 6.2) contain $\cos \nu$ and $\sin \nu$ at the degrees of freedom where the load is acting and zeros elsewhere.

A distributed load on an element, as shown in Figure 6.2, is transferred into equivalent nodal loads and then assembled into the global load vector. The global partial derivative load vector is computed by assembling the partial derivatives of the equivalent nodal load vectors of all the elements. For a uniformly distributed load, W, on the element, the partial derivative of the equivalent nodal load vector is of the form

$$\frac{\partial F_e}{\partial W} = \begin{Bmatrix} a\cos\tau + b\sin\tau \\ a\sin\tau - b\cos\tau \\ -c\cos\tau \\ a\cos\tau + b\sin\tau \\ a\sin\tau - b\cos\tau \\ -c\cos\tau \end{Bmatrix} \tag{6.44}$$

in which

$$a = \frac{L}{2}\sin\theta$$

$$b = \frac{L}{2}\cos\theta$$

$$c = \frac{L^2}{12}$$

and τ is the angle of inclination of the load as defined in Figure 6.2. Note that for this case the six nodal loads are perfectly correlated and are represented by a single random variable. Similar results can be derived for nonuniformly distributed loads. In this discussion, the distributed load is represented by a single random variable. It is more accurate to represent it by a random field, if the load has random fluctuations along the length of the element. Such a representation is developed in the next chapter.

Thus, the procedure for the computation of the partial derivative stiffness matrix and the partial derivative load vector has been developed for plane frame members. The computation of partial derivatives of the remaining matrices in Equation 6.38, namely $\frac{\partial Q}{\partial X_j}$ and $\frac{\partial S_0}{\partial X_j}$, are specific for particular limit states and will be described in the following subsections, with the help of numerical examples.

6.5.2 Limit States

Four types of limit states, commonly encountered in the practical design of steel framed structures, are discussed in detail:

1. Combined axial compression and bending of frame members
2. Pure bending of frame members
3. Side sway at the top of a plane frame
4. Vertical deflection at the midspan of a beam member in a frame

The formulation of each limit state is discussed first starting with appropriate performance criteria. Then, numerical examples are presented with a portal steel frame structure, and the results of SFEM-based reliability analysis are provided. The FORM approach is used for reliability analysis. The results have been verified by the authors using Monte Carlo simulation (both basic as well as with advanced variance reduction techniques). In the numerical examples, the sensitivities of the reliability index to different random variables are quantified. (Refer to Section 3.8 for details on sensitivity computation of the reliability index.) Random variables with low sensitivity are ignored and the gain in computational efficiency while maintaining high accuracy is illustrated.

6.5.2.1 *Combined Axial Compression and Bending Limit State* Before formulating a limit state, one has to clearly understand the difference between the actual behavior of a structural component and the design equation specified in codes of professional practice. The limit state is formulated using the actual behavior model. The code-based design equations incorporate safety factors to account for the uncertainties and to achieve the target reliability. However, for the purpose of reliability analysis, the actual behavior model should be used. This distinction is maintained in the discussion below.

Many members in a frame structure are subjected to both bending moment and axial load. According to the AISC Load and Resistance Factor (LRFD) design guidelines, the following interaction equations should be checked to satisfy the strength requirements for members in two-dimensional frames:

$$\frac{P_u}{\phi P_n} + \frac{8}{9}\frac{M_u}{\phi_b M_n} \leq 1.0 \qquad \text{if } \frac{P_u}{\phi P_n} \geq 0.2 \tag{6.45}$$

$$\frac{P_u}{2\phi P_n} + \frac{M_u}{\phi_b M_n} \leq 1.0 \qquad \text{if } \frac{P_u}{\phi P_n} < 0.2 \tag{6.46}$$

where ϕ and ϕ_b are the resistance factors, P_u is the required tensile/compressive strength, P_n is the nominal tensile/compressive strength, M_u is the required flexural strength, and M_n is the nominal flexural strength.

For reliability evaluation, the corresponding performance functions can be expressed as

$$g_1(\mathbf{R}, \mathbf{S}) = 1.0 - \left\{ \frac{P_u}{P_n} + \frac{8}{9}\frac{M_u}{M_n} \right\} \qquad \text{if } \frac{P_u}{\phi P_n} \geq 0.2 \tag{6.47}$$

$$g_2(\mathbf{R}, \mathbf{S}) = 1.0 - \left\{ \frac{P_u}{2P_n} + \frac{M_u}{M_n} \right\} \qquad \text{if } \frac{P_u}{\phi P_n} < 0.2 \tag{6.48}$$

where P_u and M_u are unfactored load effects.

Nominal values of the axial load capacity, P_n, and the bending moment capacity, M_n, can be obtained from the AISC-LRFD *Manual* (1994) as

$$P_n = A F_{cr} \qquad \text{(compression)} \tag{6.49}$$

$$P_n = A F_y \qquad \text{(tension)}$$

and

$$M_n = ZF_y \tag{6.50}$$

where

$$F_{cr} = (0.658^{\lambda_c^2})F_y \qquad \text{for } \lambda_c \le 1.5 \tag{6.51}$$

$$F_{cr} = \left\{ \frac{0.877}{\lambda_c^2} \right\} F_y \qquad \text{for } \lambda_c > 1.5 \tag{6.52}$$

$$\lambda_c = \frac{Kl}{r\pi} \sqrt{\frac{F_y}{E}} \tag{6.53}$$

and where A is the gross area of the member (in^2), F_y is the specified yield stress (ksi), Z is the plastic section modulus in the plane of bending, E is the modulus of elasticity (ksi), K is the effective length factor, l is the unbraced length of the member (in.), and r is the governing radius of gyration about the plane of buckling (in.).

To implement SFEM-based reliability analysis with these limit state equations, quantities such as $\mathbf{R}$, $\mathbf{S}$, $\mathbf{Q}$, and $\mathbf{S_0}$ need to be identified for use in Equations 6.7 to 6.9. Thus, for both performance functions (Equations 6.47 and 6.47), $\mathbf{R} = \{P_n, M_n\}^t$ and $\mathbf{S} = \{P_u, M_u\}^t$. Note that these quantities are functions of the underlying random variables such as the E, A, I, and F_y load variables.

For both performance functions, the matrix $\mathbf{Q}$ has two columns, corresponding to the two elements in the vector of load effects $\mathbf{S}$. The number of rows in $\mathbf{Q}$ is equal to the number of degrees of freedom of the structure, as is apparent from Equation 6.8. If the reliability for this limit state at local node 1 of a member is of interest, then the columns of $\mathbf{Q}$ contain the components from the first and third rows of the stiffness matrix of the member, at the global degrees of freedom corresponding to P and M. All the other elements of $\mathbf{Q}$ are zero. If the reliability at local node 2 of the member is of interest, then the columns of $\mathbf{Q}$ contain the components from the fourth and sixth rows of the stiffness matrix of the member, again at the corresponding global degrees of freedom.

Vector $\mathbf{S_0}$ consists of the fixed-end axial load and bending moment due to the distributed load on the member at the node of interest.

Example 6.3

Consider the rigid steel portal frame shown in Figure 6.3. All three members of the frame are made of the same grade of steel and have identical cross-sections (W10 × 15, using the AISC notation). The frame is subjected to uniformly distributed loads W, D, and L as shown. These loads correspond to wind load, dead load, and live load, respectively, on frames considered in practical design. The magnitudes of the loads used are only for the purpose of illustration. Eight variables are considered to be stochastic: W, the lateral load; D, the vertical (dead) load; L, the vertical (live) load; E, the Young's modulus; F_y, the yield stress; A, the area of cross section; I, the moment of inertia; and Z, the plastic section modulus. The first three variables refer to the loads, the next two to the material properties, and the last three to the sectional properties. Table 6.1 provides the statistical description of the random variables, with the means, coefficients of variation, and types of distribution.

The safety of member 3 under combined axial compression and bending is investigated in this example. The results of the reliability analysis are summarized in Table

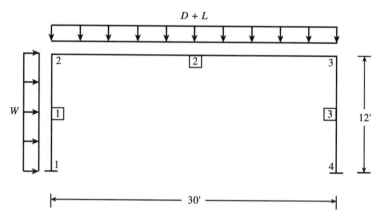

FIGURE 6.3. Rigid steel portal frame.

6.2. The checking point values of each random variable, the value of the performance function, and the reliability index are shown for each iteration. The algorithm converges to a value of $\beta = 3.088$ in just five iterations, which corresponds to a probability of failure $p_f = 0.001$.

Table 6.2 also shows the reliability sensitivity indices for the various random variables. The computation of these indices is described in Chapter 3. The signs of these indices show the direction of movement toward the design point. For the present discussion, the magnitudes of these indices are important. The safety of the member is found to be most sensitive to the stochastic variation in its yield strength, as expected. It also shows considerable sensitivity to the three loads and the area of cross section of the member. However, the sensitivity indices corresponding to the Young's modulus, area of cross section, and moment of inertia of the members are less than or equal to 0.01. The variables that show such low influence may be considered to be deterministic after the first iteration; that is, their checking point values may be fixed at the mean values. The last column of Table 6.2 shows the results of the reliability analysis using the information on the sensitivity indices. These results are not very different from those in the previous column. Thus, the use of sensitivity indices to ignore the randomness in some

Table 6.1. Rigid steel frame: description of the random variables

Variable	Units	Mean	Coefficient of Variation	Type of Distribution
D	kip/ft	0.44	0.10	Normal
L	kip/ft	0.05	0.25	Type I
W	kip	0.41	0.37	Type I
A	in^2	4.41	0.05	Normal
I	in^4	68.90	0.05	Normal
Z	in^3	16.00	0.05	Normal
E	ksi	29,000.00	0.06	Normal
F_y	ksi	39.60	0.11	Normal

Table 6.2. Reliability analysis: combined axial compression and bending of member 3 at node 3

Variable	Sensitivity Index	Initial Checking Point	Final Checking Point (without reduction)	Final Checking Point (with reduction)
D	0.47	0.44	0.48	0.48
L	0.14	0.05	0.06	0.06
W	0.25	0.41	0.48	0.48
A	−0.01	4.41	4.22	4.41
I	−0.01	68.90	65.86	68.90
Z	−0.32	16.00	15.29	15.33
E	0.00	29,000.00	29,004.60	29,000.00
F_y	−0.77	39.60	27.99	27.85
Performance function		0.39	−0.0001	−0.0001
Reliability index			3.008	3.107
Number of iterations			5	5

of the variables reduces the overall computational time without affecting the accuracy, for this problem.

6.5.2.2 Pure Bending of a Beam
The performance function for this limit state is expressed as

$$g(\mathbf{R}, \mathbf{S}) = 1.0 - \frac{M}{M_u} \qquad (6.54)$$

where M is the applied bending moment and M_u is the flexural strength of the beam. Considering the generalized behavior of a single or doubly symmetric beam bent about its major axis, the beam ultimately fails by lateral–torsional buckling, by local plate buckling, or by web buckling (Yura et al., 1978). For most rolled sections, the flange and web slenderness provided are sufficient to permit the beams to reach the plastic zone (compact sections) if the lateral bracing spacing is satisfactory. Therefore, this limit state is assumed to be determined by lateral-torsional buckling.

Lateral–torsional buckling of a beam occurs in one of the following three ranges of behavior: (1) the plastic range, (2) the inelastic range, or (3) the elastic range. For the design of a given beam, the precise value of M_u to be used in Equation 6.54 depends on the range that corresponds to the behavior of the beam; this is governed by the laterally unbraced length of the beam (L_b). Thus, for design purposes, the AISC LRFD *Manual* (1994) defines three zones of L_b as shown in Figure 6.4. The expressions for M_u corresponding to these three zones can be inferred quite easily by removing the safety factors from the design equations in the manual. For doubly symmetric I-shaped members, these are obtained as follows:

1. For $L_b < L_p$ (plastic range), the limit state corresponds to yielding of the member. Thus, the ultimate moment is

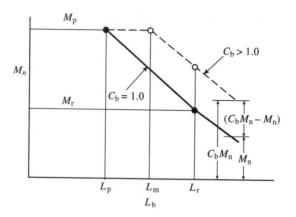

FIGURE 6.4. Moment capacity as a function of L_b.

$$M_u = M_p = Z_x F_y \tag{6.55}$$

where, for I-shaped sections,

$$L_p = 300 \frac{r_y}{\sqrt{F_{yf}}} \tag{6.56}$$

is the maximum unbraced length to reach M_p with uniform bending moment ($C_b = 1.0$), r_y is the radius of gyration about the minor axis, and F_{yf} is the yield strength of the flange.

In this case, if the performance function (Equation 6.54) is expressed in terms of the basic variables, the vectors $\mathbf{R} = \{Z_x, F_y\}^t$ and $\mathbf{S} = \{M\}$. Matrix $\mathbf{Q}$ has only one column, consisting of the components from the third or the sixth row (depending on whether the local node of interest is 1 or 2) of the member stiffness matrix at the global degrees of freedom. Vector $\mathbf{S}_0$ has a single element, the fixed-end moment at the node of interest.

2. For $L_p < L_b \leq L_r$, corresponding to inelastic behavior, the ultimate moment is computed as

$$M_u = C_b \left[M_p - (M_p - M_r) \left(\frac{L_b - L_p}{L_r - L_p} \right) \right] \leq M_p \tag{6.57}$$

where, for doubly symmetric I-shaped members,

$$L_r = \frac{r_y X_1}{F_L} \left[1 + \left(1 + X_2 F_L^2 \right)^{1/2} \right]^{1/2} \tag{6.58}$$

$$M_r = F_L S_x \tag{6.59}$$

$$X_1 = \frac{\pi}{S_x} \sqrt{\frac{EGJA}{2}} \tag{6.60}$$

$$X_2 = 4 \frac{C_w}{I_y} \left(\frac{S_x}{GJ} \right)^2 \tag{6.61}$$

where

$$G = \text{shear modulus, ksi}$$

$$J = \text{torsional constant, in}^4$$

$$S_x = \text{section modulus about the major axis, in}^3$$

$$F_L = \text{smaller of } (F_{yf} - F_r) \text{ or } F_{yw}$$

$$F_{yf} = \text{yield strength of the flange, ksi}$$

$$F_{yw} = \text{yield strength of the web, ksi}$$

$$F_r = \text{compressive residual stress in the flange, ksi}$$

$$C_w = \text{warping constant, in}^6$$

The coefficient C_b depends on the loading conditions within the unbraced length. For uniform moment, $C_b = 1.0$; for nonuniform moment diagrams, when both ends of the beam are braced, C_b is given as

$$C_b = \frac{12.5 M_{max}}{2.5 M_{max} + 3 M_A + 4 M_B + 3 M_C} \tag{6.62}$$

where

M_{max} = the absolute value of the maximum moment in the unbraced segment, kip-in

M_A = the absolute value of the moment at the quarter point of the unbraced segment

M_B = the absolute value of the moment at the centerline of the unbraced segment

M_C = the absolute value of the moment at the three-quarter point of the unbraced segment

With regard to this model (Equations 6.57 to 6.61), the vectors **R** and **S** in the performance function for this case are found to be as follows:

$$\mathbf{R} = \{E, G, A, I_y, r_y, J, Z_x, S_x, C_w, F_{yw}, F_r\}^t \tag{6.63}$$

$$\mathbf{S} = \{M\} \qquad\qquad \text{for } C_b = 1.0 \tag{6.64}$$

$$= \{M, M_1, M_2\}^t \qquad \text{for } C_b > 1.0$$

where M_1 and M_2 are the moments at the ends of an unbraced segment. For $C_b = 1.0$, **Q** and **S**$_0$ are the same as in the plastic range of behavior.

3. For $L_b > L_r$ (elastic behavior),

$$M_u = \frac{C_b S_x X_1 \sqrt{2}}{L_b / r_y} \left[1 + \frac{X_1^2 X_2}{2 \left(L_b / r_y \right)^2} \right]^{1/2} \tag{6.65}$$

In this case, the vectors **R** and **S** are determined to be

$$\mathbf{R} = \{E, G, A, I_y, r_y, J, S_x, C_w\}^t \qquad \text{for } M_u < C_b M_r \tag{6.66}$$

$$= \{S_x, F_{yw}, F_r\}^t \qquad\qquad \text{for } M_u = C_b M_r$$

$$S = \{M\} \qquad\qquad \text{for } C_b = 1.0 \qquad (6.67)$$
$$= \{M, M_1, M_2\}^t \qquad \text{for } C_b > 1.0$$

The matrix $\mathbf{Q}$ and the vector $\mathbf{S_0}$ for this case are the same as for the two cases just discussed. In this formulation, the zones of behavior defined by L_{pd}, L_p, and L_r are valid only for $C_b = 1.0$. When $C_b > 1.0$, the cutoff points for the zones are shifted as shown in Figure 6.4. Refer to the AISC LRFD *Manual* (1994) for a detailed presentation of the design criteria for lateral torsional buckling caused by bending.

Example 6.4

The bending of the beam member in the portal frame in Figure 6.3 is considered. Adequate lateral bracing is assumed ($L_b < L_p$) so that the ultimate moment capacity of the beam is equal to the plastic moment. The performance function is given in Equation 6.54. Table 6.3 shows the results of the reliability analysis. The checking point values of the random variables, the value of the performance function, and the reliability index are shown for the first and the last iterations only. The SFEM-based algorithm converges to a value of $\beta = 4.078$ in six iterations. The corresponding probability of failure is $p_f = 0.88 \times 10^{-6}$.

The reliability index is found to be sensitive to the stochastic variations in the three loads, the plastic section modulus, and the yield strength. The results of reliability analysis when we ignore the stochasticity of the other variables that have insignificant influence on the reliability are shown in the last column of Table 6.3. The results are identical in both cases.

6.5.2.3 *Side Sway at the Top of a Plane Frame* This is the simplest limit state from the point of view of SFEM-based reliability analysis. It refers to the lateral deflection

Table 6.3. Reliability analysis: pure bending of member 2 at node 2

Variable	Sensitivity Index	Initial Checking Point	Final Checking Point (without reduction)	Final Checking Point (with reduction)
D	0.39	0.44	0.48	0.48
L	0.57	0.05	0.12	0.12
W	−0.22	0.41	0.36	0.36
A	0.00	4.41	4.20	4.41
I	0.00	68.90	65.64	68.90
Z	−0.28	16.00	15.24	15.24
E	0.00	29,000.00	29,000.00	29,000.00
F_y	−0.63	39.60	26.69	26.69
Performance function		0.475	0.0018	0.0018
Reliability index			4.078	4.078
Number of iterations			6	6

at one of the top nodes of a plane frame. The performance criterion may be written as

Lateral deflection at one of the top nodes ≤ Allowable deflection

The corresponding performance function may be written as

$$g(\mathbf{R}, \mathbf{S}) = 1.0 - \frac{u_i}{u_{\lim}} \qquad (6.68)$$

where $u_{\lim}$ is the allowable lateral deflection and u_i is the lateral deflection at the ith node. The performance function is in terms of a single load effect, the nodal displacement, u_i. Hence, $\mathbf{S}$, the vector of load effects, is $\mathbf{S} = \{u_i\}$; and $\mathbf{R}$, the vector of resistance variables, is $\mathbf{R} = \{0\}$. In Equation 6.8, the displacement to load effect transformation matrix $\mathbf{Q}$ for the present case is a vector containing a value of 1 at the global degree of freedom number corresponding to u_i, and zeros elsewhere. The matrix $\mathbf{S_0} = \{0\}$.

Example 6.5

Consider once again the portal frame of Example 6.3. The limit state of the horizontal displacement at node 2 not exceeding 0.36 in (height/400) is to be evaluated for reliability.

The results of the reliability analysis are shown in Table 6.4. The SFEM-based algorithm converges to a value of $\beta = 1.187$ in just three iterations. The corresponding probability of failure is $p_f = 0.117$. For this limit state, the reliability is observed to be most sensitive to the stochastic variation in the horizontal load W as expected, followed by the Young's modulus and the moment of inertia of the cross sections. The results of the reliability analysis, ignoring the stochasticity in the other basic variables, are shown in the last column of Table 6.4. The difference between the two results is very small.

Table 6.4. Reliability analysis: horizontal displacement at node 2

Variable	Sensitivity Index	Initial Checking Point	Final Checking Point (without reduction)	Final Checking Point (with reduction)
D	0.01	0.44	0.48	0.44
L	0.00	0.05	0.05	0.05
W	0.98	0.41	0.60	0.60
A	−0.00	4.41	4.37	4.41
I	−0.14	68.90	68.33	68.34
Z	0.00	16.00	16.00	16.00
E	−0.17	29,000.00	28,654.29	28,654.09
F_y	0.00	39.60	39.60	39.60
Performance function		0.110	−0.00001	−0.00001
Reliability index			1.187	1.187
Number of iterations			3	3

6.5.2.4   *Vertical Deflection at the Midspan of a Beam*

The deflection of a floor beam under live load is one of the important criteria checked in practical design. In this case, the performance function is similar to that in Equation 6.68:

$$g(R, S) = 1.0 - \frac{v_{\text{midspan}}}{v_{\text{lim}}} \tag{6.69}$$

where v_{midspan} denotes the vertical deflection at the midspan of the beam member, and v_{lim} denotes the allowable deflection.

In the finite element scheme, the beam member may be composed of several elements. Two instances arise in the context of finite element analysis. If one of the nodes coincides with the midspan of the beam, then the midspan deflection is the vertical deflection at a node. This is similar to the previous example of lateral deflection at a node. Hence, in this case, the vector $\mathbf{R} = \{0\}$ and the vector $\mathbf{S} = \{v_{\text{midspan}}\}$. The matrix $\mathbf{Q}$ in Equation 6.8 is a vector containing the value 1 at the global degree of freedom number corresponding to $\{v_{\text{midspan}}\}$, and zeros elsewhere, while $\mathbf{S}_0 = \{0\}$.

If, however, the midspan of the beam element does not coincide with a node, then the computation of $\{v_{\text{midspan}}\}$ may involve the midspan bending moment M_0, an end moment (M_1 or M_2), and the moment of inertia I of the beam. Hence, in this case, the vector $\mathbf{R} = \{I\}$ and the vector $\mathbf{S} = \{M_0, M_1 \text{ or } M_2\}$.

Example 6.6

Consider the portal frame in Example 6.3. The vertical deflection at the midspan of the beam in the portal frame (Figure 6.3) under the load L is of interest now. The allowable deflection is 1.0 in. ($L/360$). The probability of failure corresponding to this limit state is to be estimated.

The results of SFEM-based reliability analysis are shown in Table 6.5. The checking point values of the random variables, the values of the performance function, and β are shown for the first and the last iterations only. The algorithm converges to a value of

Table 6.5. Reliability analysis: vertical deflection at the midspan of the beam

Variable	Sensitivity Index	Initial Checking Point	Final Checking Point (without reduction)	Final Checking Point (with reduction)
L	-0.95	0.05	0.36	0.36
A	0.00	4.41	4.79	4.41
I	0.20	68.90	74.86	74.81
Z	0.00	16.00	17.37	16.00
E	0.24	29,000.00	32,504.32	32,499.00
F_y	0.00	39.60	39.60	39.60
Performance function		0.837	-0.2×10^{-7}	-0.2×10^{-7}
Reliability index			8.271	8.259
Number of iterations			8	8

$\beta = 8.271$ in eight iterations. The corresponding probability of failure is $p_f \approx 0.0$. The structure is obviously very safe in this limit state. The reliability is sensitive to the stochastic variation in the load, L; Young's modulus, E; and the moment of inertia, I. The results of reliability analysis, ignoring the randomness in the other variables, are shown in the last column of Table 6.5 and are close to those obtained earlier.

6.6 CONCLUDING REMARKS

This chapter provides a detailed formulation of SFEM-based reliability analysis for structures with linear static behavior, using the displacement approach of finite element analysis. The chain rule of differentiation is found to be convenient for the computation of the gradient vector of the performance function used in the FORM algorithm. However, practical implementation for large structures with many elements is likely to create large demands on memory to store the partial derivatives of various matrices.

The latter half of the chapter provides a detailed illustration of SFEM-based reliability analysis for framed structures with linear elastic behavior. Four limit states were considered, two of them related to member strength, and two related to deflection. The purpose of these examples was to show the entire process of reliability analysis in detail: formulation of limit states, definition of random variables, identification of load-type and resistance-type variables, computation of performance function gradients, and implementation of FORM.

The four limit state examples for framed structures are found to show different levels of reliability. This is an important observation. In practical reliability-based design, it is quite common to have different limit states with different levels of reliability. Design optimization procedures have been developed to achieve the target reliabilities for different limit states (e.g., Mahadevan, 1992).

The examples presented here relate to the reliability analysis of frame members, where the stiffness matrix terms are easily computed through closed-form formulae. This makes it quite easy to analytically compute the partial derivatives during the gradient computation of the performance function. For other structures such as plate and shell problems, the stiffness matrix terms are computed using a numerical integration procedure. In that case, the partial derivatives may be computed using numerical differentiation. However, the SFEM-based reliability analysis method developed in this chapter is quite general within the context of linear elastic behavior and is applicable to frames as well as continuum problems.

7

SFEM FOR SPATIAL VARIABILITY PROBLEMS

7.1 INTRODUCTORY COMMENTS

In the reliability analysis problems of the previous chapters, the basic structural parameters were assumed to be discrete, allowing them to be represented as single-valued random variables. This assumption is valid for quantities that are concentrated at discrete points in space, such as concentrated loads and stiffnesses of joints and supports. However, most parameters in a structure are distributed in space, not concentrated at a point. Examples of such parameters are distributed loads, material properties (Young's modulus, ultimate strength, etc.) and geometric properties (width, thickness, etc.) that vary over the length of a beam, the area of a plate, and so forth. Such quantities cannot be expressed as single random variables, but only as a collection of many random variables or, more appropriately, as random processes or fields.

In this chapter, a random field description of such distributed parameters is introduced. The random field is then discretized into a set of correlated random variables for use in the SFEM-based reliability analysis as formulated in the previous chapters. Two methods of discretization of the random field are described and illustrated with the help of numerical examples. The improvement obtained by the representation of the distributed parameters as random fields is examined with the help of sensitivity measures. Similar to the finite element discretization of a structure, issues of accuracy vs. computational effort are encountered in the case of random field discretization. These are discussed and illustrated with one-dimensional and two-dimensional problems.

7.2 RANDOM PROCESSES AND RANDOM FIELDS

Several alternative and equivalent definitions for the term "random process" are available. For the purpose of this discussion, the following definition is most suitable. Figure 7.1 shows a collection of functions, $X^1(t), X^2(t), \ldots, X^n(t)$. Each function is a sample (also known as a realization) of a deterministic function, $X(t)$, where t is an independent variable.

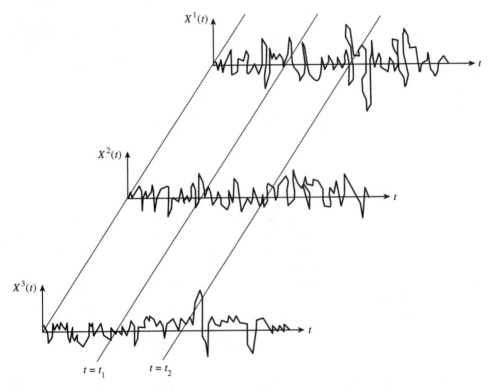

FIGURE 7.1. Ensemble of a Random Process

For example, let t denote a space coordinate such as the length of a beam and $X(t)$ denote a material property such as Young's modulus. If the entire beam is made of the same material, it is common to assume its Young's modulus to be constant over the entire length of the beam. Under such an assumption, the plot of $X(t)$ vs. t will be a horizontal straight line, corresponding to a constant value of the Young's modulus. In reality, if several "identically" constructed beams were considered, the realizations of $X(t)$ may not all be perfect straight lines, but different randomly fluctuating functions for different samples. The ensemble of all the realizations $X^i(t)$ is called a random process.

Let x_i denote the value of $X(t)$ corresponding to the location t_i. The values of x_i for each beam (at the same location t_i) will be different from each other. Therefore, x_i can be represented as a random variable. The values $x_i^1, x_i^2, \ldots, x_i^n$ are particular values of the random variable x_i for the n samples considered. Thus, for example, if the variation of the Young's modulus over the length of an ensemble of "identically" constructed beams is described as a random process, its variation at a particular cross section is a random variable.

Thus, there are two types of random variability to be considered: variation across samples and variation over space. The problems and examples in all the previous chapters have considered only the type of variability across samples; this chapter introduces additional considerations for the variability over space.

The above discussion refers to one-dimensional random processes, i.e., stochastic variation along one dimension only. Multidimensional spatially fluctuating random processes are referred to as *random fields*, and are used in the context of the Stochastic Finite Ele-ment Method. In the following discussion, the terms "random process" and "random field" are used synonymously. The mathematical treatment can be extended to random fields by considering the parameter t to be a vector of several parameters instead of a single scalar quantity. Examples of random fields commonly arise in two- and three-dimensional struc-tural problems, such as plane stress/plane strain, plate, and shell problems. For example, the thickness and Young's modulus of a rolled metal plate may vary randomly along both x and y directions, and may have to be modeled as random fields. Other examples are (1) ply orientations in a composite plate, and (2) pressure and temperature fields in a gas turbine engine.

7.3 CORRELATIVE RELATIONSHIPS IN RANDOM FIELDS

How does the spatial variability of a random quantity affect reliability analysis? Consider the example of a simple beam, where the Young's modulus, E, fluctuates over the length. Since E is random, it causes randomness in the structural response, and therefore affects the reliability of the beam. If E is considered only as a random variable, this would simply indicate variability across samples, i.e., several beams. How does one include the variabil-ity of E over the length of the beam?

The question is answered by considering several segments of the beam, as shown in Figure 7.2. The beam is divided into four segments, and four random variables are used (E_1, E_2, E_3, and E_4), one for each segment, to represent the random field E over that segment. This obviously means that for a particular value of E_i, the entire ith segment is assumed to have the same Young's modulus. The accuracy of this assumption depends on the spatial variability of the random field E and the size of the segments.

Once the random field E is replaced by random variables E_1 to E_4, then the methods of the previous chapters can be easily applied for structural and reliability analysis. Thus, a random field needs to be discretized into a set of random variables. Since these random variables are obtained from the same random field, there are statistical correlations among them. A mathematical understanding of the correlative relationships in a random field is therefore essential in order to correctly discretize the field.

As mentioned earlier, $X(t_1)$ at any particular location t_1 is a random variable with a well-defined cumulative distribution function (CDF) expressed as

$$F(x, t_1) = P\{X(t_1) \le x\} \tag{7.1}$$

Similarly, the joint cumulative distribution function of two random variables $X(t_1)$ and $X(t_2)$ at locations t_1 and t_2 is written as

$$F(x_1, t_1 x_2, t_2) = P\{X(t_1) \le x_1; X(t_2) \le x_2\} \tag{7.2}$$

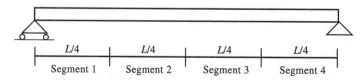

| L/4 | L/4 | L/4 | L/4 |
| Segment 1 | Segment 2 | Segment 3 | Segment 4 |

FIGURE 7.2. Simply Supported Beam

An *ensemble average* of any function of the random variables $X(t_1), X(t_2), \ldots, X(t_n)$ is defined as the expected value of such a function, given by

$$E\{g[X(t_1), X(t_2), \ldots, X(t_n)]\} \tag{7.3}$$

$$= \int_{-\infty}^{\infty} \cdots \int_{-\infty}^{\infty} g(x_1, x_2, \ldots, x_n) f_n(x_1, t_1, x_2, t_2, \ldots, x_n, t_n) \, dx_1 \, dx_2 \ldots dx_n$$

where $f_n(x_1, t_1, x_2, t_2, \ldots, x_n, t_n)$ is the joint probability density function of the random variables. The following ensemble averages are useful for further discussion.

$$\alpha_i(t) = E\left[X^i(t)\right] \tag{7.4}$$

is the ith moment of $X(t)$. Thus, the first moment—the mean—is found to be

$$\alpha_1(t) = E[X(t)] \tag{7.5}$$

Further,

$$\alpha_{ij}(t_1, t_2) = E\left[X^i(t_1)X^j(t_2)\right] \tag{7.6}$$

and

$$\mu_{ij}(t_1, t_2) = E\left\{[X(t_1) - \alpha_1(t_1)]^i \, [X(t_2) - \alpha_1(t_2)]^j\right\} \tag{7.7}$$

are second-order ensemble averages. The special case

$$\alpha_{11}(t_1, t_2) = E[X(t_1)X(t_2)] \tag{7.8}$$

is known as the *autocorrelation function* of $X(t_1)$ and $X(t_2)$, denoted as $\phi_{XX}(t_1, t_2)$. Similarly, another special case

$$\mu_{11}(t_1, t_2) = E\{[X(t_1) - \alpha_1(t_1)][X(t_2) - \alpha_1(t_2)]\} \tag{7.9}$$

is known as the *autocovariance function* of $X(t_1)$ and $X(t_2)$, denoted as $\Gamma_{XX}(t_1, t_2)$. The variance at location t_1 is observed to be

$$\text{Var}[X(t_1)] = \mu_{11}(t_1, t_1) \tag{7.10}$$

These definitions pertain to a single random field. However, in structural problems there are several random fields, which may be correlated with each other. For example, in a rolled steel beam, the area and moment of inertia may be modeled as unidimensional random fields correlated with each other. Similarly, the loads on a floor may be modeled as two-dimensional random fields correlated with each other. In such cases, the following ensemble averages are required. For two random fields $\{X(t)\}$ and $\{Y(t)\}$, the second-order averages are

$$\alpha_{ij}^{xy}(t_1, t_2) = E\left[X^i(t_1)Y^j(t_2)\right] \tag{7.11}$$

and

$$\mu_{ij}^{xy}(t_1, t_2) = E\left\{[X(t_1) - \alpha_1^x(t_1)]^i \, [Y(t_2) - \alpha_1^y(t_2)]^j\right\} \tag{7.12}$$

The special cases $\alpha_{11}^{xy}(t_1, t_2)$ and $\mu_{11}^{xy}(t_1, t_2)$ are known as the *cross-correlation function* and the *cross-covariance function*, respectively, and are denoted as $\phi_{XY}(t_1, t_2)$ and $\Gamma_{XY}(t_1, t_2)$, respectively.

7.4 HOMOGENEOUS AND WIDE-BAND RANDOM FIELDS

A random field is said to be stationary in the strict sense if its statistical properties are invariant to a shift of the origin. That is, the fields $X(t)$ and $X(t + c)$ have the same statistics for any c. Thus, the first-order density function is

$$f(x, t) = f(x, t + c) \tag{7.13}$$

for any c. This means that the first-order density function is independent of t:

$$f(x, t) = f(x) \tag{7.14}$$

Similarly the second-order density function $f(x_1, x_2, t_1 + c, t_2 + c)$ is independent of c for any c. This leads to the conclusion that

$$f(x_1, x_2, t_1, t_2) = f(x_1, x_2, \tau) \tag{7.15}$$

where $\tau = t_1 - t_2$ is the distance between two locations t_1 and t_2.

A random field is said to be stationary in the wide sense if its mean is constant:

$$E[X(t)] = \mu \tag{7.16}$$

and its autocorrelation depends only on $\tau = (t_1 - t_2)$:

$$E[X(t + \tau)X(t)] = R(\tau) \tag{7.17}$$

The term "stationary" is commonly used for one-dimensional random processes, whereas for multidimensional random fields the term "homogeneous" may be used correspondingly (Vanmarcke, 1983). For the type of problems considered in this chapter, the spatially distributed parameters are modeled as random fields that are homogeneous in the wide sense. That is, their mean values are constant over space, and their autocorrelation functions depend only on the separation between two locations.

The discussion so far has been in the context of ensemble averages, i.e., averages obtained over an ensemble of many samples or realizations:

$$E[X(t)] = \frac{1}{n}\sum_{i=1}^{n} X(t, \zeta_i) \tag{7.18}$$

There is another kind of average related to a single sample of the random process, with respect to the indexing parameter t. If t is a time coordinate, then the temporal average of the random process $X(t)$ over a time interval T may be defined as

$$\bar{X} = \frac{1}{2T}\int_{-T}^{T} X(t, \zeta)\, dt \tag{7.19}$$

If the parameter t is a space coordinate, then $\bar{X}$ is the spatial average of the random process $X(t)$ over a space interval T. Such local averages are of importance in the context of spatially distributed parameters considered in this section and are discussed in detail in Section 7.5.

The average over the entire domain is

$$Y = \lim_{T \to \infty} \frac{1}{2T} \int_{-T}^{T} X(t)\,dt \tag{7.20}$$

A stationary random process is said to be *ergodic* if its ensemble averages equal the appropriate temporal (or spatial) averages. According to the "theorem of ergodicity," any statistic of $X(t)$ can be determined from a single sample $X(t, \zeta)$ with probability 1. This is a general requirement. As a special case, a stationary random process is said to be *ergodic in the mean* if, with probability 1,

$$\bar{X} \to E[X(t)] \qquad \text{as } T \to \infty \tag{7.21}$$

Thus, for a random process $X(t)$ that is ergodic in the mean,

$$Y = \lim_{T \to \infty} \frac{1}{2T} \int_{-T}^{T} X(t)\,dt = E[X(t)] \tag{7.22}$$

Note that Y is a random variable, since any member function $X(t, \zeta_i)$ could have been chosen to evaluate the temporal (or spatial) average. However, if the random process $X(t)$ is stationary in the wide sense, then

$$E[X(t)] = \mu \tag{7.23}$$

is a constant. Thus, $Y = \mu$, a constant value, and therefore has zero variance.

For one-dimensional random fields (or random processes), the *power spectral density*, denoted as $\Phi(\omega)$, characterizes the distribution of average energy of the process in the frequency domain. For a stationary random process, the autocorrelation function $R(\tau)$ and the power spectral density $\Phi(\omega)$ form a Fourier-transform pair

$$\Phi(\omega) = \frac{1}{\pi} \int_{0}^{\infty} R(\tau) \cos(\omega\tau)\,d\tau \tag{7.24}$$

$$R(\tau) = \int_{0}^{\infty} \Phi(\omega) \sin(\omega\tau)\,d\omega \tag{7.25}$$

for even functions $R(\tau)$ and $\Phi(\omega)$. Equations 7.24 and 7.24 are the well-known Wiener–Khinchine relations between the power spectral density and the autocorrelation function of a stationary random process. These relations have been generalized to n dimensions to show that the n-dimensional Fourier transform of $R(\tau)$ equals the spectral density function $\Phi(\omega)$, where τ and ω are now regarded as n-dimensional vectors. Normalizing these functions with respect to the variance σ^2 yields another Fourier transform pair: the *correlation function* $\rho(\tau)$ and the normalized spectral density function $s(\omega)$.

A stationary random process is classified as a *wide-band* or *narrow-band* random process, depending on the nature of its power spectral density. A random process is said to be a wide-band process if its power spectral density has significant values over a wide range of

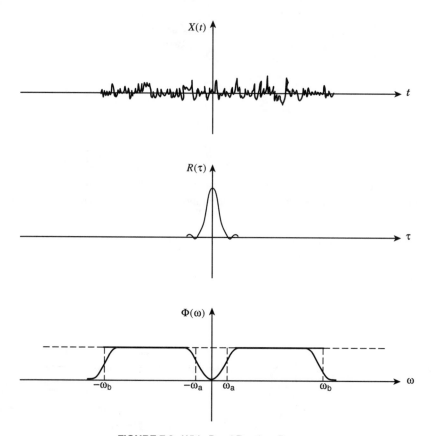

FIGURE 7.3. Wide Band Random Process

frequencies. A typical sample function, the autocorrelation function, and the power spectral density of a wide-band random process are shown in Figure 7.3. A random process is said to be a narrow-band process if its power spectral density has significant values over a narrow frequency band around a central frequency. A typical sample function, the autocorrelation function, and the power spectral density of a narrow-band random process are shown in Figure 7.4.

Extension of these definitions to multidimensional homogeneous random fields is mostly straightforward, by considering the parameters τ and w to be vectors. However, the evenness of the functions $R(\tau)$ and $\Phi(\omega)$ become questionable. In this case, Vanmarcke (1983) introduced the concept of quadrant symmetry. By definition, the autocorrelation function and the power spectral density function of quadrant symmetric homogeneous random fields (which include isotropic fields as a special class) are even functions for each component of the vectors $\tau = (\tau_1, \tau_2, \ldots, \tau_n)$ and $\omega = (\omega_1, \omega_2, \ldots, \omega_n)$, respectively.

From Figure 7.3 it is apparent that the autocorrelation function for a wide-band random field decays rapidly. This is representative of structural parameters such as distributed loads, material, and sectional properties. That is, if a random field is represented by several random variables corresponding to several locations, two random variables whose locations are separated by a shorter distance are expected to have a higher correlation than two random variables whose locations are separated by a longer distance. In the subsequent

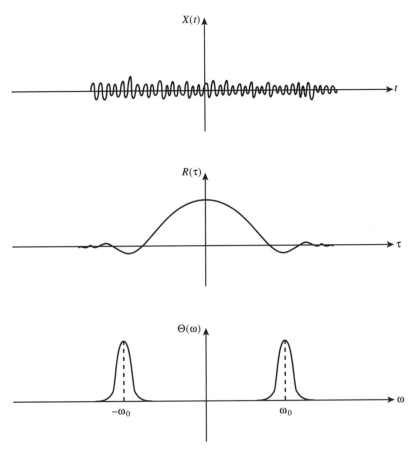

FIGURE 7.4. Narrow Band Random Process

discussion, the stochastic structural parameters are assumed to be wide-band, quadrant-symmetric random fields, homogeneous in the wide sense.

7.5   RANDOM FIELD DISCRETIZATION

The discretization of a spatially distributed random field has to be done by discretizing the structure into several elements and specifying a set of correlated random variables such that each random variable represents the random field over a particular element. Although this is similar to finite element discretization of a structure, there is an important difference between the two kinds of discretization. While finite element discretization is done to adequately represent the deterministic behavior of a structure, random field discretization is done to represent the stochastic spatial variation of any parameter over the structure. Thus, the number, size, and shape of the random field elements need not be the same as those of the finite elements. Typically, the stochastic finite element analysis of a continuum structure would involve the use of two types of meshes: (1) a finite element mesh for the deterministic part of the analysis, and (2) several random field element meshes for the stochastic part of the analysis—one for each spatially distributed stochastic parameter.

Several methods have been proposed for the discretization of random fields into random variables (Liu and Der Kiureghian, 1989). The methods can be classified into three main approaches. In the first approach, the value of the random field at a few selected locations in the element is used to define a corresponding random variable for that element. (The selected locations could be just one point—e.g., the midpoint or center of the element—which is referred to as the midpoint method. Or the random variable could be defined based on values at the nodes of the element, which is referred to as the nodal point method.) The second approach is to define a random variable for each element using the spatial average of the random field over the element (Vanmarcke and Grigoriu, 1983). The third approach is to use some form of series expansion to break a random field into a set of random variables. Two examples of this are Karhunen–Loeve expansion (Spanos and Ghanem, 1988) and Newmann expansion (Shinozuka and Deodatis, 1988). Each approach has its advantages and disadvantages and problems suitable for application. Two simple techniques, the midpoint and spatial average methods, are presented here.

In the midpoint method, the value of the random field over an element is represented by its value at the midpoint of the element. That is, the random field $X(t)$ is represented over an element i as

$$X_i = X(t_i) \tag{7.26}$$

where t_i is the location of the centroid of the element. The mean and the variance of the random variable X_i are obtained from Equations 7.5 and 7.10, while the covariance matrix of the random variables is obtained using Equation 7.9.

In the spatial averaging method, the value of the random field over an element is represented by the spatial average of the random field over the element. That is, the random field $X(t)$ is represented over an element i as

$$X_i = \frac{1}{\Omega_i} \int_{\Omega_i} X(t) \, d\Omega_i \tag{7.27}$$

where Ω_i is the length of the element if the random field is one dimensional, the area of the element if the random field is two dimensional, and the volume of the element if the random field is three dimensional. For one-dimensional random processes, the above equation may be written as

$$X_i = \frac{1}{T} \int_{-T/2}^{T/2} X(t) \, dt \tag{7.28}$$

where T is the averaging (temporal or spatial) interval.

Example 7.1

Consider the simply supported beam in Figure 7.2 with the random field E. To discretize E into several random variables, the beam is divided into four equal segments of length $L/4$, where L is the length of the entire beam. In the midpoint method, the corresponding four random variables are defined by the values of E at the midpoints of the segments, as

$$E_1 = E(0.125L)$$

$$E_2 = E(0.375L)$$

$$E_3 = E(0.625L)$$

$$E_4 = E(0.875L)$$

In the spatial averaging method, the four random variables for the four segments are defined as follows:

$$E_1 = \frac{1}{L/4} \int_0^{L/4} E(t)\, dt$$

$$E_2 = \frac{1}{L/4} \int_{L/4}^{L/2} E(t)\, dt$$

$$E_3 = \frac{1}{L/4} \int_{L/2}^{3L/4} E(t)\, dt$$

$$E_4 = \frac{1}{L/4} \int_{3L/4}^{L} E(t)\, dt$$

Numerical integration may be employed to evaluate the integral in Equation 7.27, similar to the finite element methodology. The mean value of the random variable X_i is not affected by the averaging operation. For a wide-sense homogeneous random field $X(t)$ with mean μ and variance σ^2, the mean of X_i is simply

$$E[X_i] = E[X(t)] = \mu \tag{7.29}$$

However, the variance of X_i is not obtained as a simple average and has to be expressed as

$$\mathrm{Var}[X_i] = \gamma(T)\sigma^2 \tag{7.30}$$

where $\gamma(T)$ is defined as the *variance function* of $X(t)$, which measures the reduction of the point variance σ^2 under local averaging. The variance function is dimensionless and has the following properties:

$$\gamma(T) \geq 0 \tag{7.31}$$

$$\gamma(0) = 1 \tag{7.32}$$

$$\gamma(-T) = \gamma(T) = \gamma(|T|) \tag{7.33}$$

The variance function decreases monotonically with T, its upper bound is $\gamma(0) = 1$, and for processes that are ergodic in the mean, $\gamma(\infty) = 0$ (Vanmarcke, 1983). Thus, the variance of X_i in the spatial averaging method is obtained by multiplying the point variance by a reduction factor given by the variance function, which depends on the size of the averaging interval.

The variance function can also be used to derive the coefficients of correlation between the spatially averaged random variables X_i and X_j for a one-dimensional process. Let X_i and X_j correspond to two random field elements with lengths T and T', located arbitrarily on the t axis, as shown in Figure 7.5. The coefficient of correlation is

$$\rho_{ij} = \frac{\mathrm{Cov}(X_i, X_j)}{\sigma_i \sigma_j} \tag{7.34}$$

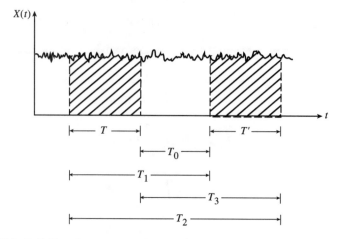

FIGURE 7.5. Definition of Distances to Compute the Covariance Between Spatial Averages

where the covariance between X_i and X_j has been found to be (Vanmarcke and Grigoriu, 1983)

$$\text{Cov}(X_i, X_j) = \frac{\sigma^2}{2TT'} \left[T_0^2 \gamma(T_0) - T_1^2 \gamma(T_1) + T_2^2 \gamma(T_2) - T_3^2 \gamma(T_3) \right] \tag{7.35}$$

The distances T_0 to T_3 are defined in Figure 7.5.

The exact pattern of decay of the spatially averaged variance depends on the correlation function $\rho(\tau)$ of the process $X(t)$:

$$\gamma(T) = \frac{2}{T} \int_0^T \left(1 - \frac{\tau}{T} \right) \rho(\tau) \, d\tau \tag{7.36}$$

Vanmarcke (1983) presented several models of the correlation function (such as triangular, exponential, and Gaussian) and the corresponding variance functions. He observed that $\gamma(T)$ becomes inversely proportional to T at large values of T for each model, and named the proportionality constant as the *scale of fluctuation*, θ:

$$\theta = \lim_{T \to \infty} T\gamma(T) \tag{7.37}$$

or

$$\gamma(T) = \frac{\theta}{T} \qquad \text{when } T \to \infty \tag{7.38}$$

This asymptotic form suggests a simple and convenient approximation to the variance function:

$$\gamma(T) = 1 \qquad T \le \theta$$

$$= \frac{\theta}{T} \qquad T > \theta \tag{7.39}$$

Vanmarcke and Grigoriu (1983) observed that the variance function resulting from this approximation agreed closely with those derived from the aforementioned exact correlation functions for a few common wide-band random processes. Thus, the scale of fluctuation is a useful parameter for obtaining simple and reasonably accurate information about the variances of local averages of random fields.

The scale of fluctuation can also be viewed as the approximate length over which strong correlation persists in the random field. This fact can be used advantageously in determining the size of the random field element mesh. Since strong correlation among the random variables obtained by discretization could cause numerical difficulties (refer to Section 7.6.2), it is advisable to use element sizes that are larger than the scale of fluctuation. A similar practical application of the scale of fluctuation was illustrated by Haldar (1984) for the statistical modeling of soil properties. When a single type of test was used at all locations, the sampling distances were chosen to be larger than the scale of fluctuation to avoid redundancy in the gathering of information. However, if different tests were to be performed at two locations, the distance between the two locations was chosen to be smaller than the scale of fluctuation for maximum effectiveness.

Example 7.2

A cantilever beam is shown in Figure 7.6. It has a uniform load W acting on it vertically downward. The Young's modulus, E, is treated as a random field. Compare the two methods of random field discretization described in this section.

Assume that the correlation coefficients between the discretized random variables of the random field E are given by the formula

$$\rho(E_i, E_j) = \exp\left(-\frac{\Delta x}{\lambda L}\right) \tag{7.40}$$

where $\rho(E_i, E_j)$ is the correlation coefficient between the random variables E_i and E_j obtained by discretizing the random field, Δx is the distance between two random field elements i and j, and L is the length of the beam. λL is referred to as the correlation length, where λ is a dimensionless constant. It characterizes the length at which the autocorrelation function decays to a specified value. In this example, the value of λ is assumed to be equal to 0.25. The resulting matrix of correlation coefficients among the variables corresponding to each random field is shown in Table 7.1 for the midpoint method as well as the method of spatial averages.

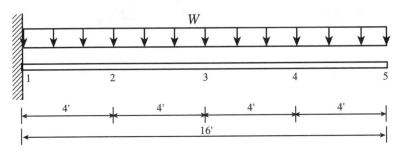

FIGURE 7.6. Cantilever Beam

Table 7.1. Correlation coefficients between the random variables for a discretized field: midpoint method and spatial averaging method

	Element 1	Element 2	Element 3	Element 4
	Midpoint Method			
Element 1	1.000	0.368	0.135	0.050
Element 2	0.368	1.000	0.368	0.135
Element 3	0.135	0.368	1.000	0.368
Element 4	0.050	0.135	0.368	1.000
	Spatial Averaging Method			
Element 1	1.000	0.738	0.272	0.100
Element 2	0.738	1.000	0.738	0.272
Element 3	0.272	0.738	1.000	0.738
Element 4	0.100	0.272	0.738	1.000

Among the two methods of discretization of random fields presented here, the midpoint discretization method (Equation 7.26) tends to overrepresent the variability within the element, whereas the spatial averaging method (Equation 7.27) tends to underrepresent the same variability. Therefore, the two methods tend to bracket the exact result for any given random field element mesh and together can be used to check the accuracy of the solution and its convergence with mesh refinement.

The discretization of a random field $X(t)$ into a set of random variables X_i also raises the question about the probability distribution functions for X_i. In the midpoint discretization method, the distribution of the random variables remains the same as that of the underlying random field. However, for the spatial averaging method, this is true only if the underlying random field is Gaussian, since the integration operator is linear. For non-Gaussian random fields, the distribution of X_i as defined by Equation 7.27 is very difficult to obtain. Der Kiureghian (1987) presented approximate descriptions of the distribution function for such cases.

Example 7.3

Consider once again the cantilever beam shown in Figure 7.6, with a downward load W. The statistical description of the random quantities of the problem is given in Table 7.2. Perform the reliability analysis of this beam using FORM for the following two failure criteria:

Table 7.2. Cantilever beam: description of the random variables

Variable	Units	Mean	Coefficient of Variation	Type of Distribution
W	kip/in	0.08	0.20	Normal
A	in^2	7.68	0.05	Normal
I	in^4	301.00	0.05	Normal
Z	in^3	44.20	0.05	Normal
E	ksi	29,000.00	0.06	Normal
F_y	ksi	39.60	0.11	Normal

1. Vertical deflection at the free end $\geq$ length/200
2. Bending moment at the support $\geq$ plastic moment capacity

The expression of the performance function and the assembly of the various matrices for the two corresponding limit states are similar to the examples described in the previous chapter.

The beam is divided into four elements. Even though the stiffness matrix of a beam element is explicitly available and an exact solution can be obtained with only one element, the discretization of the beam is done to discretize the random fields into random variables. In this example, all the random parameters have spatial variability. Therefore, each parameter can be described as a random field and discretized into sets of random variables. All the random fields are assumed to be Gaussian and uncorrelated with each other.

Two types of failures are considered for reliability analysis. The first failure criterion is concerned with the deflection of the beam, i.e.,

$$\delta_1 > \frac{L}{200} \tag{7.41}$$

where δ_1 is the vertical deflection at the free end of the beam, and L is the length of the beam. For this failure criterion, the relevant uncertain quantities are the distributed load, W; the Young's modulus, E; and the cross-section moment of inertia, I. The second type of failure is concerned with the strength of the beam. The failure criterion is

$$M_1 > ZF_y \tag{7.42}$$

where M_1 is the bending moment at the support, Z is the plastic section modulus of the beam, and F_y is the yield strength of the beam material. For this failure criterion, the relevant uncertain quantities are the distributed load, W; the plastic section modulus, Z; and the yield strength, F_y.

All five of these uncertain quantities are modeled as random fields and discretized into four random variables each, one for each finite element. The correlation coefficients among the four discretized random variables for each random field have already been computed in Table 7.1.

The FORM reliability analysis results are showed in Tables 7.3 and 7.4 for the two failure criteria. The tables show the results with two types of random field discretization: the midpoint method and the spatial averaging method. As expected, the spatial averaging method produces a higher estimate of the reliability index than the midpoint method. This is because the variance of the random variables in the midpoint method is equal to the point variance of the random field, whereas in the spatial averaging method the variance is reduced through a variance function. Naturally, the lower variance of the random variables produces a higher estimate of the reliability.

7.6 PRACTICAL IMPLEMENTATION OF RANDOM FIELD DISCRETIZATION

The previous sections developed the concepts of considering the distributed structural parameters as random fields and their use in reliability analysis. However, in the implemen-

Table 7.3. Vertical deflection of the cantilever beam: reliability analysis

		Midpoint Method		Spatial Averaging Method	
Variable	Initial Checking Point	Sensitivity Index	Final Checking Point	Sensitivity Index	Final Checking Point
W_1	0.08	−0.03	0.07	−0.03	0.07
W_2	0.08	−0.18	0.06	−0.18	0.06
W_3	0.08	−0.44	0.05	−0.44	0.05
W_4	0.08	−0.78	0.05	−0.78	0.05
E_1	29,000.00	0.29	29,745.15	0.29	29,647.14
E_2	29,000.00	0.11	29,496.54	0.11	29,583.92
E_3	29,000.00	0.03	29,230.72	0.03	29,315.00
E_4	29,000.00	0.00	29,088.00	0.00	29,133.77
I_1	301.00	0.24	306.41	0.24	305.70
I_2	301.00	0.09	304.60	0.09	305.24
I_3	301.00	0.02	302.67	0.02	303.28
I_4	301.00	0.00	301.64	0.00	301.97
Performance function			−0.0003		−0.0003
Reliability index			2.155		2.206
Number of iterations			3		3

Table 7.4. Bending moment at the fixed support of the cantilever beam: reliability analysis

		Midpoint Method		Spatial Averaging Method	
Variable	Initial Checking Point	Sensitivity Index	Final Checking Point	Sensitivity Index	Final Checking Point
W_1	0.08	0.08	0.08	0.06	0.08
W_2	0.08	0.23	0.08	0.19	0.08
W_3	0.08	0.38	0.09	0.31	0.08
W_4	0.08	0.53	0.09	0.44	0.08
Z	44.20	−0.30	43.41	−0.34	43.31
F_y	39.60	−0.66	35.89	−0.75	355.355
Performance function	0.210		0.00005		−0.00007
Reliability index	1.422		1.282		1.372
Number of iterations			3		3

tation of these concepts, the analyst has to address the following two questions of practical importance to balance efficiency and accuracy in the computations:

1. Should all the distributed parameters be considered as random fields?
2. If a parameter is considered as a random field, what is the best random field element mesh to be used?

These questions are addressed in this section.

7.6.1 Selective Consideration of Random Fields

The first question can be addressed by using sensitivity indices, which measure the relative influence of various random variables on the reliability. Refer to Chapter 3 for the definition of reliability sensitivity indices. The random field description of a stochastic parameter and subsequent discretization into several random variables significantly increases the size of the problem and the computational cost. On the other hand, the representation of a distributed parameter by a single random variable reduces the cost but implies a perfectly correlated random field, which is an approximation. Therefore, in the practical implementation of SFEM-based reliability analysis of structures with distributed parameters, the analyst needs to decide whether it is necessary to consider all the distributed parameters as random fields. It is apparent that sensitivity measures can be used to resolve this issue and enable the selective representation of only a few distributed parameters as random fields. That is, if a particular distributed parameter has a low influence on the stochastic response, then it is reasonable not to consider that parameter as a random field and discretize it into several random variables, since such increased computational effort would produce very little improvement in the final result. Such a parameter may well be modeled as a single random variable, without much loss in accuracy. Further, if its sensitivity index is very low—less than a cutoff value, say 0.05—its stochasticity may be ignored altogether as discussed earlier. To ensure economy, the analyst may first perform a trial reliability analysis considering all the stochastic parameters as random variables and select only a few of the distributed parameters to be modeled as random fields based on sensitivity information.

Notice, however, that this simple guideline, based on small values of sensitivity indices, is only to reduce the size of the problem without much loss of accuracy. Distributed parameters with large values of sensitivity indices will routinely be modeled as random fields using this guideline. Further reduction in problem size is achieved by examining the correlation length or scale of fluctuation of the random field and its effect on the statistics of the response. As mentioned earlier, the representation of a distributed parameter by a single random variable implies a perfectly correlated random field, i.e., a random field with infinite correlation length. Therefore, if a distributed parameter has a very large correlation length, it may be modeled as a random variable, despite its having a high sensitivity index.

Example 7.4

Consider the cantilever beam of Figure 7.6. In Example 7.3, four random quantities were considered and all four were treated as random fields. This example illustrates the idea of using sensitivity indices to selectively model only a few of the uncertain quantities with spatial variability as random fields. Only the reliability analysis relating to the vertical deflection of the beam at the free end is considered, and only the spatial averaging method is used in this example.

For the sake of illustration, a different type of correlation structure is assumed for the random fields now, instead of that used in Example 7.3. A triangular approximation for the decay of the variance function is used:

$$\gamma_x(T) = 1 - \frac{T}{3\theta_x} \qquad\qquad T \le \theta_x \tag{7.43}$$

$$= \frac{\theta_x}{T}\left[1 - \frac{\theta_x}{3T}\right] \qquad T > \theta_x \tag{7.44}$$

where $\gamma_x(T)$ is the variance function of the random field X, θ_x is the scale of fluctuation, and T is the distance. Using this approximation, the correlation coefficient between two discretized random variables in the ith and jth elements may be expressed as

$$\rho(x_i, x_j) = \frac{1}{2}(k-1)^2\gamma_x\left[\frac{(k-1)L}{N}\right] - 2k^2\gamma_x\left(\frac{kL}{N}\right) + (k+1)^2\gamma_x\left[\frac{(k+1)L}{N}\right] \tag{7.45}$$

in which $k = |i - j|$.

The reliability analysis of the structure is done in four steps, to clearly examine the effect of considering different distributed parameters as random fields. In the first step, all the parameters are considered as random variables, as in Example 7.3. The SFEM-based reliability analysis converges to a value of $\beta = 1.053$ in just three iterations. The load, W, has the highest sensitivity index (0.93), followed by the Young's modulus, E (0.28), and the moment of inertia, I (0.23).

In subsequent steps, the spatial variabilities of the random quantities W, E, and I are considered and the reliability analysis is repeated. In general, the description of a random parameter by a single random variable causes it to have a greater influence on the determination of the reliability index than its description as a random field. This observation can be combined with the fact that a random variable with a greater variability would have a greater influence on the reliability of a structure than if it had lesser variability. Therefore, the description of a random parameter by a single random variable tends to overrepresent its variability compared to its description by a random field (Mahadevan, 1988).

In the second step, the parameter with the highest sensitivity index, the load, W, is considered to be a random field and discretized into four random variables. As shown in Table 7.5, the reliability index jumps to a value of $\beta = 1.391$ compared to a value of 1.053 in step 1, for $\theta = 0.25L$, where θ is the scale of fluctuation, and L is the length of the beam. In the second and third steps, parameters E and I are considered to be random fields, resulting in β values equal to 1.446 and 1.488, respectively, for the same scale of fluctuation. These are much smaller changes compared to that obtained in step 2. Thus, as parameters with smaller sensitivity indices are considered as random fields, the result does not change appreciably. This indicates that the parameters E and I may well be represented by single random variables without significant loss in accuracy. Thus the computational efficiency of SFEM may be improved with the help of sensitivity indices, facilitating its application to practical problems.

Table 7.5. Cantilever beam: progressive consideration of random fields

	β Values		
θ/L	Only W as random field	W, E as random field	W, E, I as random field
0.0	1.467	1.543	1.603
0.25	1.391	1.446	1.488
0.50	1.260	1.287	1.306
0.75	1.186	1.202	1.213
1.00	1.149	1.159	1.167
1.50	1.114	1.120	1.124
2.00	1.097	1.102	1.105
2.50	1.088	1.092	1.094
3.00	1.082	1.085	1.087
∞	1.053	1.053	1.053

These three steps are repeated for various values of θ, and the results are shown in Figure 7.7. The results are similar: the consideration of W, the parameter with the highest influence, as a random field improves the estimate of β greatly, whereas consideration of the less influential E and I does not improve the estimate of β significantly. Also, as θ increases (i.e., the random field is becoming more and more correlated, thus approaching the single random variable case), the improvement by considering E and I as random fields is negligible. Thus, for the present problem, E and I may be consid-

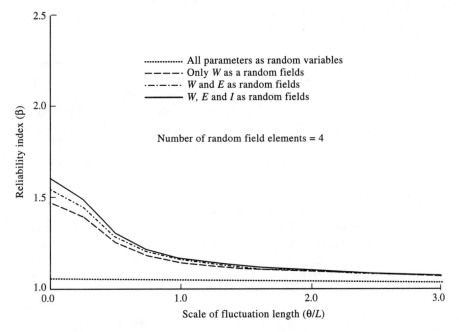

FIGURE 7.7. Cantilever Beam: Progressive Consideration of Random Fields

Table 7.6. Cantilever beam: influence of scale of fluctuation and random field mesh refinement

Number of Elements	Scale of Fluctuation, θ/L								
	0.0	0.25	0.5	0.75	1.0	1.5	2.0	2.5	3.0
	β values								
1	1.053 (3)	1.053 (3)	1.053 (3)	1.053 (3)	1.053 (3)	1.053 (3)	1.053 (3)	1.053 (3)	1.053 (3)
4	1.467 (3)	1.391 (3)	1.260 (3)	1.186 (3)	1.149 (3)	1.114 (3)	1.097 (3)	1.088 (3)	1.082 (3)
8	1.783 (3)	1.500 (3)	1.293 (3)	1.204 (3)	0.160 (3)	1.120 (3)	1.104 (3)	1.091 (3)	1.085 (3)
12	1.948 (3)	1.531 (3)	1.293 (3)	1.204 (3)	1.161 (3)	1.117 (3)	1.105 (3)	1.088 (3)	1.079 (3)
16	2.032 (11)	1.533 (5)	1.297 (5)	1.212 (3)	1.164 (5)	1.129 (3)	1.111 (3)	1.099 (3)	1.090 (3)
20	2.144 (12)	1.538 (7)	1.332 (14)	1.245 (3)	1.19 (7)	1.131 (5)	1.093 (9)	1.079 (6)	1.085 (6)
24	2.188 (7)	1.375 (30)	1.332 (4)	1.216 (4)	1.053 (4)	1.152 (4)	1.100 (5)	1.122 (11)	1.122 (10)

Note. The figures in parentheses indicate the number of iterations needed by the reliability analysis algorithm to converge to β.

ered as single random variables for $\theta \geq L$. The last row in Table 7.5 corresponds to the case when $\theta \to \infty$, which is the same as the single random variable case.

7.6.2 Stochastic Mesh Refinement

As mentioned in Section 7.5, the random field element mesh is similar to the finite element mesh in that the structure is discretized into several elements. It is apparent that the accuracy of the result of the stochastic finite element analysis depends not only on the finite element mesh, but also on the random field element meshes. Similar to the finite element methodology, refinement of the random field element meshes may be attempted to improve the results of the analysis. However, such refinement studies have to be performed with caution. The following factors merit consideration:

1. The use of a fine mesh to discretize a random field results in a large number of random variables; this greatly increases the size of the stiffness matrix and the partial derivative matrices. The analyst has to determine whether such a tremendous increase in computational effort is justified by the improvement in the results.

2. More importantly, the use of a fine mesh causes the random variables in elements close to each other to have high correlation with each other, resulting in numerical difficulties in the orthogonalization of the covariance matrix. Hence the random field element mesh size has to be chosen in order to avoid high correlations among the random variables.

3. The use of scale of fluctuation is a simple and effective means of determining the appropriate random field element mesh. As mentioned earlier, the scale of fluctuation approximately indicates the distance over which strong correlation persists within the random field. Hence it may be desirable to choose a random field element mesh in which distance between the centroids of the elements is larger than the scale of fluctuation.

Typically, the random field element meshes consist of elements that are blocks of one or more deterministic finite elements.

Example 7.5

Consider once again the cantilever beam of Figure 7.6, with the statistics shown in Table 7.2. This numerical example features the study of stochastic mesh refinement. For this purpose, the distributed load, W, is discretized over 4, 8, 12, 16, 20, and 24 elements of equal length, as shown in Table 7.6. In this table, the row showing the number of elements = 1 corresponds to the case when W is considered as a single random variable. As discussed earlier, the correlation characteristics of the random field can be expected to affect the estimated reliability and therefore the selection of the appropriate mesh for random field discretization. Thus, the reliability analysis is performed for different values of θ, the scale of fluctuation, for each random field element mesh. The results of such mesh refinement are shown in Table 7.6 and Figure 7.8.

It is observed from Figures 7.7 and 7.8 that the reliability index decreases with increasing θ. This is to be expected, since the scale of fluctuation is the length over which there is persistent strong correlation in the random field. Therefore, a higher value of θ indicates that the random variables are more strongly correlated. Thus, the limit $\theta \rightarrow \infty$ implies a perfectly correlated random field, or a single random variable, and gives the minimum value of the reliability index. As shown in Figure 7.8, for various meshes, the β values converge to the limit of $\beta = 1.053$, the result for the single random variable case.

It is also observed that a finer mesh results in a higher estimate of the reliability index, but as the mesh becomes finer, the overall trend is toward convergence to a single value. Of course, there are variations in this trend due to the interaction of the following two factors: (1) As the mesh becomes finer, the correlation between the random variables in the adjacent elements increases. (2) At the same time, the correlation between

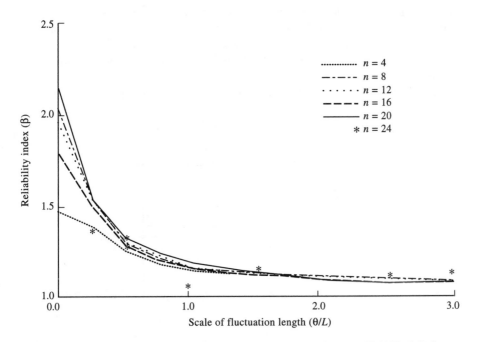

FIGURE 7.8. Cantilever Beam: Influence of Scale of Fluctuation and Random Field Mesh Refinement

random variables in elements that are far apart from each other decreases. In the present problem, the difference in β values is 0.10 or less for $\theta \geq 0.5L$. Thus the use of only four elements appears to be adequate (element length $= 0.25L$) for $\theta \geq 0.5L$. This is an important result: It indicates the minimum number of elements or the element size required to obtain the desired accuracy. For this problem, convergence to β is effectively achieved when the random field element size is one-quarter to one-half of the scale of fluctuation.

However, the results in Table 7.6 also indicate that it may not be practically desirable to use element sizes one-quarter or one-half of the scale of fluctuation when the latter is small, since convergence problems and numerical instability may be encountered as the mesh becomes finer. This is due to the high correlation between random variables in adjacent elements resulting in numerical difficulties in the orthogonalization of the covariance matrix, as discussed earlier. Even if there is convergence, it takes a larger number of iterations to achieve this as the mesh becomes finer, as shown in Table 7.6.

Example 7.6

Consider the portal frame shown in Figure 7.9. Six random parameters are considered: dead load, D; live load, L; wind load, W; modulus of elasticity, E; area of cross section, A; and moment of inertia, I. All three members of the frame have identical cross sections. The statistical description of the random parameters is given in Table 7.7. Note that this example considers non-Gaussian random fields. The limit state considered for FORM reliability analysis is

$$\text{Vertical deflection at midspan of the beam} \leq \frac{\text{Length}}{300}$$

Two random variables—wind load, W, and area of cross section, A—were found to have an insignificant influence on the reliability index during the first trial (considering all stochastic parameters as random variables), and are henceforth not considered in this discussion. Table 7.8 shows this result, along with the sensitivity indices corresponding to other random variables. The remaining four variables, D, L, E, and I, are considered as random fields one by one as in the previous example. Each member is divided

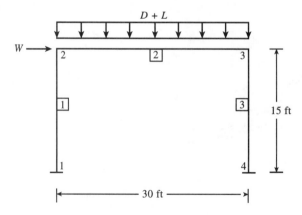

FIGURE 7.9. Portal Frame

Table 7.7. Portal frame: description of random variables

Variable	Units	Nominal Value	Mean Nominal Value	Coefficient of Variation	Type of Distribution
D	kip/ft	3.0	1.05	0.10	Lognormal
L	kip/ft	1.1	1.00	0.25	Type I
W	k	6.5	0.78	0.37	Type I
A	in^2	19.7	1.00	0.05	Lognormal
I	in^4	954.0	1.00	0.05	Lognormal
Z	in^3	130.00	1.00	0.05	Lognormal
E	ksi	29,000.00	1.00	0.06	Lognormal
F_y	ksi	36.0	1.05	0.10	Lognormal

Note. The information in the last three columns is based on Ellingwood et al. (1980).

into four elements for random field discretization. The results are shown in Figure 7.10. Unlike the cantilever beam in Examples 7.3 and 7.4, all four variables have comparable values of sensitivity indices. Therefore, the change in reliability index due to the consideration of each additional variable as a random field is significant. Thus, all four quantities merit treatment as random fields.

As in the previous example, random field element meshes are refined for each of the four parameters, and the results are shown in Figures 7.11 to 7.14. It is observed that mesh refinement has no significant effect beyond $\theta = 0.5L$ for D and L, and beyond $\theta = L$ for E and I. That is, for D and L, random field element size $= 0.5L$ appears to be sufficient beyond $\theta = 0.5L$. For E and I, random field element size $= 0.5L$ appears to be sufficient beyond $\theta = L$. Thus, given the scale of fluctuation, there is a lower limit on the element size below which there is no improvement in results.

Compare the results of this example with those of Example 7.5. The results are qualitatively similar, but they also indicate that the actual relationship between scale of fluctuation and optimum random field element size depends on the structure and the random field considered. For smaller values of θ, the load variables D and L have a larger variation and therefore require smaller elements compared to the resistance variables E and I. In general, element size equal to the scale of fluctuation may be considered adequate.

Table 7.8. Reliability analysis of portal frame

Quantity	Units	Sensitivity Index	Initial Checking Point	Final Checking Point
D	kip/ft	−0.59	3.15	3.36
L	kip/ft	−0.51	1.10	1.32
W	k	0.00	5.07	4.76
A	in^2	0.01	19.7	19.66
E	ksi	0.48	29,000.00	27,942.39
I	in^4	0.39	954.00	930.23
$g(X)$			−0.16	0.00006
β				1.247

Note. Number of iterations $= 3$.

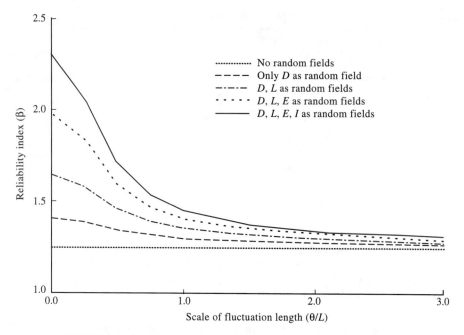

FIGURE 7.10. Portal Frame: Progressive Consideration of Random Fields

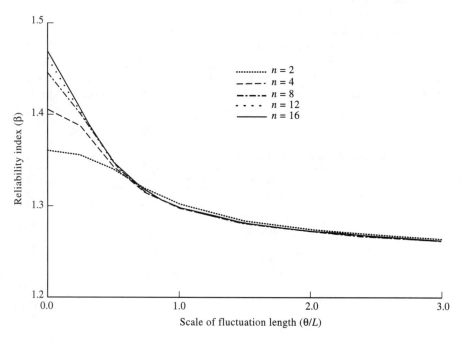

FIGURE 7.11. Portal Frame: Influence of Scale of Fluctuation and Random Field Discretization of *D*

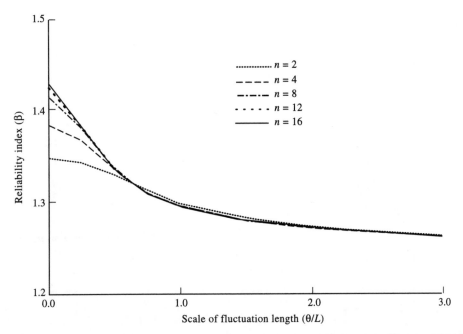

FIGURE 7.12. Portal Frame: Influence of Scale of Fluctuation and Mesh Refinement of Random Field L

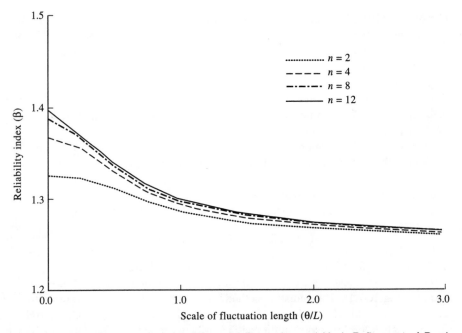

FIGURE 7.13. Portal Frame: Influence of Scale of Fluctuation and Mesh Refinement of Random Field E

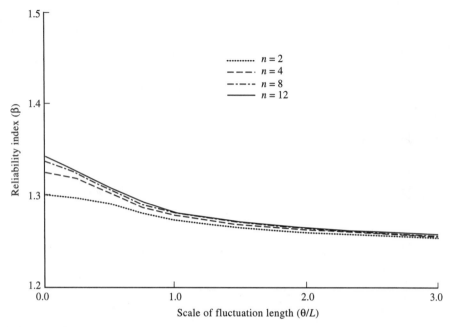

FIGURE 7.14. Portal Frame: Influence of Scale of Fluctuation and Mesh Refinement of Random Field *I*

Example 7.7

Consider a 24×24-in^2 steel plate, 1-inch thick, with a 4-inch-diameter circular hole. One quarter of the plate is shown in Figure 7.15, with a finite element mesh. The plate is acted upon by distributed edge loads as shown. The stochastic parameters for this problem are Young's modulus, E; thickness of the plate, t; and the two distributed edge loads, W_1 and W_2. Their statistical description is given in Table 7.9. Note that the distributed parameters need to be represented as two-dimensional random fields. What is the appropriate random field element mesh to discretize the random field E (Young's modulus) into random variables?

In this problem, isoparametric 4-node quadrilateral elements are used for the finite element analysis. Deterministic analysis with the finite element mesh using $W_1 = W_2 = 12.0$ kip/in results in the computation of the stress σ_{xx} at node 31 (Figure 7.15) to be equal to 24.075 ksi, which agrees very closely with the theoretical value of 24.0 ksi.

The following strength limit state is considered for the reliability analysis:

$$\sigma_{xx} \text{ at node } 31 \leq F_2, \text{ the yield strength}$$

The yield strength is considered to be deterministic, equal to 36.0 ksi.

Four different meshes are considered for the discretization of the random field E, as shown in Figure 7.15. The midpoint method is used in this example for the discretization of the random field. The random field, E, is assumed to have an exponential correlation function. Therefore, the correlation coefficient between any two random variables E_i and E_j obtained by the discretization of E is computed using

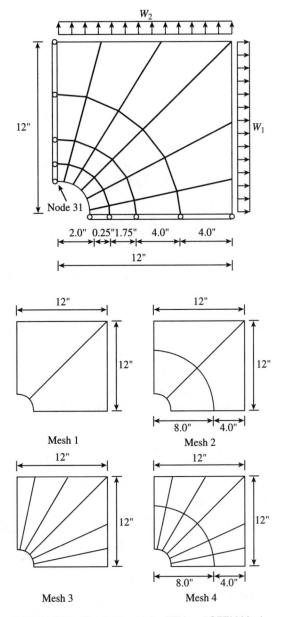

FIGURE 7.15. Plate with a Hole: FEM and SFEM Meshes

$$\rho(E_i, E_j) = \exp\left[-\left(\frac{\Delta x}{\lambda_x L}\right)^2 - \left(\frac{\Delta y}{\lambda_y L}\right)^2\right] \qquad (7.46)$$

where Δx and Δy are the differences between the x and y coordinates of the midpoints of the elements i and j in the random field discretization, $L = 12$ in, and $\lambda_x L$ and $\lambda_y L$ are the correlation lengths in the x and y directions. λ_x and λ_y are dimensionless

Table 7.9. Plate with a hole: description of random variables

Variable	Units	Mean	Coefficient of Variation	Type of Distribution
W_1	kip/in	12.00	0.20	Normal
W_2	kip/in	12.00	0.20	Normal
E	ksi	29,000.00	0.10	Normal
t	in	1.00	0.20	Normal

constants; they characterize the length over which the autocorrelation function decays to a prespecified value. In this example, it is assumed that $\lambda_x = \lambda_y = \lambda$, which implies an isotropic correlation structure.

Different values are assumed for λ, and their influence on the results obtained from the various random field meshes is examined. Table 7.10 shows the results of the reliability analysis, considering as a first step all the stochastic parameters as random variables only. As illustrated in Examples 7.3 and 7.4, the examination of the sensitivity indices for the various parameters in this step should determine which distributed parameters are to be considered as random fields. However, for the sake of this example, only one parameter, E, is modeled as a random field.

The results of reliability analysis with different random field element meshes are shown for different values of λ in Table 7.11. The first observation is that the modeling of E as a random field has resulted in higher values of β than that in Table 7.10. Second, the observation along the rows of Table 7.11 shows that for the same random field element mesh, as the value of λ decreases, the value of β increases. From Equation 7.46, it is apparent that as the correlation length decreases, the decay of correlation becomes rapid at short distances. Thus, the reliability of the structure is high for a weakly correlated random field (i.e., low correlation length). On the other hand, the representation of a random field by a single random variable implies the assumption of a perfectly correlated random field, i.e., $\lambda \to \infty$, and results in the determination of a lower value of the reliability index. The reliability of a structure would be lower if a stochastic parameter had greater variability than if the same parameter had lower variability. This leads to the conclusion that the modeling of a stochastic parameter as a random variable tends to overrepresent its variability, whereas its modeling as a random field underrepresents its variability, especially with the decrease in correlation length.

Table 7.10. Reliability analysis of plate with a hole

Variable	Sensitivity Index	Initial Checking Point	Final Checking Point
W_1	0.16	12.00	12.78
W_2	−0.05	12.00	11.74
E	−0.67	29,000.00	25,700.00
t	−0.72	1.00	0.87
Performance function		0.331	0.048
Reliability index		0.263	1.745

Note. Number of interations = 13.

Table 7.11. Plate with a hole: influence of correlation length and random field discretization

| | Correlation Length Parameter, λ | | | |
	1.0	0.75	0.50	0.25
	β values			
Mesh 1	2.380	2.819	3.925	
Mesh 2	2.602	3.056	3.344	
Mesh 3	Numerical instability	2.700	3.222	Very slow convergence ($>$100 interations)
Mesh 4	Numerical instability	Numerical instability	2.775	

The observation along the columns in Table 7.11 shows that for larger correlation lengths, numerical instability occurs earlier in the process of stochastic mesh refinement. This is because larger correlation length implies higher correlation coefficients between the random variables in elements close to each other, resulting in numerical instability in the orthogonalization of the covariance matrix. As the correlation length decreases, the algorithm is stable for more refined meshes. However, for $\lambda = 0.25$ the algorithm showed very slow convergence (more than 100 iterations) even for the first mesh; therefore, computations for this case were not pursued further.

The observation of the reliability indices for the various meshes for $\lambda = 0.5$ shows that as the number of elements increases, the reliability index decreases. However, the results for other values of λ show no such trend. It should be noted that the correlation coefficient between the variables in any two elements is dependent on the quantities $\Delta x / \lambda$ and $\Delta y / \lambda$. The matrix of correlation coefficients is thus different for different combinations of the correlation length and the random field element mesh. The reliability index, β, appears to be sensitive to these combinations in the present problem.

Thus, this numerical example with a two-dimensional random field demonstrates that the correlation characteristics of the random field have a strong influence not only on the estimated β values but also on the convergence and numerical feasibility of the reliability analysis for various meshes. The choice of the appropriate random field element size appears to be dependent on the structure. Therefore, it is necessary to have such investigation for particular types of structures to enable the choice of an appropriate mesh for random field discretization.

7.7 CONCLUDING REMARKS

This chapter considered problems with uncertain quantities that have spatial variability. Such quantities may be represented as random fields, which are then discretized into sets of correlated random variables for the sake of SFEM-based reliability analysis.

Several practical issues in the implementation arise in the use of random fields in reliability analysis to consider trade-offs between accuracy and efficiency. Two issues are presented: (1) the selective consideration of a few stochastic parameters as random fields, and (2) the choice of the appropriate mesh for random field discretization. The sensitivity indices for the stochastic parameters are very useful in selectively modeling only a few of the distributed parameters as random fields, resulting in computational efficiency while ensuring sufficient accuracy.

The examples in this chapter all considered structures with linear elastic behavior. The displacement-based finite element method was used for structural analysis. However, the concepts of random field representation and discretization are quite general and are equally applicable to nonlinear structural analysis and to other finite element analysis methods.

8

SFEM-BASED RELIABILITY EVALUATION OF NONLINEAR TWO- AND THREE-DIMENSIONAL STRUCTURES

8.1 INTRODUCTORY COMMENTS

Safety evaluation of frames considering their realistic behavior is becoming an important challenge to the engineering profession. A properly designed frame is expected to behave linearly when subjected to design loads. This was discussed in Chapter 6. However, even when the load level is low, assuring linear elastic behavior, it may not behave linearly due to the presence of partially restrained (PR) connections and less than ideal support conditions.

Most structures consist of many structural elements that are joined together by various types and forms of connections. As discussed in Section 8.2.3, for steel structures these connections are usually modeled as fully restrained (FR) and partially restrained (PR) or semirigid. Despite these classifications, almost all steel connections used in practice are, in fact, essentially PR connections with different rigidities.

Furthermore, the structures are supported on foundations. For analysis purposes, the support conditions are usually idealized as fixed, pinned, or roller. However, in reality, they are usually not perfectly fixed or pinned. They can be considered to be partially fixed with different rigidities.

If the load level is high enough to force the frame to behave inelastically, an evaluation of the frame's safety must consider the geometric and material nonlinearities in addition to the nonlinearities resulting from the PR connections and support conditions. If the failure of the frame is imminent, these sources of uncertainty cannot be avoided. Depending on the load combination used to estimate the failure probability, the governing performance of interest, i.e., strength or serviceability, could be different if the different sources of nonlinearities are considered appropriately. In spite of significant advances in analytical capabilities and design philosophies, the realistic safety evaluation of frames has yet to be undertaken comprehensively, and needs further attention from the profession.

To evaluate the safety of complicated frames in the presence of different sources of nonlinearity, a finite element-based formulation is desirable, since it is also the first step in a conventional deterministic analysis. In this way, different sources of nonlinearity in the structural systems can be incorporated easily. As is described in detail later, the deterministic finite element method (FEM) consists of an iterative procedure to capture the nonlinear behavior of the structure. The efficiency of the deterministic FEM is important for the success of the stochastic finite element method (SFEM). As discussed in Chapters 6 and 7, since the proposed SFEM algorithm is based on tracking the uncertainty propagating through the steps of deterministic analysis, this efficiency is very important.

Deterministic evaluation of the nonlinear responses of frame structures using either displacement-based or stress-based FEMs has been discussed extensively in the literature. In displacement-based FEM, shape functions are used to describe the displacements at the node of the elements. The nature of these formulations requires a large number of elements to model a member with large deformation. The need for a large number of elements makes this approach computationally uneconomical. Although explicit formulations can be obtained to evaluate the integrals of the stiffness matrix, the drawback of using a large number of elements is still unavoidable. Conceptually, the most efficient way to analyze a nonlinear problem is to use fewer elements and express the tangent stiffness matrix explicitly.

The discussions in Chapters 6 and 7 utilize displacement-based FEM. The unified SFEM algorithm proposed here does not require that a specific type of deterministic FEM be used. The assumed stress-based FEM can also be used in the algorithm, and is the subject of this chapter.

It has been reported extensively in the literature that the assumed stress-based FEM has several advantages over the displacement-based FEM, especially for frame structures. In the assumed stress-based FEM, the tangent stiffness can be expressed in explicit form, the stresses of an element can be obtained directly, fewer elements (sometimes as low as half the elements required by the other method) are required in describing a large deformation configuration, and integration is not required to obtain the tangent stiffness. It is very accurate and efficient in analyzing the nonlinear responses of frame structures.

Since SFEM-based study is expected to be computationally more demanding than conventional deterministic FEM, the desirable features of the assumed stress-based FEM need to be exploited to the fullest extent. An efficient SFEM is discussed in this chapter using the assumed stress-based FEM and considering the nonlinearities due to geometry, material, partially restrained (PR) connections, and support conditions. The discussion is specifically applicable to steel frame structures, but the method can be easily modified to study other types of structures. The theoretical background and the application of this method is discussed in detail for two-dimensional (2D) and three-dimensional (3D) structures later in this chapter.

8.2 SOURCES OF NONLINEARITY

The purpose of this section is to briefly introduce the readers to the sources of nonlinearity and the need to incorporate them in any structural analysis. The discussion is not exhaustive but is necessary to formulate the nonlinear SFEM concept discussed later.

It is important to identify the sources of nonlinearity in a typical frame structure. Since steel frames are made of slender members and the ends are connected differently than in frames made of concrete, additional sources of nonlinearity are expected to be present in steel frames. In addition to the geometric and material nonlinearities, the major sources of nonlinearity expected to be present in a typical steel frame are the flexibility in the connections and the modeling of support conditions. All these nonlinearities will cause the configuration of a structure to deviate quite noticeably from its undeformed configuration. In such cases, linear analysis is obviously inappropriate, and nonlinear structural analysis should be used. Each source of nonlinearity is discussed very briefly below.

8.2.1 Geometric Nonlinearity

Nonlinear behavior can be produced from changes in the geometry of the frame, and is usually referred to in the literature as the geometric nonlinearity. The effects of geometric nonlinearity give rise to the secondary moments, which are caused by the axial force acting either through the lateral displacement of the member relative to its chord $(P - \delta)$ or through the relative lateral displacement of the two ends of the member $(P - \Delta)$, and the finite rigid body deformation with small to moderate relative rotation of a member.

In general, a frame structure can be represented by beam–column elements, that is, elements subjected to both bending moment and axial load simultaneously. Realistic representation of their behavior is the essence of nonlinear analysis. Either the beam–column approach or the finite element concept can be used to formulate a geometrically nonlinear beam–column element. In the beam–column approach, a local coordinate is usually used as the member coordinate in developing the element stiffness matrix. The member deformations and joint displacements are separated by attaching a set of local coordinates to the member in the deformed position. The nonlinear behavior of an element can be represented by updating its tangent stiffness. In this approach, numerical integration is usually needed to obtain the tangent stiffness matrix; however, with symbolic manipulation procedures, an explicit form of the approximate tangent stiffness matrix can be obtained.

In the finite element approach, the effect of geometric nonlinearity is accounted for by inclusion of higher order terms in the strain–displacement relationship. Conventionally, either the total Lagrangian (TL) or updated Lagrangian (UL) formulation is used in this approach. In the total Lagrangian (TL) formulation, all static and kinetic variables are referred to the initial configuration at time zero. In the updated Lagrangian (UL) formulation, the same stress and strain decomposition as in the TL formulation are employed, but all variables are referred to the last known configuration. When consistently developed, the response prediction using the TL or UL method should have the same results. Both formulations were reviewed by Bathe and Bolourchi (1979), who found that the UL formulation was computationally more effective.

The mathematical aspect of the deterministic FEM formulation to consider geometric nonlinearity is discussed in Section 8.3.1 for 2D problems, and in Section 8.12.2 for 3D problems.

8.2.2 Material Nonlinearity

Material nonlinearity arises when yielding occurs or if the stress–strain behavior exhibits a nonlinear constitutive relationship. For steel structures, material nonlinearity arises when yielding spreads through the cross section (plastification) and along the member length (plastic zone) as the moment in the cross section increases from the initial yield moment M_y to the full plastic moment M_p. Depending on the desired degree of accuracy, two models are available to consider the material nonlinearity in steel frame analysis. The first, known as the concentrated plasticity (plastic hinge) model, ignores the progressive yielding that takes place in the cross section and in the member. The second and more sophisticated model, known as the distributed plasticity (plastic zone) model, takes into consideration the spread of yield in the cross section and along the member length. Considering common engineering practice, only the plastic hinge model is used here.

The essential computational problem of material nonlinearity is that equilibrium equations must be written using material properties that depend on strains, but the strains are not known in advance. The key to solving material nonlinearity problems is to properly define a constitutive relationship and the yield criterion in tracing the stress–strain path. In the analysis of steel structures, the three most common assumptions for the material behavior are elastic–perfectly plastic, isotropic strain hardening, and kinematic strain hardening models. Among these models, the elastic–perfectly plastic material behavior is used widely for steel structures in the engineering profession because of its simplicity. Elastic–perfectly plastic behavior is characterized by an initial elastic material response, on which a plastic deformation is superimposed after a certain level of stress has been reached. In essence, plastic deformation is designated by an irreversible plastic strain that is not time-dependent and that can be sustained only once a certain level of stress has been reached. In this chapter, the topic is restricted to elastoplastic material nonlinearity.

Various yield criteria have been suggested for metals. The simple and frequently used yield criteria are the Tresca criterion and Von Mises criterion (Owen and Hinton, 1980). The Tresca yield criterion states that yielding begins when the maximum shear stress reaches the value of the maximum shear stress occurring under simple tension. Von Mises suggested that yielding occurs when the distortion energy equals the distortion energy at yield in simple tension. The Von Mises concept in the form of interaction equations is used in this study, and is discussed in Sections 8.7.1 and 8.14.1.

The mathematical aspects of the deterministic FEM formulation to consider material nonlinearity are discussed in Section 8.3.2 for 2D problems, and in Section 8.12.3 for 3D problems.

8.2.3 Flexibility of Connections

Nonlinear structural behavior can be produced by the flexibility of connections. In the past, the American Institute of Steel Construction (AISC), representing the steel industry in the United States, defined three basic types of connections:

1. Type I, commonly designated as rigid frame (continuous frame), assumes that beam-to-column connections have sufficient rigidity to hold the original angles between intersecting members virtually unchanged.
2. Type II, commonly designated as simple frame (unrestrained, pinned), assumes that the ends of the beams and girders are connected for shear only, and are free to rotate under the gravity load.
3. Type III, commonly designated as semirigid or partially restrained (PR), assumes that the connections of beams and girders possess a quantifiable and known moment capacity intermediate in degree between the rigidity of Type I and the flexibility of Type II.

Currently, the Load and Resistance Factor Design (LRFD) code of the AISC (1994) suggests two types of models for connections: fully restrained (FR) and partially restrained (PR). Conventional analysis and design of steel frames is based on the assumption that beam-to-column connections are either fully restrained or perfectly pinned (PP). These two connection models give the upper and lower bounds for the effect of the connection flexibility on the structural response. Despite these assumptions, almost all steel connections used in practice are, in fact, essentially partially restrained connections with different rigidities. It has been established both theoretically and experimentally that these connections exhibit nonlinear response characteristics, even when the applied loads are very small. The concept of FR or PP connections is really just an assumption made to simplify calculations, and is a major weakness in current analytical procedures. In reality, FR connections possess some flexibility and PP connections possess some rigidity. The consideration of connection rigidity is essential in the analysis and design of large deformed frame structures. Although much research work has been conducted on the moment–rotation behavior of steel framing connections, only a few attempts have been made to incorporate connection deformation into the structural analysis of a frame.

The properties of a flexible connection are generally represented by a relationship between the applied moment and the relative rotation of the connection, and are commonly represented by M–Θ curves, as shown in Figure 8.1. Since connections are complicated, consisting of several elements in which there are many contact points where the yielding and/or local buckling usually occurs, modes of contact may change during deformation, and they behave nonlinearly even when the applied moment is small. Furthermore, since it is very difficult to represent a connection using any accurate analytical model, experimental results are usually used to model the flexibility of a connection. Several important static test results for connection rigidity have been reported in the literature and can be used as the basis for analysis. Some typical moment-relative rotation (M–Θ) curves obtained from experiments are shown in Figure 8.2. The corresponding connection geometries are shown in Figure 8.3. It can be seen that all moment–rotation curves are nonlinear right from the start of loading. This is discussed further in Section 8.3.3.

8.2.4 Support Conditions

The modeling uncertainty associated with support conditions is a major source of uncertainty, which has not received the attention it deserves. Structures are supported on foundations. For analysis purposes, the support conditions are usually idealized as fixed, pinned, or roller. In reality, they are usually not perfectly fixed or pinned. If the supports are not

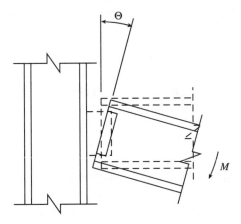

FIGURE 8.1. Moment and relative rotation of a PR connection.

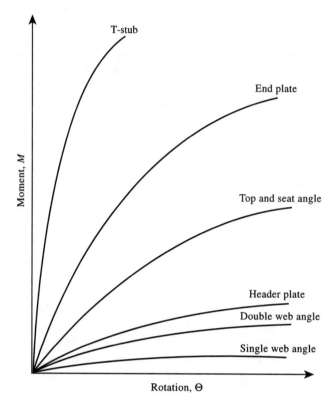

FIGURE 8.2. M–Θ curves for commonly used connections.

perfectly fixed, they will introduce different amounts of flexibility and the design load effects in terms of location and magnitude will be changed. Most importantly, the lateral stiffness of the structure will be reduced. This may cause the controlling performance limit state of the structure to be lateral deflection rather than the strength limit state.

It is expected that a more realistic modeling of the structural support conditions would considerably increase the amount of nonlinearity and uncertainty in the problem. Supports can be modeled as partially restrained with different rigidities. Thus, the discussion on the modeling of PR connections is also applicable to the modeling of supports.

8.3 DETERMINISTIC NONLINEAR FEM FORMULATION (TWO-DIMENSIONAL PROBLEMS)

8.3.1 Geometric Nonlinearity

The essential steps of the deterministic nonlinear FEM required for the proposed SFEM are described very briefly in the following sections. The details of the assumed stress-based nonlinear FEM formulation can be found in the literature (Haldar and Nee, 1989; Kondoh and Atluri, 1987; Shi and Atluri, 1988, 1989). For the analysis of elastic, geometrically nonlinear beam–column elements, the following assumptions are made:

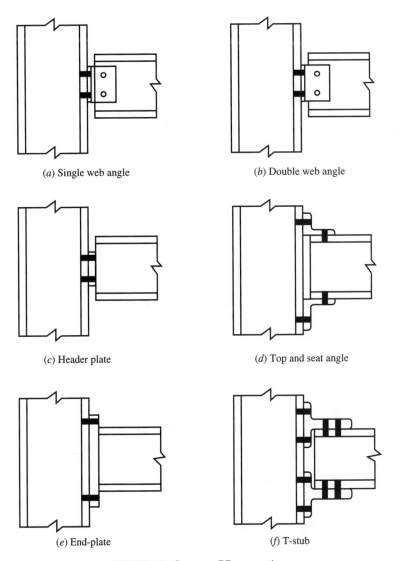

(a) Single web angle

(b) Double web angle

(c) Header plate

(d) Top and seat angle

(e) End-plate

(f) T-stub

FIGURE 8.3. Common PR connections.

1. The transverse section of an element remains plane during the deformation, and the shear deformation and the warping effects are considered to be negligible.

2. The element can undergo arbitrary large, rigid rotation but small axial stretch and relative rotation.

3. The material is assumed to be linearly elastic.

A prismatic 2D beam–column element with initial length l lying along the X_1 axis and deforming in the X_1, X_2 plane is shown in Figure 8.4. X_i ($i = 1, 2,$ and 3) denotes the global coordinates and $\hat{X}_i$ ($i = 1, 2,$ and 3) denotes the deformed local coordinates.

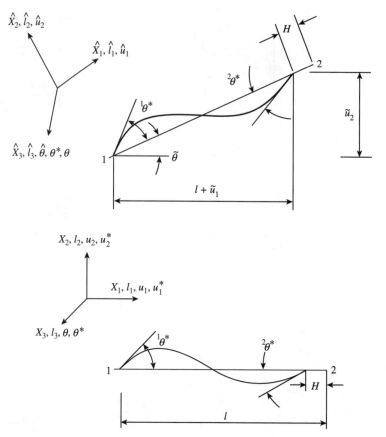

FIGURE 8.4. Kinematic relationships between deformed local and global displacements for a 2D beam–column element.

Figure 8.4 shows the kinematic relationship between the deformed local and global displacements of a 2D beam–column element.

Assuming that the large deformation of a 2D beam–column element subjected to both axial force and bending moment can be considered as the superposition of components (Mallett and Marcal, 1968) with arbitrary large rigid rotation and small axial stretch and relative rotation, the total rotation angle, θ, at each node of the element can be expressed as

$$\theta = \theta^* + \tilde{\theta} \tag{8.1}$$

where θ^* is the small relative rotation, and $\tilde{\theta}$ is the arbitrary large rigid rotation. Under the assumption of small relative rotation, the total axial deformation of the member H can be shown to be approximately

$$H \approx \left[(l + \tilde{u}_1)^2 + (\tilde{u}_2)^2\right]^{1/2} - l \tag{8.2}$$

where l is the original length of a member, $\tilde{u}_1$ and $\tilde{u}_2$ are the relative displacements of an element in the X_1 and X_2 global coordinates, respectively, that is, $\tilde{u}_i = {}^2u_i - {}^1u_i$, and ${}^2u_i$ and ${}^1u_i$ are the nodal displacements of the two nodes of an element in the X_i global coordinate. All these parameters are shown in Figure 8.4.

The rigid rotation $(\tilde{\theta})$ can be expressed as the function of the relative nodal displacements in the X_1 and X_2 coordinates as

$$\tan \tilde{\theta} = \frac{\tilde{u}_2}{l + \tilde{u}_1} \tag{8.3}$$

Referring to Figure 8.5, the element's axial force, $\hat{N}$, and moment, $\hat{M}$, can be shown to be

$$\left\{ \begin{matrix} \hat{N} \\ \hat{M} \end{matrix} \right\} = \begin{bmatrix} 1 & 0 & 0 \\ 0 & 1 - \dfrac{\hat{X}_1}{l} & \dfrac{\hat{X}_1}{l} \end{bmatrix} \left\{ \begin{matrix} n \\ m_1 \\ m_2 \end{matrix} \right\} \tag{8.4}$$

where n is the nodal axial force, and m_1 and m_2 are the moments at node 1 and 2 of the element, respectively. They can be expressed in matrix form as

$$\sigma = \{ n \quad m_1 \quad m_2 \}^{\text{t}} \tag{8.5}$$

The nodal forces for a 2D beam-element are shown in Figure 8.5.

The commonly used linear iterative strategy for solving nonlinear structural problems can be expressed as

$$\mathbf{K}^{(n)} \Delta \mathbf{D}^{(n)} = \mathbf{F}^{(n)} - \mathbf{R}^{(n-1)} \tag{8.6}$$

where $\mathbf{K}^{(n)}$, $\Delta \mathbf{D}^{(n)}$ and $\mathbf{F}^{(n)}$ are the global tangent stiffness matrix, the incremental displacement vector, and the external load vector of the nth iteration, respectively, and $\mathbf{R}^{(n-1)}$ is the internal force vector of the $(n - 1)$th iteration. It is important to formulate all the quantities in Equation 8.6 according to the assumed stress-based finite element method.

Referring to Figure 8.5, the tangent stiffness matrix of a 2D beam–column element can be expressed as

$$\mathbf{K} = \mathbf{A}_{\sigma d0}^{\text{t}} \mathbf{A}_{\sigma\sigma}^{-1} \mathbf{A}_{\sigma d0} + \mathbf{A}_{dd0} \tag{8.7}$$

where $\mathbf{A}_{\sigma\sigma}$ is the elastic property matrix and can be expressed as

$$\mathbf{A}_{\sigma\sigma} = \begin{bmatrix} \dfrac{l}{EA} & 0 & 0 \\ 0 & \dfrac{l}{3EI} & \dfrac{l}{6EI} \\ 0 & \dfrac{l}{6EI} & \dfrac{l}{3EI} \end{bmatrix} \tag{8.8}$$

where A is the cross-sectional area, E is the Young's modulus, and I is the moment of inertia of the cross section. $\mathbf{A}_{\sigma d0}^{\text{t}}$ is the transformation matrix and can be shown to be

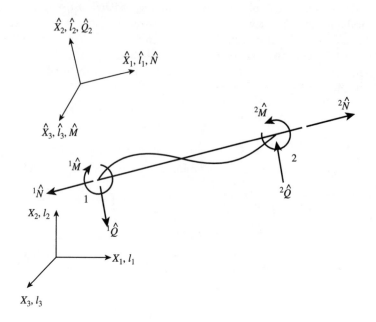

(a) Nodal forces of a 2D beam–column element

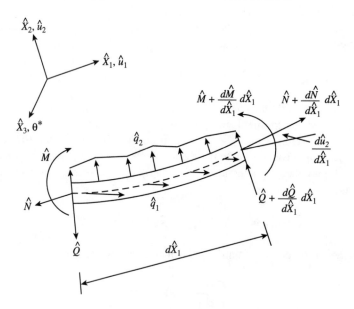

(b) Free-body diagram of a deformed 2D element

FIGURE 8.5. Nodal forces and the free-body diagram of a 2D beam–column element.

$$
\mathbf{A}^t_{\sigma d0} =
\begin{bmatrix}
-\cos\tilde{\theta} & \dfrac{\sin\tilde{\theta}}{l} & -\dfrac{\sin\tilde{\theta}}{l} \\[2ex]
-\sin\tilde{\theta} & -\dfrac{\cos\tilde{\theta}}{l} & \dfrac{\cos\tilde{\theta}}{l} \\[2ex]
0 & -1 & 0 \\[2ex]
\cos\tilde{\theta} & -\dfrac{\sin\tilde{\theta}}{l} & \dfrac{\sin\tilde{\theta}}{l} \\[2ex]
\sin\tilde{\theta} & \dfrac{\cos\tilde{\theta}}{l} & -\dfrac{\cos\tilde{\theta}}{l} \\[2ex]
0 & 0 & 1
\end{bmatrix}
\tag{8.9}
$$

$\mathbf{A}_{dd0}$ is the geometric stiffness matrix, which can be expressed as

$$
\mathbf{A}_{dd0} =
\begin{bmatrix}
c_1 & c_2 & 0 & -c_1 & -c_2 & 0 \\
c_2 & d_1 & 0 & -c_2 & -d_1 & 0 \\
0 & 0 & 0 & 0 & 0 & 0 \\
-c_1 & -c_2 & 0 & c_1 & c_2 & 0 \\
-c_2 & -d_1 & 0 & c_2 & d_1 & 0 \\
0 & 0 & 0 & 0 & 0 & 0
\end{bmatrix}
\tag{8.10}
$$

where

$$
c_1 = n\frac{(\sin\tilde{\theta})^2}{l} + \frac{m_2 - m_1}{l}\frac{-\sin\tilde{\theta}\cos\tilde{\theta}}{l}
\tag{8.11}
$$

$$
c_2 = n\frac{-\sin\tilde{\theta}\cos\tilde{\theta}}{l} - \frac{m_2 - m_1}{l}\frac{(\cos\tilde{\theta})^2}{l}
\tag{8.12}
$$

$$
d_1 = n\frac{(\cos\tilde{\theta})^2}{l} - \frac{m_2 - m_1}{l}\frac{\sin\tilde{\theta}\cos\tilde{\theta}}{l}
\tag{8.13}
$$

The internal force vector $\mathbf{R}$ in Equation 8.6 for a 2D beam–column element can be expressed as

$$
\mathbf{R} = -\mathbf{A}^t_{\sigma d0}\mathbf{A}^{-1}_{\sigma\sigma}\mathbf{R}_\sigma + \mathbf{R}_{d0}
\tag{8.14}
$$

where

$$
\mathbf{R}_\sigma =
\begin{Bmatrix}
\dfrac{nl}{EA} - \left\{[(l+\tilde{u}_1)^2 + (\tilde{u}_2)^2]^{1/2} - l\right\} \\[2ex]
\dfrac{1}{EI}\left(\dfrac{m_1 l}{3} + \dfrac{m_2 l}{6}\right) + \left[{}^1\theta - \tan^{-1}\left(\dfrac{\hat{u}_2}{l+\tilde{u}_1}\right)\right] \\[2ex]
\dfrac{1}{EI}\left(\dfrac{m_1 l}{6} + \dfrac{m_2 l}{3}\right) - \left[{}^2\theta - \tan^{-1}\left(\dfrac{\tilde{u}_2}{l+\tilde{u}_1}\right)\right]
\end{Bmatrix}
\tag{8.15}
$$

and

$$
\mathbf{R}_{d0} = \left\{
\begin{array}{c}
-n\cos\tilde{\theta} - \dfrac{m_2 - m_1}{l}\sin\tilde{\theta} \\[2mm]
-n\sin\tilde{\theta} + \dfrac{m_2 - m_1}{l}\cos\tilde{\theta} \\[2mm]
-m_1 \\[2mm]
n\cos\tilde{\theta} + \dfrac{m_2 - m_1}{l}\sin\tilde{\theta} \\[2mm]
n\sin\tilde{\theta} - \dfrac{m_2 - m_1}{l}\cos\tilde{\theta} \\[2mm]
m_2
\end{array}
\right\}
\tag{8.16}
$$

By substituting Equations 8.8, 8.9, 8.10, 8.15, and 8.16 into Equations 8.7 and 8.14, we can obtain the tangent stiffness matrix and the internal force vector in explicit form. Using the linear iterative solution algorithm of Equation 8.6, geometrically nonlinear problems can be solved efficiently. In this study, the modified Newton–Raphson method with arc-length procedure is used to solve Equation 8.6.

8.3.2 Elastoplastic Material Nonlinearity

The geometric nonlinearity and the material nonlinearity both need to be incorporated in the finite element formulation. As mentioned in Section 8.2.2, the elastoplastic material nonlinearity is used here for the mathematical formulation. Plastic hinges are assumed to form when the yield condition is satisfied. Thus, when the combined axial force and bending moments satisfy a prescribed yield function at a node of an element, a plastic hinge is assumed to occur instantaneously at that location. The yield function used in this study for a beam–column element has the following general form:

$$
f(\hat{N}, \hat{M}, \sigma_y) = 0 \qquad \text{at} \quad \hat{X}_1 = l_p
\tag{8.17}
$$

where $\hat{N}$ and $\hat{M}$ are the axial force and bending moment, respectively, σ_y is the yield stress of an element, and l_p is the location of a plastic hinge along the length of an element.

The presence of plastic hinges in a frame is expected to increase the axial and rotational deformation. The elastoplastic tangent stiffness, $\mathbf{K}_p$, and the elastoplastic internal force, $\mathbf{R}_p$, of the element can be shown to be (Kondoh and Atluri, 1987)

$$
\mathbf{K}_p = \mathbf{K} - \mathbf{A}_{\sigma d0}^t \mathbf{A}_{\sigma\sigma}^{-1} \mathbf{V}_p \mathbf{C}_p^t \mathbf{A}_{\sigma d0}
\tag{8.18}
$$

and

$$
\mathbf{R}_p = \mathbf{A}_{\sigma d0}^t (\mathbf{A}_{\sigma\sigma}^{-1} \mathbf{V}_p \mathbf{C}_p^t - \mathbf{A}_{\sigma\sigma}^{-1}) \hat{\mathbf{R}}_\sigma + \mathbf{R}_{d0}
\tag{8.19}
$$

where

$$
\mathbf{V}_p = \left\{
\begin{array}{c}
-\dfrac{\partial f}{\partial \hat{N}} \\[10pt]
-\dfrac{\partial f}{\partial \hat{M}} \left(1 - \dfrac{\hat{X}_1}{l} \right) \\[10pt]
-\dfrac{\partial f}{\partial \hat{M}} \left(\dfrac{\hat{X}_1}{l} \right)
\end{array}
\right\}
\tag{8.20}
$$

$$
\mathbf{C}_p^t = (\mathbf{V}_p^t \mathbf{A}_{\sigma\sigma}^{-1} \mathbf{V}_p)^{-1} \mathbf{V}_p^t \mathbf{A}_{\sigma\sigma}^{-1}
\tag{8.21}
$$

and

$$
\hat{\mathbf{R}}_\sigma = \mathbf{R}_\sigma + \left\{
\begin{array}{c}
H_p \\[10pt]
\theta_p^* \left(1 - \dfrac{\hat{X}_1}{l} \right) \\[10pt]
\theta_p^* \left(\dfrac{\hat{X}_1}{l} \right)
\end{array}
\right\}
\quad \text{at} \quad \hat{X}_1 = l_p
\tag{8.22}
$$

In Equation 8.22, H_p and θ_p^* are the additional axial elongation and additional relative rotation due to the plastic hinges. All other parameters were defined earlier.

Considering either Equations 8.7 and 8.14 or Equations 8.18 and 8.19, depending on the elastic or inelastic state of the material, the effect of the material nonlinearity can be incorporated in the analysis. The basic simplicity of the algorithm is not lost even when the material and geometric nonlinearities are considered simultaneously.

The basic concept is that the assumed stress-based tangent stiffness matrix and the internal force vector can be developed in a conventional way using the geometric and material properties of an element. Since $\mathbf{K}$ and $\mathbf{R}$ can be expressed in explicit form, geometrically nonlinear problems can be solved efficiently using Equation 8.6. The algorithm has been verified extensively in the literature (Haldar and Nee, 1989; Kondoh and Atluri, 1987; Shi and Atluri 1988, 1989) in terms of its accuracy, efficiency, and robustness.

8.3.3 Partially Restrained or Flexible Connections

The PR connections need to be incorporated in the analysis at this stage. Connections can be modeled as structural elements that transmit forces between beams and columns. In general, these forces may be axial force, shear, torsion, or bending moment. For two-dimensional problems, the torsional effect on the connection deformation can be neglected. Furthermore, it has been shown that the effects of shear and axial forces are small in comparison to the bending moment, and can be neglected. Thus, only the effect of the bending moment needs to be considered.

The important effect of the bending moment is the relative rotation (Θ) of the connection. It represents the change in angle between the beam and the column from its original configuration. As discussed earlier, the comprehensive properties of the PR connection are represented by an $M-\Theta$ curve. For a deterministic analysis, there is a unique $M-\Theta$ curve for a connection. The $M-\Theta$ relationship for a particular connection can be obtained from experiments, but test results are available only at some discrete points. To include the effect of connection flexibility in the analytical procedure, formulas that predict a continuous

M–Θ relationship must be used to properly model the connection. Several analytical expressions have been proposed to represent the M–Θ curve. These include the piecewise linear model (Razzaq, 1983), the polynomial model (Frye and Morris, 1975), the exponential model (Liu, 1985), the B-spline model (Jones et al., 1982), and the Richard model (Richard and Abbott, 1975; Ramberg and Osgood, 1943).

Based on the above discussion, it is clear that for nonlinear or limit state analysis, a linear model of the M–Θ curve is unacceptable. Some of the nonlinear representations of PR connections suggested in the literature are shown in Figure 8.6. The bilinear and piecewise linear models are simple to use, but the connection stiffness represented by these models can change abruptly, which may cause numerical difficulties.

The polynomial and exponential models are efficient and simple to use in practical problems, especially in dealing with complex structures with an analysis procedure using the finite element method. The polynomial model can be expressed in series form as

$$M(\Theta) = \sum_{i=0}^{m} \alpha_i \, f_i(\Theta) \tag{8.23}$$

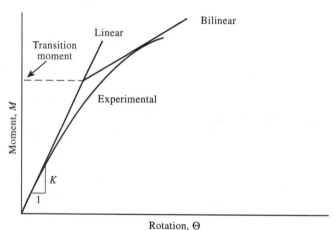

(a) Linear and bilinear models

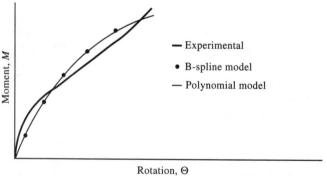

(b) B-spline and polynomials models

FIGURE 8.6. Analytical models for M–Θ curves.

where $M(\Theta)$ is the moment capacity of a connection as a function of Θ, m is the highest power of the polynomial (it can be of the order of up to $n - 1$, where n is the quantity of actual moment rotation data); and α_i's are the constants to be estimated by minimizing the total least-squares error. $f_i(\Theta)$ is a deterministic shape function. For a polynomial model, it can be expressed as

$$f_i(\Theta) = \Theta^i \qquad (8.24)$$

One form of exponential function commonly used to represent an $M-\Theta$ curve is

$$M(\Theta) = C_0 + \sum_{i=1}^{n} C_i (1 - e^{-\Theta/2\alpha_i}) + C_{n+1}\Theta \qquad (8.25)$$

where α is the scaling factor and C_i's are the connection model parameters. The constants C_i can be determined by curve-fitting to actual moment–rotation data. The polynomial model has the undesirable characteristic that negative connection stiffness may be encountered. The derivative of the exponential function represented by Equation 8.25 is always positive. That means no negative stiffness of the connection will be encountered at any stage of loading. However, it is easier and equally accurate to use a carefully chosen polynomial function.

The B-spline model is similar to the polynomial model, but a cubic polynomial model is used to fit the segments of a curve. A cubic function is used within each segment, so the first and second derivatives are maintained between adjacent segments. This method gives a very good approximation of the $M-\Theta$ curve shown in Figure 8.6. The main disadvantage of this model is that a large number of parameters are needed to define a curve.

The Richard model is also frequently used because of its generality and applicability to any type of connections, including those with strain-softening behavior. This model represents the observed experimental data very well and is convenient to implement in a computer program. Its four parameters are derived from a rational interpretation of response. The four-parameter Richard model can be expressed as

$$M(\Theta) = \frac{(k - k_p)\Theta}{\left(1 + \left|(k - k_p)\Theta/M_0\right|^N\right)^{1/N}} + k_p\Theta \qquad (8.26)$$

where k is the initial or elastic stiffness, k_p is the plastic stiffness, M_0 is the reference moment, and N is the curve shape parameter. These parameters are shown in Figure 8.7.

Two methods can be used to incorporate the flexible connections in the finite element analysis procedure: the connection element method and the matrix transformation method.

8.3.3.1 Method 1: Connection Element Approach

The connection element approach is equivalent to adding a rotational spring at the member's end. The stiffness of the structure is obtained using the standard assembly method by considering the presence of the spring. This will add the extra degrees of freedom due to the rotational spring. In this approach, the flexible connection is modeled as a kind of beam element. The difference between this beam element and ordinary beam elements of the structure is that the equivalent Young's modulus for the connection beam element is used instead of the Young's modulus for the ordinary beam element. The stiffness and the current Young's modulus

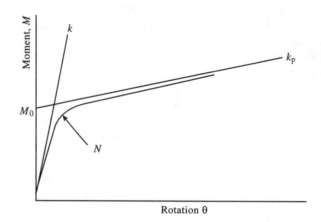

FIGURE 8.7. A typical M–Θ curve according to the Richard model.

of an element are functions of the relative rotation Θ; thus, they need to be continuously updated.

The stiffness of a connection can be represented by the following expression

$$k(\Theta) = E(\Theta)\frac{I}{l} \tag{8.27}$$

where $E(\Theta)$ and $k(\Theta)$ are the current Young's modulus and the current stiffness as a function of Θ, and I and l are the moment of inertia and length of a connection beam element, respectively. $k(\Theta)$ is the slope of the M–Θ curves, and can be expressed as

$$k(\Theta) = \frac{dM(\Theta)}{d\Theta} \tag{8.28}$$

For the polynomial model, using Equation 8.23, $k(\Theta)$ can be shown to be

$$k(\Theta) = \sum_{i=0}^{m} \alpha_i \frac{df_i(\Theta)}{d\Theta} \tag{8.29}$$

For the exponential model, using Equation 8.25, $k(\Theta)$ becomes

$$k(\Theta) = \left(\sum_{i=1}^{n} \frac{C_i}{2\alpha_i} e^{-\Theta/2\alpha_i} \right) + C_{n+1} \tag{8.30}$$

For the Richard model, using Equation 8.26, $k(\Theta)$ can be shown to be

$$k(\Theta) = \frac{k - k_{\mathrm{p}}}{\left[1 + \left| \frac{(k - k_{\mathrm{p}})\theta}{M_0} \right|^{N} \right]^{(N+1)/N}} + k_{\mathrm{p}} \tag{8.31}$$

Thus, knowing $k(\Theta)$, I, and l, the current Young's modulus $E(\Theta)$ can be evaluated by Equation 8.27. By substituting $E(\Theta)$ for E in Equation 8.8 for the connection element, the tangent stiffness matrix of the structure can be obtained at every iteration step. This approach is used here.

8.3.3.2 *Method 2: Matrix Transformation Approach*   This method considers the elements of flexibly connected frames as a kind of beam–column element with rotational springs at the ends. It generates the stiffness matrix of an element with a flexible connection. Then the static condensation method is used to obtain a modified element stiffness matrix that incorporates the additional stiffness contributed by the rotational spring.

The same M–Θ relationships that are used for the connection element method can be used here, and the connection rigidity can be expressed as in Equation 8.28. The difference between this method and the connection element method is that the tangent stiffness is reformulated considering the presence of rotational springs instead of just changing the equivalent Young's modulus. This method can be tedious but does not increase the number of degrees of freedom.

Using the assumed stress method, the tangent stiffness matrix for the beam–column–spring element can be expressed as (Shi and Atluri, 1989)

$$\mathbf{K} = \mathbf{A}^t_{\sigma d0}\tilde{\mathbf{A}}^{-1}_{\sigma\sigma}\mathbf{A}_{\sigma d0} + \mathbf{A}_{dd0} \tag{8.32}$$

$\mathbf{A}_{\sigma d0}$ and $\mathbf{A}_{dd0}$ are the same as in Equations 8.9 and 8.10, respectively, and the matrix $\tilde{\mathbf{A}}^{-1}_{\sigma\sigma}$ is

$$\tilde{\mathbf{A}}^{-1}_{\sigma\sigma} = \begin{bmatrix} \dfrac{EA}{l} & 0 & 0 \\ 0 & \xi(2+b) & -\xi \\ 0 & -\xi & \xi(2+a) \end{bmatrix} \tag{8.33}$$

where

$$\xi = \frac{6EI}{l[(2+a)(2+b)-1]} \tag{8.34}$$

and a and b are modification terms of the rotational spring at the ends. They can be expressed as

$$a = \frac{6EI}{^1K(\Theta)l} \tag{8.35}$$

and

$$b = \frac{6EI}{^2K(\Theta)l} \tag{8.36}$$

Therefore, two types of element stiffness matrices are needed in the formualtion of the global stiffness matrix: an element without the rotational spring, and an element with the rotational spring. For the element without the rotational spring, the stiffness matrix can be obtained using Equation 8.7. For the element with rotational springs at the ends, Equation 8.32 should be used. The element stiffness matrix has to be updated in every iteration step corresponding to the instantaneous M–Θ relationship.

As mentioned earlier, the connection element approach is used here. It must be emphasized that the procedure to obtain the tangent stiffness does not change even with the existence of flexible connections. The only difference is that the Young's modulus of the beam–column element used to represent the PR connection needs to be updated at each iteration. The tangent stiffness of the structure with PR connections can still be expressed in explicit form. With this method, the degrees of freedom of a structure are increased by the number of PR connections added, but this is not a great disadvantage if a computer is used. The advantage, of course, lies in the fact that it is no longer necessary to compute the modified fixed-end moments and the modified member stiffness matrices.

8.3.4 Support Conditions

As mentioned earlier, support conditions can be modeled as PR connections with different rigidities. When the Richard model and the connection element approach are used, the support conditions can be easily incorporated in the formulation, so the discussion on PR connections is also applicable to support conditions. Thus, a deterministic finite element model that considers nonlinearities due to geometry, material, PR connections, and support conditions is now available.

8.4 GENERAL STRATEGY TO SOLVE NONLINEAR STRUCTURAL PROBLEMS

No matter which type of nonlinearity is being considered, a set of nonlinear equations has to be solved in the context of the finite element method. Similar solution techniques can be applied to different types of nonlinearities. The basic concept behind numerical solution of nonlinear equations is linearization. In the linearization procedure, the nonlinear problem is changed to a set of linear problems. This is because the deformed geometry of the structure is not known during the formulation of the equilibrium and kinetic relationships. The deformed geometry of the structure obtained from a preceding step or cycle of calculations is used as the basis for formulating the equilibrium and kinetic relationships for the current cycle of calculations. The stiffness matrix must be updated at every cycle of calculations to take the nonlinear effects into account. Since the equilibrium configuration of the structure changes at every cycle of calculations, the solution of a nonlinear problem is obtained using a series of linearized analyses.

Different algorithms have been proposed to obtain approximate linear solutions to nonlinear problems. The two most commonly used nonlinear solution techniques for a finite element-based approach are the increment procedure and the iteration procedure. In the increment procedure, the load acting on a deformable body can be considered as being applied in increments until a desired level is reached. This method is very simple to implement and is computationally fast. If the incremental step chosen is small enough, the solution may be convergent. However, this method is known to drift from the true solution, and it is impossible to estimate the error from the true solution since no attempt is made to ensure that a particular solution is in equilibrium.

Iteration methods include the Newton–Raphson, modified Newton–Raphson, and simplified Newton–Raphson methods. The advantages of the iteration method are that it is simple to use and the equilibrium is checked at every iteration step. In spite of its many desirable features, it suffers from severe disadvantages when the stiffness matrix deteriorates. Numerical difficulties may be encountered and convergence may not be achieved for problems that exhibit limit point, snapthrough, and snapback behavior. To circumvent this problem, the modified arc-length method can be used. The arc-length method was originally developed by Riks (1979) and Wempner (1971), and later modified by Crisfield (1983) and Ramm (1981). When the arc-length method is used in the Newton–Raphson algorithm, the iteration procedure for solving nonlinear equations is greatly improved.

8.5 UNCERTAINTY ANALYSIS

The modeling of uncertainties in a problem is the main objective of this book. It is discussed in the following sections.

8.5.1 Uncertainty in Basic Random Variables

The uncertainties associated with basic random variables, that is, the load- and resistance-related variables, have been studied extensively in the literature. Several procedures with various degrees of complexity were described by Haldar and Mahadevan (2000). The uncertainty in most of the random variables to be used in subsequent discussions has already been quantified and documented. No additional discussion is necessary at this time. The uncertainty in random variables is identified as needed in the examples considered later. In general, once the uncertainty in a variable is quantified, the information can be used in the subsequent analysis.

8.5.2 Uncertainty in PR Connections and Support Conditions

The quatification of uncertainties in the PR connections and the support conditions is relatively new and needs some additional discussion here. Experimental data are always dependent on test conditions, and the $M-\Theta$ relationship obtained from such test results is generally accepted in a statistical sense. Obviously, the uncertainties in the $M-\Theta$ relationship need to be explicitly addressed in any acceptable analytical procedure. To incorporate the actual behavior of frame structures with PR connections or support conditions or both into the design and analysis, the nonlinearity and the randomness associated with it should be considered appropriately.

A considerable amount of uncertainty is expected in $M-\Theta$ curves. The uncertainty in the $M-\Theta$ curve comes from two major sources: the uncertainty arising during the fabrication of the connections and the supports, and the modeling uncertainty. The fabrication uncertainty is due to various random factors involved during the manufacturing and assembly processes. The fabrication of a connection or support is a complex process. The main steps of such a procedure include fabricating the components into required forms, putting them in place, and assembling them by bolting or welding. For large structures, it is impossible to carry out all these steps under identical conditions. So connections or supports with the same nominal design parameters may not have exactly the same mechanical properties; instead, the mechanical properties are uncertain. The modeling uncertainty comes mainly from the following three sources:

1. Because of the complexity of the connections and supports, the mechanical properties, namely the $M-\Theta$ relationships, can be obtained only through experiments. The

test procedure and the statistical evaluation of the test results will introduce uncertainty.

2. Many different types of connections and supports are used in practice, but only a few representative types have been tested and have available $M-\Theta$ curves. Analysts and designers have to approximate the moment–rotation relationships for other types of connections based on experience, and approximation introduces uncertainty.

3. Several moment–rotation-predicting formulas are available for modeling a connection. Each formula is obtained from a particular theory and set of test data. It is impossible to have similar modeling error for all kinds of connections if they are represented by the same formula. This introduces another source of uncertainty in modeling the connection. Since supports are being modeled as connections here, this also adds to the modeling uncertainty.

In general, since the $M-\Theta$ relationships of PR connections are obtained through tests, the uncertainty in their mechanical properties must be considered in a realistic manner. Due to these uncertainties, one cannot precisely predict the mechanical properties of the connections for given design parameters. The probabilistic method can be used to include the uncertain properties in the PR connection and support conditions in an analysis.

Rauscher and Gerstle (1992) reported a study that provides statistical information on the $M-\Theta$ curves for slip critical and bearing-type bolted connection. Nominally identical framing connection (double-web angle connection) specimens from different sources were tested under identical conditions by one or more fabricators. A considerable amount of scatter in the $M-\Theta$ curves was observed.

In reliability analysis, a computational model is necessary to account properly for the scatter in the connection behavior. A stochastic computational model is presented here to account for the scatter in the connection behavior by considering the four parameters in the Richard model, k, k_p, M_0, and N, to be random variables. The statistical properties of these four parameters can be obtained through experiments and practice. This concept is illustrated in Figure 8.8. Each of the four parameters has its own statistical characteristics.

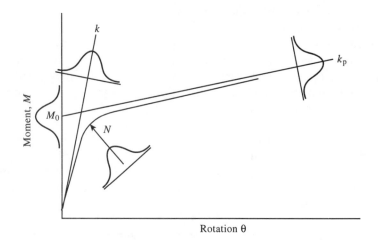

FIGURE 8.8. Uncertainty in the Richard model.

These parameters can now be added to the list of other basic random variables, and should be treated similarly. This is discussed in Section 8.9.

8.6   A UNIFIED STOCHASTIC FINITE ELEMENT METHOD

The basic deterministic nonlinear finite element algorithm and the uncertainty associated with the parameters in the algorithm have been presented in the previous sections. How to incorporate uncertainty in the nonlinear finite element algorithm to evaluate reliability is presented next.

The reliability analysis procedure presented here is based on the first-order reliability method (FORM) and necessitates the definition of a limit state function, $G(\mathbf{x}, \mathbf{u}, \mathbf{s})$, where vector $\mathbf{x}$ denotes the set of basic random variables pertaining to a structure (e.g., loads, material properties and structural geometry), vector $\mathbf{u}$ denotes the set of displacements involved in the limit state function, and vector $\mathbf{s}$ denotes the set of load effects (except the displacement) involved in the limit state function (e.g., stresses, internal forces). The displacement $\mathbf{u}$ can be expressed as $\mathbf{u} = \mathbf{QD}$, where $\mathbf{D}$ is the global displacement vector and $\mathbf{Q}$ is a transformation matrix. In general, $\mathbf{x}$, $\mathbf{u}$, and $\mathbf{s}$ are related in an algorithmic sense, for example, a finite element code. The algorithm evaluates the performance function deterministically, with the corresponding gradients at each iteration point. It converges to the most probable failure point (or checking point or design point) and calculates the corresponding reliability index β. The following iteration scheme is used to find the checking point:

$$\mathbf{y}_{i+1} = \left[\mathbf{y}_i^t \boldsymbol{\alpha}_i + \frac{G(\mathbf{y}_i)}{|\nabla G(\mathbf{y}_i)|} \right] \boldsymbol{\alpha}_i \qquad (8.37)$$

where

$$\nabla G(\mathbf{y}) = \left[\frac{\partial G(\mathbf{y})}{\partial y_1}, \ldots, \frac{\partial G(\mathbf{y})}{\partial y_n} \right]^t \qquad (8.38)$$

$$\boldsymbol{\alpha}_i = -\frac{\nabla G(\mathbf{y}_i)}{|\nabla G(\mathbf{y}_i)|} \qquad (8.39)$$

and

$$\nabla G = \left[\frac{\partial G}{\partial \mathbf{s}} \mathbf{J}_{s,x} + \left(\mathbf{Q} \frac{\partial G}{\partial \mathbf{u}} + \frac{\partial G}{\partial \mathbf{s}} \mathbf{J}_{s,D} \right) \mathbf{J}_{D,x} + \frac{\partial G}{\partial \mathbf{x}} \right] \mathbf{J}_{y,x}^{-1} \qquad (8.40)$$

In Equation 8.40, $\mathbf{J}_{i,j}$ are the Jacobians of transformation and $y_1, y_2, \ldots, y_n$ are statistically independent random variables in the standard normal space. The evaluation of the quantities in Equation 8.40 will depend on the problem under consideration (linear or nonlinear, 2D or 3D, etc.) and the performance functions used. The essential numerical aspect of SFEM is the evaluation of three partial derivatives, $\partial G/\partial \mathbf{s}$, $\partial G/\partial \mathbf{u}$, and $\partial G/\partial \mathbf{x}$, and four Jacobians, $\mathbf{J}_{s,x}$, $\mathbf{J}_{s,D}$, $\mathbf{J}_{D,x}$, and $\mathbf{J}_{y,x}$. These are evaluated in the following sections.

8.7 PERFORMANCE FUNCTIONS AND PARTIAL DIFFERENTIALS

The safety of a structure needs to be evaluated with respect to predetermined performance criteria. The performance criteria are usually expressed in the form of limit state functions, which are functional relationships among all the load effects and resistance-related parameters. Two types of limit state functions are commonly used in engineering profession: the limit state function of strength, which defines safety against extreme loads during the intended life of the structure, and the limit state function of serviceability, which defines the functional requirements.

8.7.1 Strength Performance Functions

Many members in a framed structure are subjected to both bending moment and axial load. According to the AISC's Load and Resistance Factor Design (LRFD) design guidelines, the following interaction equations should be checked to satisfy the strength requirements for 2D members

$$\frac{P_u}{\phi P_n} + \frac{8}{9}\left(\frac{M_{ux}}{\phi_b M_{nx}}\right) \leq 1.0 \qquad \text{if} \quad \frac{P_u}{\phi P_n} \geq 0.2 \tag{8.41}$$

$$\frac{P_u}{2\phi P_n} + \frac{M_{ux}}{\phi_b M_{nx}} \leq 1.0 \qquad \text{if} \quad \frac{P_u}{\phi P_n} < 0.2 \tag{8.42}$$

where ϕ and ϕ_b are the resistance factors, P_u is the required tensile/compressive strength, P_n is the nominal tensile/compressive strength, M_{ux} is the required flexural strength, and M_{nx} is the nominal flexural strength.

For reliability evaluation, the strength limit state function can be expressed as

$$G(\mathbf{x}, \mathbf{u}, \mathbf{s}) = 1.0 - \left(\frac{P_u}{P_n} + \frac{8}{9}\frac{M_{ux}}{M_{nx}}\right) \qquad \text{if} \quad \frac{P}{\phi P_n} \geq 0.2 \tag{8.43}$$

$$G(\mathbf{x}, \mathbf{u}, \mathbf{s}) = 1.0 - \left(\frac{P_u}{2P_n} + \frac{M_{ux}}{M_{nx}}\right) \qquad \text{if} \quad \frac{P_u}{\phi P_n} < 0.2 \tag{8.44}$$

where P_u and M_{ux} in Equations 8.43 and 8.44 are unfactored load effects.

Expressions of $\partial G/\partial \mathbf{x}$, $\partial G/\partial \mathbf{u}$, and $\partial G/\partial \mathbf{s}$ can be derived from Equations 8.43 and 8.44. Neither G function contains any explicit displacement component. Therefore, $\partial G/\partial \mathbf{u} = 0$ for both Equations 8.43 and 8.44. To calculate $\partial G/\partial \mathbf{x}$, the basic random variables need to be defined. The Young's modulus, E; area, A; yield stress, F_y; plastic modulus, Z_x; and the moment of inertia of a cross section I with the external load F are considered to be basic random variables here. Therefore, the following expression applies:

$$\frac{\partial G}{\partial \mathbf{x}} = \left\{\frac{\partial G}{\partial E} \quad \frac{\partial G}{\partial A} \quad \frac{\partial G}{\partial I} \quad \frac{\partial G}{\partial Z_x} \quad \frac{\partial G}{\partial F_y}\right\} \tag{8.45}$$

From the AISC-LRFD *Manual* (1994), one obtains

$$P_n = A F_{cr} \text{ (compression)} \quad \text{or} \quad P_n = A F_y \text{ (tension)} \tag{8.46}$$

and

$$M_{nx} = Z_x F_y \tag{8.47}$$

where

$$F_{\mathrm{cr}} = (0.658^{\lambda_c^2}) F_y \qquad \text{when} \quad \lambda_c \leq 1.5 \tag{8.48}$$

$$F_{\mathrm{cr}} = \left(\frac{0.877}{\lambda_c^2} \right) F_y \qquad \text{when} \quad \lambda_c > 1.5 \tag{8.49}$$

and

$$\lambda_c = \frac{Kl}{r\pi} \sqrt{\frac{F_y}{E}} \tag{8.50}$$

where A is the gross area of member (in^2), F_y is the specified yield stress (ksi), E is the modulus of elasticity (ksi), K is the effective length factor, l is the unbraced length of member (inches), and r is the governing radius of gyration about the plane of buckling (inches).

From Equations 8.46–8.50 and the strength limit state functions given by Equations 8.43 and 8.44, the following results can be obtained.

8.7.1.1 *Compression Members*   When $P_u/\phi P_n \geq 0.2$, and if $\lambda_c \leq 1.5$,

$$\frac{\partial G}{\partial E} = \frac{P_u}{P_n E} \left(-\lambda_c^2 \ln 0.658 \right)$$

$$\frac{\partial G}{\partial A} = \frac{P_u}{A P_n} \left(1 + \lambda_c^2 \ln 0.658 \right)$$

$$\frac{\partial G}{\partial I} = \frac{-P_u \lambda_c^2}{P_n I} \ln 0.658$$

$$\frac{\partial G}{\partial Z_x} = \frac{8}{9} \frac{M_{ux}}{M_{nx} Z_x}$$

and

$$\frac{\partial G}{\partial F_y} = \frac{P_u}{P_n F_{\mathrm{cr}}} 0.658^{\lambda_c^2} \left(1 + \lambda_c^2 \ln 0.658 \right) + \frac{8}{9} \frac{M_{ux}}{M_{nx} F_y}$$

When $P_u/\phi P_n \geq 0.2$, and $\lambda_c > 1.5$,

$$\frac{\partial G}{\partial E} = \frac{P_u}{P_n E}$$

$$\frac{\partial G}{\partial A} = 0$$

$$\frac{\partial G}{\partial I} = \frac{P_u}{P_n I}$$

$$\frac{\partial G}{\partial Z_x} = \frac{8}{9} \frac{M_{ux}}{M_{nx} Z_x}$$

and

$$\frac{\partial G}{\partial F_y} = \frac{8}{9} \frac{M_{ux}}{M_{nx} F_y}$$

When $P_u/\phi P_n < 0.2$, and $\lambda_c \leq 1.5$,

$$\frac{\partial G}{\partial E} = \frac{P_u}{2P_n E} \left(-\lambda_c^2 \ln 0.658 \right)$$

$$\frac{\partial G}{\partial A} = \frac{P_u}{2P_n A} \left(1 + \lambda_c^2 \ln 0.658 \right)$$

$$\frac{\partial G}{\partial I} = \frac{P_n}{2P_n I} \left(-\lambda_c^2 \ln 0.658 \right)$$

$$\frac{\partial G}{\partial Z_x} = \frac{M_{ux}}{M_{nx} Z_x}$$

and

$$\frac{\partial G}{\partial F_y} = \frac{P_u}{2P_n F_{cr}} 0.658^{\lambda_c^2} \left(1 + \lambda_c^2 \ln 0.658 \right) + \frac{M_{ux}}{M_{nx} F_y}$$

When $P_u/\phi P_n < 0.2$ and $\lambda_c > 1.5$,

$$\frac{\partial G}{\partial E} = \frac{P_u}{2P_n E}$$

$$\frac{\partial G}{\partial A} = 0$$

$$\frac{\partial G}{\partial I} = \frac{P_u}{2P_n I}$$

$$\frac{\partial G}{\partial Z_x} = \frac{M_{ux}}{M_{nx} Z_x}$$

and

$$\frac{\partial G}{\partial F_y} = \frac{M_{ux}}{M_{nx} F_y}$$

8.7.1.2 Tension Members When $P_u/\phi P_n \geq 0.2$,

$$\frac{\partial G}{\partial E} = 0$$

$$\frac{\partial G}{\partial A} = \frac{P_u}{A P_n}$$

$$\frac{\partial G}{\partial I} = 0$$

$$\frac{\partial G}{\partial Z_x} = \frac{8}{9} \frac{M_{ux}}{M_{nx} Z_x}$$

and

$$\frac{\partial G}{\partial F_y} = \frac{P_u}{P_n F_y} + \frac{8}{9} \frac{M_{ux}}{M_{nx} F_y}$$

When $P_u/\phi P_n < 0.2$,

$$\frac{\partial G}{\partial E} = 0$$

$$\frac{\partial G}{\partial A} = \frac{P_u}{2A P_n}$$

$$\frac{\partial G}{\partial I} = 0$$

$$\frac{\partial G}{\partial Z_x} = \frac{M_{ux}}{M_{nx} Z_x}$$

and

$$\frac{\partial G}{\partial F_y} = \frac{P_u}{2P_n F_y} + \frac{M_{ux}}{M_{nx} F_y}$$

To calculate the partial differential $\partial G/\partial \mathbf{s}$, the following expression can be used:

$$\frac{\partial G}{\partial \mathbf{s}} = \left[\frac{\partial G}{\partial P_u} \frac{\partial G}{\partial M_u} \right] \tag{8.51}$$

when $P_u/\phi P_n \geq 0.2$,

$$\frac{\partial G}{\partial P_u} = -\frac{1}{P_n} \quad \text{and} \quad \frac{\partial G}{\partial M_u} = -\frac{8}{9M_n}$$

When $P_u/\phi P_n < 0.2$,

$$\frac{\partial G}{\partial P_u} = -\frac{1}{2P_n} \quad \text{and} \quad \frac{\partial G}{\partial M_u} = -\frac{1}{M_n}$$

8.7.2 Serviceability Performance Function

For the serviceability criterion, the following limit state function is used:

$$G(\mathbf{x}, \mathbf{u}, \mathbf{s}) = 1.0 - \frac{\delta}{\delta_{limit}} \tag{8.52}$$

where δ is the calculated displacement component and δ_{limit} is the allowable maximum value of the displacement component.

From Equation 8.52, one obtains

$$\frac{\partial G}{\partial \mathbf{x}} = \frac{\partial G}{\partial \mathbf{s}} = 0 \tag{8.53}$$

and

$$\frac{\partial G}{\partial \mathbf{u}} = \begin{bmatrix} \dfrac{\partial G}{\partial \delta} & \mathbf{0} \end{bmatrix} \tag{8.54}$$

$$\text{where} \quad \frac{\partial G}{\partial \delta} = -\frac{1}{\delta_{\text{limit}}}$$

Although the actual δ_{limit} values chosen in design offices vary within a certain range, the following two serviceability design criteria are commonly used:

1. Vertical deflections at the midspan of the beam under live load should be less than or equal to $1/360$ of the span of the beam. In this case, $\delta_{\text{limit}}^{\text{deflection}} = l/360$.
2. Side sway at the top of the frame should be less than or equal to $1/400$ of the height of the frame. In this case, $\delta_{\text{limit}}^{\text{drift}} = h/400$.

Both criteria are used in this study.

8.8   EVALUATION OF JACOBIANS AND ADJOINT VARIABLE METHOD

To evaluate the gradient ∇G, the evaluation of the three partial derivatives on the right-hand side of Equation 8.40 is necessary. They are easy to compute since $G(\mathbf{x}, \mathbf{u}, \mathbf{s})$ is an explicit function of $\mathbf{x}$, $\mathbf{u}$, and $\mathbf{s}$, as discussed in the previous section. The four Jacobians in Equation 8.40 need to be computed properly now. Because of the triangular nature of the transformation, $\mathbf{J}_{y,x}$ and its inverse are easy to compute. Since $\mathbf{s}$ is not an explicit function of the basic random variables $\mathbf{x}$, $\mathbf{J}_{s,x} = 0$. The Jacobians of the transformation $\mathbf{J}_{s,D}$ and $\mathbf{J}_{D,x}$, however, are not easy to compute since $\mathbf{s}$, $\mathbf{D}$, and $\mathbf{x}$ are implicit functions of each other. The adjoint variable method (Ryu et al., 1985) is used here to compute the product of the second term in Equation 8.40 directly, instead of evaluating its constituent parts. It is accurate and computationally efficient when a large number of basic random variables are involved in a problem.

An adjoint vector λ can be introduced such that

$$\lambda^t \mathbf{K} = \mathbf{Q} \frac{\partial G}{\partial \mathbf{u}} + \frac{\partial G}{\partial \mathbf{s}} \mathbf{J}_{s,D} \tag{8.55}$$

To implement the adjoint variable method, the adjoint vector λ needs to be evaluated at this stage. It depends on the limit state function being considered. If it is a serviceability limit state function represented by Equation 8.52, then $\partial G/\partial \mathbf{x} = \partial G/\partial \mathbf{s} = \mathbf{0}$ and $\partial G/\partial \mathbf{u} = [\partial G/\partial \delta \quad \mathbf{0}]$, and $\partial G/\partial \delta = -1/\delta_{\text{limit}}$. In Equation 8.55, all the parameters except λ are known, and it can be evaluated easily.

For the strength limit state functions represented by Equations 8.43 and 8.44, $\partial G/\partial \mathbf{u} = 0$, and $\partial G/\partial \mathbf{s}$ is known from Equation 8.51; however, $\mathbf{J}_{s,D}$ needs to be evaluated to estimate λ. Normally, when the strength performance functions are considered, the internal force vector σ is the only contribution to the load effect $\mathbf{s}$ and can be expressed as $\mathbf{s} = \mathbf{A}\sigma$, where $\mathbf{A}$ is the transformation matrix with constant elements. Thus, one can obtain

$$\mathbf{J}_{s,D} = \frac{\partial \mathbf{s}}{\partial \mathbf{D}} = \mathbf{A} \frac{\partial \sigma}{\partial \mathbf{D}} = \mathbf{A} \begin{bmatrix} \dfrac{\delta \sigma}{\partial \mathbf{d}} & \mathbf{0} \end{bmatrix} \tag{8.56}$$

where $\mathbf{d}$ is the nodal displacement vector in the global coordinate for the element and can be defined as

$$\mathbf{d} = \{^1u_1 \quad ^1u_2 \quad ^1\theta \quad ^2u_1 \quad ^2u_2 \quad ^2\theta\} \tag{8.57}$$

The relationship between the increment of internal nodal force $\Delta\sigma$ and the increment of global displacement $\Delta\mathbf{D}$ can be expressed as

$$\Delta\sigma = \mathbf{A}_{\sigma\sigma}^{-1}(\mathbf{A}_{\sigma d0}\Delta\mathbf{D} - \mathbf{R}_\sigma) \tag{8.58}$$

Equation 8.58 needs to be solved iteratively. Once the algorithm converges, $\Delta\mathbf{D}$ and $\Delta\sigma$ will be almost zero vectors, and Equation 8.58 becomes $\mathbf{R}_\sigma = 0$. Then the relationship between σ and $\mathbf{d}$ can be shown to be

$$\sigma = \begin{Bmatrix} n \\ m_1 \\ m_2 \end{Bmatrix} = \frac{E}{l} \begin{Bmatrix} AH \\ -2I(2^1\theta^* + {}^2\theta^*) \\ 2I({}^1\theta^* + 2^2\theta^*) \end{Bmatrix} \tag{8.59}$$

where

$$^i\theta^* = {}^i\theta - \tan^{-1}\left(\frac{\tilde{u}_2}{l + \tilde{u}_1}\right) \tag{8.60}$$

All the parameters in Equation 8.60 were defined in Equations 8.1 and 8.3. The derivative of the internal forces with respect to the displacements becomes

$$\frac{\partial\sigma}{\partial\mathbf{d}} = \left[\frac{\partial n}{\partial\mathbf{d}} \quad \frac{\partial m_1}{\partial\mathbf{d}} \quad \frac{\partial m_2}{\partial\mathbf{d}}\right]^t \tag{8.61}$$

where

$$\frac{\partial n}{\partial\mathbf{d}} = \frac{EA}{l} \left\{ \frac{-(l+\tilde{u}_1)}{L} \quad \frac{-\tilde{u}_2}{L} \quad 0 \quad \frac{(l+\tilde{u}_1)}{L} \quad \frac{\tilde{u}_2}{L} \quad 0 \right\} \tag{8.62}$$

$$\frac{\partial m_1}{\partial\mathbf{d}} = \frac{2EI}{l} \left\{ \frac{3\tilde{u}_2}{L^2} \quad \frac{-3(l+\tilde{u}_1)}{L^2} \quad -2 \quad \frac{-3\tilde{u}_2}{L^2} \quad \frac{3(l+\tilde{u}_1)}{L^2} \quad -1 \right\} \tag{8.63}$$

$$\frac{\partial m_2}{\partial\mathbf{d}} = \frac{2EI}{l} \left\{ \frac{-3\tilde{u}_2}{L^2} \quad \frac{3(l+\tilde{u}_1)}{L^2} \quad 1 \quad \frac{3\tilde{u}_2}{L^2} \quad \frac{-3(l+\tilde{u}_1)}{L^2} \quad 2 \right\} \tag{8.64}$$

where L is the deformed length and can be calculated as $L = \sqrt{(l+\tilde{u}_1)^2 + (\tilde{u}_2)^2}$. Thus, for the strength limit state, all the parameters in Equation 8.55 are known except λ, and it can be evaluated very easily.

From Equation 8.55, it can also be shown that

$$\lambda^t\mathbf{K}^{(n)}\mathbf{J}_{D,x} = \left(\mathbf{Q}\frac{\partial G}{\partial\mathbf{u}} + \frac{\partial G}{\partial\mathbf{s}}\mathbf{J}_{s,D}\right)\mathbf{J}_{D,x} = \lambda^t\left(\frac{\partial\mathbf{F}}{\partial\mathbf{x}} - \frac{\partial\mathbf{R}^{(n-1)}}{\partial\mathbf{x}}\right) \tag{8.65}$$

where $\partial\mathbf{F}/\partial\mathbf{x}$ is easy to obtain since the explicit dependence of $\mathbf{F}$ on the basic variables is known, assuming the external load is not affected by the structural response, and $\partial\mathbf{R}/\partial\mathbf{x}$ can be derived as discussed below.

To calculate $\partial \mathbf{R}/\partial \mathbf{x}$, Equation 8.14 can be written as

$$\mathbf{R} = -\mathbf{A}_{\sigma d0}^{t}\mathbf{R}_{A\sigma} + \mathbf{R}_{d0} \tag{8.66}$$

where $\mathbf{R}_{A\sigma} = \mathbf{A}_{\sigma\sigma}^{-1}\mathbf{R}_{\sigma}$ and can be expressed as

$$\mathbf{R}_{A\sigma} = \left\{ \begin{array}{c} n - \dfrac{EAH}{l} \\[3mm] m_1 + \dfrac{2EI}{l}(2\,{}^1\theta^* + {}^2\theta^*) \\[3mm] m_2 - \dfrac{2EI}{l}({}^1\theta^* + 2\,{}^2\theta^*) \end{array} \right\} \tag{8.67}$$

Since $\mathbf{R}_{d0}$ and $\mathbf{A}_{\sigma d0}^{t}$ are not functions of the basic variables, the derivative of $\mathbf{R}$ with respect to $\mathbf{x}$ can be expressed as

$$\left.\frac{\partial \mathbf{R}}{\partial \mathbf{x}}\right|_{D,\sigma} = -\mathbf{A}_{\sigma d0}^{t}\frac{\partial \mathbf{R}_{A\sigma}}{\partial \mathbf{x}} \tag{8.68}$$

To evaluate Equation 8.68, the basic random variables need to be identified. The practical implementation of the SFEM-based reliability analysis needs to be done in a computer environment. The amount of memory required for the computer grows rapidly with the number of basic random variables. As mentioned in Section 3.8, sensitivity analysis can be used to reduce the number of basic random variables. Among the structural parameters, only the Young's modulus, E; area, A; and the moments of inertia of a cross section, I, with the external load, $\mathbf{F}$, are considered to be basic random variables in this section. The partial differentiation on the right-hand side of Equation 8.68 can be calculated by considering the elastic and elastoplastic nonlinear cases discussed in the following sections.

8.8.1 Elastic Nonlinear Case

In the elastic nonlinear case, $\partial \mathbf{R}_{A\sigma}/\partial \mathbf{x}$ for a beam–column element can be shown to be

$$\frac{\partial \mathbf{R}_{A\sigma}}{\partial \mathbf{x}} = \left[\begin{array}{cccc} \dfrac{\partial \mathbf{R}_{A\sigma}}{\partial E} & \dfrac{\partial \mathbf{R}_{A\sigma}}{\partial A} & \dfrac{\partial \mathbf{R}_{A\sigma}}{\partial I} & 0 \end{array} \right] \tag{8.69}$$

where

$$\frac{\partial \mathbf{R}_{A\sigma}}{\partial E} = \left\{ -\frac{AH}{l} \quad \frac{2I}{l}(2\,{}^1\theta + {}^2\theta^*) \quad -\frac{2I}{l}({}^1\theta^* + 2\,{}^2\theta^*) \right\}^{t} \tag{8.70}$$

$$\frac{\partial \mathbf{R}_{A\sigma}}{\partial A} = \left\{ -\frac{EH}{l} \quad 0 \quad 0 \right\}^{t} \tag{8.71}$$

$$\frac{\partial \mathbf{R}_{A\sigma}}{\partial I} = \left\{ 0 \quad \frac{2E}{l}(2\,{}^1\theta^* + {}^2\theta^*) \quad -\frac{2E}{l}(\theta^* + 2\,{}^2\theta^*) \right\}^{t} \tag{8.72}$$

8.8.2 Elastoplastic Nonlinear Case

In this book, when the combined axial force and bending moments of an elastic–perfectly plastic material model satisfy a prescribed yield function at a node of an element, a plastic hinge is assumed to occur instantaneously at that location. $\mathbf{V}_p \mathbf{C}_p^t$ becomes unity and $\mathbf{K}_p$ becomes zero. Therefore,

$$\frac{\partial \mathbf{R}}{\partial \mathbf{x}} = 0 \qquad (8.73)$$

So far, all the information required for the computation of the gradient of the performance function (Equation 8.40) is available for the elastic and elastoplastic nonlinear cases.

8.9 CONSIDERATION OF UNCERTAINTIES IN CONNECTION MODEL

In general, the response of a frame structure is greatly influenced by the mechanical properties of the connections. So the uncertainties in the connections will contribute to the uncertainty in the structural response. Since connections are important sources of uncertainty, they should be considered explicitly in the structural reliability analysis. In this section, a methodology is developed to incorporate the uncertainties in PR connections in the SFEM. The uncertainties in the PR connection were quantified in Section 8.5.2.

In practice, the parameters in a typical moment–rotation curve are estimated from experimental results using a curve-fitting technique. Therefore, the deterministic moment–rotation curves do not account for the scatter in the connection behavior. In this book, the scatter in the connection behavior is modeled by considering the four parameters in the Richard model, k, k_p, M_0, and N, to be the basic random variables as shown in Figure 8.8. For connection elements to incorporate uncertainties in the connection behavior, the following additional items need to be considered

$$\frac{\partial \mathbf{R}_{A\sigma}}{\partial \mathbf{x}} = \left[\frac{\partial \mathbf{R}_{A\sigma}}{\partial k} \quad \frac{\partial \mathbf{R}_{A\sigma}}{\partial k_p} \quad \frac{\partial \mathbf{R}_{A\sigma}}{\partial M_0} \quad \frac{\partial \mathbf{R}_{A\sigma}}{\partial N} \quad 0 \right] \qquad (8.74)$$

Equations 8.26, 8.27, and 8.28 can be used to calculate some of the terms in Equation 8.74.

Considering the major axis rotational flexibility of a connection, the components in Equation 8.74 can be expressed as

$$\frac{\partial \mathbf{R}_{A\sigma}}{\partial \xi_i} = 2 \frac{\partial K(\theta)}{\partial \xi_i} [0 \quad (2\,^1\theta^* + \,^2\theta^*) \quad -(^1\theta^* + 2\,^2\theta^*)]^t \qquad (8.75)$$

where $\xi_i = k$, k_p, M_0, and N.

Using Equation 8.31, Equation 8.75 can be evaluated as

$$\frac{\partial K(\theta)}{\partial k} = \frac{1 - Na^N}{(1 - a^N)^{(2N+1)/N}} \qquad (8.76)$$

$$\frac{\partial K(\theta)}{\partial k_p} = \frac{-(N+2)a^N - 1}{(1 + a^N)^{(2N+1)/N}} + 1 \qquad (8.77)$$

$$\frac{\partial K(\theta)}{\partial M_0} = \frac{(N+1)a^N}{(1 + a^N)^{(2N+1)/N}} \frac{k - k_p}{M_0} \qquad (8.78)$$

$$\frac{\partial K(\theta)}{\partial N} = \left[-\frac{a^N(1+N)\log a}{N(1+a^N)^{(1+2N)/N}} + \frac{-N + (1+N)\log(1+a^N)}{N^2(1+a^N)^{(1+N)/N}} \right](k - k_p) \qquad (8.79)$$

where

$$a = \left| \frac{(k - k_{\mathrm{p}})\theta}{M_0} \right| \tag{8.80}$$

All the quantities required to compute ∇G in Equation 8.40 are now available in simple explicit forms considering the geometric nonlinearity, elastoplastic material behavior, and flexibility of connections. Equation 8.37 can be used to compute the design point, and the corresponding reliability index can also be calculated. To improve the stability of the non-linear finite-element analysis, the Newton–Raphson method with the arc-length procedure is used here.

As mentioned earlier, the support conditions can be modeled as flexible connections. Thus, the discussion on flexible connections is also applicable to model support conditions.

8.10 LINEAR ELASTIC CASE

The discussion so far in this chapter has considered the nonlinear behavior of structures. However, it may be of interest to estimate the reliability of structures assuming linear behavior. In this situation, the initial stiffness of the elements will obviously remain constant throughout the analysis. Also, the stresses in the elements cannot exceed the yield strength of the material, and the flexible connections and support conditions cannot be used. The problem is simple and can be formulated in the following way.

The displacement response vector $\mathbf{D}$ of a linear structure under static loads can be written as

$$\mathbf{KD} = \mathbf{F} \tag{8.81}$$

where $\mathbf{K}$ is the linear global stiffness matrix and $\mathbf{F}$ is the global load vector.

The discussion in Section 8.6 is valid for linear SFEM, except that the evaluation of Equation 8.65 needs some modification. It can be rewritten as

$$\lambda^{\mathrm{t}} \mathbf{KJ}_{D,x} = \left(\mathbf{Q} \frac{\partial G}{\partial \mathbf{u}} + \frac{\partial G}{\partial \mathbf{s}} \mathbf{J}_{s,D} \right) \mathbf{J}_{D,x} = \lambda^{\mathrm{t}} \bar{\mathbf{R}} \tag{8.82}$$

and

$$\bar{\mathbf{R}} = \frac{\partial \mathbf{F}}{\partial \mathbf{x}} - \frac{\partial [\mathbf{K}\tilde{\mathbf{D}}]}{\partial \mathbf{x}} \tag{8.83}$$

where a tilde ($\sim$) over a variable indicates that it is to remain constant during partial differentiation operations.

$\mathbf{J}_{s,D}$ in Equation 8.82 can be calculated using Equation 8.56 and Equations 8.61–8.64. If externally applied loads are not included in the basic random variable vector, the term $\partial \mathbf{F}/\partial \mathbf{x}$ in Equation 8.83 is a zero matrix. Otherwise, $\partial \mathbf{F}/\partial \mathbf{x}$ can be calculated since the explicit dependence of $\mathbf{F}$ on the basic random variables is known. The term $\partial [\mathbf{K}\tilde{\mathbf{D}}]/\partial \mathbf{x}$ can be expressed explicitly as

$$\frac{\partial[\mathbf{K\tilde{D}}]}{\partial\mathbf{x}} = \left[\frac{\partial[\mathbf{K\tilde{D}}]}{\partial x_1} \quad \frac{\partial[\mathbf{K\tilde{D}}]}{\partial x_2} \quad \cdots \quad \frac{\partial[\mathbf{K\tilde{D}}]}{\partial x_m} \right] \tag{8.84}$$

where m is the total number of basic random variables. For any variable x_i, the following equation can be obtained:

$$\frac{\partial[\mathbf{K\tilde{D}}]}{\partial x_i} = \frac{\partial\mathbf{K}}{\partial x_i}\mathbf{D} \tag{8.85}$$

$\partial\mathbf{K}/\partial x_i$ in Equation 8.85 can be calculated at the element level, and then assembled at the global level. Since the stiffness matrix for an element can be expressed by Equation 8.7, the following expression can be obtained:

$$\frac{\partial\mathbf{K}}{\partial x_i} = \sum_{j=1}^{n}\frac{\partial\mathbf{K}_j^e}{\partial x_i} = \sum_{j=1}^{n}(\mathbf{A}_{\sigma d0}^t)_j\frac{\partial(\mathbf{A}_{\sigma\sigma}^{-1})_j}{\partial x_i}(\mathbf{A}_{\sigma d0})_j \tag{8.86}$$

where n is the total number of elements in the system.

For a 2D element, $\mathbf{A}_{\sigma\sigma}^{-1}$ is the inverse of Equation 8.8, which can be shown to be

$$\mathbf{A}_{\sigma\sigma}^{-1} = \frac{1}{l}\begin{bmatrix} EA & 0 & 0 \\ 0 & 4EI & -2EI \\ 0 & -2EI & 4EI \end{bmatrix} \tag{8.87}$$

If E, A, and I are included in the basic random variable vector, the following expressions can be derived:

$$\frac{\partial\mathbf{A}_{\sigma\sigma}^{-1}}{\partial E} = \frac{1}{l}\begin{bmatrix} A & 0 & 0 \\ 0 & 4I & -2I \\ 0 & -2I & 4I \end{bmatrix} \tag{8.88}$$

$$\frac{\partial\mathbf{A}_{\sigma\sigma}^{-1}}{\partial A} = \frac{1}{l}\begin{bmatrix} E & 0 & 0 \\ 0 & 0 & 0 \\ 0 & 0 & 0 \end{bmatrix} \tag{8.89}$$

and

$$\frac{\partial\mathbf{A}_{\sigma\sigma}^{-1}}{\partial I} = \frac{1}{l}\begin{bmatrix} 0 & 0 & 0 \\ 0 & 4E & -2E \\ 0 & -2E & 4E \end{bmatrix} \tag{8.90}$$

Once $\partial\mathbf{A}_{\sigma\sigma}^{-1}/\partial x_i$'s are determined, $\partial\mathbf{K}/\partial x_i$ can be evaluated using Equation 8.86. Its assembly procedure will be very similar to the assembly procedure of the global stiffness matrix in regular FEM. By using Equations 8.84 and 8.85, the right-hand side of Equation 8.83 can be calculated. By substituting Equation 8.83 into Equation 8.40, the gradient of the performance function can be obtained, and reliability analysis of linear structures can be performed. Thus, the SFEM procedure outlined in Section 8.5 for nonlinear structures can also be used for reliability analysis of linear structures without difficulty.

8.11 NUMERICAL EXAMPLES—RELIABILITY ANALYSIS OF GEOMETRICALLY NONLINEAR 2D FRAMES WITH FLEXIBLE CONNECTIONS AND SUPPORTS

The methodology developed in the previous sections needs verification in terms of its accuracy, efficiency, and robustness. Several examples are given in the following sections for this purpose. A computer program is written to implement the linear and nonlinear SFEM algorithm. This program is used in estimating risk or reliability of all the structures considered next. Monte Carlo simulation is used to verify the algorithm.

Example 8.1 *One-Story Frame*

A frame, shown in Figure 8.9, is considered in this example. The geometric dimensions of the frame are shown in the figure. The following load combinations suggested in the LRFD guidelines are used to select the size of the members: (1) $1.4D$; (2) $1.2D + 1.6L$; (3) $1.2D + 0.5L + 1.3W$; and (4) $0.9D - 1.3W$, where D, L, and W are the dead, live, and wind load, respectively. Beams are assumed to have adequate lateral supports to develop the plastic moment capacity of the member.

Using the nominal values of the loads given in Table 8.1, the frame is analyzed and designed, and W16 × 67 steel is found to be adequate for all three members of the frame.

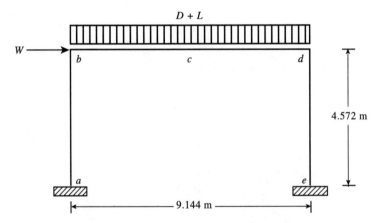

FIGURE 8.9. One-story frame in Example 8.1.

Table 8.1. Description of basic random variables in Example 8.1

Variables	Nominal Values	Mean/Nominal	COV	Distribution
E (MPa)	199,948.04	1.00	0.06	Lognormal
A (cm^2)	127.10	1.00	0.05	Lognormal
I (cm^4)	39,708.48	1.00	0.05	Lognormal
Z_x (cm^3)	2,130.32	1.00	0.05	Lognormal
F_y (MPa)	248.21	1.05	0.10	Lognormal
D (kN/m)	43.78	1.05	0.10	Lognormal
L (kN/m)	16.05	1.00	0.25	Type I
W (kN)	28.91	0.78	0.37	Type I

The probabilistic descriptions of the basic variables required for the reliability analysis are listed in Table 8.1. The reliability of the frame is evaluated using the SFEM method. Different load combinations are considered for the strength limit states. For the drift serviceability limit state, as discussed in Section 8.7.2, $\delta_{\text{limit}}^{\text{drift}}$ is equal to 1.14 cm for this example. Similarly, for the serviceability limit state of deflection at c, the midspan of the beam, $\delta_{\text{limit}}^{\text{deflection}}$ is considered to be 2.54 cm for this example under unfactored live load.

Example 8.2 *One-Story Frame With FR Connections (Linear Case)*

First, the frame is analyzed assuming all the members are connected rigidly. The reliability indexes of the frame corresponding to the strength and serviceability limit states are calculated using SFEM considering the linear and nonlinear behavior of the frame. The results for linear behavior are listed in Tables 8.2–8.5. Monte Carlo simulation is also conducted to verify the analytical results. Considering the magnitude of the failure

Table 8.2. Linear analysis of the FR frame in Example 8.2, strength limit state (beam at *d*), load: *D* + *L*

Variables	Sensitivity Index	Initial Point	Final Point
E (MPa)	−0.0005	199,948.04	199,570.62
A (cm^2)	−0.0084	127.10	126.77
I (cm^4)	−0.0015	39,708.48	39,650.21
Z_x (cm^3)	−0.3649	2,130.32	2,011.02
F_y (MPa)	−0.7474	260.62	205.95
D (kN/m)	0.3673	45.97	51.22
L (kN/m)	0.4163	16.05	21.33
Performance function		0.3610	0.0008
Reliability index		4.39	3.10
Monte Carlo simulation			2.82

Note. Number of iterations = 4.

Table 8.3. Linear analysis of FR frame in Example 8.2, strength limit state (column at *d*), load: *D* + *L*

Variables	Sensitivity Index	Initial Point	Final Point
E (MPa)	−0.0011	199,948.04	199,552.14
A (cm^2)	−0.0192	127.10	126.64
I (cm^4)	−0.0017	39,708.48	39,649.37
Z_x (cm^3)	−0.3019	2,130.32	2,040.19
Fy (MPa)	−0.6426	260.62	216.98
D (kN/m)	0.4496	45.97	51.82
L (kN/m)	0.5417	16.05	22.66
Performance function		0.3337	−0.0001
Reliability index		3.49	2.78
Monte Carlo simulation			2.88

Note. Number of iterations = 4.

Table 8.4. Linear analysis of FR frame in Example 8.2, serviceability (drift at *b*), load: *D* + *L* + *W*

Variables	Sensitivity Index	Initial Point	Final Point
E (MPa)	−0.2183	199,948.04	190,490.57
A (cm^2)	−0.0051	127.10	126.97
I (cm^4)	−0.1769	39,708.48	38,545.94
Z_x (cm^3)			
F_y (MPa)			
D (kN/m)	0.0044	45.97	46.08
L (kN/m)	0.0035	16.05	16.25
W (kN)	0.9597	22.55	143.81
Performance function		0.8071	−0.0009
Reliability index		12.84	5.96

Note: Number of iterations = 4.

Table 8.5. Linear analysis of FR frame in Example 8.2, serviceability (deflection at *c*), live load only

Variables	Sensitivity Index	Initial Point	Final Point
E (MPa)	−0.2099	199,948.04	188,098.43
A (cm^2)	−0.0041	127.10	126.84
I (cm^4)	−0.1709	39,708.48	38,094.33
Z_x (cm^3)			
F_y (MPa)			
L (kN/m)	0.9627	45.97	54.25
Performance function		0.7326	−0.0003
Reliability index		11.06	4.71

Note. Number of iterations = 4.

Table 8.6. Summary of safety indexes for FR frame in Examples 8.2 and 8.3

Limit State	Location	β (Linear Analysis)	β (Nonlinear Analysis)
Strength	Beam at d	3.10	3.31
	Column at d	2.78	3.08
Serviceability	Drift at b	5.96	5.47
	Deflection at c	4.71	4.77

probability, 10,000 Monte Carlo simulation cycles are used for verification. The results of the Monte Carlo simulation are close to those of SFEM. For comparison purposes, the reliability indexes at various locations are summarized in Table 8.6.

Example 8.3 *One-Story Frame With FR Connections (Nonlinear Case)*

The same frame with FR connections is analyzed considering its nonlinear behavior. The results are given in Tables 8.7–8.10. The results for both linear and nonlinear behavior of the frame are summarized in Table 8.6. For the simple one-story frame struc-

Table 8.7. Nonlinear analysis of FR frame in Example 8.3, strength limit state (Beam at *d*), load: *D* + *L*

Variables	Sensitivity Index	Initial Point	Final Point
E (MPa)	−0.2907	199,948.04	188,403.25
A (cm^2)	−0.0168	127.10	126.58
I (cm^4)	−0.2342	39,708.48	38,152.60
Z_x (cm^3)	−0.3384	2,130.32	2,011.84
F_y (MPa)	−0.6928	260.62	206.29
D (kN/m)	0.3412	45.97	51.19
L (kN/m)	0.3865	16.05	21.28
Performance function		0.3593	−0.0004
Reliability index		4.02	3.31
Monte Carlo simulation			3.09

Note. Number of iterations = 5.

Table 8.8. Nonlinear analysis of FR frame in Example 8.3, strength limit state (Column at *d*), load: *D* + *L*

Variables	Sensitivity Index	Initial Point	Final Point
E (MPa)	−0.3487	199,948.04	187,138.95
A (cm^2)	−0.0359	127.10	126.26
I (cm^4)	−0.2727	39,708.48	38,028.57
Z_x (cm^3)	−0.2695	2,130.32	2,041.17
F_y (MPa)	−0.5736	260.62	217.39
D (kN/m)	0.4031	45.97	51.77
L (kN/m)	0.4885	16.05	22.64
Performance function		0.3322	−0.0001
Reliability index		3.04	3.08
Monte Carlo simulation			2.91

Note. Number of iterations = 5.

Table 8.9. Nonlinear analysis of FR frame in Example 8.3, serviceability (drift at *b*), load: *D* + *L* + *W*

Variables	Sensitivity Index	Initial Point	Final Point
E (MPa)	−0.0275	199,948.04	195,550.36
A (cm^2)	−0.0007	127.10	126.97
I (cm^4)	−0.1609	39,708.48	38,055.62
Z_x (cm^3)			
F_y (MPa)			
D (kN/m)	0.0086	45.97	46.11
L (kN/m)	0.0068	16.05	16.74
W (kN)	0.9865	22.55	130.96
Performance function		0.7989	0.0008
Reliability index		12.54	5.47

Note. Number of iterations = 6.

Table 8.10. Nonlinear analysis of FR frame in Example 8.3, serviceability (deflection at *c*), live load only

Variables	Sensitivity Index	Initial Point	Final Point
E (MPa)	−0.0444	199,948.04	197,071.96
A (cm^2)	−0.0009	127.10	126.90
I (cm^4)	−0.1569	39,708.48	38,204.63
Z_x (cm^3)			
F_y (MPa)			
L (kN/m)	0.9866	45.97	56.65
Performance function		0.7322	−0.0009
Reliability index		11.02	4.77

Note. Number of iterations = 5.

ture in this example, the linear and nonlinear behaviors are not expected to be very different. The reliability indexes are very similar for the linear and nonlinear analyses. However, for the strength performance function, the reliability indexes corresponding to the nonlinear analyis are greater than those in the linear case, since the nonlinear analysis results in lower internal stresses. The reliability indexes corresponding to the serviceability limit state are much higher than those for the strength limit state. This indicates that the strength limit state is the governing limit state, and the structure is expected to develop strength failure first. The structure is stiff enough not to develop an excessive drift or deflection problem.

Example 8.4 *One-Story Frame With PR Connections*

The flexibility of connections is considered next. Both beam-to-column connections at *b* and *d* in Figure 8.9 are considered to be partially restrained. Three M–Θ curves for the connections, shown in Figure 8.10, are considered, representing different amounts of rigidity in the connections. Curve 1 represents high rigidity, curve 2 represents intermediate rigidity, and curve 3 represents very flexible behavior. The probabilistic descriptions of the four parameters of the Richard model are listed in Table 8.11 for all three cases. The frame with two flexible connections is analyzed again using the algorithm discussed earlier. The reliability indexes for the strength and serviceability limit states using curves 1, 2, and 3 are listed as β_1, β_2, and β_3, respectively, in Table 8.12.

The behavior of the frame with rigid and PR connections can be compared at this stage by examining the results listed in Tables 8.6 and 8.12. When PR connections are

Table 8.11. Statistical description of Richard parameters for PR connections

Random Variables	Mean			COV	Distribution
	Curve 1	Curve 2	Curve 3		
k (kN-m/rad)	1.1298×10^6	1.4688×10^5	5.6492×10^4	0.15	Normal
k_p (kN-m/rd)	1.1298×10^5	1.1298×10^4	1.1298×10^3	0.15	Normal
M_0 (kN-m)	508.43	451.94	338.95	0.15	Normal
N	0.50	1.0	1.5	0.05	Normal

Table 8.12. Summary of safety indexes for PR frame in Example 8.4

Limit State	Location	β_1	β_2	β_3	β_4
Strength	Beam*	3.55	3.42	2.60	3.07
	Column at d	3.19	3.89	5.18	3.12
Serviceability	Drift at b	5.17	4.47	4.06	2.86
	Deflection at c	4.61	4.16	3.66	3.91

*Location for the beam: at d for the first two cases and at c for others.

considered, the reliability indexes for both the strength and serviceability limit states changed significantly. This is expected for the strength limit state because a redistribution of moments is expected due to the presence of the flexible connection. In this particular case, the bending moment at c increased, while the bending moment at d decreased because of the PR connections. The design moment for the beam shifted from d to c as the rigidity of the connections gradually decreased. As the rigidity of the connections decreased, the reliability indexes for the strength limit state decreased for the beam and increased for the column, making the beam more prone to failure than the columns. Thus, for the frame under consideration, the lower rigidity in the connections has a beneficial effect on the column and a detrimental effect on the beam. The increase in the spread of the reliability indexes between beam and columns indicates inefficiency in the design as the rigidity in the connections is lowered. However, the reliability indexes according to the serviceability criterion started decreasing as the rigidity of the PR connections decreased. This is also expected since the PR connections lowered the overall stiffness of the frame, and the serviceability criterion became as important as the strength criterion. If the flexibility in the connections is significant, the frame needs to be reanalyzed and redesigned since significant changes are expected in the design loads.

Using curve 3, the frame is redesigned; W16 × 89 steel is selected for the beam and W14 × 53 steel is selected for the columns. The reliability of the frame is then estimated for both limit states and is shown in Table 8.12 as β_4. In this case, the serviceability became the controlling limit state. All four reliability indexes are very similar, indicating an efficient design.

The example demonstrates that the frame must be redesigned if the flexibility in the PR connections is significant. The routine assumption in the profession that the connections are rigid is not appropriate. Furthermore, the example amplifies the point that serviceability capacities of frames with PR connections should be thoroughly investigated.

Example 8.5 *One-Story Frame With Flexible Supports*

Next, the connections at locations b and d are assumed to be rigid, but the two supports are considered to be flexible, represented by the three curves shown in Figure 8.10. The corresponding reliability indexes for various limit states are evaluated and shown in Table 8.13. In this case, the reliability indexes for the strength limit state increased for both the beam and columns compared to the fixed support condition. The reliability indexes for the vertical deflection of the beam at the midspan remained virtually the same, but decreased significantly for the lateral deflection, compared to the fixed support condition.

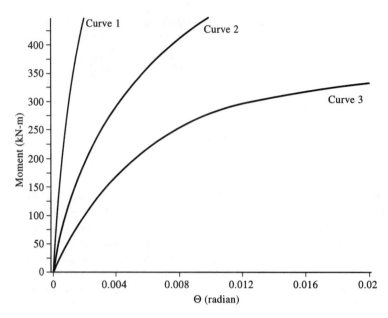

FIGURE 8.10. M–Θ curves for connections.

Table 8.13. Safety indexes for flexible supports at *a* and *e* in Example 8.5

Limit State	Location	β_1	β_2	β_3
Strength	Beam at d	3.33	3.45	3.55
	Column at d	3.09	3.18	3.35
Serviceability	Drift at b	5.06	4.25	3.48
	Deflection at c	4.74	4.65	4.58

Example 8.6 *One-Story Frame With PR Connections and Flexible Supports*

Finally, all the connections and supports are considered to be flexible, represented by the three curves considered earlier. The corresponding reliability indexes are given in Table 8.14.

The compounding effect of the flexible supports and the PR connections can be observed by comparing Tables 8.6 and 8.14. In this case, the reliability indexes for both the strength and serviceability criteria went below 3.0. If either the supports or con-

Table 8.14. Reliability indexes for flexible supports at *a* and *e*, and PR connections at *b* and *d* in Example 8.6

Limit State	Location	β_1	β_2	β_3
Strength	Beam at d	3.58	3.31	2.29
	Column at d	3.20	4.15	7.15
Serviceability	Drift at b	4.93	3.76	2.40
	Deflection at c	4.68	4.15	3.63

nections are very flexible, or if both the supports and connections are very flexible, the frame needs to be redesigned. This observation is expected to be more important for multistory frames with relatively lower lateral stiffness.

Example 8.7 *Two-Story Symmetrical Frame*

A two-story frame, shown in Figure 8.11, is considered in this example. The geometric dimensions of the frame are shown in the figure and the nominal values of the loads are given in Table 8.15. Using the LRFD code, W18 × 55 steel is selected for the beams and W14 × 68 steel is selected for the columns. The probabilistic properties of all the basic variables are given in Table 8.15.

Both the strength and serviceability limit states are considered in this example. For the serviceability limit state, the permissible lateral displacement at the top of the frame is again considered not to exceed $h/400$, i.e., 1.83 cm for this example. And the allowable vertical deflection at midspan of the beam is again considered to be $l/360$ under unfactored live load, i.e., 2.03 cm for this example.

First, the frame is analyzed assuming all the members are connected rigidly. The reliability indexes of the frame corresponding to the strength and serviceability limit states are summarized in Table 8.16, considering the linear and nonlinear behavior of the frame. This case is similar to the previous example of a one-story frame, in which the reliability indexes according to the strength limit state for the nonlinear case appear to be higher than the linear case, indicating that the nonlinear analysis results in lower internal

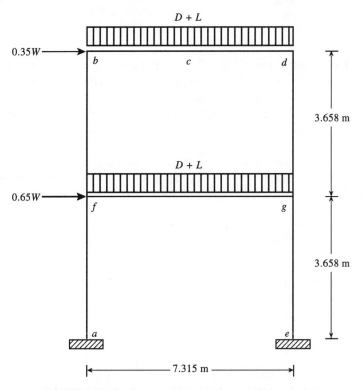

FIGURE 8.11. Two-story symmetrical frame in Example 8.7.

Table 8.15. Description of basic random variables in Example 8.7

Variables	Nominal Values	Mean/Nominal	COV	Distribution
E (MPa)	199,948.04	1.00	0.06	Lognormal
A^b (cm^2)	104.52	1.00	0.05	Lognormal
I^b (cm^4)	37,044.60	1.00	0.05	Lognormal
Z_x^b (cm^3)	1,835.35	1.00	0.05	Lognormal
A^c (cm^2)	129.03	1.00	0.05	Lognormal
I^c (cm^4)	30,093.53	1.00	0.05	Lognormal
Z_x^c (cm^3)	1,884.51	1.00	0.05	Lognormal
F_y (MPa)	248.21	1.05	0.10	Lognormal
D (kN/m)	43.78	1.05	0.10	Lognormal
L (kN/m)	16.05	1.00	0.25	Type I
W (kN)	57.03	0.78	0.37	Type I

Note. Superscript b represents beam and superscript c represents column.

Table 8.16. Summary of safety indexes for the FR frame in Example 8.7

Limit State	Location	β (Linear Analysis)	β (Nonlinear Analysis)
Strength	Beam at d	3.46	3.92
	Column at d	3.30	3.56
Serviceability	Drift at b	1.93	1.88
	Deflection at c	4.40	4.41

stresses. The reliability indexes for the linear and nonlinear cases for the serviceability limit state are also found to be very similar since the structural deformation may not be significantly different in both cases. For simplicity, all the columns are assumed to be the same size. The example indicates that the lower column size may need to be increased to satisfy the serviceability requirement, although it is adequate from the strength point of view.

The flexibility of connections is considered next. All four beam-to-column connections at b, d, f, and g are considered to be partially restrained. For this example, curve 3 in Figure 8.10 is considered to represent the $M-\Theta$ curve. The frame with PR connections is again analyzed using the algorithm. The reliability indexes for the strength and serviceability limit states are listed in Table 8.17. The reliability indexes for both limit states changed significantly. For the strength limit state, the PR connections reduce the

Table 8.17. Summary of safety indexes for the PR frame in Example 8.7

Limit State	Location	β
Strength	Beam at c	2.99
	Column at d	5.34
Serviceability	Drift at b	0.68
	Deflection at c	3.53

bending moments at the connections and increase the bending moments at the midspan of the beams. This is expected because of the redistribution of bending moments. The reliability indexes for the strength limit state decreased for the beams and increased for the columns, making the beams more prone to failure than the columns.

The reliability indexes for the serviceability limit state went down significantly. This is expected since the PR connections lower the overall stiffness of the frame. However, the importance of the PR connections in the design of flexible multistory buildings is quite evident from the drop in the reliability indexes for the serviceability limit state. This simple example clearly indicates that the flexibility in the PR connections must be considered appropriately, particularly for the design of flexible structures. The serviceability could be the most important limit state in this case. The structure may need to be reanalyzed and redesigned considering the PR connections.

Example 8.8 *Two-Story Unsymmetrical Frame*

An unsymmetrical two-story, two-bay rectangular frame as shown in Figure 8.12 is considered next. This is a more general frame from a practical application point of view. The lateral loads can be considered as wind or seismic loads. All the basic random variables are shown in Table 8.18a. Initially, all the connections are considered to be FR-type.

The strength limit state functions (Equations 8.43 and 8.44) are considered first. The results of the reliability analysis are summarized in Table 8.18b. The serviceability limit state function (Equation 8.52) is considered next. The allowable lateral deflection at node 1 is assumed to be 2.54 cm. The results are given in Table 8.19.

The structure with 6 flexible connection elements is considered next. The descriptions of the structural parameters and loads are shown in Table 8.18. The Richard model is used to represent the flexible connections. Three $M-\Theta$ curves shown in Figure 8.10 are considered. The results are shown in Table 8.20.

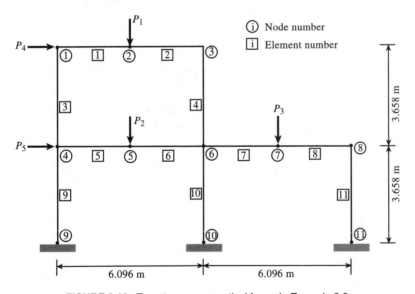

FIGURE 8.12. Two-story unsymmetical frame in Example 8.8.

Table 8.18a. Statistical descriptions of all the random variables in the two-story unsymmetrical frame in Example 8.8

Variable	Unit	Mean Value	COV	Distribution	Final Checking Point	Sensitivity Index
E	MPa	199,948.04	0.06	Lognormal	190,042.34	−0.271995
Beam						
A	cm^2	76.13	0.05	Lognormal	75.97	−0.005313
I_x	cm^4	21,519.16	0.05	Lognormal	20,855.07	−0.200367
Z_x	cm^3	819.35	0.05	Lognormal	791.23	−0.224213
Column						
A	cm^2	114.19	0.05	Lognormal	114.07	0.001224
I_x	cm^4	14,318.36	0.05	Lognormal	14,247.85	−0.024560
Z_x	cm^3	983.22	0.05	Lognormal	983.22	0.000000
F_y	MPa	260.62	0.10	Lognormal	226.45	−0.452156
Dead load						
P_1	kN	44.48	0.10	Lognormal	44.27	0.000941
P_2	kN	88.96	0.10	Lognormal	93.02	0.165162
P_3	kN	88.96	0.10	Lognormal	89.93	0.052567
Wind load						
P_4	kN	44.48	0.37	Type I	96.01	0.771541
P_5	kN	22.24	0.37	Type I	22.57	0.072145

Table 8.18b. Reliability analysis of the frame in Example 8.8 (strength limit state for element 6)

	Nonlinear	Linear
Reliability index	$\beta = 3.006$	$\beta = 3.017$

Table 8.19a. Safety indexes for the serviceability limit state at Node 1; $\delta_{limit} = 2.54$ cm, for the two-story unsymmetrical frame in Example 8.8

Variable	Unit	Mean Value	COV	Distribution	Final Checking Point	Sensitivity Index
E	MPa	199,948.04	0.06	Lognormal	199,477.13	−0.004125
Beam						
A	cm^2	76.13	0.05	Lognormal	76.03	−0.000035
I_x	cm^4	21,519.16	0.05	Lognormal	21,488.74	−0.001464
Column						
A	cm^2	114.19	0.05	Lognormal	114.05	−0.000021
I_x	cm^4	14,318.36	0.05	Lognormal	14,162.82	−0.085153
Dead load						
P_1	kN	44.48	0.10	Lognormal	44.25	−0.000995
P_2	kN	88.96	0.10	Lognormal	88.69	0.008263
P_3	kN	88.96	0.10	Lognormal	88.60	0.004068
Wind load						
P_4	kN	44.48	0.37	Type I	93.90	0.992914
P_5	kN	22.24	0.37	Type I	22.33	0.082252

Table 8.19b. Reliability analysis of the frame in Example 8.8 (serviceability limit state)

	Nonlinear	Linear
Reliability index	$\beta = 2.274$	$\beta = 2.283$

Table 8.20. Summary of safety indexes for different PR connections in Example 8.8

Connection Type	Safety Index (Nonlinear)
FR connections	2.274
PR connections	
Curve 1	2.233
Curve 2	1.927
Curve 3	1.483

Example 8.9 *Truss*

The truss shown in Figure 8.13 is considered next. Its geometric description is given in the figure. The basic random variables in the problem are summarized in Table 8.21. The maximum allowable deflection at the midspan is considered to be 7.62 cm. Using the serviceability limit state represented by Equation 8.52, and considering the linear and nonlinear behavior, the reliability of the truss is evaluated. The results are shown in Table 8.21. The reliability indexes for linear and nonlinear analyses are very different. For trusses, the effect of geometric nonlinearity is expected to be significant.

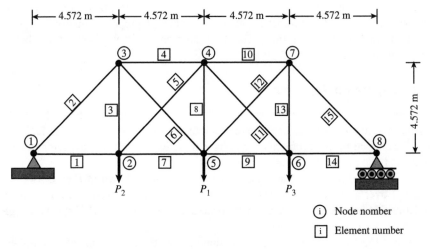

FIGURE 8.13. Truss in Example 8.9.

Table 8.21a. Reliability analysis of a truss in Example 8.9, serviceability limit state, allowable vertical deflection at node 5 = 7.62 cm

Variable	Unit	Mean Value (Initial Checking Point)	COV	Distribution	Final Checking Point Linear	Final Checking Point Nonlinear	Sensitivity Index Linear	Sensitivity Index Nonlinear
E	MPa	199,948.04	0.06	Lognormal	199,013.80	199,257.87	−0.00818	−0.00860
$A_1, A_2,$ $A_3, A_{13},$ A_{14}, A_{15}	cm^2	10.32	0.05	Lognormal	10.30	10.31	−0.00189	−0.00224
$A_4, A_5,$ $A_7, A_6,$ $A_8, A_9,$ $A_{10}, A_{11},$ A_{12}	cm^2	6.45	0.05	Lognormal	5.59	5.96	−0.48251	−0.48643
P_1	kN	266.89	0.10	Lognormal	399.29	338.93	0.84244	0.83740
P_2	kN	88.96	0.10	Lognormal	97.83	94.02	0.16918	0.17617
P_3	kN	88.96	0.10	Lognormal	97.85	94.02	0.16965	0.17617

Table 8.21b. Reliability analysis for the truss in Example 8.9

	Nonlinear	Linear
Reliability index	$\beta = 3.223$	$\beta = 5.889$

8.12 THREE-DIMENSIONAL PROBLEMS

The two-dimensional SFEM presented in the previous sections is extended next to consider three-dimensional (3D) problems. It will be denoted hereafter the 3D SFEM algorithm.

8.12.1 Deterministic Nonlinear FEM Formulation

As in the two-dimensional case, the essential steps of the deterministic nonlinear FEM are necessary considering the geometric properties, material properties, partially restrained connections, and support conditions. The discussion on 2D frames is still valid, but needs to be extended to 3D problems.

8.12.2 Geometric Nonlinearity

The commonly used linear iterative strategy represented by Equation 8.6 to solve nonlinear problems is still applicable, except that explicit expressions for the variables $\mathbf{K}$, $\mathbf{F}$, and $\mathbf{R}$ need to be defined for 3D problems. Obviously, these expressions are expected to be more complicated, but not really different than the 2D case.

A 3D beam–column element in an arbitrary position is shown in Figure 8.14, where a caret (^) over a variable indicates that it is a deformed coordinate. From Figure 8.14, the relationship between the member undeformed local coordinates $(\mathbf{X}_i', \mathbf{e}_i')$ and the global coordinates $(\mathbf{X}_i, \mathbf{e}_i)$ can be derived.

Figures 8.15a and b show the free-body diagrams of the projections of an infinitesimal length of a large deformed element on the $\hat{X}_1 - \hat{X}_2$ and $\hat{X}_1 - \hat{X}_3$ planes, respectively. The

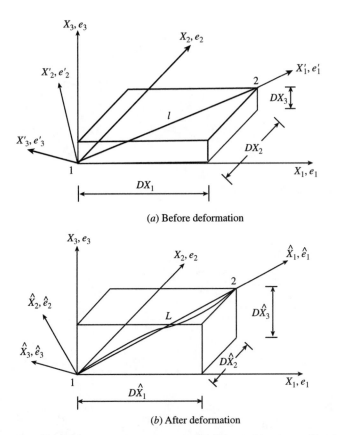

FIGURE 8.14. Kinematic relationships of undeformed and deformed local coordinates to the global coordinates for a 3D beam–column element in arbitrary position.

tangent stiffness matrix for the 3D beam–column element can be estimated using Equation 8.7. The expressions for the corresponding matrixes are given below.

The elastic property matrix $\mathbf{A}_{\sigma\sigma}$ can be shown to be

$$
\mathbf{A}_{\sigma\sigma} =
\begin{bmatrix}
\dfrac{l}{EA} & 0 & 0 & 0 & 0 & 0 \\[2mm]
0 & \dfrac{l}{GJ} & 0 & 0 & 0 & 0 \\[2mm]
0 & 0 & \dfrac{l}{3EI_2} & \dfrac{l}{6EI_2} & 0 & 0 \\[2mm]
0 & 0 & \dfrac{l}{6EI_2} & \dfrac{l}{3EI_2} & 0 & 0 \\[2mm]
0 & 0 & 0 & 0 & \dfrac{l}{3EI_3} & \dfrac{l}{6EI_3} \\[2mm]
0 & 0 & 0 & 0 & \dfrac{l}{6EI_3} & \dfrac{l}{3EI_3}
\end{bmatrix}
\tag{8.91}
$$

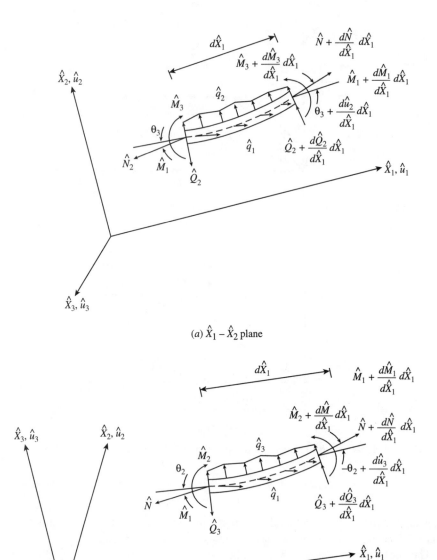

(a) $\hat{X}_1 - \hat{X}_2$ plane

(b) $\hat{X}_1 - \hat{X}_3$ plane

FIGURE 8.15. Projection of a free-body diagram of a 3D beam–column element.

where A is the cross-sectional area, I is the moment of inertia, J is the polar moment of inertia, E is the Young's modulus, and G is the shear modulus. $\mathbf{A}^t_{\sigma d0}$ is the transposed matrix. If the element is not parallel to the X_3 axis, $\mathbf{A}^t_{\sigma d0}$ can be expressed as

$$\mathbf{A}^t_{\sigma d0} = \begin{bmatrix} -\dfrac{\partial H}{\partial\tilde u_1} & 0 & \dfrac{-1}{l}\dfrac{\partial H}{\partial\tilde u_3}\dfrac{\partial L}{\partial\tilde u_1} & \dfrac{1}{l}\dfrac{\partial H}{\partial\tilde u_3}\dfrac{\partial L_1}{\partial\tilde u_1} & \dfrac{1}{l}\dfrac{\partial L_1}{\partial\tilde u_2} & \dfrac{-1}{l}\dfrac{\partial L_1}{\partial\tilde u_2} \\[2ex] -\dfrac{\partial H}{\partial\tilde u_2} & 0 & \dfrac{-1}{l}\dfrac{\partial H}{\partial\tilde u_3}\dfrac{\partial L_1}{\partial\tilde u_2} & \dfrac{1}{l}\dfrac{\partial H}{\partial\tilde u_3}\dfrac{\partial L_1}{\partial\tilde u_2} & \dfrac{-1}{l}\dfrac{\partial L_1}{\partial\tilde u_1} & \dfrac{1}{l}\dfrac{\partial L_1}{\partial\tilde u_1} \\[2ex] -\dfrac{\partial H}{\partial\tilde u_3} & 0 & \dfrac{1}{l}\dfrac{L_1}{L} & \dfrac{-1}{l}\dfrac{L_1}{L} & 0 & 0 \\[2ex] 0 & -\dfrac{\partial H}{\partial\tilde u_1} & \dfrac{\partial L_1}{\partial\tilde u_2} & 0 & \dfrac{\partial H}{\partial\tilde u_3}\dfrac{\partial L_1}{\partial\tilde u_1} & 0 \\[2ex] 0 & -\dfrac{\partial H}{\partial\tilde u_2} & -\dfrac{\partial L_1}{\partial\tilde u_1} & 0 & \dfrac{\partial H}{\partial\tilde u_3}\dfrac{\partial L_1}{\partial\tilde u_2} & 0 \\[2ex] 0 & -\dfrac{\partial H}{\partial\tilde u_3} & 0 & 0 & \dfrac{-L_1}{L} & 0 \\[2ex] \dfrac{\partial H}{\partial\tilde u_1} & 0 & \dfrac{1}{l}\dfrac{\partial H}{\partial\tilde u_3}\dfrac{\partial L_1}{\partial\tilde u_1} & \dfrac{-1}{l}\dfrac{\partial H}{\partial\tilde u_3}\dfrac{\partial L_1}{\partial\tilde u_1} & \dfrac{-1}{l}\dfrac{\partial L_1}{\partial\tilde u_2} & \dfrac{1}{l}\dfrac{\partial L_1}{\partial\tilde u_2} \\[2ex] \dfrac{\partial H}{\partial\tilde u_2} & 0 & \dfrac{1}{l}\dfrac{\partial H}{\partial\tilde u_3}\dfrac{\partial L_1}{\partial\tilde u_2} & \dfrac{-1}{l}\dfrac{\partial H}{\partial\tilde u_3}\dfrac{\partial L_1}{\partial\tilde u_2} & \dfrac{1}{l}\dfrac{\partial L_1}{\partial\tilde u_1} & \dfrac{-1}{l}\dfrac{\partial L_1}{\partial\tilde u_1} \\[2ex] \dfrac{\partial H}{\partial\tilde u_3} & 0 & \dfrac{-1}{l}\dfrac{L_1}{L} & \dfrac{1}{l}\dfrac{L_1}{L} & 0 & 0 \\[2ex] 0 & \dfrac{\partial H}{\partial\tilde u_1} & 0 & \dfrac{-\partial L_1}{\partial\tilde u_2} & 0 & \dfrac{-\partial H}{\partial\tilde u_3}\dfrac{\partial L_1}{\partial\tilde u_1} \\[2ex] 0 & \dfrac{\partial H}{\partial\tilde u_2} & 0 & \dfrac{\partial L_1}{\partial\tilde u_1} & 0 & \dfrac{-\partial H}{\partial\tilde u_3}\dfrac{\partial L_1}{\partial\tilde u_2} \\[2ex] 0 & \dfrac{\partial H}{\partial\tilde u_3} & 0 & 0 & 0 & \dfrac{L_1}{L} \end{bmatrix} \tag{8.92}$$

where l is the initial undeformed length of an element, L is the deformed length, H is the total axial elongation, and $\tilde u_i$ is the relative displacement of one node with respect to the other of an element in the global coordinate. These parameters are shown in Figures 8.14 and 8.15. If an element is parallel to the X_3 coordinate, a similar relationship can be obtained for $\mathbf{A}^t_{\sigma d0}$.

$\mathbf{A}_{dd0}$ can be derived from

$$\Delta \mathbf{T}^t \mathbf{P} = \mathbf{A}_{dd0}\Delta\mathbf{d} \tag{8.93}$$

where $\mathbf{P}$ is an internal nodal force vector for each element and can be expressed as

$$\mathbf{P} = \{\hat N \quad {}^1\hat Q_2 \quad {}^1\hat Q_3 \quad {}^1\hat M_1 \quad {}^1\hat M_2 \quad {}^1\hat M_3 \quad \hat N \quad {}^2\hat Q_2 \quad {}^2\hat Q_3 \quad {}^2\hat M_1 \quad {}^2\hat M_2 \quad {}^2\hat M_3\}^t \tag{8.94}$$

The internal axial force ($\hat N$), transverse shear forces ($\hat Q_2$, $\hat Q_3$), and moments ($\hat M_1$, $\hat M_2$, $\hat M_3$) are the six forces at each node of a 3D beam–column element.

The nodal displacement vector **d** in the global coordinate for each element is denoted as

$$\mathbf{d} = \{{}^{1}u_1 \quad {}^{1}u_2 \quad {}^{1}u_3 \quad {}^{1}\theta_1 \quad {}^{1}\theta_2 \quad {}^{1}\theta_3 \quad {}^{2}u_1 \quad {}^{2}u_2 \quad {}^{2}u_3 \quad {}^{2}\theta_1 \quad {}^{2}\theta_2 \quad {}^{2}\theta_3\}^{t} \quad (8.95)$$

The left superscripts in Equations 8.94 and 8.95 represent the node number and the right subscripts represent the directions of the global coordinates.

The rotational matrix **T** relates the deformed local and global unit basis. Assuming that the initial position of an element is arbitrary but not parallel to the X_3 coordinate, the first three rows of $\Delta \mathbf{T}^t \mathbf{P}$ in Equation 8.93 can be shown to be

$$\Delta \mathbf{T}^t \mathbf{P} = \Delta \begin{bmatrix} \dfrac{\partial H}{\partial \tilde{u}_1} & \dfrac{-\partial L_1}{\partial \tilde{u}_2} & \dfrac{-\partial H}{\partial \tilde{u}_3}\dfrac{\partial L_1}{\partial \tilde{u}_1} \\[2ex] \dfrac{\partial H}{\partial \tilde{u}_2} & \dfrac{\partial L_1}{\partial \tilde{u}_1} & \dfrac{-\partial H}{\partial \tilde{u}_3}\dfrac{\partial L_1}{\partial \tilde{u}_2} \\[2ex] \dfrac{\partial H}{\partial \tilde{u}_3} & 0 & \dfrac{L_1}{L} \end{bmatrix} \begin{Bmatrix} \hat{N} \\[1ex] {}^{1}\hat{Q}_2 \\[1ex] {}^{1}\hat{Q}_3 \end{Bmatrix} = \begin{bmatrix} A_{11} & A_{12} & A_{13} \\ A_{21} & A_{22} & A_{23} \\ A_{31} & A_{32} & A_{33} \end{bmatrix} \begin{Bmatrix} \Delta \tilde{u}_1 \\ \Delta \tilde{u}_2 \\ \Delta \tilde{u}_3 \end{Bmatrix}$$

$$(8.96)$$

where

$$\begin{Bmatrix} \Delta \tilde{u}_1 \\ \Delta \tilde{u}_2 \\ \Delta \tilde{u}_3 \end{Bmatrix} = \begin{bmatrix} -1 & 0 & 0 & 0 & 0 & 0 & 1 & 0 & 0 & 0 & 0 & 0 \\ 0 & -1 & 0 & 0 & 0 & 0 & 0 & 1 & 0 & 0 & 0 & 0 \\ 0 & 0 & -1 & 0 & 0 & 0 & 0 & 0 & 1 & 0 & 0 & 0 \end{bmatrix} \Delta \begin{Bmatrix} {}^{1}u_1 \\ {}^{1}u_2 \\ {}^{1}u_3 \\ {}^{1}\theta_1 \\ {}^{1}\theta_2 \\ {}^{1}\theta_3 \\ {}^{2}u_1 \\ {}^{2}u_2 \\ {}^{2}u_3 \\ {}^{2}\theta_1 \\ {}^{2}\theta_2 \\ {}^{2}\theta_3 \end{Bmatrix} \quad (8.97)$$

and

$$A_{11} = \frac{\partial^2 H}{\partial \tilde{u}_1^2}\hat{N} - \frac{\partial^2 L_1}{\partial \tilde{u}_1 \partial \tilde{u}_2}{}^{1}\hat{Q}_2 - \left(\frac{\partial^2 H}{\partial \tilde{u}_3 \partial \tilde{u}_1}\frac{\partial L_1}{\partial \tilde{u}_1} + \frac{\partial H}{\partial \tilde{u}_3}\frac{\partial^2 L_1}{\partial \tilde{u}_1^2} \right){}^{1}\hat{Q}_3$$

$$A_{12} = \frac{\partial^2 H}{\partial \tilde{u}_1 \partial \tilde{u}_2}\hat{N} - \frac{\partial^2 L_1}{\partial \tilde{u}_2^2}{}^{1}\hat{Q}_2 - \left(\frac{\partial^2 H}{\partial \tilde{u}_3 \partial \tilde{u}_2}\frac{\partial L_1}{\partial \tilde{u}_1} + \frac{\partial H}{\partial \tilde{u}_3}\frac{\partial^2 L_1}{\partial \tilde{u}_1 \partial \tilde{u}_2} \right){}^{1}\hat{Q}_3$$

$$A_{13} = \frac{\partial^2 H}{\partial \tilde{u}_1 \partial \tilde{u}_3}\hat{N} - \left(\frac{\partial^2 H}{\partial \tilde{u}_3^2}\frac{\partial L_1}{\partial \tilde{u}_1} \right){}^{1}\hat{Q}_3$$

$$A_{21} = \frac{\partial^2 H}{\partial \tilde{u}_2 \partial \tilde{u}_1}\hat{N} + \frac{\partial^2 L_1}{\partial \tilde{u}_1^2}{}^{1}\hat{Q}_2 - \left(\frac{\partial^2 H}{\partial \tilde{u}_1 \partial \tilde{u}_3}\frac{\partial L_1}{\partial \tilde{u}_2} + \frac{\partial H}{\partial \tilde{u}_3}\frac{\partial^2 L_1}{\partial \tilde{u}_1 \partial \tilde{u}_2} \right){}^{1}Q_3$$

$$A_{22} = \frac{\partial^2 H}{\partial \tilde{u}_2^2} \hat{N} + \frac{\partial^2 L_1}{\partial \tilde{u}_1 \partial \tilde{u}_2} {}^1 \hat{Q}_2 - \left(\frac{\partial^2 H}{\partial \tilde{u}_2 \partial \tilde{u}_3} \frac{\partial L_1}{\partial \tilde{u}_2} + \frac{\partial H}{\partial \tilde{u}_3} \frac{\partial^2 L_1}{\partial \tilde{u}_2^2} \right) {}^1 \hat{Q}_3$$

$$A_{23} = \frac{\partial^2 H}{\partial \tilde{u}_2 \partial \tilde{u}_3} \hat{N} - \left(\frac{\partial^2 H}{\partial \tilde{u}_3^2} \frac{\partial L_1}{\partial \tilde{u}_2} \right) {}^1 \hat{Q}_3$$

$$A_{31} = \frac{\partial^2 H}{\partial \tilde{u}_1 \partial \tilde{u}_3} \hat{N} + \left(\frac{\partial L_1}{\partial \tilde{u}_1} \frac{1}{L} - \frac{L_1}{L^2} \frac{\partial H}{\partial \tilde{u}_1} \right) {}^1 \hat{Q}_3$$

$$A_{32} = \frac{\partial^2 H}{\partial \tilde{u}_2 \partial \tilde{u}_3} \hat{N} + \left(\frac{\partial L_1}{\partial \tilde{u}_2} \frac{1}{L} - \frac{L_1}{L^2} \frac{\partial H}{\partial \tilde{u}_2} \right) {}^1 \hat{Q}_3$$

$$A_{33} = \frac{\partial^2 H}{\partial \tilde{u}_3^2} \hat{N} - \left(\frac{L_1}{L^2} \frac{\partial H}{\partial \tilde{u}_3} \right) {}^1 \hat{Q}_3$$

Thus, $\mathbf{A}_{dd0}$ can be calculated using Equation 8.93.

The internal force vector $\mathbf{R}$ in Equation 8.6 can be expressed by Equation 8.14. The corresponding terms can be estimated as discussed below. $\mathbf{R}\sigma$ can be shown to be

$$\mathbf{R}_\sigma = \left\{ \begin{array}{c} \dfrac{nl}{EA} - H \\[2mm] \dfrac{m_1 l}{GJ} - ({}^2\theta_1^* - {}^1\theta_1^*) \\[2mm] \dfrac{1}{EI_2} \left(\dfrac{{}^1 m_2 l}{3} + \dfrac{{}^2 m_2 l}{6} \right) + {}^1\theta_2^* \\[2mm] \dfrac{1}{EI_2} \left(\dfrac{{}^1 m_2 l}{6} + \dfrac{{}^2 m_2 l}{3} \right) - {}^2\theta_2^* \\[2mm] \dfrac{1}{EI_3} \left(\dfrac{{}^1 m_3 l}{3} + \dfrac{{}^2 m_3 l}{6} \right) + {}^1\theta_3^* \\[2mm] \dfrac{1}{EI_3} \left(\dfrac{{}^1 m_3 l}{6} + \dfrac{{}^2 m_3 l}{3} \right) - {}^2\theta_3^* \end{array} \right\} \tag{8.98}$$

$\mathbf{R}_{d0}$ is the homogeneous portion of the internal nodal force vector, and can be shown to be

$$\mathbf{R}_{d0} = \mathbf{T}^t \mathbf{P} \tag{8.99}$$

All the terms were defined earlier.

In Equation 8.6, $\mathbf{F}$ is an external nodal force vector for each element and can be written as

$$\mathbf{F} = \{ {}^1 F_1 \quad {}^1 F_2 \quad {}^1 F_3 \quad {}^1 M_1 \quad {}^1 M_2 \quad {}^1 M_3 \quad {}^2 F_1 \quad {}^2 F_2 \quad {}^2 F_3 \quad {}^2 M_1 \quad {}^2 M_2 \quad {}^2 M_3 \}^t$$

$$\tag{8.100}$$

All the terms in Equation 8.6 can be assembled in incremental form. Both the standard and modified Newton–Raphson methods with arc-length procedure can be used to solve Equation 8.6. Since the tangent stiffness matrix and the internal force vector are in explicit form, Equation 8.6 can be solved efficiently.

8.12.3 Elastoplastic Material Nonlinearity

The material nonlinearity also needs to be addressed to evaluate the reliability index. As in the 2D case, the elastoplastic material nonlinearity is considered here. The yield function for 3D problems used in this book has the following general form:

$$f(\hat{N}, \quad \hat{M}_1, \quad \hat{M}_2, \quad \hat{M}_3, \quad \sigma_y) = 0 \qquad \text{at} \quad \hat{X}_1 = l_\text{p} \qquad (8.101)$$

where $\hat{N}$ and $\hat{M}$ are the axial force and bending moments, respectively; σ_y is the yield stress of an element; and l_p is the location of a plastic hinge along the length of an element.

The presence of plastic hinges in a frame is expected to increase the axial and rotational deformation. The elastoplastic tangent stiffness $\mathbf{K}_\text{p}$ and the elastoplastic internal force $\mathbf{R}_\text{p}$ of the element can be evaluated using Equations 8.18 and 8.19. The evaluation procedures for all the 3D problem parameters in these equations except $\mathbf{V}_\text{p}$, $\mathbf{C}_\text{p}$, and $\mathbf{R}_\sigma$ were explained earlier. These three parameters can be expressed as

$$\mathbf{V}_\text{p} = \begin{Bmatrix} -\dfrac{\partial f}{\partial \hat{N}} \\[2mm] -\dfrac{\partial f}{\partial M_1} \\[2mm] -\dfrac{\partial f}{\partial \hat{M}_2}\left(1 - \dfrac{\hat{X}_1}{l}\right) \\[2mm] -\dfrac{\partial f}{\partial \hat{M}_2}\left(\dfrac{\hat{X}_1}{l}\right) \\[2mm] -\dfrac{\partial f}{\partial \hat{M}_3}\left(1 - \dfrac{\hat{X}_1}{l}\right) \\[2mm] -\dfrac{\partial f}{\partial \hat{M}_3}\left(\dfrac{\hat{X}_1}{l}\right) \end{Bmatrix} \qquad (8.102)$$

$$\mathbf{C}_\text{p}^\text{t} = (\mathbf{V}_\text{p}^\text{t} \mathbf{A}_{\sigma\sigma}^{-1} \mathbf{V}_\text{p})^{-1} \mathbf{V}_\text{p}^\text{t} \mathbf{A}_{\sigma\sigma}^{-1} \qquad (8.103)$$

and

$$\hat{\mathbf{R}}_\sigma = \mathbf{R}_\sigma + \begin{Bmatrix} H_\text{p} \\[2mm] \theta_{\text{p}1}^* \\[2mm] \theta_{\text{p}2}^*\left(1 - \dfrac{\hat{X}_1}{l}\right) \\[2mm] \theta_{\text{p}2}^*\left(\dfrac{\hat{X}_1}{l}\right) \\[2mm] \theta_{\text{p}3}^*\left(1 - \dfrac{\hat{X}_1}{l}\right) \\[2mm] \theta_{\text{p}3}^*\left(\dfrac{\hat{X}_1}{l}\right) \end{Bmatrix} \qquad \text{at} \quad \hat{X}_1 = l_\text{p} \qquad (8.104)$$

where H_p and θ_p^* are the additional axial elongation and additional relative rotation due to the plastic hinges. All other parameters were defined earlier.

Considering either Equations 8.7 and 8.14 or Equations 8.18 and 8.19, depending on the elastic or inelastic state of the material, the effect of the material nonlinearity can be incorporated in the analysis. The basic simplicity of the algorithm is not lost even when the material and geometric nonlinearities are considered simultaneously.

The basic concept is that the assumed stress-based tangent stiffness matrix and the internal force vector can be developed conventionally using the geometric and material properties of an element. Since $\mathbf{K}$ and $\mathbf{R}$ can be expressed in explicit form, geometrically nonlinear problems can be solved efficiently using Equation 8.6.

8.12.4 Partially Restrained or Flexible Connections

The PR connections need to be incorporated in the algorithm at this stage. As discussed for 2D problems, the comprehensive properties of the flexible connection are represented by an $M-\Theta$ curve. As in the 2D problems, the Richard model represented by Equation 8.26 is used for 3D problems.

Equation 8.26 represents the in-plane behavior of the connections. For 3D frames, the out-of-plane behavior of the connections needs to be considered properly. Almost all the experimental and analytical methods available in the literature consider only the in-plane or major axis behavior of PR connections. In a three-dimensional representation of a real structure, the rotational flexibility of both the major and minor axes of PR connections needs to be considered appropriately. As discussed in Section 8.3.3.1, the current major axis Young's modulus $E(\Theta)$ can be evaluated using Equation 8.27. To evaluate the current minor axis Young's modulus, the same equation can be used, except that the minor axis $K(\Theta)$ and I values are used instead. For the major axis rotational flexibility of a connection element, Equation 8.26 can still be used, where k, k_p, M_0, and N represent the four parameters of the Richard model about the major axis. The same Richard model can also be used to represent the minor axis rotational flexibility, except that four other parameters, denoted as k', k_p', M_0', and N', need to be evaluated with respect to the minor axis.

8.12.5 Support Conditions

As in 2D problems, support conditions can be modeled as PR connections with different rigidities. Using the Richard model and the connection element approach, the support conditions can be easily incorporated in the formulation.

Thus, a 3D deterministic finite-element model that considers nonlinearities due to geometry, material, PR connections, and flexible support conditions is now available.

8.13 SFEM FOR 3D PROBLEMS

The gradient vector in Equation 8.40 needs to be evaluated for 3D problems. The same three partial derivatives and four Jacobians in Equation 8.40 need to be evaluated considering the performance criteria and the basic random variables in the problem.

8.14 PERFORMANCE FUNCTIONS AND PARTIAL DIFFERENTIALS

As in the 2D case, the safety of a structure needs to be evaluated with respect to predetermined performance criteria. Both the strength and serviceability performance criteria are considered here.

8.14.1 Strength Performance Functions

For 3D problems, the strength limit state function can be expressed as

$$G(\mathbf{x}, \mathbf{u}, \mathbf{s}) = 1.0 - \frac{P_u}{P_n} - \frac{8}{9}\left(\frac{M_{ux}}{M_{nx}} + \frac{M_{uy}}{M_{ny}}\right) \qquad \text{if} \quad \frac{P_u}{\phi P_n} \geq 0.2 \qquad (8.105)$$

$$G(\mathbf{x}, \mathbf{u}, \mathbf{s}) = \left(\frac{P_u}{2P_n} + \frac{M_{ux}}{M_{nx}} + \frac{M_{uy}}{M_{ny}}\right) \qquad \text{if} \quad \frac{P_u}{\phi P_n} < 0.2 \qquad (8.106)$$

where ϕ is the resistance factor, P_u is the required tensile/compressive strength, P_n is the nominal tensile/compressive strength, M_{ux} and M_{uy} are the required flexural strength for the x axis and y axis, respectively, and M_{nx} and M_{ny} are the nominal flexural strength for the x axis and y axis, respectively. As stated in the LRFD *Manual* (AISC, 1994), P_n can be evaluated as $P_n = AF_{cr}$ (compression) or $P_n = AF_y$ (tension), $M_{nx} = Z_x F_y$ and $M_{ny} = Z_y F_y$. Please refer to the LRFD *Manual* for further detail.

One can derive $\partial G/\partial \mathbf{x}$, $\partial G/\partial \mathbf{u}$, and $\partial G/\partial \mathbf{s}$ from Equations 8.105 and 8.106. Since neither G function contains an explicit displacement component, $\partial G/\partial \mathbf{u} = 0$. To calculate $\partial G/\partial \mathbf{x}$, the basic random variables are defined here as the Young's modulus, E; area, A; yield stress, F_y; plastic modulus, Z_x and Z_y; and the moments of inertia of a cross section, J, I_x, and I_y with external load F. Therefore,

$$\frac{\partial G}{\partial \mathbf{x}} = \left\{ \frac{\partial G}{\partial E} \quad \frac{\partial G}{\partial A} \quad \frac{\partial G}{\partial J} \quad \frac{\partial G}{\partial I_x} \quad \frac{\partial G}{\partial I_y} \quad \frac{\partial G}{\partial Z_x} \quad \frac{\partial G}{\partial Z_y} \quad \frac{\partial G}{\partial F_y} \right\} \qquad (8.107)$$

Considering either Equation 8.105 or 8.106, the corresponding terms in Equation 8.107 can be calculated as discussed below. Refer to Equations 8.46 to 8.50 for all other variables required to evaluate Equation 8.107.

8.14.1.1 *Compression Members*   Assume that the x axis is the governing direction. The formulations for the case in which the y axis is the governing direction can be derived accordingly. When $P_u/\phi P_n \geq 0.2$, and $\lambda_c \leq 1.5$,

$$\frac{\partial G}{\partial E} = \frac{P_u}{P_n E}(-\lambda_c^2 \ln 0.658)$$

$$\frac{\partial G}{\partial A} = \frac{P_u}{A P_n}(1 + \lambda_c^2 \ln 0.658)$$

$$\frac{\partial G}{\partial J} = 0$$

$$\frac{\partial G}{\partial I_x} = \frac{-P_u \lambda_c^2}{P_n I_x}\ln 0.658$$

$$\frac{\partial G}{\partial I_y} = 0$$

$$\frac{\partial G}{\partial Z_x} = \frac{8}{9} \frac{M_{ux}}{M_{nx} Z_x}$$

$$\frac{\partial G}{\partial Z_y} = \frac{8}{9} \frac{M_{uy} Z_y}{M_{ny} Z_y}$$

$$\frac{\partial G}{\partial F_y} = \frac{P_u}{P_n F_{cr}} 0.658^{\lambda_c^2} (1 + \lambda_c^2 \ln 0.658) + \frac{8}{9 F_y} \left(\frac{M_{ux}}{M_{nx}} + \frac{M_{uy}}{M_{ny}} \right)$$

When $P_u / \phi P_n \geq 0.2$, and $\lambda_c > 1.5$,

$$\frac{\partial G}{\partial E} = \frac{P_u}{P_n E}$$

$$\frac{\partial G}{\partial A} = 0$$

$$\frac{\partial G}{\partial J} = 0$$

$$\frac{\partial G}{\partial I_x} = \frac{P_u}{P_n I_x}$$

$$\frac{\partial G}{\partial I_y} = 0$$

$$\frac{\partial G}{\partial Z_x} = \frac{8}{9} \frac{M_{ux}}{M_{nx} Z_x}$$

$$\frac{\partial G}{\partial Z_y} = \frac{8}{9} \frac{M_{uy}}{M_{ny} Z_y}$$

$$\frac{\partial G}{\partial F_y} = \frac{8}{9 F_y} \left(\frac{M_{ux}}{M_{nx}} + \frac{M_{uy}}{M_{ny}} \right)$$

When $P_u / \phi P_n < 0.2$, and $\lambda_c \leq 1.5$,

$$\frac{\partial G}{\partial E} = \frac{P_u}{2 P_n E} (-\lambda_c^2 \ln 0.658)$$

$$\frac{\partial G}{\partial A} = \frac{P_u}{2 P_n A} (1 + \lambda_c^2 \ln 0.658)$$

$$\frac{\partial G}{\partial J} = 0$$

$$\frac{\partial G}{\partial I_x} = \frac{-P_u \lambda_c^2}{2 P_n I_x} \ln 0.658$$

$$\frac{\partial G}{\partial I_y} = 0$$

$$\frac{\partial G}{\partial Z_x} = \frac{M_{ux}}{M_{nx} Z_x}$$

$$\frac{\partial G}{\partial Z_y} = \frac{M_{uy}}{M_{ny} Z_y}$$

$$\frac{\partial G}{\partial F_y} = \frac{P_u}{2 P_n F_{cr}} 0.658^{\lambda_c^2} (1 + \lambda_c^2 \ln 0.658) + \frac{M_{ux}}{M_{nx} F_y} + \frac{M_{uy}}{M_{ny} F_y}$$

When $P_u / \phi P_n < 0.2$, and $\lambda_c > 1.5$,

$$\frac{\partial G}{\partial E} = \frac{P_u}{2 P_n E}$$

$$\frac{\partial G}{\partial A} = 0$$

$$\frac{\partial G}{\partial J} = 0$$

$$\frac{\partial G}{\partial I_x} = \frac{P_u}{2 P_n I_x}$$

$$\frac{\partial G}{\partial I_y} = 0$$

$$\frac{\partial G}{\partial Z_x} = \frac{M_{ux}}{M_{nx} Z_x}$$

$$\frac{\partial G}{\partial Z_y} = \frac{M_{uy}}{M_{ny} Z_y}$$

$$\frac{\partial G}{\partial F_y} = \frac{M_{ux}}{M_{nx} F_y} + \frac{M_{uy}}{M_{ny} F_y}$$

8.14.1.2 Tension Members When $P_u / \phi P_n \geq 0.2$,

$$\frac{\partial G}{\partial E} = 0$$

$$\frac{\partial G}{\partial A} = \frac{P_u}{A P_n}$$

$$\frac{\partial G}{\partial J} = 0$$

$$\frac{\partial G}{\partial I_x} = 0$$

$$\frac{\partial G}{\partial I_y} = 0$$

$$\frac{\partial G}{\partial Z_x} = \frac{8}{9} \frac{M_{ux}}{M_{nx} Z_x}$$

$$\frac{\partial G}{\partial Z_y} = \frac{8}{9} \frac{M_{uy}}{M_{ny} Z_y}$$

$$\frac{\partial G}{\partial F_y} = \frac{P_u}{P_n F_y} + \frac{8}{9F_y}\left(\frac{M_{ux}}{M_{nx}} + \frac{M_{uy}}{M_{ny}}\right)$$

When $P_u/\phi P_n < 0.2$,

$$\frac{\partial G}{\partial E} = 0$$

$$\frac{\partial G}{\partial A} = \frac{P_u}{2A P_n}$$

$$\frac{\partial G}{\partial J} = 0$$

$$\frac{\partial G}{\partial I_x} = 0$$

$$\frac{\partial G}{\partial I_y} = 0$$

$$\frac{\partial G}{\partial Z_x} = \frac{M_{ux}}{M_{nx} Z_x}$$

$$\frac{\partial G}{\partial Z_y} = \frac{M_{uy}}{M_{ny} Z_y}$$

$$\frac{\partial G}{\partial F_y} = \frac{P_u}{2P_n F_y} + \frac{1}{F_y}\left(\frac{M_{ux}}{M_{nx}} + \frac{M_{uy}}{M_{ny}}\right)$$

The following expression is used to calculate the partial derivative $\partial G/\partial s$:

$$\frac{\partial G}{\partial s} = \left[\frac{\partial G}{\partial P_u} \quad \frac{\partial G}{\partial M_{ux}} \quad \frac{\partial G}{\partial M_{uy}}\right] \tag{8.108}$$

For $P_u/\phi P_n \geq 0.2$,

$$\frac{\partial G}{\partial P_u} = -\frac{1}{P_n}, \quad \frac{\partial G}{\partial M_{ux}} = -\frac{8}{9M_{nx}}, \quad \text{and} \quad \frac{\partial G}{\partial M_{uy}} = -\frac{8}{9M_{ny}}$$

When $P_u/\phi P_n < 0.2$,

$$\frac{\partial G}{\partial P_u} = -\frac{1}{2P_n}, \quad \frac{\partial G}{\partial M_{ux}} = -\frac{1}{M_{nx}}, \quad \text{and} \quad \frac{\partial G}{\partial M_{uy}} = -\frac{1}{M_{ny}}$$

8.14.2 Serviceability Performance Function

The discussion in Section 8.6.2 for 2D structures is valid for 3D structures also.

8.15 EVALUATION OF JACOBIANS USING THE ADJOINT VARIABLE METHOD FOR 3D PROBLEMS

As in the 2D case, the implementation of FORM in the reliability estimation requires the evaluation of the gradient of the performance function represented by Equation 8.40. Only the procedures to estimate the four Jacobians, $\mathbf{J}_{s,x}$, $\mathbf{J}_{y,x}$, $\mathbf{J}_{s,D}$, and $\mathbf{J}_{D,x}$ are necessary at this stage. Since $\mathbf{s}$ is not an explicit function of the basic random variables $\mathbf{x}$, $\mathbf{J}_{s,x} = 0$. The Jacobians $\mathbf{J}_{s,D}$ and $\mathbf{J}_{D,x}$ are evaluated with the help of adjoint vector λ, as discussed in Section 8.7.

Using Equation 8.55, λ can be estimated by considering the limit state functions. For the serviceability limit state function represented by Equation 8.52, λ is very easy to calculate, as described for 2D problems. For the strength limit state functions represented by Equations 8.43 and 8.44, the Jacobians $\mathbf{J}_{s,D}$ represented by Equation 8.56 need to be evaluated to estimate λ. The nodal displacement vector in the global coordinate for a 3D element is defined by Equation 8.95, and the internal force vector σ becomes

$$\sigma = \left\{ n \quad m_1 \quad {}^1m_2 \quad {}^2m_2 \quad {}^1m_3 \quad {}^2m_3 \right\}^t \tag{8.109}$$

The relationship between σ and $\mathbf{d}$ for 3D problems can be derived as

$$\sigma = \left\{ \begin{array}{c} n \\ m_1 \\ {}^1m_2 \\ {}^2m_2 \\ {}^1m_3 \\ {}^2m_3 \end{array} \right\} = \frac{1}{l} \left\{ \begin{array}{c} EAH \\ GJ({}^2\theta_1^* - {}^1\theta_1^*) \\ -2EI_2(2\,{}^1\theta_2^* + {}^2\theta_2^*) \\ 2EI_2({}^1\theta_2^* + 2\,{}^2\theta_2^*) \\ -2EI_3(2\,{}^1\theta_3^* + {}^2\theta_3^*) \\ 2EI_3({}^1\theta_3^* + 2\,{}^2\theta_3^*) \end{array} \right\} \tag{8.110}$$

where ${}^\alpha\theta_i^*$ is the relative rotation of node α ($\alpha = 1$ or 2) given by Equation 8.60.

To evaluate Equation 8.56, the derivative of the internal forces with respect to the displacement becomes

$$\frac{\partial \sigma}{\partial \mathbf{d}} = \left\{ \frac{\partial n}{\partial \mathbf{d}} \quad \frac{\partial m_1}{\partial \mathbf{d}} \quad \frac{\partial {}^1m_2}{\partial \mathbf{d}} \quad \frac{\partial {}^2m_2}{\partial \mathbf{d}} \quad \frac{\partial {}^1m_3}{\partial \mathbf{d}} \quad \frac{\partial {}^2m_3}{\partial \mathbf{d}} \right\}^t \tag{8.111}$$

where

$$\frac{\partial n}{\partial \mathbf{d}} = \frac{AE}{Ll} \left\{ \begin{array}{c} -(DX_1 + \tilde{u}_1) \\ -(DX_2 + \tilde{u}_2) \\ -(DX_3 + \tilde{u}_3) \\ 0 \\ 0 \\ 0 \\ (DX_1 + \tilde{u}_1) \\ (DX_2 + \tilde{u}_2) \\ (DX_3 + \tilde{u}_3) \\ 0 \\ 0 \\ 0 \end{array} \right\} \tag{8.112}$$

$$\frac{\partial m_1}{\partial \mathbf{d}} = \frac{GJ}{l} \begin{Bmatrix} 0 \\ 0 \\ 0 \\ -(T_0)_{11} \\ -(T_0)_{12} \\ -(T_0)_{13} \\ 0 \\ 0 \\ 0 \\ (T_0)_{11} \\ (T_0)_{12} \\ (T_0)_{13} \end{Bmatrix} \tag{8.113}$$

$$\frac{\partial^1 m_2}{\partial \mathbf{d}} = \frac{-2EI_2}{l} \begin{Bmatrix} 3\dfrac{(T_0)_{11}\xi_1 - (T_0)_{31}\xi_2}{\xi_1^2 + \xi_2^2} \\[2mm] 3\dfrac{(T_0)_{12}\xi_1 - (T_0)_{32}\xi_2}{\xi_1^2 + \xi_2^2} \\[2mm] 3\dfrac{(T_0)_{13}\xi_1 - (T_0)_{33}\xi_2}{\xi_1^2 + \xi_2^2} \\[2mm] 2(T_0)_{21} \\ 2(T_0)_{22} \\ 2(T_0)_{23} \\[1mm] -3\dfrac{(T_0)_{11}\xi_1 - (T_0)_{31}\xi_2}{\xi_1^2 + \xi_2^2} \\[2mm] -3\dfrac{(T_0)_{12}\xi_1 - (T_0)_{32}\xi_2}{\xi_2^2 + \xi_2^2} \\[2mm] -3\dfrac{(T_0)_{13}\xi_1 - (T_0)_{33}\xi_2}{\xi_1^2 + \xi_2^2} \\[2mm] (T_0)_{21} \\ (T_0)_{22} \\ (T_0)_{23} \end{Bmatrix} \tag{8.114}$$

$$\frac{\partial^2 m_2}{\partial \mathbf{d}} = \frac{2EI_2}{l} \left\{ \begin{array}{c} 3\dfrac{(T_0)_{11}\xi_1 - (T_0)_{31}\xi_2}{\xi_1^2 + \xi_2^2} \\[2ex] 3\dfrac{(T_0)_{12}\xi_1 - (T_0)_{32}\xi_2}{\xi_1^2 + \xi_2^2} \\[2ex] 3\dfrac{(T_0)_{13}\xi_1 - (T_0)_{33}\xi_2}{\xi_1^2 + \xi_2^2} \\[1.5ex] (T_0)_{21} \\ (T_0)_{22} \\ (T_0)_{23} \\[1ex] -3\dfrac{(T_0)_{11}\xi_1 - (T_0)_{31}\xi_2}{\xi_1^2 + \xi_2^2} \\[2ex] -3\dfrac{(T_0)_{12}\xi_1 - (T_0)_{32}\xi_2}{\xi_1^2 + \xi_2^2} \\[2ex] -3\dfrac{(T_0)_{13}\xi_1 - (T_0)_{33}\xi_2}{\xi_1^2 + \xi_2^2} \\[1.5ex] 2(T_0)_{21} \\ 2(T_0)_{22} \\ 2(T_0)_{23} \end{array} \right\} \tag{8.115}$$

$$\frac{\partial^1 m_3}{\partial \mathbf{d}} = \frac{-2EI_3}{l} \left\{ \begin{array}{c} 3\dfrac{(T_0)_{21}\xi_2 - (T_0)_{11}\xi_3}{\xi_2^2 + \xi_3^2} \\[2ex] 3\dfrac{(T_0)_{22}\xi_2 - (T_0)_{12}\xi_3}{\xi_2^2 + \xi_3^2} \\[2ex] 3\dfrac{(T_0)_{23}\xi_2 - (T_0)_{13}\xi_3}{\xi_2^2 + \xi_3^2} \\[1.5ex] 2(T_0)_{31} \\ 2(T_0)_{32} \\ 2(T_0)_{33} \\[1ex] -3\dfrac{(T_0)_{21}\xi_2 - (T_0)_{11}\xi_3}{\xi_2^2 + \xi_3^2} \\[2ex] -3\dfrac{(T_0)_{22}\xi_2 - (T_0)_{12}\xi_3}{\xi_2^2 + \xi_3^2} \\[2ex] -3\dfrac{(T_0)_{23}\xi_2 - (T_0)_{13}\xi_3}{\xi_2^2 + \xi_3^2} \\[1.5ex] (T_0)_{31} \\ (T_0)_{32} \\ (T_0)_{33} \end{array} \right\} \tag{8.116}$$

$$\frac{\partial^2 m_3}{\partial \mathbf{d}} = \frac{2EI_3}{l} \left\{ \begin{array}{c} 3\dfrac{(T_0)_{21}\xi_2 - (T_0)_{11}\xi_3}{\xi_2^2 + \xi_3^2} \\[3mm] 3\dfrac{(T_0)_{22}\xi_2 - (T_0)_{12}\xi_3}{\xi_2^2 + \xi_3^2} \\[3mm] 3\dfrac{(T_0)_{23}\xi_2 - (T_0)_{13}\xi_3}{\xi_2^2 + \xi_3^2} \\[3mm] (T_0)_{31} \\ (T_0)_{32} \\ (T)_{33} \\[2mm] -3\dfrac{(T_0)_{21}\xi_2 - (T_0)_{11}\xi_3}{\xi_2^2 + \xi_3^2} \\[3mm] -3\dfrac{(T_0)_{22}\xi_2 - (T_0)_{12}\xi_3}{\xi_2^2 + \xi_3^2} \\[3mm] -3\dfrac{(T_0)_{23}\xi_2 - (T_0)_{13}\xi_3}{\xi_2^2 + \xi_3^2} \\[3mm] 2(T_0)_{31} \\ 2(T_0)_{32} \\ 2(T_0)_{33} \end{array} \right\} \tag{8.117}$$

where

$$\xi_1 = (T_0)_{31}\tilde{u}_1 + (T_0)_{32}\tilde{u}_2 + (T_0)_{33}\tilde{u}_3$$

$$\xi_2 = (T_0)_{11}\tilde{u}_1 + (T_0)_{12}\tilde{u}_2 + (T_0)_{13}\tilde{u}_3 + l$$

$$\xi_3 = (T_0)_{21}\tilde{u}_1 + (T_0)_{22}\tilde{u}_2$$

$\tilde{u}_i$ is the relative displacement of node 2 and node 1 of an element in the global coordinate and T_0 is a rotational matrix that relates the global (X_i, e_i) and undeformed local coordinates (X_i', e_i'), as shown in Figure 8.14a. It can be shown to be

$$\mathbf{T}_0 = \begin{bmatrix} \dfrac{DX_1}{l} & \dfrac{DX_2}{l} & \dfrac{DX_3}{l} \\[3mm] -\dfrac{DX_2}{l_1} & \dfrac{DX_1}{l_1} & 0 \\[3mm] -\dfrac{DX_1 DX_3}{ll_1} & -\dfrac{DX_2 DX_3}{ll_1} & \dfrac{l_1}{l} \end{bmatrix} \tag{8.118}$$

where DX_i's are the coordinate difference between node 2 and node 1 in the global coordinate X_i, and l is the initial length of an element. It can be estimated as

$$l = \left[(DX_1)^2 + (DX_2)^3 + (DX_3)^2 \right]^{1/2} \tag{8.119}$$

and

$$l_1 = \left[(DX_1)^2 + (DX_2)^2\right]^{1/2} \tag{8.120}$$

Equation 8.66 can be used to calculate $\partial R/\partial x$. $R_{A\sigma}$ for 3D problems becomes

$$R_{A\sigma} = \begin{Bmatrix} n - \dfrac{EAH}{l} \\[2mm] m_1 - \dfrac{GJ}{l}({}^2\theta_1^* - {}^1\theta_1^*) \\[2mm] {}^1m_2 + \dfrac{2EI_2}{l}(2\,{}^1\theta_2^* + {}^2\theta_2^*) \\[2mm] {}^2m_2 - \dfrac{2EI_2}{l}({}^1\theta_2^* + 2\,{}^2\theta_2^*) \\[2mm] {}^1m_3 + \dfrac{2EI_3}{l}(2\,{}^1\theta_3^* + {}^2\theta_3^*) \\[2mm] {}^2m_3 - \dfrac{2EI_3}{l}({}^1\theta_3^* + 2\,{}^2\theta_3^*) \end{Bmatrix} \tag{8.121}$$

Since R_{d0} and $A_{\sigma d0}^t$ are not functions of basic variables, the derivative of R with respect to x can be evaluated using Equation 8.68. As mentioned earlier, the evaluation of Equation 8.68 requires identification of the basic random variables. Among the structural parameters, only the Young's modulus, E; area, A; and the moments of inertia of a cross section, J, I_2, I_3, with the external load, F, are considered to be basic random variables. The partial differentiation on the right-hand side of Equation 8.68 can be calculated for the elastic and elastoplastic nonlinear cases, as discussed in the following sections.

8.15.1 Elastic Nonlinear Case

In this case, $\partial R_{A\sigma}/\partial x$ for a 3D beam–column element can be shown to be

$$\frac{\partial R_{A\sigma}}{\partial x} = \left[\begin{array}{cccccc} \dfrac{\partial R_{A\sigma}}{\partial E} & \dfrac{\partial R_{A\sigma}}{\partial A} & \dfrac{\partial R_{A\sigma}}{\partial J} & \dfrac{\partial R_{A\sigma}}{\partial I_2} & \dfrac{\partial R_{A\sigma}}{\partial I_3} & 0 \end{array}\right] \tag{8.122}$$

where

$$\frac{\partial R_{A\sigma}}{\partial E} = \frac{1}{l}\begin{Bmatrix} -AH \\[2mm] -\dfrac{J}{2(1+\mu)}({}^2\theta_1^* - {}^1\theta_1^*) \\[2mm] 2I_2(2\,{}^1\theta_2 + {}^2\theta_2^*) \\[2mm] -2I_2({}^1\theta_2^* + 2\,{}^2\theta_2^*) \\[2mm] 2I_3(2\,{}^1\theta_3^* + {}^2\theta_3^*) \\[2mm] -2I_3({}^1\theta_3^* + 2\,{}^2\theta_3^*) \end{Bmatrix} \tag{8.123}$$

$$\frac{\partial \mathbf{R}_{A\sigma}}{\partial A} = \frac{1}{l} \begin{Bmatrix} -EH \\ 0 \\ 0 \\ 0 \\ 0 \\ 0 \end{Bmatrix} \tag{8.124}$$

$$\frac{\partial \mathbf{R}_{A\sigma}}{\partial J} = \frac{1}{l} \begin{Bmatrix} 0 \\ -G(^2\theta_1^* - {}^1\theta_1^*) \\ 0 \\ 0 \\ 0 \\ 0 \end{Bmatrix} \tag{8.125}$$

$$\frac{\partial \mathbf{R}_{A\sigma}}{\partial I_2} = \frac{1}{l} \begin{Bmatrix} 0 \\ 0 \\ 2E(2\,{}^1\theta_2^* + {}^2\theta_2^*) \\ -2E(^1\theta_2^* + 2\,{}^2\theta_2^*) \\ 0 \\ 0 \end{Bmatrix} \tag{8.126}$$

$$\frac{\partial \mathbf{R}_{A\sigma}}{\partial l_3} = \frac{1}{l} \begin{Bmatrix} 0 \\ 0 \\ 0 \\ 0 \\ 2E(2\,{}^1\theta_3^* + {}^2\theta_3^*) \\ -2E(^1\theta_3^* + 2\,{}^2\theta_3^*) \end{Bmatrix} \tag{8.127}$$

8.15.2 Elastoplastic Nonlinear Case

The plastic hinge model is used in this section. When the combined axial force and bending moments of an elastic–perfectly-plastic material model satisfy a prescribed yield function at a node of an element, a plastic hinge is assumed to occur instantaneously at that location. $\mathbf{V}_p\mathbf{C}_p^t$ becomes unity and $\mathbf{K}_p$ becomes zero. Thus, $\partial \mathbf{R}/\partial \mathbf{x}$ becomes zero.

The information is now available to evaluate the gradient of a performance function using Equation 8.40 for the elastic and elastoplastic nonlinear cases.

8.16 CONSIDERATION OF UNCERTAINTIES IN THE CONNECTION MODEL

For 3D frames, the out-of-plane behavior of the connections needs to be considered properly. Both major and minor axis rotational flexibility of connections are considered here. Denoting k, k_p, M_0, and N as the four parameters representing the major axis rotational flexibility of a connection element, and k', k_p', M_0', and N' as the four parameters representing the minor axis rotational flexibility of the connection element, the stochastic information on both major and minor axes can be included in Equation 8.122 as follows:

$$\frac{\partial \mathbf{R}_{A\sigma}}{\partial \mathbf{x}} = \begin{bmatrix} \dfrac{\partial \mathbf{R}_{A\sigma}}{\partial k} & \dfrac{\partial \mathbf{R}_{A\sigma}}{\partial k_p} & \dfrac{\partial \mathbf{R}_{A\sigma}}{\partial M_0} & \dfrac{\partial \mathbf{R}_{A\sigma}}{\partial N} & \dfrac{\partial \mathbf{R}_{A\sigma}}{\partial k'} & \dfrac{\partial \mathbf{R}_{A\sigma}}{\partial k'_p} & \dfrac{\partial \mathbf{R}_{A\sigma}}{\partial M'_0} & \dfrac{\partial \mathbf{R}_{A\sigma}}{\partial N'} & 0 \end{bmatrix}$$

(8.128)

The components in the major axis direction in Equation 8.128 can be shown to be

$$\frac{\partial \mathbf{R}_{A\sigma}}{\partial \xi_i} = 2\frac{\partial \mathbf{K}(\Theta)}{\partial \xi_i} \{0 \quad 0 \quad (2\,{}^1\theta_2^* + {}^2\theta_2^*) \quad -({}^1\theta_2^* + 2\,{}^2\theta_2^*) \quad 0 \quad 0\}^t$$

(8.129)

where $\xi_i = k$, k_p, M_0, and N. The corresponding components in the minor axis direction are

$$\frac{\partial \mathbf{R}_{A\sigma}}{\partial \xi'_i} = 2\frac{\partial \mathbf{K}(\Theta)}{\partial \xi'_i} \{0 \quad 0 \quad 0 \quad 0 \quad (2\,{}^1\theta_2^* + {}^2\theta_2^*) \quad -({}^1\theta_2^* + 2\,{}^2\theta_2^*)\}^t$$

(8.130)

where $\xi'_i = k'$, k'_p, M'_0, and N', and $\partial K(\Theta)/\partial \xi_i$ and $\partial K(\Theta)/\partial \xi'_i$ can be estimated using Equations 8.75 to 8.79.

The quantities required for the computation of ∇G in Equation 8.40 for a 3D frame are now available in a simple explicit form considering the geometric nonlinearity, elastoplastic material behavior, and partially restrained connections.

8.17 LINEAR ELASTIC CASE

For the linear elastic case, the discussion on 2D problems is still valid. To calculate $\mathbf{A}_{\sigma\sigma}^{-1}$ in Equation 8.86, it can be shown that

$$\mathbf{A}_{\sigma\sigma}^{-1} = \frac{1}{l}\begin{bmatrix} EA & 0 & 0 & 0 & 0 & 0 \\ 0 & GJ & 0 & 0 & 0 & 0 \\ 0 & 0 & 4EI_2 & -2EL_2 & 0 & 0 \\ 0 & 0 & -2EI_2 & 4EI_2 & 0 & 0 \\ 0 & 0 & 0 & 0 & 4EI_3 & -2EI_3 \\ 0 & 0 & 0 & 0 & -2EI_3 & 4EI_3 \end{bmatrix}$$

(8.131)

If E, A, J, I_2, and I_3 are included in the basic random variable vector, the following expressions can be derived:

$$\frac{\partial \mathbf{A}_{\sigma\sigma}^{-1}}{\partial E} = \frac{1}{l}\begin{bmatrix} A & 0 & 0 & 0 & 0 & 0 \\ 0 & \dfrac{J}{2(1+\mu)} & 0 & 0 & 0 & 0 \\ 0 & 0 & 4I_2 & -2I_2 & 0 & 0 \\ 0 & 0 & -2I_2 & 4I_2 & 0 & 0 \\ 0 & 0 & 0 & 0 & 4I_3 & -2I_3 \\ 0 & 0 & 0 & 0 & -2I_3 & 4I_3 \end{bmatrix}$$

(8.132)

$$\frac{\partial \mathbf{A}_{\sigma\sigma}^{-1}}{\partial A} = \frac{1}{l}\begin{bmatrix} E & 0 & 0 & 0 & 0 & 0 \\ 0 & 0 & 0 & 0 & 0 & 0 \\ 0 & 0 & 0 & 0 & 0 & 0 \\ 0 & 0 & 0 & 0 & 0 & 0 \\ 0 & 0 & 0 & 0 & 0 & 0 \\ 0 & 0 & 0 & 0 & 0 & 0 \end{bmatrix}$$

(8.133)

$$\frac{\partial \mathbf{A}_{\sigma\sigma}^{-1}}{\partial J} = \frac{1}{l} \begin{bmatrix} 0 & 0 & 0 & 0 & 0 & 0 \\ 0 & G & 0 & 0 & 0 & 0 \\ 0 & 0 & 0 & 0 & 0 & 0 \\ 0 & 0 & 0 & 0 & 0 & 0 \\ 0 & 0 & 0 & 0 & 0 & 0 \\ 0 & 0 & 0 & 0 & 0 & 0 \end{bmatrix} \tag{8.134}$$

$$\frac{\partial \mathbf{A}_{\sigma\sigma}^{-1}}{\partial I_2} = \frac{1}{l} \begin{bmatrix} 0 & 0 & 0 & 0 & 0 & 0 \\ 0 & 0 & 0 & 0 & 0 & 0 \\ 0 & 0 & 4E & -2E & 0 & 0 \\ 0 & 0 & -2E & 4E & 0 & 0 \\ 0 & 0 & 0 & 0 & 0 & 0 \\ 0 & 0 & 0 & 0 & 0 & 0 \end{bmatrix} \tag{8.135}$$

$$\frac{\partial \mathbf{A}_{\sigma\sigma}^{-1}}{\partial I_3} = \frac{1}{l} \begin{bmatrix} 0 & 0 & 0 & 0 & 0 & 0 \\ 0 & 0 & 0 & 0 & 0 & 0 \\ 0 & 0 & 0 & 0 & 0 & 0 \\ 0 & 0 & 0 & 0 & 0 & 0 \\ 0 & 0 & 0 & 0 & 4E & -2E \\ 0 & 0 & 0 & 0 & -2E & 4E \end{bmatrix} \tag{8.136}$$

Again, substituting Equation 8.83 into Equation 8.40, the gradient of the performance function can be obtained for linear problems. Once the gradient is known, the SFEM procedure will give the corresponding reliability.

8.18 NUMERICAL EXAMPLE

A computer program was written to implement the methodology discussed in the previous sections. The methodology is clarified with the help of an example.

Example 8.10 *(Three-Dimensional Structure)*

A two story, three-dimensional space frame structure is considered. The geometric dimensions of the frame and the sizes of the beams and columns are shown in Figure 8.16. The probabilistic description of the basic variables required for the analysis of the structure is given in Table 8.22. The M–Θ curves for the major and minor axes are assumed to be the same for all connections. They are shown in Figure 8.17. The probabilistic descriptions of the four parameters of the Richard model are given in Table 8.23.

Both the strength and serviceability limit states are considered in the examples discussed below. The allowable vertical deflection at the midspan of the beam under live load is assumed to be span/360. For the side sway, the allowable lateral movement at the top of the frame is assumed to be height/400. The strength limit states are given by Equations 8.105 and 8.106.

Assuming that all the members are connected rigidly, the reliability indexes for the strength and serviceability limit states are calculated using SFEM. Reliability indexes for beam 1 and column 2, shown in Figure 8.16, are calculated and shown in Table 8.24. Reliability indexes for the drift at the top of the column at location A and

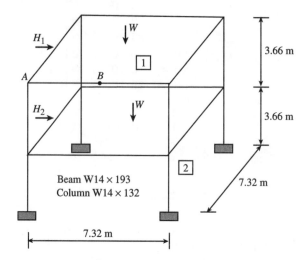

FIGURE 8.16. Two-story 3D structure.

Table 8.22. Description of basic random variables in the 3D frame

Random Variables	Beam		
	Mean	COV	Distribution
A^b (cm^2)	366.45	0.05	Lognormal
J^b (cm^4)	1,448.49	0.05	Lognormal
I_2^b (cm^4)	99,895.54	0.05	Lognormal
I_3^b (cm^4)	38,751.15	0.05	Lognormal
Z_x^b (cm^3)	5,817.41	0.05	Lognormal
Z_y^b (cm^3)	2,949.67	0.05	Lognormal
A^c (cm^2)	250.32	0.05	Lognormal
J^c (cm^4)	511.96	0.05	Lognormal
I_2^c (cm^4)	63,683.41	0.05	Lognormal
I_3^c (cm^4)	22,809.48	0.05	Lognormal
Z_x^c (cm^3)	3,834.57	0.05	Lognormal
Z_y^c (cm^3)	1,851.74	0.05	Lognormal
F_y (MPa)	260.62	0.10	Lognormal
E (MPa)	199,948.04	0.06	Lognormal
W (kN/m^2)	6.225	0.20	Lognormal
H_1 (kN)	81.398	0.30	Type I
H_2 (kN)	40.477	0.30	Type I

Note: Superscript b represents beam and superscript c represents column.

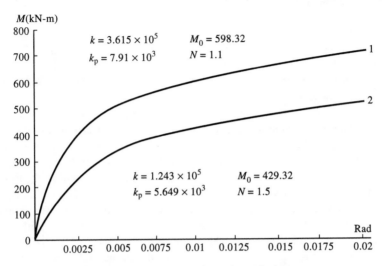

FIGURE 8.17. M–Θ curves for major and minor axes.

Table 8.23. Statistical description of the parameters of the Richard model

Random Variable	Mean	COV	Distribution
	Major Axis		
k (kN-m/rad)	3.615×10^5	0.15	Normal
k_p (kN-m/rad)	7.910×10^3	0.15	Normal
M_0 (kN-m)	598.79	0.15	Normal
N	1.1	0.05	Normal
	Minor Axis		
k' (kN-m/rad)	1.243×10^5	0.15	Normal
k_p' (kN-m/rad)	5.649×10^3	0.15	Normal
M_0' (kN-m)	429.32	0.15	Normal
N'	1.5	0.05	Normal

Table 8.24. Summary of safety indexes for the 3D frame

Limit State	Location	β FR Connection	β PR Connection
Strength	Beam [1]	4.26	3.78
	Column [2]	3.35	3.62
Serviceability	Drift [A]	5.84	4.15
	Deflection [B]	6.29	5.54

vertical deflection at the midspan of the top floor beam at location B are also given in Table 8.24.

The reliability indexes corresponding to the serviceability limit state are much higher than those for the strength limit state. This indicates that the strength limit state is the governing limit state, and the structure is expected to develop strength failure first. The structure is stiff enough to not develop excessive drift or deflection. A properly designed structure is expected to behave this way. This has always been the intent of design codes.

The connections are then assumed to be PR, as shown in Figure 8.17, and the corresponding reliabilities for the strength and serviceability limit states are evaluated; the results are summarized in Table 8.24. The reliability indexes for both limit states changed significantly. For the strength limit state, the PR connections reduce the bending moments at the connections and the bending moments at the midspan of the beams. This is expected because of the redistribution of the bending moments. The reliability indexes for the strength limit state decreased for the beams and increased for the columns.

The reliability indexes for the serviceability limit state decreased significantly. This is expected since the PR connections lower the overall stiffness of the frame. However, the importance of PR connections in the design of flexible multistory frames is quite evident from the drop in the reliability indexes for the serviceability limit state.

This simple example indicates that the flexibility in the PR connections must be considered appropriately, particularly for the design of flexible structures. The serviceability could be the most critical limit state in this case. The structure may have to be reanalyzed and redesigned considering the PR connections.

8.19 CONCLUDING REMARKS

A stochastic finite element-based algorithm has been developed in this chapter for the reliability analysis of complicated nonlinear two- and three-dimensional structures in which the performance function is not available as a closed-form expression in terms of the basic random variables. The assumed stress-based FEM is used to improve the efficiency of the algorithm. The corresponding matrix form of the First Order Reliability Method was derived. The algorithm does not depend on the explicit solution of the limit state equation; instead, it uses the information about the gradient vector of the performance function to converge to the design point. The basic principle of the Stochastic Finite Element Method is used to formulate the computation of the gradient vector, through the computation of the partial derivatives of the stiffness matrix, load vector, and so forth. The method is applied to the reliability analysis of 2D and 3D frame structures. Several important limit states have been considered, and the results of the numerical examples show the desirability of the proposed method. The use of sensitivity indexes improves the computational efficiency. Finally, a procedure using the algorithm has been developed for the reliability analysis of frames with uncertain connection rigidity.

The unified stochastic finite element method developed in this chapter can be used for both the displacement-based and stress-based finite element methods. Several examples with intermediate steps were given in this chapter to verify the basic concept. The algorithm has the potential for routine use in the reliability analysis of real linear and nonlinear structures.

9

STRUCTURES UNDER
DYNAMIC LOADING

9.1 INTRODUCTORY COMMENTS

Chapters 6 to 8 presented the implementation of reliability analysis methods for structures under static, time-invariant loads. Many engineering systems face time-varying demands and loads. This variation with time is of two types: over a long duration, and over a short duration. For example, one could consider the increase in traffic loads on a bridge over a number of years (long duration), or the fluctuations in bending moments, stresses, and strains as a single load moves from one end to another (short duration). Another example is of a water supply system. One type of planning considers the demands on the system due to population growth over a number of years (long duration), while another type of analysis and design considers the fluctuating demands on the water supply during a day (short duration). The analysis methods and the issues in the two types of duration are very different. The variation of reliability over the long-term life of an engineering system is currently an active area of research, involving the development of analytical and simulation methods for reliability estimation, inspection planning strategies, etc.

This chapter focuses on reliability analysis under short-duration, time-variant loads on a system, such as loads experienced by a structure within a few seconds during an earthquake. Even here, two types of analysis can be used: time domain and frequency domain. This chapter considers load description in the time domain. In the first part of this chapter, linear behavior of structures is assumed and the displacement approach of finite element analysis is used. The overall approach is the same as in Chapter 6 and is in the context of the first-order reliability method (FORM). The performance function $g(\mathbf{X})$ is evaluated using linear dynamic analysis. The chain rule of differentiation is used to compute the gradient of $g(\mathbf{X})$. These two pieces of information are used at each step during the search for the minimum distance point in the FORM algorithm. In the second part of this chapter, reliability evaluation of nonlinear structures subjected to short duration loading using stress-based finite element analysis is discussed.

It should be carefully noted that randomness in dynamic loads is appropriately addressed by random vibration techniques, which are beyond the scope of this book. This chapter presents only the application of the SFEM approach to a simple dynamic problem where only the amplitude of the loading is considered to be a random variable. The shape of the loading history is assumed to be deterministic. The state of the art in SFEM-based reliability analysis under dynamic loads is still an active area of research and it is not ready for everyday use. The material presented here is therefore only a simple extension of the concepts in Chapters 6 and 8.

Under these limitations, SFEM-based reliability analysis of linear structures for dynamic loads may be divided into three parts:

1. *Dynamic Analysis* Computes the structural response (displacement, stress, etc.) due to the applied dynamic load. The performance function for reliability analysis is formulated in terms of these response quantities. In the simple case being considered in this chapter, the performance function is written in terms of the peak response.

2. *Sensitivity Analysis* Computes the partial derivatives of the performance function with respect to the basic random variables using the chain rule of differentiation through all the steps of the dynamic analysis.

3. *Reliability Analysis* Uses the Rackwitz–Fiessler algorithm to search for the most probable point on the limit state. The dynamic analysis is performed for each iteration of the reliability analysis using the current values of all the random variables, and the partial derivatives are also calculated at each step. Once the most probable point is found, a first-order or second-order approximation to the limit state is used to compute the failure probability.

These three steps are described in detail in the following sections.

9.2 DYNAMIC ANALYSIS OF LINEAR STRUCTURES

A brief overview of linear dynamic analysis is presented in this section to facilitate the sensitivity computation in the next section. Consider a structure, such as that shown in Figure 9.1, with a dynamic load such as that shown in Figure 9.2. The variation of the structural displacement with time is computed through the following equation of motion as

$$[\mathbf{M}]\{\ddot{\mathbf{y}}\} + [\mathbf{C}]\{\dot{\mathbf{y}}\} + [\mathbf{K}]\{\mathbf{y}\} = \{\mathbf{f}\} \tag{9.1}$$

where $[\mathbf{M}]$ is the mass matrix, $[\mathbf{C}]$ is the damping matrix, $[\mathbf{K}]$ is the stiffness matrix, $\{\ddot{\mathbf{y}}\}$ is the vector of accelerations, $\{\dot{\mathbf{y}}\}$ is the vector of velocities, $\{\mathbf{y}\}$ is the vector of displacements, and $\{\mathbf{f}\}$ is the vector of forces acting on the system. The acceleration acting on all the masses is the same for a ground-excited system; therefore, $\{\mathbf{f}\} = [\mathbf{M}]\{\mathbf{1}\}\ddot{\mathbf{y}}_s$, where $\ddot{\mathbf{y}}_s$ is the ground acceleration.

For linear dynamic analysis, $[\mathbf{K}]$, $[\mathbf{M}]$, and $[\mathbf{C}]$ are independent of time. For a deterministic history, $\{\mathbf{f}\}$ is a known function of time and initial values of $\{\mathbf{y}\}$ and $\{\dot{\mathbf{y}}\}$ are prescribed. The modal superposition method (Paz, 1997) transforms Equation 9.1 so that $\{\mathbf{y}\}$, $\{\dot{\mathbf{y}}\}$, and

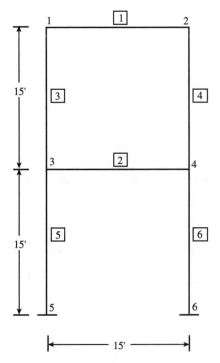

FIGURE 9.1. Two-story, one-bay frame.

$\{\ddot{y}\}$ are replaced by $\{q\}$, $\{\dot{q}\}$, and $\{\ddot{q}\}$, where $\{q\}$ is a vector of modal amplitudes. Thus,

$$\{y\} = [\Phi]\{q\} \tag{9.2}$$

where Φ is the matrix of mode shape vectors Φ_i. Since $[K]$ and $[M]$ are orthogonal with respect to $[\Phi]$

$$\{\Phi_i\}^t[M]\{\Phi_j\} = \{\Phi_i\}^t[K]\{\Phi_j\} = 0 \qquad \text{if} \quad i \neq j \tag{9.3}$$

If $[\Phi]$ is normalized such that $\{\Phi_i\}^t[M]\{\Phi_i\} = 1$, then

$$[\Phi]^t[M][\Phi] = [I] \quad \text{and} \quad [\Phi]^t[K][\Phi] = [\omega^2] \tag{9.4}$$

where $[I]$ is a unit matrix and $[\omega^2]$ is the spectral matrix, a diagonal matrix composed of the squares of the natural frequencies of the structure. By substituting Equation 9.2 in Equation 9.1, premultiplying by $\{\Phi_i\}^t$ and simplifying, the equation of motion of the system can be written as

$$\ddot{q}_i + 2\xi_i \omega_i \dot{q}_i + \omega_i^2 q_i = \{\Phi_i\}^t\{f\} \tag{9.5}$$

where ξ_i is the modal damping ratio (assuming viscous damping) and ω_i is the natural frequency of vibration of mode i. The damping matrix is also assumed to be orthogonal with respect to $[\Phi]$.

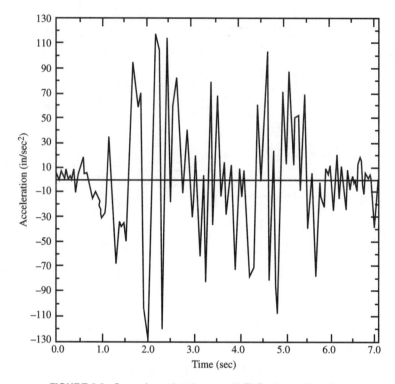

FIGURE 9.2. Ground acceleration record: El Centro earthquake.

Equation 9.5 represents a set of uncoupled equations, one for each degree of freedom, which can be solved for q_i as a function of time. For any general loading, this can be done by using Duhamel's integral:

$$q_i(t) = \frac{1}{m\omega_{D_i}} \int_{\tau=0}^{\tau=t} F(\tau)e^{-\xi\omega_i(t-\tau)} \sin\omega_{D_i}(t-\tau)d\tau \qquad (9.6)$$

where $F(\tau)$ is the modal load history. For any general loading, $F(\tau) = \{\Phi\}^t\{f\}$ and for ground excitation, $F(\tau) = \{\Phi\}^t[M]\{1\}\ddot{y}_s$. q_i can then be used to calculate the displacements $\{y\}$ using Equation 9.2. Numerical integration has to be used for evaluating Equation 9.6 in most cases, since the solution cannot be obtained in closed form for most real loadings like earthquake ground excitation.

9.3 SENSITIVITY AND RELIABILITY ANALYSIS

The algorithm to search for the most probable point on the limit state requires the derivatives of the performance function (and, therefore, the structural response) with respect to the random variables. As formulated in the earlier chapters, sensitivity analysis is incorporated into the structural analysis to compute these derivatives. This sensitivity analysis procedure for the case of dynamic loading is discussed below. Rigid plane frames are used as an example. The same procedure may be adopted for other types of structures as well. In the following discussion, steps 1 to 4 relate to sensitivity analysis, and steps 5 and 6 relate to reliability analysis.

Step 1 The first step in the dynamic analysis of frames is the computation of the stiffness and mass matrices. Their partial derivatives with respect to the basic random variables are also computed. This computation is quite simple, since in the case of the plane frame element used here, the element matrices can be written explicitly in terms of the basic variables and numerical integration is not necessary.

   The element stiffness matrix $\mathbf{k}_e$ and its partial derivatives are the same as in the case of static analysis in Chapter 6. The element mass matrices can be computed using either the consistent mass formulation or the lumped mass formulation. In the case of consistent mass formulation, the element mass matrix $\mathbf{m}_e$ and its derivative with respect to dead load DL and live load LL are of the form

$$\mathbf{m}_e, \frac{\partial \mathbf{m}_e}{\partial \mathrm{DL}}, \frac{\partial \mathbf{m}_e}{\partial \mathrm{LL}} = \frac{\mu L}{420} \begin{bmatrix} a & b & c & d & -b & e \\ b & f & g & -b & h & i \\ c & g & j & -e & -i & k \\ d & -b & -e & a & b & -c \\ -b & h & -i & b & f & -g \\ e & i & k & -c & -g & j \end{bmatrix} \tag{9.7}$$

where

$$a = 156 \sin^2 \theta + 140 \cos^2 \theta$$

$$b = -16 \sin \theta \cos \theta$$

$$c = -22L \sin \theta$$

$$d = 54 \sin^2 \theta + 70 \cos^2 \theta$$

$$e = 13L \sin \theta$$

$$f = 140 \sin^2 \theta + 156 \cos^2 \theta$$

$$g = 22L \cos \theta$$

$$h = 70 \sin^2 \theta + 54 \cos^2 \theta$$

$$i = 13L \cos \theta$$

$$j = 4L^2$$

$$k = -3L^2$$

In the case of lumped mass formulation,

$$\mathbf{m}_e, \frac{\partial \mathbf{m}_e}{\partial \mathrm{DL}}, \frac{\partial \mathbf{m}_e}{\partial \mathrm{LL}} = \frac{\mu L}{2} \begin{bmatrix} 1 & 0 & 0 & 0 & 0 & 0 \\ 0 & 1 & 0 & 0 & 0 & 0 \\ 0 & 0 & 0 & 0 & 0 & 0 \\ 0 & 0 & 0 & 1 & 0 & 0 \\ 0 & 0 & 0 & 0 & 1 & 0 \\ 0 & 0 & 0 & 0 & 0 & 0 \end{bmatrix} \tag{9.8}$$

For $\mathbf{m}_e$, $\mu = DL + LL$. For $\partial \mathbf{m}_e / \partial DL$ and $\partial \mathbf{m}_e / \partial LL$, $\mu = 1$. Note that the third and the sixth row and column have zeros. These correspond to rotational degrees of freedom. Thus, the only masses considered in this formulation are those that have translational degrees of freedom.

The global stiffness matrix $[\mathbf{K}]$ and mass matrix $[\mathbf{M}]$ can be assembled using the above element matrices. Using the same procedure, global partial derivative matrices $\partial[\mathbf{K}]/\partial\mathbf{X}$ and $\partial[\mathbf{M}]/\partial\mathbf{X}$ can also be assembled.

Step 2   The next step is to calculate eigenvectors and eigenvalues, which is done as follows:

- Compute the eigenvalues λ_i by solving

$$|[\mathbf{K}] - \lambda_i[\mathbf{M}]| = 0 \qquad (9.9)$$

This yields one eigenvalue for each degree of freedom.

- Compute the eigenvectors Φ_i by solving

$$([\mathbf{K}] - \lambda_i[\mathbf{M}])\Phi_i = 0 \qquad (9.10)$$

for each value of λ_i.

Following the basic principle of SFEM, the partial derivatives of eigenvalues and eigenvectors have to be computed at each iteration. The partial derivatives of eigenvalues and eigenvectors have been derived by Fox and Kapoor (1968) and Hasselman and Hart (1972) as follows.

Eigenvalues   Multiply Equation 9.10 by Φ_i^t

$$\Phi_i^t([\mathbf{K}] - \lambda_i[\mathbf{M}])\Phi_i = 0 \qquad (9.11)$$

Differentiate the above equation with respect to X, a basic random variable:

$$\frac{\partial \Phi_i^t}{\partial \mathbf{X}}\left([\mathbf{K}] - \lambda_i[\mathbf{M}]\right)\Phi_i + \Phi_i^t([\mathbf{K}] - \lambda_i[\mathbf{M}])\frac{\partial \Phi_i}{\partial \mathbf{X}} \qquad (9.12)$$

$$+ \Phi_i^t\left(\frac{\partial[\mathbf{K}]}{\partial \mathbf{X}} - \lambda_i\frac{\partial[\mathbf{M}]}{\partial \mathbf{X}} - \frac{\partial\lambda_i}{\partial \mathbf{X}}[\mathbf{M}]\right)\Phi_i = 0$$

Since $\Phi_i \neq 0$, $([\mathbf{K}] - \lambda_i[\mathbf{M}]) = 0$, then

$$\frac{\partial\lambda_i}{\partial \mathbf{X}}\Phi_i^t[\mathbf{M}]\Phi_i = \Phi_i^t\left(\frac{\partial[\mathbf{K}]}{\partial \mathbf{X}} - \lambda_i\frac{\partial[\mathbf{M}]}{\partial \mathbf{X}}\right)\Phi_i \qquad (9.13)$$

Since Φ_i are orthonormal with respect to $[\mathbf{M}]$,

$$\frac{\partial\lambda_i}{\partial \mathbf{X}} = \Phi_i^t\left(\frac{\partial[\mathbf{K}]}{\partial \mathbf{X}} - \lambda_i\frac{\partial[\mathbf{M}]}{\partial \mathbf{X}}\right)\Phi_i \qquad (9.14)$$

Since the natural frequencies $\omega_i = \sqrt{\lambda_i}$,

$$\frac{\partial\omega_i}{\partial \mathbf{X}} = -\frac{1}{2\sqrt{\lambda_i}}\frac{\partial\lambda_i}{\partial \mathbf{X}} \qquad (9.15)$$

Eigenvectors Since the eigenvectors form a complete set of vectors, any n-component vector can be represented by their linear combination:

$$\frac{\partial \Phi_i}{\partial \mathbf{X}} = \sum_{j=1}^{n} a_{ij} \Phi_j \tag{9.16}$$

Differentiating Equation 9.10 with respect to X, substituting Equation 9.16 and rearranging the terms, we get

$$([\mathbf{K}] - \lambda_i [\mathbf{M}]) \sum_{j=1}^{n} a_{ij} \Phi_j = \left(\frac{\partial [\mathbf{K}]}{\partial \mathbf{X}} - \lambda_i \frac{\partial [\mathbf{M}]}{\partial \mathbf{X}} - \frac{\partial \lambda_i}{\partial \mathbf{X}} [\mathbf{M}] \right) \Phi_i \tag{9.17}$$

Multiplying both sides of the equation by Φ_k^t gives

$$\sum_{j=1}^{n} a_{ij} \Phi_k^t ([\mathbf{K}] - \lambda_i [\mathbf{M}]) \Phi_j = \Phi_k^t \left(\frac{\partial [\mathbf{K}]}{\partial \mathbf{X}} - \lambda_i \frac{\partial [\mathbf{M}]}{\partial \mathbf{X}} - \frac{\partial \lambda_i}{\partial \mathbf{X}} [\mathbf{M}] \right) \Phi_i \tag{9.18}$$

Since the mode shapes are orthonormal, $k = j$, $\Phi_k^t [\mathbf{K}] \Phi_j = \lambda_j$, and $\Phi_k^t [\mathbf{M}] \Phi_j = 1$ on the left-hand side. On the right-hand side, if $i \neq j$, then $\Phi_k^t [\mathbf{M}] \Phi_i = 0$, and

$$a_{ij} = \frac{\Phi_j^t \left(\frac{\partial [\mathbf{K}]}{\partial \mathbf{X}} - \lambda_i \frac{\partial [\mathbf{M}]}{\partial \mathbf{X}} \right) \Phi_i}{\lambda_i - \lambda_j}, \qquad i \neq j \tag{9.19}$$

If $i = j = k$, then differentiating $\Phi_i^t [\mathbf{M}] \Phi_i = 1$ gives

$$2 \Phi_i^t [\mathbf{M}] \frac{\partial \Phi_i}{\partial \mathbf{X}} = -\Phi_i^t \frac{\partial [\mathbf{M}]}{\partial \mathbf{X}} \Phi_i \tag{9.20}$$

Equation 9.16 gives

$$a_{ii} = -\frac{\Phi_i^t \frac{\partial [\mathbf{M}]}{\partial \mathbf{X}} \Phi_i}{2} \tag{9.21}$$

Step 3   The Duhamel integral in Equation 9.6 for many practical loadings needs to be evaluated numerically since the loading data are available only at discrete time values and not as a continuous variable. Several numerical integration techniques are available for this purpose, such as the trapezoidal rule and Simpson's rule. These methods replace the integral by a suitable summation. The response obtained in such cases is always approximate.

An alternative approach is based on obtaining the exact analytical solution of the integral for the linear loading function. This method does not introduce any approximations for loadings like ground excitation, which are inherently segmentally linear. The equations for computing the displacements at each time step t_i are given by Paz (1997) and are reproduced below:

$$q(t_i) = \frac{e^{-\xi \omega t_i}}{m \omega_D} \{ A_D(t_i) \sin \omega_D t_i - B_D(t_i) \cos \omega_D t_i \} \tag{9.22}$$

For small values of modal damping ξ, the damped natural frequency of vibration ω_D can be assumed to be equal to ω, the natural frequency of vibration. Also, since the mode shape vectors are orthonormal with respect to the mass matrix, $m = \{\phi\}^t[M]\{\phi\} = 1$:

$$A_D(t_i) = A_D(t_{i-1}) + \left[F(t_{i-1}) - t_{i-1}\frac{\Delta F_i}{\Delta t_i}\right]I_1 + \frac{\Delta F_i}{\Delta t_i}I_4 \qquad (9.23)$$

$$B_D(t_i) = B_D(t_{i-1}) + \left[F(t_{i-1}) - t_{i-1}\frac{\Delta F_i}{\Delta t_i}\right]I_2 + \frac{\Delta F_i}{\Delta t_i}I_3 \qquad (9.24)$$

$$I_1 = \left.\frac{e^{\xi\omega\tau}}{\omega^2}(\xi\omega\cos\omega\tau + \omega\sin\omega\tau)\right|_{t_{i-1}}^{t_i} \qquad (9.25)$$

$$I_2 = \left.\frac{e^{\xi\omega\tau}}{\omega^2}(\xi\omega\sin\omega\tau - \omega\cos\omega\tau)\right|_{t_{i-1}}^{t_i} \qquad (9.26)$$

$$I_3 = \left.\left(\tau - \frac{\xi}{\omega}\right)I_2' + \frac{1}{\omega}I_1'\right|_{t_{i-1}}^{t_i} \qquad (9.27)$$

$$I_4 = \left.\left(\tau - \frac{\xi}{\omega}\right)I_1' - \frac{1}{\omega}I_2'\right|_{t_{i-1}}^{t_i} \qquad (9.28)$$

where I_1' and I_2' are I_1 and I_2 before their evaluation at their limits and

$$F(t_i) = \{\Phi\}^t\{f(t_i)\}, \text{ for any general loading} \qquad (9.29)$$

$$= \{\Phi\}^t[M]\{1\}\ddot{y}_s(t_i), \text{ for ground excitation}$$

$$\Delta F_i = F(t_i) - F(t_{i-1}) \qquad (9.30)$$

$$\Delta t_i = t_i - t_{i-1} \qquad (9.31)$$

All of the above equations are differentiated with respect to the basic random variables and the chain rule is used to get the partial derivatives of q. The partial derivatives of the displacements y can then be computed by differentiating Equation 9.2 with respect to the random variables X:

$$\frac{\partial y}{\partial X} = \left(\frac{\partial[\Phi]}{\partial X}q + [\Phi]\frac{\partial q}{\partial X}\right) \qquad (9.32)$$

Note that in time-domain analysis, the response of the structure is computed for many short time intervals. However, the performance function for the reliability problem in this case is formulated only in terms of the peak response, such as peak displacement and peak stress. Since the shape of the loading history is assumed to be fixed, the peak response is assumed to occur at the same instant of time. Only its magnitude is a random variable, so the reliability analysis requires only the peak response and its derivatives with respect to the random variables, and not at every interval. Therefore, it is computationally efficient to first perform a deterministic analysis and identify the instant when peak response occurs. Then compute the partial derivatives only at this instant when performing reliability analysis.

Step 4   Calculate the desired response $\mathbf{S}$ using computation of the form $\mathbf{S} = \mathbf{Q}^t\mathbf{y}$, where $\mathbf{Q}$ is the displacement-to-load-effect transformation matrix. The partial derivatives of $\mathbf{S}$ with respect to the basic random variables can be calculated using

$$\frac{\partial \mathbf{S}}{\partial \mathbf{X}} = \frac{\partial \mathbf{Q}^t}{\partial \mathbf{X}}\mathbf{y} + \mathbf{Q}^t\frac{\partial \mathbf{y}}{\partial \mathbf{X}} \tag{9.33}$$

Note here that the symbol $\mathbf{y}$ refers to the displacement response of the structure. The nature of $\mathbf{S}$ depends on the limit state being considered, as described below.

Step 5   Compute the performance function

$$g(\mathbf{X}) = g\{\mathbf{R}, \mathbf{S}\} \tag{9.34}$$

where $\mathbf{R}$ and $\mathbf{S}$ are vectors of resistance and response variables occuring in the performance function. For example, consider the limit states relating to displacement and flexural failure, similar to the ones considered in Chapter 6 with static analysis. The performance function for interstory drift may be written as

$$g(\mathbf{X}) = 1 - \frac{u_i - u_{i-1}}{d_p} \tag{9.35}$$

where u_i and u_{i-1} are the peak horizontal deformations at the top and bottom of a story, respectively, and d_p is the maximum permissible interstory drift. Here, $\mathbf{R} = \{0\}$ and $\mathbf{S} = \{u_i - u_{i-1}\}$. For the limit state corresponding to flexural failure of a member,

$$g(\mathbf{X}) = 1 - \frac{M_x}{Z F_y} \tag{9.36}$$

where M_x is the maximum bending moment about the major axis, and Z and F_y are the plastic section modulus and the yield stress respectively. Here, $\mathbf{R} = \{Z, F_y\}$ and $\mathbf{S} = \{M_x\}$.

Step 6   Once the performance function and its derivatives are computed, the reliability analysis procedure is the same as that followed in the earlier chapters. An iteration in the search for the minimum distance point on the limit state in the standard normal space is performed. At the next iteration point, all of the above steps are repeated. This continues until convergence to the minimum distance point.

Example 9.1

The two-story frame shown in Figure 9.1, subjected to a dynamic load in the form of a ground acceleration record from the El Centro earthquake, is analyzed for two limit states in this example (Mehta, 1991). Details of the earthquake ground acceleration history are shown in Figure 9.2. In this example, the shape of the acceleration history is assumed to be deterministic. Only its magnitude is assumed to be random, with the peak value having a mean of 128.29 in/sec^2 and a coefficient of variation of 1.38. (Such a large coefficient of variation is typical of earthquake load modeling. See ASCE 7-95

(1996) for details.) This random variable is assumed to have a Type II extreme value distribution. Since the history is assumed to be deterministic, when the peak value is changed by a certain factor, the y coordinate of the entire acceleration history is changed by the same factor. The modal damping ratio for this frame is also assumed to be a random variable, with a mean value of 0.05, coefficient of variation 0.65, and a lognormal distribution.

The frame is also subjected to static dead load (D), and live load (L). Details of these two random variables are shown in Table 9.1, along with other random variables relating to the sectional and material properties of the frame members. The dead load and the live load are taken into account by converting them to equivalent uniformly distributed masses.

The performance function for reliability analysis is written in terms of the bending moment at node 5 in member 5 as

$$g(X) = 1.0 - \frac{M}{M_p} \tag{9.37}$$

where M is the maximum moment at node 5 due to the dynamic load, and $M_p = ZF_y$ is the plastic moment capacity of member 5, Z is the plastic section modulus, and F_y is the yield strength.

The dynamic analysis of the frame and the SFEM-based reliability analysis are implemented using the steps described earlier. The reliability index is estimated to be 0.845, and the analysis converges in four iterations, as shown in Table 9.2. Note the sensitivity indices of the random variables. Among the 15 random variables considered, only 6 have a sensitivity index greater than 0.05. In particular, the earthquake acceleration has a very high sensitivity index of 0.92, which is typical of reliability analysis under earthquake loading. Using the first-order approximation to the limit state, the

Table 9.1. Two-story frame: description of random variables

Variable	Units	Mean	COV	Distribution
D	k/ft	3.00	0.10	Lognormal
L	k/ft	0.235	0.25	Type I
f	in/s^2	128.29	1.38	Type II
ξ		0.05	0.65	Lognormal
A_{tb}	in^2	8.80	0.05	Lognormal
I_{tb}	in^4	170.00	0.05	Lognormal
Z_{tb}	in^3	36.60	0.05	Lognormal
A_{lb}	in^2	10.00	0.05	Lognormal
I_{lb}	in^4	340.00	0.05	Lognormal
Z_{lb}	in^3	54.60	0.05	Lognormal
A_c	in^2	14.70	0.05	Lognormal
I_c	in^4	394.00	0.05	Lognormal
Z_c	in^3	72.40	0.05	Lognormal
E	ksi	29,000.00	0.06	Lognormal
F_y	ksi	39.60	0.11	Lognormal

Note. 1 kip = 4.55 kN, 1 in = 0.0254 m. tb, top beam; lb, lower beam; c, column.

Table 9.2. Two-story frame: bending of member 5 at node 5

Variable	Sensitivity Index	Initial Checking Point	Final Checking Point
D	0.164	3.00	3.03
L	0.051	0.235	0.23
f	0.920	128.29	151.75
ξ	−0.282	0.05	0.038
A_{tb}	0.000	8.80	8.8
I_{tb}	−0.001	170.00	170.0
Z_{tb}	0.000	36.60	36.60
A_{lb}	0.000	10.00	10.00
I_{lb}	−0.001	340.00	340.00
Z_{lb}	0.000	54.60	54.60
A_c	−0.000	14.70	14.70
I_c	0.022	394.00	394.00
Z_c	−0.087	72.10	72.10
E	0.010	29,000.00	29,000.00
F_y	−0.190	39.60	38.71
Performance function		0.2364	−0.0003
Reliability index (in 4 iterations)			0.8450
Probability of failure—FORM			0.1991
Monte Carlo estimate			0.2156

Note. tb, top beam; lb, lower beam; c, column.

probability of failure is estimated as 0.1991. This result is within 10% of the Monte Carlo simulation estimate of 0.2156.

Example 9.2

In this example, a six-story, two-bay frame is analyzed (Mehta, 1991) with 54 degrees of freedom. The dimensions and properties of this frame are taken from Narayanan (1985) and are reproduced in Figure 9.3 and Table 9.3. The reliability analysis is performed for the bending limit state of member 29 at node 20, similar to the limit state in the previous example. Thus, the performance function is written as

$$g(X) = 1.0 - \frac{M}{M_p} \tag{9.38}$$

where M is now the maximum moment at node 20 due to the dynamic load, and $M_p = Z F_y$ is the plastic moment capacity of member 29, where Z is the plastic section modulus and F_y is the yield strength.

This structure is subjected to the same ground acceleration history from an El Centro record as shown in the previous example. The results of the SFEM-based reliability analysis are summarized in Table 9.4. Thirty-five random variables are initially considered in this problem. However, when the reliability sensitivity indices are examined, only five random variables (f, ξ, D_2, E, and F_y) are found to be significant (greater than 0.05) for the limit state considered, as seen in Table 9.4. All the remaining vari-

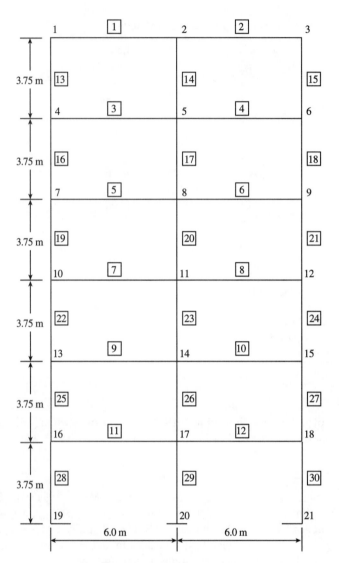

FIGURE 9.3. Six-story, two-bay frame.

ables are treated as deterministic at their mean values after the first iteration. This results in considerable savings in computation time.

Example 9.3

The six-story frame of Example 9.2 is considered again, but with a different type of dynamic loading. A horizontal excitation with a half-sine pulse forcing function is applied at each story (Figure 9.4), with the amplitude F_z which varies with height z as

Table 9.3. Six-story frame: description of random variables

Variable	Units	Mean	COV	Distribution
Dead load on roof, D_1	kN/m	16.875	0.10	Lognormal
Live load on roof, L_1	kN/m	1.688	0.25	Type I
Dead load on floor, D_2	kN/m	21.600	0.10	Lognormal
Live load on floor, L_2	kN/m	3.938	0.25	Type I
Ground acceleration, f	m/s^2	3.948	1.38	Type II
Half-sine nodal load, W	kN	16.950	0.37	Type I
Modal damping ratio, ξ		0.05	0.65	Lognormal
A_1 (1,2)	cm^2	39.90	0.05	Lognormal
I_1 (1,2)	cm^4	4,427.00	0.05	Lognormal
Z_1 (1,2)	cm^3	394.80	0.05	Lognormal
A_2 (3,4,5,6)	cm^2	47.40	0.05	Lognormal
I_2 (3,4,5,6)	cm^4	7,143.00	0.05	Lognormal
Z_2 (3,4,5,6)	cm^3	539.30	0.05	Lognormal
A_3 (7,8)	cm^2	56.90	0.05	Lognormal
I_3 (7,8)	cm^4	12,052.00	0.05	Lognormal
Z_3 (7,8)	cm^3	771.70	0.05	Lognormal
A_4 (9,10,11,12)	cm^2	68.30	0.05	Lognormal
I_4 (9,10,11,12)	cm^4	18,576.00	0.05	Lognormal
Z_4 (9,10,11,12)	cm^3	1,046.00	0.05	Lognormal
A_5 (13,15,16,18)	cm^2	58.80	0.05	Lognormal
I_5 (13,15,16,18)	cm^4	4,564.00	0.05	Lognormal
Z_5 (13,15,16,18)	cm^3	497.40	0.05	Lognormal
A_6 (14,17,19,21,22,24)	cm^2	75.80	0.05	Lognormal
I_6 (14,17,19,21,22,24)	cm^4	6,088.00	0.05	Lognormal
Z_6 (14,17,19,21,22,24)	cm^3	652.00	0.05	Lognormal
A_7 (20,23)	cm^2	114.00	0.05	Lognormal
I_7 (20,23)	cm^4	14,307.00	0.05	Lognormal
Z_7 (20,23)	cm^3	1,288.00	0.05	Lognormal
A_8 (25,27,28,30)	cm^2	75.80	0.05	Lognormal
I_8 (25,27,28,20)	cm^4	6,088.00	0.05	Lognormal
Z_8 (25,27,28,20)	cm^3	652.00	0.05	Lognormal
A_9 (26,29)	cm^2	136.60	0.05	Lognormal
I_9 (26,29)	cm^4	17,510.00	0.05	Lognormal
Z_9 (26,29)	cm^3	1,485.00	0.05	Lognormal
E	GPa	205.00	0.06	Lognormal
F_y	MPa	280.00	0.11	Lognormal

Note. The numbers in parentheses represent the applicable member numbers.

$$F_z = F_{10} \qquad \text{for} \quad z \leq 10 \text{ m} \qquad (9.39)$$

$$F_z = F_{10} \left(\frac{z}{10}\right)^{0.8} \qquad \text{for} \quad z > 10 \text{ m} \qquad (9.40)$$

where F_{10} is the mean load at 10 m above the ground and z is the height above the ground in meters. This type of loading may be considered to be similar to a wind load,

Table 9.4. Six-story frame: El Centro earthquake loading

Variable	Sensitivity Index	Initial Checking Point	Final Checking Point
D_1	0.040	16.875	16.875
L_1	0.010	1.688	1.688
D_2	0.103	21.600	21.686
L_2	0.045	3.938	3.938
f	0.983	3.948	4.548
ξ	−0.097	0.05	0.04
A_1	0.000	39.90	39.90
I_1	−0.003	4,427.00	4,427.00
Z_1	0.000	394.80	394.80
A_2	0.000	47.40	47.40
I_2	−0.014	7,143.00	7,143.00
Z_2	0.000	539.30	539.30
A_3	0.000	56.90	56.90
I_3	−0.005	12,052.00	12,052.00
Z_3	0.000	771.70	771.70
A_4	−0.000	68.30	68.30
I_4	−0.010	18,576.00	18,576.00
Z_4	0.000	1,046.00	1,046.00
A_5	−0.000	58.80	58.80
I_5	−0.005	4,564.00	4,564.00
Z_5	0.000	497.40	497.40
A_6	−0.000	75.80	75.80
I_6	−0.007	6,088.00	6,088.00
Z_6	0.000	652.00	652.00
A_7	0.000	114.00	114.00
I_7	−0.001	1,4307.00	1,4307.00
Z_7	0.000	1,288.00	1,288.00
A_8	−0.000	75.80	75.80
I_8	−0.013	11,360.00	11,360.00
Z_8	0.000	988.50	988.50
A_9	0.000	136.60	136.60
I_9	0.012	17,510.00	17,510.00
Z_9	−0.031	1,485.00	1,485.00
E	−0.056	205.00	204.05
F_y	−0.067	280.00	276.43
Performance function		0.3584	0.0007
Reliability index (in 13 iterations)			1.0861
Probability of failure—FORM			0.1379
Monte Carlo estimate			0.1013

where the wind pressure on a building multiplied by an appropriate area is applied as a horizontal nodal force on each story. For the sake of simplicity, analysis for a small duration with a single half-sine pulse is illustrated. Once again, the shape of the load history is deterministic; only the magnitude is random.

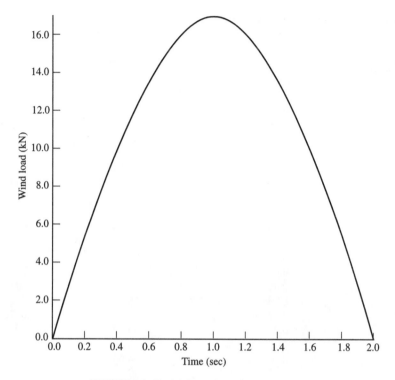

FIGURE 9.4. Nodal dynamic excitation history.

The results of SFEM-based reliability analysis are summarized in Table 9.5. The SFEM reliability analysis initially has 35 random variables. However, when the reliability sensitivity indices are examined, only four (W, ξ, F_y, and Z_9) are found to be significant, i.e., greater than 0.05, for the limit state considered, as seen in Table 9.5. Similar to the previous example, all the remaining variables are treated as deterministic at their mean values after the first iteration.

In all three examples presented above, the failure probabilities are quite large. In practical design, such large probabilities would be unacceptable. The examples were chosen with large failure probabilities to facilitate easy verification with Monte Carlo simulation. When the failure probability is large, a smaller number of Monte Carlo samples are adequate, saving computational effort. In realistic structural analysis and design, the failure probabilities will be much smaller, and Monte Carlo simulation will be time-consuming.

9.4 MATRIX CONDENSATION FOR LARGE STRUCTURES

The dynamic analysis described in earlier sections is done through numerical integration in the time domain. This is quite time-consuming and makes the analysis of structures with a large number of degrees of freedom uneconomical. Therefore, matrix condensation procedures have been used in the dynamic analysis of large structures. This section describes matrix condensation and the method to integrate it with SFEM-based dynamic reliability analysis.

Table 9.5. Six-story frame: sinusoidal loading

Variable	Sensitivity Index	Initial Checking Point	Final Checking Point
D_1	−0.006	16.875	16.875
L_1	−0.003	1.688	1.688
D_2	0.014	21.600	21.600
L_2	0.012	3.938	3.938
W	0.928	16.950	36.019
ξ	−0.100	0.05	0.042
A_1	0.000	39.90	39.90
I_1	0.000	4,427.00	4,427.00
Z_1	0.000	394.80	394.80
A_2	0.000	47.40	47.40
I_2	0.002	7,143.00	7,143.00
Z_2	0.000	539.30	539.30
A_3	0.000	56.90	56.90
I_3	−0.000	12,052.00	12,052.00
Z_3	0.000	771.70	771.70
A_4	0.000	68.30	68.30
I_4	−0.028	1,8576.00	1,8576.00
Z_4	0.000	1,046.00	1,046.00
A_5	0.000	58.80	58.80
I_5	0.000	4,564.00	4,564.00
Z_5	0.000	497.40	497.40
A_6	0.000	75.80	75.80
I_6	−0.000	6,088.00	6,088.00
Z_6	0.000	652.00	652.00
A_7	0.000	114.00	114.00
I_7	0.002	14,307.00	14,307.00
Z_7	0.000	1,288.00	1,288.00
A_8	−0.000	75.80	75.80
I_8	−0.048	11,360.00	11,360.00
Z_8	0.000	988.50	988.50
A_9	0.000	136.60	136.60
I_9	0.066	17,510.00	17,480.00
Z_9	−0.145	1,485.00	1,482.40
E	−0.008	205.00	205.00
F_y	−0.317	280.00	277.13
Performance function		0.5440	0.0100
Reliability index (in 21 iterations)			2.2820
Probability of failure—FORM			0.0113
Monte Carlo estimate			0.0108

The size of the problem grows by an order of magnitude in the case of SFEM-based reliability analysis, since (1) the partial derivatives of matrices [**K**] and [**M**] have to be computed and stored at each iteration of the reliability analysis, and (2) several iterations need to be performed in searching for the minimum distance point. Therefore, it is useful to incorporate matrix condensation procedures in SFEM to keep the computations within manageable limits for large structures.

The condensation algorithms assume that the displacements of a few degrees of freedom, called the master degrees of freedom, govern the response of the structure. This is particularly true of building structures. The number of degrees of freedom can be reduced by eliminating the other degrees of freedom, called the slave degrees of freedom, using matrix condensation. Therefore, the most important step in condensation is the selection of the master degrees of freedom. In the case of framed structures subjected to predominantly horizontal excitation, the master degrees of freedom can be selected by following the guidelines given below:

- Since the floors are very stiff in the horizontal plane, all the nodes at each floor level may be assumed to have the same horizontal displacement; therefore, the structure may be assumed to have only one master horizontal degree of freedom at each floor level and the other horizontal degrees of freedom may be condensed.

- Vertical deformations are produced only by bending of the structure and are very small compared to the horizontal displacements; therefore, the vertical degrees of freedom may be condensed.

- Similarly, the rotations of nodes for predominantly horizontal excitations are generally very small; therefore, the rotational degrees of freedom may also be condensed.

The above assumptions give an equivalent structure with only one degree of freedom per floor; therefore, dynamic analysis of large structures can be carried out economically. The remainder of this section develops this idea for dynamic reliability analysis with SFEM. The stiffness matrix is condensed using static condensation, while the mass matrix is condensed using Guyan's approach (Guyan, 1965).

9.4.1 Static Condensation

The system equations are written as

$$\begin{bmatrix} [\mathbf{K}_{ss}] & [\mathbf{K}_{sm}] \\ [\mathbf{K}_{ms}] & [\mathbf{K}_{mm}] \end{bmatrix} \begin{Bmatrix} \{\mathbf{y}_s\} \\ \{\mathbf{y}_m\} \end{Bmatrix} = \begin{Bmatrix} \{\mathbf{f}_s\} \\ \{\mathbf{f}_m\} \end{Bmatrix} \tag{9.41}$$

where subscripts s and m specify slave and master degrees of freedom respectively. Assuming that the forces acting on the slave degrees of freedom $\{\mathbf{f}_s\}$ are zero, the reduced structure equation can be obtained as

$$[\overline{\mathbf{K}}] = [\mathbf{K}_{mm}] - [\mathbf{K}_{ms}][\mathbf{K}_{ss}]^{-1}[\mathbf{K}_{sm}] \tag{9.42}$$

where $[\overline{\mathbf{K}}]$ is used in $\{\mathbf{f}_m\} = [\overline{\mathbf{K}}]\{\mathbf{y}_m\}$. Equation 9.41, which represents the relationship between s and m degrees of freedom, can be modified to give $\{\mathbf{y}\}$ using

$$\{\mathbf{y}\} = [\mathbf{T}]\{\mathbf{y}_m\} \tag{9.43}$$

where

$$\{\mathbf{y}\} = \begin{Bmatrix} \{\mathbf{y}_s\} \\ \{\mathbf{y}_m\} \end{Bmatrix} \quad \text{and} \quad [\mathbf{T}] = \begin{bmatrix} [\overline{\mathbf{T}}] \\ [\mathbf{I}] \end{bmatrix}$$

Substituting the above equations in Equation 9.42 gives

$$[\overline{\mathbf{K}}] = [\mathbf{T}]^t[\mathbf{K}][\mathbf{T}] \tag{9.44}$$

In practice $[\overline{\mathbf{K}}]$ can be obtained by applying Gauss–Jordan elimination (Paz, 1997) on the stiffness matrix arranged as above. This results in

$$\begin{bmatrix} [\mathbf{I}] & -[\overline{\mathbf{T}}] \\ [0] & [\overline{\mathbf{K}}] \end{bmatrix} \begin{Bmatrix} \{\mathbf{y}_s\} \\ \{\mathbf{y}_m\} \end{Bmatrix} = \begin{Bmatrix} \{\mathbf{f}_s\} \\ \{\mathbf{f}_m\} \end{Bmatrix} \tag{9.45}$$

which gives the transformation matrix $[\overline{\mathbf{T}}]$ as well as the condensed stiffness matrix $[\overline{\mathbf{K}}]$. These transformations do not introduce any approximations into the calculations, since they involve only Gauss–Jordan elimination of the stiffness matrix, which does not make any assumptions about the behavior of the structure or the properties of the structural matrices.

9.4.2 Guyan Condensation

The eigenvalue problem for any system can be arranged as in the last section as follows:

$$\begin{bmatrix} [\mathbf{K}_{ss}] & [\mathbf{K}_{sm}] \\ [\mathbf{K}_{ms}] & [\mathbf{K}_{mm}] \end{bmatrix} \begin{Bmatrix} \{\Phi_s\} \\ \{\Phi_m\} \end{Bmatrix} = \lambda \begin{bmatrix} [\mathbf{M}_{ss}] & [\mathbf{M}_{sm}] \\ [\mathbf{M}_{ms}] & [\mathbf{M}_{mm}] \end{bmatrix} \begin{Bmatrix} \{\Phi_s\} \\ \{\Phi_m\} \end{Bmatrix} \tag{9.46}$$

It is assumed that $[\mathbf{M}_{ss}]$, $[\mathbf{M}_{sm}]$, and $[\mathbf{M}_{ms}]$ are negligible compared to $[\mathbf{M}_{mm}]$, which makes the transformations developed for static case valid for the dynamic case also. This is an approximation, and the accuracy of the solution is dependent on the selection of the master degrees of freedom and the accuracy of the assumption. For example, in the case of lumped mass formulation, all the rotational degrees of freedom are massless, and if those degrees of freedom are condensed, only $[\mathbf{M}_{mm}]$ is nonzero and the transformation is exact.

In any general case, the condensed mass matrix $[\overline{\mathbf{M}}]$ is obtained by a transformation similar to $[\overline{\mathbf{K}}]$:

$$[\overline{\mathbf{M}}] = [\mathbf{T}]^t[\mathbf{M}][\mathbf{T}] \tag{9.47}$$

9.5 SENSITIVITY ANALYSIS OF CONDENSED STRUCTURE

The first step in the SFEM analysis of condensed systems is to calculate the partial derivatives of the condensed stiffness and mass matrices. The differentiation of Equation 9.44 gives

$$\frac{\partial[\overline{\mathbf{K}}]}{\partial \mathbf{X}} = \frac{\partial[\mathbf{T}]^t}{\partial \mathbf{X}}[\mathbf{K}][\mathbf{T}] + [\mathbf{T}]^t\frac{\partial[\mathbf{K}]}{\partial \mathbf{X}}[\mathbf{T}] + [\mathbf{T}]^t[\mathbf{K}]\frac{\partial[\mathbf{T}]}{\partial \mathbf{X}} \tag{9.48}$$

A similar formula may be developed for the mass matrix. This formula may be simplified by an approximation neglecting the terms involving the partial derivatives of $[\mathbf{T}]$, under the assumption that their contribution to the partial derivatives of the eigenvalues and eigenvectors is small (Hasselman and Hart, 1972). (The effect of this approximation will be examined later, in Examples 9.4 and 9.5.) Thus, the partial derivatives of the condensed stiffness and mass matrices are approximated as

$$\frac{\partial [\overline{K}]}{\partial X} = [T]^t \frac{\partial [K]}{\partial X} [T] \tag{9.49}$$

$$\frac{\partial [\overline{M}]}{\partial X} = [T]^t \frac{\partial [M]}{\partial X} [T] \tag{9.50}$$

The next step in SFEM is to compute the partial derivatives of the eigenvalues and eigenvectors using the partial derivatives of the condensed stiffness and mass matrices. Substitution of Equation 9.43 in Equation 9.14 gives

$$\frac{\partial \lambda_i}{\partial X} = \Phi_{m_i}^t [T]^t \left(\frac{\partial [K]}{\partial X} - \lambda_i \frac{\partial [M]}{\partial X} \right) [T] \Phi_{m_i} \tag{9.51}$$

$$\frac{\partial \lambda_i}{\partial X} = \Phi_{m_i}^t \left(\frac{\partial [\overline{K}]}{\partial X} - \lambda_i \frac{\partial [\overline{M}]}{\partial X} \right) \Phi_{m_i} \tag{9.52}$$

where Φ_{m_i} is the ith mode of vibration for condensed system. Substituting Equation 9.43 in Equation 9.19 gives

$$a_{ij} = \frac{\Phi_{m_j}^t [T]^t \left(\frac{\partial [K]}{\partial X} - \lambda_i \frac{\partial [M]}{\partial X} \right) [T] \Phi_{m_i}}{\lambda_i - \lambda_j} \qquad i \neq j \tag{9.53}$$

$$a_{ij} = \frac{\Phi_{m_j}^t \left(\frac{\partial [\overline{K}]}{\partial X} - \lambda_i \frac{\partial [\overline{M}]}{\partial X} \right) \Phi_{m_i}}{\lambda_i - \lambda_j} \qquad i \neq j \tag{9.54}$$

Substituting Equation 9.43 in Equation 9.21 gives

$$a_{ii} = -\frac{\Phi_{m_i}^t [T]^t \frac{\partial [M]}{\partial X} [T] \Phi_{m_i}}{2} \tag{9.55}$$

or

$$a_{ii} = -\frac{\Phi_{m_i}^t \frac{\partial [\overline{M}]}{\partial X} \Phi_{m_i}}{2} \tag{9.56}$$

Once the matrices, eigenvalues, and eigenvectors for the condensed system are available, modal superposition analysis can be performed on the condensed system to evaluate the displacements of the master degrees of freedom $\{y_m\}$. The displacements of the whole structure $\{y\}$ can then be computed using Equation 9.43. Differentiation of Equation 9.43 gives

$$\frac{\partial y}{\partial X} = \frac{\partial [T]}{\partial X} y_m + [T] \frac{\partial y_m}{\partial X} \tag{9.57}$$

Since $\partial [T]/\partial X$ is assumed to be zero, the partial derivatives of the displacement can be written as

$$\frac{\partial \mathbf{y}}{\partial \mathbf{X}} = [\mathbf{T}]\frac{\partial \mathbf{y}_m}{\partial \mathbf{X}} \tag{9.58}$$

The performance function and response statistics can then be calculated in the same way as for the original structure, and the Rackwitz–Fiessler algorithm can be used to search for the minimum distance point on the limit state and to estimate the reliability as explained earlier.

Example 9.4

The reliability analysis problem of Example 9.2 (six-story frame with a ground acceleration) is repeated here with condensation. The same random variable data and the limit state are used for the sake of comparison. The results of SFEM-based reliability analysis with matrix condensation are shown in Table 9.6.

Example 9.5

The reliability analysis problem of Example 9.3 (six-story frame with a half-sine pulse load) is repeated here with condensation. The same random variable data and the limit state are used for the sake of comparison. The results of SFEM-based reliability analysis with matrix condensation are shown in Table 9.7.

In Examples 9.4 and 9.5, the eigenvalues and eigenvectors obtained after matrix condensation match closely with the corresponding values of the uncondensed structure in Examples 9.2 and 9.3, respectively. However, in Example 9.4 (ground excitation), the estimates of displacements have errors of the order of 10 to 20%, thus significantly affecting the estimation of reliability. The origin of this error is in the condensed mass matrix $\overline{\mathbf{M}}$. Since the error in the evaluation of eigenvalues and eigenvectors using condensation is negligible, the left-hand side of the equation of motion in modal amplitudes, Equation 9.5,

$$\ddot{q}_i + 2\xi_i\omega_i\dot{q}_i + \omega_i^2 q_i = \{\Phi_i\}^{\mathrm{t}}\{\mathbf{f}\} \tag{9.59}$$

remains unaffected. However, the evaluation of the modal force vector $\{\mathbf{f}\}$ involves the mass matrix in the case of ground excitation. That is,

$$
\begin{aligned}
\{\mathbf{f}\} &= \{\Phi\}^{\mathrm{t}}\{f\}, &&\text{for any general loading} \tag{9.60}\\
&= \{\Phi\}^{\mathrm{t}}[\overline{\mathbf{M}}]\{1\}\ddot{y}_s(t_i), &&\text{for ground excitation}
\end{aligned}
$$

The modal masses $\{\Phi\}^{t}[\overline{\mathbf{M}}]\{1\}$ are higher by about 10 to 20% because of the error in the computation of $\overline{\mathbf{M}}$. This results in increased modal forces and therefore increased deflections. The displacements in the condensed structure are always larger than the original structure so that the results obtained for the probability of failure are always on the conservative side. However, the mass matrix is not involved in the case of the half-sine pulse loading at the nodes and the results with condensation are within 3% of the results for the original structure. Thus, the assumption of negligible masses for the slave degrees of freedom critically affects the accuracy of the results.

In addition, there is a significant difference between the results of SFEM and Monte Carlo simulation when condensation is used. This is due to another approximation in-

Table 9.6. Six-story frame with condensation: El Centro earthquake loading

Variable	Sensitivity Index	Initial Checking Point	Final Checking Point
D_1	0.030	16.875	16.875
L_1	−0.007	1.688	1.688
D_2	0.084	21.600	21.639
L_2	0.037	3.938	3.938
f	0.986	16.950	40.776
ξ	−0.101	0.05	0.040
A_1	−0.000	39.90	39.90
I_1	−0.001	4,427.00	4,427.00
Z_1	0.000	394.80	394.80
A_2	0.000	47.40	47.40
I_2	−0.011	7,143.00	7,143.00
Z_2	0.000	539.30	539.30
A_3	0.000	56.90	56.90
I_3	−0.004	12,052.00	12,052.00
Z_3	0.000	771.70	771.70
A_4	−0.000	68.30	68.30
I_4	−0.013	18,576.00	18,547.10
Z_4	0.000	1,046.00	1,046.00
A_5	−0.000	58.80	58.80
I_5	−0.003	4,564.00	4,564.00
Z_5	0.000	497.40	497.40
A_6	−0.000	75.80	75.80
I_6	−0.006	6,088.00	6,088.00
Z_6	0.000	652.00	652.00
A_7	0.000	114.00	114.00
I_7	−0.002	1,4307.00	1,4307.00
Z_7	0.000	1,288.00	1,288.00
A_8	−0.000	75.80	75.80
I_8	−0.013	11,360.00	11,360.00
Z_8	0.000	988.50	988.50
A_9	0.000	136.60	136.60
I_9	0.019	17,510.00	17,510.00
Z_9	−0.031	1,485.00	1,482.70
E	−0.041	205.00	205.00
F_y	−0.067	280.00	276.602
Performance function		0.2788	0.0009
Reliability index (in 12 iterations)			0.9331
Probability of failure—FORM			0.1736
Monte Carlo estimate			0.1210

troduced by the assumption that $\partial[\mathbf{T}]/\partial\mathbf{X} = 0$ while computing the response sensitivity of the condensed structure for SFEM analysis. Other factors that affect the difference between SFEM and Monte Carlo results include nonlinearity of the limit state (AFOSM uses only a linear approximation), and the highly nonnormal extreme value distributions relating to earthquake acceleration amplitude and the modal damping ratio, which also have high coefficients of variation.

Table 9.7. Six-story frame with condensation: sinusoidal loading

Variable	Sensitivity Index	Initial Checking Point	Final Checking Point
D_1	0.005	16.875	16.875
L_1	−0.003	1.688	1.688
D_2	0.014	21.600	21.600
L_2	0.012	3.938	3.938
W	0.924	16.950	35.869
ξ	−0.099	0.05	0.042
A_1	0.000	39.90	39.90
I_1	0.000	4,427.00	4,427.00
Z_1	0.000	394.80	394.80
A_2	0.000	47.40	47.40
I_2	0.002	7,143.00	7,143.00
Z_2	0.000	539.30	539.30
A_3	−0.000	56.90	56.90
I_3	−0.000	12,052.00	12,052.00
Z_3	0.000	771.70	771.70
A_4	−0.000	68.30	68.30
I_4	−0.053	1,8576.00	1,8547.10
Z_4	0.000	1,046.00	1,046.00
A_5	0.000	58.80	58.80
I_5	0.000	4,564.00	4,564.00
Z_5	0.000	497.40	497.40
A_6	0.000	75.80	75.80
I_6	0.000	6,088.00	6,088.00
Z_6	0.000	652.00	652.00
A_7	0.000	114.00	114.00
I_7	0.000	14,307.00	14,307.00
Z_7	0.000	1,288.00	1,288.00
A_8	−0.001	75.80	75.80
I_8	−0.051	11,360.00	11,360.00
Z_8	0.000	988.50	988.50
A_9	0.000	136.60	136.60
I_9	0.097	17,510.00	17,510.00
Z_9	−0.144	1,485.00	1,482.70
E	−0.007	205.00	205.00
F_y	−0.316	280.00	277.207
Performance function			0.0098
		0.5419	
Reliability index (in 22 iterations)			2.2687
Probability of failure—FORM			0.0116
Monte Carlo estimate			0.0111

Table 9.8 highlights the great savings in computational time due to the use of condensation in SFEM. When condensation is not used, SFEM takes a lot of time, especially when multiple limit states are considered. This is due to the computation of large partial derivative matrices of **K** and **M** in SFEM. In the case of the six-story frame with 54 degrees of freedom, **K** and **M** have $54 \times 54 = 2904$ elements each. Their partial derivative matrices for 5 random variables have $2{,}904 \times 5 = 14{,}520$ elements each, which

Table 9.8. Effect of condensation on computational accuracy and efficiency

Example	Original				Condensed			
	DOF	Itrn.	P_f	CPU (sec)	DOF	Itrn.	P_f	CPU (sec)
9.2 and 9.4	54	13	.1379	8381	6	12	.1736	1385
9.3 and 9.5	54	21	.0113	2879	6	22	.0116	201

have to be computed along with **K** and **M** for each iteration of the reliability analysis. It can be seen that the amount of computational effort grows rapidly with the number of degrees of freedom and number of random variables. On the other hand, when condensation is used, only a few degrees of freedom are considered, thus greatly reducing the computational effort in the SFEM sensitivity analysis.

Realistic structures with small failure probabilities will require a large number of Monte Carlo simulations, even with variance reduction techniques. Therefore, the performance efficiency of SFEM would compare much more favorably with Monte Carlo simulation for practical structures, especially when condensation is incorporated.

9.6 DYNAMIC ANALYSIS OF NONLINEAR STRUCTURES

In the previous sections of this chapter, we presented the reliability analysis of linear structures subjected to short-duration dynamic loading using the displacement-based finite element method. In this section we present the reliability analysis of nonlinear structures using the stress-based finite element method. In both linear and nonlinear reliability analyses, the uncertainty in the amplitude of the dynamic loading is specifically addressed.

In general, the reliability analysis of nonlinear structures in the time domain is very difficult. Recently, Huh (1999) suggested a method. It is briefly discussed here to provoke further discussion and development. The algorithm intelligently integrates the concept of the response surface method (RSM), the finite element method and FORM. Since the performance function of a nonlinear dynamic structural system is implicit, RSM is used to approximately generate the performance function and FORM is used to calculate the corresponding reliability index, coordinates of the checking point, and the sensitivity indexes for the random variables involved in the problem. The salient features of the algorithm are discussed next.

9.6.1 Response Surface Method

RSM is an important element of the algorithm and needs to be introduced more formally. The primary purpose of RSM in reliability analysis is to approximately construct a polynomial in explicit form to represent the implicit performance function $g(\mathbf{X})$ (e.g., first-order or second-order), where $\mathbf{X}$ is a vector containing all the load and resistance-related input random variables. Essentially, RSM is a set of statistical techniques designed to find the "best" value of the response considering the uncertainty or variations in the values of input variables (Khuri and Cornell, 1996). Initially, it was developed for models in biology and agriculture (Box et al., 1978). However, it has recently been used for a variety of problems.

RSM can be used to approximately generate the implicit performance function or the structural response statistics, that is, the mean and coefficient of variation of the response. In the latter case, some additional approximations are necessary to approximate the performance function from the information on the response statistics. Therefore, it may be reasonable to use RSM to approximate the performance function in an explicit form for the nonlinear dynamic problem considered here. However, an iterative strategy needs to be employed for its implementation.

The essential components of basic RSM are: (1) the degree of polynomial to be used to represent the performance function, (2) the experimental region and codified variables, (3) selection of sampling or design points, known as the experimental design, and (4) the determination of the center point. They are briefly discussed next.

9.6.1.1 Degree of Polynomial in Performance Function

In Chapter 5, Section 5.3, a linear performance function $g(\mathbf{X})$ was represented by Equation 5.3. $g(\mathbf{X})$ is called the true performance function, but it is unknown. It is assumed to be a continuous function of the individual random variable X_i. Considering the nonlinear dynamic response behavior of structures in the context of RSM, a second-order polynomial can be used to represent the performance function.

A second-order polynomial can be represented without cross terms as

$$\hat{g}(\mathbf{X}) = b_0 + \sum_{i=1}^{k} b_i X_i + \sum_{i=1}^{k} b_{ii} X_i^2 \tag{9.61}$$

and with cross terms as

$$\hat{g}(\mathbf{X}) = b_0 + \sum_{i=1}^{k} b_i X_i + \sum_{i=1}^{k} b_{ii} X_i^2 + \sum_{i=1}^{k-1} \sum_{j>1}^{k} b_{ij} X_i X_j \tag{9.62}$$

where $Xi(i = 1, 2, \ldots, k)$ is the ith random variable, k is the total number of random variables present in the problem, b_0, b_i, b_{ii}, and b_{ij} are the unknown coefficients to be determined, and $\hat{g}(\mathbf{X})$ is an approximate representation of $g(\mathbf{X})$. The number of coefficients, p, to be estimated for Equations 9.61 and 9.62, are $p = 2k + 1$ and $p = (k + 1)(k + 2)/2$, respectively.

9.6.1.2 Experimental Region and Coded Variables

For the efficient and accurate construction of a performance function, the experimental region should be kept to a minimum (Khuri and Cornell, 1996). The uncertainty in the random variables can be used to define the experimental region. The region or bounds for each input variable X_i can be specified as

$$X_i^{\text{region}} = X_i^C \pm h_i \sigma_{X_i} x_i \tag{9.63}$$

where X_i^c and σ_{X_i} are the center point (to be discussed in Section 9.6.1.4) and standard deviation of a random variable X_i, respectively, h_i is an arbitrary factor used to define the region, and x_i is the coded variable. The coded variable can be defined as

$$x_i = \frac{X_i - X_i^C}{h_i \sigma_{X_i}} i = 1, 2, \ldots, k \tag{9.64}$$

All the terms in Equation 9.64 have already been defined. The value of h_i is generally considered to be between 1 and 3. From Equation 9.64, it is clear that the value of a coded variable at the center point is zero. The transformation of random variables to coded variables facilitates the evaluation of the coefficients in Equations 9.61 and 9.62 (Snee, 1973).

9.6.1.3 *Experimental Design*

In experimental design, the sets of values of the random variables are selected within the experimental region to analyze a structure deterministically to evaluate the coefficients of the performance function $g(\mathbf{X})$. The number of design points should be kept to a minimum to increase the computational efficiency but must be at least equal to the number of coefficients p needed to define a performance function. Efficient location of design points around the center point, discussed next, is essential in the accurate construction of the performance function.

To construct performance functions containing second-order polynomials, the available procedures for experimental design can be divided into two categories: classical design and saturated design.

Classical Design: Central Composite Design In the classical design approach, responses are first calculated at specified design points and then a regression analysis is carried out to formulate the performance function. In this approach, one of the conceptually simpler designs is factorial design, in which the response values are estimated for each variable sampled at equal intervals. In order to fit a second order surface for k input variables, the sampling points must have at least three levels for each variable, leading to 3^k factorial design.

Box and Wilson (1951) introduced a more efficient approach, known as central composite design (CCD), for fitting second order surfaces. This design can be used for polynomials with cross terms as in Equation 9.62. It consists of a complete 2^k factorial design, a center point, and two axial points on the axis of each random variable at a distance α from the center point where $\alpha = \sqrt[4]{2^k}$ in order to make the design rotatable. The total number of design points in CCD is $N = 2^k + 2k + 1$; N is much more than the number of coefficients $p = (k + 1)(k + 2)/2$ to be estimated. Regression analysis can be used to estimate the coefficients of the performance function. Other experimental design methods, such as Box–Behnken design (Fox, 1993) have also been used in response surface-based reliability analysis.

Saturated Design Saturated design consists of only as many design points as the total number of coefficients necessary to define a polynomial representing the performance function. The unknown coefficients are obtained by solving a set of linear equations, without conducting any regression analysis. The design can be used for second-order polynomials with and without cross terms. For second-order polynomials without cross terms, the total number of required design points is $2k + 1$, where k is the number of random variables. For second-order polynomials with cross terms, the total number of required design points is $(k + 1)(k + 2)/2$.

Considering the capabilities of saturated design and CCD, three experimental design models to select design points are considered in this section: (1) saturated design using a second-order polynomial without cross terms, (2) saturated design using a second-order polynomial with cross terms, and (3) central composite design with a second-order polynomial with cross terms. They are denoted hereafter as Models 1, 2, and 3, respectively.

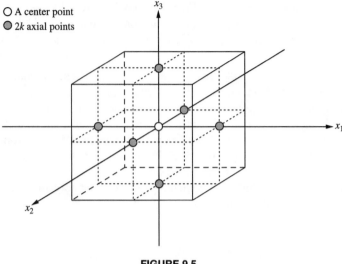

FIGURE 9.5.

The design points for these three models for $k = 3$ are shown in Figures 9.5, 9.6, and 9.7, respectively, in the coded variables space. They are compared in Table 9.9 in terms of the number of coefficients p to be estimated and the number of design points N required as a function of the number of random variables k.

Since computational efficiency depends on the number of design points, it can be observed from Table 9.9 that Model 1 is the most efficient and Model 3 is the least efficient. However, as will be discussed later, Model 3 is the most accurate of the three models. Con-

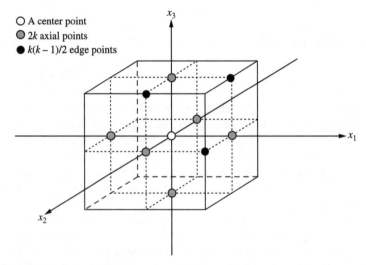

FIGURE 9.6. Model 2: Saturated Design using a full second order polynomial with cross terms in the coded variables space for $k = 3$.

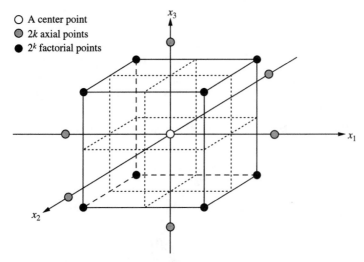

FIGURE 9.7. Model 3: Central Composite Design using a full second order polynomial with cross terms in the coded variables space for $k = 3$.

sidering both computational efficiency and accuracy, a mixture of the three models may be desirable.

9.6.1.4 *Determination of the Center Point*

The center point around which the design points are selected plays an important role in the construction of the performance function. Determining the location of the center point is not a simple task, and a systematic iterative approach is necessary. Information on the reliability index and the coordinates of the corresponding checking point, obtained from the FORM analysis, can be used to select the center point. Bucher and Bourgund (1990) suggested an iterative linear interpolation scheme for this purpose, as discussed next.

Since the location of the center point is unknown at the beginning of the iteration, the center point is initially assumed to be located at the mean values of the random variables

Table 9.9. Comparison of the three models

Model	Number of Coefficients	k	Number of Design Points	
Model 1	$p = 2k + 1$	7 17 25	3 8 12	7 17 25	$N = 2k + 1$
Model 2	$p = \dfrac{(k+1)(k+2)}{2}$	10 45 91	3 8 12	10 45 91	$N = \dfrac{(k+1)(k+2)}{2}$
Model 3		10 45 91	3 8 12	15 273 4142	$N = 2^k + 2k + 1$

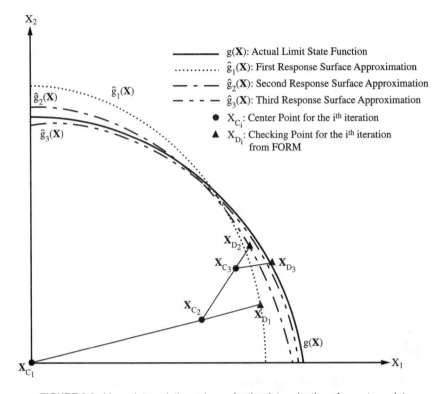

FIGURE 9.8. Linear interpolation scheme for the determination of a center point.

X_i's. It is denoted as $\mathbf{x}_{C_1}$ in Figure 9.8. This will produce a performance function $\hat{g}_1(\mathbf{X})$ as shown in Figure 9.8. Then, a linear interpolation scheme can be used to select the next center point as

$$\mathbf{x}_{C_2} = \mathbf{x}_{C_1} + \left(\mathbf{x}_{D_1} - \mathbf{x}_{C_1}\right) \times \frac{g\left(\mathbf{x}_{C_1}\right)}{g\left(\mathbf{x}_{C_1}\right) - g\left(\mathbf{x}_{D_1}\right)} \qquad \text{if} \quad g\left(\mathbf{x}_{D_1}\right) \geq g\left(\mathbf{x}_{C_1}\right) \qquad (9.65)$$

or

$$\mathbf{x}_{C_2} = \mathbf{x}_{D_1} + \left(\mathbf{x}_{C_1} - \mathbf{X}_{D_1}\right) \times \frac{g\left(\mathbf{x}_{D_1}\right)}{g\left(\mathbf{x}_{D_1}\right) - g\left(\mathbf{x}_{C_1}\right)} \qquad \text{if} \quad g\left(\mathbf{x}_{D_1}\right) < g\left(\mathbf{x}_{C_1}\right) \qquad (9.66)$$

Another performance function can be constructed by considering design points around the new center point $\mathbf{x}_{C_2}$. The iteration process needs to be continued until all the random variables at the center point converge to a preselected criterion. For each random variable, the difference between 2 successive iterations can be limited, for example to 1%, giving the required convergence criterion. It generally takes only 2 to 3 iterations to satisfy this convergence criterion (Huh and Haldar, 1999). The iteration scheme is shown graphically in Figure 9.8.

Comments The concept behind the construction of the performance function using RSM is simple. However, considering computational efficiency and accuracy, the implementation of a basic RSM scheme in the reliability evaluation of nonlinear dynamic system could be impractical. If an RSM scheme needs to be implemented, the number of random variables needed to construct the response surface should be kept to a minimum. This can be done using the sensitivity analysis discussed in Section 3.8. Furthermore, a mixture of 3 models, discussed in Section 9.6.1.3, may need to be developed for the construction of an RSM. A computationally efficient model, such as Model 1, can be used in the first $(n-1)$ iterations, and a more accurate model, such as Model 3, can be used at the last iteration.

9.6.2 Reliability Evaluation

As discussed in Section 8.7, reliability is always calculated with respect to a performance function. The RSM-based algorithm is capable of estimating reliability considering both the strength and serviceability limit states. The essential steps required for estimating risk or reliability are summarized next.

Step 1   Set the initial center point to start the iteration at the mean values of all the random variables.

Step 2   Select an experimental design model considering the type of polynomial, number of random variables, and computational efficiency.

Step 3   Analyze the structure deterministically at all the design points according to the model selected in Step 2 using the nonlinear finite element method discussed in Chapter 8.

Step 4   Generate the performance function by evaluating the coefficients in Equation 9.61 or 9.62.

Step 5   Using the performance function thus generated, estimate the reliability index satisfying a convergence criterion as discussed in Section 3.5 and the corresponding coordinates of the checking point using FORM .

Step 6   Using the information on the sensitivity indexes obtained in Step 5, consider less sensitive random variables to be deterministic constants at their mean values.

Step 7   Find the coordinates of the new center point to be used for the next iteration using the linear interpolation scheme given in Equation 9.65 or 9.66.

Step 8   Repeat Steps 2 to 7 until the coordinates of the center point of all random variables converge to a preselected tolerance criterion. This completes the first $(n-1)$ iterations.

Step 9   Select a type of polynomial and corresponding experimental design model considering accuracy for the final iteration.

Step 10   Construct the final response surface.

Step 11   Calculate the reliability index and the corresponding coordinates of the checking point.

9.6.3 Numerical Example

A computer program was developed to implement the procedure discussed in the previous section. The procedure is clarified with the help of an example.

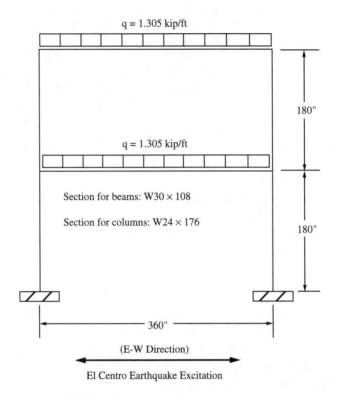

FIGURE 9.9. A two story steel frame structure.

Example 9.6

A two-story one-bay frame shown in Figure 9.9 is considered to describe the algorithm just discussed. All the beams are made of W30 × 108 and all the columns are made of W24 × 176. A36 steel is used for this example. The frame is excited by the first 5.12 sec of the East-West component of the 1940 El Centro earthquake, as shown in Figure 9.10. The amplitude of the earthquake is considered to be a random variable, represented by a Type I extreme value distribution. Information on all the random variables considered in the formulation is shown in Table 9.10.

To increase the efficiency of the RSM algorithm, 3 models identified in Table 9.9 are grouped into 6 schemes, as shown in Table 9.11. For this example, only the serviceability limit state is considered. The allowable maximum lateral displacement at the top of the frame is considered to be 0.9 inches. The corresponding performance function is

$$g(\mathbf{X}) = 0.9 - y_{\max}(\mathbf{X}) \qquad (9.67)$$

Using the algorithm, the frame is analyzed using an SGI Origin 2000 supercomputer. After a sensitivity analysis, it was observed that the 9 random variables shown in Table 9.10 are not equally important in the reliability analysis. To increase computational efficiency, 3 cases are considered: (1) all 9 variables are considered to be random, (2) seven variables are considered to be random; the section modulus of beams and columns are considered to be constants, and (3) five of them are considered to be ran-

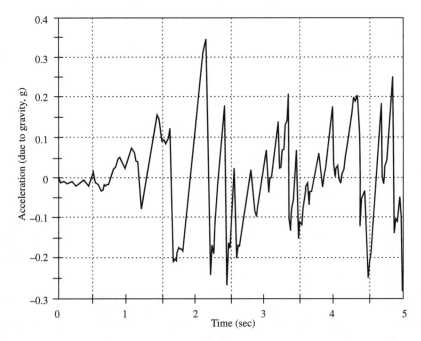

FIGURE 9.10. El Centro earthquake time history for 5.12 seconds (East-West component).

dom; the areas and section modulus of beams and columns are considered to constants. They are denoted as Cases 1 to 3 in Table 9.12.

The probability of failure and the error associated with the estimation, the CPU time, and the total number of design points for all 3 cases and 6 schemes are summarized in Table 9.12. The probability of failure and the CPU time for 100,000 cycles of Monte Carlo simulation for each case are shown in Table 9.12. The error in the probability of failure using the algorithm is estimated using the simulation results. In each case, the

Table 9.10. Statistical description of the random variables for Cases 1, 2, and 3

Random Variable	Mean value	Case 1		Case 2		Case 3	
		C.O.V	Distribution	C.O.V	Distribution	C.O.V	Distribution
E (ksi)	29,000	0.06	Lognormal	0.06	Lognormal	0.06	Lognormal
A^b (in^2)	31.7	0.05	Lognormal	0.05	Lognormal	—	Constant
I_x^b (in^4)	4,470	0.05	Lognormal	0.05	Lognormal	0.05	Lognormal
Z_x^b (in^3)	346	0.05	Lognormal	—	Constant	—	Constant
A^c (in^2)	51.7	0.05	Lognormal	0.05	Lognormal	—	Constant
I_x^c (in^4)	5,680	0.05	Lognormal	0.05	Lognormal	0.05	Lognormal
Z_x^c (in^3)	511	0.05	Lognormal	—	Constant	—	Constant
ξ (damping)	0.02	0.15	Lognormal	0.15	Lognormal	0.15	Lognormal
g_e	1.0	0.20	Type I	0.20	Type I	0.20	Type I

g_e is a factor to consider uncertainty in the amplitude of actual seismic acceleration
Superscript b represents beam and superscript c represents column

Table 9.11. Six schemes for the proposed algorithm

Schemes	Intermediate Iteration	Final Iteration
Scheme 1		Model 1
Scheme 2		Model 2
Scheme 3		Model 3
Scheme 4	Model 1	Model 2
Scheme 5	Model 1	Model 3
Scheme 6	Model 2	Model 3

Table 9.12. Reliability analysis for three cases

	Case	RSM						MCS*
		Scheme 1	Scheme 2	Scheme 3	Scheme 4	Scheme 5	Scheme 6	
1	P_f	0.03824	0.03750	0.03583	0.03482	0.03592	0.03579	0.03627
	error (%)	−5.43	−3.39	1.21	4.00	0.96	1.32	
	CPU$^\Delta$	54 sec	144 sec	1340 sec	84 sec	476 sec	536 sec	83479 sec
	TNDP$^\bullet$	57	165	1593	93	569	641	
2	P_f	0.03816	0.03691	0.03598	0.03424	0.03592	0.03583	0.03606
	error (%)	5.82	2.36	−0.22	−5.05	−0.39	−0.64	
	CPU$^\Delta$	43 sec	91 sec	357 sec	62 sec	148 sec	173 sec	81890 sec
	TNDP$^\bullet$	45	108	429	66	173	215	
3	P_f	0.03796	0.03688	0.03546	0.03416	0.03608	0.03598	0.03583
	error (%)	5.94	2.93	−1.03	−4.66	0.70	0.42	
	CPU$^\Delta$	32 sec	57 sec	111 sec	39 sec	59 sec	76 sec	81414 sec
	TNDP$^\bullet$	33	63	129	43	65	85	

$^i h = 1.5$ (for intermediate iterations) and $^f h = 1.0$ (for the final iteration)
*100,000 cycles of Monte Carlo simulation
$^\bullet$Total number of design points, i.e., total number of deterministic dynamic analyses
$^\Delta$SGI Origin 2000 supercomputer

value of h in Equation 9.63 is considered to be 1.5 for intermediate iterations, and 1.0 for the final iteration.

Several interesting observations can be made from the results shown in Table 9.12. The probability of failure for all 3 cases is very similar to the simulation results, indicating that the algorithm is viable. In all 3 cases, Scheme 1 gave the maximum error in the estimation of the probability of failure, but required the least number of design points and thus the least amount of time. Scheme 5, that is, the performance function constructed using saturated design with a second-order polynomial without cross terms for all intermediate iterations and central composite design with a second-order polynomial with cross terms at the final iteration, appears to be most desirable considering both computational efficiency and accuracy. Although it is not emphasized here, the value of h plays an important role in the accuracy of the failure probability estimation. By comparing Scheme 5 for all 3 cases, it can be observed that the probability of failure estimated using only 5 sensitive random variables (Case 3) is not significantly different than when all 9 variables are considered to be random (Case 1). However, Case 1 took almost 10 times more CPU time than Case 3, indicating the significance of the sensitivity analysis.

The CPU times shown in Table 9.12 are for an SGI Origin 2000 supercomputer. For a simple frame excited by only 5.12 sec of real earthquake loading, 100,000 cycles of Monte Carlo simulation required over 23 hours of supercomputer time. Thus, simulation may not be a practical option for estimating the reliability of real nonlinear structures excited by short duration dynamic loading. However, various modifications of the algorithm presented here can be attempted, resulting in significant improvement in computational efficiency (81414 sec for simulation versus 59 sec for Case 3–Scheme 5) without sacrificing any accuracy.

9.7 CONCLUDING REMARKS

This chapter presented the application of SFEM-based reliability analysis to structures subject to dynamic loading. The dynamic loading was assumed to have a deterministic shape, and only the magnitude was considered to be random. In Sections 9.1 to 9.5, linear elastic structural behavior was assumed, and the analysis was done in the time domain. Formulations were presented for the computation of response sensitivity in the case of plane frames. For such structures (and linear analysis), the mass and stiffness matrices are analytically available. Therefore, it is easy to compute the response sensitivity to be used in the reliability analysis. In the case of nonlinear behavior, the response sensitivities will need to be computed using an iterative perturbation analysis. Since dynamic analysis in the time domain is quite time-consuming, the use of condensation and its effect on the SFEM-based reliability analysis were presented. A realistic six-story frame structure was used for illustration.

In Section 9.6, the reliability of nonlinear structures under short-duration dynamic loading was presented. The method integrates the concept of the response surface method, the finite element method, and FORM. The method is illustrated with the help of an example consisting of a simple frame excited by an actual earthquake loading. A supercomputer was used to implement the algorithm.

In this chapter the randomness in the dynamic loading was simply represented by treating only the magnitude as a random variable. A more realistic representation of the randomness in dynamic loading is through a random process description. In that case, the structural response is computed through the methods of random vibration, which are beyond the scope of this book. Considerable research is still underway to combine random vibration and SFEM-based reliability analysis. This chapter presented only a simple extension of the SFEM concept to structures under dynamic loading using time-domain analysis.

As mentioned at the beginning of this chapter, the reliability analysis of structures and systems under time-dependent loading is an evolving research topic. Research is being done on computational modeling of fatigue and fracture reliability, long-term durability modeling, inspection and repair planning, and so forth. Empirical methods of reliability engineering that are based on testing and statistics are being used in mechanical and electrical systems to address such issues. However, for large systems where full-scale testing to failure is impossible, efficient computational methods have to be developed to estimate the reliability. The SFEM-based approach presented in this book, which combines finite element analysis and reliability analysis, appears quite promising in achieving this objective.

APPENDIX 1

TABLE OF CUMULATIVE STANDARD NORMAL DISTRIBUTION

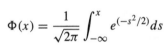
$$\Phi(x) = \frac{1}{\sqrt{2\pi}} \int_{-\infty}^{x} e^{(-s^2/2)} ds$$

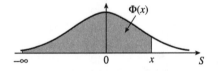

x	$\Phi(x)$	x	$\Phi(x)$	x	$\Phi(x)$	x	$\Phi(x)$
0.00	0.50000	0.25	0.59871	0.50	0.69146	0.75	0.77337
0.01	0.50399	0.26	0.60257	0.51	0.69497	0.76	0.77637
0.02	0.50798	0.27	0.60642	0.52	0.69847	0.77	0.77935
0.03	0.51197	0.28	0.61026	0.53	0.70194	0.78	0.78230
0.04	0.51595	0.29	0.61409	0.54	0.70540	0.79	0.78524
0.05	0.51994	0.30	0.61791	0.55	0.70884	0.80	0.78814
0.06	0.52392	0.31	0.62172	0.56	0.71226	0.81	0.79103
0.07	0.52790	0.32	0.62552	0.57	0.71566	0.82	0.79389
0.08	0.53188	0.33	0.62930	0.58	0.71904	0.83	0.79673
0.09	0.53586	0.34	0.63307	0.59	0.72240	0.84	0.79955
0.10	0.53983	0.35	0.63683	0.60	0.72575	0.85	0.80234
0.11	0.54380	0.36	0.64058	0.61	0.72907	0.86	0.80511
0.12	0.54776	0.37	0.64431	0.62	0.73237	0.87	0.80785
0.13	0.55172	0.38	0.64803	0.63	0.73565	0.88	0.81057
0.14	0.55567	0.39	0.65173	0.64	0.73891	0.89	0.81327
0.15	0.55962	0.40	0.65542	0.65	0.74215	0.90	0.81594
0.16	0.56356	0.41	0.65910	0.66	0.74537	0.91	0.81859
0.17	0.56749	0.42	0.66276	0.67	0.74857	0.92	0.82121
0.18	0.57142	0.43	0.66640	0.68	0.75175	0.93	0.82381
0.19	0.57535	0.44	0.67003	0.69	0.75490	0.94	0.82639
0.20	0.57926	0.45	0.67364	0.70	0.75804	0.95	0.82894
0.21	0.58317	0.46	0.67724	0.71	0.76115	0.96	0.83147
0.22	0.58706	0.47	0.68082	0.72	0.76424	0.97	0.83398
0.23	0.59095	0.48	0.68439	0.73	0.76730	0.98	0.83646
0.24	0.59483	0.49	0.68793	0.74	0.77035	0.99	0.83891

x	$\Phi(x)$	x	$\Phi(x)$	x	$\Phi(x)$	x	$\Phi(x)$
1.00	0.84134	1.50	0.93319	2.00	0.97725	2.50	0.99379
1.01	0.84375	1.51	0.93440	2.01	0.97778	2.51	0.99396
1.02	0.84614	1.52	0.93574	2.02	0.97831	2.52	0.99413
1.03	0.84849	1.53	0.93699	2.03	0.97882	2.53	0.99430
1.04	0.85083	1.54	0.93822	2.04	0.97932	2.54	0.99446
1.05	0.85314	0.55	0.93943	2.05	0.97982	2.55	0.99461
1.06	0.85543	1.56	0.94062	2.06	0.98030	2.56	0.99477
1.07	0.85769	1.57	0.94179	2.07	0.98077	2.57	0.99492
1.08	0.85993	1.58	0.94295	2.08	0.98124	2.58	0.99506
1.09	0.86214	1.59	0.94408	2.09	0.98169	2.59	0.99520
1.10	0.86433	1.60	0.94520	2.10	0.98214	2.60	0.99534
1.11	0.86650	1.61	0.94630	2.11	0.98257	2.61	0.99547
1.12	0.86864	1.62	0.94738	2.12	0.98300	2.62	0.99560
1.13	0.87076	1.63	0.94845	2.13	0.98341	2.63	0.99573
1.14	0.87286	1.64	0.94950	2.14	0.98382	2.64	0.99585
1.15	0.87493	1.65	0.95053	2.15	0.98422	2.65	0.99598
1.16	0.87698	1.66	0.95154	2.16	0.98461	2.66	0.99609
1.17	0.87900	1.67	0.95254	2.17	0.98500	2.67	0.99621
1.18	0.88100	1.68	0.95352	2.18	0.98537	2.68	0.99632
1.19	0.88298	1.69	0.95449	2.19	0.98574	2.69	0.99643
1.20	0.88493	1.70	0.95543	2.20	0.98610	2.70	0.99653
1.21	0.88686	1.71	0.95637	2.21	0.98645	2.71	0.99664
1.22	0.88877	1.72	0.95728	2.22	0.98679	2.72	0.99674
1.23	0.89065	1.73	0.95818	2.23	0.98713	2.73	0.99683
1.24	0.89251	1.74	0.95907	2.24	0.98745	2.74	0.99693
1.25	0.89435	1.75	0.95994	2.25	0.98778	2.75	0.99702
1.26	0.89617	1.76	0.96080	2.26	0.98809	2.76	0.99711
1.27	0.89796	1.77	0.96164	2.27	0.98840	2.77	0.99720
1.28	0.89973	1.78	0.96246	2.28	0.98870	2.78	0.99728
1.29	0.90147	1.79	0.96327	2.29	0.98899	2.79	0.99736
1.30	0.90320	1.80	0.96407	2.30	0.98928	2.80	0.99744
1.31	0.90490	1.81	0.96485	2.31	0.98956	2.81	0.99752
1.32	0.90658	1.82	0.96562	2.32	0.98983	2.82	0.99760
1.33	0.90824	1.83	0.96638	2.33	0.99010	2.83	0.99767
1.34	0.90988	1.84	0.96712	2.34	0.99036	2.84	0.99774
1.35	0.91149	1.85	0.96784	2.35	0.99061	2.85	0.99781
1.36	0.91308	1.86	0.96856	2.36	0.99086	2.86	0.99788
1.37	0.91466	1.87	0.96926	2.37	0.99111	2.87	0.99795
1.38	0.91621	1.88	0.96995	2.38	0.99134	2.88	0.99801
1.39	0.91774	1.89	0.97062	2.39	0.99158	0.28	0.99807
1.40	0.91924	1.90	0.97128	2.40	0.99180	2.90	0.99813
1.41	0.92073	1.91	0.97193	2.41	0.99202	2.91	0.99819
1.42	0.92220	1.92	0.97257	2.42	0.99224	2.92	0.99825
1.43	0.92364	1.93	0.97320	2.43	0.99245	2.93	0.99831
1.44	0.92507	1.94	0.97381	2.44	0.99266	2.94	0.99836
1.45	0.92647	1.95	0.97441	2.45	0.99286	2.95	0.99841
1.46	0.92785	1.96	0.97500	2.46	0.99305	2.96	0.99846
1.47	0.92922	1.97	0.97558	2.47	0.99324	2.97	0.99851
1.48	0.93056	1.98	0.97615	2.48	0.99343	2.98	0.99856
1.49	0.93189	1.99	0.97670	2.49	0.99361	2.99	0.99861

x	$\Phi(x)$	x	$\Phi(x)$	x	$1 - \Phi(x)$
3.00	0.99865	3.50	0.99977	4.00	3.1686e−05
3.01	0.99869	3.51	0.99978	4.05	2.5622e−05
3.02	0.99874	3.52	0.99978	4.10	2.0669e−05
3.03	0.99878	3.53	0.99979	4.15	1.6633e−05
3.04	0.99882	3.54	0.99980	4.20	1.3354e−05
3.05	0.99886	3.55	0.99981	4.25	1.0696e−05
3.06	0.99889	3.56	0.99981	4.30	8.5460e−06
3.07	0.99893	3.57	0.99982	4.35	6.8121e−06
3.08	0.99896	3.58	0.99983	4.40	5.4170e−06
3.09	0.99900	3.59	0.99983	4.45	4.2972e−06
3.10	0.99903	3.60	0.99984	4.50	3.4008e−06
3.11	0.99906	3.61	0.99985	4.55	2.6849e−06
3.12	0.99910	3.62	0.99985	4.60	2.1146e−06
3.13	0.99913	3.63	0.99986	4.65	1.6615e−06
3.14	0.99916	3.64	0.99986	4.70	1.3023e−06
3.15	0.99918	3.65	0.99987	4.75	1.0183e−06
3.16	0.99921	3.66	0.99987	4.80	7.9435e−07
3.17	0.99924	3.67	0.99988	4.85	6.1815e−07
3.18	0.99926	3.68	0.99988	4.90	4.7987e−07
3.19	0.99929	3.69	0.99989	4.95	3.7163e−07
3.20	0.99931	3.70	0.99989	5.00	2.8710e−07
3.21	0.99934	3.71	0.99990	5.10	1.7012e−07
3.22	0.99936	3.72	0.99990	5.20	9.9834e−08
3.23	0.99938	3.73	0.99990	5.30	5.8022e−08
3.24	0.99940	3.74	0.99991	5.40	3.3396e−08
3.25	0.99942	3.75	0.99991	5.50	1.9036e−08
3.26	0.99944	3.76	0.99992	5.60	1.0746e−08
3.27	0.99946	3.77	0.99992	5.70	6.0077e−09
3.28	0.99948	3.78	0.99992	5.80	3.3261e−09
3.29	0.99950	3.79	0.99992	5.90	1.8236e−09
3.30	0.99952	3.80	0.99993	6.00	9.9012e−10
3.31	0.99953	3.81	0.99993	6.10	5.3238e−10
3.32	0.99955	3.82	0.99993	6.20	2.8347e−10
3.33	0.99957	3.83	0.99994	6.30	1.4947e−10
3.34	0.99958	3.84	0.99994	6.40	7.8049e−11
3.35	0.99960	3.85	0.99994	6.50	4.0358e−11
3.36	0.99961	3.86	0.99994	6.60	2.0665e−11
3.37	0.99962	3.87	0.99995	6.70	1.0479e−11
3.38	0.99964	3.88	0.99995	6.80	5.2616e−12
3.39	0.99965	3.89	0.99995	6.90	2.6161e−12
3.40	0.99966	3.90	0.99995	7.00	1.2881e−12
3.41	0.99968	3.91	0.99995	7.10	6.2805e−13
3.42	0.99969	3.92	0.99996	7.20	3.0320e−13
3.43	0.99970	3.93	0.99996	7.30	1.4500e−13
3.44	0.99971	3.94	0.99996	7.40	6.8612e−14
3.45	0.99972	3.95	0.99996	7.50	3.2196e−14
3.46	0.99973	3.96	0.99996	7.60	1.4988e−14
3.47	0.99974	3.97	0.99996	7.70	6.8834e−15
3.48	0.99975	3.98	0.99997	7.80	3.1086e−15
3.49	0.99976	3.99	0.99997	7.90	1.4433e−15

APPENDIX 2

EVALUATION OF GAMMA FUNCTION

- Exact Evaluation (Euler's integral);

$$\Gamma(z) = \int_0^\infty t^{z-1} e^{-t} dt$$

- Approximate Evaluation of Gamma Function
1. Integer Values, x

   $\Gamma(1+x) = x\Gamma(x) = x!$

2. Noninteger Values, x

   (a) For $1 \leq z \leq 2$ (i.e., $0 \leq x \leq 1$), $\Gamma(z) = \Gamma(1+x)$

   Polynomial Approximation for Gamma Function*

   (i) $\Gamma(1+x) = 1 + a_1 x + a_2 x^2 + a_3 x^3 + a_4 x^4 + a_5 x^5 + \varepsilon(x)$, $|\varepsilon(x)| \leq 5 \times 10^{-5}$

   $a_1 = -0.57486\,46$, $a_2 = 0.95123\,63$, $a_3 = -0.69985\,88$
   $a_4 = 0.42455\,49$, $a_5 = -0.10106\,78$

   (ii) $\Gamma(1+x) = 1 + b_1 x + b_2 x^2 + b_3 x^3 + b_4 x^4 + b_5 x^5 + b_6 x^6 + b_7 x^7 + b_8 x^8 + \varepsilon(x)$,
   $|\varepsilon(x)| \leq 3 \times 10^{-7}$

   $b_1 = -0.57719\,1652$, $b_2 = 0.98820\,5891$, $b_3 = -0.89705\,6937$,
   $b_4 = 0.91820\,6857$, $b_5 = -0.75670\,4078$, $b_6 = 0.4821\,9394$,
   $b_7 = -0.1935\,7818$, $b_8 = 0.03586\,8343$

Example: $\Gamma(\frac{11}{8}) = \Gamma(1 + \frac{3}{8}) = 1 - 0.577191652 \times (\frac{3}{8}) + 0.988205891 \times (\frac{3}{8})^2 + \cdots + 0.035868343 \times (\frac{3}{8})^8 = 0.888913365$

   (b) For $0 < z \leq 1$, (i.e., $0 \leq x < 1$) $\Gamma(z) = \Gamma(1-x)$, use reflection formula

$$\Gamma(z) = \Gamma(1-x) = \frac{\pi x}{\Gamma(1+x)\sin(\pi x)}$$

---

*Abramowitz, M., and Stegun, I. A. *Handbook of Mathematical Functions with Formulas, Graphs, and Mathematical Tables*. U.S. Department of Commerce, National Bureau of Standards, Applied Mathematics Series 55, Washington, DC, 1964.

Example: $\Gamma(\frac{5}{8}) = \Gamma(1 - \frac{3}{8}) = \pi(\frac{3}{8})/[\Gamma(1 + \frac{3}{8})\sin(3\pi/8)] = \pi(\frac{3}{8})/[0.888913365\sin(3\pi/8)] = 1.434519178$

(c) For $z > 2$ (i.e., $x > 1$), $\Gamma(z) = \Gamma(1 + x)$, use recurrence formula.

$$\Gamma(z) = \Gamma(1 + x) = \left[\prod_{n-1}^{k}(z - n)\right]\Gamma(1 + \alpha)$$

where $0 \leq \alpha < 1$, integer value $k = (x - \alpha)$, and $\Gamma(1 + \alpha)$ can be obtained from equations (i) or (ii).

Example: $\Gamma(5.2) = \Gamma(1 + 4.2) = (5.2 - 1) \times (5.2 - 2) \times (5.2 - 3) \times (5.2 - 4) \times \Gamma(1 + 0.2)$
$= 35.4816 \times 0.918168911 = 32.578102$,
where $z = 5.2$, $x = 4.2$, $\alpha = 0.2$, $k = (4.2 - 0.2) = 4$

- Evaluation of Gamma Function Using Spreadsheets

  Microsoft Excel—EXP(GAMMALN(z))
  EXP(GAMMALN($\frac{11}{8}$)) = 0.888913569,
  EXP(GAMMALN($\frac{5}{8}$)) = 1.434518848,
  EXP(GAMMALN(5.2)) = 32.57809604.

  Quattro Pro—@EXP(@GAMMALN(X))
  @EXP(@GAMMALN($\frac{11}{8}$)) = 0.88891336,
  @EXP(@GAMMALN($\frac{5}{8}$)) = 1.43451904,
  @EXP(@GAMMALN(5.2)) = 32.57809603.

GRAM–SCHMIDT ORTHOGONALIZATION

Consider a matrix $\mathbf{R}_0$, with row vectors $\mathbf{r}_{01}, \mathbf{r}_{02}, \ldots, \mathbf{r}_{0n}$. This has to be transformed to a matrix $\mathbf{R}$, whose row vectors $\mathbf{r}_1, \mathbf{r}_2, \ldots, \mathbf{r}_n$ are orthogonal to each other, with the nth row the same as in matrix $\mathbf{R}_0$, that is, $\mathbf{r}_n = \mathbf{r}_{0n}$.

The Gram–Schmidt (G-S) method to achieve this may be written as follows. The nth row vector of matrix $\mathbf{R}$ is simply, $\mathbf{r}_n = \mathbf{r}_{0n}$. The other rows of matrix $\mathbf{R}$ are computed (going from the $(n-1)$th row to the first row, in that order) using the formula

$$\mathbf{r}_k = \mathbf{r}_{0k} - \sum_{j=k+1}^{n} \frac{\mathbf{r}_j \mathbf{r}_{0k}^t}{\mathbf{r}_j \mathbf{r}_j^t} \mathbf{r}_j \tag{A3.1}$$

where the superscript t implies the transpose of the row vector. Note that the rows of $\mathbf{R}$ have to be computed in the reverse order, from n to 1.

If the rows of $\mathbf{R}$ are required to be orthonormal (i.e., the rows are orthogonal to each other, and each row vector is of unit length), then each row should be normalized separately at the end.

Example (Application of the G-S Method to Structural Reliability) Consider the example in Section 3.6. The direction cosines of the β vector are $\alpha_1 = 0.867$, $\alpha_2 = 0.498$. The coordinate system (Y_1, Y_2) has to be transformed to another orthogonal coordinate system (Y_1', Y_2') such that Y_2' is along the β vector. To do this, a transformation matrix $\mathbf{R}$ needs to be constructed such that $\mathbf{Y}' = \mathbf{R}\mathbf{Y}$.

First, a matrix $\mathbf{R}_0$ is selected as

$$\mathbf{R}_0 = \begin{bmatrix} 1 & 0 \\ 0.867 & 0.498 \end{bmatrix}$$

The rows of the transformed matrix $\mathbf{R}$ are obtained using the G-S method as follows. The second row is simply, $\mathbf{r}_2 = \{0.867 \quad 0.498\}$. To compute the first row, use Equation A3.1.

In this formula, $k = 1$ and $n = 2$. Therefore,

$$\mathbf{r}_1 = \{1 \quad 0\} - \frac{\{0.867 \quad 0.498\} \begin{Bmatrix} 1 \\ 0 \end{Bmatrix}}{\{0.867 \quad 0.498\} \begin{Bmatrix} 0.867 \\ 0.498 \end{Bmatrix}} \{0.867 \quad 0.498\}$$

$$= \{0.248 \quad -0.432\}$$

When the elements of $\mathbf{r}_1$ are normalized to produce a unit vector, then $\mathbf{r}_1 = \{0.498 \quad -0.867\}$. Thus, the matrix $\mathbf{R}$ is obtained as

$$\mathbf{R}_0 = \begin{bmatrix} 0.498 & -0.867 \\ 0.867 & 0.498 \end{bmatrix}$$

This is identical to the result obtained with the special equation for two variables in Equation 3.62.

APPENDIX 4

CONVERSION FACTORS

		Customary to SI
inches (in.)	meters (m)	0.0254
inches (in.)	centimeters (cm)	2.54
inches (in.)	millimeters (mm)	25.4
feet (ft)	meters (m)	0.305
yards (yd)	meters (m)	0.914
miles (miles)	kilometers (km)	1.609
degrees (°)	radians (rad)	0.0174
acres (acre)	hectares (ha)	0.405
acre-feet (acre-ft)	cubic meters (m^3)	1233
gallons (gal)	cubic meters (m^3)	3.79×10^{-3}
gallons (gal)	liters (l)	3.79
pounds (lb)	kilograms (kg)	0.4536
tons (ton, 2000 lb)	kilograms (kg)	907.2
pound force (lbf)	newtons (N)	4.448
pounds per sq in. (psi)	newtons per sq m (N/m^2)	6895
pounds per sq ft (psf)	newtons per sq m (N/m^2)	47.88
foot-pounds (ft-lb)	joules (J)	1.356
horsepowers (hp)	watts (W)	746
British thermal units (BTU)	joules (J)	1055
British thermal units (BTU)	kilowatt-hours (kwh)	2.93×10^{-4}

DEFINITIONS

newton: force that will give a 1-kg mass an acceleration of 1 m/sec^2
joule: work done by a force of 1 N over a displacement of 1 m
1 newton per sq m (N/m^2) = 1 pascal
1 kilogram force (kgf) = 9.807 N
1 gravity acceleration (g) = 9.807 m/sec^2
1 are (a) = 100 m^2
1 hectare (ha) = 10,000 m^2
1 kip (kip) = 1000 lb

REFERENCES

Abramowitz, M., and Stegun, I. A. (Eds.) *Handbook of Mathematical Functions with Formulas, Graphs, and Mathematical Tables*. U.S. Department of Commerce, National Bureau of Standards, Applied Mathematics Series 55, Washington, DC, 1964.

American Concrete Institute. *Building Code Requirements for Reinforced Concrete*, ACI 318-95, 1995.

American Institute of Steel Construction. *Manual of Steel Construction Load & Resistance Factor Design*, 1st ed., 1986.

American Institute of Steel Construction. *Manual of Steel Construction Load & Resistance Factor Design*, 2d ed., 1994.

American Society of Civil Engineers, ASCE 7-95. *Minimum Design Loads for Buildings and Other Structures*, New York, 1996.

Ang, A. H-S., and Cornell, C. A. Reliability bases of structural safety and design. *Journal of Structural Division, ASCE*, 100(ST9): 1755–1769, 1974.

Ang, A. H-S., and Tang, W. H. *Probability Concepts in Engineering Design*, Vol. I: *Basic Principles*. Wiley, New York, 1975.

Ang, A. H-S., and Tang, W. H.. *Probability Concepts in Engineering Design*, Vol. II: *Decision, Risk and Reliability*. Wiley, New York, 1984.

Ayyub, B. M. The nature of uncertainty in structural engineering. *Uncertainty Modelling and Analysis: Theory and Applications*, B. M. Ayyub and M. M. Gupta (Eds.). Elsevier Science, Amsterdam, 1994.

Ayyub, B. M., and Haldar, A. Practical structural reliability techniques. *Journal of Structural Engineering, ASCE*, 110(8): 1707–1724, 1984.

Ayyub, B. M., and Haldar, A. Decisions in construction operation. *Journal of the Construction Engineering and Management Division, ASCE*, 111(4): 343–357, 1985.

Ayyub, B. M., and McCuen, H. *Probability, Statistics, & Reliability for Engineers*, CRC Press, New York, 1997.

Ayyub, B. M., Guran, A., and Haldar, A. (Eds.) *Uncertainty Modeling in Vibration, Control, and Fuzzy Analysis of Systems*, World Scientific, River Edge, NJ, 1997.

Baecher, G. B., and Ingra, T.S. Stochastic FEM in settlement predictions. *Journal of the Geotechnical Engineering Division, ASCE*, 107(GT4): 449–463, 1981.

Bathe, K. J., and Bolourchi, S. Large displacement analysis of three-dimensional beam structures. *International Journal of Mechanics*, 14: 961–986, 1979.

Baybutt, P., and Kurth, R. E. Methodology for uncertainty analysis of light water reactor meltdown accident consequences: methodology development. *Topical Report*, Battelle Columbus Laboratories, Columbus, OH, 1978.

Beard, V. D. Actual applications of special structural steels. *Transactions of the American Society of Civil Engineers*, 102: 1308–1332, 1937.

Benjamin, J. R., and Cornell, C. A. *Probability, Statistics, and Decision for Civil Engineers*. McGraw-Hill, New York, 1970.

Bergman, L. A. Numerical solutions to the first passage problem in stochastic structural dynamics. In: *Computational Mechanics of Probabilistic and Reliability Analysis*, W. K. Liu and T. Belytschko (Eds.). PlmePress International, Lausanne, pp. 379–408, 1989.

Bennett, R. M., and Ang, A. H-S. *Investigation of Methods for Structural System Reliability*, Structural Research Series No. 510. University of Illinois, Urbana, 1983.

Bickel, P. J., and Doksum, K. A. *Mathematical Statistics: Basic Ideas and Selected Topics*. Holden–Day, San Francisco, 1977.

Bleistein, N., and Handelsman, R. A. *Asymptotic Expansions of Integrals*. Holt, Rinehart & Winston, New York. 1975.

Bjorhovde, R., Galambos, T. V., and Ravindra, M. K. LRFD criteria for steel beam–columns. *Journal of the Structural Division, ASCE*, 104(ST9): 1371–1388, 1978.

Bourgund, U., Ouypornprasert, W., and Prenninger, P. H. W. Advanced Simulation Methods for the Estimation of System Reliability. *Internal Working Report No. 19*, Institut fur Mechanik, Universitat Innsbruck, Innsbruck, Austria, 1986.

Box, G. E. P., and Behnken, D. W. Some new three level designs for the study of quantitative variables. *Technometrics*, 2(4): 455–475, 1960.

Box, G. E. P., and Cox, D. R. An analysis of transformation. *Journal of the Royal Statistical Society, Series B*, 26: 211–252, 1964.

Box, G. E. P., and Muller, M. E. A note on the generation of random normal deviates. *Annals of Mathematical Statistics*, 29: 610–611, 1958.

Box, G. E. P., and Wilson, K. B. On the experimental attainment of optimum conditions. *Journal of the Royal Statistical Society*, B-13: 1–38, 1951.

Breitung, K. Asymptotic approximations for multinormal integrals. *Journal of Engineering Mechanics, ASCE*, 110(3): 357–366, 1984.

Briggs, G. A. *Plume Rise*. U.S. Atomic Energy Commission, Washington, DC, 1969.

Bucher, C. G. Adaptive sampling: an iterative fast Monte Carlo procedure. *Structural Safety*, 5: 119–126, 1988.

Bucher, C. G., and Bourgund, U. Efficient use of response surface methods. *Report No. 9-87*, Institut fur Mechanik, Universitat Innsbruck, Innsbruck, Austria, 1987.

Bucher, C. G., and Bourgund, U. A fast and efficient response surface approach for structural reliability problems. *Structural Safety*, 7:57–66, 1990.

Bucher, C. G., Chen, Y. M., and Schueller, G. I. Time variant reliability analysis utilizing response surface approach. *Reliability and Optimization of Structural Systems '88: 2nd IFIP WG7.5*, Springer-Verlag, Berlin, Germany, 1989.

Burden, R. L., and Faires, J. D. *Numerical Analysis*, 6th ed. Brooks/Cole, Pacific Grove, CA, 1997.

Cai, G. Q., and Elishakoff, I. Refined second-order reliability analysis. *Structural Safety*, 14: 267–276, 1994.

Canadian Standards Association, Standards for the Design of Cold-Formed Steel Members in Buildings, CSA S-137, 1974.

Casciati, F., and Faravelli, L., *Fragility Analysis of Complex Structural Systems*, Research Studies Press, Somerset, England, 1991.

Castillo, E. *Extreme Value Theory in Engineering*. Academic, San Diego, 1988.

Chen, X., and Lind, N. C. Fast probability integration by three-parameter normal tail approximation. *Structural Safety*, 1: 269–276, 1983.

Chowdhury, M. R., Wang, D., and Haldar, A. Reliability assessment of pile-supported structural systems. *Journal of Structural Engineering, ASCE*, 120(1): 80–88, 1998.

Clough, R. W., and Penzien, J. *Dynamics of Structures*, 2nd Ed. McGraw-Hill, London, 1993.

Cochran, W. G., and Cox, M. C. *Experimental Designs*. Wiley, New York, 1957.

Comité European du Béton, Joint Committtee on Structural Safety CEB-CECM-IABSE-IASS-RILEM. First order reliability concepts for design codes. *CEB Bulletin No. 112*, 1976.

Connor, W. S., and Zelen, M. Fractional factorial experimental designs for factors at three levels. *National Bureau of Standards*, Washington, DC, Applied Mathematics Series No. 54, 1959.

Cook, R. D. *Finite Element Modeling for Stress Analysis*. Wiley, New York, 1995.

Cornell, C. A. Bounds on the reliability of structural systems. *Journal of the Structural Division, ASCE*, 93(ST1): 171–200, 1967.

Cornell, C. A. A probability-based structural code. *Journal of the American Concrete Institute*, 66(12): 974–985, 1969.

Crisfield, M. A. An arc-length method including line searches and accelerations. *International Journal of Mechanics*, 19: 267–278, 1983.

Cruse, T. A., Burnside, O. H., Wu, Y-T., Polch, E. Z., and Dias, J. B. Probabilistic structural analysis methods for select space propulsion system structural components (PSAM). *Computers and Structures*, 29(5): 891–901, 1988.

Cruse, T. A., Chamis, C. C., and Millwater, H. R. An overview of the NASA(LeRC)-SwRI probabilistic structural analysis (PSAM) program. *Proceedings, 5th International Conference on Structural Safety and Reliability (ICOSSAR)*, San Francisco, pp. 2267–2274, 1989.

Cruse, T. A., Mahadevan, S., Huang, Q., and Mehta, S. Mechanical system reliability and risk assessment. *AIAA Journal*, 32(11): 2249–2259, 1994.

Das, M. M., *Principles of Geotechnical Engineering*, 4th ed., PWS, Boston, 1998.

DebChaudhury, A., Rajagopal, K. R., Ho, H., and Newell, J. F. A probabilistic approach to the dynamic analysis of ducts subjected to multibase harmonic and random excitation. *Proceedings, 31st AIAA/ASME/ASCE/AHS/ASC Structures, Structural Dynamics and Materials Conference*, Long Beach, California, 1990.

Deodatis, G. Weighted integral method. I: stochastic stiffness matrix. *Journal of Engineering Mechanics, ASCE*, 117(8): 1851–1864, 1991.

Der Kiureghian, A. Multivariate distribution models for structural reliability. *Proceedings, 9th International Conference on Structural Mechanics in Reactor Technology*, Lausanne, Switzerland, 1987.

Der Kiureghian, A., and Ke, J-B. Finite-element based reliability analysis of frame structures. *Proceedings, International Conference on Structural Safety and Reliability (ICOSSAR)*, Kobe, Japan, Vol. 1, pp. 395–404, 1985.

Der Kiureghian, A., and Ke, J-B. The stochastic finite element method in structural reliability. In: *Lecture Notes in Engineering: Stochastic Structural Mechanics*, Y. K. Lin and G. I. Schueller (Eds.). Springer, New York. Vol. 31, pp. 66–83, 1987.

Der Kiureghian, A., and Ke, J-B. The stochastic finite element method in structural reliability. *Probabilistic Engineering Mechanics*, 3(2): 83–91, 1988.

Der Kiureghian, A., and Liu, P. L. Structural reliability under incomplete probability information. *Research Report No. UCB/SESM-85/01*. University of California, Berkeley, 1985.

Der Kiureghian, A., Lin, H. Z., and Hwang, S. J. Second-order reliability approximations. *Journal of Engineering Mechanics, ASCE*, 113(8): 1208–1225, 1987.

Dey, A., and Mahadevan, S. Ductile system reliability analysis using adaptive importance sampling. *Structural Safety*, 20: 137–154, 1998.

Dias, J. B. Probabilistic finite element methods for problems in solid mechanics. Thesis, Stanford University, Palo Alto, CA, 1990.

Dias, J. B., and Nagtegaal, J. C. Efficient algorithims for use in probabilistic finite element analysis. In: *Advances in Aerospace Structural Analysis, AD-09*, O. H. Burnside and C. H. Parr (Eds.), ASME, pp. 37–50, 1985.

Ditlevsen, O. Structural reliability and the invariance problem. *Research Report No. 22*, Solid Mechanics Division, University of Waterloo, Waterloo, Canada, 1973.

Ditlevsen, O. Generalized second moment reliability index. *Journal of Structural Mechanics*, 7(4): 435–451, 1979a.

Ditlevsen, O. Narrow reliability bounds for structural systems. *Journal of Structural Mechanics*, 3: 453–472, 1979b.

Ditlevsen, O. Principle of normal tail approximation. *Journal of the Engineering Mechanics Division, ASCE*, 107(EM6): 1191–1208, 1981.

Ditlevsen, O., and Bjerager, P. Reliability of highly redundant plastic structures. *Journal of Engineering Mechanics, ASCE*, 110(5): 671–693, 1984.

Ellingwood, B. R., and Mori, Y. Reliability-based condition assessment of concrete structures. *Proceedings, International Conference on Structural Safety and Reliability (ICOSSAR)*, Innsbruck, Austria, pp. 1779–1786, 1993.

Ellingwood, B. R., Galambos, T. V., MacGregor, J. G., and Cornell, C. A. Development of a probability based load criterion for American National Standard A58. *NBS Special Publication 577*, U.S. Department of Commerce, Washington, DC, 1980.

Ellingwood, B. R., MacGregor, J. G., Galambos, T. V., and Cornell, C. A. Probability based load criteria: load factors and load combinations. *Journal of Structural Engineering, ASCE*, 108(ST5): 978–997, 1982.

Elms, D. G. System health approach for risk management and design. *Proceedings, International Conference on Structural Safety and Reliability (ICOSSAR)*, Kyoto, Japan, pp. 271–277, 1998.

Faravelli, L. Response surface approach for reliability analysis. *Journal of Engineering Mechanics, ASCE*, 115(12): 2763–2781, 1989.

Fiessler, B., Neumann, H. J., and Rackwitz, R. Quadratic limit states in structural reliability. *Journal of Engineering Mechanics, ASCE*, 105(4): 661–676, 1979.

Fisher, A. D., and Chou, K. C. Reliability investigation of prestressed bulb-tee bridge beams. *Proceedings, International Conference on Structural Safety and Reliability (ICOSSAR)*, Kyoto, Japan, pp. 1921–1926, 1998.

Foschi, R. O., Isaacson, M., Allyn, N., and Yee, S. Combined wave-iceberg loading on offshore structures. *Canadian Journal of Civil Engineering*, 23: 1099–1110, 1996.

Fox, E. P. Methods of integrating probabilistic design within an organization's design system using Box–Behnken matrices. *Proceedings, 34th AIAA/ASME/ASCE/AHS/ASC Structures, Structural Dynamics and Materials Conference*, Paper No. AIAA-93-1380-CP, 1993.

Fox, R. L., and Kapoor, M. P., Rates of change of eigenvalues and eigenvectors. *AIAA Journal*, 6: 2426–2427, 1968.

Frangopol, D. M., and Moses, F. Reliability-based structural optimization. *Advances in Design Optimization*, Chapman & Hall, London, pp. 492–570, 1994.

Freudenthal, A. M. Safety and the probability of structural failure. *ASCE Transactions*, 121: 1337–1397, 1956.

Freudenthal, A. M., Garrelts, J. M., and Shinozuka, M. The analysis of structural safety. *Journal of Structural Engineering, ASCE*, 92(ST1): 267–324, 1966.

Frye, M. J., and Morris, G. A. Analysis of flexibly connected steel frames. *Canadian Journal of Civil Engineering*, 2(3): 280–291, 1975.

Fu, G. Reliability models for assessing highway bridge safety. *Proceedings, International Conference on Structural Safety and Reliability (ICOSSAR),* Kyoto, Japan, pp. 1883–1888, 1998.

Fu, G., and Moses, F. Importance sampling in structural system reliability. *Proceedings, ASCE Joint Specialty Conference on Probabilistic Methods,* Blacksburg, Virginia, pp. 340–343, 1988.

Galambos, T. V. *Guide to Stability Design Criteria for Metal Structures,* 4th ed. Structural Stability Research Council, Wiley, New York, 1988.

Gao, L. Stochastic finite element method for the reliability analysis of nonlinear frames with PR connections. Thesis, Department of Civil Engineering and Engineering Mechanics, University of Arizona, Tucson, 1994.

Gao, L., and Haldar, A. Nonlinear SFEM-based reliability for space structures. *Proceedings, Sixth International Conference on Structural Safety and Reliability (ICOSSAR '93),* Vol. 1, pp. 325–332, 1993.

Gao, L., and Haldar, A. Safety evaluation of frames with PR connections. *Journal of Structural Engineering, ASCE,* 121(7): 1101–1109, 1995.

Gao, L., Haldar, A., and Shome, N. Strength and serviceability requirements in seismic design using nonlinear SFEM. In: *IUTAM Symposium on Advances in Nonlinear Stochastic Mechanics,* A. Naess and S. Krenk (Eds.). Kluwer Academic, pp. 179–188, 1996.

Ghanem, R., and Spanos, P. D. Galerkin-based response surface approach for reliability analysis. *Proceedings, 5th International Conference on Structural Safety and Reliability, ICOSSAR '90,* San Francisco, pp. 1081–1088, 1990.

Ghanem, R., and Spanos, P. D. *Stochastic Finite Elements: A Spectral Approach.* Springer, Berlin, 1991.

Gollwitzer, S., and Rackwitz, R. An efficient numerical solution to the multinormal integral. *Probabilistic Engineering Mechanics,* 3(2): 98–101, 1988.

Grandhi, R., and Wang, L. Higher order failure probability calculation using nonlinear approximations. *Proceedings, 37th AIAA/ASME/ASCE/AHS/ASC Conference on Structures,* Structural Dynamics and Materials, Salt Lake City, Utah, pp. 1292–1306, 1996.

Grigoriu, M., and Khater, M. Finite difference analysis of stochastic beams on elastic foundation. *Report 85-7,* Department of Structural Engineering, Cornell University, Ithaca, NY, 1985.

Gumbel, E. J. Statistical theory of extreme values and some practical applications. *Applied Mathematics Series 33,* National Bureau of Standards, Washington, DC, February 1954.

Gumbel, E. J. *Statistics of Extremes.* Columbia University Press, New York, 1958.

Gurley, K., Tognarelli, M., and Kareem, A. Analysis and simulation tools for wind engineering. *Probabilistic Engineering Mechanics,* 12(1): 9–31, 1997.

Guyan, R. J. Reduction of stiffness and mass matrices. *AIAA Journal,* 3: 380, 1965.

Hadjian, A. H., Smith, C. B., Haldar, A., and Ibanez, P. Variability in engineering aspects of structural modeling. *Proceedings, Sixth World Conference on Earthquake Engineering,* Vol. III, Paper No. 9-31, pp. 2729–2734, New Delhi, India, 1977.

Haldar, A. Liquefaction study: A decision analysis framework. *Journal of the Geotechnical Engineering Division, ASCE,* 106(GT12): 1297–1312, 1980.

Haldar, A. Probabilistic evaluation of construction deficiencies. *Journal of the Construction Engineering Division, ASCE,* 107(CO1): 107–119, 1981.

Haldar, A. Probabilistic evaluation of welded structures. *Journal of the Structural Engineering Division, ASCE,* 108(ST9): 1943–1955, 1982.

Haldar, A. Statistical Site Characterization. *Fourth Australia-New Zealand Conference on Geomechanics,* Perth, Australia, Vo. 2, pp. 530–534, 1984.

Haldar, A., and Chern, S. Uncertainty in dynamic anisotropic strength of sand. *Journal of the Geotechnical Engineering Division, ASCE,* 113(5): 528–533, 1987.

Haldar, A., and Gao, L. Reliability analysis of nonlinear 3-D structures by SFEM. In: *IUTAM Symposium on Probabilistic Structural Mechanics: Advances in Structural Reliability Methods*, P. D. Spanos and Y.-T. Wu (Eds.). Springer, New York, pp. 189–204, 1993.

Haldar, A., and Huh, J. Reliability analysis of structures subjected to dynamic loadings using nonlinear SFEM. *International Union of Theoretical and Applied Mechanics (IUTAM), Nonlinearity and Stochastic Structural Dynamics*, 1999.

Haldar, A., and Mahadevan, S. A design-oriented stochastic einite element method. *Proceedings, Ninth International Conference on Structural Mechanics in Reactor Technology (SMiRT)*, Lausanne, Switzerland, Vol. M, pp. 307–312, 1987.

Haldar, A., and Mahadevan, S. First-order/second-order reliability methods (FORM/SORM). In: *Probabilistic Structural Mechanics Handbook, C. Sundararajan* (Ed.). Chapman & Hall, New York, pp. 27–52, 1993.

Haldar, A., and Mahadevan, S. *Probability, Reliability, and Statistical Methods in Engineering Design*, Wiley, New York, 2000.

Haldar, A., and Miller, F. J. Statistical estimation of relative density. *Journal of the Geotechnical Engineering Division, ASCE*, 110(4): 525–530, 1984.

Haldar, A., and Nee, K-M. Elasto-plastic large deformation analysis of PR steel frames for LRFD. *Computers and Structures*, 34(5): 811–823, 1989.

Haldar, A., and Tang, W. H. Uncertainty analysis in relative density. *Journal of the Geotechnical Engineering Division, ASCE*, 105(GT7): 899–904, 1979.

Haldar, A., and Zhou, Y. An efficient SFEM algorithm for nonlinear structures. In: *Computational Stochastic Mechanics*, P. D. Spanos and C. A. Brebbia (Eds.). Elsevier, New York, pp. 851–862, 1991a.

Haldar, A., and Zhou, Y. SFEM based reliability analysis of nonlinear structures with flexible connections. *Proceedings, Sixth International Conference on Applications of Statistics and Probability in Civil Engineering (ICASP 6)*, Vol. 1, pp. 354–361, 1991b.

Haldar, A., and Zhou, Y. Reliability of geometrically nonlinear PR frames. *Journal of Engineering Mechanics, ASCE*, 118(10): 2148–2155, 1992.

Haldar, A., Guran, A., and Ayyub, B. M. (Eds.). *Uncertainty Modeling in Finite Element, Fatigue, and Stability of Systems*, World Scientific, River Edge, NJ, 1997.

Handa K., and Anderson, K. Application of finite element methods in stochastic analysis of structures. *Proceedings, 3rd International Conference on Structural Safety and Reliability, ICOSSAR '81*, Trondheim, Norway, pp. 395–408, 1981.

Harbitz, A. An efficient sampling method for probability of failure calculation. *Structural Safety*, 3(2): 109–115, 1986.

Hasofer, A. M. and Lind, N. C. Exact and invariant second moment code format. *Journal of the Engineering Mechanics Division, ASCE*, 100(EM1): 111–121, 1974.

Hasselman, T. K., and Hart, G. C. Modal analysis of random structural systems. *Journal of the Engineering Mechanics Division, ASCE*, 98: 561–579, 1972.

Hisada T., and Nakagiri, S. Role of the stochastic finite element method in structural safety and reliability. *Proceedings, 4th International Conference on Structural Safety and Reliability, ICOSSAR '85*, Kobe, Japan, Vol. 1, pp. 385–394, 1985.

Hoel, P. G. *Introduction to Mathematical Statistics*, 3d ed. Wiley, New York, 1962.

Hohenbichler, M., and Rackwitz, R. Non-normal dependent vectors in structural safety. *Journal of the Engineering Mechanics Division, ASCE*, 107(EM6): 1227–1238, 1981.

Hohenbichler, M., and Rackwitz, R. First-order concepts in system reliability. *Structural Safety*, 1: 177–188, 1983.

Hohenbichler, M., Gollwitzer, S., Kruse, W., and Rackwitz, R. New light on first- and second-order reliability methods. *Structural Safety*, 4: 267–284, 1987.

Huh, J. Dynamic reliability analysis for nonlinear structures using stochastic finite element method. Thesis, Department of Civil Engineering and Mechanics, University of Arizona, Tucson, 1999.

Hwang, H., Jaw, J.-W., and Shau, H.-J. Seismic performance assessment of code-designed structures. *Technical Report NCEER-88-0007*, National Center for Earthquake Engineering Research, State University of New York, Buffalo, 1988.

Jiang, M., Corotis, R. B., and Ellis, J. H. Reliability-based inspection and repair management strategies. *Proceedings, International Conference on Structural Safety and Reliability (ICOSSAR)*, Kyoto, Japan, pp. 143–150, 1998.

Johnson, N. I., and Kotz, S. *Distributions in Statistics: Continuous Multivariate Distributions*. Wiley, New York, 1975.

Jones, S. W., Kirby, P. A., and Nethercot, D. A., Column with semi-rigid joints, *Journal of Structural Engineering, ASCE,* 108(ST2): 361–372, 1982.

Kanegaonkar, H. B., Haldar, A., and Ramesh, C. K. Fatigue analysis of offshore platforms with uncertainty in foundation conditions. *Structural Safety*, 3(2): 117–134, 1986.

Karamchandani, A. Structural system reliability analysis methods. *Report No. 83*, John A. Blume Earthquake Engineering Center, Stanford University, Stanford, CA, 1987.

Karamchandani, A., Bjerager, P., and Cornell, A. C. Adaptive importance sampling. *Proceedings, International Conference on Structural Safety and Reliability (ICOSSAR)*, San Francisco, CA, pp. 855–862, 1989.

Kareem, A. Wind effects on structures: a probabilistic viewpoint. *Probabilistic Engineering Mechanics*, 2(4): 166–200, 1987.

Khaleel, M. A., and Simonen, F. A. A probabilistic model for reactor pressure vessels using latin hypercube sampling method. *Proceedings, International Conference on Structural Safety and Reliability (ICOSSAR)*, Kyoto, Japan, pp. 1321–1324, 1998.

Khuri, A., and Cornell, J. A. *Response Surfaces: Designs and Analyses*. Dekker, New York, 1987.

Kiely, G. *Environmental Engineering*. McGraw-Hill International (UK), Berkshire, England, 1997.

Kim, S. H., and Na, S. W. Response surface method using vector projected sampling points. *Structural Safety*. 19(1): 3–19, 1997.

Kondoh, K., and Atluri, S. N. Large deformation, elasto-plastic analysis of frames under nonconservative loading, using explicit derived tangent stiffness based on assumed stress. *Computational Mechanics*, 2(1): 1–25, 1987.

Koyluglu, H. U., and Nielsen, S. R. K. New approximations for SORM integrals. *Structural Safety*, 13(4): 235–246, 1994.

Kwak, K.-H., Cho, H-N., and Choi, Y-M. Reliability-based integrity assessment of segmental PC box girder bridges. *Proceedings, International Conference on Structural Safety and Reliability (ICOSSAR)*, Kyoto, Japan, pp. 1857–1865, 1998.

Laplace, P. S. M. *A Philosophical Essay on Probabilities*. Translated from the sixth French edition by F.W. Truscott and F. L. Emory. Dover, New York, 1951.

Lawrence, M. A. A basis random variable approach to stochastic structural analysis. Thesis, University of Illinois, Urbana, 1986.

Lin, Y. K. *Probabilistic Theory of Structural Dynamics*. McGraw-Hill, New York, 1967.

Lind, N. C. The design of structural design norms. *Journal of Structural Mechanics*, 1: 357–370, 1973.

Liu, E. M. *Effects of connection flexibility and panel zone deformation on the behavior of plane steel frames*. Thesis, Purdue University, West LaFayette, IN, 1985.

Liu, P.-L., and Der Kiureghian, A. Optimization algorithms for structural reliability analysis. *Report UCB/SESM-86/09*. Dept. of Civil Engineering, University of California, Berkeley, 1986.

Liu, P.-L., and Der Kiureghian, A. Finite-element reliability methods for geometrically nonlinear stochastic structures. *Research Report No. UCB/SEMM-89/05*, University of California, Berkeley, California, 1989.

Liu, W. K., Belytschko, T., and Mani, A. Probabilistic finite elements for transient analysis in nonlinear continua. In: *Advances in Aerospace Analysis* O. H. Burnside and C. H. Parr (Eds.). ASME Winter Annual Meeting, Miami, Florida, pp. 9–24, 1985.

Liu, W. K., Belytschko, T., and Mani, A. Probabilistic finite elements for nonlinear structural dynamics. *Computer Methods in Applied Mechanics and Engineering*, 56: 61–86, 1986a.

Liu, W. K., Belytschko, T., and Mani, A. Random field finite elements. *International Journal for Numerical Methods in Engineering*, 23: 1831–1845, 1986b.

Loeve, M., *Probability Theory*, 4th ed. Springer, New York, 1977.

Low, B. K., and Tang, W. H. Efficient reliability evaluation using spreadsheet. *Journal of Engineering Mechanics, ASCE*, 123(7): 749–752, 1997.

Lutes, L. D., and Sarkani, S. *Stochastic Analysis of Structural and Mechanical Vibrations*. Prentice Hall, Upper Saddle River, NJ, 1997.

Lui, E. M. Effects of connection flexibility and panel zone deformation on the behavior of plane steel frames. Thesis, Department of Civil Engineering, Purdue University, West Lafayette, IN, 1985.

Madsen, H. O. Krenk, S., and Lind, N. C. *Methods of Structural Safety*, Prentice-Hall, Englewood Cliffs, NJ, 1986.

Maes, M. A., and Breitung, K. W. Direct approximation of the extreme value distribution of nonhomogeneous Gaussian random fields. *Proceedings of the 15th International Conference on Offshore Mechanics and Arctic Engineering*, Vol. 2, pp. 103–109, 1996.

Mahadevan, S. Stochastic finite element-based structural reliability analysis and optimization. Thesis, Georgia Institute of Technology, Atlanta, 1988.

Mahadevan, S. Probabilistic optimum design of framed structures. *Computers and Structures*, 42(3): 365–374, 1992.

Mahadevan, S. and Chamis, C. C. Structural system reliability under multiple failure modes. *34th AIAA/ASME/ASCE/AHS/ASC Structures, Structural Dynamics, and Materials Conference (SDM)*, La Jolla, California, pp. 707–713, April 1993.

Mahadevan, S., and Cruse, T. A. An advanced first-order method for system reliability. *Proceedings, Sixth ASCE Joint Specialty Conference on Probabilistic Mechanics and Structural and Geotechnical Reliability*, Denver, Colorado, pp. 487–490, July 1992.

Mahadevan, S., Cruse, T. A., Huang, Q., and Mehta, S. Structural reanalysis for system reliability computation. *Reliability Technology 1992, AD-Vol. 28*, ASME Winter Annual Meeting, Anaheim, California, pp. 169–187, November 1992.

Mahadevan, S., and Dey, A. Adaptive Monte Carlo simulation for time-dependent reliability analysis. *AIAA Journal*, 35(2): 321–326, 1997.

Mahadevan, S., and Haldar, A. Practical random field discretization in stochastic finite element analysis. *Structural Safety*, 9: 283–304, 1991a.

Mahadevan, S., and Haldar, A., Stochastic FEM-based validation of LRFD, *Journal of Structural Engineering, ASCE,* 117(5): 1393–1412, 1991b.

Mahadevan, S., and Liu, X. Probabilistic optimum design of composite structures. *Journal of Composite Materials*, 32(1): 68–82, 1998.

Mahadevan, S., Liu, X., and Xiao, Q. A probabilistic progessive failure model for composite laminates. *Journal of Reinforced Plastics and Composites*, 16(11): 1020–1038, 1997.

Mahadevan, S., and Mehta, S. Dynamic reliability of large frames. *Computers and Structures*, 47(1): 57–67, 1993.

Mahadevan, S., Mehta, S., Tryon, R. G., and Cruse, T. A. System reliability design analysis of propulsion structures. *ASME International Gas Turbine Conference*, Cincinnati, Ohio, May 1993.

Mahadevan, S., and Xiao, Q. Large structural systems reliability. *Proceedings, 2nd International Symposium on Uncertainty Modeling and Analysis*, College Park, Maryland, pp. 560–566, April 1993.

Mahadevan, S., and Xiao, Q. Stochastic finite element-based reliability analysis of earthquake loaded structures. *Proceedings of the Sixth International Conference on Applications of Statistics and Probability in Civil Engineering*, Mexico City, Mexico, pp. 399–406, 1991.

Mallet, R. H. and Marcal, P. V. Finite analysis of nonlinear structures. *Journal of Structural Engineering, ASCE*, 96(ST9): 2081–2105, 1968.

Marek, P., Gustar, M., and Anagnos, T. *Simulation-based Reliability Assessment for Structural Engineers*. CRC Press, Boca Raton, FL, 1995.

Mehta, S. Stochastic finite element-based reliability analysis of dynamically loaded frames. Thesis, Vanderbilt University, Nashville, TN, 1991.

Melchers, R. E. *Structural Reliability: analysis and prediction*. Ellis Horwood, Chichester, UK, 1987.

Melchers, R. E. Improved importance sampling methods for structural system reliability calculation. *Proceedings, 5th International Conference on Structural Safety and Reliability, ICOSSAR '89*, San Francisco, California, pp. 1185–1192, 1989.

Moan, T. Reliability and risk analysis for design and operations planning of offshore structures. *Proceedings, International Conference on Structural Safety and Reliability (ICOSSAR)*, Innsbruck, Austria, pp. 21–43, 1994.

Morgenstern, D. Einfache Beispiele Zweidimensionaler Verteilungen. *Mitteilingsblatt fur Mathematische Statistik*, 8: 234–235, 1956.

Murotsu, Y., Okada, M., Yonezawa, M., and Taguchi, K. Reliability assessment of redundant structures. *Proceedings, 3rd International Conference on Structural Safety and Reliability, ICOSSAR'81*, Trondheim, Norway, pp. 315–329, 1981.

Myers, R. H., and Montgomery, D. C. *Response Surface Methodology: Process and Product Optimization Using Designed Experiments*. Wiley, New York, 1995.

Naess, A. A method for extrapolation of extreme value data. *Proceedings, 11th ASCE Engineering Mechanics Conference*, Ft. Lauderdale, Florida, Vol. 1, pp. 273–276, 1996.

Narayanan, R. *Steel Framed Structures: Stability and Strength*. Elsevier Applied Science, London, 1985.

Nataf, A. Determination des distribution dont les marges sont donees. *Comptes Rendus de l'Academic des Sciences*, 225: 42–43, 1962.

Newell, J. F., Rajagopal, K. R., Ho, H., and Cunniff, J. M. Probabilistic structural analysis of space propulsion systems LOX post. *Proceedings, 31st AIAA/ASME/ASCE/AHS/ASC Structures*, Structural Dynamics and Materials Conference, Long Beach, California, 1990.

Nigam, N. C., *Introduction to Random Vibrations*, MIT Press, Cambridge, MA, 1983.

Nowak, A. S. *Calibration of LRFD Bridge Design Code, Final Report*. prepared for NCHRP, TRB, National Research Council, Department of Civil Engineering, University of Michigan, Ann Arbor, MI, 1993.

Owen, O. R., and Hinton, E. *Finite Element in Plasticity: Theory and Practice*. Pineridge Press, Swansea, UK, 1980.

Paloheimo, E. Eine Bemessugsmethode, die Sich auf Variierende Fraktilen Grundet. Arbeitstagung des Deutschen Betonvereins, Sicherheit Von Betonbauten, Berlin, 1973.

Papoulis, A. *Probability, Random Variables, and Stochastic Processes*. McGraw-Hill, New York, 1965.

Paz, M. *Structural Dynamics: Theory and Computation*, 4th ed, Chapman & Hall, New York, 1997.

Petersen, R. G. *Design and Analysis of Experiments*. Dekker, New York, 1985.

Pradlwarter, H. J., and Schueller, G. I. Stochastic structural dynamics: a primer for practical applications. In: *Uncertainty Modeling in Vibration and Fuzzy Analysis of Structural Systems,* B.M. Ayyub, A. Guran, and A. Haldar (Eds.). World Scientific, River Edge, NJ, pp. 1–27, 1997.

Quek, S-T., and Ang, A. H-S. *Structural System Reliability by the Method of Stable Configuration.* Structural Research Series No. 529. University of Illinois, Urbana, 1986.

Rackwitz, R. Practical probabilistic approach to design. *Bulletin No. 112,* Comité European du Béton, Paris, 1976.

Rackwitz, R., and Fiessler, B. Note on discrete safety checking when using non-normal stochastic models for basic variables. *Load Project Working Session,* MIT, Cambridge, MA, June 1976.

Rackwitz, R., and Fiessler, B. Structural reliability under combined random load sequences. *Computers and Structures,* 9(5): 484–494, 1978.

Rahman, S., and Leis, B. Risk-based consideraions in developing strategies to ensure pipeline integrity, Part I: theory; Part II: applications. *ASME Journal of Pressure Vessel Technology,* 116(3): 278–283, and 284–289, 1994.

Rajagopal, K. R., Debchaudhury, A., and Newell, J. F. Verification of NESSUS code on space propulsion components. *Proceedings, 5th International Conference on Structural Safety and Reliability, ICOSSAR '89,* San Francisco, pp. 2299–2306, 1989.

Rajashekhar, M. R., and Ellingwood, B. R. A new look at the response surface approach for reliability analysis. *Structural Safety,* 12: 205–220, 1993.

Ramberg, R., and Osgood, W.R. Description of stress-strain curves by three parameters. *Technical Note No. 902,* National Advisory Committee for Aeronautics, Washington, DC, 1943.

Ramm, E. Strategies for tracing the nonlinear response near limit points. *Nonlinear Finite Element Analysis in Structural Mechanics,* W. Wunderlich, E. Stein, and K. J. Bathe (Eds.). pp. 63–89, Springer, Berlin, 1981.

Rauscher, T. R., and Gerstle, K. H. Reliability of rotational behavior of framing connections. *Engineering Journal, AISC,* 29(1): 12–19, 1992.

Razzaq, Z. End restraint effect on steel column strength. *Journal of Structural Engineering, ASCE,* 109(ST2): 314–334, 1983.

Reid, S. G. Load testing for reliability based design of culverts. *Proceedings of the 7th International Conference on Applications of Statistics and Probability in Civil Engineering (ICASP7),* Paris, pp. 903–909, 1995.

Richard, R. M., and Abbott, B. J., Versatile elastic–plastic stress–strain formula, *Journal of Engineering Mechanics, ASCE,* 101(EM4): 511–515, 1975.

Riks, E. An incremental approach to the solution of snapping and buckling problems. *International Journal of Solids and Structures,* 15: 529–551, 1979.

Rosenblatt, M. Remarks on a multivariate transformation. *Annals of Mathematical Statistics,* 23(3): 470–472, 1952.

Rosenbleuth, E., and Esteva, L. Reliability bases for some Mexican codes. *ACI Publication SP-31,* pp. 1–41, 1972.

Rosowsky, D. V., and Fridley, K. J. Effect of discrete member size on reliability of wood beams. *Journal of Structural Engineering, ASCE,* 123(6): 831–835, 1997.

Ross, S. M. *A Course in Simulation.* Macmillan, New York, 1990.

Ryu, Y. S., Haririan, M., Wu, C. C., and Arora, J. S. Structural design sensitivity analysis of nonlinear response. *Computers and Structures,* 21(1/2): 245–255, 1985.

Schueller, G. I. Structural reliability: recent advances. Freudenthal Lecture, *Proceedings, International Conference on Structural Safety and Reliability (ICOSSAR),* Kyoto, Japan, pp. 3–35, 1998.

Schueller, G. I., Bucher, C. G., Bougund, U., and Ouypornprasert, W. On efficient computational schemes to calculate structural failure probabilities. *Lecture Notes in Engineering (31): Stochastic Structural Mechanics,* Y. K. Lin and G. I. Schueller (Eds.). Springer, New York, pp. 388–410, 1987.

Schueller, G. I., and Stix, R. A critical appraisal of methods to determine failure probabilities. *Structural Safety,* 4,(4): 193–309, 1987.

Sexsmith, R., Leal, J., and Bartlett, M. Yukon highway bridges: reliability and retrofit for heavy ore-haul. *Proceedings, International Conference on Structural Safety and Reliability (ICOSSAR),* Kyoto, Japan, pp. 137–141, 1998.

Shi, G., and Atluri, S. N. Elasto-plastic large deformation analysis of space-frames. *International Journal for Numerical Methods in Engineering,* 26: 589–615, 1988.

Shi, G., and Atluri, S. N. Static and dynamic analysis of space frames with non-linear flexible connections. *International Journal for Numerical Methods in Engineering,* 28: 2635–2650, 1989.

Shinozuka, M. Basic analysis of structural safety. *Journal of the Structural Division, ASCE,* 109(3): 721–740, 1983.

Shinozuka, M. Stochastic fields and their digital simulation. In: *Stochastic Methods in Structural Dynamics, G. I. Schueller and M. Shinozuka (Eds.).,* pp. 93–133, Martinus Nijhoff Publishers, 1987.

Shinozuka, M., and Deodatis, G. Response variability of stochastic finite element systems. *Journal of Engineering Mechanics, ASCE,* 114(3): 499–519, 1988.

Shinozuka, M., and Nomoto, T. Response variability due to special randomness of material properties. *Technical Report,* Department of Civil Engineering, Columbia University, New York, 1980.

Shooman, M. L. *Probabilistic reliability: an engineering approach.* McGraw-Hill, New York, 1968.

Singhal, A., and Kiremidjian, A. S. A method for probabilistic evaluation of seismic structural damage. *Journal of Structural Engineering, ASCE,* 122(12): 1459–1467, 1996.

Snee, R. D. Some aspects of nonorthogonal data analysis: Part I, developing prediction equations. *Journal of Quality Tech.,* 5:67–79, 1973.

Soares, C. G. *Risk and Reliability in Marine Technology.* A.A. Balkema, Rotterdam, 1998.

Sorensen, J. D., and Thoft-Christensen, P. Integrated reliability-based optimal design of structures. In: *Reliability and Optimization of Structural Systems, IFIP WG 7.5,* P. Thoft-Christensen (Ed.), Springer, Berlin, pp. 385–398, 1987.

Southwest Research Institute. A short course on probabilistic structural analysis methods and NESSUS workshop. Vols. 1–3. San Antonio, TX, 1990.

Southwest Research Institute. Nonlinear evaluation of stochastic structures under stress (NESSUS). In: *User's Manual.* San Antonio, TX, 1991.

Spanos, P. D., and Ghanem, R. Stochastic finite element expansion for random media. *Journal of Engineering Mechanics, ASCE,* 115(EM4): 1035–1053, 1989.

Spanos, P. D., and Zeldin, B. A. Indirect sampling method for stochastic mechanics problems. *Proceedings, International Conference on Structural Safety and Reliability (ICOSSAR),* Innsbruck, Austria, pp. 1755–1761, 1993.

Stewart, M. G. Serviceability reliability analysis of reinforced concrete structures. *Journal of Structural Engineering, ASCE,* 122(7): 794–803, 1996.

Stewart, M. G., and Melchers, R. E. *Probabilistic Risk Assessment of Engineering Systems.* Chapman & Hall, London, 1997.

Terzaghi, K. *Theoretical Soil Mechanics.* Wiley, New York, 1943.

Thoft-Christensen, P., and Murotsu, Y. *Application of Structural Systems Reliability Theory.* Springer, Berlin, 1986.

Torng, T. Y., Wu, Y-T., and Millwater, H. R. Structural system reliability calculation using probabilistic fault tree analysis method. *Proceedings, the 33rd AIAA/ASME/ASCE/AHS/ASC Conference on Structures, Structural Dynamics and Materials*, Dallas, pp. 603–613, 1992.

Tvedt, L. *Two Second-Order Approximations to the Failure Probability*, Report No. RDIV/20-004-83. Det Norske Veritas, Hovik, Norway, 1983.

Tvedt, L. Distribution of quadratic forms in normal space: application to structural reliability. *Journal of Engineering Mechanics, ASCE*, 116(6): 1183–1197, 1990.

Vanderplaats, G. N. *Numerical Optimization Techniques for Engineering Design: With Applications*. McGraw-Hill, New York, 1984.

Vanmarcke, E. H. *Random Fields: Analysis and Synthesis*. MIT Press, Cambridge, MA, 1983.

Vanmarcke, E. H., and Grigoriu, M. Stochastic finite element analysis of simple beams. *Journal of the Engineering Mechanics, ASCE*, 109(5): 1203–1214, 1983.

Wang, D., Chowdhury, M., and Haldar, A. System reliability evaluation considering strength and serviceability requirements. *Computers and Structures*, 62(5): 883–896, 1997.

Wang, L. P., and Grandhi, R. V. Intervening variables and constraint approximation in safety index and failure probability calculations. *Structural Optimization*, 10(1): 2–8, 1995.

Wempner, G. A. Discrete Approximations related to nonlinear theories of solids. *International Journal of Solids and Structures*, 7: 1581–1599, 1971.

Wen, Y-K. *Structural Load Modeling and Combination for Performance and Safety Evaluation*. Elsevier, Amsterdam, 1990.

Wu, Y. T. Efficient methods for mechanical and structural reliability analysis and design. Thesis, University of Arizona, Tucson, 1984.

Wu, Y-T. An adaptive importance sampling method for structural system reliability analysis, reliability technology 1992, Vol. AD-28, T. A. Cruse (Ed.). *Proceedings, ASME Winter Annual Meeting*, Anaheim, California, pp. 217–231, 1992.

Wu, Y-T., Burnside, O. H., and Cruse, T. A. Probabilistic methods for structural response analysis. *Computational Mechanics of Probabilistic and Reliability Analysis*, W. K. Liu and T. Belytschko (Eds.). Elmpress, Lausanne, pp. 182–195, 1987.

Wu, Y-T., Millwater, H. R., and Cruse, T. A. An advanced probabilistic structural analysis method for implicit performance functions. *AIAA Journal*, 28(9): 1663–1669, 1990.

Wu, Y-T., and Wirsching, P. H. New algorithm for structural reliability estimation. *Journal of Engineering Mechanics, ASCE*, 113(9): 1319–1336, 1987.

Xiao, Q., and Mahadevan, S. Fast failure mode identification for ductile structural system reliability. *Structural Safety*, 13(4): 207–226, 1994a.

Xiao, Q., and Mahadevan, S. Plasticity effects on frame member reliability. *Structural Safety*, 16: 201–214, 1994b.

Xiao, Q., and Mahadevan, S. Second-Order upper bounds on probability of intersection of failure events. *Journal of Engineering Mechanics, ASCE*, 120(3): 670–675, 1994c.

Yamazaki, F., Shinozuka, M., and Dasgupta, G. Newmann expansion for stochastic finite element analysis. *Journal of Engineering Mechanics, ASCE*, 114(8): 1335–1354, 1988.

Yang, J. N. Application of reliability methods to fatigue, quality assurance and maintenance. Freudenthal Lecture, *Proceedings, International Conference on Structural Safety and Reliability (ICOSSAR)*, Innsbruck, Austria, pp. 3–18, 1993.

Yao, J. T. P. *Safety and Reliability of Existing Structures*. Pitman Advanced Publishing Program, Boston, 1985.

Yao, J. T. P. Reliability issues in civil infrastructure systems. *Proceedings, International Conference on Structural Safety and Reliability (ICOSSAR)*, Kyoto, Japan, pp. 49–57, 1998.

Yao, Th. H.-J., and Wen, Y. K. Response surface method for time-variant reliability analysis. *Journal of Structural Engineering, ASCE*, 122(2): 193–201.

Yura, J. A., Galambos, T. V., and Ravindra, M. K. The bending resistance of steel beams. *Journal of Structural Division, ASCE*, 104(ST9): 1355–1370, 1978.

Zeldin, B. A., and Spanos, P. D. On random field discretization in stochastic finite elements. *Journal of Applied Mechanics*, 65: 320–327, 1998.

Zerva, A., and Zhang, O. Correlation patterns in characteristics of spatially variable seismic ground motions. *Earthquake Engineering and Structural Dynamics*, 11: 19–39, 1997.

Zhang, R. Statistical estimation of earthquake ground motion characteristics. *Proceedings, 11th World Conference on Earthquake Engineering,* Acapulco, Mexico [CD-ROM].

Zhao, Z., and Haldar, A. Fatigue damage evaluation and updating using nondestructive inspections. *Journal of Engineering Fracture Mechanics,* 53(5): 775–788, 1996.

Zhao, Z., Haldar, A., and Breen, F. L. Fatigue reliability evaluation of steel bridges. *Journal of Structural Division, ASCE,* 120(5): 1608–1623, 1994a.

Zhao, Z., Haldar, A., and Breen, F. L. Fatigue reliability updating through inspections for bridges. *Journal of Structural Division, ASCE,* 120(5): 1624–1642, 1994b.

Zhou, Y. Efficient stochastic finite element method for the reliability analysis of nonlinear frame structures. Thesis, Department of Civil Engineering and Engineering Mechanics, University of Arizona, Tucson, 1991.

INDEX

An engineering approach to linear algebra

An engineering approach to linear algebra

W. W. Sawyer

Professor jointly to
the Department of Mathematics
and the College of Education
University of Toronto

Cambridge
at the University Press
1972

Published by the Syndics of the Cambridge University Press
Bentley House, 200 Euston Road, London NW1 2DB
American Branch: 32 East 57th Street, New York, N.Y.10022

© Cambridge University Press 1972

Library of Congress Catalogue Card Number: 70–184143

ISBN: 0 521 08476 8

Typeset by William Clowes & Sons, Limited, London, Beccles, and Colchester

Printed in the United States of America

Contents

[v]   218436

Preface

This book first took form as duplicated notes handed out to first year engineering students at the University of Toronto. The preparation of this material was undertaken because no published book seemed to meet the needs of first year engineers. Some books were mathematically overweight; in order to prove every statement made, long chains of propositions were included, which served only to exhaust and antagonize the students. Other books, emphasizing applications to engineering, involved mathematics beyond the student's comprehension. This was naturally so, for while linear algebra has a wealth of applications to engineering and science, many of these lie outside the range of the first year student, both in their mathematical and in their scientific content.

Thus problems of treatment arose both on the mathematical and the engineering side.

In some discussions of mathematics teaching, it seems to be assumed that only two courses are open: either the full rigorous proof of every statement from a set of axioms, or a cookbook approach in which the student memorizes some rigmarole. The end result of these two approaches is much the same; in neither case does the average student have any understanding of what he is doing. A third course is in fact possible, in which the student learns to picture or to imagine the things he is dealing with, and then, because his imagination is working, he is able to reason about these things. The objective of this book is to let the student achieve such insight and reasoning ability. I would strenuously oppose any suggestion that this book is less mathematical because it steers in the direction of this third course. Imagination is the source of all mathematics; a mathematician begins with an idea; the details of proofs come later. All the explanations, all the arguments, all the justifications brought forward in this book could, if one desired, be put in a rigorous form that would satisfy a professional mathematician, but it would not help the first year student if this were done.

If a mathematics course for engineers is to succeed, it must give the student the feeling that it will be of practical value in enabling him to master his profession. No mathematical concept has been included in this book without a careful examination of its value for applications, and the right of the student to demand evidence of such applicability has throughout been respected. The difficulty that most actual applications are very complicated has been met by discussing simplified situations, but indicating the existence of more realistic problems.

So far as I know, no other book uses the approach of the opening sections, which start with the real world and show how real situations lead to the basic concepts of linear algebra. The exercises at the end of §3 are intended to show the variety of situations in which these concepts are involved. Students should be warned that they will find these exercises hard. They should be prepared to bypass them to some extent, to go ahead with the rest of the book, and to come back to these exercises from time to time.

Some books are written to provide problems and extra reading for students whose main instruction is derived from a series of lectures. That is not the purpose of this book. It would be interesting to apply the spirit of operational research to the institution of the university lecture. What purpose are lectures supposed to achieve, and to what extent do they in fact achieve it? For several years now thoughtful teachers have raised this question. Stephen Potter in his book *The Muse in Chains* maintains that universities have simply not realized that printing has been invented, and that the lecture has survived from the time when the lecturer had a handwritten copy of Aristotle which he read at dictation speed in order that each student might possess a similar manuscript. Professor Crowther in New Zealand used to maintain that precise communication was possible only through the written or printed word. When we speak, we improvise. When we write, we examine what we have written and amend it many times until it expresses truly what we wish to say. It was with such remarks in mind that this book was composed. The printed material was handed to the students to read. In the class periods and tutorials, students raised difficulties and these were discussed – between students themselves, between students and graduate assistants, and between students and the 'lecturer'. It was in such discussions that the most effective learning was done. A clear explanation is a useful starting point, but it is hardly ever the end. Students are liable to misinterpret or fail to understand the explanation in the most individual and unpredictable ways. It is only through discussion that such misunderstandings can be discovered and cleared up.

September 1971   W. W. S.

1 *Mathematics and engineers*

Is it justified to have a course in mathematics for engineers? Would it not be better to provide a course in electrical calculations for electrical engineers, chemical calculations for chemical engineers and so on through all the departments? The reason for teaching mathematics is that mathematics is concerned not with particular situations but with patterns that occur again and again. This is most obvious in elementary mathematics. In arithmetic, the number 40 may occur as $40, 40 horsepower, 40 tons, 40 feet, 40 atoms, 40 ohms and so on indefinitely. It would be most wasteful if we decided not to teach a child the general idea of 40, but left this idea to be explained in every activity involving counting or measurement.

Advanced mathematics cannot claim the universal relevance that arithmetic has. There are mathematical topics vital for aerodynamics that leave the production engineer cold. An engineer cannot simply decide to learn mathematics. He must judge wisely what mathematics will serve him best. His aim is not only to find mathematics that will help him frequently now, but also to guess what mathematics is most likely to help in industries and processes still to be invented.

Linear algebra qualifies on both counts; it is already used in most branches of engineering, and has every prospect of continuing to be.

Very many situations in engineering involve the mathematical concept of mapping. We shall restrict ourselves to the simplest class of these, namely *linear mappings*. We shall explain what these are and how to recognize an engineering situation in which they occur. And again in the interests of simplicity, we will not consider linear mappings in general, but restrict ourselves to linear mappings involving only a finite number of dimensions.

2 *Mappings*

Obviously we can find many applications of the scheme

$$\text{input} \xrightarrow{\text{Process}} \text{output}.$$

The process may involve actual material, as when certain components or ingredients (input) are used to produce some manufactured article (output). Or the process may

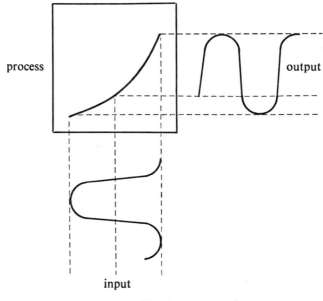

process

output

input

Fig. 1

be one of calculation, the input certain data, the output an answer. Another type of example would be a public address system; the input being the sounds made by a speaker or performer, the process being one of amplification and perhaps distortion, the output the sounds or noise emerging.

Diagrammatically, this last example might be shown as in Fig. 1.

Here distortion is occurring. The input is supposed to be a pure sine wave; the output certainly is not.

This is an example of a *mapping*, but it would not seem sensible to describe it as a linear mapping. We notice, for instance, that the graph representing the process is *not* a straight line.

There are two ways in which we might go about getting a straight-line graph. One way would probably be expensive – to use extremely good apparatus, which would yield a straight-line graph. The other way to avoid distortion is simpler and cheaper – to turn down the volume control. This means that we operate with a very small part of the curve. It usually happens that a small part of a curve is nearly (though of course not exactly) straight.

In fact, most of the examples we think of, that seem to arise in the first way (a graph that is really straight) are probably instances of the second way (using a small piece of a curve). Here we are thinking not only of our particular example – sound reproduction – but of all the cases in science where a straight-line law seems to hold. Someone might suggest, say, Hooke's Law: that for a stretched spring, force is proportional to deformation. But if we allow sufficiently large forces to act, the graph connecting force and deformation looks as is shown in Fig. 2. It seems reasonable to suppose that, if we

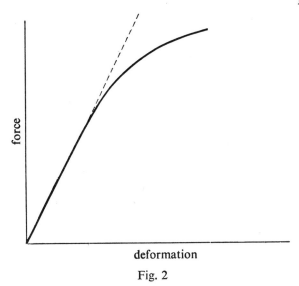

deformation

Fig. 2

could make very precise measurements, we would find a slight curvature in every part of the graph. But Hooke's formula is very convenient: it is easy to use and, for forces of moderate size, it gives all the accuracy we need.

In the same way, the formula $V = gt$ for a body falling from rest is linear, and is useful in many small-scale situations, but would be entirely misleading if used by astronauts. Ohm's Law, giving current and voltage proportional, is linear, but heavy currents are liable to produce changes in resistance and ultimately of course to melt the conductor.

Linear algebra then, like calculus, relies on the fact that, very often, a small part of a curve can be efficiently approximated by a straight line, a small part of a surface by a plane, and so on. (Question for discussion: what does 'and so on' mean here?) The qualification 'very often' is necessary; mathematicians have studied curves that are infinitely wriggly; for these no part, however small, can be approximated by a line. The functions corresponding to such graphs could appear only in quite sophisticated engineering problems.

The utility of linear algebra thus depends on a very general consideration – the tendency of functions to have good linear approximations. This argument is not tied to any branch of science, or to any particular department of engineering, or to the present state of the engineer's art; even if the whole of our present technology became obsolete, the argument would, in all probability, retain its force.

Sometimes of course an engineer has to deal with disturbances of such a scale that he cannot regard all the operations involved as effectively linear. He is then confronted with *non-linear problems*, which are much harder. As mentioned earlier, we do not plan to discuss these.

The illustrations so far given are not sufficient to make precise just what we mean by *linearity*. This idea we shall have to discuss for quite a while yet. Before we go into this it may be wise to remove a possible source of confusion. A linear mapping is a generalization of a type of function we meet in elementary algebra, namely $x \to y$, where $y = mx$ and m is constant. In elementary work, the function defined by $y = mx + b$ is usually called linear. It is important to realize that, in the language used by university mathematicians, $x \to mx + b$ would *not* be described as a linear function when $b \neq 0$. Our scheme 'input, process, output' may suggest why this is reasonable. We always suppose that if nothing goes in, nothing comes out. Zero input produces zero output. With $y = mx$, taking $x = 0$ makes $y = 0$, which is satisfactory. However, with $y = mx + b$, putting $x = 0$ gives $y = b$, which is not zero when $b \neq 0$. It has been agreed not to attach the label 'linear mapping' in this case.

3   *The nature of the generalization*

When $x \to y$ occurs in elementary work, it is understood that both the input x and the output y are real numbers. But the input–output idea is capable of vast generalization. Input and output could be almost anything: all we require is that the input in some way determines the output. For example, in designing an apartment building, we might be concerned about the effects of winds; the input would then be a specification of the wind acting on the building, the output perhaps the extra forces acting on the foundations as the result of such wind pressure. For an electronic computer, the programme fed in could be the input, its response the output. In an automated factory, the materials supplied could constitute the input, the goods produced the output.

It would be easy to produce many more examples; all that is necessary is that x, *what goes in* must determine y, *what comes out*; x and y can stand for extremely complicated objects or collections of data.

Any realistic person will realize that our scheme for a process or mapping, $x \to y$, is so general that very little information can be derived from it as it stands. Being told only that we have a mapping $x \to y$ is like being confronted with a machine having many controls and being told only that the machine is consistent – it always reacts in the same way to any particular setting of the controls. It may be comforting to know this, but it is not much to go on.

It is here that the restriction to *linear* mapping is helpful. But what do we mean by *linear* when we are dealing not with 'number → number' but with 'anything → anything'? Indeed, how has it come about that people have thought of using the term *linear* in such a vague and general situation?

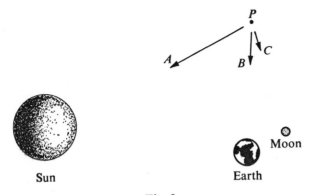

Fig. 3

To answer these questions we have to realize that this terminology arose only after many years' experience in a variety of subjects. In many sciences we meet something called a Principle of Superposition. In the theory of gravitation, for instance, we may want to find the force on an object at the point P, due to the attractions of the Sun, Earth and Moon. Let A be the force the object would experience if the Sun alone acted on it, B the force that would be produced by the Earth alone, and C that due to the Moon alone (Fig. 3). The principle of superposition states that the force due to the Sun, Earth and Moon acting simultaneously is the combined effect of the forces A, B, C. Thus one complex problem is broken into three simpler problems.

Again, in the theory of gravitation, a principle of proportion can be used. If a mass M exerts a force F on an object, then – provided the positions are unchanged – a mass of $3.7M$ will exert a force $3.7F$ on the object (Fig. 4). A mass kM would exert a force kF.

We should not think of the principles of superposition and proportion as applying to everything. For example, they do not apply to traffic.

In the situation shown in Fig. 5, it is possible that 20 cars a minute arriving from the north might pass without delay, if none came from the west. Also 30 cars a minute from the west might pass without delay if none came from the north. It would be rash to assume that, if both streams of traffic were arriving, 50 cars a minute would emerge

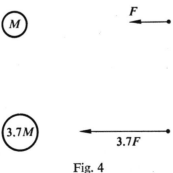

Fig. 4

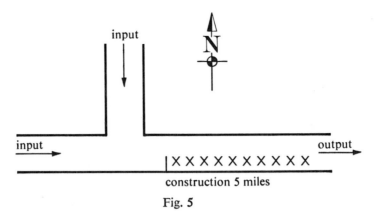

construction 5 miles

Fig. 5

beyond the construction area. Superposition does not apply. Nor does proportionality. If 10 times as many cars arrive per minute we do not expect to find 10 times as many emerging; rather we expect the rush-hour crawl.

Informally we may define a linear mapping as one arising in a situation to which the principles of superposition and proportionality apply.

Notice that, to make this explanation clear, we have to state carefully what we mean by superposition and proportionality both for input and output. In the gravitation question above, the input consists of a specification of the bodies exerting attraction. To superpose the three situations in which the Sun alone, the Earth alone and the Moon alone act on an object, we consider the Sun, Earth and Moon acting simultaneously. The output is concerned with the force experienced by the object; to apply superposition we must know how to find the combined effect of the three forces A, B and C. Someone unfamiliar with mechanics might be at a loss how to find this combined effect.

A number of situations are given below for consideration. In each case, it is necessary to spell out reasonable interpretations of superposition and proportionality both for the input and for the output. Then one has to consider whether principles of superposition and proportionality would apply (as in the gravitation problem) or would fail (as in the traffic problem).

Students may wish to select the situations that are most familiar to them.

Situations

(1) Purchasing articles at fixed prices (no reduction for quantity).
  Input. Specification of articles purchased.
  Output. Cost.

(2) Sand, or other heavy material, rests on a light beam (i.e. one of negligible weight). Spring balances measure the reactions at the ends (Fig. 6).

Fig. 6

Input. The distribution of weight along the beam.
Output. The readings of the spring balances.

(3) An electrical system of the type shown in Fig. 7.

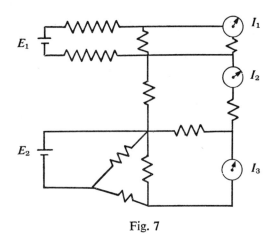

Fig. 7

Input. Voltages E_1, E_2.
Output. Currents I_1, I_2, I_3.

(4) Manufacturing.
   Input. Materials required.
   Output. Articles manufactured.

(5) Would it make sense to reverse input and output in Situation (4) i.e. to consider?
   Input. Articles manufactured.
   Output. Materials required.

(6) In order to obtain crude estimates of how quickly the temperature of a furnace is rising, the temperature (T) is measured at unit intervals of time, and the changes (ΔT) are calculated; for example

T	100		400		650		770		840
ΔT		300		250		120		70	

Input. The numbers in the T row.

Output. The numbers in the ΔT row.

(7) A student is given problems in differentiation. For example, given

$$f(x) = x^2 + 5x + 2,$$

he writes $f'(x) = 2x + 5$. (He does not make any mistakes.)

Input. The function $f(x)$.

Output. The function $f'(x)$.

(8) In an experiment, the mileage gone and the speedometer readings of a car are recorded at various times.

Input. Figures for mileage at these times.

Output. Speedometer readings at these times.

(9) In a radio programme listeners are able to phone in their opinions. In order to exclude obscenity, blasphemy, slander, sedition, etc. it is arranged that their remarks are heard on the radio ten seconds after they are spoken.

Input. The sounds made into the telephone.

Output. The sounds heard on the radio, on an occasion when nothing is censored.

(10) A number of pianos are available. A tape recorder deals faithfully with a certain range of notes on the piano. Notes above that it reproduces at half strength, and notes below that not at all.

Input. The sounds made by the pianos.

Output. The sounds as reproduced by the tape recorder.

(11) In pulse code telephony, the voice produces certain vibrations, shown in Fig. 8 as a graph. 8000 times a second a computer measures the ordinates (i.e. the y-values) shown here as upright lines.

Input. The voice graph.

Output. The values measured by the computer.

(12) Some function f, reals to reals is specified. Its values $f(0), f(1), f(2)$ are calculated.

Input. The function f.

Output. The numbers $f(0), f(1), f(2)$.

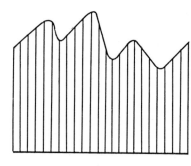

Fig. 8

(13) A function f, reals to reals, is specified. From it we calculate another function, g.
Input. The function f.
Output. The function g.

Discuss the following cases.

(a) $g(x) = f(2x)$.
(b) $g(x) = f(x - 1)$.
(c) $g(x) = f(x^2)$.
(d) $g(x) = [f(x)]^2$.

4 Symbolic conditions for linearity

Superposition and proportionality lend themselves to a very convenient representation by mathematical symbols.

In the gravitation problem, we wanted to find the attraction due to the Sun *and* the Earth *and* the Moon. In beginning arithmetic we associate 'and' with $+$. So it is natural to represent the superposition of Sun, Earth and Moon by $S + E + M$. The forces produced by Sun, Earth and Moon acting separately were indicated by A, B and C; it is natural to represent their combined effect by $A + B + C$, and indeed this notation will already be familiar to most students as vector addition of forces.

Thus from
$$S \rightarrow A$$
$$E \rightarrow B$$
$$M \rightarrow C$$

we conclude
$$S + E + M \rightarrow A + B + C.$$

Note that it is a subtle question of language whether we should say that the plus signs on the two sides of this statement have the same or different meanings. Someone could maintain that the meanings were the same because $S + E + M$ stands for 'the combined effect of S and E and M', while $A + B + C$ stands for 'the combined effect of A and B and C'. An opponent of this view would point out that the detailed procedure by which we find the combined effect of three forces is very different from that of considering three massive bodies existing together.

It will be seen from the list of situations given above that superposition applies to a great variety of problems. It would be very confusing to have a different sign for superposition in each of these cases. As we discussed right at the outset, mathematics is concerned with the similarities between different things, so that we learn one idea which can be used in many different circumstances. In order to bring out similarities, we use the same sign, +, whenever superposition is involved. But we must bear in mind the different systems that exist: we must never fall into the error of writing an expression that indicates the addition of a number to a force, or a mass to a voltage.

Proportionality can also be represented very simply. S standing for the Sun, $10S$ would indicate a body having the same position as the Sun but 10 times its mass. Again $10A$ would represent a force having the same direction as A but 10 times its magnitude.

The principle of proportionality shows that from the given fact

$$S \to A,$$

we may conclude

$$10S \to 10A.$$

We may now sum up our explanation of linearity.

We suppose we have an input, the elements of which may be denoted by $u, v, w, \ldots$ and an output with elements $u^*, v^*, w^*, \ldots$ In both the input and the output, we have sensible definitions of addition and of multiplication by a number, so we know what is meant by $u + v$ and ku, where k is any real number, and also what is meant by $u^* + v^*$ and ku^*. We have a mapping M which makes $u \to u^*$, $v \to v^*$, $w \to w^*$. This mapping will be called linear if for any two elements, u and v of the input, we find $u + v \to u^* + v^*$, and also that, for any number k and for any u, $ku \to ku^*$.

This is illustrated in Fig. 9.

The statement $u + v \to u^* + v^*$ means that we have two ways of finding where the mapping M sends $u + v$. First there is the obvious way, that works for any mapping. Starting with u and v in the input, we find $u + v$ which is also an element of the input. We then follow the arrow and find which element of the output we land on. The second method, which is not allowable for an arbitrary mapping but works with a linear

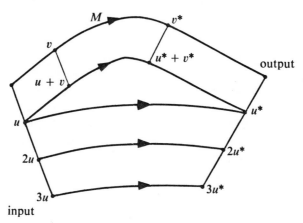

Fig. 9

mapping, is to begin with u and v, follow the arrows from each to u^* and v^*, and *then* add to find $u^* + v^*$. This is an element of the output and is found by the addition rules appropriate to the output.

In the same way, if we want to find the effect of the mapping on an element of the type ku, there are two ways of doing it. Consider for example $k=3$; what does the mapping do to $3u$? We may begin with u, find $3u$, and then follow the arrow. Alternatively, we may begin with u, follow the arrow to u^*, and thence find $3u^*$. That is, it does not matter whether we first multiply by 3 and then map, or first map and then multiply by 3. Similarly for addition, it does not matter whether we first add and then map, or first map and then add.

When we are dealing with a general mapping, knowing $u \rightarrow u^*$ does not enable us to say where any other element goes. But with a linear mapping, the information that $u \rightarrow u^*$ tells us immediately the fate of infinitely many other points. We know $2u$ must go to $2u^*$, that $3u \rightarrow 3u^*$, $7u \rightarrow 7u^*$, $0.51u \rightarrow 0.51u^*$, $-8u \rightarrow -8u^*$ and so on. The fate of any element of the form ku is determined once we know the fate of u.

If we know what a linear mapping does to two points, say $u \rightarrow u^*$ and $v \rightarrow v^*$, we can make even more extensive deductions. We know, for example, $5u \rightarrow 5u^*$ and $3v \rightarrow 3v^*$, by proportionality. Superposing these, we deduce $5u + 3v \rightarrow 5u^* + 3y^*$. Obviously, there is nothing special here about the numbers 5 and 3. For any numbers a and b we could deduce $au + bv \rightarrow au^* + by^*$.

From the strictly mathematical point of view, our explanation is still incomplete. In summing up, above, we said that 'in both the input and the output, we have sensible definitions of addition and of multiplication by a number'. But how do we tell that a definition is sensible? What are the requirements? These questions we must answer later on. We do not deal with them immediately, as we do not wish to distract attention from the important and central idea of a linear mapping, expressed verbally by reference to superposition and proportionality and symbolically by the conditions, if $u \rightarrow u^*$ and $v \rightarrow v^*$, then $u + v \rightarrow u^* + v^*$ and $ku \rightarrow ku^*$.

The following two little exercises are intended to fix these conditions in mind, and also to call attention to forms in which they may occur in the literature.

Exercises

1   A mapping is a function. If we denote it by the symbol F, the element to which u goes, instead of being written u^*, will be written $F(u)$. Express the conditions $u + v \rightarrow u^* + v^*$, $ku \rightarrow ku^*$ in this notation, that is, using F and not making any use of the arrow, $\rightarrow$, or the star, $*$.

2   This hardly differs from Exercise 1. If we denote the mapping by M, we may instead of u^* write simply Mu. Express the two conditions for linearity in this notation. When so written, they resemble equations in elementary algebra. With what laws of elementary algebra do you associate the resulting equations?

5   *Graphical representation*

Some of the conclusions of §4 are not at all surprising if we remember the first example of §3, under the heading 'Situations' – 'purchasing articles at fixed prices (no reduction for quantity)'. If in a store you are told that an article costs $10, the salesman will be inclined to think you are feeble-minded if you then proceed to ask 'What would two of them cost? How much would three cost?' If article A costs $10 and article B costs $20, you are expected to understand for yourself what 5 of A and 3 of B would cost.

The purchasing situation is in many ways typical of situations involving linear mappings, and the type of argument just used for finding the cost of $5A + 3B$ can be applied to *any* linear mapping. This is emphasized here, as experience shows that students often fail to appreciate this idea and to use it freely.

Of course the purchasing situation is special in one respect; the output is always simply a sum in dollars; it can be specified by giving a single number. It is easy to devise a situation where this special feature is removed. Suppose we wish not only to purchase certain articles but also to transport them to some rather inaccessible place. We may then be concerned to know not only the price but also the weight. We now need two numbers to specify the output, – price in dollars of the purchase and weight in pounds. But our procedure is not essentially affected. If we know the price and weight of A, and have similar information for B, we can work out the price and the weight for 5 of A and 3 of B. What we are using – though we are hardly conscious of this – are the principles of proportionality and superposition.

It has been said that an engineer thinks by drawing pictures. Pictorial thinking is certainly important for linear algebra, as may be guessed from the very name of the subject; while the second word, 'algebra', suggests calculation, the first word, 'linear' (having to do with straight lines) clearly comes from geometry. Most students will cope most successfully with linear algebra if they use ideas that are suggested by pictures and checked by calculation. There are certain limitations to what we can do graphically. We live in a world of 3 dimensions. We can deal with problems involving 2 numbers by means of graph paper; when 3 numbers occur, we can use or imagine solid models. Beyond that we are stuck.

The only solution to this difficulty seems to be to become thoroughly familiar with those situations which we can draw or represent physically. These help us to remember the essential ideas, and will often suggest an approach to the problems involving more than 3 numbers. Fortunately, many of the results in linear algebra do not depend essentially on the number of dimensions involved. There are strong analogies between the situations in 2 or 3 dimensions, which we can show graphically, and those in higher dimensions which we cannot.

For this reason, it is useful to see how our present topic looks graphically, in the very convenient case, a mapping from 2 dimensions to 2 dimensions. This is the purpose of the questions given below. Each question calls for points on one piece of graph paper to be mapped onto another piece. The investigation is complete when a student has mapped enough points to see the emergence of a pattern. There is no point in undertaking the drudgery of going past this stage. On the other hand, you should not stop until you have reached the stage at which you can predict, *without calculation*, to what positions further points would map. Each question is based on the illustration we used earlier, where we are concerned with the price and weight of purchases involving two articles, *A* and *B*.

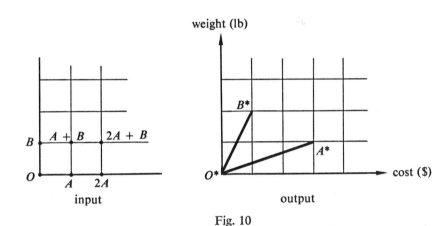

Fig. 10

Questions

1 In Fig. 10 $A \to A^*$; this is intended to indicate that article *A* costs $3 and weighs 1 pound. Similarly from $B \to B^*$ we read off that *B* costs $1 and weighs 2 pounds. On the input diagram mark the purchases that other points represent (some are already marked), and on the output diagram mark the points that give the corresponding cost and weight.

2 As in question 1, except that *A* now costs $1 and weighs 2 pounds, while *B* costs $2 and weighs 1 pound.

3 Now *A* costs $1 and has negligible weight, *B* costs $1 and weighs 1 pound.

4 *A* costs $1 and weighs 1 pound, *B* costs $2 and weighs 2 pounds.

5 *A* costs $1 and has negligible weight, and exactly the same is true of *B*.

In Questions 4 and 5, it will be found that several points in the input may go to the same point in the output. It is of interest to mark on the input diagram the set of points that go to some selected point on the output diagram.

In all these questions it will be noticed that we have only used the first quadrants of the graph paper. This is due to the nature of our illustration; the number of articles purchased, their cost and their weight are naturally taken to be non-negative. In most engineering applications we are concerned with quantities that may be positive or negative. If you care to experiment with situations in which the price is negative (e.g. garbage, which you pay people to take) or the weight (balloons filled with hydrogen or helium), you will find this does not bring in any essentially new feature. The network, or pattern of points, that appears in the output diagram still has much the same character as before.

In drawing such diagrams, it is convenient to use integers, since these correspond to the grid of the graph paper. In most applications, fractional values are perfectly possible, and points with fractional co-ordinates are subject to the mapping. However, a diagram such as Fig. 11 does effectively convey to our minds what is happening to *every* point of the plane, not only to the points of the grid that are actually drawn.

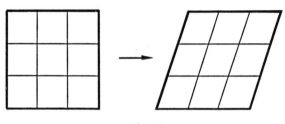

Fig. 11

For discussion

Which of the following are true for the mapping in Questions 1, 2, 3 above? How do your answers have to be modified if we allow also the mappings in Questions 4 and 5?

(1) Every point goes to a point.

(2) Distinct points go to distinct points.

(3) Every point in the output plane arises from some point of the input plane. (*Note.* We suppose throughout that the diagrams have been extended by allowing fractional and negative numbers.)

(4) Points in line go to points in line.

(5) Squares go to squares.

(6) Rectangles go to rectangles.

(7) Parallelograms go to parallelograms.

(8) Right angles go to right angles.

(9) The mapping does not change the sizes of angles.

(10) A circle goes to a circle.

(11) An ellipse goes to an ellipse (a circle being accepted as a special case of an ellipse).

(12) A hyperbola goes to a hyperbola.

(13) A parabola goes to a parabola.

6 Vectors in a plane

Most students have already met vectors and know that it is possible to define operations on them purely geometrically, without reference to any system of co-ordinates. Accordingly, we give these definitions without any introductory explanation.

We suppose a plane given, in which a special point O has been marked. If P and Q are any two points, by $P + Q$ we understand the point that forms a parallelogram with O, P and Q, as in Fig. 12.

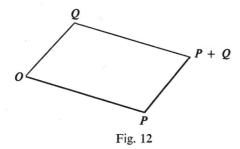

Fig. 12

Again, for any point P, by kP we understand the point that lies in the line OP, and is k times as far from O as P is.

A convention is understood here. If $k > 0$, then kP is on the same side of O as P; if $k < 0$, P and kP are separated by O. If $k = 0$, of course kP is at O. Fig. 13 shows the cases $k = 3, k = -2$. (Students sometimes fall into the error of thinking that we get to $3P$ by taking 3 intervals *beyond* P; this is not so, we reach $3P$ by measuring 3 intervals from O, the first interval being from O to P.)

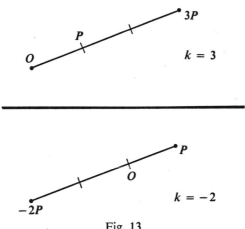

Fig. 13

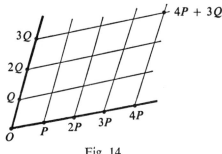

Fig. 14

In mechanics we tend to speak of 'the vector OP', and to connect the points O and P by an arrow symbol. In graphical work, these arrows may complicate and obscure the diagram. As the point O is laid down once and for all, it is perfectly satisfactory to do as we did in our first definition and speak of 'the point P', 'the point Q', 'the point $P + Q$'.

There is no mathematical difference involved here. It is simply a question of whether the diagram is clearer when the points O and P are connected by an arrow, or when they are left as they are. In any situation, we are free to use whichever we think conveys a better picture to the mind.

For instance, it seems much better not to draw the arrows in the situation now to be discussed. Suppose P and Q are any two points of the plane, such that O, P, Q are not in line. We mark the multiples of P, say $2P, 3P, 4P, \ldots$, and those of Q, such as $2Q, 3Q, \ldots$ By drawing parallel lines, we obtain a grid, as in Fig. 14.

The point at the top right of this diagram forms a parallelogram with O, $4P$ and $3Q$ and hence must be the point $4P + 3Q$.

The grid in this diagram could be used as graph paper. The point we have just considered would then be specified by its co-ordinates $(4, 3)$, since it is reached from O by 4 steps in the direction P, followed by 3 steps in the direction Q.

This point thus can be specified either as the point $(4, 3)$ in the co-ordinate system with basis $[P, Q]$ or, in the vector form, as $4P + 3Q$.

Similarly, the other points in the diagram can be specified in these two ways. (Consider some of them.)

Conventional graph paper is rather restricted. Being based on squares, the axes are perpendicular, and the intervals on both axes are the same length. The system based on two vectors, P and Q is much more flexible. We can choose P and Q so as to fit the system neatly to a problem in which the most important lines are not perpendicular.

Again it often happens that a problem is stated in terms of axes which are later found not to be the most convenient. In order to keep the work from becoming impossibly difficult, it is necessary to change to a new system. Change of axes is often thought of as being a very difficult procedure. However, we have seen that in the vector form a point can be specified by a very simple expression such as $4P + 3Q$. Manipulating such expressions should not be difficult and in fact, by the vector approach, change of axes can be done very easily, as we shall see in §9.

Note on a superstition

If we compare the co-ordinate specification, $(4, 3)$, with the vector specification, $4P + 3Q$, we see that the same pair of numbers appears in each. This effect is quite general, and we naturally use algebra to state it; if x stands for the first number and y for the second, we may assert that, in the P, Q system, the point with co-ordinates (x, y) is the point $xP + yQ$.

On many occasions, students have been asked in tests to give the vector expression for the point (x, y) and have written the answer $aP + bQ$. Some teacher must have given them the idea that x and y had some special, sacred quality, that these letters always stood for unknowns, and could not appear in an answer. This is a complete misconception. To go back to a first exercise in algebra, the correct answer to the question 'A room contains x boys and y girls, how many children in all?' is $x + y$. This could be regarded as a formula. It is a general result, in so far as x and y may stand for *any* whole numbers. Whether x and y are unknowns or not depends on the circumstances – whether we have seen the room and have had time to count the boys and girls. We cannot use the formula until we have been given the values of x and y.

In the same way, the statement 'the point (x, y) may be specified as $xP + yQ$' can be regarded as a formula or rule. If we are asked to apply it to the point $(8, 5)$, we know x has the value 8 and y the value 5, so $(8, 5)$ may be specified as $8P + 5Q$.

The assertion 'the point (x, y) may be specified as $aP + bQ$' cannot be so used. If we are told that x and y have the values 8 and 5, we still need to ask, 'What are the values of a and b? By what procedure are a and b to be derived from x and y?' Really, such an assertion does not make sense as it stands.

7   Bases

In §6 we considered the grid built on two vectors P and Q. Only part of this grid was shown in our last diagram, but it is clear that, in that diagram, the grid could be extended to cover the whole plane. Now any point covered by the grid can be expressed in the form $xP + yQ$, by reading off its co-ordinates x, y. (These of course may be fractional or negative.) Accordingly, in the situation illustrated, every point of the plane can be specified in the form $xP + yQ$. Also, this can be done in only one way.

For example we can get the point $4P + 3Q$ only by taking $x = 4, y = 3$. Any other choice of x and y will land us somewhere else.

An expression such as $4P + 3Q$ is called a *linear combination* of P and Q. More informally, we may refer to it as a *mixture* of P and Q. We can have linear combinations, or mixtures, of any number of elements. If in some system we have elements A, B, C, D, and the expression $5A - 2B + 9.1C + 0.3D$ is meaningful, then that expression is a mixture of A, B, C and D, as would be any expression of the form $xA + yB + zC + tD$, with real numbers x, y, z, t.

The points P and Q are said to form a *basis* of the plane, since they have the following properties: (1) every point of the plane is a mixture of P and Q, (2) each point can be specified as such a mixture in only one way.

It will be seen that a basis for the plane can be chosen in many ways. Almost any two points will do for P and Q. What does 'almost' mean here? First of all, neither P nor Q must be a zero vector, i.e. neither point may be at O. Secondly, P and Q must not be in line with O. If they were, every point $xP + yQ$ would lie on that line; the grid would not fill the whole plane.

Thus there are infinitely many ways of choosing a basis with 2 points (or vectors) in it. Would it be possible to have a basis with 1 or 3 points in it?

If we had only one point, say R, we could only form expressions of the form xR. As we saw at the beginning of §6, xR is defined as a point on the line OR. Thus we would not be able to find an expression to specify the points of the plane not on OR.

What about a basis with 3 vectors? Consider for example the three points, A, B, C in Fig. 15. They certainly have the first property. The point (x, y) of the graph paper is $xA + yB$, so we can get it without using C. We may write it as $xA + yB + 0 \cdot C$ if we want to bring out that it has the required form for a mixture of A, B and C. But this expression is one of many. For example, the point J may be written as $2A + 3B$, or $A + 2B + C$, or $B + 2C$, or $-A + 3C$, or in many other ways.

The vector C is redundant, for $C = A + B$. C is already a mixture of A and B. By mixing A, B and C we get nothing beyond what we could get by mixing A and B. If we

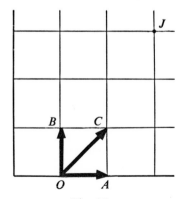

Fig. 15

interpret *A* as a gallon of red paint and *B* as a gallon of white paint, *C* would be a mixture of 1 gallon of red and 1 gallon of white paint. It does not extend in any way the range of tints we can produce.

A collection of vectors is said to be *linearly dependent* if one vector is a mixture of the remaining vectors. Thus *A*, *B* and *C* are linearly dependent. One can see that a linearly dependent collection of vectors is no good as a basis. Property (2) will always fail for it. A basis must consist of linearly independent vectors.

In the plane, as we have seen, it seems that every basis contains the same number of vectors, namely 2. For this reason a plane is said to be 'of 2 dimensions'. Further any 3 vectors in a plane are bound to be linearly dependent. In a space of 2 dimensions it is impossible to find 3 linearly independent vectors.

In a strictly mathematical development, these statements would be proved, and it would be shown that corresponding statements can be made in spaces of 3, 4, 5, . . ., dimensions. The proofs are somewhat tedious, and the time spent on them is somewhat unrewarding, since the results are what we would in any case guess or expect to be true. Later on, we shall return to some of the points involved, in a rather more formal way. For the present we will restrict ourselves to giving a definition of vector space and of basis, and a statement, without proof, of the theorems.

In §2 and §3 we considered mappings input → output. In all the situations listed at the end of §3, it was possible to define addition and multiplication by *k* 'in a sensible way'. Thus each input and each output was an example of a vector space. But in many of the situations it was not possible to specify what was coming in or what was coming out by a finite collection of numbers. Such inputs or outputs were examples of vector spaces of infinite dimension, which we do not intend to discuss further. But there were cases in which the input, or the output, or both, could be specified by a finite list of numbers. (For which inputs was this true? For which outputs?) In several cases we can adapt the situation described so as to make it finitely specifiable. For instance, in Situation 2, instead of putting sand on the light beam, we may suppose 3 weights hung from it, as in Fig. 16.

The input is now specified by giving the 3 weights W_1, W_2, W_3; the inputs form a space of 3 dimensions. The output is specified by giving the 2 reactions; the possible outputs form a space of 2 dimensions. (Strictly speaking, we ought to allow for these

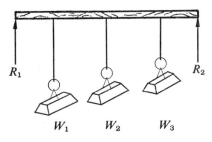

Fig. 16   Situation 2*a*.

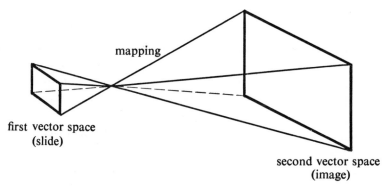

first vector space
(slide)

second vector space
(image)

Fig. 17

quantities becoming negative: a weight could be replaced by an upward pull, and it might be necessary to hold an end down rather than to support it. If we are only interested in actual weights and in positive reactions we are mapping *part* of a 3-dimensional space to *part* of a 2-dimensional space.)

In reading the formal definitions below, we should have in mind not only the plane and our everyday space of 3-D, but also such systems as occur in Situation 2*a* just discussed, or in other suitable examples resembling the situations given at the end of §3. It should be remembered that a linear mapping connects the input and the output. When we are looking for an example of a vector space we look at *the input alone* or the *output alone*. For instance, when a slide is projected onto a screen, the points of the slide *or* the points of the screen may help us to visualize a space of 2 dimensions. The projection, by which the little picture on the slide produces the big picture on the screen, is then a mapping from one vector space to the other (Fig. 17). It might well be a linear mapping.

We now come to our formal definitions. We suppose we have some system consisting of various objects, situations, ideas or other elements. In this system addition is defined; given any two elements, a and b, there is a definite element c, called the sum of a and b, and written $c = a + b$. Also if k is any real number, we know what is meant by ka. Further, we suppose it is possible to choose a basis within the system, that is, n elements $v_1, v_2, \ldots, v_n$ such that each element can be expressed as a linear combination (or mixture) of them, for example $a = a_1v_1 + a_2v_2 + \cdots + a_nv_n$ where $a_1, a_2, \ldots, a_n$ are real numbers; we also require that this can be done in one way only. Further, if $c = a + b$, then $c_1 = a_1 + b_1$, $c_2 = a_2 + b_2, \ldots, c_n = a_n + b_n$. Also we require that the numbers occurring in ka be $ka_1, ka_2, \ldots, ka_n$.

(In strict logic, it is necessary to supplement these requirements by statements which many students will feel are too obvious to need mentioning, such as, for example that v_1 and $1 \cdot v_1 + 0 \cdot v_2 + 0 \cdot v_3 + \cdots + 0 \cdot v_n$ mean the same thing.)

In these circumstances, we are dealing with a *vector space of finite dimension.*

We naturally expect to say that the space is in fact of dimension n. But a point has to be cleared up. In a vector space we expect to find many ways of choosing a basis.

For instance in the plane any pair of vectors (with different directions) will do, and all these choices are on an equal footing. Suppose we were to find one basis with the n vectors $v_1, \ldots, v_n$, and another basis, with a different number, m, of vectors, $u_1, \ldots, u_m$. The first basis would lead us to assign n dimensions to the space, the second would indicate m dimensions. We would be very surprised if such a thing did happen – for example, if it turned out that a plane could be the same thing as a space of 3 dimensions. And in fact there is a theorem – all the bases in a vector space contain the same number of vectors. That number we define to be the dimension of the space. This theorem is proved with the help of another theorem – in space of n dimensions it is impossible to have $(n + 1)$ linearly independent vectors.

In some books you will find a vector defined as something that has length and direction. These were in fact the first objects to be considered as vectors – such things as force, velocity, acceleration. But now, as we have seen, 'vector spaces' comprise a great variety of systems. In present usage, a vector means simply an element of such a system.

Questions

1 Usually 3 vectors in 3 dimensions form a basis of the space. In what cases do they not do so? (This question is most easily considered in the everyday, geometrical space of 3 dimensions.)

2 A drawer contains a certain collection of nuts, bolts, washers and nails. Can such a collection be regarded as an element in a vector space? If so, discuss how $a + b$ and ka are defined; the dimension of the space; and a simple example of a basis.

3 For a collection of vectors to form a basis, the collection must satisfy two conditions, (i) every point of space is representable as a mixture of vectors in the collection, (ii) this representation can be done in only one way (i.e. the vectors in the collection must be linearly independent).

For each collection below, state whether it is a basis or not. If it is not, state which condition or conditions it fails to satisfy.

Space of 2 dimensions

   (a) (1, 0)   (b) (3, 1), (3, −1)   (c) (2, 0), (2, 1), (2, 2)   (d) (2, 3), (4, 6)

Space of 3 dimensions

   (e) (1, 0, 0), (0, 1, 0)   (f) (1, 0, 0), (0, 1, 0), (0, 0, 1), (1, 1, 1)

   (g) (1, 1, 0), (1, −1, 0), (3, 1, 5) (h) (1, 1, 1), (1, 2, 3), (2, 3, 4)

4 Find a simple way of choosing a basis for any input or output, at the end of §3, that is a space of finite dimension; also for any space of finite dimension that can be obtained by modifying the description of an infinite dimensional space occurring in that list of situations.

8 Calculations in a vector space

In a formal mathematical treatment, the definition of vector space would be followed by a number of theorems about the properties of such spaces: it would be shown that addition was commutative and associative, that subtraction could be defined, that a distributive law held, and so forth. Subsequent calculations would be justified by citing these theorems. It is not a difficult exercise to check that these various properties do hold.

There is a common sense consideration that enables us to see how calculations are made in vector spaces and to feel confident that these calculations will have properties closely resembling those of elementary arithmetic.

Arithmetic, as taught to young children, considers situations involving one kind of article. Children are asked to add 2 apples to 3 apples or $2\frac{1}{2}$ gallons to $3\frac{1}{4}$ gallons. It is only a very small extension of this to consider several articles – add 2 nuts and 3 bolts to 4 nuts and 5 bolts. But now we are essentially dealing with a vector space of 2 dimensions. We have seen this on a number of occasions – the input to Situation 1 at the end of §3, various remarks in §5, and Question 2 at the end of §7. Now if we add a_1 nuts and b_1 bolts to a_2 nuts and b_2 bolts we get $(a_1 + a_2)$ nuts and $(b_1 + b_2)$ bolts. We have only to replace 'nuts' by v_1 and 'bolts' by v_2 to arrive at the formal definition of addition in a vector space of 2 dimensions. Multiplication by k can be arrived at similarly. If instead of 2 articles we have n articles, our statements take longer to write out, but no new idea is involved. This means that work in n dimensions proceeds along much the same lines as work in 2 dimensions. Thus in vector theory we do not often have to refer to the number of dimensions of the space; there are many theorems that hold for a space of any finite number of dimensions. In vector calculations we are using a very simple extension of arithmetic; an expression such as $a_1v_1 + a_2v_2$ can arise in many contexts, but if, when computing with it, we imagine it to mean 'a_1 articles v_1 together with a_2 articles v_2' we shall obtain the correct answer.

There is one important distinction to bear in mind however. If, on ordinary graph paper, we use the point (2, 3) to signify '2 nuts and 3 bolts', then the horizontal and vertical axes acquire a special significance. Any point on the horizontal axis represents a collection of nuts alone; any point on the vertical axis represents bolts alone. All other points represent collections in which both articles are present. Now in a vector space, no direction has any special significance. We may picture a vector space of 2 dimensions as a plane on which the point O has been marked. But nothing else is marked; in particular, no line through O has special significance. One line through O is as good as another. Any two distinct lines through O will serve as axes, and vectors in them will constitute a basis. Any basis is as good as any other basis.

In §6 we met a grid composed of parallelograms and saw that such a grid could be used as graph paper. Most students have experience only of conventional graph paper, based on squares. It is therefore desirable to do some work with this more general kind of graph paper, to acquire familiarity with it and to discover which results remain true for it.

In the worked example below, three approaches are used for dealing with certain straight lines. In the first approach, we simply guess the equations of the lines. In the second approach, vector notation gives a way of specifying the lines. In the third approach, the vector results are used to prove the correctness of our earlier guesses.

Worked example

Investigate the lines *OFMV* and *PGNW* shown in Fig. 18, the co-ordinates x and y being measured parallel to the lines *OE* and *OS*, with *OP* and *OQ* as units in these directions.

First approach The points O, F, M, V have co-ordinates $(0, 0)$, $(1, 1)$, $(2, 2)$, $(3, 3)$. In each of these, the x and y co-ordinates are equal. This *suggests* that *OFMV* has the equation $y = x$.

The points P, G, N, W have co-ordinates $(1, 0)$, $(2, 1)$, $(3, 2)$, $(4, 3)$. The equation $y = x - 1$ is suggested for the line *PGNW*.

Second approach *OFMV* is the line joining O to F. By definition, the vector tF lies on this line, and is t times as far from O as F is. If t runs through all real numbers, positive, zero or negative, the point tF will take every possible position on the line *OFMV*. Thus this line may be specified as consisting of all points of the form tF.

The line *PGNW* results if every point of the line *OFMV* experiences a displacement equal to the vector *OP*. For example, the point W arises when V is given such a displacement. We observe that *OVWP* is a parallelogram, so $W = V + P$. Similar considerations apply to every point of the line *PGNW*; each point can be obtained by adding P to the vector for a point on the line *OFMV*. As the general point of *OFMV* is tF, the general point of *PGNW* is of the form $P + tF$, where t is any real number.

An alternative way of seeing this result is the following; if a mass starts at position P and moves with velocity *OF* for t seconds, it will reach a point on the line *PGNW*.

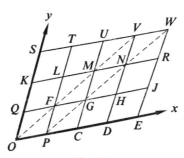

Fig. 18

The velocity OF may be written simply as F, since single letters are understood to denote vectors starting at the origin. Thus after t seconds the mass will be at the position $P + tF$. As time passes, the moving point describes the entire line. (Negative times must be considered to get the points below P.)

Third approach   We have $F = P + Q$, so $tF = tP + tQ$. Accordingly tF has the co-ordinates (t, t) by the argument of §6 in which we identified (x, y) with $xP + yQ$. Thus for any point of the line $OFMV$ we have $x = t, y = t$ for some t; hence $y = x$.

The line $PGNW$ consists of all points of the form $P + tF$. As $F = P + Q$, we have $P + tF = P + t(P + Q) = (t + 1)P + tQ$. Thus for any point of $PGNW$ we have $x = t + 1, y = t$ for some real number t. Hence $y = x - 1$ is the equation of the line, as we guessed earlier.

In the work of the second approach, it should be noted that $P + tF$ is not the only correct answer. The moving mass does not have to be at P when $t = 0$; it could equally well be at any other point of the line $PGNW$. Again, we are free to choose the speed with which it describes the line; the velocity could be, for instance $\frac{1}{2}F$ or $-3F$. Thus $G + \frac{1}{2}tF$ and $W - 3F$ are among the many possible correct ways of specifying the general point of this line. We may speak of any such expression as a *vector parametric specification* of the line, with parameter t.

Exercises

(All questions relate to Fig. 18.)

1   For the lines listed below, carry out the procedure just described, that is (i) guess the equation, (ii) find a vector parametric specification, (iii) from your answer to (ii) deduce the equation of the line.

(a) CHR (b) QLU (c) $SLGD$

(*Hint.* This line is parallel to QP. The vector that goes from Q to P is $P - Q$.)

(d) $TMHE$ (e) $OQKS$ (f) $PFLT$ (g) $KLMNR$

2   The position of a moving point at time t is given by $Q + tP$. State (a) where the point is at $t = 0$, (b) its velocity, (c) the line in which it moves, (d) its co-ordinates at time t, (e) the equation of the line it moves along.

3   Identify on Fig. 18 the lines with the following equations:

(a) $x + y = 3$ (b) $y = x + 1$ (c) $y = \frac{1}{2}x + 1$ (d) $y = 3 - \frac{1}{2}x$ (e) $y = 2 - \frac{1}{2}x$.

9   Change of axes

It very often happens that a problem can be immensely simplified by changing to a new set of axes. It is therefore important to know how such a change is made. Fortunately with vector notation changing axes is very simple.

Fig. 19 shows a plane on which two grids have been marked, one based on the points A, B, the other on the points P, Q. The point H has the co-ordinates (2, 1) in the P, Q system and (10, 5) in the A, B system. How could we, by calculation, pass from one set of co-ordinates to the other? More generally, if a point has co-ordinates (X, Y) in the P, Q system, how shall we calculate its co-ordinates, (x, y), in the A, B system?

Vector notation gives us a simple way for making such calculations. It also gives us a much shorter and more convenient way of indicating which system we are using, and also avoids the dangers of error due to mistaking co-ordinates in one system for co-ordinates in the other. Our initial information, that H is (2, 1) in the P, Q system can be written quite shortly as $H = 2P + Q$. Now, looking at P and Q, we see that $P = 4A + B$ and $Q = 2A + 3B$. Substituting these in our first equation, we find $H = 2(4A + B) + (2A + 3B)$, that is $H = 10A + 5B$, from which we can, if we wish, read off the co-ordinates (10, 5) in the A, B system.

We can go through the same steps with any point. Quite generally, suppose the

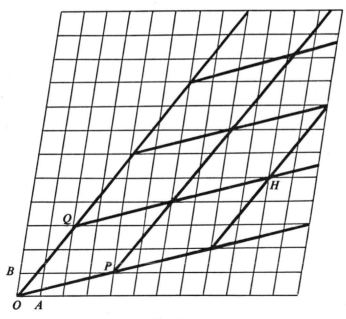

Fig. 19

point S has the co-ordinates (X, Y) in the P, Q system. Then $S = XP + YQ$. Substituting, as before, for P and Q, we find $S = X(4A + B) + Y(2A + 3B)$ so

$$S = (4X + 2Y)A + (X + 3Y)B.$$

Now if S is (x, y) in the A, B system, this means $S = xA + yB$. Comparing the two equations, we obtain

$$x = 4X + 2Y$$
$$y = X + 3Y.$$

These equations tell us how the co-ordinates in the two systems are connected. We can solve these equations, to find X and Y in terms of x and y. We find

$$X = 0.3x - 0.2y$$
$$Y = -0.1x + 0.4y.$$

Thus, given the co-ordinates of any point in one system we can calculate its co-ordinates in the other system.

In the diagram both grids contain what are sometimes called oblique axes – that is, the axes are not at right angles. This did not cause any difficulty in the calculations; in fact, right angles have nothing to do with our present work. In pure vector theory they are not mentioned and are not even defined.

Note that what we called 'change of axes' could equally well have been called 'change of basis'. In the A, B system every point is shown in the form $xA + yB$, that is, we use A, B as a basis. In the P, Q system, every point is shown as $XP + YQ$, a mixture of P and Q; here we are using P, Q as a basis. (See the definition of *basis* in §7.)

Questions

1 On ordinary graph paper mark the following points; $A = (1, 0)$, $B = (0, 1)$, $C = (2, 1)$, $D = (1, 2)$. Draw a reasonable amount of the C, D grid, in all quadrants. Complete the table below: on each line should appear four different ways of specifying one and the same point.

A, B system		C, D system	
Co-ordinates	Vector form	Vector form	Co-ordinates
(2, 1)	2A + B	C	(1, 0)
		D	
		C + D	
		2C + D	
		2C	(2, 0)
			(2, −1)
(2, −2)			
(0, −3)			
(−2, −4)			

If $XC + YD = xA + yB$, find equations giving x, y in terms of X, Y and also equations giving X, Y in terms of x, y. Check for accuracy by using the data read off from your diagram and recorded in the table above.

2   In the same way, consider the A, B system and the E, F system where E is $(1, 1)$ and F is $(-1, 1)$ on the original graph paper.

3   We sometimes wish to introduce only one new axis. Do the same work using the A, B system and the A, E system.

4   Co-ordinates x, y are understood to be in the A, B system.
Find
(*a*) The equation in the C, D system that corresponds to $5x^2 - 8xy + 5y^2 = 9$.
(*b*) The equation in the E, F system corresponding to $x^2 - xy + y^2 = 1$.
(*c*) The equation in the A, E system corresponding to $x^2 - 2xy + 3y^2 = 1$.

5   Let $P = 4A + B, Q = 2A + 3B$. Find the equations in the P, Q system of the lines whose equations in the A, B system are $x + y = 10$ and $3y = 2x$. Check your result by drawing part of the P, Q grid, with the help of ordinary graph paper.
A diagram in §9 shows A, B, P, Q but with the axes OA and OB not at right angles. Could this diagram be used to check these results?

6   'Quadratic expressions of the type $ax^2 + bx + c$ form a vector space.' Justify this statement and give any definitions needed to make its meaning clear. Of how many dimensions is this space? What is the most obvious basis for it? Do the three quadratics $(x - 1)^2, x^2, (x + 1)^2$ form a basis? Give an example of changing from one basis to another in this space, and obtain the relevant equations.
In questions 7, 8 and 9 a 3-dimensional co-ordinate system (x, y, z) is based on vectors A, B, C and a system (X, Y, Z) on P, Q, R.

7   If $P = 3B + 5C, Q = A + 6C, R = 2A + 4B$, find expressions giving the co-ordinates (x, y, z) in the A, B, C system of the point whose co-ordinates in the P, Q, R system are (X, Y, Z).
The points for which $6x - 3y - z = 0$ form a plane. Find the equation of this plane in the X, Y, Z system.

8   Find the expressions for the (x, y, z) co-ordinates of a point in terms of its (X, Y, Z) co-ordinates if $P = A + 10B + 5C, Q = A + 20B - 5C, R = A - 10B + 15C$.
Find the equations involving X, Y, Z that correspond to
(*a*) $15x - y - z = 0$,   (*b*) $20x - y - 2z = 0$,   (*c*) $5x - y - z = 0$.

9   Express (x, y, z) in terms of X, Y, Z when, in the A, B, C system of co-ordinates, P is $(1, 1, 1)$, Q is $(1, 2, 3)$ and R is $(1, 4, 9)$.
Find the equations involving X, Y, Z that correspond to
(*a*) $8y = 5x + 3z$,   (*b*) $3x - 3y + z = 0$.
What equation in x, y, z corresponds to $Z = 0$?

Worked example

A complicated metal structure is specified by giving the co-ordinates of all the corners of the plates from which it is built. The points $O = (0, 0, 0)$, $C = (-1, 2, 2)$, $D = (2, -1, 2)$, $E = (3, 0, 6)$, $F = (0, 3, 6)$ are known to be the corners of a flat plate bounded by the polygon $ODEFC$. It is also known that the sides OC and OD are perpendicular and of equal length. A scale drawing, from which this plate could be manufactured, is required.

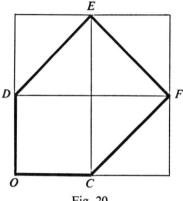

Fig. 20

Solution   We are told that the points E and F lie in the plane OCD. This means that each of them must be expressible in the form $XC + YD$.

If $E = XC + YD$, this means $(3, 0, 6) = X(-1, 2, 2) + Y(2, -1, 2)$, that is $3 = -X + 2Y, 0 = 2X - Y, 6 = 2X + 2Y$. From the first two equations we find $X = 1, Y = 2$. The third equation is satisfied by these values. (If it were not, it would mean that E did not lie in the plane OCD, contrary to the information provided.) Thus $E = C + 2D$. Similarly we may find $F = 2C + D$.

Accordingly, if we take C and D as the basis for conventional graph paper, E is $(1, 2)$ and F is $(2, 1)$. The shape of the plate is thus as shown in Fig. 20.

The points O, C, D having the same meanings as in the worked example above, make drawings of the following shape (all are polygons):

10 $OCMD$, where $M = (1, 1, 4)$.

11 $OCGD$, where $G = (2, 2, 8)$.

12 $OCHD$, where $H = (-1, -1, -4)$.

13 $OCJHK$, where $J = (-3, 3, 0)$, $H = (-1, -1, -4)$, $K = (3, -3, 0)$.

10   Specification of a linear mapping

In §3 and §4 we discussed what was meant by a mapping being linear. In §3 we characterized a mapping as being linear if it enjoyed the principles of proportionality and superposition. In §4 we gave this a more algebraic form; if a linear mapping sends $u \rightarrow u^*, v \rightarrow v^*$ then $c_1u + c_2v \rightarrow c_1u^* + c_2v^*$; that is, if we know what a linear mapping M does to any two vectors u and v, we know what it does to any mixture of u and v.

Suppose now the input is a vector space of n dimensions. This means that every vector is a mixture of the vectors $v_1, \ldots, v_n$ in a basis of the input space. Thus for any vector a of the input we may write $a = a_1v_1 + \cdots + a_nv_n$. Suppose now we know the fate of the basis vectors, that is, we know $v_1 \to v_1^*, v_2 \to v_2^*, \ldots, v_n \to v_n^*$. By proportionality, we know $a_1v_1 \to a_1v_1^*, a_2v_2 \to a_2v_2^*, \ldots, a_nv_n \to a_nv_n^*$. By superposition, we can now deduce

$$a_1v_1 + a_2v_2 + \cdots + a_nv_n \to a_1v_1^* + a_2v_2^* + \cdots + a_nv_n^*.$$

Now all the quantities on the right-hand side are known; we assumed $v_1^*, v_2^*, \ldots, v_n^*$, the point to which the basis vectors go, to be given, and the other quantities, $a_1, a_2, \ldots, a_n$, are the co-ordinates of the point a that is under consideration. Thus, with a linear mapping, if we know what happens to the basis vectors, we know what happens to any vector in the space.

The fate of any basis determines the fate of the entire input space, when the mapping is linear.

The investigations in §5 were intended to prepare for this idea. The questions there were concerned with a plane having A, B and a basis. The diagram for Question 1 shows A^* and B^* in the output space. Thus we knew, in this question, what the mapping did to the basis vectors A and B. What happened to other points, such as $A + B$, $2A + B$, could be determined in the first place by calculation. The graphical work, however, suggested that, in general, the input grid mapped to a very similar grid in the output plane. Thus a very simple geometrical construction enabled us to see the effect of the mapping, once the points A^*, B^* were given. It may help to fix this idea in mind if, at this stage, one or two questions, of the type given in §5 are answered by *purely geometrical means.*

Worked example

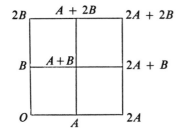 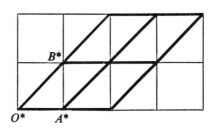

Fig. 21

On one piece of squared paper $A = (1, 0)$ and $B = (0, 1)$. This piece of paper is then mapped onto another piece in such a way that $A^* = (1, 0)$ and $B^* = (1, 1)$ as in Fig. 21.

Without any calculations, by eye alone, draw the network to which the squares of the first piece map, and read off the co-ordinates of the points to which the points $2A$, $A + B$, $2A + B$, $2B$, $A + 2B$ and $2A + 2B$ map.

Solution   The parallelogram with sides O^*A^* and O^*B^* is completed. The required network of cells congruent to this parallelogram is easily drawn by eye. We now read off $A + B \rightarrow (2, 1)$, $2A \rightarrow (2, 0)$, $2A + B \rightarrow (3, 1)$, $2B \rightarrow (2, 2)$, $A + 2B \rightarrow (3, 2)$, $2A + 2B \rightarrow (4, 2)$.

Exercises

1 Carry out the procedure of the worked example for the following cases.
  (a) $A^* = (1, 1)$, $B^* = (0, 1)$ (b) $A^* = (1, 1)$, $B^* = (-1, 1)$
  (c) $A^* = (1, 0)$, $B^* = (0, -1)$ (d) $A^* = (2, 0)$, $B^* = (1, -1)$
  (e) $A^* = (2, 1)$, $B^* = (-1, 1)$

2 Check in the worked example above and in your answers to Question 1 that the geometrical method there used leads to the same results as the algebraic argument, e.g. that the point to which $A + 2B$ maps is in fact the same point as $A^* + 2B^*$.

Note   Our theorem states that the fate of *any* basis settles the fate of the whole plane. It is not necessary for the basis to be the obvious one consisting of A and B. For example, if we are told P^* and Q^*, where $P = A + B$ and $Q = 2A + 3B$, we can deduce A^* and B^*. For $A = 3P - Q$, so $A^* = 3P^* - Q^*$; similarly $B = Q - 2P$, so $B^* = Q^* - 2P^*$.

3 Find A^* and B^* from the information given in each of the following cases.
  (a) $A \rightarrow A$, $A + B \rightarrow 2A + B$ (b) $A + B \rightarrow A + B$, $A - B \rightarrow B - A$
  (c) $A + B \rightarrow B - A$, $2A + B \rightarrow 2B - A$ (d) $A \rightarrow A + B$, $A + B \rightarrow B$

4 An examination question asks, 'If a linear mapping sends $(1, 0)$ to $(3, 4)$ and $(0, 1)$ to $(5, 6)$ where does it send $(1, 1)$?' A student, observing the sequence of numbers 3, 4, 5, 6 in the question, writes '$(7, 8)$' for his answer. This is simply a guess; it makes no use of the information that the mapping is linear, and in fact the answer is incorrect. What should the answer be, and what argument, using the properties of linear mappings, allows us to arrive logically at the correct answer?

5 A student is asked, 'If a linear mapping sends $(1, 0)$ to $(2, 1)$ and $(0, 1)$ to $(1, 2)$, where does it send $(3, 5)$?' He observes that the mapping increases each co-ordinate of the points mentioned by 1; in effect he assumes the mapping is $(x, y) \rightarrow (x + 1, y + 1)$ and answers '$(3, 5) \rightarrow (4, 6)$'. Why is his argument certainly incorrect? What is the correct answer? (*Hint*, if needed; see the end of §2.)

A student learning calculus is, without being aware of it, making use of the fact that differentiation is a linear operation. Suppose he knows that, for differentiation, $x^3 \rightarrow 3x^2$, $x^2 \rightarrow 2x$, $x \rightarrow 1$, $1 \rightarrow 0$. Now $x^3, x^2, x, 1$ form a basis for cubics; any cubic is a mixture of these. The linearity of the operation of differentiating now allows the student to differentiate any cubic, for

$$a_3 x^3 + a_2 x^2 + a_1 x + a_0 \rightarrow a_3(3x^2) + a_2(2x) + a_1(1) + a_0(0).$$

Some applications of the idea explained in this section will be found in the questions below. Before proceeding to these questions, we will consider an application to mechanics. This application is to such a simple situation that it has no claim to practical value; the problem could be solved equally well without any mention of linearity; it serves simply as an illustration of the basis-mapping principle in a concrete situation.

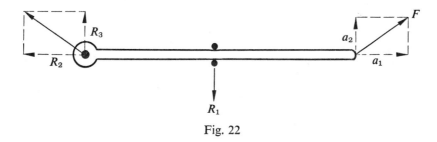

Fig. 22

We suppose a light rod has one end fastened by a smooth pin joint. It is prevented from rotating by smooth nails on either side of its midpoint (see Fig. 22). A force F acts on its free end. We are interested in the reactions produced by this force. The input here is the force F, with horizontal and vertical components a_1, a_2. The output is specified by R_1, R_2, R_3; here R_1 is the reaction at the midpoint, R_2 and R_3 are the horizontal and vertical components of the reaction at the pin joint.

We consider two situations. In the first situation unit, horizontal force acts at the free end (Fig. 23). It is obvious that this will be balanced by a unit horizontal force at

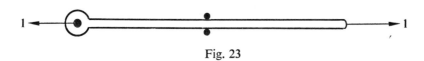

Fig. 23

the pin joint, and that there will be no reaction from the nails. Thus the output is $R_1 = 0$, $R_2 = 1$, $R_3 = 0$.

In the second situation, the free end experiences unit vertical force. It is easily seen that the forces on the rod will now be as shown in Fig. 24. That is, the output is $R_1 = 2$, $R_2 = 0$, $R_3 = 1$.

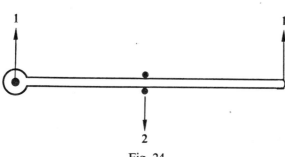

Fig. 24

Now this problem belongs to a large class of statical questions which are known to involve a linear mapping. The unit horizontal and vertical forces form a basis for the input force; the general F is obtained by adding a_1 times the horizontal unit to a_2 times the vertical unit. Taking these multiples of the reactions in the two special cases, we find that the force (a_1, a_2) produces the reactions $R_1 = 2a_2$, $R_2 = a_1$, $R_3 = a_2$.

As already mentioned, this solution could be obtained very easily by the usual methods of statics. It has been included only as a very elementary illustration of the idea that the solution of a complex problem can be found by combining the solutions of several simple cases, provided the problem is one involving a linear mapping from data to solution.

Questions

1   In Situation 2a of §7, the input is specified by the values of W_1, W_2, W_3. What would be a simple basis for the input? To what special situations would the vectors of the basis correspond?

Suppose that in Situation 2a it is impossible to measure any of the lengths involved, but that we can arrange for any loading W_1, W_2, W_3 that we wish and measure the reactions R_1, R_2 produced. Explain what series of experiments you would conduct in order to be able to calculate the reactions caused by any loading W_1, W_2, W_3. (The weights are always hung from the same 3 points.) Give explicit instructions that would enable a person ignorant of mechanics to calculate the reactions after performing the appropriate experiments. (The person may be assumed to understand the use of a formula.)

2   A piece of apparatus has two dials that may be set to any desired values. These settings constitute the input. Three meters indicate the output (Fig. 25). Nothing is known about the apparatus

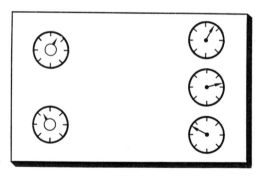

Fig. 25

except that intput $\rightarrow$ output is linear. Describe what experiments and measurements you would make in order to predict the meter readings corresponding to any dial settings. Embody your method in a formula. Design an electrical circuit to illustrate this problem.

3   In numerical work we meet tables of differences such as the following:

$$
\begin{array}{ccccccccccccc}
0 & & 1 & & 6 & & 18 & & 40 & & 75 & & 126 \\
& 1 & & 5 & & 12 & & 22 & & 35 & & 51 & \\
& & 4 & & 7 & & 10 & & 13 & & 16 & & \\
& & & 3 & & 3 & & 3 & & 3 & & &
\end{array}
$$

The numbers in each row represent the differences of adjacent terms in the row above.

Let p, q, r, s represent the first numbers in the first, second, third and fourth row respectively, and let every number in the fourth row be s. (In our example, $p = 0, q = 1, r = 4, s = 3$.)

It is known that when $p = 1, q = r = s = 0$, every number in the top row is 1. When $p = 0$, $q = 1, r = 0, s = 0$, the numbers in the top row are 0, 1, 2, 3, ... that is, they are given by the expression x.

When $p = 0, q = 0, r = 1, s = 0$ the numbers in the top row are given by the expression $\frac{1}{2}x(x - 1)$, in which the values 0, 1, 2, 3, ... are substituted in turn for x.

When $p = 0, q = 0, r = 0, s = 1$ the numbers in the top row are given by the expression $\frac{1}{6}x(x - 1)(x - 2)$.

It may be assumed that the mapping from the data p, q, r, s to the expression for the top row is linear.

Find the expression that corresponds to arbitrary values of p, q, r, s. Use this result to find the expression corresponding to 0, 1, 6, 18, 40, 75, 126, the numbers in the top row of our example above.

4   In a scientific experiment it is known that there is a law of the form $y = mx + b$. In order to determine m and b, the values of y are measured for two fixed values of x; $x = s$ gives $y = p$ and $x = t$ gives $y = q$. Is the mapping $(p, q) \rightarrow (m, b)$ linear?

When $p = 1, q = 0$ we find $y = (x - t)/(s - t)$.

When $p = 0, q = 1$ we find $y = (x - s)/(t - s)$.

What formula for y corresponds to arbitrary values of p and q?

5   It is known that x and y are related by an equation of the form $y = ax^2 + bx + c$. The values of y corresponding to $x = s$, $x = t$ and $x = u$ are p, q, r.

Find the equation for y corresponding to $p = 1, q = 0, r = 0$. (*Hint.* The Factor Theorem is relevant.)

Write the equations corresponding to $p = 0, q = 1, r = 0$ and to $p = 0, q = 0, r = 1$.

Deduce the equation corresponding to arbitrary values p, q, r.

11   *Transformations*

In most of the mappings we have considered so far, the output has been different in nature from the input. In the mapping merchandise $\rightarrow$ cost, the input is a certain collection of articles, the output a certain amount of money. In one of the statics problems, the input was a specification of 3 weights, the output was 2 reactions. In an electrical problem, the input was 2 voltages, the output was 3 currents in amperes.

It can happen, however, that the input and output are of the same character. For instance, in the process of differentiation such as $x^2 + 5x + 2 \rightarrow 2x + 5$, the input is a polynomial and so is the output. In a single-transistor amplifier, the input may be specified by means of a voltage and a current, the output also.

Another example occurs in the numerical solution of equations. Suppose we wish to solve an equation $f(x) = 0$ where the graph $y = f(x)$ has the appearance shown in Fig. 26 in a certain interval.

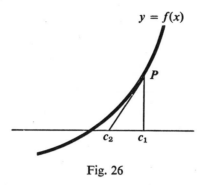

Fig. 26

Let c_1 be any value of x in the interval. Newton's method proceeds as follows. Let P be the point on the graph $y = f(x)$ corresponding to $x = c_1$. Draw the tangent at P, and let this tangent cut the horizontal axis at the point where $x = c_2$. Then c_2 will be a better approximation to the solution of $f(x) = 0$ than c_1.

Here the input is c_1, an approximation to the solution. The output is c_2, also an approximation to the solution (and in fact an improved approximation). Thus in the mapping $c_1 \rightarrow c_2$, the input and the output are of the same nature.

The advantage of such a situation is that we can repeat the operation. If we now take c_2 as our input, we can obtain a still better approximation c_3 by the same process. This type of calculation is particularly suitable for automatic computing. We programme the computer for the operation $c_1 \rightarrow c_2$, and tell it to keep on taking the output and feeding it back in as an input. In this way the computer produces a sequence of better and better approximations $c_1, c_2, c_3, c_4, \ldots$ It is told to stop when c_n and c_{n+1} differ by a sufficiently small amount. The process may be illustrated as in Fig. 27. *Note:* the

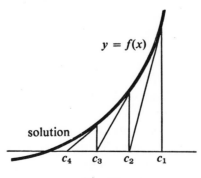

Fig. 27

sequence $c_1, c_2, c_3, \ldots$ does not always converge if the graph $y = f(x)$ is of a different type from that illustrated here.

The method of solution described above is an example of *iteration*. In this example, at each stage the input to the computer consists of a single number. But in general an iteration process may involve several numbers. For example, the initial input might consist of 10 numbers. From these the computer would obtain an output of 10 numbers, which would then form the input for the next stage of the process.

It is obvious that, for iteration to be possible, the output and the input must involve the same number of numbers. If the input is specified by (a_1, a_2, a_3) and the output by (b_1, b_2), we cannot feed the output back into the computer. A computer will not accept 2 numbers if its programme calls for an input of 3 numbers. But for iteration to be meaningful, it is not only necessary for the number of co-ordinates in the input and the output to be equal; they must also have the same significance. For instance, there are machines into which you can put 10 cents and obtain a fluid which is alleged to be coffee. So we have a mapping $10x$ cents produce x cups of coffee: $10x \to x$. But we cannot iterate this mapping; if we invest 100 cents and get 10 cups of coffee, we cannot pour the 10 cups of coffee back into the machine and get anything at all. We have a mapping from the vector space of cents to the vector space of cups of coffee. The amount of cents and the amount of coffee are both specified by a single number; both spaces have the same dimension, namely 1, but this agreement is not enough to permit iteration.

Arrows go from the cents space to the coffee space; arrows do not go from the coffee space anywhere (Fig. 28).

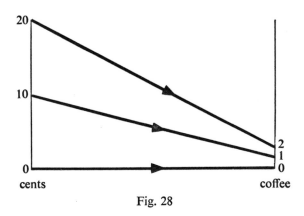

Fig. 28

For iteration to be possible, mapping must be from a vector space *to itself*.

A mapping from a space to itself is called a *transformation*.

A familiar example of a transformation is a rotation. Let T represent the operation of rotating the points of a plane through 30° about the origin. If P is any point of the plane, the transformation T sends P to a point P^*, which is also a point of the same plane (Fig. 29).

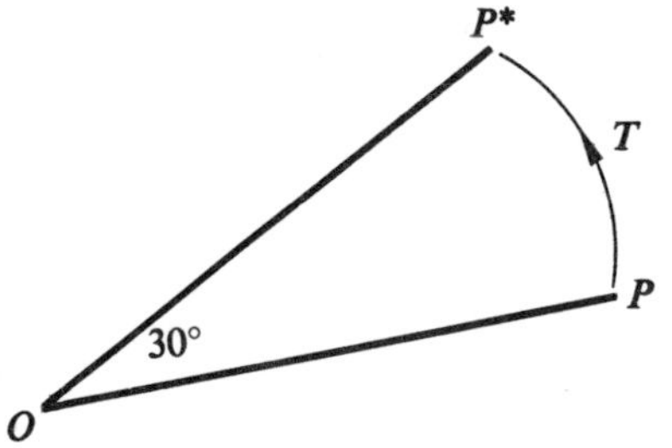

Fig. 29

We can iterate the transformation T if we wish. If we start with any point P_0, applying T again and again we would obtain points $P_1, P_2, P_3 \ldots$ spaced at $30°$ around a circle. P_{12} would coincide with P_0. We may write $P_1 = TP_0$, $P_2 = TP_1$, $P_3 = TP_2$ and so on. These equations also suggest that we might write $P_2 = TTP_0$, $P_3 = TTTP_0$, so that repeated application of the transformation is naturally written in a way that reminds us of repeated multiplication in elementary algebra. We use the same abbreviations as in elementary algebra, T^2 for TT, T^3 for TTT, and so on. As a rotation of $30°$, done 12 times, brings us back to our original position, we have $T^{12} = I$, where I represents the identity transformation.

Of course, the fact that we can iterate T does not mean that we are compelled to. There may be occasions when we are interested in a rotation through $30°$, done once and for all. But the fact that powers $T^2, T^3, \ldots$ can be defined is helpful, and can often be exploited to our advantage. *Any* transformation can be raised to a power in this way. Transformations thus differ from more general types of mapping; if M is a mapping from one vector space to another then (as with the cents to coffee mapping) the symbol M^2 is entirely meaningless.

Questions

1   Each of the transformations listed below is periodic, that is, if repeated a certain number of times it brings the plane back to its original position. Thus for some natural number n the transformation T satisfies $T^n = I$. Find the smallest n for each of the following transformations:

(*a*) a rotation of $180°$ about the origin,
(*b*) a rotation of $60°$ about the origin,
(*c*) reflection in the horizontal axis.

2   Find T^2 and T^3 for each transformation, T, of the real numbers listed here:

(*a*) doubling every number, that is, $x \to 2x$,
(*b*) adding 10 to every number, $x \to x + 10$,
(*c*) squaring every number, $x \to x^2$.

3   Which of the following transformations, T, of the real numbers are periodic?

(*a*) $x \to -x$ (*b*) $x \to x + 1$ (*c*) $x \to 10 - x$
(*d*) $x \to 1/x$ (*e*) $x \to \frac{1}{2}(x + 1)$ (*f*) $x \to (x - 1)/x$

4   A transformation T acts on every point of the plane except the origin, and sends the point (x, y) to the point with co-ordinates $x/(x^2 + y^2)$, $y/(x^2 + y^2)$. Find the effect of T^2.

5 Would a convergent sequence be obtained if we applied iteration procedure

(a) to the transformation $x \to x/2$,
(b) to the transformation $x \to \frac{1}{2}(1 + x)$,
(c) to the transformation $x \to -2x$?

6 A well known procedure for finding $\sqrt{a}$ is to iterate the transformation $x \to \frac{1}{2}x + \frac{1}{2}(a/x)$. What would happen if we tried to find $\sqrt{(-1)}$ by applying this procedure with $a = -1$?

Worked example

The mapping considered in the worked example on page 29 can be regarded as a transformation if we suppose the two pieces of squared paper to coincide. If this transformation is called T, find T^2 and T^3. What is the effect of T^n?

Solution Transformations are sufficiently specified by their effect on the basis vectors, A and B. As $A^* = A$, the transformation T leaves A where it was, and however many times T may be repeated this will remain so. T sends B to $A + B$. When T acts again, it sends $A + B$ to the point (2, 1), as we found in the earlier worked example; (2, 1) is the point $2A + B$. Thus $T^2A = 2A + B$. When T acts for the third time, $2A + B$ is sent to (3, 1) as we saw earlier, that is, to $3A + B$, so $T^3B = 3A + B$.

Thus T^2 is specified by $A \to A$, $B \to 2A + B$ and T^3 by $A \to A$, $B \to 3A + B$.

T represents a shearing action, which can be demonstrated by pressing suitably on the side of a pack of cards. Points on the horizontal axis, $y = 0$, do not move. Each time T acts, the points on the line $y = 1$ are pushed one further unit to the right. It will be seen that, if T acts n times, points on $y = 1$ move n units to the right, so T^n is specified by $A \to A$, $B \to nA + B$.

7 Find T^2 and T^3 for the transformations corresponding to Question 1, (a), (b) and (c) on page 36. Consider the effect of T^n in these cases.

8 A transformation T acts on polynomials. If the input is $P(x)$ the output is $1 + xP(x)$. Investigate the sequence of polynomials obtained by repeatedly applying T, the initial input being the constant polynomial 1.

9 A transformation T maps the polynomial $P(x)$ to the polynomial $1 + \int_0^x P(t)\, dt$. Discuss the sequence of polynomials obtained by applying T repeatedly with initial input 1. To what well known function is it related? *Note.* The transformation in this question, as also in several earlier questions, is not linear.

12   *Choice of basis*

We noted in §10 that any linear mapping could be specified by stating what it did to the vectors in any basis. Transformations being a particular kind of mapping, it follows that this remark applies to linear transformations.

While a linear transformation *can* be specified by means of its effect on *any* basis, it is not a matter of indifference which basis is used. One basis may give us a very simple specification, that is easy to calculate with and that allows us to see exactly what the transformation does, while another basis may give a specification that is complicated and difficult to imagine.

If we are given a transformation, there is a procedure for finding a basis in which it will have the simplest form, and later we shall discuss such procedures. For the moment we are merely concerned to give examples which will show the effect that choice of basis can have on the specification of a transformation. How these examples have been selected or constructed need not be considered at this stage.

We begin with an example of a transformation specified in a convenient manner. It is a transformation of a plane, and we use as basis the points A, B on ordinary graph paper, where $A = (1, 0)$ and $B = (0, 1)$. The transformation is completely specified by the information $A \rightarrow A^*$, $B \rightarrow B^*$ where $A^* = (2, 0)$ and $B^* = (0, -1)$.

In order to avoid a confused diagram, Fig. 30 shows separately a number of points

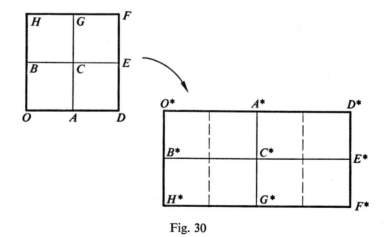

Fig. 30

$O, A, B, C, D, E, F, G, H$ and the points $O^*, A^*, B^*, C^*, D^*, E^*, F^*, G^*, H^*$ to which these go. However, it should be remembered that the transformation maps the plane to itself: we should imagine the right-hand grid drawn on a transparent sheet, which is then placed over the left-hand grid in such a way that O^* covers O and A^* covers D.

Here it is understood that O^*A^* is twice as long as OA, and that O^*B^* is as long as OB, but in the opposite direction.

The convenience of this specification arises from the fact that each axis is transformed into itself. If P is any point on the horizontal axis, then P^* is also on the horizontal axis. In fact, in vector notation, $P^* = 2P$; the scale of the horizontal axis is doubled. Similarly, if Q is on the vertical axis, Q^* will also be on the vertical axis. We have $Q^* = -Q$. The vertical axis undergoes a reversal of direction with unaltered scale.

Thus the problem of seeing what the transformation does is broken down. We can begin by forgetting all the points except those on the horizontal axis; these are rearranged among themselves. We then consider those on the vertical axis, and these also are rearranged among themselves. Finally, knowing what happens to the points on the axes, we can deduce what happens to any point. Let R be any point. Drop perpendiculars from it to the axes, to give the rectangle $OPRQ$ (Fig. 31).

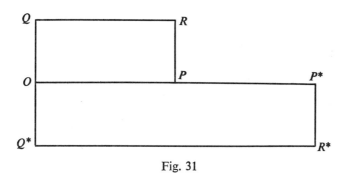

Fig. 31

We know $P \rightarrow P^*$ and $Q \rightarrow Q^*$ as shown here. Since $R = P + Q$, and this transformation is linear, $R^* = P^* + Q^*$ and R^* is in the position shown.

We will now consider a second transformation. This transformation is quite as simple as the transformation just considered, to which indeed it bears a very strong resemblance. This simplicity, however, is obscured by our choice of basis; we shall again use the points A, B of ordinary graph paper. This is an unsuitable choice, the transformation does not fit neatly into it. This transformation sends $A \rightarrow A^*$, $B \rightarrow B^*$ where $A^* = (1, 1)$ and $B^* = (2, 0)$.

Fig. 32 shows 8 points of the plane, O, A, B, C, D, E, J, K and the points $O^*, A^*, B^*,$ C^*, D^*, E^*, J^*, K^* to which the transformation sends them. As before, we have to imagine the two parts of the diagram superposed; O^* should cover O, while B^* falls on D and A^* on C.

Now of course by extending the two grids in this diagram it is possible to determine where any point of the plane goes, but it will be agreed that this diagram does not give us any simple and immediate way of feeling what the transformation does to the plane, or seeing quickly where any particular point goes. However, a way of doing this can be disentangled from the information at our disposal.

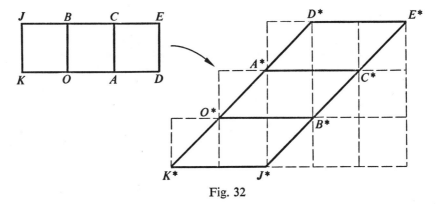

Fig. 32

It may be remembered that in our first example each axis transformed into itself. This suggests that we look at our diagram and see if we can find any line that transforms into itself.

It seems reasonable that there should be such a line. The line segment OA transforms into O^*A^*. That is to say, the transformation changes its direction by $+45°$; it swings this line in the counterclockwise direction. On the other hand OB is transformed to O^*B^*, an angle change of $-90°$, a clockwise swing. Imagine a whole lot of lines sprouting from O in directions lying between OA and OB. Those close to OB will experience a direction change close to $-90°$; those close to OA will experience a direction change of close to $+45°$ when the transformation acts on them. Somewhere between OB and OA we expect to find the place where the minus sign changes to plus; at this point there should be a direction unchanged by the transformation. We might try OC, but this does not quite work; OC is transformed to O^*C^*. This is still a clockwise swing, but it is much smaller than $90°$ in magnitude. We are getting closer to the line we want. If our next guess is OE, we have arrived at the desired place, for O^*E^* is in the same direction as OE and indeed is exactly twice as long. In vector rotation, $E^* = 2E$.

Further search on the diagram will uncover a second line that transforms into itself. It is the line through O and J, for J, J^* and the origin are in line. In fact, $J^* = -J$.

Compare the equations just found with those we had in our first example. Here we have $E^* = 2E, J^* = -J$. In our first example we had $P^* = 2P, Q^* = -Q$. In each case one vector is doubled and one vector is reversed.

In the first example, any horizontal vector was doubled, any vertical vector was reversed. In the second example, any vector lying in the line OE is doubled, any vector lying in the line OJ is reversed (Fig. 33).

This allows us to tell easily where the transformation sends any point S. We draw lines SU and SV parallel to OJ and OE, and take U on the line OE and V on the line OJ. Then $U \to U^*$ and $V \to V^*$, as shown in the diagram and as $S = U + V$, we can find $S^* = U^* + V^*$ by completing a parallelogram.

It should be noted that this construction is modelled, step for step, on the construction we used for the first transformation. The first transformation fitted neatly to the

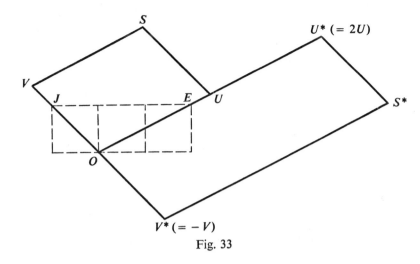

Fig. 33

axes OA and OB, which happened to be perpendicular. The second transformation fitted axes OU and OV, which happened not to be perpendicular. This did not in any way affect our ability to carry through the construction. In our first example, it was purely accidental that $OPRQ$ was a rectangle; all that mattered was that it was a parallelogram. The only use we made of this shape was to write $R = P + Q$, and for such an equation we need only to know that we are dealing with a parallelogram.

Technical terms

The examples we have just considered involve ideas that are fundamental for work with transformations. Naturally, a certain number of words and phrases have been coined, so that these ideas can be referred to without long explanations.

A line that transforms into itself is called an *invariant line*. 'Invariant' means 'unchanging'. *We shall apply this term only to lines through O.*

In our first example, we had a point P for which $P^* = 2P$, and a point Q for which $Q^* = (-1)Q$. Thus the vector P gets doubled when the transformation acts on it, and the vector Q gets multiplied by -1. A vector that gets multiplied by a number is called an *eigenvector* of the transformation. Thus, if V is an eigenvector, $V^* = \lambda V$ for some number λ; the number λ is called the *eigenvalue*. Thus, in our first example, P is an eigenvector with eigenvalue 2, since $P^* = 2P$. Similarly Q is an eigenvector with eigenvalue -1, since $Q^* = (-1)Q$. The symbol λ is nearly always used when an eigenvalue is discussed.

Any point on an invariant line, except the origin, must give an eigenvector. Suppose V is such a point on an invariant line. Then V^* also lies on the line, for the line transforms into itself. But every point of the line is of the form kV. So, for some k, $V^* = kV$; this means that V is an eigenvector with eigenvalue k.

We exclude the origin, O, when we are speaking of eigenvectors. Under any linear transformation, $O \rightarrow O$, so knowing $O \rightarrow O$ does not tell us anything about the transformation. Further, an eigenvector gives us the direction of an invariant line; if V is an eigenvector, the line OV is invariant. Here again, it is no use taking O for the vector; 'the line OO' is meaningless. So we agree that the term 'eigenvector' is not to cover the origin, O.

Note, however, that while an eigen*vector* V is, by definition, not 0, it is perfectly acceptable for the eigen*value* λ to be zero. In that case, $V^* = \lambda V = (0)V = 0$. But the *line OV* still has a perfectly good meaning; we can use it as one of our axes, and it will help to give a simple specification of the transformation. In this case the term 'invariant line' may seem strange, for every point of the line then goes to the origin; it seems that the transformation does change the line – the line shrinks to the single point O. However, the definition of *invariant line* is so framed that this case is not excluded; the definition requires that every point of the line maps to some point of the line. We may put it negatively – no point of the line maps to a point outside the line. So to speak, the transformation, so far as the points of the line are concerned, is a domestic matter; we can see where they go without knowing anything about points outside the line. If they all happen to land on the same point, O, which is in the line, this does not conflict with the requirement. The definition, of course, is framed in this way because this is found most convenient in the light of the calculations we have to make. An invariant line gives us an axis that is useful if we are trying to find a simple specification for the transformation.

Students sometimes become confused because they think of V^* as the eigenvector. As a rule this does not matter; V^* lies on the same invariant line as V, and every point of this line is an eigenvector, as we have seen. But there is trouble in the case where $\lambda = 0$, for then $V^* = 0$, and V^* is thus not acceptable as an eigenvector. Yet V, which is not 0, is still perfectly good; OV is an invariant line and a useful axis; V is accepted as an eigenvector.

These remarks should be borne in mind when answering questions.

If V is an eigenvector, every multiple kV of V (where $k \neq 0$) is also an eigenvector. For if $V \rightarrow \lambda V$, by proportionality $kV \rightarrow k\lambda V$, so the effect of the transformation on kV is to multiply it by λ. If we are asked to find the eigenvectors of a linear transformation, we often take this principle as understood. Thus, for the first transformation of this section we may say that the eigenvectors are A (with $\lambda = 2$) and B (with $\lambda = -1$). We do not bother to say 'A or any non-zero multiple of A'; we may even speak of the transformation as having 2 eigenvectors. Actually it has an infinity of eigenvectors; what we really mean is that it has 2 linearly independent eigenvectors, or 2 invariant lines.

Questions

In all these questions, the points O, A, B, C, J have the meanings attached to them in the previous section; that is, on conventional graph paper, O is $(0, 0)$, $A = (1, 0)$, $B = (0, 1)$, $C = (1, 1)$, $J = (-1, 1)$. Rotations are given with the angle measured in the usual sense; a clockwise rotation of $30°$ thus receives angle $-30°$. The term 'rotation' should be used only for movements that can be carried out within the plane of the paper. A transformation that sends any point (x, y) to $(x, -y)$

will be called 'a reflection in the horizontal' or (as some students prefer) 'a flip about the horizontal'. At a later stage, it will sometimes be necessary to classify movements as rotations or reflections; it will be confusing if the word rotation is applied to both (though of course in fact a reflection can be produced by rotating through 180° about an axis lying in the plane of the paper).

All rotations are understood to be about the origin, and all reflections in lines through the origin.

1 In the following lists, some linear transformations of the plane are described geometrically, and some are specified in vector notation. In some cases, the same transformation may occur in two or more forms. Group the descriptions and specifications in such a way as to make clear when this happens.

(a) rotation through 90° (b) rotation through 180° (c) rotation through 45°
(d) reflection in OA (e) reflection in OB (f) reflection in OC
(g) reflection in OJ (h) $A \to A, B \to -B$ (i) $A \to B, B \to -A$
(j) $A \to -A, B \to -B$ (k) $C \to -C, J \to -J$ (l) $A \to -A, C \to -C$
(m) $C \to J, J \to -C$ (n) $A \to B, C \to J$ (o) $B \to -B, C \to -J$
(p) $A \to -A, C \to J$ (q) $C \to C, J \to -J$ (r) $A \to B, B \to A$
(s) $C \to J, J \to C$ (t) $A \to -B, B \to -A$ (u) $A \to C, C \to B$
(v) $A \to C, B \to J$

2 Investigate the invariant lines of the transformations given in Question 1. Do any of these have no invariant line at all? Do any have an infinity of invariant lines? Discuss the eigenvalues and eigenvectors of these transformations.

3 In a dilation, or dilatation, the origin stays fixed, but the scale of the diagram is changed in the ratio $1:k$, in the manner suggested by Fig. 34. What invariant lines have such a transformation? Consider the cases $k = 2, k = 0.1, k = 0, k = -1$.

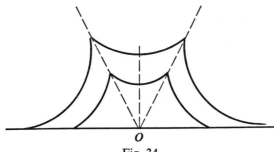

O

Fig. 34

4 In projection onto the horizontal axis, any point P goes to P^*, the point where the ordinate through P meets OA (Fig. 35). Discuss the invariant lines, eigenvectors and eigenvalues.

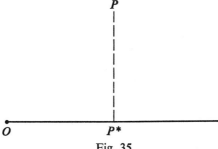

O   *P**

Fig. 35

5   Fig. 36 illustrates a particular case of oblique projection onto the horizontal axis. Discuss invariant lines, eigenvectors and eigenvalues for this projection.

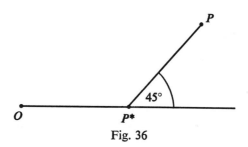

Fig. 36

6   A linear transformation S sends $A \rightarrow A$, $B \rightarrow C$. Make a sketch to show the effect of this transformation. What are the co-ordinates of the points to which S sends the points that were originally at $(1, 1)$, $(2, 1)$, $(3, 1)$? To what point does S send $(n, 1)$? Suppose that lines of slopes 1, $\frac{1}{2}$ and $\frac{1}{3}$ through the origin are drawn before the transformation S is applied. What will be the slopes of the lines to which S sends these? Are there any invariant lines in the plane for this transformation? If so, how many and which lines?

7   A linear transformation sends $(1, 0)$ to $(2, 1)$ and $(0, 1)$ to $(1, 2)$. By trial and error, find the invariant lines, eigenvectors and eigenvalues.

8   A linear transformation T sends $A \rightarrow C$, $C \rightarrow B$. Where does T send B? Investigate the transformations T^3 and T^6.

9   A transformation is obtained by first applying a rotation of $90°$ about the origin, and then reflecting in the horizontal axis; the transformation is given by the final effect of this. Is there any single transformation mentioned in Question 1 that coincides with this transformation? (Suggested approach: consider what the transformation does to A and B.)

10   In question 9, suppose the order of the operations reversed, that is, suppose the reflection applied first, and then the rotation. Answer this amended question.

11   Any line through the origin, all of whose points map to the origin, O, is an invariant line, and its points (other than O) are eigenvectors with $\lambda = 0$. Find the invariant lines that are mapped entirely into the origin, and the eigenvectors with $\lambda = 0$, for the following transformations of the plane.

(*a*) $A \rightarrow O$, $B \rightarrow B$ (*b*) $A \rightarrow \frac{1}{2}(A + B)$, $B \rightarrow \frac{1}{2}(A + B)$
(*c*) $(x, y) \rightarrow (x, x)$ (*d*) $(x, y) \rightarrow (y, y)$
(*e*) $(x, y) \rightarrow (x + y, 0)$ (*f*) $(x, y) \rightarrow (x - y, 0)$

Make sketches, showing the effect of these transformations, and describe the transformations in geometrical terms.

13 Complex numbers

We now come to consider complex numbers. This subject is important to us in two ways. First, complex numbers are used in many engineering calculations and give a great increase in mathematical power. They play a decisive role in all questions concerned with vibrations, whether mechanical, structural or electrical. They are used in the theory of alternating electrical currents. They play an essential part in quantum theory, which is the foundation of all work on atoms and nuclei, and hence of chemical and nuclear engineering. Thus complex numbers are directly used on many occasions, and this is our first reason for being concerned with them. Our second reason is that they throw considerable light on linear algebra. We are going to develop an algebraic theory of linear transformations. Now this seems a very peculiar thing to do. It seems complete nonsense to talk of a reflection being subtracted from a rotation. One naturally asks – how did anyone ever arrive at this strange idea? And once the idea has arisen that we can define addition and multiplication for transformations, the question still remains – how shall we choose suitable definitions? Now complex numbers were developed a couple of centuries before linear algebra, and provided a stimulus, a model and the essential guiding ideas for this later development. When this background is known, it is much easier to see linear algebra as a reasonable subject and to remember how it works. Accordingly we have two aims – to consider how calculations are made with complex numbers, and to see how complex numbers suggested the ideas of linear algebra.

The story begins in Italy, around 1575. A little earlier, a formula had been found for the solution of cubic equations. It was a rather complicated formula, involving two square roots and two cube roots. For example, applied to the equation $x^3 = 15x + 4$ it gave

$$x = \sqrt[3]{2 + \sqrt{-121}} + \sqrt[3]{2 - \sqrt{-121}}.$$

This is particularly unsatisfactory, since this equation has in fact a very simple solution. If we try $x = 4$, we find $x^3 = 64$ and $15x + 4 = 64$, so 4 is a solution. The solution given by the formula is not merely complicated; it contains in two places the 'impossible' symbol, $\sqrt{-121}$. A mathematician called Bombelli decided to try to work with this impossible number. He accepted $\sqrt{-1}$ as a possible number, and applied the usual procedures of algebra to it. We will use modern notation to explain what he did. Today most mathematicians write i for $\sqrt{-1}$; most engineers write j. So, if this number is accepted, we can write $\sqrt{-121} = 11\sqrt{-1} = 11j$. Bombelli then found, no doubt after considerable trial and error, that it was possible to extract the two cube roots. For consider the cube of $2 + j$. We have $(2 + j)^2 = 4 + 4j + j^2 = 3 + 4j$ since $j^2 = -1$. Multiplying again by $2 + j$ we find $(3 + 4j)(2 + j) = 6 + 11j + 4j^2 = 2 + 11j$. This is exactly what stands under the first cube root sign. Accordingly, we can extract this first cube root, and obtain $2 + j$. Similarly, in the second cube root we can check that

$\sqrt[3]{2 - 11j} = 2 - j$. Accordingly the whole formula for the solution in this case boils down to $(2 + j) + (2 - j) = 4$. We know this is what it should be. The final answer does not contain j, so we do not have to know what j means to use it.

For more than two centuries things continued more or less like this. In problem after problem it was found that if you accepted the symbol j and applied the usual rules of algebra to it, you got correct results. Yet no one was able to say why this happened.

Around the year 1800 a new device was brought in which aided calculation with complex numbers and which eventually helped to explain them. As often happens with new discoveries, the same idea occurred at about the same time to men in widely separated places – to Gauss in Germany, to Wessel in Scandinavia and to Argand in France. Wessel's work remained unknown for many years; Gauss was an eminent mathematician whose name was associated with many discoveries; the new idea became known as the Argand diagram. Argand was, in fact, the first to publish this idea. It is essentially a graphical procedure for dealing with complex numbers. If we have two complex numbers, c and z, where $c = a + jb$ and $z = x + jy$, the usual procedures of algebra tell us what we are to understand by $c + z$ and cz, namely

$$c + z = (a + x) + j(b + y),$$
$$cz = (ax - by) + j(bx + ay).$$

To these definitions Argand brought the idea that we could represent complex numbers by means of dots on ordinary graph paper. A dot on the point $(3, 4)$ would indicate that we had in mind $3 + 4j$; a dot on (a, b) would indicate $a + jb$; a dot on (x, y) would indicate $x + jy$.

So far, of course, we have not made any mathematical advance; we have simply devised a kind of signalling code for indicating what complex number we have in mind. The question then arises – will the algebraic definitions of addition and multiplication have simple geometrical constructions corresponding to them when the Argand diagram is used? Addition certainly does. As Fig. 37 shows, the points c, z, $c + z$ and the origin form a parallelogram. Thus complex numbers are added by the usual process of vector addition. It must be remembered that in 1800 the idea of vector was still unknown. The

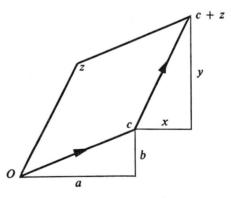

Fig. 37

concept of vector addition must have been formed partly from experience with adding complex numbers, and partly from adding forces in mechanics.

For the geometrical interpretation of multiplication it is best to proceed by stages and consider particular cases.

First, suppose $x + jy$ is to be multiplied by a real number a. Algebraically,

$$a(x + jy) = ax + jay.$$

Thus multiplication by a sends the point (x, y) on the Argand diagram to the point (ax, ay). This point is on the line from the origin to (x, y), but is a times as far away from the origin as (x, y) (Fig. 38(a)). (This clearly suggests the later definition for multiplying a vector by a number.)

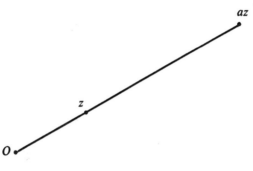

Fig. 38(a)

Next consider j multiplying $x + jy$. By algebra $j(x + jy) = jx + j^2y = -y + jx$. Thus multiplication by j sends (x, y) to $(-y, x)$. By observing the triangles in Fig. 38(b), we see that the second could be obtained from the first by rotating through 90°. Thus multiplication by j does not affect the length of a vector, but rotates the vector through a right angle (Fig. 38(c)).

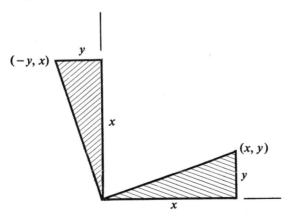

Fig. 38(b)

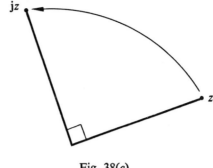

Fig. 38(*c*)

We can now see the effect of multiplication by j*b*. For j*bz* can be obtained from *z* by multiplying by *b* and then multiplying by j (Fig. 38(*d*)).

Now finally we can see what multiplication by $a + jb$ does. For $(a + jb)z = az + jbz$. So given *z*, we construct the vectors (or points) for *az* and j*bz* and then apply vector addition to these.

This construction has been carried out in Fig. 39. If *z* is at a distance *L* from the origin, $(a + jb)z$ is at a distance $L\sqrt{a^2 + b^2}$. Thus, when we apply the transformation $z \rightarrow (a + jb)z$, the length of every vector gets multiplied by the same number, $\sqrt{a^2 + b^2}$. Further every vector gets turned through the same angle θ, for in the figure

$$\tan \theta = (bL)/(aL) = b/a,$$

which does not depend on *z*.

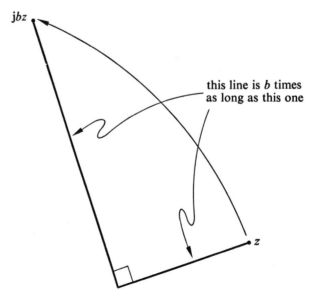

this line is *b* times as long as this one

Fig. 38(*d*)

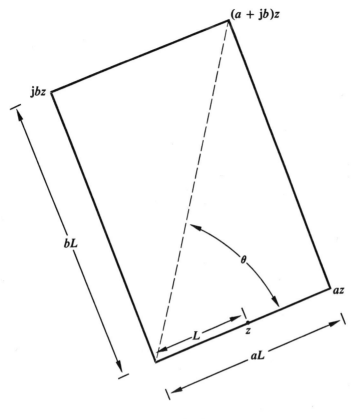

Fig. 39

If we consider the polar co-ordinates of the point $a + jb$, we see that $r = \sqrt{a^2 + b^2}$ and $\tan \theta = b/a$, Fig. 40. Thus, by observing the position of $a + jb$ on the Argand diagram we can immediately read off the angle through which multiplication by $a + jb$ rotates every vector and the change of scale it produces.

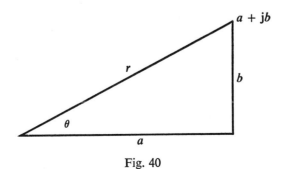

Fig. 40

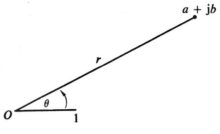

Fig. 41

It is not surprising that r and θ should occur in this connection. For on the Argand diagram the number 1, which is $1 + j \cdot 0$, appears at the point $(1, 0)$. Now

$$(a + jb) \cdot 1 = a + jb.$$

So multiplication by $a + jb$ must produce that rotation and that change of scale that is needed to bring the point $(1, 0)$ to the position (a, b) (Fig. 41).

Earlier, we used the fact that $\tan \theta = b/a$, and students sometimes point out that there is more than one solution of this equation. The argument just given shows that we need not be bothered by this; $\tan \theta = b/a$ is only part of what we know about θ. From the diagrams above we can read off $\sin \theta$ and $\cos \theta$ also if we wish. There remains of course the possibility of replacing θ by $\theta + 2\pi n$, where n is an integer, but this does not affect our work. We are only interested in the mapping $z \to (a + jb)z$; we do not mind whether people imagine the plane rotating several times before it comes to rest; we are only concerned with the final result.

Our line of thought so far has been the following. The experience of several centuries suggests that the usual rules of algebra, applied to complex numbers, lead to correct results. If we agree to let the point (x, y) be used as a graphical representation of the complex number, $x + jy$, we find that both addition and multiplication of complex numbers can be carried out by simple geometrical operations. Addition is done by drawing parallelograms; multiplication by means of rotation and change of scale, as just explained.

It is desirable to practise these geometrical operations in the Argand diagram until they have become fixed in the mind. A student can make up routine exercises of this kind for himself; samples are given here.

Exercises

1 Plot the following pairs of numbers on the Argand diagram. By geometrical considerations, predict where the point representing their sum should lie. Check this by applying the usual processes of arithmetic to the complex numbers.

(*a*) 1 and j   (*b*) $1 + j$ and $1 - j$ (*c*) $2 + j$ and $1 + 2j$
(*d*) 1 and -1 (*e*) j and $-j$ (*f*) $2 + j$ and $-2 + j$
(*g*) $2 + 2j$ and $-1 - j$ (*h*) $-1 + j$ and 1

2 In the same way, for the pairs below, predict geometrically where the product should be and check arithmetically. For each complex number occurring it will of course be necessary to consider its distance, r, from the origin and the angle, θ, at which it occurs.

(*a*) $1 + j$ and j (*b*) $2j$ and j (*c*) $1 + j$ and $1 + j$
(*d*) $1 + 2j$ and $2 + j$ (*e*) $3 + 4j$ and $4 + 3j$ (*f*) $2 + j$ and $2 - j$
(*g*) $2 + j$ and $-2 + j$

3 On the Argand diagram the powers of a number, $1, z, z^2, z^3, \ldots$, display a characteristic pattern. Calculate and plot the powers of z for the values given below. Check that their arrangement agrees with that expected on geometrical grounds.

(*a*) j (*b*) $-j$ (*c*) $1 + j$
(*d*) $(1/\sqrt{2}) + j(1/\sqrt{2})$ (*e*) $(\tfrac{1}{2}) + j(\tfrac{1}{2})$ (*f*) $(\sqrt{3}) + j$
Where would you expect the negative powers $z^{-1}, z^{-2}, z^{-3}, \ldots$ to lie?

4 We know that, if c lies at angle θ and distance r, multiplication by c has a simple geometrical effect. What about division by c?

Division by c is multiplication by c^{-1}. Where would you expect to find c^{-1} on the Argand diagram?

What is the product of $\sqrt{3} + j$ and $(\tfrac{1}{4}\sqrt{3}) - j(\tfrac{1}{4})$? Relate this result to the earlier parts of this question

5 Express in the form $a + jb$ the point of the Argand diagram that lies at angle θ and distance r from the origin.

6 To obtain a square root of -1 we have to introduce a new symbol j. To get the square root of j do we need to introduce another new symbol?

---

14   *Calculations with complex numbers*

In work with complex numbers we have two approaches at our disposal – by algebra and by geometry. At each stage of the work we should consider which approach will be simpler and more effective. If, for instance, we wish to calculate a product such as $(2 + 3j)(4 + 5j)$ the direct arithmetical treatment is much shorter and easier than the geometrical; by geometry we would have to determine the angles corresponding to $2 + 3j$ and $4 + 5j$ and then deal with their sum, which would clearly involve much more work than simply multiplying out the original expression.

It is quite different if we wish to extract a root of a complex number. A certain type of linear differential equation, with important engineering applications, leads to algebraic equations with complex roots. Thus we may find ourselves confronted with an equation such as $m^4 = -1$ or $m^3 = j$. By algebra we would probably begin by writing

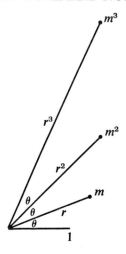

Fig. 42

$m = x + jy$; this would lead to an awkward pair of simultaneous equations. By geometry it is possible to solve these equations mentally after a little practice. The reason for this is the existence of a simple geometrical pattern formed by the positions of the powers $m, m^2, m^3, \ldots$ To solve, say, $m^3 = j$, we consider where we would have to put m to ensure that m^3 landed on j.

If m is at angle θ and distance r from O, multiplication by m rotates the plane through θ and enlarges the scale r times. If we multiply by m twice, the effect is to rotate through 2θ and to enlarge r^2 times; similarly, multiplication three times by m gives a rotation 3θ and an enlargement r^3 times. Suppose we apply these operations to the point 1; we obtain the points shown in Fig. 42.

Thus m^3 appears as the point at angle 3θ and distance r^3. Now we wish to solve $m^3 = j$, that is, to choose r and θ in such a way that angle 3θ and distance r^3 will land us on j. Now j is at distance 1 from O, so evidently we must choose $r = 1$; for $r > 1$ makes m^3 lie outside the unit circle and $r < 1$ makes m^3 lie inside the unit circle. (By the unit circle we mean the circle centre O, radius 1.) Now j lies on the vertical axis, thus at an angle $\frac{1}{2}\pi$. Thus our first thought is to take $\theta = \frac{1}{6}\pi$ and produce the situation shown in Fig. 43.* We can see that the co-ordinates of the point m are $\frac{1}{2}\sqrt{3}$, $\frac{1}{2}$, thus this point represents the complex number $(\frac{1}{2}\sqrt{3}) + j(\frac{1}{2})$, which we may write $(\sqrt{3} + j)/2$. It can be verified arithmetically that the cube of this number is in fact j.

In solving an equation we need to find *all* the solutions. Now $(\sqrt{3} + j)/2$ is a solution, but it is not the only solution, as may be easily seen. For consider $(-j)^3$. We have $(-j)^3 = -j^3 = -j^2 \cdot j = j$ since $j^2 = -1$. Thus $-j$ satisfies $m^3 = j$. In geometrical terms, how does it manage to do it? For $-j$ we have $r = 1$, $\theta = -\pi/2$. The cube involves r^3 and 3θ; thus $(-j)^3$ will be at distance 1, angle $-3\pi/2$, and this is an alternative

---

* For theoretical results such as $\sin \theta = \theta - (\theta^3/6) + \cdots$ it is essential for θ to be in radians, not in degrees. For description of a geometrical figure, such as those used here, it is possible to specify by degrees; j at 90°, m at 30°.

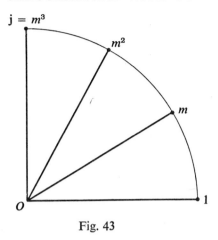

Fig. 43

way of describing where j lies. Thus to find the cube roots of a number, we have not only to consider the obvious specification of its position, but to take into account every angle that is associated with its position. Thus for j we have to consider every angle of the form $(\pi/2) + 2n\pi$, where n is an integer. If $3\theta = (\pi/2) + 2n\pi$ we have

$$\theta = (\pi/6) + (2n\pi/3).$$

If various integral values of n are tried, it will be found that we obtain only 3 distinct positions for m, as shown in Fig. 44.

It is not surprising that we cannot get more than 3 solutions, for it can be proved that in any field a cubic equation has at most 3 solutions. However there is perhaps some novelty in a number having 3 cube roots. When we are working with real numbers, we think of 2 as being *the* cube root of 8. However, if we write the equation $m^3 = 8$ this leads us to $m^3 - 8 = 0$, which may be written $(m - 2)(m^2 + 2m + 4) = 0$. The quadratic factor leads to complex solutions, $-1 + j\sqrt{3}$ and $-1 - j\sqrt{3}$, so when complex numbers are accepted we must regard 8 as having 3 cube roots.

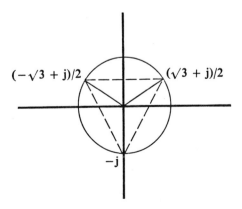

Fig. 44

With real numbers there is some uncertainty about how many solutions an equation will have. A quadratic may have 2 solutions, like $x^2 - 5x + 6 = 0$, or none, like

$$x^2 + 1 = 0.$$

A cubic may have 3, like $x^3 - x = 0$, or only 1, like $x^3 - x + 10 = 0$. When we work with complex numbers, this uncertainty disappears; *an equation of the nth degree always has its full set of n solutions*. This statement has to be interpreted in a certain way, for there is still the possibility of repeated roots. For example, even with complex numbers we cannot find any solution of $(x - 1)^3 = 0$ other than 1. As the equation may be written $(x - 1)(x - 1)(x - 1) = 0$ we sometimes say the solution 1 occurs 3 times. Our earlier statement may be clarified by expressing it in terms of factors; *with complex numbers, any polynomial of degree n can be expressed as the product of n linear factors.* These factors need not be all different.

If the cube roots of 8 are plotted on the Argand diagram, it will be seen that they form an equilateral triangle. We saw the same thing with the cube roots of j. (Does a similar remark apply to the 4th roots of any complex number? To the *n*th roots?)

We have shown in some detail how to find the cube root of a number by geometrical means. The same idea is easily adapted to finding a square root, a 4th root, or any other root.

Exercises

Solve the following equations, and plot the solutions on the Argand diagram.

(*a*) $m^2 = j$ (*b*) $m^2 = -1$ (*c*) $m^2 = -j$ (*d*) $m^3 = -1$
(*e*) $m^3 = 1$ (*f*) $m^4 = 1$ (*g*) $m^4 = -1$ (*h*) $m^6 = 1$
(*i*) $m^2 = 2 + j \cdot 2\sqrt{3}$ (*j*) $m^3 = -2 + 2j$ (*k*) $m^3 = -8$ (*l*) $m^4 = -324$

Fractions

The simplification of fractions is as a rule more convenient by the arithmetical or algebraic approach than by geometry. If we had to deal with a fraction such as

$$(1 + \sqrt{3})/(5 + 2\sqrt{3})$$

we would use the trick of multiplying above and below by $5 - 2\sqrt{3}$. The new denominator would then be $5^2 - (2\sqrt{3})^2$ or 13, in which $\sqrt{3}$ does not appear. As j stands for $\sqrt{-1}$, the same trick can be applied here. Thus to simplify $(1 + j)/(5 + 2j)$ we might proceed

$$\frac{1 + j}{5 + 2j} = \frac{(1 + j)(5 - 2j)}{(5 + 2j)(5 - 2j)} = \frac{7 + 3j}{25 - 4j^2} = \frac{7 + 3j}{29}.$$

If we write this last fraction as $(7/29) + j(3/29)$ we have it in the standard form $a + jb$.

Exercises

Find in standard form

(*a*) $(1 + j)/(2 + 3j)$ (*b*) $5/(3 + 4j)$ (*c*) $(1 - j)/(1 + j)$
(*d*) $1/(1 + j)$ (*e*) $1/(\cos \theta - j \sin \theta)$ (*f*) $1/(1 + j \tan \theta)$

15   Complex numbers and trigonometry

In this section we shall show how several results in trigonometry can be obtained very simply by using complex numbers. Much of this section may be already familiar. It is put here in order to illustrate the power that the use of complex numbers gives us. We shall obtain an even better impression of that power if we remember that the present section does not represent the most important application of complex numbers, and in fact constitutes only a very small part of what can be done with them.

All our results flow directly from the following three obvious remarks concerning rotations about the origin.

(1) The inverse of a rotation through θ is a rotation through $-\theta$.

(2) A rotation through θ combined with a rotation through ϕ gives a rotation though $\theta + \phi$.

(3) A rotation through θ done n times gives a rotation through $n\theta$.

To express these statements in terms of complex numbers, we have only to recall that the mapping $z \to cz$ corresponds to a rotation through θ and a change of scale in the ratio $r{:}1$. If $r = 1$, this mapping will be a pure rotation. Under the mapping $z \to cz$, the point 1 goes to c. Thus, we can find the value of c for a given rotation by seeing where that rotation sends the point 1.

A rotation through θ sends 1 to the point P whose co-ordinates are $(\cos \theta, \sin \theta)$ (see Fig. 45). Thus P represents the complex number $\cos \theta + \mathrm{j} \sin \theta$. We note that $r = 1$ for P. Thus multiplication by $\cos \theta + \mathrm{j} \sin \theta$ gives a rotation through θ.

A rotation through $-\theta$ would bring 1 to Q with co-ordinates $(\cos \theta, -\sin \theta)$. Thus Q corresponds to $\cos \theta - \mathrm{j} \sin \theta$, and multiplication by this complex number gives rotation through $-\theta$.

We are now in a position to translate our three statements about rotations into statements about complex numbers.

The first statement gives a very familiar result. Rotation through θ combined with

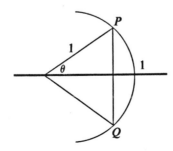

Fig. 45

rotation through $-\theta$ leaves us where we started. Leaving us where we started corresponds to multiplication by 1. So we have

$$\left.\begin{array}{r}(\cos\theta + j\sin\theta)(\cos\theta - j\sin\theta) = 1 \\ \cos^2\theta - j^2\sin^2\theta = 1 \\ \cos^2\theta + \sin^2\theta = 1.\end{array}\right\} \tag{1}$$

This of course we could prove quite easily by other methods. We have not yet demonstrated any increase of power. But statements (2) and (3) are more rewarding.

Statement (2) considers a rotation through θ combined with a rotation through ϕ to produce a rotation through $\theta + \phi$. This gives

$$(\cos\theta + j\sin\theta)(\cos\phi + j\sin\phi) = \cos(\theta + \phi) + j\sin(\theta + \phi). \tag{2}$$

Multiplying out we find

$$\cos(\theta + \phi) + j\sin(\theta + \phi) = (\cos\theta\cos\phi - \sin\theta\sin\phi) + j(\cos\theta\sin\phi + \sin\theta\cos\phi).$$

The equality sign indicates that both expressions give the same point on the Argand diagram. Now points coincide only when both co-ordinates agree. If (a, b) coincides with (x, y) we must have $x = a$ and $y = b$. Thus from $a + jb = x + jy$ we can conclude $x = a$ and $y = b$. This is often referred to as 'equating real and imaginary parts', an expression that goes back to the time when j was considered an 'imaginary' or 'impossible' number.

Accordingly we have

$$\cos(\theta + \phi) = \cos\theta\cos\phi - \sin\theta\sin\phi$$
$$\sin(\theta + \phi) = \cos\theta\sin\phi + \sin\theta\cos\phi.$$

Thus we have obtained the addition formulas for cosine and sine.

Students may have met this development in the reverse order – the addition formulas being proved (usually rather awkwardly), and then equation (2) above, deduced from these. But considerations of complex numbers show that equation (2) is obvious, and can be used to give the addition formulas quickly and easily.

Statement (3) leads to a result usually associated with the name of De Moivre, although in fact De Moivre never gave it in the form we know. Using the same ideas as we did earlier, we find it gives directly

$$(\cos\theta + j\sin\theta)^n = \cos n\theta + j\sin n\theta.$$

This can be useful for finding the trigonometric functions of multiple angles. For example, if we wish to recall $\cos 3\theta$ and $\sin 3\theta$ we may write $n = 3$. It is convenient to use the abbreviation c for $\cos\theta$ and s for $\sin\theta$. Then

$$\cos 3\theta + j\sin 3\theta = (c + js)^3 = c^3 + 3jc^2s - 3cs^2 - js^3 = (c^3 - 3cs^2) + j(3c^2s - s^3).$$

From this we read off $\cos 3\theta = c^3 - 3cs^2$ and $\sin 3\theta = 3c^2s - s^3$. If one wishes, these may be put in other forms by using $c^2 + s^2 = 1$.

This treatment of trigonometry can be taken a stage further. There is a remarkable result, which makes trigonometry simply a branch of algebra. To derive this result in a strictly logical manner is a considerable undertaking, and would demand more time and knowledge than can reasonably be expected from an engineering student at this stage. The result itself is valuable. Accordingly in the next section an account of it will be given, and the considerations that lead to it will be sketched, but no claim will be made that this constitutes rigorous deduction.

16   Trigonometry and exponentials

In elementary mathematics we meet exponentials, such as 2^x, 10^x and later e^x, and trigonometric functions such as sine and cosine. The two types of function differ both in origin and behaviour. Exponentials arise from arithmetic and can be handled with the help of a few simple formulas, such as $a^m \cdot a^n = a^{m+n}$ and $(a^m)^n = a^{mn}$. Trigonometric functions, on the other hand, are introduced with the help of geometrical ideas (lengths, right angles, circles) and involve a large number of rather complicated formulas. It was therefore a matter for considerable surprise when Euler, around the year 1750, showed that sines and cosines were simply exponentials in disguise, and that the complicated results of trigonometry could be deduced from the simple properties of exponentials.

Yet engineers and scientists might well have suspected some link between the two. For the engineer can visualize the meaning of sine and exponential by associating these with certain physical situations. If a mass is attached to a spring, it vibrates to and fro. The graph of its motion would be a curve that repeats again and again a rise and fall. One would naturally guess that the function belonging to this graph would be the sine, and so indeed it is. Sines we associate with simple harmonic motion – a mass on a spring, or a pendulum.

Exponentials we associate with unstable equilibrium. We may with difficulty persuade a pole to balance in a vertical position above its support. But the slightest disturbance will cause it to lean, and the forces acting on it will cause it to move more and more rapidly away from its position of equilibrium. A good approximation to the early stages of this motion is provided by the exponential function. Another illustration would be a particle resting on a rotating disc. If it is at the exact centre, it can remain there indefinitely. But if it moves ever so little it will experience a centrifuge effect and be moved outwards with ever increasing violence, its distance from the centre and its velocity increasing exponentially.

Now stable and unstable equilibria are not two totally separated things. It is possible to begin with a stable system and then alter it gradually until it becomes unstable, like a foolish captain who takes on so much cargo that his ship capsizes.

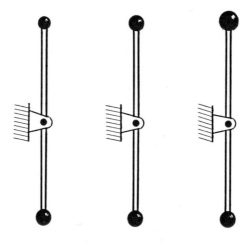

Fig. 46

For instance, we may imagine a bar with a large weight at its lower end and a small weight at the top (Fig. 46). With a pin at its midpoint, it could oscillate like a pendulum. If we gradually increase the weight at the top, the oscillation will become slower and slower and finally a point will be reached where it ceases to be a small oscillation at all; the overloaded top plunges towards the depths.

It is characteristic of the unstable situation that the acceleration carries the system away from equilibrium, and the further the system goes, the greater the acceleration is. Since d^2s/dt^2 or $\ddot{s}$, the second derivative, represents acceleration, it is not surprising that we frequently meet the equation $\ddot{s} = k^2s$ in connection with unstable systems. For such a system, $s = e^{kt}$ is a possible motion.

In a stable system, we usually find acceleration proportional to displacement but in this case the acceleration is pulling the object back towards equilibrium, and the equation is $\ddot{s} = -c^2s$; a possible motion is $s = \sin ct$.

Both the equations above are of the form $\ddot{s} = ms$ where m is a constant. We pass from unstable to stable as m passes from positive to negative. Now $\ddot{s} = -c^2s$ would become identical with $\ddot{s} = k^2s$ if we were allowed to write $k^2 = -c^2$ or $k = jc$. Thus $s = e^{kt}$ would be replaced by $s = e^{jct}$, and we would be led to conjecture, on engineering grounds, that if our calculations led to an exponential with an 'imaginary' exponent, this indicated that we were dealing with some kind of oscillation. This idea plays a large part in investigations of stability, and is among the most important applications of complex numbers to practical problems.

The connection between oscillations and imaginary exponents was in fact discovered by the mathematician Euler around 1750. At that time mathematicians were very much interested in physical applications, and the argument just given may well have occurred to him at some stage of his work. The connection, however, can be demonstrated by purely mathematical formalisms, which we now proceed to indicate.

One way of arriving at this connection is by means of the series for e^x, $\sin x$ and $\cos x$. By quite elementary calculus, using only integration, we can obtain the results

$$e^x = 1 + x + (x^2/2!) + (x^3/3!) + (x^4/4!) + (x^5/5!) + \cdots$$
$$\cos x = 1 \qquad - (x^2/2!) \qquad\quad + (x^4/4!) \qquad\qquad - \cdots$$
$$\sin x = \qquad x \qquad\quad - (x^3/3!) \qquad\qquad + (x^5/5!) - \cdots$$

These results are obtained of course for real values of x. In looking at these results one can hardly fail to be struck by the fact that e^x seems to be built from the same ingredients as $\cos x$ and $\sin x$; e^x contains all the terms of the form $x^n/n!$, $\cos x$ contains the even ones of these and $\sin x$ the odd ones; in e^x all the signs are positive, while in $\cos x$ and $\sin x$ the signs are plus and minus alternately. We could simply observe this as an interesting coincidence and leave it at that. But for a scientist or a mathematician an unexplained coincidence is an indication that some new law or theorem is waiting to be discovered.

If the signs in $\cos x$ and $\sin x$ were all positive we would simply add these series, for $\cos x + \sin x$ agrees with e^x apart from the presence of minus signs. If we are to get genuine agreement, we must find some way of bringing minus signs into the exponential series. If we compare the signs in $\cos x$ with the even terms of the series for e^x, we find that $\cos x$ has a minus sign in the terms containing x^2, x^6, x^{10}, ..., that is, the odd powers of x^2. In fact the signs in the cosine series are as in the sequence $-x^2$, $(-x^2)^2$, $(-x^2)^3$, $(-x^2)^4$, ... This might suggest putting $y^2 = -x^2$, or $x = jy$. And in fact if we put $x = jy$ in the series for e^x, we immediately achieve our goal; the calculation leads directly to the equation $e^{jy} = \cos y + j \sin y$.

Earlier we saw that $\cos \theta + j \sin \theta$ was associated with rotation through θ; now we see that it may be written in the compact from $e^{j\theta}$, and instead of using statements (1), (2), (3) about rotations we can use the algebraic properties of exponents.

Another way of looking at e^x leads to the same conclusion. The original idea of e^x was related to compound interest, and from this point of view it was found natural to define e as the limit of $[1 + (1/n)]^n$ as $n \to \infty$, and then to prove $e^x = \lim [1 + (x/n)]^n$ as $n \to \infty$. If x is a complex number, this last definition is still workable. For any whole number n we can find $[1 + (x/n)]^n$; this merely involves multiplication of complex numbers. We can then see whether the result approaches some fixed point when n is made very large.

So, in order to see what $e^{j\theta}$ should mean, we first consider $[1 + (j\theta/n)]^n$. Multiplication by $1 + (j\theta/n)$ brings 1 to the point with co-ordinates $(1, \theta/n)$. If we keep multiplying by $1 + (j\theta/n)$, we shall obtain a succession of points lying on a spiral, as in Fig. 47. In this figure $OA = 1$ and $AP_1 = \theta/n$. Thus, by Pythagoras,

$$r = \sqrt{1 + \frac{\theta^2}{n^2}} \quad \text{and} \quad \tan \alpha = \theta/n.$$

In the Argand diagram, A represents 1, P_1 represents $1 + (j\theta/n)$, and the number we are interested in is the nth power of $1 + (j\theta/n)$. Thus it occurs at distance r^n and angle $n\alpha$.

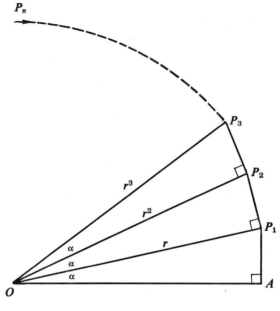

Fig. 47

Now r is slightly larger than 1, and so r^n also must be larger than 1. However, in the expression for r we see $1 + (\theta^2/n^2)$, and the n^2 in the denominator has a more powerful effect than the exponent n in r^n. It can be proved that, as n grows, r^n approaches 1. Thus the *limit* point is at a distance 1 from O, and lies on the unit circle.

For a small angle ϕ, $\tan \phi$ and ϕ are approximately equal. In fact ϕ is slightly smaller than $\tan \phi$. Thus from the equation $\tan \alpha = \theta/n$ we know that α is very nearly equal to θ/n, but is just a little less. Accordingly $n\alpha$ is nearly θ, but is slightly less. Here again it can be proved that $n\alpha$, which is $n \tan^{-1}(\theta/n)$, does tend to θ when $n \to \infty$.

Thus the *limit* of $[1 + (j\theta/n)]^n$ as $n \to \infty$ is at distance 1 and angle θ. Its co-ordinates are accordingly $(\cos \theta, \sin \theta)$ and the complex number it represents is $\cos \theta + j \sin \theta$. But this limit is, by definition, what we mean by $e^{j\theta}$. Thus once more we are led to the conclusion

$$e^{j\theta} = \cos \theta + j \sin \theta.$$

Both arguments above may in places go beyond a student's knowledge. In the second argument, the procedure for calculating the limits may not be known. In the first argument, the actual calculations are not difficult but questions of interpretation arise: what do we mean by an infinite series of complex numbers? When does it converge? How do we define e^z, $\sin z$ and $\cos z$ if $z = x + jy$, a complex number? A whole course might be devoted to answering these questions thoroughly. For the present we are not trying to go beyond the position of the eighteenth-century mathematician – that it seems very reasonable to identify $e^{j\theta}$ with $\cos \theta + j \sin \theta$, and that correct and extremely useful results follow when we do so.

We have already mentioned one important application – the study of oscillations and the solution of the differential equations arising in connection with oscillations. This topic belongs to calculus and will not be further developed here.

It was mentioned at the beginning of this section that all the formulas of trigonometry could be deduced from the simple properties of exponentials. This remark will now be explained.

In our first argument we obtained the equation $e^{j\theta} = \cos\theta + j\sin\theta$ by putting $x = j\theta$ in the series for e^x. If we put $x = -j\theta$ we are led to the equation

$$e^{-j\theta} = \cos\theta - j\sin\theta.$$

We thus have two equations involving $\cos\theta$ and $\sin\theta$, and we may regard these as simultaneous equations with $\cos\theta$ and $\sin\theta$ as unknowns. Solving these we find $\cos\theta$ and $\sin\theta$ in terms of $e^{j\theta}$ and $e^{-j\theta}$. We have

$$\cos\theta + j\sin\theta = e^{j\theta}, \tag{1}$$

$$\cos\theta - j\sin\theta = e^{-j\theta}. \tag{2}$$

$$\tfrac{1}{2}(1) + \tfrac{1}{2}(2) \cdots \cos\theta = \tfrac{1}{2}(e^{j\theta} + e^{-j\theta}), \tag{3}$$

$$\frac{1}{2j}(1) - \frac{1}{2j}(2) \cdots \sin\theta = \frac{1}{2j}(e^{j\theta} - e^{-j\theta}). \tag{4}$$

Students are sometimes worried by the fact that j appears in the denominator in equation (4); they think this means that $\sin\theta$ will not be a real number. But the quantity $e^{j\theta} - e^{-j\theta}$ in the bracket contains a factor j, as may be seen by substituting in the series for e^x. If we did not have j in the denominator to cancel this factor we would indeed be confronted by a paradox.

Equations (3) and (4) fulfil our promise; they define cosine and sine purely in terms of exponential functions.

An abbreviation will prove convenient. For this section only, we introduce a convention. Given quantities $A, B, C, D, \ldots$ denoted by capital letters we shall understand by the corresponding small letters $a, b, c, d, \ldots$ the following:

$$a = e^{jA}, \qquad b = e^{jB}, \qquad c = e^{jC}, \qquad d = e^{jD}.$$

Then we have $e^{-jA} = 1/a$ so we can obtain purely algebraic expressions for $\cos A$ and $\sin A$, namely

$$\cos A = \frac{1}{2}\left(a + \frac{1}{a}\right), \qquad \sin A = \frac{1}{2j}\left(a - \frac{1}{a}\right).$$

Thus the trigonometric identity $\cos^2 A + \sin^2 A = 1$ may be regarded as a consequence of the algebraic identity

$$\frac{1}{4}\left(a + \frac{1}{a}\right)^2 - \frac{1}{4}\left(a - \frac{1}{a}\right)^2 = 1.$$

In the exercises below, students are asked to translate into algebra trigonometric expressions of gradually increasing complexity. It will be found that even an expression such as cos $(A - 2B + 3C + 4D)$ has a fairly simple algebraic equivalent.

In the nineteenth century, students were expected to achieve great virtuosity in the handling of trigonometric expressions. Today there is rightly less stress on such manipulative skill, but there are still occasions where it is useful for an engineer to understand some argument involving trigonometric formulas. This may be because engineers deal with objects that have actual shapes, so that trigonometry arises in its primitive role; it may also be in some situation where waves or vibrations are involved; again it may be that trigonometric functions arise in the middle of some mathematical process. The exercises below lead to a sort of dictionary, allowing any statement about sines and cosines to be translated into a purely algebraic equation. Algebra is usually easier to handle than trigonometry; it is more systematic. However translating trigonometry into algebra does not automatically remove all difficulties; sometimes the resulting algebraic equation is quite hard to verify or prove.

Exercises

1 By using the convention explained above, translate the expressions below into algebraic expressions involving only the small letters a, b, c, d.

(a) cos$(A + B)$	(b) cos$(A - B)$	(c) sin$(A + B)$
(d) sin$(A - B)$	(e) cos $2A$	(f) sin $2A$
(g) cos $3A$	(h) sin $3A$	(i) cos $4A$
(j) sin $4A$	(k) cos$(A + 2B)$	(l) sin$(A + 2B)$
(m) cos$(A - 2B)$	(n) sin$(A - 2B)$	(o) cos$(A - 2B + 3C)$
(p) sin$(A - 2B + 3C)$	(q) cos$(A - 2B + 3C - 4D)$	(r) sin$(A - 2B + 3C - 4D)$

2 By means of the results found in Exercise 1 above, translate the following equations into algebra, investigate whether they are correct or incorrect.

(a) $1 + \cos 2A = 2 \cos^2 A$.

(b) $2 \cos A \cos B = \cos(A + B) + \cos(A - B)$.

(c) $\cos(A + B) = \cos A \cos B + \sin A \sin B$.

(d) $\cos(A - B) = \cos A \cos B + \sin A \sin B$.

(e) $\cos 3A = 3 \cos^3 A - 4 \cos A$.

(f) $\cos 3A = 4 \cos^3 A - 3 \cos A$.

17 Complex numbers and convergence

One of the arguments used in §16 depended on the use of infinite series. The development of calculus 300 years ago led rapidly to the discovery that very many functions could be represented as infinite series and this has remained a very powerful method for dealing with them. However, an infinite series, as one might guess, is a dangerous thing to use; if you do not understand it properly it can lead to entirely incorrect results. In this section we shall state (but not prove) a very simple result that enables us to distinguish correct from incorrect usage of infinite series. This result depends on complex numbers and the Argand diagram, and could not be stated or understood without the use of complex numbers.

The dangers of infinite series can be seen from the simplest example, one that occurs already in secondary school work. The series $1 + x + x^2 + x^3 + \cdots$ is a geometrical progression; its sum to infinity, calculated from the formula $a/(1 - r)$ is $1/(1 - x)$. This series can also be reached by starting with $1/(1 - x)$ and doing long division, or using the binomial theorem with exponent -1.

We will now compare the behaviour of the series and the fraction $1/(1 - x)$. In Fig. 48, the graph of $y = 1/(1 - x)$ is shown by a curve, the rectangular hyperbola. The value

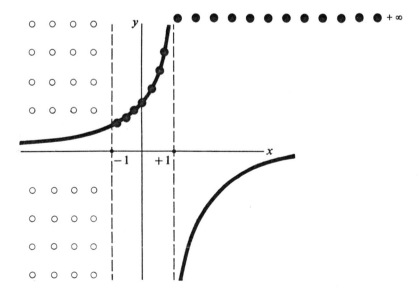

Fig. 48

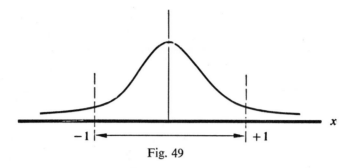

Fig. 49

of the series is indicated by large dots. For $-1 < x < 1$, as is well known, the series agrees perfectly with $1/(1 - x)$; the dots lie on the curve. When $x > 1$, this agreement totally disappears. For example, when $x = 2$, the series gives $1 + 2 + 4 + 8 + \cdots$ with sum $+\infty$. The fraction gives -1. Any value of x larger than 1 gives $+\infty$ for the series. We indicate this, rather symbolically, by dots along the 'ceiling' of the graph. For $x < -1$, the series is unsatisfactory in another way; if we put, for example, $x = -2$, the series gives $1 - 2 + 4 - 8 + 16 - 32 \cdots$; it oscillates more and more wildly and does not settle down to any number at all. The diagram tries to indicate this by dots scattered at random. Thus the series is a true and faithful servant so long as x lies between -1 and $+1$; outside this region it gives results such as $-1 = +\infty$ and is totally unreliable. In this its behaviour is typical for power series.

If we examine the graph, we can see why something unusual should be expected as x approaches $+1$, for here the graph goes off to infinity. But why should the series begin to misbehave when we pass -1? When $x = -1$, the fraction has the value $\frac{1}{2}$, and the graph is perfectly smooth and well behaved.

This question becomes even more acute if we consider the fraction $1/(1 + x^2)$ and the corresponding series $1 - x^2 + x^4 - x^6 + \cdots$ One could hardly ask for a tamer and better behaved graph than that of $1/(1 + x^2)$, as shown in Fig. 49. There is nothing in this graph to suggest that the interval -1 to $+1$ has any special significance, yet here again the series behaves perfectly within that interval; it becomes meaningless outside it.

The situation changes entirely as soon as we begin to consider complex values. The fraction $1/(1 + x^2)$ will be undefined if $x^2 + 1 = 0$, that is, if $x = j$ or $x = -j$. Thus on the Argand diagram we naturally mark the points j and $-j$ as danger points. Imagine pins put at these points and a balloon centred on the origin being gradually blown up, (Fig. 50). Suppose we blow it up, but are careful not to let it reach the pins. Then in

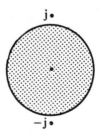

Fig. 50

the part of the Argand diagram covered by the balloon, all will be well; the series will sum to its correct value. But this is not all; within this region, you can apply any formal procedure that any sane person would think of; you can integrate the series, differentiate it, rearrange the terms, multiply it by itself or by another well behaved series, and in each case arrive at the result you want.

The danger points are known as *singularities*. The series for a complex function, $f(z)$, will behave well in any circle, with centre O, that does not reach as far as the nearest singularity. The functions e^z, sin z, cos z do not have a singularity for any finite value of z. Thus the series for them can be used with confidence for every value of z. As we have seen, $1/(1 - z)$ and $1/(1 + z^2)$ have singularities at unit distance from the origin. They are well behaved in any circle with centre O and radius less than 1.

A function can have a singularity without itself becoming infinite. For example $\sqrt{1 - z}$ has a singularity at $z = 1$ although its value there is zero. However, its derivative is infinite when $z = 1$.

For most engineering applications, it is sufficient to know that a function has a singularity where it, or any derivative, becomes infinite or undefined. The circle, with centre O, that goes through the nearest singularity, is called the *circle of convergence*. Inside this circle, all is well. On it, or outside it, anything may happen.

18   Complex numbers: terminology

This section is quite brief. Its aim is simply to explain certain names and symbols the student may meet in books or lectures.

When we were dealing with multiplication of complex numbers, a great role was played by r, the distance of a point from the origin, and θ, the angle at which the point occurs. It is not surprising that names have been coined for these.

For any real number x, the sign $|x|$ denotes its distance from 0. For complex numbers the same symbol is used for the distance from O in the Argand diagram. Thus if z is a complex number at distance r from O, we write $|z| = r$. This quantity is referred to as the *modulus* or *absolute value* of z.

The angle θ at which z occurs is called the *argument* of z, or arg z for short.

It can often happen that we are dealing with a complex number, $z = x + jy$, but we are only interested in one of the numbers x, y. There are names which allow us to indicate this; x is called the *real part* of $x + jy$, and y the *imaginary part*. Abbreviations are used, such as $x = \text{Re}(x + jy)$ or $y = \text{Im}(x + jy)$. These may vary a little from author to author. Note that the imaginary part of $7 + 13j$ indicates the real number 13; it does *not* indicate 13j. This is obviously more convenient; in engineering, any measurement we make involves *real numbers*. For purposes of calculation it may be convenient to know that the number we want turns up as the *coefficient* of j in some complicated expression. If the opposite convention were used, we would have to keep pointing out that we did not want jy, the part of $x + jy$ that contains $\sqrt{-1}$, but only y, the number that appears multiplying j in this part of the expression.

Routine exercises

1 Write down $|z|$ for the following values of z:

(*a*) j   (*b*) −j   (*c*) 4 + 3j
(*d*) 4 − 3j   (*e*) 1 + j   (*f*) −3j
(*g*) $a + b$j   (*h*) $\cos\theta + j\sin\theta$ (*i*) $e^{j\theta}$

2 Give arg z for:

(*a*) j   (*b*) −3   (*c*) −j
(*d*) 1 + j   (*e*) −2 + 2j   (*f*) $(\sqrt{3}) + j$
(*g*) 1 − j$\sqrt{3}$ (*h*) $\cos\theta + j\sin\theta$ (*i*) $e^{j\theta}$

3 Re and Im indicate real and imaginary parts. Give the following:

(*a*) Re(2 + 3j)   (*b*) Im(2 + 3j)   (*c*) Re($a + b$j)   (*d*) Im($a + b$j)
(*e*) Re 1/(1 + j)   (*f*) Im 1/(1 + j)   (*g*) Re $e^{j\theta}$ (*h*) Im $e^{j\theta}$
(*i*) Re $e^{-j\theta}$ (*j*) Im $e^{-j\theta}$

19   The logic of complex numbers

We have now followed the development of complex numbers through two stages. The first stage lasted about two centuries; in it mathematicians applied the rules of algebra to complex numbers and obtained correct and useful results, but always with some uncertainty as to what complex numbers were and why they worked so well. In the second stage, it was found that the Argand diagram gave a useful geometrical method for calculating with complex numbers.

We now come to the third stage, in which a logical account of complex numbers is sought, and a fourth stage in which mathematicians pass beyond complex numbers and see if there are yet other systems ('hypercomplex numbers') that can be profitably studied.

The Argand diagram is a kind of analogue computer. If we are given a complex number, we know where to put a dot to represent it. Conversely, given a dot on the diagram, we know how to read off the complex number it represents. Given two dots, p and q, we know geometrical constructions for marking the dots $p + q$ and pq. By continuing these constructions we can arrive at points corresponding to longer algebraic expressions, such as for example $(p + q)^2$ and $p^2 + 2pq + q^2$. (A little care is necessary here to specify the order in which the operations are done.) In the algebra of real numbers, the two expressions just mentioned would always be equal. But in building up a logical theory of complex numbers we are not allowed to make use of the guess that complex numbers obey the same laws as real ones. Rather, we define the geometrical operations that give sum and product. Starting from any two points p and q in the plane, a sequence of additions and multiplications leads us to $(p + q)^2$. So we can construct the point $(p + q)^2$. Similarly, the constructions corresponding to a sequence of additions and multiplications will lead us to $p^2 + 2pq + q^2$. When these constructions are carried out, it may be that $(p + q)^2$ and $p^2 + 2pq + q^2$ will always coincide, or they may not – so far as we know at present. But the matter is outside our control. We have defined the process, and have to carry it out according to the definitions. We may experiment and find in many particular cases that the points do coincide, or we may be able to prove that they always will.

Thus the Argand diagram becomes something like a machine we have set up, into which complex numbers can be fed and operations carried out. A statement such as 'For all p and q, $(p + q)^2 = p^2 + 2pq + q^2$' is to be regarded as true if the dots that correspond to the two constructions always coincide. On the basis of long experience, we expect that every formula valid in elementary algebra will be valid for this machine. Can we prove this?

Here we find ourselves up against a difficulty. There are infinitely many equations that hold in elementary algebra. We cannot deal with each one individually. How shall we know when we have proved enough to make sure that all the rest hold?

Mathematicians put this question to themselves in the years 1800–40, and it was in these years that the terms 'commutative', 'associative' and 'distributive' (which so obsess the writers of recent school textbooks) were coined. The idea was that every formula in elementary algebra is a logical consequence of these properties. Accordingly, if we can demonstrate that our Argand diagram 'machine' obeys $p + q = q + p$, $p(qr) = (pq)r$, $p(q + r) = pq + pr$ and so forth, then we know that operations with polynomials will work just the same for complex numbers as for real numbers. There are other properties that ought to be checked, if the full analogy with arithmetic and algebra is to be established – that subtraction and division (except by zero) are always possible; that complex numbers playing the roles of 1 and 0 exist; that $pq = 0$ only if $p = 0$ or $q = 0$. In all, not more than a dozen properties have to be checked. These properties are known as the axioms of a field. A field may informally be defined as a system in which there are operations called addition and multiplication, which can be handled just as if we were dealing with our usual arithmetic.

In the nineteenth century it was proved that complex numbers were such a system, and that we had logical justification for applying the usual procedures of arithmetic and algebra to them.

In many ways, for an engineer and even for a mathematician, the conclusion reached is more important than the details of the proof. There are many occasions on which all we need to remember is that we can apply the usual methods of arithmetic, algebra and calculus to complex numbers. It can hinder you, in the course of such a calculation, if you start thinking of every complex number as a point in a plane, that is as a vector. The Argand diagram should be used when it is appropriate. For some calculations, as we have seen, it is extremely useful. It also can be used as one way of showing that work with complex numbers has a logical justification. But there are also occasions when the wisest thing is to forget all about it and return to the simple faith that expressions like $(2 + 3j)/(4 + 5j)$ and $e^{j\theta}$ do have some meaning and can be handled just as if they were real numbers.

Exercises

1 Investigate the properties of commutativity, associativity and distributivity for operations with complex numbers.

2 Discuss subtraction for complex numbers.

3 Investigate when division by $a + jb$ is possible.

20   The algebra of transformations

One of the most persistent habits of mathematicians is generalizing. A particular device has proved useful for solving a certain type of problem. Almost the first question a mathematician asks is whether this device can be amended in some way so that it can be used in a wider class of problems.

In the Argand diagram multiplication by a complex number, $z \to cz$, appears as a particular kind of transformation, involving rotation and change of scale. If p and q are two complex numbers, we find ourselves considering the transformation $z \to (p + q)z$ and the transformation $z \to (pq)z$. Thus, for these very special transformations addition and multiplication are defined. It will only be a matter of time before some mathematician asks, 'Using our experience with complex numbers as a guide, can we define addition and multiplication for *any* transformations?'

A definition of multiplication is easily found. To multiply by pq, we can first multiply by q and then multiply the result by p. Thus, if $qz = z^*$ and $pz^* = z^{**}$, we have $z^{**} = (pq)z$. Now this last statement remains meaningful if p and q are thought of as representing *any* transformation of some space.

Thus, if S and T are transformations of any space, we define ST as having the effect of the transformation T followed by the transformation S. This definition is quite in keeping with the usual way of writing functions of a real variable; for instance, log sin x could be found by first looking up sin x in a table of sines, and then finding the logarithm of that number. Similarly, given tables of sines and logarithms, we could construct a table for sin log x, by doing the operations in the reverse order.

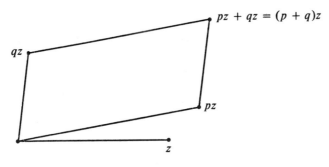

Fig. 51

Addition calls for a little more thought. In the Argand diagram we know all the laws of algebra hold. So, if we want to find $(p + q)z$ we know we can write this as $pz + qz$. The addition here will be seen in the Argand diagram as vector addition. We shall have the diagram of Fig. 51.

This suggests that we should define the transformation $S + T$ as follows; to find $(S + T)z$, first find Sz and Tz, and then add these to give $Sz + Tz$. This assumes we are dealing with a space in which addition has been defined. As we are going to be concerned only with transformations of vector spaces, in all our work this condition will be met.

Students sometimes raise the objection – we set out to define $S + T$ but we have not succeeded in doing that. We have only defined $(S + T)z$. How do we get rid of this z?

In fact there is no need to get rid of z. A transformation sends a vector z to a vector z^*. The transformation is completely defined when we know where it sends each vector. Now in the diagram and construction above, z represents *any* vector. So we can use the construction to find where each vector goes; that is to say, the transformation $S + T$ is completely defined by the construction above.

For example, suppose we are dealing with a plane. Let I denote the identity transformation, M_1 reflection in the horizontal axis and M_2 reflection in the vertical axis.

What is $I + M_1$? Consider any point z. By definition, $(I + M_1)z$ means $Iz + M_1z$. Now Iz is simply z, while M_1z is its reflection as shown in Fig. 52. We can find

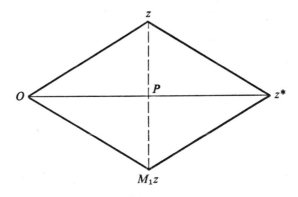

Fig. 52

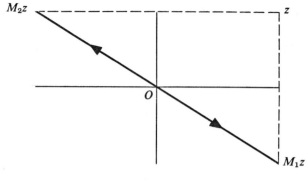

Fig. 53

$z + M_1z$ by drawing a parallelogram. This gives us z^*, the point to which $I + M_1$ sends z. If P is the projection of z on the horizontal axis, z^* is twice as far from O as P. We now know exactly what the transformation $I + M_1$ does to each point of the plane. The transformation is fully defined; we have a complete specification of it.

In the same way, we can find $M_1 + M_2$. Fig. 53 shows any point z, and its reflections in the axes. It will be seen that M_1z and M_2z are equal and opposite vectors; their sum is thus the zero vector, 0. Thus the transformation $M_1 + M_2$ sucks every point of the plane into the origin. This transformation is usually thought of as the zero transformation; we may denote it by 0, and write $M_1 + M_2 = 0$.

Note that we have three zeros in our work, the number 0, the vector 0 and the transformation 0. It will usually be clear which of these we have in mind; for instance, in the equation $M_1 + M_2 = 0$, as the sum of two transformations is a transformation, not a vector or a number, it is clear that the 0 in this equation must stand for the zero transformation.

If we write $0v$, where v is a vector, it may not be clear whether we mean the vector v multiplied by the *number* 0, or the result of applying the *transformation* 0 to the vector v. Fortunately both interpretations give the same result – the zero vector, 0.

You may meet some books in which different signs are used to distinguish the three zeros; the sign θ (Greek theta) is sometimes used for one of them.

If S and T are linear transformations, the product ST will also be linear, and so will the sum $S + T$. (How do we know this?)

It will be remembered from §10 that for any linear transformation, the fate of the entire space is determined by the fate of the vectors in any basis. This is a possible method for finding what ST and $S + T$ are, when S and T are given. It is not always the best method.

Worked example

Let M_1 be as above, and J represent rotation through 90° about the origin. Find M_1J. In Fig. 54, consider what M_1J does to A and B. J sends A to B, and M_1 then sends B to

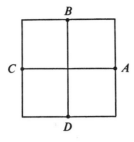

Fig. 54

D. Thus $(M_1J)A = D$. J sends B to C and M_1 leaves C where it was. So $(M_1J)B = C$.
Thus for M_1J we see $A \to D$, $B \to C$.

Thus M_1J is reflection in the line with equation $x + y = 0$.

Multiplying a transformation by a number

We have explained how transformations are added. There is nothing to stop us adding
a transformation to itself. Thus, given T, we can find $T + T$. If we add T again we can
find $T + T + T$. Naturally we abbreviate these to $2T$ and $3T$. Let v be any vector, and
let $Tv = v^*$. Then, by definition, $(T + T)v = Tv + Tv = v^* + v^* = 2v^*$. Thus
$(2T)v = 2v^*$. In the same way, $(3T)v = 3v^*$.

These results suggest that, if k is any number (not necessarily a natural number), we
should define kT so that $(kT)v = kv^*$. Now v^* is short for Tv, so our equation means
$(kT)v = k(Tv)$.

We accordingly *define* kT by the equation just given. To anyone used to elementary
algebra, it may be hard to see that this equation says anything: the two sides seem to be
exactly the same thing.

Suppose however this equation had not been given but that instead students suddenly
found an expression such as $3.5T$ occurring in a problem. Surely someone would ask,
'What do you mean by multiplying a transformation by 3.5?' And the student would
be quite right to ask this: it is not obvious what, for instance, 3.5 times a reflection
means.

The right-hand side of the equation $(kT)v = k(Tv)$ is meaningful. The transformation
T sends the vector v to a vector represented by Tv. Thus Tv is a vector, and we know
what multiplying a vector by k means. So $k(Tv)$ has a clear meaning. On the left, we
have kT, a symbol that has not been used before. It is up to us to explain what we mean
by it. A person who introduces a new symbol is entitled to say what he uses it to stand
for. This seems to give him a lot of freedom; he could say kT stood for a pound of
cheese. But in practice, his freedom is not so great. The symbol kT looks like multiplication in elementary algebra. If we defined it in such a way that it did not behave, more or
less, like the multiplications we did at school, it would be misleading. People would be
influenced by their old habits and would continually be making mistakes and getting

wrong results. This would certainly happen if we defined our new symbol kT in such a way that the equation $(kT)v = k(Tv)$ did *not* hold. So we are practically forced to try this definition. Now of course it might turn out that even this definition did not work out well; it might lead to complications because of some other definition we wanted to adopt. This would mean that the behaviour of transformations was essentially different from that of numbers. If this were so, we would be wise to use a symbolism for transformations utterly unlike that for numbers; this would be the best way to avoid confusion. However, this possibility is not realized in fact. There is a very strong analogy between the behaviour of transformations and that of numbers. It does pay us to use the symbols of algebra for transformations; more often than not, the results we expect will be true. We can learn, in a reasonably short time, what processes in elementary algebra we must *not* apply to transformations. It is really remarkable that transformations and numbers should have properties as similar as they do.

The equation $(kT)v = k(Tv)$ that defines the transformation kT can be put into words. It says that the transformation kT can be carried out as follows: apply the transformation T, then change the scale in the ratio $k:1$.

Exercises

I, M_1, M_2, and J have the meanings already defined in this section.

1   Find a geometrical description, as simple as circumstances permit, for each of the transformations listed below. In each case, say what are the invariant lines (if any); state the eigenvectors and eigenvalues (if any).

(a) M_2J (b) JM_2 (c) JM_1
(d) M_1M_2 (e) M_2M_1 (f) M_1^2, that is, M_1M_1
(g) M_2^2 (h) J^2 (i) $M_2J^2M_1$
(j) $I + M_2$ (k) $I + J$ (l) $I + M_1 + M_2$
(m) $M_2J + JM_2$ (Use your answers to (a) and (b) above.)
(n) $JM_1 + JM_2$

2   Which of the following statements are true and which false?

(a) $JM_1 = M_1J$ (b) $JM_2 = M_2J$ (c) $M_1M_2 = M_2M_1$
(d) $JM_1 = M_2J$ (e) $JM_2 = M_1J$

3   In the light of your work with the earlier questions, would you expect the statement $ST = TS$, where S and T are any two transformations to be true always, sometimes or never? What about the statement $S + T = T + S$?

4 $JM_1 + JM_2$ was found in question 1(n). Would we get the same result if we found $J(M_1 + M_2)$?

5   (a) In the text $I + M_1$ was found. How would you describe $\frac{1}{2}(I + M_1)$ most simply as a geometrical operation?
(b) Similarly describe $\frac{1}{2}(I + M_2)$.
(c) What is $(1/\sqrt{2})(I + J)$? Have we met essentially this result in any earlier section?

21 Subtraction of transformations

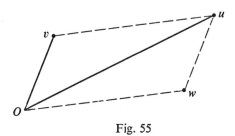

Fig. 55

In §20 we defined addition and multiplication of transformations, and by repeated use of these definitions we can build expressions such as $STS + 2S + 3T$ or $S^2 + 5S + 6$. However, we cannot yet attach a meaning to an expression involving subtraction such as $S - T$ or even $-T$. Such expressions do arise naturally. For instance, in §20 we found $M_1 + M_2 = 0$ for the reflections in the axes. This suggests $M_2 = -M_1$, if we can find some way of explaining what the minus sign means in this connection.

We deal with this question in much the same way as a teacher might in Grade 1. If a child does not know $7 - 4$ the teacher may ask '4 and what make 7?' Thus a child does not have to learn a 'subtraction table'. Knowing the addition table gives the subtraction results. 'What is $7 - 4$?' and '4 and what makes 7?' are equivalent questions. If we replace 'what?' by an algebraic symbol, we may say that $x = 7 - 4$ and $4 + x = 7$ are equivalent.

We know that $7 - 4$ indicates a definite number, 3. This means that there is some number denoted by $7 - 4$, and that there is only one such number – the equation $4 + x = 7$ has a solution, and (unlike, say, a quadratic) it has only one solution.

When we are dealing with transformations, or even with vectors, subtraction is defined along essentially the same lines.

If u and v are vectors, by $u - v$ we mean that vector which, added to v, will give u.

Suppose $w = u - v$, that is, suppose $w + v = u$. Then u must be the corner of a parallelogram formed when w and v are added. Given u and v, we can construct w, as shown in Fig. 55. There is only one possible position for w. It will be noticed that the path from O to w is parallel to the path from v to u, and of equal length. Thus the magnitude and direction of the vector $u - v$ can be seen, by considering the path 'from v to u'.

Usually $O - v$, the difference between v and O, is shortened to $-v$. Thus $-v$ means (in accordance with our definition) the vector that has to be added to v to give O. This of course, is the vector equal and opposite to v, as in Fig. 56.

Now of course $u - v$ and $u + (-v)$ ought to mean the same thing, and you can check that in fact they do. In Fig. 55, you may notice that from u to w is a displacement equal and opposite to that from O to v.

Fig. 56

Now we return to the subject of transformations. If we write $W = S - T$, then W means that transformation which, added to T, gives S. We have to check that some transformation does this, and that only one does.

If $S = W + T$, for any vector v, $Sv = Wv + Tv$, from the definition of addition in §20. Thus $Wv = Sv - Tv$. Sv and Tv are both vectors – they are the points to which v is sent under the transformations S and T. $Sv - Tv$ is the difference of these vectors. As we have just seen, you can subtract any vector from any vector and a definite vector results. Thus there will always be one, and only one, Wv that satisfies the required condition.

If we like, we can define $-T$, and then get $S - T$ by taking $S + (-T)$. This is often the most convenient way to do things. By $-T$ we understand of course $0 - T$, the transformation that, added to T, gives 0. This means $T + (-T) = 0$. So for any vector v, $Tv + (-T)v = 0v$. Now 0 is the transformation that sucks everything into the origin. Whatever v, $0v = 0$. (Notice how this looks right, for someone used to elementary algebra. What kinds of things are the two zeros here – transformations, vectors, numbers?)

So $Tv + (-T)v = 0$, that is $(-T)v$ is the vector equal and opposite to Tv. Again this looks right; we could write it $(-T)v = -(Tv)$.

So we can find the effect of $-T$ as follows; see where T sends each point, then reverse all the vectors.

For example, consider $-J$, where J as before denotes rotation through 90°. J sends P to Q (Fig. 57). Reversing OQ, we find the point P^*. Thus $-J$ is a rotation through $-90°$ for $-J$ sends P to P^*.

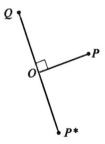

Fig. 57

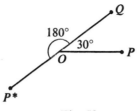

Fig. 58

Do not jump to the conclusion that, if R is a rotation through θ, $-R$ is a rotation through $-\theta$. This is *not* true; 90° is a very special angle. Suppose for instance R denotes rotation through 30°. Then R sends P to Q as shown in Fig. 58, and $-Q$ gives P^*. We do not get from P to P^* by a rotation of $-30°$.

It will be useful to consider $-I$. The identity transformation I leaves any point P unchanged. Thus $-I$ sends P to P^* as in Fig. 59. Thus $-I$ denotes a rotation through 180°.

If we now look back at the figure where R denoted rotation through 30°, we see that $-R$ denoted rotation through 30° followed by rotation through 180°. As rotation through 30° is denoted by R and through 180° by $-I$, the previous sentence can be boiled down into the equation $-R = (-I)R$, which again looks very plausible as a piece of algebra.

In §20, we noticed $M_1 + M_2 = 0$ which we may now write as $M_2 = -M_1$. It is useful at this stage to look back at the diagram showing M_1z and M_2z, and to see how this particular case illustrates what we have just been considering. We see that indeed $M_2 = (-I)M_1$; the effect of reflection in the horizontal followed by a rotation of 180° is a reflection in the vertical.

Now that addition, subtraction and multiplication have been defined we can build expressions that look very much like those at the beginning of an elementary algebra book, such as $(S - T)^2$; $(S - T)(S + T)$; $(S + T)(S^2 - ST + T^2)$. We have seen above that many true statements about transformations look exactly like statements in elementary algebra. *However, as the exercises below will show, equations that are true for numbers are not always true for transformations.* So, for the present, we have to check each time, by going back to the meaning of the statement, whether it is true for transformations or not. Later we shall try to find a more systematic way of deciding which equations work for transformations and which do not.

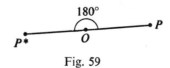

Fig. 59

Co-ordinate Methods

So far, our discussion of transformations has been based entirely on *geometrical diagrams*. This has been done because drawing is one of the main ways in which an engineer

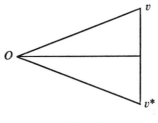

Fig. 60

thinks and communicates. It is not enough for an engineer to calculate; he must be able to *see* what he is doing. In North American education this aspect of mathematics tends to be neglected. However, it is often helpful to calculate; though, whenever possible, calculation should be illustrated by a diagram. Many results that we have found earlier by purely geometrical thinking can be found or checked by co-ordinate methods. The following worked examples illustrate the procedure.

(1) Where does I send the point (x, y)? I leaves it unchanged; so I makes $(x, y) \rightarrow (x, y)$.

(2) Where does M_1 send (x, y)? M_1 is a reflection in the horizontal: it sends v to v^* in Fig. 60. Thus the x co-ordinate is unchanged, but the y component changes sign. So we have $M_1: (x, y) \rightarrow (x, -y)$.

(3) Where does $I + M_1$ send (x, y)? By the definition of addition,

$$(I + M_1)v = Iv + M_1v = v + v^*$$

in the figure just used. Now $v = (x, y)$ and $v^* = (x, -y)$ as we found above. Superposing these, we find $(x, y) \rightarrow (2x, 0)$ when $I + M_1$ acts. This agrees with the conclusion reached in §20, and gives a more compact way of recording the result. In words we can express the effect of $I + M_1$ as follows; the x co-ordinate is doubled, the y co-ordinate becomes zero.

(4) Where does M_1J send (x, y)? By the definition of the product of transformations, we must let J act, and then see what M_1 does to the result. Fig. 61 shows v, and we have

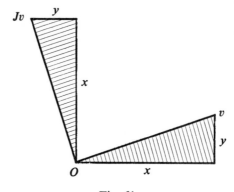

Fig. 61

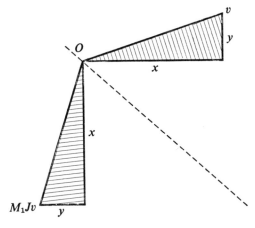

Fig. 62

shaded in a certain triangle. J rotates everything through 90°. If we imagine it picking up the shaded triangle, of which v is one corner, it will take it to the shaded triangle of which Jv is a corner. We see that the co-ordinates of Jv are $(-y, x)$. This of course is a standard result and the argument just used should be familiar to the student.

M_1 changes the sign of the second co-ordinate. So M_1 acting on Jv gives $(-y, -x)$, as in Fig. 62. This again agrees with our conclusion found in §20, that M_1J is a reflection in the line with slope -1.

The co-ordinate approach may be found helpful in some of the following exercises.

It is important that students should work right through the following questions. There may seem to be a lot of questions here, but each is quite short, and each is really a small part of an investigation. For instance, the first five questions form a single topic. The questions are not merely exercises on the definitions, but are designed to show what the algebra of transformations is like. A student who worked Exercises (1) to (12) only and then stopped would be left with an entirely misleading impression. These questions show an effect that happens *sometimes*; the later questions show that it does *not* happen always.

Exercises

1 Where does $2M_1$ send (x, y)?

2 Where does M_1^2 send (x, y)? What is the simplest symbol for M_1^2?

3 Where does $I + 2M_1 + M_1^2$ send (x, y)?

4 Where does $(I + M_1)^2$ send (x, y)? (The effect of $I + M_1$ has been discussed in the text.)

5 Elementary algebra suggests that $(I + M_1)^2$ and $I + 2M_1 + M_1^2$ might be the same transformation. Do the answers to questions 3 and 4 confirm this or not?

6 Where does $-I$ send (x, y)?

7 Where does M_2 send (x, y)?

8   Investigate $M_1 M_2$ and $M_2 M_1$. Are they equal? Is there any simpler way of writing them?

9   Where does $M_1 + M_2$ send (x, y)? What is the simplest symbol for $M_1 + M_2$?

10   Where does $(I + M_2)$ send (x, y)?

11   Where does $(I + M_1)(I + M_2)$ send (x, y)? What is the simplest symbol for this transformation?

12   Elementary algebra suggests that $(I + M_1)(I + M_2)$ and $I + M_1 + M_2 + M_1 M_2$ might be the same transformation. Investigate this with the help of your answers to questions 8, 9 and 11.

13   The effects of J and M_1 have been given in the text. Where does $J + M_1$ send (x, y)?

14   Where does $J + M_1$ send $(7, 3)$? Where does it send $(4, 4)$? Where does $(J + M_1)^2$ send $(7, 3)$?

15   Choose any point, e.g. $(3, 2)$, and see where $(J + M_1)^2$ sends it. Experiment a little with different points. Investigate by algebra the effect of $(J + M_1)^2$ on any point (x, y).

16   Where does JM_1 send (x, y)? Does JM_1 have the same effect as $M_1 J$, which was considered in the Worked example 4 above?

17   Simplify $J^2 + 2JM_1 + M_1^2$, or find its effects on (x, y), whichever you prefer. Is

$$J^2 + 2JM_1 + M_1^2$$

the same transformation as $(J + M_1)^2$?

18   Is $(J + M_1)^2$ the same as either of the following?
  (a) $J^2 + 2M_1 J + M_1^2$ (b) $J^2 + M_1 J + JM_1 + M_1^2$

19   In elementary algebra, a quadratic has only two solutions. For transformations, the quadratic $X^2 = I$ has the solutions $X = I$ and $X = -I$. Has it any other solutions? If so, how many?

20   In elementary algebra, $ab = 0$ only if $a = 0$ or $b = 0$. Is it true for transformations that $ST = 0$ only if $S = 0$ or $T = 0$? (Some of your answers to earlier questions provide evidence relevant to this.)

22   Matrix notation

In working the exercises of §21, you probably found that the use of the co-ordinates to specify points made the work easier. This idea can be carried further. By a very simple device, we can specify linear transformations in terms of numbers. This allows us to specify a transformation in a convenient form, and it also allows us to calculate the numbers that specify transformations such as $S + T$ and ST, when we are given the numbers that specify S and T.

In §10, 'Specification of a linear mapping' we met the doctrine that the fate of a basis determines the fate of the whole space. In the plane, the points $A = (1, 0)$ and $B = (0, 1)$ constitute a basis. If a linear transformation T sends these points, say, to $A^* = (4, 1)$

and $B^* = (2, 3)$, this fixes what happens to every point of the plane. As A^* has 2 co-ordinates and B^* also has 2 co-ordinates, it follows that any linear transformation of the plane can be specified by giving 4 numbers, the co-ordinates of these two points. Thus we could specify T by writing $(1, 0) \rightarrow (4, 1)$, $(0, 1) \rightarrow (2, 3)$. However, a shorter and more convenient symbolism can be found.

In this section, we are concerned to translate into algebra definitions which have already been given in geometrical form. Much of this work is quite straightforward. Students will find it improves their confidence and understanding of the definitions if they carry out the details of this work for themselves, rather than simply following an exposition of it. Accordingly this section is presented as a series of questions. After each question, there will be a row of stars. Such a row of stars is an invitation to the student to turn away from the book and find the answer for himself. Below the stars, the answer to the question will be explained.

For the particular transformation T considered above we have the information $(1, 0) \rightarrow (4, 1)$, $(0, 1) \rightarrow (2, 3)$, and we know that this fixes where every point of the plane goes. It ought to be possible, then, to derive a formula showing where any given point, (x, y), goes. If T sends (x, y) to (x^*, y^*), what equations give x^* and y^* in terms of (x, y)? We know T is linear, so we can use the principles of proportionality and superposition.

$$* \qquad * \qquad *$$

We use proportionality first. If the input $(1, 0)$ is multiplied by any number x, the output $(4, 1)$ will be multiplied by that same number. So $(x, 0) \rightarrow (4x, x)$. By the same principle, applying multiplication by y to the statement $(0, 1) \rightarrow (2, 3)$, we find $(0, y) \rightarrow (2y, 3y)$ for any number y. We now know where any point on either axis goes. Superposing $(x, 0)$ and $(0, y)$ gives (x, y). So we conclude $(x, y) \rightarrow (4x + 2y, x + 3y)$ and we may write

$$x^* = 4x + 2y,$$
$$y^* = x + 3y.$$

So, appropriately enough, a linear transformation, T, is expressed by a pair of linear equations. We would obtain much the same kind of result if we took any other numbers in place of $(4, 1)$ and $(2, 3)$. If we replace the particular numbers by algebraic symbols and carry through exactly the same argument, we find that $(1, 0) \rightarrow (a, c)$, $(0, 1) \rightarrow (b, d)$ lead to the equations

$$\left. \begin{array}{l} x^* = ax + by, \\ y^* = cx + dy. \end{array} \right\} \tag{1}$$

These equations show why we made the apparently strange choice of symbols in $(1, 0) \rightarrow (a, c)$. By having $(1, 0) \rightarrow (a, c)$ rather than $(1, 0) \rightarrow (a, b)$ we get the letters a, b, c, d in their natural order in the equations (1).

Here again four numbers are used to specify a transformation. Matrix notation is based on one very simple idea, that of arranging the four numbers in the positions they occupy in the equations above. Thus we would specify the transformation T by writing

$$\begin{pmatrix} 4 & 2 \\ 1 & 3 \end{pmatrix}$$

while the general linear transformation, corresponding to equations (1), would be written

$$\begin{pmatrix} a & b \\ c & d \end{pmatrix}.$$

In many of the transformations we have considered in our examples, some of the terms seem to be missing. For instance M_1, reflection in the horizontal, has $x^* = x$, $y^* = -y$. In order to exhibit this in the pattern of equations (1), we insert the coefficients 0 and 1 implied by these equations. Thus, for M_1

$$x^* = 1x + 0y,$$
$$y^* = 0x - 1y.$$

and so, in matrix notation

$$M_1 = \begin{pmatrix} 1 & 0 \\ 0 & -1 \end{pmatrix}.$$

In the same way, if we meet the symbol

$$\begin{pmatrix} 0 & -1 \\ 1 & 0 \end{pmatrix}$$

this tells us that in equations (1) we are to substitute $a = 0$, $b = -1$, $c = 1$, $d = 0$. We then find $x^* = -y$, $y^* = x$, that is $(x, y) \rightarrow (-y, x)$, which near the end of §21 we saw to be the transformation J, rotation through 90°.

One advantage of using the matrix symbolism is that we can easily write the effect of successive transformations. Thus M_1J is immediately written as

$$\begin{pmatrix} 1 & 0 \\ 0 & -1 \end{pmatrix} \begin{pmatrix} 0 & -1 \\ 1 & 0 \end{pmatrix}.$$

Much more writing would be needed if we had to express this combined transformation by means of equations.

Successive transformations frequently arise in engineering. There are obvious examples such as transistor amplifiers in which each stage takes a signal and operates on it. The final output is the result of all these operations in turn applied to the original input. But in any branch of engineering less obvious examples arise. In a computing process, certain data may be taken, a calculation made with them giving another set of numbers, which then become the starting point for a further calculation or subroutine. In numerical analysis, frequent use is made of iteration, the process mentioned in §11, by which the same transformation is applied again and again. In structural analysis, we find, if not successive transformations, at any rate successive mappings, which can be represented by matrices. Thus in the direct stiffness method, we start with the loads applied to a structure, use a matrix to deduce the displacements, and from these displacements, with the help of another matrix, we find the stresses.

Suppose we have two transformations, H and N, specified in matrix form by

$$N = \begin{pmatrix} p & q \\ r & s \end{pmatrix}, \qquad H = \begin{pmatrix} a & b \\ c & d \end{pmatrix}.$$

The combined transformation NH has been defined in §20; it is found by applying H and then applying N to the result. Thus if H sends (x, y) to (x^*, y^*) and N then sends (x^*, y^*) to (x^{**}, y^{**}), the transformation NH sends (x, y) to (x^{**}, y^{**}). We have the equations

$$\left. \begin{array}{l} x^* = ax + by, \\ y^* = cx + dy, \end{array} \right\} \quad (1) \qquad \left. \begin{array}{l} x^{**} = px^* + qy^*, \\ y^{**} = rx^* + sy^*. \end{array} \right\} \quad (2)$$

These correspond to the matrices given for H and N.

NH is the transformation $(x, y) \rightarrow (x^{**}, y^{**})$. To find the matrix specification for NH, we have to answer two questions, neither of which calls for any special trick or ingenuity. What equations give (x^{**}, y^{**}) in terms of (x, y)? What is the matrix corresponding to these equations?

*　　*　　*

The first question can be answered immediately by substituting in equations (2) the values of (x^*, y^*) given in equations (1). For example, for x^{**} we find

$$p(ax + by) + q(cx + dy).$$

Collecting together the x and y terms, we find the first equation shown in (3) below. The second equation below is found by applying a similar process to y^{**}. We obtain

$$\left. \begin{array}{l} x^{**} = (pa + qc)x + (pb + qd)y, \\ y^{**} = (ra + sc)x + (rb + sd)y. \end{array} \right\} \quad (3)$$

The matrix form is obtained by writing the coefficients here in the appropriate places, as in the matrix for NH below. We may show the result of combining N and H to obtain NH by the scheme below.

$$\begin{pmatrix} p & q \\ r & s \end{pmatrix} \begin{pmatrix} a & b \\ c & d \end{pmatrix} = \begin{pmatrix} pa + qc & pb + qd \\ ra + sc & rb + sd \end{pmatrix}. \quad (4)$$

There is a mechanical process for obtaining this result. It we examine, for example, $pb + qd$, the number that occurs in the first row and second column of NH, we see that it involves the numbers p, q in the first row of N, and the numbers b, d in the second column of H. Corresponding numbers in these two sets are multiplied (giving pb and

qd) and the results added (giving $pb + qd$). It is the custom to perform these calculations at first with the help of finger movements. A finger of the left-hand moves across the first row of N; a finger of the right-hand moves down the second column in H. Our attention is thus concentrated on the entries we have to deal with.

Similar considerations apply to the other entries in NH.

In writing out a matrix such as that for NH care should be taken to keep the four expressions involved clearly separated. One should remember that $pa + qc$ represents a single number, and that it is not added or multiplied to any of the other numbers appearing in the scheme. It simply records the coefficient of x in the equation for x^{**}.

It may be helpful to consider a numerical example. If we take $p = 10, q = 3, r = 40$, $s = 2, a = 20, b = 30, c = 1, d = 3$, equations (1) and (2) become

$$\left. \begin{array}{l} x^* = 20x + 30y \\ y^* = x + 3y \end{array} \right\} \quad (1a) \qquad \left. \begin{array}{l} x^{**} = 10x^* + 3y^*, \\ y^{**} = 40x^* + 2y^*. \end{array} \right\} \quad (2a)$$

From these we find

$$\left. \begin{array}{l} x^{**} = 203x + 309y, \\ y^{**} = 802x + 1206y. \end{array} \right\} \qquad (3a)$$

In matrix shorthand this is expressed by

$$\begin{pmatrix} 10 & 3 \\ 40 & 2 \end{pmatrix} \begin{pmatrix} 20 & 30 \\ 1 & 3 \end{pmatrix} = \begin{pmatrix} 203 & 309 \\ 802 & 1206 \end{pmatrix}.$$

The numbers used here have been chosen so that the algebraic formula behind each can be seen. Thus 1206 corresponds to $rb + sd$ in the general formula. If 1206 is read as 12 hundred and 6, the '12 hundred' corresponds to rb, namely 40×30, and the 6 to sd, namely 2×3. In the other entries similarly the 'hundreds' correspond to one term in the formula, the 'units' to another term.

This comment may seem unnecessarily elementary. However, experience shows that errors arise in students' work, owing to algebraic symbols becoming detached from their correct position, and getting mixed up with symbols in some other column of the matrix.

The columns in a matrix

In §10 we saw that the fate of a basis under a linear mapping determines the fate of the whole space. As we mentioned at the beginning of this section, in the plane an obvious basis consists of the vectors A and B, where A is $(1, 0)$ and B is $(0, 1)$. What happens to these under the transformation specified by

$$\begin{pmatrix} a & b \\ c & d \end{pmatrix}?$$

To answer this question, we go back to the equations indicated by this matrix, namely $x^* = ax + by, y^* = cx + dy$.

To find the effect of the transformation on A, we substitute the co-ordinates of A, that is, $x = 1$, $y = 0$. We find $x^* = a$, $y^* = c$.

To find the effect on B we substitute $x = 0$, $y = 1$ and find $x^* = b$, $y^* = d$.

Thus, for this transformation, $(1, 0) \rightarrow (a, c)$ and $(0, 1) \rightarrow (b, d)$.

If we look at the matrix above, we notice that a and c, the co-ordinates of A^*, occur in the first column, while b and d, the co-ordinates of B^*, occur in the second column.

This is a very useful result. It means that, given a matrix, we can see from it immediately, without any calculation, where it sends the points A and B. We simply read off the columns of the matrix.

For instance, early in this section we showed that a certain transformation T was given by the matrix

$$\begin{pmatrix} 4 & 2 \\ 1 & 3 \end{pmatrix}.$$

Reading off the figures in the columns we see that, for T, we have $A^* = (4, 1)$ and $B^* = (2, 3)$. In this particular case, this is not new information, since T was originally defined by specifying A^* and B^*. But the general doctrine is valuable: if we meet any matrix, we can get some idea of its geometrical effect simply by reading off the columns of the matrix.

Conversely, if for some transformation we know A^* and B^*, we can rapidly write down the matrix for this transformation. For instance, suppose we need the matrix for J, rotation through $90°$. J sends A to $(0, 1)$, so in the first column we must write 0 and 1, in that order, starting at the top. J sends B to $(-1, 0)$ so -1 and 0 must be written, in that order, in the second column. Thus we find

$$J = \begin{pmatrix} 0 & -1 \\ 1 & 0 \end{pmatrix}.$$

We assume here that conventional graph paper is being used, with perpendicular axes and a grid composed of squares.

This procedure, while simple, is useful, and to fix it in mind a student should take some transformations known to him, – reflections, rotations, shears, etc. – and write down the matrices for them by this method. In particular, the matrix for rotation through θ, which occurs in many engineering problems, should be found by this method.

Students with some skill in manipulation may find the matrix for reflection in the line $y = mx$ (or the line $y = m \tan A$) by this method.

Vectors as columns

When we are specifying a transformation we may want to show, not only the coefficients a, b, c, d in the equations

$$x^* = ax + by,$$
$$y^* = cx + dy,$$

but also the fact that this transformation has input (x, y) and output (x^*, y^*). Now x and y occur on the same level, while x^* occurs above y^* in these equations. Yet (x, y) and (x^*, y^*) represent the same kind of object; they both specify points (or vectors) in the plane. Should we show them horizontally or vertically?

There are two arguments to show why, in matrix work, it is convenient to write vectors as columns $\begin{pmatrix} x \\ y \end{pmatrix}$ rather than as rows (x, y).

The first argument comes from our paragraphs above on the columns of a matrix. These columns show the co-ordinates of A^* and B^*. If we form the habit of writing vectors as columns, it will be easy for us to split the matrix $\begin{pmatrix} a & b \\ c & d \end{pmatrix}$ and read off

$$A^* = \begin{pmatrix} a \\ c \end{pmatrix}, \qquad B^* = \begin{pmatrix} b \\ d \end{pmatrix}.$$

If we do not form this habit, we are liable to make the mistake of reading off the rows (a, b) and (c, d) which of course do *not* tell us where A^* and B^* are.

A second reason comes from the mechanical procedure for matrix multiplication. We saw by studying equation (4) that the entries in the matrix product involved numbers from a row in the first matrix and a column in the second:

$$(\rightarrow) \ (\downarrow).$$

If we write a vector as a column we can use the same mechanical procedure to work out Mv, a *matrix* acting on a *vector*. Thus we find

$$\begin{pmatrix} a & b \\ c & d \end{pmatrix} \begin{pmatrix} x \\ y \end{pmatrix} = \begin{pmatrix} ax + by \\ cx + dy \end{pmatrix} = \begin{pmatrix} x^* \\ y^* \end{pmatrix}.$$

The product has 2 rows, because the first factor, the matrix, has 2 rows; it has 1 column because the second factor, the vector, has 1 column. It will be seen that both the input and the output vectors now appear as columns.

It will be found that writing vectors as columns proves a very convenient convention for matrix work. It is unfortunate that in co-ordinate geometry, invented long before matrices, we use the arrangement (x, y) rather than the column. Some effort is needed if one is not to become confused.

Exercises

1   Write in the usual notation of elementary algebra the equations expressed by

$$\begin{pmatrix} x^* \\ y^* \end{pmatrix} = \begin{pmatrix} 2 & 5 \\ 8 & 3 \end{pmatrix} \begin{pmatrix} x \\ y \end{pmatrix}.$$

2   Write in matrix notation (as in question 1)

$$x^* = 4x + 9y,$$
$$y^* = 16x + 25y.$$

3 Vectors u and v are specified by $\acute{u} = \begin{pmatrix} 1 \\ 3 \end{pmatrix}$, $v = \begin{pmatrix} 20 \\ 10 \end{pmatrix}$. Write, in column form, the following vectors:

(a) $u + v$ (b) $2u$ (c) $3v$ (d) $2u + 3v$
(e) ku (f) mv (g) $ku + mv$ (h) $u - v$.

4 Work out the following matrix products.

(a) $\begin{pmatrix} 2 & 1 \\ 5 & 3 \end{pmatrix} \begin{pmatrix} 3 & -1 \\ -5 & 2 \end{pmatrix}$ (b) $\begin{pmatrix} 1 & 2 \\ 2 & 4 \end{pmatrix} \begin{pmatrix} 2 & -4 \\ -1 & 2 \end{pmatrix}$

(c) $\begin{pmatrix} \cos\theta & -\sin\theta \\ \sin\theta & \cos\theta \end{pmatrix} \begin{pmatrix} \cos\theta & \sin\theta \\ -\sin\theta & \cos\theta \end{pmatrix}$ (d) $\begin{pmatrix} 1 & 1 \\ 0 & 1 \end{pmatrix} \begin{pmatrix} a & b \\ c & d \end{pmatrix}$

(e) $\begin{pmatrix} a & b \\ c & d \end{pmatrix} \begin{pmatrix} 1 & 1 \\ 0 & 1 \end{pmatrix}$ (f) $\begin{pmatrix} a & b \\ c & d \end{pmatrix} \begin{pmatrix} d & -b \\ -c & a \end{pmatrix}$

(g) $\begin{pmatrix} 1 & 2 \\ 3 & 4 \end{pmatrix} \begin{pmatrix} a & b \\ c & d \end{pmatrix}$ (h) $\begin{pmatrix} a & b \\ c & d \end{pmatrix} \begin{pmatrix} 1 & 2 \\ 3 & 4 \end{pmatrix}$

Optional puzzle; when are the answers to (g) and (h) equal?

5 Work out the squares of the following matrices.

(a) $\begin{pmatrix} 0 & 1 \\ 1 & 0 \end{pmatrix}$ (b) $\begin{pmatrix} 2 & 0 \\ 0 & 3 \end{pmatrix}$ (c) $\begin{pmatrix} 0 & -1 \\ 1 & 0 \end{pmatrix}$

(d) $\begin{pmatrix} 1 & 0 \\ 0 & -1 \end{pmatrix}$ (e) $\begin{pmatrix} 1 & -2 \\ 0 & -1 \end{pmatrix}$ (f) $\begin{pmatrix} 1 & b \\ c & -1 \end{pmatrix}$

(g) $\begin{pmatrix} 1 & 1 \\ -1 & -1 \end{pmatrix}$ (h) $\begin{pmatrix} 0 & 1 \\ 0 & 0 \end{pmatrix}$ (i) $\begin{pmatrix} 0 & 2 \\ 0 & 0 \end{pmatrix}$

6 The transformation I leaves every point where it was; the transformation 0 maps every point to the origin. Write the matrices for I and 0.

7 Write the matrices for M_1, M_2 and J, the transformations involved in the exercises at the end of §20. Use matrix multiplication to find the products in (a) to (i) of question 1, and all parts of question 2 in those exercises.

8 *An investigation.* In elementary algebra, the equation $x^2 = 1$ has only 2 solutions, and $x^2 = 0$ only 1 solution. The calculations in question 5 above suggest that the matrix equations $M^2 = I$ and $M^2 = 0$ may behave very differently. Find all 2×2 matrices that satisfy $M^2 = I$; try to interpret your results geometrically. Do the same for $M^2 = 0$.

Worked example

Find the matrix that represents a rotation of $45°$ about the origin, O. It is understood that the co-ordinates refer to conventional squared graph paper.

Solution The columns of the required matrix represent the points to which $(1, 0)$ and $(0, 1)$ are mapped by this rotation. A rotation of $45°$ sends $(1, 0)$ to $(1/\sqrt{2}, 1/\sqrt{2})$ and $(0, 1)$ to $(-1/\sqrt{2}, 1/\sqrt{2})$. We write this information, using column vectors:

$$\begin{pmatrix} 1 \\ 0 \end{pmatrix} \rightarrow \begin{pmatrix} 1/\sqrt{2} \\ 1/\sqrt{2} \end{pmatrix} \qquad \begin{pmatrix} 0 \\ 1 \end{pmatrix} \rightarrow \begin{pmatrix} -1/\sqrt{2} \\ 1/\sqrt{2} \end{pmatrix}.$$

The required matrix is now obtained by putting the two output columns side by side:

$$\begin{pmatrix} 1/\sqrt{2} & -1/\sqrt{2} \\ 1/\sqrt{2} & 1/\sqrt{2} \end{pmatrix}.$$

9 By the method of the worked example, or otherwise, find matrices that represent the following transformations of the plane.

(a) reflection in the horizontal axis

(b) reflection in the vertical axis

(c) reflection in the line $y = x$

(d) rotation of 180° about O

(e) rotation of 90° about O

(f) rotation of −90° about O

(g) rotation of −45° about O

(h) rotation of 30° about O

(i) rotation through any given angle α about O

(j) reflection in the line $y = x \tan \alpha$

Addition and subtraction of matrices

In §20 and §21 we saw that the sum and difference of two transformations, say H and N, were defined in such a way that for any vector v we would have the equations,

$$(H + N)v = Hv + Nv \quad \text{and} \quad (H - N)v = Hv - Nv.$$

Suppose we have the matrix specification of H and N, namely

$$H = \begin{pmatrix} a & b \\ c & d \end{pmatrix}, \qquad M = \begin{pmatrix} p & q \\ r & s \end{pmatrix}.$$

The statements in the first paragraph show us the procedure by which we can find the matrix for $H + N$. Here again, it is good if the student can derive this result for himself, by following the strategy now to be indicated.

First, we suppose the vector v to have arbitrary co-ordinates x and y. The student may work with the traditional co-ordinate form (x, y) or with the column vector form $\begin{pmatrix} x \\ y \end{pmatrix}$. If he can work with the latter, he will be forming a useful habit; if, however, he finds this unfamiliar format inhibits his thinking, he may be wise to use the more familiar arrangement.

In order to find $H + N$, we have to answer the following questions:

(1) What is the vector Hv in terms of x and y?

(2) What is Nv?

(3) What is $Hv + Nv$, found by adding our answers to (1) and (2)?

By definition, our answer to (3) gives $(H + N)v$. This answer involves x and y. To write the matrix for $H + N$ we have to pick out the coefficients in this answer, and write them in appropriate positions, as explained earlier in this section, when matrix notation was first introduced.

No ingenuity is required to carry this programme through. The work is left to the student; it is essential that the two exercises below should be worked, as the results to which they lead are essential for all later work.

Exercises

1 Carry out the instructions above for finding the matrix $H + N$.

2 By a similar procedure, find the matrix $H - N$.

Multiplying a matrix by a number

In §20, we defined kT, the product of a number k and a transformation T, by the equation $(kT)v = k(Tv)$. This told us what the transformation kT did to any vector v.

We now wish to express this definition in matrix notation.

Let us consider the transformation H, specified by the matrix $\begin{pmatrix} a & b \\ c & d \end{pmatrix}$. What matrix will specify kH ?

By definition, $(kH)v = k(Hv)$. We will suppose v to be $\begin{pmatrix} x \\ y \end{pmatrix}$, and work out the expression $k(Hv)$ on the right-hand side of the equation:

$$Hv = \begin{pmatrix} a & b \\ c & d \end{pmatrix} \begin{pmatrix} x \\ y \end{pmatrix} = \begin{pmatrix} ax + by \\ cx + dy \end{pmatrix}.$$

The last expression represents a vector in the plane. It has two numbers in it, $ax + by$ and $cx + dy$, which are the two co-ordinates of the vector Hv.

To obtain $k(Hv)$ we must multiply the vector Hv by the number k. This means that each co-ordinate must be multiplied by k. Accordingly

$$k(Hv) = \begin{pmatrix} kax + kby \\ kcx + kdy \end{pmatrix}.$$

This last vector gives $(kH)v$, the output of the transformation kH when the input is v. Let x^* and y^* be the co-ordinates of this output vector. Then the effect of the transformation kH is shown by the equations,

$$x^* = kax + kby,$$
$$y^* = kcx + kdy.$$

To specify the transformation kH in vector form we need only to pick out the coefficients in these equations. We find

$$kH = \begin{pmatrix} ka & kb \\ kc & kd \end{pmatrix}.$$

Thus the rule for passing from the matrix H to the matrix kH is very simple – multiply each entry by k.

Summary

We now have routines by which matrices may be added, subtracted and multiplied. Division by a matrix is not always possible; this question will be discussed later.

An electronic computer is capable of dealing with a calculation that has been reduced to routine. In this section we have written only 2×2 matrices – matrices with 2 rows and 2 columns. For these the input vector, v, is specified by 2 numbers, and the output vector v^* also by 2 numbers. One convenience of matrix notation is that we can use the same symbolism $v^* = Mv$ when the input and the output each consist of 1000 numbers.

The routines for $n \times n$ matrices, where n is any natural number, are essentially the same as for 2×2 matrices. Thus we can instruct a computer to carry out a certain sequence of operations, and these instructions will apply to a whole class of problems. In one problem there may be 50 unknowns, in another 500. Of course, in each problem it will be necessary to feed into the computer data for that particular problem, so the computer knows how many unknowns are involved and what transformations have to be applied to these. But the general procedure, the strategy to be followed, can be laid down once and for all, and matrix notation gives a very convenient way of doing this.

In this section we have introduced matrix notation by considering transformations. Transformations map a space to itself; if the space is of n dimensions, the input and output each involve n numbers. This is in many ways the most fruitful and interesting case. However, other mappings can be of engineering importance. We have met examples in which the input consisted of 3 forces and the output of 2 reactions, or the input of 2 voltages and the output 3 current measurements. Such mappings lead to rectangular rather than square matrices, and we do not then have the same freedom to add, subtract and multiply as in this section. However, when such operations are possible, the formalisms are very similar to those just considered, and we shall come to them in due course.

Exercises on addition, subtraction and multiplication of matrices

1   Find $\begin{pmatrix} 1 & 2 \\ 3 & 4 \end{pmatrix} + \begin{pmatrix} 80 & 60 \\ 70 & 50 \end{pmatrix}$.

2   Find $\begin{pmatrix} 10 & 20 \\ 30 & 40 \end{pmatrix} - \begin{pmatrix} 5 & 6 \\ 7 & 8 \end{pmatrix}$.

Questions 1 and 2 serve simply to check that the student has understood the procedure for addition and subtraction of matrices. If further routine examples of this kind are needed, they can be produced simply by varying the numbers involved.

For questions 3 to 11, the following symbols are used:

$$A = \begin{pmatrix} 1 & 1 \\ 0 & 1 \end{pmatrix}, \quad B = \begin{pmatrix} 1 & 0 \\ 1 & 1 \end{pmatrix}, \quad C = \begin{pmatrix} 0 & 1 \\ 1 & 0 \end{pmatrix}, \quad I = \begin{pmatrix} 1 & 0 \\ 0 & 1 \end{pmatrix}.$$

3   Find $A + B - C$.

4   Find $A^2 - 2A + I$ and $(A - I)^2$. Are they equal?

5   Find $(A + B)(A - B)$; $(A - B)(A + B)$; $A^2 - B^2$. Are any two of these equal?

6   Find $B^2 - 2B + I$ and $(B - I)^2$. Are they equal?

7   If, for some matrix M we find $M^2 = 0$, does it follow that $M = 0$? Justify your answer.

8   Are $(A + B)(I + C)$ and $(I + C)(A + B)$ equal?

9   Find $(I + C)(I - C)$.

10 Find $AC - B$ and C^2.

11 Elementary algebra suggests $(AC - B)C = AC^2 - BC$. With the help of your answers to question 10 check whether this equation is correct or not.

12 Let $D = \begin{pmatrix} 2 & 1 \\ 3 & 2 \end{pmatrix}$. Write $2D, 3D, 4D, 5D$.

13 Is there any number k for which $D^2 + I = kD$?

14 Let $E = \begin{pmatrix} 1 & 2 \\ 3 & 4 \end{pmatrix}$. Calculate $E^2 - 5E - 2I$.

15 Let $F = \begin{pmatrix} 0 & 1 \\ 2 & 3 \end{pmatrix}$. Is it possible to find numbers p, q, for which $F^2 - pF + qI = 0$?

16 We have seen that some 2×2 matrices satisfy quadratic equations. Do they all do so? Given the matrix $M = \begin{pmatrix} a & b \\ c & d \end{pmatrix}$, can we be sure there will exist numbers p and q such that

$$M^2 - pM + qI = 0?$$

23   An application of matrix multiplication

The remarks at the end of §22 about the engineering applications of matrix algebra can be illustrated by an electrical example that does not call for any specialized knowledge.

We consider boxes, each having 4 terminals. On the input side, we suppose a current i flows in at one terminal and out at the other. Similarly on the output side a current I flows as shown in Fig. 63. Let v be the voltage difference between the input terminals,

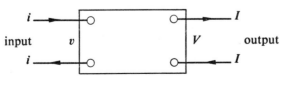

Fig. 63

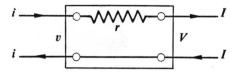

Fig. 64

V between the output terminals. The vector $w = \begin{pmatrix} v \\ i \end{pmatrix}$ measures the input, while $W = \begin{pmatrix} V \\ I \end{pmatrix}$ measures the output. In the cases we shall consider $w \to W$ will be a linear transformation.

We now consider certain special boxes, which ultimately will be put together to form more complicated circuits. Box (1) is very simple. It contains a resistance of r ohms, as shown in Fig. 64.

Clearly $I = i$ since the current has no chance to do anything but flow straight through. The current i flowing through resistance r produces a potential drop ri, so $V = v - ri$. Thus

$$V = v - ri,$$
$$I = i.$$

Picking out the coefficients of v and i, we see that the output of this box is related to the input by the equation $W = Aw$, where

$$A = \begin{pmatrix} 1 & -r \\ 0 & 1 \end{pmatrix}.$$

Box (2) contains a resistance R connected as shown in Fig. 65. Since the input and output terminals are directly connected, there can be no change of voltage. We have $V = v$. However, by Ohm's Law, a current v/R flows down the resistance, and there will be this much less current to pass to the output. So $I = i - (v/R)$.

Accordingly, being careful to write the equations and symbols in the correct order, we obtain the equations

$$V = v,$$
$$I = -(1/R)v + i.$$

So, for this box, $W = Bw$ where

$$B = \begin{pmatrix} 1 & 0 \\ -1/R & 1 \end{pmatrix}.$$

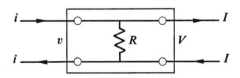

Fig. 65

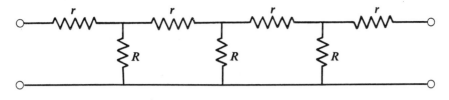

Fig. 66

Now consider two stations connected by a cable which is imperfectly insulated. After all, no insulation is perfect. We might approximate the behaviour of such a cable by a circuit such as that shown in Fig. 66.

The current flows along the cable through the resistances r, but leaks away to earth through the resistances R. Now this circuit can be constructed by connecting a number of boxes (1) and (2), as in Fig. 67.

Let w_1 represent the input to the first box; w_2, the output of that box, is also the input of the second box; w_3 the output of the second box is the input of the third box, and so we continue until we reach the final output w_8. We have $w_2 = Aw_1$, $w_3 = Bw_2$, $w_4 = Aw_3$, $w_5 = Bw_4$, $w_6 = Aw_5$, $w_7 = Bw_6$, $w_8 = Aw_7$.

It follows that $w_8 = ABABABAw_1$. Thus the effect of such a leaky cable can be found by a routine process of matrix multiplication. It is clear that with the help of an electronic computer we could find the effect of such a circuit with thousands of boxes so connected.

Alternating current circuits involving resistance, capacity and inductance can be dealt with in a somewhat similar way, by using matrices in which complex numbers occur.

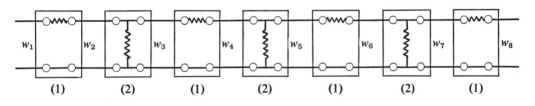

Fig. 67

Exercises

1 Calculate the matrix $ABABABA$ for the case $r = 1$, $R = 1$.

2 The circuit shown is symmetrical, so that interchanging input and output should not make any difference. Check that this is so for the matrix you found in question 1. (*Note.* The condition that a matrix represents a symmetrical circuit needs to be formulated with some care, as there is a small point that is easily overlooked. It may help to consider a simpler matrix such as ABA or even A; if the circuits corresponding to these are drawn it will be seen that they too are symmetrical.)

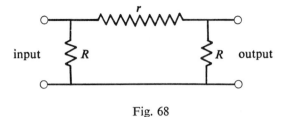

Fig. 68

3   Find the matrix corresponding to the circuit shown in Fig. 68.

4   Draw the circuits corresponding to the matrices AB and BA. Are they electrically equivalent?

24   An application of linearity

In §23, a fairly complex circuit was built up from very simple components and its behaviour was deduced by a routine procedure. The processes to be described in this section differ greatly from those used in §23, but they have the same general purpose – to cope with a complex situation by relating it to simple ones.

In this section, we will use certain vibration problems as illustrations, but it should be realized that the methods are perfectly general; they can be used in many situations that have nothing to do with vibration.

Our present considerations are built on the ideas of two earlier sections. In §10 we met the idea that the fate of any vector was determined once the fate of a basis had been ascertained. In §12 we saw that a basis could be chosen in many ways and that a suitable choice of basis could lead to a great simplification.

The simplest vibration problem of all is that of a mass attached to a spring. Let x, v, a denote the displacement, velocity and acceleration of the mass. For simplicity we suppose we have unit mass and that the constant for the spring is unity. Then $a = -x$. We know $v = dx/dt$ and $a = dv/dt$, since d/dt means 'the rate of change of'. It is very convenient to represent differentiation by a dot, so dx/dt is written $\dot{x}$ and d^2x/dt^2 becomes $\ddot{x}$. We then may write $v = \dot{x}$, $a = \dot{v} = \ddot{x}$ and $a = -x$ takes the form $\ddot{x} = -x$.

If we consider the graph drawn by a vibrating body on a surface that moves steadily past it, we are immediately reminded of the graphs of sine or cosine. If we try $x = \sin t$, on differentiating twice we find $\dot{x} = \cos t$ and $\ddot{x} = -\sin t$, which means $\ddot{x} = -x$ for every time t. Thus $x = \sin t$ is a possible motion for our mass on a spring. Similarly, we can check that $x = \cos t$ is a possible motion.

Now if we have a mass on the end of the spring, we can displace it any small distance we like and send it off with a speed chosen by ourselves. Once that is done, the process develops automatically. At every moment, the amount the spring is stretched determines the acceleration of the mass. So the initial position and initial velocity determine the subsequent motion.

The initial position and initial velocity are specified by a pair of numbers. This is our input. The formula for the subsequent motion contains a function of time, t. This is our output.

For the motion $x = \sin t$ we have, when $t = 0$, $x = 0$ and $v = 1$. Our input is $(0, 1)$. We may write $(0, 1) \rightarrow \sin t$.

For the motion $x = \cos t$, when $t = 0$ we have $x = 1$, $v = 0$. Our input here is $(1, 0)$ and so $(1, 0) \rightarrow \cos t$.

It can be shown that the mapping, input $\rightarrow$ output, is linear. As $(1, 0)$ and $(0, 1)$ form a basis for the input, it follows from the doctrine of §10 that the output for any input is now known.

By proportionality, since $(1, 0) \rightarrow \cos t$, multiplying by a constant A, we find $(A, 0) \rightarrow A \cos t$. (The A here is simply a number. It has nothing to do with the matrix A in §23.)

Similarly, since $(0, 1) \rightarrow \sin t$, multiplying by a constant B, we find

$$(0, B) \rightarrow B \sin t.$$

Superposing these two results, we find that $(A, B) \rightarrow A \cos t + B \sin t$, so now we know that $x = A \cos t + B \sin t$ if the mass starts from position A with velocity B. This covers every possible motion.

In a calculus course, $x = A \cos t + B \sin t$ is shown to be the most general solution of the differential equation $\ddot{x} + x = 0$. The language may differ somewhat from that used here, but the essential content is the same.

We will now consider an arrangement, as shown in Fig. 69, with two unit masses and three springs. The central spring is stronger than the others; it has constant 4 while they each have constant 1. The ends are fixed and the arrangement is symmetrical. If we imagine the masses displaced distances x and y, as shown, with $y > x$, the central spring will be in tension and exerting forces of magnitude $4(y - x)$ at each end. The left spring will exert a tension of x, and the right-hand spring will be compressed and exerting a force y. The forces on the masses will thus be $-x + 4(y - x)$ and $-4(y - x) - y$ respectively. Thus the equations of motion (found from force = mass × acceleration) will be $\ddot{x} = -5x + 4y$ and $\ddot{y} = 4x - 5y$.

Now we can start the system moving by giving an arbitrary displacement and velocity to each of the particles, so we have 4 constants at our disposal. We do not want to work

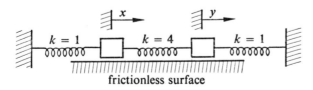

frictionless surface

Fig. 69

in 4 dimensions, so let us agree that both masses start from the equilibrium position. From that position we give each mass a bang to start it moving. This reduces the number of constants to 2; we can choose an initial velocity for each mass. Thus by an input of, say, (3, 8), we understand that we have given the left-hand mass an initial velocity of 3 units, and the right-hand mass an initial velocity of 8 units, both in the positive direction.

Our first example might suggest that we try to see the motions that result from inputs (1, 0) and (0, 1). But it is not easy to see what happens in these cases. Input (1, 0) for instance means that we leave the right-hand mass alone, but start the left-hand mass moving towards the right with velocity 1. The central spring will become compressed; it will bring the right-hand mass into motion, and it is not simple to predict how the later movements will develop.

We are dealing with a symmetrical system. This suggests that, instead of considering unsymmetrical inputs like (1, 0) and (0, 1), we might do better to put the masses on an equal footing and consider the input (1, 1). Input (1, 1) means that both masses begin moving to the right with velocity 1. Since they have the same velocity, at the beginning of the motion the distance between them is not changing, and the middle spring is being neither stretched nor compressed. This may suggest to us the question – can this state of affairs continue? Can the masses continue to move exactly in step, so the middle spring exerts no force? It is fairly easy to see that this can happen. Imagine the middle spring removed. The two end springs have the same stiffness constants, so the masses can perform identical oscillations. If they are sent off, each with unit velocity, we shall have after time t the equations $x = \sin t$, $y = \sin t$, as we saw earlier. If, now we replace the central spring, it will make no difference, since $y - x = 0$ at all times. This spring will exert no force. (We are assuming the masses of the springs to be negligible, so replacing the middle spring does not affect the inertia of the system.)

Accordingly we have identified a possible motion:

$$\text{input } (1, 1) \to \text{output } (\sin t, \sin t). \tag{1}$$

The two entries in the output are the formula for x and the formula for y, at time t.

A rather more formal way of seeing that this motion is possible is the following. We pose the question, is it possible to have a motion in which at all times $x = y$? If we substitute $x = y$ in the equations of motion $\ddot{x} = -5x + 4y$, $\ddot{y} = 4x - 5y$ *both* equations lead to $\ddot{x} = -x$. Thus $x = y = \sin t$ satisfies all requirements.

There is another symmetrical motion. Our system has mirror symmetry about its centre. If we started the masses off with unit velocity towards the centre, we would expect this kind of symmetry to continue. The system would appear balanced; the displacement of one mass to the right would equal the displacement of the other to the left.

This leads us to ask – is there a motion in which at all times $y = -x$? If we substitute $-x$ for y in the equations of motion we obtain the equations $\ddot{x} = -9x$ and $-\ddot{x} = 9x$. The second of these is equivalent to the first. Thus all conditions will be met if $y = -x$ and $\ddot{x} = -9x$. Now $\ddot{x} = -9x$ has the solution $x = \sin 3t$. Differentiating gives

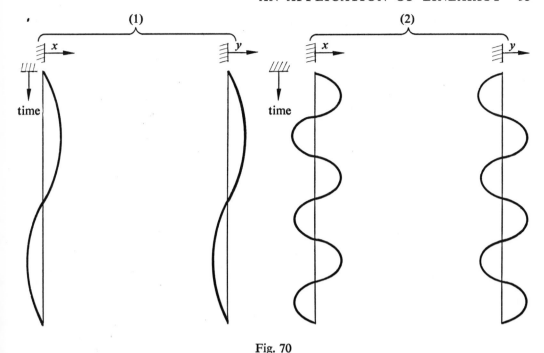

Fig. 70

$\dot{x} = 3 \cos 3t$, so when $t = 0$, $\dot{x}$, the velocity of the first mass, has the value 3. Since $y = -x$, we know $y = -\sin 3t$ at all times, and the initial velocity of the second mass is -3.

Thus

$$(3, -3) \rightarrow (\sin 3t, -\sin 3t). \qquad (2)$$

If the vibrating masses were made to record their positions on a moving strip, the results would be as in Fig. 70, in which (1) shows the motion $(\sin t, \sin t)$ and (2) shows $(\sin 3t, -\sin 3t)$, in which the vibration is 3 times as rapid as in (1). Both motions are easy to imagine.

It can be proved that for this system the mapping input $\rightarrow$ output is linear. Accordingly we may use the principles of superposition and proportionality. Since

$$(1, 1) \rightarrow (\sin t, \sin t),$$

for any constant number A, $(A, A) \rightarrow (A \sin t, A \sin t)$. Similarly, for any constant number B, $(3B, -3B) \rightarrow (B \sin 3t, -B \sin 3t)$. Superposing we see

$$(A + 3B, A - 3B) \rightarrow (A \sin t + B \sin 3t, A \sin t - B \sin 3t).$$

Thus, for any numbers A and B, a possible vibration is given by the equations

$$x = A \sin t + B \sin 3t, \, y = A \sin t - B \sin 3t.$$

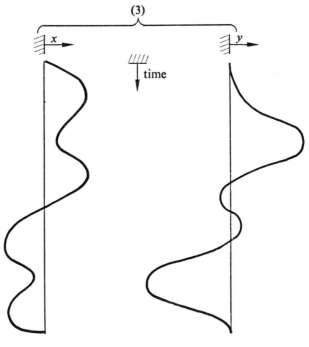

Fig. 71

For example, we may take $A = 2, B = 1$. This gives

$$x = 2 \sin t + \sin 3t, y = 2 \sin t - \sin 3t. \tag{3}$$

The corresponding motion is indicated by the graphs in Fig. 71. This motion is anything but simple to imagine. The two simple vibrations with different periods have combined to produce an unsymmetrical and rather irregular motion, which would still, however, be periodic. The graphs above would repeat indefinitely, at any rate in the idealized situation where frictional damping is ignored.

If the time scale were suitably chosen, these vibrations would produce audible

sounds. If vibration 1 gave ♭𝄢 vibration 2 would give 𝄞

A person with a good musical ear, when exposed to vibration 3, would hear both the notes, bass C and treble G. It is interesting that our ears (or brains) spontaneously analyse the complex vibration 3 into the two simple vibrations 1 and 2, of which it is composed.

The simple vibrations, in which only one frequency or musical note occurs, are known as the *normal modes* of vibration. The simple system of masses and springs that we have considered is not exceptional. In a subject known as general dynamics a theory of vibrations is developed which shows that similar features are to be expected in the vibrations of any structure.

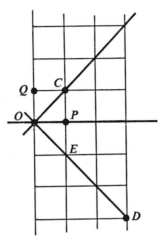

Fig. 72

Inputs, such as $(1, 0)$, $(0, 1)$, $(1, 1)$, $(3, -3)$, involve only 2 numbers each and can be plotted on graph paper. In Fig. 72 these four cases are represented by the points P, Q, C and D. As we saw, the basis A, B was not helpful in our problem, but the vectors C and D, which also form a basis, gave us a simple treatment. In this particular problem we were led to consider C and D by considerations of symmetry. Now symmetry is an important idea in mathematical work but we cannot expect always to find it. For instance, in our vibrating system if the masses had differed in magnitude by even 1 per cent, symmetry would have been destroyed. Yet surely physically this would have made very little difference to the system; surely it would still produce two notes, and by setting it suitably in motion we could make it produce just one of these notes. Can we then find some way of showing the significance of the vectors C and D without appealing to symmetry? If we look at our equations of motion

$$\ddot{x} = -5x + 4y$$
$$\ddot{y} = 4x - 5y$$

we see that they embody the matrix

$$\begin{pmatrix} -5 & 4 \\ 4 & -5 \end{pmatrix} = M, \quad \text{say.}$$

Are the vectors C and D specially related to this matrix? Let us see what it does to them. If we let this transformation act on C we get

$$\begin{pmatrix} -5 & 4 \\ 4 & -5 \end{pmatrix} \begin{pmatrix} 1 \\ 1 \end{pmatrix} = \begin{pmatrix} -1 \\ -1 \end{pmatrix}.$$

When it acts on D we get

$$\begin{pmatrix} -5 & 4 \\ 4 & -5 \end{pmatrix} \begin{pmatrix} 3 \\ -3 \end{pmatrix} = \begin{pmatrix} -27 \\ 27 \end{pmatrix}.$$

We notice that C^* is a multiple of C, and D^* is a multiple of D. In fact $MC = -C$ and $MD = -9D$. Thus C and D are eigenvectors and the lines OC and OD are invariant lines.

In §12 we saw that simplification could often be achieved by choosing invariant lines as axes, and in §9 we had a simple procedure for changing axes. Accordingly let us see what happens to our equations of motion if we take OC and OD as axes. As base vectors we can choose any points in the invariant lines (one point in each line, of course). In the line OD it will be simpler to choose E, with co-ordinates $(1, -1)$ rather than D with co-ordinates $(3, -3)$. Taking C and E as basis, the point with co-ordinates (X, Y) is given by the vector $XC + YE$. Now $C = P + Q$ and $E = P - Q$ so

$$XC + YE = X(P + Q) + Y(P - Q) = (X + Y)P + (X - Y)Q.$$

We are thus led to put $x = X + Y, y = X - Y$. Substituting these values in the equations of motion we find $\ddot{X} + \ddot{Y} = -X - 9Y$ and $\ddot{X} - \ddot{Y} = -X + 9Y$. Solving for $\ddot{X}$ and $\ddot{Y}$ we obtain $\ddot{X} = -X, \ddot{Y} = -9Y$.

At the beginning we had equations such as $\ddot{x} = -5x + 4y$ in which the x and y were mixed up together, and a student might well not know what to do. By going over to the invariant lines as axes, we obtain the equation $\ddot{X} = -X$ in which X alone appears, and which we recognize as being just like the equation for a single mass on the end of a spring. Similarly, we obtain the equation $\ddot{Y} = -9Y$, another simple vibration equation, involving Y alone.

It is thus suggested that, if *we choose the invariant lines as axes, the problem will split into simpler problems.*

We can see why this should be so. If (x, y) are the co-ordinates of a moving point, then $(\ddot{x}, \ddot{y})$ are the co-ordinates of the vector that represents its acceleration. The point (x, y) is of course only a graphical device to show the displacements of the masses on the spring. As the system vibrates, the point (x, y) will move around in the plane. The vector $(\ddot{x}, \ddot{y})$ shows how this point is accelerated at any time.

Fig. 73 shows the direction of the acceleration for various positions of the point (x, y). It will be seen that if the point (x, y) starts on one of the invariant lines, its acceleration will be such as to keep it in that line. If, however, the point starts not on either of these lines, the acceleration will cause it to move on some curve, the nature of which we cannot readily predict. Students may be familiar with the Lissajous curves that can be produced by combining simple oscillations. In general the path followed by the moving point (x, y) will be such a curve.

What was said earlier in this section may now be repeated. The method used here is not restricted to vibration problems; it would produce a simplification in any problem where the matrix $\begin{pmatrix} -5 & 4 \\ 4 & -5 \end{pmatrix}$, that we have been considering, occurred. And of course it would be equally useful for many other matrices.

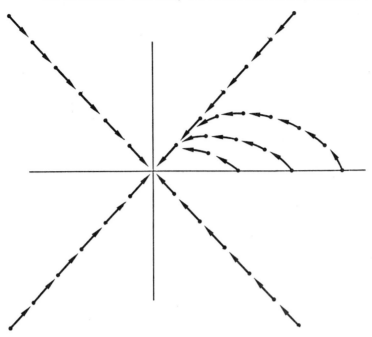

Fig. 73   In this diagram the dot represents a position (x, y). The arrow gives the direction of the acceleration $(\ddot{x}, \ddot{y})$ corresponding to that position.

Having seen that problems are simplified when we change our axes to the invariant lines, the question naturally arises – how do we find the invariant lines? This question will be considered in the next section.

25   Procedure for finding invariant lines, eigenvectors and eigenvalues

The procedure for finding invariant lines, eigenvectors and eigenvalues falls into two stages. In Stage 1, the eigenvalues are found from a certain equation. In Stage 2 the eigenvectors belonging to each eigenvalue are found; this immediately gives the invariant lines.

For understanding the procedure, it is best to work backwards. We will suppose that, somehow, Stage 1 has been completed. Someone, say, has told us what the eigenvalues are. How would we then find the eigenvectors?

Suppose, for instance, we are dealing with the matrix $\begin{pmatrix} 1 & 2 \\ -1 & 4 \end{pmatrix}$, and we have received the information that the eigenvalues are 2 and 3. This means that there is one vector which gets doubled when the transformation acts on it; there is another vector that gets tripled. Our task is to find these vectors. It is not at all difficult.

Which vector gets doubled? Suppose its co-ordinates are x and y. It gets doubled if

$$\begin{pmatrix} 1 & 2 \\ -1 & 4 \end{pmatrix} \begin{pmatrix} x \\ y \end{pmatrix} = 2\begin{pmatrix} x \\ y \end{pmatrix}. \tag{1}$$

We interpret this matrix equation in the language of elementary algebra. This gives the simultaneous equations

$$x + 2y = 2x,$$
$$-x + 4y = 2y.$$

If we simplify these equations, we find that each of them boils down to the same equation $2y = x$. Students are sometimes worried by this; they ask, 'How can I find two unknowns from one equation?' In fact something would be seriously wrong if they could find a definite solution for x and y. We saw in §12 that, if v is an eigenvector, so is kv for any number $k \neq 0$. Any point on an invariant line (except the origin, which is not acceptable for this purpose) gives an eigenvector. There must therefore be an infinity of solutions; the equations cannot possibly fix x and y.

In our example, the algebra leads us to the single equation $2y = x$. This is the only condition the vector has to meet. It does not matter what y is (barring always $y = 0$) provided x is twice as large. You can check that equation (1) is satisfied if $x = 2$, $y = 1$ or if $x = 100$, $y = 50$ or if $x = -0.3$, $y = -0.15$. We can if we like give the general solution, $\begin{pmatrix} 2t \\ t \end{pmatrix}$ where t can be any number except 0, or we may pick out a particular solution $\begin{pmatrix} 2 \\ 1 \end{pmatrix}$ on the understanding that any non-zero multiple of this will do equally well (see the paragraph in §12 immediately before the questions at the end of that section).

As was mentioned in §12, the symbol λ is widely used for eigenvalues. Thus our result may be recorded as $\begin{pmatrix} 2 \\ 1 \end{pmatrix}$, $\lambda = 2$. This means that the vector $\begin{pmatrix} 2 \\ 1 \end{pmatrix}$ gets multiplied by 2 when the transformation acts on it.

We were also told that 3 was an eigenvalue. So, searching for a vector that gets tripled, we write the equation

$$\begin{pmatrix} 1 & 2 \\ -1 & 4 \end{pmatrix} \begin{pmatrix} x \\ y \end{pmatrix} = 3\begin{pmatrix} x \\ y \end{pmatrix}. \tag{2}$$

This gives us $x + 2y = 3x$ and $-x + 4y = 3y$, and these are equivalent to the single condition $x = y$. Thus any vector $\begin{pmatrix} t \\ t \end{pmatrix}$ is tripled by the transformation, and has $\lambda = 3$.

The invariant lines are $2y = x$ for $\lambda = 2$, and $y = x$ for $\lambda = 3$.

If we introduce new co-ordinates (X, Y) with the invariant lines as axes the transformation will take the form $(X, Y) \rightarrow (X^*, Y^*)$ with $X^* = 2X$, $Y^* = 3Y$, since the vectors in the first line are doubled and in the second line are tripled. The matrix expressing the transformation relative to the new axes is thus (by reading off the coefficients in the two equations) $\begin{pmatrix} 2 & 0 \\ 0 & 3 \end{pmatrix}$.

Note that we are still dealing with the same transformation. What happens to each point is exactly the same throughout. But this transformation is expressed by a different matrix when we introduce a new system of axes.

The matrix $\begin{pmatrix} 2 & 0 \\ 0 & 3 \end{pmatrix}$ is said to be in *diagonal form* since the entries not on the main diagonal are zero. Note that the entries in the main diagonal are the eigenvalues. This always happens. The order in which the eigenvalues occur is arbitrary. If we had taken $y = x$ for our first axis and $2y = x$ for our second axis, the transformation would have been specified by the matrix $\begin{pmatrix} 3 & 0 \\ 0 & 2 \end{pmatrix}$.

Exercises

In the following questions, students should check that they have found the correct eigenvectors, by applying the transformation to the vectors they have found and seeing if in fact these get multiplied by the numbers specified.

1 Find the eigenvectors, invariant lines, and matrix resulting when the invariant lines are taken as axes, for each of the following transformations.

(a) $\begin{pmatrix} 2 & -3 \\ 1 & -2 \end{pmatrix}$ which has eigenvalues 1 and -1

(b) $\begin{pmatrix} -4 & 2 \\ -15 & 7 \end{pmatrix}$ with eigenvalues 1 and 2

(c) $\begin{pmatrix} 3 & -1 \\ 0 & 2 \end{pmatrix}$ with eigenvalues 2 and 3

(d) $\begin{pmatrix} 2 & 1 \\ 1 & 2 \end{pmatrix}$ with eigenvalues 1 and 3

(e) $\begin{pmatrix} 2 & 2 \\ 2 & -1 \end{pmatrix}$ with eigenvalues 3 and -2

(f) $\begin{pmatrix} 1 & 1 \\ 1 & 1 \end{pmatrix}$ with eigenvalues 2 and 0

(g) $\begin{pmatrix} 1 & -1 \\ 0 & 0 \end{pmatrix}$ with eigenvalues 0 and 1

2 The matrix $\begin{pmatrix} a & b \\ c & d \end{pmatrix}$ is called *symmetric* when $b = c$, as in questions (d), (e), (f) above. What do you notice about the invariant lines in these questions?

Finding eigenvalues

In practice, of course, we are rarely in the position of being told the eigenvalues of a matrix. Before we can do the type of work in the exercises just given, we have to find out for ourselves what the eigenvalues are.

Equations (1) and (2) at the beginning of this section were formed when we looked for vectors that were multiplied by 2 and by 3 when the transformation $\begin{pmatrix} 1 & 2 \\ -1 & 4 \end{pmatrix}$ acted.

Equation (1) was equivalent to the pair of equations $x + 2y = 2x$, $-x + 4y = 2y$ and these turned out to be equivalent to the single equation $2y = x$. It followed that $x = 2$, $y = 1$ would do as a solution.

What would have happened if our informant had been mistaken and had asked us to search for a vector that got multiplied by 5 when the transformation was applied? We would have written

$$x + 2y = 5x, \\ -x + 4y = 5x. \tag{3}$$

The first equation simplifies to $y = 2x$, the second to $2y = 3x$. The only solution is $x = 0$, $y = 0$. Now under *any* linear transformation, the origin goes to the origin. Knowing this fact does not in any way help us to find new axes in which the transformation will appear simpler. Thus looking for a vector that gets enlarged 5 times does not help us.

If we are looking for a vector that gets multiplied by any number λ, we write the equations

$$x + 2y = \lambda x, \\ -x + 4y = \lambda y,$$

which may be put in the form

$$(1 - \lambda)x + 2y = 0, \\ -x + (4 - \lambda)y = 0. \tag{4}$$

Both equations represent straight lines through the origin. If we choose an unsuitable value for λ, such as $\lambda = 5$, we get two different lines through the origin as happened for equations (3). If we take a suitable value of λ, such as $\lambda = 3$ used in equations (2), we find we get the same line twice. In fact $\lambda = 3$ gives $-2x + 2y = 0$ and $-x + y = 0$. These equations do not appear identical, but their graphs coincide. For $\lambda = 5$ there is only one point that lies on both lines, and so satisfies both equations (3). It is the origin which, as we have seen, is no use to us. For $\lambda = 3$, any point of the line $y = x$ satisfies both equations, and thus qualifies as an eigenvector.

Thus to find a suitable value of λ, we have to arrange for the two lines in equations (4) to coincide.

Let us look at the general question of when two lines through the origin coincide. Suppose we have two lines with equations $px + qy = 0$, $rx + sy = 0$. What is the condition for them to coincide? Students often produce the following solution. The first line has slope $-p/q$, the second $-r/s$. For the lines to coincide these slopes must be equal. After clearing fractions the condition $ps - qr = 0$ is found.

Now this result is in fact perfectly correct. The lines will coincide if $ps - qr = 0$, and will not coincide if $ps - qr \neq 0$. But the manner of proof is open to some objections. We have divided by q and by s; what if q or s or both should happen to be zero?

We can tidy up the argument so as to avoid any question of dividing by zero. We use a procedure that is sometimes used to solve equations. We can get rid of y by taking s times the first equation minus q times the second. This gives us $(ps - qr)x = 0$. Similarly, we can get an equation not involving x if we take p times the second equation minus r times the first. This gives $(ps - qr)y = 0$. Thus, if $ps - qr \neq 0$, we must have $x = 0$ and $y = 0$. The lines in this case intersect only at the origin.

We still have to show that if $ps - qr$ is zero, then the equations have a solution other than $x = 0$, $y = 0$. This is easy. The equation $px + qy = 0$ is automatically satisfied by $x = -q$, $y = p$. These values make $rx + sy$ equal to $ps - qr$, which is given to be zero. So we have a point that satisfies both equations. But there is one last snag. If $p = q = 0$, this solution will not be other than $x = 0$, $y = 0$. However, we can clear this up. If $p = q = 0$, the first equation is $0x + 0y = 0$ and *any* values of x and y will satisfy this. We have only to pick a point, other than the origin, to satisfy the second equation $rx + sy = 0$ and we have done what is required.

A special sign exists for $ps - qr$. It is called the *determinant* and is written $\begin{vmatrix} p & q \\ r & s \end{vmatrix}$

The determinant is a single number – not a vector or a matrix. Thus we have a theorem. *The equations*

$$px + qy = 0,$$
$$rx + sy = 0,$$

have a solution other than the trivial solution $x = 0$, $y = 0$ when, and only when, the determinant $\begin{vmatrix} p & q \\ r & s \end{vmatrix} = 0.$

We now return to our particular problem. If we did not already know the eigenvalues, how would we determine λ to make the equations (4) have a solution other than $x = 0$, $y = 0$? We write the determinant, replacing p, q, r, s by the coefficients that occur in equations (4).

This gives us

$$\begin{vmatrix} 1 - \lambda & 2 \\ -1 & 4 - \lambda \end{vmatrix} = 0$$

which is the same as saying

$$(1 - \lambda)(4 - \lambda) - (-1)(2) = 0.$$

This reduces to $\lambda^2 - 5\lambda + 6 = 0$ with solutions $\lambda = 2$ and $\lambda = 3$.

We have required a fairly long explanation to arrive at this procedure, but once the idea has been understood, the actual calculation is quite short. This constitutes Stage 1, mentioned at the beginning of the section.

Worked example

Find the eigenvalues of $\begin{pmatrix} 1 & 4 \\ 2 & 3 \end{pmatrix}$.

Solution   If λ is an eigenvalue,

$$\begin{pmatrix} 1 & 4 \\ 2 & 3 \end{pmatrix} \begin{pmatrix} x \\ y \end{pmatrix} = \lambda \begin{pmatrix} x \\ y \end{pmatrix}$$

will have a non-trivial solution.

This leads to the equations

$$\left. \begin{array}{c} (1 - \lambda)x + ry = 0, \\ 2x + (3 - \lambda)y = 0. \end{array} \right\} \tag{5}$$

These will have a non-trivial solution if and only if

$$\begin{vmatrix} 1 - \lambda & 4 \\ 2 & 3 - \lambda \end{vmatrix} = 0,$$

that is

$$(1 - \lambda)(3 - \lambda) - (2)(4) = 0.$$

Whence $\lambda^2 - 4\lambda - 5 = 0$, with solutions $\lambda = -1$, $\lambda = 5$. The correctness of this calculation is automatically checked at Stage 2, when we come to find the eigenvectors. Substituting $\lambda = -1$ in equations (5) above, we obtain $2x + 4y = 0$, $2x + 4y = 0$, so $x = 2$, $y = -1$ for instance will do as an eigenvector. Similarly $\lambda = 5$ leads to $-4x + 4y = 0$, $2x - 2y = 0$ with solution $x = 1$, $y = 1$ for an eigenvector.

If, when we reach Stage 2, we find things going wrong – that is to say, we find $x = 0$, $y = 0$ is the only solution of our equations – this means that we must have made a mistake in Stage 1.

Exercises

Find eigenvalues and eigenvectors for the following matrices:

(a) $\begin{pmatrix} 3 & 1 \\ 1 & 3 \end{pmatrix}$ (b) $\begin{pmatrix} 4 & 1 \\ 1 & 4 \end{pmatrix}$ (c) $\begin{pmatrix} 0 & 2 \\ -2 & 5 \end{pmatrix}$

(d) $\begin{pmatrix} -2 & 2 \\ -3 & 5 \end{pmatrix}$ (e) $\begin{pmatrix} 1 & 1 \\ 2 & 2 \end{pmatrix}$ (f) $\begin{pmatrix} 2 & 1 \\ 0 & 3 \end{pmatrix}$

(g) $\begin{pmatrix} a & b \\ b & a \end{pmatrix}$ where a, b are any pair of numbers.

(h) What difficulties arise in trying to find eigenvalues and eigenvectors for $\begin{pmatrix} a & -b \\ b & a \end{pmatrix}$? Are there any values a, b for which this matrix has eigenvectors? (We are concerned only with real numbers.)

26 Determinant and inverse

We have now met two expressions that look somewhat similar but have very different meanings. They are M and Δ where

$$M = \begin{pmatrix} p & q \\ r & s \end{pmatrix} \quad \text{and} \quad \Delta = \begin{vmatrix} p & q \\ r & s \end{vmatrix}.$$

In each we see the same four numbers, p, q, r, s; in M these numbers are enclosed in curved brackets, in Δ in straight lines. It is important to observe carefully this small distinction in notation. It indicates that M is a matrix, which specifies a transformation, while Δ is a determinant, a single number. We speak of Δ as 'the determinant of the matrix M', and we may write $\Delta = \det M$. Thus Δ is a single number associated with the transformation M. It is natural to ask what is the meaning of the number Δ; what does it tell us about the transformation?

If

$$A = \begin{pmatrix} 1 \\ 0 \end{pmatrix} \quad \text{and} \quad B = \begin{pmatrix} 0 \\ 1 \end{pmatrix},$$

we know that A^* and B^* can be read off from the columns of M (see §22). We have

$$A^* = \begin{pmatrix} p \\ r \end{pmatrix}, \qquad B^* = \begin{pmatrix} q \\ s \end{pmatrix}.$$

Let us consider the area of the parallelogram $O^*A^*C^*B^*$ in Fig. 74.

It will help us if we drop perpendiculars from A^* and C^* onto the x-axis and from B^* and C^* onto the y-axis as in Fig. 75. The parallelogram is now inside a rectangle, and its area can be found by subtracting from the area of the whole rectangle the area of the shaded parts. This gives us $(p + q)(r + s) - pr - qs - 2qr$ which simplifies to $ps - qr$, the determinant Δ.

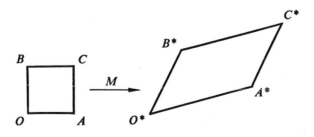

Fig. 74

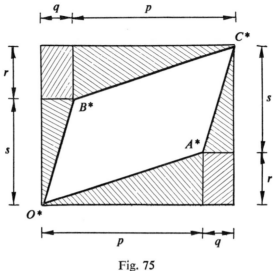

Fig. 75

The vectors A^*, B^* are the columns of the matrix M. We could indicate this by writing $M = (A^*, B^*)$, and our result could then be stated – the area of the parallelogram determined by the vectors A^* and B^* is $\det(A^*, B^*)$. Incidentally we have a result here that gives a method for finding the area of a parallelogram; it could be used by someone who did not know anything about the idea of a linear transformation.

This result, however, is very significant in connection with the transformation M. Imagine any region in the original plane that can be broken up into a number of unit squares such as $OACB$. When the transformation M acts, this region will be changed into a number of parallelograms, each congruent to $O^*A^*C^*B^*$. Thus, if the original region contained N unit squares, the transformed region would have area $N\Delta$. From this we can be led to the conclusion that, when the transformation M acts on the plane, all areas are changed in the ratio $1:\Delta$. *The determinant of M thus measures the ratio in which areas change when M acts on the plane.*

This gives us a way of generalizing the idea of determinant. Transformations in 3 dimensions are specified by a matrix with 3 rows and 3 columns. The transformation will change the unit cube into a region with parallel plane faces, and all volumes will be changed in the same ratio. We *define* the determinant as being the number that specifies this ratio. In 4, 5 or more dimensions similar results apply: there are 'boxes' which we cannot easily imagine that correspond to squares in 2 dimensions or cubes in 3 dimensions. We can specify the 'volume' of a region by breaking it up into boxes. Mathematicians have carried out the details of this in a subject known as Measure Theory.

One point should be noted. Suppose we consider the matrix $\begin{pmatrix} 0 & 1 \\ 1 & 0 \end{pmatrix}$ which represents a reflection in the line $y = x$. If we substitute these values in $\Delta = ps - qr$, we get the value -1. The size of the answer is correct – in a reflection areas are unchanged; they change in the ratio $1:1$, so we might have expected to find Δ equal to 1. The negative

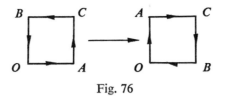

Fig. 76

sign is an indication that the plane has been turned over; anticlockwise has been changed to clockwise.

So the determinant gives us an extra piece of information. By its sign it tells us whether such a reversal has taken place or not.

If we have two 2 × 2 matrices S and T, the product ST indicates that we first apply T to the plane, and then apply S to the result. When T is applied, all areas will be multiplied by det T. When S acts, they will again be multiplied by det S. Thus altogether areas get multiplied by (det S)·(det T). Now det(ST) indicates the ratio in which areas change under the transformation ST. Accordingly we must have

$$\det(ST) = (\det S)\cdot(\det T).$$

The determinant of a product equals the product of the determinants.

For example, we have by matrix multiplication

$$\begin{pmatrix} 2 & 3 \\ 1 & 4 \end{pmatrix} \begin{pmatrix} 20 & 10 \\ 3 & 2 \end{pmatrix} = \begin{pmatrix} 49 & 26 \\ 32 & 18 \end{pmatrix}.$$

The determinants of the matrices on the left-hand side are

$$\begin{vmatrix} 2 & 3 \\ 1 & 4 \end{vmatrix} = 5 \quad \text{and} \quad \begin{vmatrix} 20 & 10 \\ 3 & 2 \end{vmatrix} = 10.$$

Accordingly we expect the determinant of the matrix on the right-hand side to be 50. And in fact

$$\begin{vmatrix} 49 & 26 \\ 32 & 18 \end{vmatrix} = 882 - 832 = 50.$$

This last result holds in any number of dimensions. To prove it in 3 dimensions we replace the word 'area' by 'volume'; in higher dimensions we use the corresponding measure of what a box holds.

Inverse matrix

The matrix M introduced at the beginning of this section corresponds to the equations

$$x^* = px + qy, \tag{1}$$

$$y^* = rx + sy. \tag{2}$$

If we know the input, $\begin{pmatrix} x \\ y \end{pmatrix}$, these equations tell us the output vector $\begin{pmatrix} x^* \\ y^* \end{pmatrix}$. Occasionally we need to proceed in the opposite direction. We require a certain output; what input will give it? In this case, x^* and y^* are given; we want to find x and y. So we have to solve equations (1) and (2) for x and y. We form the equations $s(1) - q(2)$. (This means, we multiply (1) by s, (2) by q, and subtract.) We also form the equation $p(2) - r(1)$. We thus obtain

$$sx^* - qy^* = (ps - qr)x, \tag{3}$$

$$-rx^* + py^* = (ps - qr)y. \tag{4}$$

We notice that both x and y have the coefficient Δ, the determinant of M, and we want to divide by this. As a rule, we shall be able to do so, but there is the exceptional case, in which $\Delta = 0$. These cases we must consider separately.

(1) General case, $\Delta \neq 0$ In this case, we can divide equations (3) and (4) by Δ and obtain

$$x = (s/\Delta)x^* - (q/\Delta)y^*, \tag{5}$$

$$y = -(r/\Delta)x^* + (p/\Delta)y^*. \tag{6}$$

Equations (1) and (2) tell us that, for any input, there is just one possible output. Equations (5) and (6) tell us that, for any output, there is just one possible input. We have what is known as a one-to-one mapping. We may represent it diagrammatically as in Fig. 77.

We began with $v \rightarrow v^*$, where $v^* = Mv$. If we start with v^* and want to get back to v, it is natural to write $v = M^{-1}v^*$. M^{-1} is called the *inverse matrix*: it specifies the inverse transformation. Some authors prefer M^I or ^IM, with I here indicating 'inverse'.

The product $M^{-1}M$ indicates that we start with (say) v, cross to v^*, and return to v. Thus $M^{-1}Mv = v$ for any vector v. Thus $M^{-1}M$ is the identity transformation; we may write $M^{-1}M = I$. Products usually depend on order; ST and TS are not always the same. So we had better look at MM^{-1}. We start with v^* in the output; M^{-1} sends it back to v; then M sends it forward to v^*. Thus MM^{-1} gives $v^* \rightarrow v^*$. It also is an identity transformation. Accordingly we have $MM^{-1} = M^{-1}M = I$.

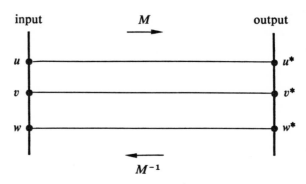

Fig. 77

Definition If we have two matrices, M and M^{-1}, such that $MM^{-1} = M^{-1}M = I$, then M^{-1} is called the inverse of M.

(2) Special case, $\Delta = 0$ Our algebra above showed that this case led to difficulties, since in equations (3) and (4) the coefficients of x and y become zero. It is impossible to divide and obtain equations (5) and (6).

Geometrically, $\Delta = 0$ means that the parallelogram $O*A*C*B*$, used earlier in this section, has area zero. How can this arise? There are two possibilities. One of the sides, $O*A*$ or $O*B*$, may be of zero length, or the angle $B*O*A*$ may be zero.

The first possibility would arise, for example, with the matrix $\begin{pmatrix} 1 & 0 \\ 0 & 0 \end{pmatrix}$. For this matrix $x* = x$, $y* = 0$. Every point is projected onto the x-axis.

The second possibility arises with $\begin{pmatrix} 1 & 2 \\ 1 & 2 \end{pmatrix}$. For this matrix,

$$A* = \begin{pmatrix} 1 \\ 1 \end{pmatrix}, \qquad B* = \begin{pmatrix} 2 \\ 2 \end{pmatrix}.$$

Both points lie on the line through the origin from south-west to north-east.

In both cases, the plane is crushed into a line. Thus every area becomes zero. There is an even more extreme possibility, represented by the matrix 0, for which every point maps to the origin.

The equation $\Delta = 0$ indicates that the space has been transformed into a space of lower dimension. We have been considering the example of the plane, but this remark applies equally well to spaces of any dimension. For instance, in 3 dimensions, if every volume is multiplied by 0, this can only mean that space has been crushed down into a plane, a line or a point.

We can now see why there is difficulty in forming the inverse transformation. This point is sufficiently illustrated by the projection $x* = x$, $y* = 0$ mentioned above. In this transformation, every point P goes to the point $P*$ on the x-axis immediately below it (Fig. 78).

If we now try to prescribe an output $(x*, y*)$ and go back from it to an input (x, y) we run into two kinds of difficulty – one a famine and one a glut. If we want an output

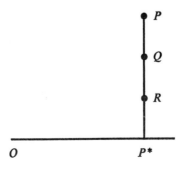

Fig. 78

such as (3, 2), a point that does not lie on the *x*-axis, we cannot get it. No point goes to this position. On the other hand if we take a point P^* on the *x*-axis, as in Fig. 78, it is true $P \to P^*$, but also $Q \to P^*$ and $R \to P^*$ and so do an infinity of other points. We do not know which to take for $M^{-1}P^*$. Someone might suggest: 'Choose one of them at random'. But this will not work out right. Suppose we say, let $M^{-1}P^* = P$. Now $M^{-1}M = I$, when an inverse exists. Apply this equation to Q. $M^{-1}MQ = IQ = Q$. Now $MQ = P^*$, so $M^{-1}MQ = M^{-1}P^* = P$ by what we agreed just now. We have obtained $Q = P$. But this is wrong, P and Q are not the same point. The glut of candidates for the position $M^{-1}P^*$ when P^* is on the *x*-axis is just as embarrassing as the famine, the absence of any candidate, when P^* is off that axis.

These things work out in essentially the same way in any number of dimensions. For convenience of language, we will discuss this question in terms of 3-dimensional space, on the understanding that essentially the same arguments and results apply in any number of dimensions.

When $\Delta = 0$, the unit cube transforms to a region whose volume is zero. This can happen only if A^*, B^*, C^* (corresponding to the unit vectors in the axes, A, B, C) lie in a plane through the origin, O^*. (In the more extreme cases, when A^*, B^*, C^* lie in a line through O^*, or even coincide with O^*, this condition still holds.) Then M^{-1} cannot exist, for we cannot find an input that gives a point outside this plane.

When $\Delta = 0$, there is a complete line, every point of which transforms into the origin. For the three vectors, A^*, B^*, C^* lie in a plane. By this we mean that the lines O^*A^*, O^*B^*, O^*C^* lie in a plane. As we agreed in §7, three vectors in a plane cannot be linearly independent; one of them, say C^*, must be a combination of the other two. Suppose then $C^* = hA^* + kB^*$. This means $hA^* + kB^* - C^* = 0$. Now in a linear mapping, $xA + yB + zC \to xA^* + yB^* + zC^*$. Putting $x = h$, $y = k$, $z = -1$, it follows $hA + kB - C \to hA^* + kB^* - C^* = 0$. So $hA + kB - C$ is a vector that maps to 0. This vector cannot itself be 0, since it has $z = -1$. Call this vector v_1. Then $v_1 \to 0$ and, by proportionality, $tv_1 \to 0$ for any number t. The points of the form tv_1, the multiples of v_1, fill a line. Q.E.D.

Let u be any vector. Since $u \to u^*$ and $tv_1 \to 0$, $u + tv_1 \to u^*$, by superposition. Since t can be any number, we see that there are infinitely many points that map to u^*.

This shows that quite generally when $\Delta = 0$ we have the alternative of famine or glut. Either an output is unobtainable from any input, or it arises from infinitely many different inputs. On both counts, such a transformation cannot possess an inverse.

Let us look at the general case, $\Delta \neq 0$, for 3-dimensional space. In this case, A^*, B^*, C^* cannot be linearly dependent; for if they were, the vectors would lie in a plane, and volumes would be sent to zero. It looks reasonable, on geometrical grounds, and it can in fact be proved, that any three linearly independent vectors in 3 dimensions form a basis for the space – we can build a co-ordinate framework on them. Thus every point P^* can be represented in just one way as $xA^* + yB^* + zC^*$. Now

$$xA + yB + zC \to xA^* + yB^* + zC^*,$$

so we certainly can find a point P that maps to P^*. And there can only be one such point.

For suppose $x_1 A + y_1 B + z_1 C$ and $x_2 A + y_2 B + z_2 C$ both mapped to P^*. We would have $P^* = x_1 A^* + y_1 B^* + z_1 C^* = x_2 A^* + y_2 B^* + z_2 C^*$, so P^* would have co-ordinates (x_1, y_1, z_1) and (x_2, y_2, z_2) in the A^*, B^*, C^* system. But a point can only have one set of co-ordinates, so $x_1 = x_2$, $y_1 = y_2$, $z_1 = z_2$; this means the points mapping to P^* must coincide. So when $\Delta \neq 0$, every point P^* can be got by choosing a suitable input, and only one such input exists. We have a one-to-one correspondence, and the inverse transformation M^{-1} exists.

It will be a linear transformation, since $u + v \to u^* + v^*$ and $ku \to ku^*$ for M. For M^{-1}, we reverse the arrows.

We may sum this up as follows.

Theorem When det $M \neq 0$, there exists an inverse matrix, M^{-1}, such that

$$MM^{-1} = M^{-1}M = I.$$

The equation $Mv = w$ then has the unique solution $v = M^{-1}w$.

When det $M = 0$, no inverse exists. The equation $Mv = w$ will have either 0 or ∞ solutions. There will always be some non-zero vector v_1 such that $Mv_1 = 0$.

Note on the role of determinants

In the nineteenth century great attention was paid to determinants and their properties, and determinants were used to solve systems of equations. In this century, computers have been used to solve problems in which hundreds of variables may be involved. For such work determinants are completely *un*-suitable. It is useful to recognize determinants when we are working in 2 or 3 dimensions. For work involving several variables, determinants occur in theoretical arguments, such as those just used, and in the procedure for finding eigenvalues (§25).

This procedure gives interest to the theorem above, which states that $Mv = 0$ has a non-zero solution when det $M = 0$. The first part of the theorem shows that, when det $M \neq 0$, the only solution of $Mv = 0$ is $v = 0$; for $v = 0$ obviously is a solution, and the transformation being one-to-one, there cannot be any other solution.

In the eigenvector–eigenvalue search, we are trying to find λ and non-zero v so that $Mv = \lambda v$. This equation may be written $Mv = \lambda Iv$ or $(M - \lambda I)v = 0$. By the theorem, a non-zero solution will exist when $\det(M - \lambda I) = 0$. The equations used in the final stages of §25 were of this form. For instance, in the worked example, we had

$$M = \begin{pmatrix} 1 & 4 \\ 2 & 3 \end{pmatrix}.$$

Since

$$\lambda I = \begin{pmatrix} \lambda & 0 \\ 0 & \lambda \end{pmatrix}, \qquad M - \lambda I = \begin{pmatrix} 1 - \lambda & 4 \\ 2 & 3 - \lambda \end{pmatrix}.$$

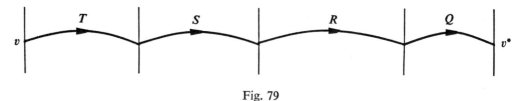

Fig. 79

In the worked example, it will be seen that the determinant of this matrix was equated to zero, so that suitable values of λ could be found. When these values of λ are substituted, $\det(M - \lambda I) = 0$ and so it is certain that we shall be able to satisfy $(M - \lambda I)v = 0$ by some non-zero vector v, that is, a vector that is acceptable as an eigenvector.

Terminology   A matrix that has no inverse is said to be *singular*.

The equation $\det(M - \lambda I) = 0$ is called the *characteristic equation* of M. The roots of the characteristic equation, as we have seen, are the *eigenvalues* of M. Some authors refer to the eigenvalues as *characteristic* numbers or *latent roots*.

A matrix that has an inverse is sometimes called *invertible*, sometimes *non-singular*.

Inverse of a product

It is sometimes necessary to consider the inverse of a product of matrices, such as $QRST$ say. Here we suppose that each of the matrices Q, R, S and T has an inverse. We may represent the product transformation graphically by Fig. 79, where

$$v^* = QRSTv.$$

The inverse transformation brings us back from v^* to v. Moving back, we find we perform Q^{-1}, R^{-1}, S^{-1}, T^{-1} in that order – the reverse of the order we followed when we went from v to v^*. Thus the inverse of $QRST$ is $T^{-1}S^{-1}R^{-1}Q^{-1}$.

We can verify this formally, for we have

$$(QRST)(T^{-1}S^{-1}R^{-1}Q^{-1}) = QRS(TT^{-1})S^{-1}R^{-1}Q^{-1}$$
$$= QRS \cdot I \cdot S^{-1}R^{-1}Q^{-1} = QR(SS^{-1})R^{-1}Q^{-1} = QR \cdot I \cdot R^{-1}Q^{-1}$$
$$= Q(RR^{-1})Q^{-1} = QIQ^{-1} = QQ^{-1} = I.$$

At several stages of this argument we have omitted a factor I. This is legitimate, since I means 'leave things as they were'. In the same way in arithmetic we can omit the factors 1 in a product such as $3 \times 7 \times 1 \times 2 \times 1 \times 9$.

If we wish, we can use the same procedure to check formally that the product is still I when we put $T^{-1}S^{-1}R^{-1}Q^{-1}$ to the left of $QRST$ instead of the right; that is, we can verify that $(T^{-1}S^{-1}R^{-1}Q^{-1})(QRST) = I$.

Questions

1   For each matrix listed below, draw a diagram to show its effect on the square $OACB$, where $O = (0, 0)$, $A = (1, 0)$, $B = (0, 1)$ and $C = (1, 1)$. (As explained in §22, the columns of the matrix provide a quick way of doing this, in cases where the geometrical meaning of the transformation is not immediately obvious.) Find the area of the resulting region, and compare it with the determinant of the matrix. In which cases does the plane experience an effect of turning over, so that clockwise and counterclockwise circuits are interchanged?

(a) $\begin{pmatrix} 2 & 0 \\ 0 & 1 \end{pmatrix}$ (b) $\begin{pmatrix} 1 & 0 \\ 0 & 3 \end{pmatrix}$ (c) $\begin{pmatrix} 2 & 0 \\ 0 & 3 \end{pmatrix}$ (d) $\begin{pmatrix} 2 & 0 \\ 0 & \frac{1}{2} \end{pmatrix}$ (e) $\begin{pmatrix} 1 & 0 \\ 0 & -1 \end{pmatrix}$ (f) $\begin{pmatrix} 1 & -1 \\ 1 & 1 \end{pmatrix}$

(g) $\begin{pmatrix} 1 & 1 \\ 1 & -1 \end{pmatrix}$ (h) $\begin{pmatrix} \cos \alpha & -\sin \alpha \\ \sin \alpha & \cos \alpha \end{pmatrix}$ (i) $\begin{pmatrix} \cos \alpha & \sin \alpha \\ \sin \alpha & -\cos \alpha \end{pmatrix}$.

2   For complex numbers $(4 + j3)(x + jy) = x^* + jy^*$ with $x^* = 4x - 3y$ and $y^* = 3x + 4y$, so multiplication by $4 + j3$ is associated with the matrix

$$M = \begin{pmatrix} 4 & -3 \\ 3 & 4 \end{pmatrix}.$$

If every complex number in the Argand diagram is multiplied by $4 + j3$, in what way does this transformation affect (a) lengths, (b) areas? Show that your answers agree with those that would be obtained by considering the value of det M.

Replace $4 + j3$ by the general number $a + jb$ and carry out the same investigation.
Does the value of the determinant in this general case throw any light on the question – could a complex number exist, such that multiplication by it produced a reflection in the Argand diagram?

3   Let

$$S = \begin{pmatrix} a & -b \\ b & a \end{pmatrix}, \qquad T = \begin{pmatrix} x & -y \\ y & x \end{pmatrix}.$$

Calculate the product matrix ST. Find the determinants of S, T and ST. Verify that

$$\det S \cdot \det T = \det(ST).$$

4   Which of the following matrices do *not* possess inverses?

(a) $\begin{pmatrix} 1 & 0 \\ 0 & 0 \end{pmatrix}$ (b) $\begin{pmatrix} 1 & 2 \\ 3 & 4 \end{pmatrix}$ (c) $\begin{pmatrix} 1 & 2 \\ 3 & 6 \end{pmatrix}$ (d) $\begin{pmatrix} 0 & 2 \\ 3 & 0 \end{pmatrix}$

(e) $\begin{pmatrix} 1 & 1 \\ 0 & 1 \end{pmatrix}$ (f) $\begin{pmatrix} 1 & 1 \\ 1 & 1 \end{pmatrix}$ (g) $\begin{pmatrix} 8 & 10 \\ 12 & 15 \end{pmatrix}$.

5   Let

$$S = \begin{pmatrix} 2 & 1 \\ 3 & 2 \end{pmatrix}, \qquad T = \begin{pmatrix} 1 & 2 \\ 3 & 7 \end{pmatrix}.$$

Calculate ST. Find det S, det T and $\det(ST)$. Find S^{-1}, T^{-1} and $(ST)^{-1}$. Of the matrices $(ST)^{-1}$ $S^{-1}T^{-1}$ and $T^{-1}S^{-1}$ which, if any, would you expect to be equal? Check your answer by direct calculation of these matrices.

6   (a) Make a sketch to show the effect of the matrix $\begin{pmatrix} 2 & 1 \\ 1 & 2 \end{pmatrix}$ on the square with corners $O = (0, 0)$, $C = (1, 1)$, $D = (0, 2)$ and $E = (-1, 1)$. In what ratios are areas changed? What is the determinant of the matrix? What are the eigenvectors and eigenvalues of this matrix? (These can be seen from the diagram; it is not necessary to use the routine of §25.)

(b) Answer the same questions as in (a), but with the matrix $\begin{pmatrix} 2.5 & 0.5 \\ 0.5 & 2.5 \end{pmatrix}$.

(c) Do the same, but with the matrix $\begin{pmatrix} a & b \\ b & a \end{pmatrix}$.

(d) Does there appear to be any connection between the eigenvalues and the value of the determinant? Is there any logical reason for expecting such a connection?

7 Each matrix below has determinant zero. According to §26, it should map the whole plane into a certain line. There should be another line, every point of which maps to the origin. For each matrix, find these two lines and indicate by a sketch how the plane is transformed by the matrix. Each matrix has two invariant lines; show these on your sketch and label each of them with the appropriate eigenvalue, λ.

(a) $\begin{pmatrix} 1 & 0 \\ 0 & 0 \end{pmatrix}$ (b) $\begin{pmatrix} 1 & 0 \\ 1 & 0 \end{pmatrix}$ (c) $\begin{pmatrix} \frac{1}{2} & \frac{1}{2} \\ \frac{1}{2} & \frac{1}{2} \end{pmatrix}$ (d) $\begin{pmatrix} 1 & 1 \\ 1 & 1 \end{pmatrix}$ (e) $\begin{pmatrix} 0.2 & 0.4 \\ 0.4 & 0.8 \end{pmatrix}$.

8 Let

$$S = \begin{pmatrix} 1 & a \\ 0 & 1 \end{pmatrix}, \quad T = \begin{pmatrix} 1 & 0 \\ b & 1 \end{pmatrix}.$$

Verify that the inverses of S and T are

$$S^{-1} = \begin{pmatrix} 1 & -a \\ 0 & 1 \end{pmatrix}, \quad T^{-1} = \begin{pmatrix} 1 & 0 \\ -b & 1 \end{pmatrix}.$$

Calculate the matrices ST and $S^{-1}T^{-1}$. Would you expect $S^{-1}T^{-1}$ to be the inverse of ST? Would you expect the product of ST and $S^{-1}T^{-1}$ to be a matrix of determinant 1? Check your answers by calculation.

Would the answer to either question be altered if we replaced $S^{-1}T^{-1}$ throughout by $T^{-1}S^{-1}$?

27   *Properties of determinants*

We have defined determinants geometrically. If u and v are the columns of a matrix in 2 dimensions, the determinant of that matrix is the area of the parallelogram with sides u and v. For a 3×3 matrix, with columns u, v, w, the determinant equals the volume for the cell* with sides u, v, w, and for higher dimensions there are corresponding definitions. This leaves determinants a little up in the air. We would like to know the explicit formulas for these quantities, and their properties for occasions when we need to calculate with them.

In §26 we found an explicit formula for $\det(u, v)$, the area of the parallelogram with sides u and v. For $u = \begin{pmatrix} p \\ r \end{pmatrix}$, $v = \begin{pmatrix} q \\ s \end{pmatrix}$ we found

$$\det(u, v) = \begin{vmatrix} p & q \\ r & s \end{vmatrix} = ps - qr.$$

---

* The traditional name for the figure in 3 dimensions corresponding to the parallelogram is 'parallelepiped', an awful word both to spell and to pronounce. I use 'cell' instead, and understand by this word the figure in any number of dimensions that corresponds to the parallelogram in 2 dimensions.

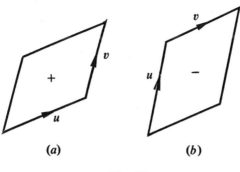

(a)   (b)

Fig. 80

It was pointed out in that section that this quantity may be positive or negative. It is useful to indicate u and v as in Fig. 80, with the second vector drawn from the end of the first. In case (a), where the arrows run round the parallelogram in a counterclockwise direction, $\det(u, v)$ is positive; in case (b), where the arrows run in a clockwise sense, $\det(u, v)$ is negative.

This might seem to be a complication, but in fact it makes the work easier. This feature appears already in one dimension. Suppose we have points on a line, and we know B is 10 units from A, and C is 3 units from B. Then the distance AC may be either 7 or 13 units, depending on the directions of AB and BC. But if we use directed distances, and write $AB = +10$, $BC = +3$ to mean that B is 10 units to the right of A and C is 3 to the right of B, we are sure that $AC = +13$. With this system, if A, B and C have co-ordinates a, b, c we write $AB = b - a$, $BC = c - b$, $AC = c - a$ and we can be sure $AB + BC = AC$. Now of course AB, BC and AC may be positive or negative depending on the relative positions of the points A, B, C.

With areas it helps in many ways if we allow $+$ and $-$ signs. Fig. 81 illustrates $\det(u, v)$ and $\det(3u, v)$. Obviously, $\det(3u, v)$ is 3 times $\det(u, v)$. This suggests that, for any number k we ought to have $\det(ku, v) = k\det(u, v)$. But what about $k = -1$?

If the formula is to hold, we must have $\det(-u, v) = -\det(u, v)$, so the parallelogram of Fig. 82 should have area equal to that of (a) in size, but opposite in sign.

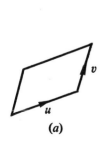

(a)

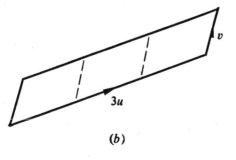

(b)

Fig. 81   Fig. 82

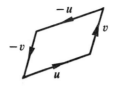

Fig. 83

It does no harm if we put arrows on all the sides of the parallelogram, as in Fig. 83. Any two consecutive sides can be used to describe the area. Thus we could call the area $\det(v, -u)$ or $\det(-u, -v)$ or $\det(-v, u)$. All of these are equal to $\det(u, v)$ as you can check by using the formula $ps - qr$.

The sign convention is particularly handy when we want to join regions together. Thus, if we take two counterclockwise parallelograms such as (a) and (b) in Fig. 81 we can put them together to make the parallelogram in Fig. 84. You will notice that the arrows on the common boundary are in opposite directions. We can regard this part as cancelling out, and we are simply left with the parallelogram involving $4u$ and v. This agrees with the result $\det(3u, v) + \det(u, v) = \det(4u, v)$.

If we want to put together a clockwise parallelogram as in Fig. 82 and an anticlockwise one like (b), to get this cancelling of common boundaries we have to arrange them as in Fig. 85.

After cancelling the parts with opposing arrows, we are left with the shaded parallelogram, counterclockwise, in agreement with the result

$$\det(3u, v) + \det(-u, v) = \det(2u, v),$$

given by the determinant formula.

Note that if we superposed Fig. 81 (a) and Fig. 82, everything would cancel. This agrees with $\det(u, v) + \det(-u, v) = 0$.

A familiar idea is that the area of a parallelogram, such as $OCED$ in Fig. 86 is not altered if we chop off a triangle ODF from one side and replace it by a congruent triangle CEG on the other side.

The vector $CG = CE + EG$. Now EG has the same direction as u, and so must be ku for some k. Thus $CG = v + ku$. The statement area $OCED$ = area $OCGF$ may

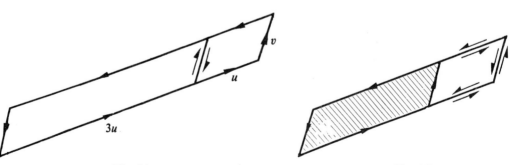

Fig. 84   Fig. 85

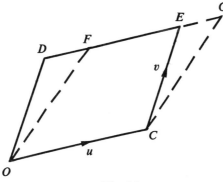

Fig. 86

accordingly be written $\det(u, v) = \det(u, v + ku)$ for any number k. Thus the determinant is unaltered in value if we add a multiple of one column to another.

This works in any number of dimensions. For 3 dimensions we can draw the picture in Fig. 87, which shows $\det(u, v, w) = \det(u, v + ku, w)$. For dimensions higher than 3 we cannot draw pictures, but we can explain by means of co-ordinate geometry or vector notation what set of points is replaced by a congruent set.

Of course the three vectors u, v, w are on an equal footing. We have, for instance, $\det(u, v, w) = \det(u + kw, v, w)$. We may add a multiple of any column to any other column.

So far we have two principles.

(I) Enlarging one vector k times enlarges the volume of the cell k times. For instance, $\det(ku, v, w) = k \det(u, v, w)$.

(II) Adding a multiple of one vector to another makes no difference to the determinant. For instance $\det(u, v, w) = \det(u, v, w + kv)$.

From these two principles we can deduce a useful result:

(III) Interchanging two vectors changes the sign of the determinant.

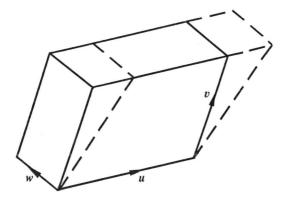

Fig. 87

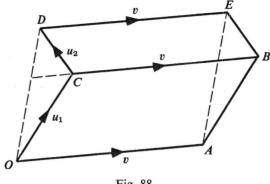

Fig. 88

Proof; $\det(u, v) = \det(u, u + v)$ by (II) $= \det(-v, u + v)$ by (II); we have added -1 times second vector to first.

Use (II) again, add first vector to second; this gives $\det(-v, u)$ which by (I), with $k = -1$, is $-\det(v, u)$ as required.

If we are in three or more dimensions, we carry out the same steps. For example $\det(u, v, w) = \det(u, u + v, w) = \det(-v, u + v, w) = \det(-v, u, w) = -\det(v, u, w)$. The presence of w makes no difference to the argument.

We can of course apply (II) more than once to obtain results such as

$$\det(u, v, w) = \det(u, v, w + au) = \det(u, v, w + au + bv)$$

for any two numbers a, b. Geometrically, this is equivalent to a statement about two cells with the same base (sides u, v) and the same height.

In Fig. 88 we have two parallelograms $OABC$ and $CBED$ lying in a plane. It is easily seen that area $OABC$ + area $CBED$ = area $OAED$. As $OD = u_1 + u_2$, we may write this result as $\det(u_1 + u_2, v) = \det(u_1, v) + \det(u_2, v)$. In 3 dimensions we have a similar result, namely $\det(u_1 + u_2, v, w) = \det(u_1, v, w) + \det(u_2, v, w)$. This follows from the fact that cells with the same base and same height have equal volumes. So we have:

(IV) $\det(u_1 + u_2, v, w) = \det(u_1, v, w) + \det(u_2, v, w)$, with similar results involving the vectors v and w.

This again can be proved for any number of dimensions.

Suppose we fix v and w, and consider u alone as variable. (I) and (IV) taken together show that the mapping $u \rightarrow \det(u, v, w)$ is linear. (I) shows that proportionality applies, and (IV) shows that superposition applies. Similarly, $\det(u, v, w)$ is linear in v and is linear in w.

This does not mean that determinants are linear functions when we consider all the vectors varying. For instance, in 2 dimensions we have had the formula $ps - qr$ for the determinant. If, for example, we put $q = 3, s = 2$, this reduces to $2p - 3r$, an expression linear in the entries in the first column. If we put constants in the first column, say $p = 5, r = 6$, we get $5s - 6q$, again a linear expression. But if we consider p, q, r, s as all variable the determinant has to be regarded as of the second degree, just as for instance in co-ordinate geometry the hyperbola $xy = 1$ is a conic, and is classified with the ellipse and parabola, equations of the second degree, although it is linear in x alone (y fixed) or in y alone (x fixed).

· A fifth principle is sometimes referred to as 'skinning a determinant'. Consider the determinant

$$\begin{vmatrix} 1 & 0 & 0 \\ 0 & p & q \\ 0 & r & s \end{vmatrix}$$

If we call the three columns here u, v and w, u is the unit vector along the x-axis, while v and w are in the plane perpendicular to the x-axis (Fig. 89). The volume of the cell they determine is thus the area of the base (the parallelogram with sides v, w) multiplied by the height (which is 1). The area of the base is $\begin{vmatrix} p & q \\ r & s \end{vmatrix}$. We know this from our study of areas in the plane. Accordingly we have

(V)

$$\begin{vmatrix} 1 & 0 & 0 \\ 0 & p & q \\ 0 & r & s \end{vmatrix} = \begin{vmatrix} p & q \\ r & s \end{vmatrix}.$$

The name 'skinning the determinant' is given because we here strip off the top row and the first column.

Now of course the x-axis is no better than the y-axis or the z-axis, and there must be corresponding results when the first column is

$$\begin{pmatrix} 0 \\ 1 \\ 0 \end{pmatrix} \text{ or } \begin{pmatrix} 0 \\ 0 \\ 1 \end{pmatrix}.$$

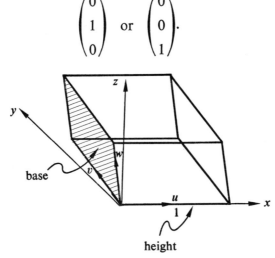

Fig. 89

In these cases, some care is needed to see whether the sign is + or −. There is a rule covering this, and perhaps the following is as good a way as any to obtain it.

Matrix multiplication gives the following equation:

$$\begin{pmatrix} 0 & 1 & 0 \\ 1 & 0 & 0 \\ 0 & 0 & 1 \end{pmatrix} \begin{pmatrix} 1 & 0 & 0 \\ 0 & p & q \\ 0 & r & s \end{pmatrix} = \begin{pmatrix} 0 & p & q \\ 1 & 0 & 0 \\ 0 & r & s \end{pmatrix}.$$

We saw in §26 that, for any two matrices S, T, we have $\det(ST) = (\det S) \cdot (\det T)$.

What is the determinant of the first matrix in the equation above? If we interchanged the first two columns, the matrix would become

$$\begin{pmatrix} 1 & 0 & 0 \\ 0 & 1 & 0 \\ 0 & 0 & 1 \end{pmatrix},$$

the identity matrix, I. The identity transformation leaves all volumes unchanged; it multiplies them by 1. So det $I = 1$. By (III), interchanging two columns changes the sign of the determinant. So the determinant of the first matrix is −1. (We might have guessed this. The matrix represents a reflection in the plane $y = x$. This is why we brought it in – to relate the unit vector on the y-axis to the unit vector on the x-axis.)

Accordingly, taking the determinants of the matrices in the equation we find

$$-\begin{vmatrix} 1 & 0 & 0 \\ 0 & p & q \\ 0 & r & s \end{vmatrix} = \begin{vmatrix} 0 & p & q \\ 1 & 0 & 0 \\ 0 & r & s \end{vmatrix}.$$

Thus

$$\begin{vmatrix} 0 & p & q \\ 1 & 0 & 0 \\ 0 & r & s \end{vmatrix} = -\begin{vmatrix} p & q \\ r & s \end{vmatrix}.$$

In this case, we pay for striking out the first column and the second row by introducing a minus sign.

We can get the 1 from the second row into the bottom row if we apply a reflection in the plane $y = z$.

Accordingly we write the equation

$$\begin{pmatrix} 1 & 0 & 0 \\ 0 & 0 & 1 \\ 0 & 1 & 0 \end{pmatrix} \begin{pmatrix} 0 & p & q \\ 1 & 0 & 0 \\ 0 & r & s \end{pmatrix} = \begin{pmatrix} 0 & p & q \\ 0 & r & s \\ 1 & 0 & 0 \end{pmatrix}$$

which again comes by matrix multiplication. Here again, the first matrix has determinant -1, and, taking determinants we find

$$\begin{vmatrix} 0 & p & q \\ 0 & r & s \\ 1 & 0 & 0 \end{vmatrix} = - \begin{vmatrix} 0 & p & q \\ 1 & 0 & 0 \\ 0 & r & s \end{vmatrix} = + \begin{vmatrix} p & q \\ r & s \end{vmatrix}.$$

Thus, in the three cases considered, the required signs are $+$, $-$, $+$. In higher dimensions the signs alternate as we continue in this way. Thus, for example, in a 6×6 determinant, if 1 occurs somewhere in the first column, and zeros fill the rest of that column and the rest of the row in which 1 appears, we may strike out that column and row, and affix a sign which is $+$, $-$, $+$, $-$, $+$, $-$ according as the row deleted is the 1st, 2nd, 3rd, 4th, 5th or 6th.

We now have the machinery needed for giving the explicit formula for a determinant. We will carry it through for 3 dimensions; a similar method can be used for any number of dimensions.

Suppose then we wish to find the determinant

$$\begin{vmatrix} a & d & g \\ b & e & h \\ c & f & i \end{vmatrix}.$$

The determinant, as we saw, is linear in each of the vectors involved. The vector in the first column is

$$\begin{pmatrix} a \\ b \\ c \end{pmatrix} \quad \text{which is} \quad \begin{pmatrix} a \\ 0 \\ 0 \end{pmatrix} + \begin{pmatrix} 0 \\ b \\ 0 \end{pmatrix} + \begin{pmatrix} 0 \\ 0 \\ c \end{pmatrix}.$$

By superposition then

$$\begin{vmatrix} a & d & g \\ b & e & h \\ c & f & i \end{vmatrix} = \begin{vmatrix} a & d & g \\ 0 & e & h \\ 0 & f & i \end{vmatrix} + \begin{vmatrix} 0 & d & g \\ b & e & h \\ 0 & f & i \end{vmatrix} + \begin{vmatrix} 0 & d & g \\ 0 & e & h \\ c & f & i \end{vmatrix}.$$

We consider the three parts separately. By (I), the first determinant equals

$$a \begin{vmatrix} 1 & d & g \\ 0 & e & h \\ 0 & f & i \end{vmatrix}.$$

We apply (II) twice. Add $-d$ times the first column to the second; then add $-g$ times the first column to the third column. This gives

$$a \begin{vmatrix} 1 & 0 & 0 \\ 0 & e & h \\ 0 & f & i \end{vmatrix} = a \begin{vmatrix} e & h \\ f & i \end{vmatrix}, \quad \text{by skinning.}$$

Apply similar procedures to the other two determinants, with care to remember the signs required. We find the original determinant equals

$$a \begin{vmatrix} e & h \\ f & i \end{vmatrix} - b \begin{vmatrix} d & g \\ f & i \end{vmatrix} + c \begin{vmatrix} d & g \\ e & h \end{vmatrix}.$$

This result is easily remembered. The coefficient of a is the result of crossing out the row and column containing a. For the coefficient of b we cross out the row and column containing b, but require a minus sign. Similar considerations apply to the coefficient of c, but the sign returns to plus.

The 2×2 determinants are given by the formula introduced in §25. We use this and by multiplying out obtain the result

$$\begin{vmatrix} a & d & g \\ b & e & h \\ c & f & i \end{vmatrix} = a(ei - fh) - b(di - fg) + c(dh - eg)$$

$$= aei + bfg + cdh - afh - bdi - ceg.$$

It will be noticed that there are 6 terms. We never find two letters in the same row or the same column appearing in a product. The numerical coefficients are all either $+1$ or -1.

It is possible to check the sign of any term as follows. Suppose we wish to check for cdh. We put $c = d = h = 1$, and all the other entries zero. This gives

$$\begin{vmatrix} 0 & 1 & 0 \\ 0 & 0 & 1 \\ 1 & 0 & 0 \end{vmatrix} = +1.$$

We then see how many interchanges of columns are needed to bring the determinant to the form det I. We find, by (III):

$$\begin{vmatrix} 0 & 1 & 0 \\ 0 & 0 & 1 \\ 1 & 0 & 0 \end{vmatrix} = - \begin{vmatrix} 1 & 0 & 0 \\ 0 & 0 & 1 \\ 0 & 1 & 0 \end{vmatrix} = + \begin{vmatrix} 1 & 0 & 0 \\ 0 & 1 & 0 \\ 0 & 0 & 1 \end{vmatrix} = +1.$$

This is in agreement with our earlier equation.

A 2×2 determinant has 2 terms when multiplied out in the form $ps - qr$. When we calculate a 3×3 determinant, as we did above, each entry in the first column has a

2×2 determinant as its coefficient. Thus there are 3×2 terms in the complete expression. If we apply the same procedure to a 4×4 determinant, there are 4 entries in the first column, and each of these has for its coefficient a 3×3 determinant with 3×2 terms. Thus the complete expression for a 4×4 determinant has $4 \times 3 \times 2$ terms, that is 24 terms. A 5×5 determinant has $5 \times 24 = 120$ terms and quite generally an $n \times n$ determinant has $n!$ terms. So a 10×10 determinant already has 3 628 800 terms.

As engineers may need to handle problems with hundreds or thousands of variables, it is easily understood why determinants are usually avoided as a means of computation. Thus given a matrix M there is a formula for its inverse, M^{-1}, in terms of determinants but it is rarely wise to use it. It is better to solve the equations $Mv = w$ directly, since, when an inverse exists, $v = M^{-1}w$.

Determinants can be useful, both for theoretical results and for particular calculations, when the number of dimensions is small, as for instance in geometrical, mechanical and electrical problems about our everyday physical space of 3 dimensions.

The eigenvalues of a matrix are usually of physical and engineering significance. Whole conferences are held on how to calculate them, but here again the methods used do not involve determinants. Yet it is useful to know that the eigenvalues of a matrix are given by an equation $\det(M - \lambda I) = 0$. If M is an $n \times n$ matrix, this equation is of the nth degree and we can classify the possible situations, by considering whether the roots of this equation are real or complex, distinct or repeated. In the same way, it may be useful to know that there is a single condition, $\det M = 0$, for a matrix M to possess no inverse.

The present position seems to be that an engineer does not need the expertise in manipulating determinants that was required of nineteenth century schoolboys, but that the properties of determinants should be part of his general knowledge. He may encounter them at any time in his reading.

In the exercises below, Questions 2 and 3 bring out properties of determinants that should be known to the student. These questions are in effect part of the theory covered in this section.

In the exercises below, questions 1(c), 1(d), 1(e), 4, 5, 6 and 7 could be slogged out by multiplying the determinants out completely. This, however, would be unwise and in some cases extremely laborious. These questions are intended to illustrate the properties of determinants (such as (I), (II), (III), (IV) in §27). The worked examples below show how such questions can be answered without heavy calculations.

Worked example 1

Calculate

$$\begin{vmatrix} 10 & 4 & 7 \\ 20 & 5 & 8 \\ 30 & 6 & 9 \end{vmatrix}.$$

Solution   The first column contains the factor 10. By (l) the determinant equals

$$10 \begin{vmatrix} 1 & 4 & 7 \\ 2 & 5 & 8 \\ 3 & 6 & 9 \end{vmatrix}.$$

To the second column add -1 times the first (by (II)). This gives

$$10 \begin{vmatrix} 1 & 3 & 7 \\ 2 & 3 & 8 \\ 3 & 3 & 9 \end{vmatrix}.$$

Now add -1 times the first column to the third. We obtain

$$10 \begin{vmatrix} 1 & 3 & 6 \\ 2 & 3 & 6 \\ 3 & 3 & 6 \end{vmatrix}.$$

Finally, add -2 times the second column to the third, to give

$$10 \begin{vmatrix} 1 & 3 & 0 \\ 2 & 3 & 0 \\ 3 & 3 & 0 \end{vmatrix}.$$

The third column now contains the zero vector; hence the determinant is zero.

Worked example 2

Simplify $\det(v + w, u + w, u + v)$ where u, v, w are any 3 vectors in 3 dimensions.

$$
\begin{aligned}
\det(v + w, u + w, u + v) &= \det(v + w, u - v, u + v) & \text{by (column 2)} - \text{(column 1),} \\
&= \det(v + w, u - v, 2u) & \text{by (column 3)} + \text{(column 2),} \\
&= 2 \det (v + w, u - v, u) & \text{from factor 2 in column 3,} \\
&= 2 \det(v + w, -v, u) & \text{by (column 2)} - \text{(column 3),} \\
&= 2 \det(w, -v, u) & \text{by (column 1)} + \text{(column 2),} \\
&= -2 \det(w, v, u) & \text{from factor } -1 \text{ in column 2,} \\
&= 2 \det(u, v, w) & \text{on interchanging columns 1 and} \\
& & \text{3, with resulting change of sign.}
\end{aligned}
$$

Note   Fallacies can result from unwise attempts to carry out several steps at once, as is shown by the following example.

A fallacious procedure   Let u, v, w be any 3 vectors in 3 dimensions. Then

$$\det(u, v, w) = \det(u - v, v - w, w - u)$$

by subtracting column 2 from column 1, column 3 from column 2 and column 1 from column 3. Add column 3 to column 2; the result is $\det(u - v, v - u, w - u)$. Add column 2 to column 1. This gives $\det(0, v - u, w - u)$. As the first column now contains the zero vector, the determinant must be zero.

The argument above purports to show that every 3×3 determinant is zero. The fallacy lies in the first step. If the steps are carried out one at a time, it will be found that no combination of (I), (II), (III), (IV) will lead from $\det(u, v, w)$ to

$$\det(u - v, v - w, w - u).$$

In order to avoid fallacies, it is wise to perform one step at a time, and to indicate the principle involved. This helps considerably if you need at a later time to check the correctness of the work. A detailed explanation, as in Worked example 1 above, is very wearisome to write. In Worked example 2, the commentary is much more concise, and it is clear that further abbreviations are possible.

Needless to say, the numbering (I), (II), (III), (IV) is purely for reference within this book, and this numbering need not be memorized.

Exercises on Determinants

1 Calculate the following:

(a) $\begin{vmatrix} 0 & 0 & 1 \\ 0 & 1 & 0 \\ 1 & 0 & 0 \end{vmatrix}$ (b) $\begin{vmatrix} 0 & 0 & 1 \\ 1 & 0 & 0 \\ 0 & 1 & 0 \end{vmatrix}$ (c) $\begin{vmatrix} 1 & 0 & 0 \\ 0 & 1 & 1 \\ 0 & 0 & 0 \end{vmatrix}$

(d) $\begin{vmatrix} a & d & d \\ b & e & e \\ c & f & f \end{vmatrix}$ (e) $\begin{vmatrix} a & d & a+d \\ b & e & b+e \\ c & f & c+f \end{vmatrix}$.

2 Let

$$A = \begin{pmatrix} k & 0 & 0 \\ 0 & 1 & 0 \\ 0 & 0 & 1 \end{pmatrix}, \quad B = \begin{pmatrix} 1 & k & 0 \\ 0 & 1 & 0 \\ 0 & 0 & 1 \end{pmatrix}, \quad C = \begin{pmatrix} 1 & 0 & 0 \\ k & 1 & 0 \\ 0 & 0 & 1 \end{pmatrix}, \quad M = \begin{pmatrix} a & d & g \\ b & e & h \\ c & f & i \end{pmatrix}.$$

Some of the calculations below provide alternative proof of principles already stated; identify the principles. Others lead to new principles; which should be noted.

First work out determinant A, determinant B and determinant C. Substitute these values in the equations below, after working out any matrix multiplications appearing in these equations.

(a) $\det(MA) = (\det M)(\det A)$
(b) $\det(AM) = (\det A)(\det M)$
(c) $\det(MB) = (\det M)(\det B)$
(d) $\det(MC) = (\det M)(\det C)$
(e) $\det(BM) = (\det B)(\det M)$
(f) $\det(CM) = (\det C)(\det M)$.

3   By M^T we understand the transpose of M, that is to say, the matrix obtained when the columns of M are changed into rows, that is

$$M^T = \begin{pmatrix} a & b & c \\ d & e & f \\ g & h & i \end{pmatrix}.$$

Work out det M^T and compare it with the expression for det M given in the text. What do you notice? What general conclusion is suggested?

4   (*a*) Simplify

$$\begin{vmatrix} a & d & 5a + 7d \\ b & e & 5b + 7e \\ c & f & 5c + 7f \end{vmatrix}.$$

(*b*) Find the value of

$$\begin{vmatrix} 1 & 4 & 4001 \\ 2 & 5 & 5002 \\ 3 & 6 & 6003 \end{vmatrix}.$$

5   For 3 particular vectors, u, v, w, it is known that $\det(u, v, w) = 10$. Find the values of the following determinants:

(*a*) $\det(2u, 3v, 4w)$ (*b*) $\det(v, u, w)$ (*c*) $\det(w, u, v)$
(*d*) $\det(u, v, 3u + 2v)$ (*e*) $\det(u + v + w, v + w, w)$.

6   Calculate

$$(a) \begin{vmatrix} 1 & 17 & 1017 \\ 2 & 18 & 2018 \\ 3 & 19 & 3019 \end{vmatrix} \qquad (b) \begin{vmatrix} 108 & 107 \\ 106 & 105 \end{vmatrix}.$$

7   Simplify

(*a*) $\det(2u + 3w, 3w, 2u + 5v) \div \det(u, v, w)$
(*b*) $\det(u + w, v + w, -u - v - 2w)$
(*c*) $\det(u - 4v + 6w, -3u + 5v - 3w, 2u - v - 3w)$.

8   The following matrices have determinant zero, so each matrix must map some non-zero vector to the origin. For each matrix, find some non-zero vector it maps to the origin.

$$(a) \begin{pmatrix} 0 & 1 & -1 \\ -1 & 0 & 1 \\ 1 & -1 & 0 \end{pmatrix} \quad (b) \begin{pmatrix} 1 & -2 & 1 \\ 2 & -1 & 0 \\ 3 & 0 & -1 \end{pmatrix} \quad (c) \begin{pmatrix} 1 & 2 & 3 \\ -2 & -1 & 0 \\ 1 & 0 & -1 \end{pmatrix}$$

$$(d) \begin{pmatrix} 1 & 2 & 3 \\ 4 & 5 & 6 \\ 7 & 8 & 9 \end{pmatrix} \quad (e) \begin{pmatrix} 1 & 4 & 7 \\ 2 & 5 & 8 \\ 3 & 6 & 9 \end{pmatrix}.$$

9   The following matrices have determinant zero and each maps a plane to the origin. For each matrix find any two independent vectors that it maps to the origin.

$$(a) \begin{pmatrix} 1 & 0 & 0 \\ 0 & 0 & 0 \\ 0 & 0 & 0 \end{pmatrix} \quad (b) \begin{pmatrix} 1 & 1 & 1 \\ 1 & 1 & 1 \\ 1 & 1 & 1 \end{pmatrix} \quad (c) \begin{pmatrix} 1 & 1 & 1 \\ 2 & 2 & 2 \\ 3 & 3 & 3 \end{pmatrix} \quad (d) \begin{pmatrix} 1 & 2 & 3 \\ 1 & 2 & 3 \\ 1 & 2 & 3 \end{pmatrix} \quad (e) \begin{pmatrix} 1 & 2 & 3 \\ 2 & 4 & 6 \\ 3 & 6 & 9 \end{pmatrix}.$$

10 Let

$$M = \begin{pmatrix} 7 & 2 & 0 \\ 2 & 6 & -2 \\ 0 & -2 & 5 \end{pmatrix}.$$

Verify that $\det(M - \lambda I) = 0$ for $\lambda = 3$, 6 or 9. Find non-zero vectors u, v, w such that

$$(M - 3I)u = 0, \quad (M - 6I)v = 0, \quad (M - 9I)w = 0.$$

(The vectors so found are eigenvectors of M, corresponding to eigenvalues 3, 6 and 9 respectively.)

11 Verify that, for

$$M = \begin{pmatrix} 0 & \sqrt{2} & 0 \\ \sqrt{2} & 0 & \sqrt{2} \\ 0 & \sqrt{2} & 0 \end{pmatrix}$$

$\det(M - \lambda I) = -\lambda^3 + 4\lambda$. For what values of λ is this determinant zero? Find the corresponding eigenvectors. (*Note*. An eigenvalue can be zero; an eigenvector cannot.)

12 Show that for

$$M = \begin{pmatrix} -1 & -2 & 0 \\ -2 & 0 & 2 \\ 0 & 2 & 1 \end{pmatrix}$$

we have $\det(M - \lambda I) = -\lambda^3 + 9\lambda$. Find the eigenvalues and eigenvectors of M.

13 Show that for

$$M = \begin{pmatrix} 2 & 2 & -3 \\ 2 & 2 & 3 \\ -3 & 3 & 3 \end{pmatrix}$$

the characteristic equation $\det(M - \lambda I) = 0$ is equivalent to $\lambda^3 - 7\lambda^2 - 6\lambda + 72 = 0$. Solve this equation, it being given that one solution is $\lambda = -3$. Find the eigenvalues and eigenvectors of M.

14 Find the eigenvalues and eigenvectors of the matrix

$$\begin{pmatrix} 3 & 4 & -4 \\ 4 & 5 & 0 \\ -4 & 0 & 1 \end{pmatrix}.$$

15 Let

$$M = \begin{pmatrix} \frac{1}{3} & \frac{1}{3} & \frac{1}{3} \\ \frac{1}{3} & \frac{1}{3} & \frac{1}{3} \\ \frac{1}{3} & \frac{1}{3} & \frac{1}{3} \end{pmatrix}.$$

For this matrix both $\det M$ and $\det(M - I)$ are zero. Show that $M - I$ maps a non-zero vector to the origin (as in question 8) while M itself maps an entire plane to the origin. Describe the eigenvalues and corresponding eigenvectors of the matrix M.

16 Show that the matrix

$$\begin{pmatrix} \frac{1}{6} & \frac{1}{6} & \frac{1}{3} \\ \frac{1}{6} & \frac{1}{6} & \frac{1}{3} \\ \frac{1}{3} & \frac{1}{3} & \frac{2}{3} \end{pmatrix}$$

has eigenvalues 0 and 1. Investigate the eigenvectors belonging to these values.

17 Investigate the eigenvalues and corresponding eigenvectors of the matrix

$$\begin{pmatrix} \frac{2}{3} & -\frac{1}{3} & -\frac{1}{3} \\ -\frac{1}{3} & \frac{2}{3} & -\frac{1}{3} \\ -\frac{1}{3} & -\frac{1}{3} & \frac{2}{3} \end{pmatrix}.$$

In what way does the answer resemble, and in what way does it differ from, the answer to question 15?

18 The matrix

$$\begin{pmatrix} \frac{5}{6} & -\frac{1}{6} & -\frac{1}{3} \\ -\frac{1}{6} & \frac{5}{6} & -\frac{1}{3} \\ -\frac{1}{3} & -\frac{1}{3} & \frac{1}{3} \end{pmatrix}$$

has eigenvalues 0 and 1. Investigate the eigenvectors corresponding to these values. Compare and contrast the answer to this question with the answer to question 16.

Cramer's rule

Suppose we have two equations

$$ax + by + cz = 0, \tag{1}$$

$$dx + ey + fz = 0. \tag{2}$$

Here we have 2 equations for 3 unknowns, so we cannot expect to fix x, y, z completely. In fact, we can see that if x, y, z, is any solution, so is kx, ky, kz. The most we can hope to do is to determine the ratios $x:y:z$.

Geometrically, equations (1) and (2) represent two planes through the origin, in space of 3 dimensions; their intersection will be a line. At least, this will be so in general. There will be exceptional cases; for instance we may have $a = b = c = 0$, in which case the first equation tells us nothing at all; or the two planes may coincide. We will suppose we are dealing with the general situation.

If we take $f(1) - c(2)$ we get rid of z and find $x(af - cd) + y(bf - ce) = 0$. This will be satisfied if $x = k(bf - ce)$ and $y = -k(af - cd)$, where k is any number.

If we get rid of x by taking $-d(1) + a(2)$ we obtain $y(ae - bd) + z(af - cd) = 0$. This will be satisfied if we use the value already found for y and take $z = k(ae - bd)$.

It may be checked that the values of x, y, z just found do in fact satisfy equations (1) and (2).

It will be noticed that the expressions found have the form of determinants. We may write our solution in the form

$$x = k \begin{vmatrix} b & c \\ e & f \end{vmatrix}, \quad y = -k \begin{vmatrix} a & c \\ d & f \end{vmatrix}, \quad z = k \begin{vmatrix} a & b \\ d & e \end{vmatrix}.$$

The equations (1) and (2) had the coefficient scheme

$$\begin{array}{ccc} x & y & z \\ a & b & c \\ d & e & f \end{array}$$

In the solution, the determinant associated with x can be formed by striking out the symbols underneath x in the coefficient scheme. Similar rules apply for y and z.

With these determinants we must put the signs $+$, $-$, $+$, alternating just as they did when we found the formula for 3×3 determinant. There is also of course the arbitrary multiplier k.

This procedure, known as Cramer's rule, can be extended to the general case of n linear equations in $(n + 1)$ unknowns, but its practical value is so small that it will not be discussed here

28 Matrices other than square; partitions

So far we have used only square matrices. These correspond to transformations in which a space is mapped onto itself, or possibly to cases in which a space is mapped onto another space having the same number of dimensions.

Students sometimes seem to feel that it is wrong to have a map involving spaces with different dimensions, but in fact this is perfectly in order. It is indeed an everyday experience. If we buy some pencils and notebooks, the price of a pencil and of a notebook being fixed, the mapping, purchase $\rightarrow$ payment, is from 2 dimensions to 1 dimension. For we need two numbers, x, y, to specify the purchase, x pencils and y notebooks, but only one number to specify the money paid. In an electrical circuit the voltages supplied by 2 batteries may determine the currents in 3 resistances, or in 32, or in 3007. In a pin jointed framework, the small displacements of the pins determine the tensions in the bars. There is no reason to expect that the number of co-ordinates needed to specify the positions of the pins will equal the number of the bars. Thus it is in no way unusual for an input, specified by m numbers to lead to an output specified by n numbers, with $m \neq n$. Thus a mapping from m dimensions to n dimensions is in no way strange or impermissible.

A linear mapping $(x, y, z) \rightarrow (s, t)$ would be represented by equations

$$\left.\begin{array}{l} ax + by + cz = s, \\ dx + ey + fz = t. \end{array}\right\} \tag{1}$$

In matrix form this would appear as

$$\begin{pmatrix} a & b & c \\ d & e & f \end{pmatrix} \begin{pmatrix} x \\ y \\ z \end{pmatrix} = \begin{pmatrix} s \\ t \end{pmatrix}. \tag{2}$$

The mapping here is represented by a 2×3 matrix – that is, 2 rows and 3 columns. Note that 2 gives the dimensions of the output, 3 the dimensions of the input.

When we form a product ST it is understood the output of T becomes the input of S. It will be impossible to form the product ST if the output of T is not acceptable as an input for S; that is to say, if it is of different dimension. We can form this product if S is an $m \times n$ matrix and T an $n \times p$ matrix. For instance, S could be 3×2 and T could be 2×4, as here:

$$\left. \begin{aligned} \xi &= aX + bY, \\ \eta &= cX + dY, \\ \zeta &= eX + fY, \end{aligned} \right\} \tag{3}$$

$$\left. \begin{aligned} X &= gx + hy + iz + jt, \\ Y &= kx + my + nz + pt. \end{aligned} \right\} \tag{4}$$

Here X, Y form a link between x, y, z, t and ξ, η, ζ. If we substitute for X, Y we can obtain equations for the mapping $(x, y, z, t) \rightarrow (\xi, \eta, \zeta)$. If this is done, and all equations are then expressed in matrix form, we arrive at the result:

$$\begin{pmatrix} a & b \\ c & d \\ e & f \end{pmatrix} \begin{pmatrix} g & h & i & j \\ k & m & n & p \end{pmatrix} = \begin{pmatrix} ag + bk & ah + bm & ai + bn & aj + bp \\ cg + dk & ch + dm & ci + dn & cj + dp \\ eg + fk & eh + fm & ei + fn & ej + fp \end{pmatrix}. \tag{5}$$

It will be seen that this resembles very closely the procedure for matrix multiplication that we have been using until now. We go across the rows in the first matrix, down the columns of the second matrix.

If we try to form an impossible product, we soon notice it. Combining a row with a column, we find that we run out of letters in one before the other is finished, and we are at a loss how to proceed. Thus we automatically receive warning that we are writing nonsense.

If S is the matrix corresponding to equations (3), and T to equations (4), the sum $S + T$ is meaningless. We cannot work from the definition $(S + T)v = Sv + Tv$. For S accepts only vectors v from a 2-dimensional space and T accepts only vectors from a 4-dimensional space. Thus we cannot find an input v that makes both Sv and Tv intelligible. Nor could it help us if we could. For Sv, an output from S, is a vector in 3 dimensions, while Tv, an output from T is in 2 dimensions. We cannot add vectors of different dimensionality. For $S + T$ to be meaningful S and T must have the same input space (otherwise we cannot find v to feed in) and the same output space (otherwise we cannot add Sv and Tv).

Thus $S + T$ can be defined if both S and T are 2×3 matrices. If

$$S = \begin{pmatrix} a & b & c \\ d & e & f \end{pmatrix} \qquad T = \begin{pmatrix} g & h & i \\ j & k & l \end{pmatrix}$$

then

$$S + T = \begin{pmatrix} a + g & b + h & c + i \\ d + j & e + k & f + l \end{pmatrix}.$$

Here again the rule closely resembles what we have already been doing. We have only to remember that addition is not always possible, and this is fairly easy since here too the procedure becomes impossible if we apply it to a meaningless situation.

Multiplication by a number k remains exactly as before. For instance, with the matrix S as in the last example

$$kS = \begin{pmatrix} ka & kb & kc \\ kd & ke & kf \end{pmatrix}.$$

This operation can always be done. The mapping kS means that we apply the mapping S, and then multiply the resulting vector by k. There is no obstacle to carrying this out.

The inverse matrix is not defined when the matrix is not square. For instance, with the matrix equation (2), corresponding to the equations (1), we can go from x, y, z to s, t but not in the reverse direction. Given s and t, we cannot find the 3 unknowns x, y, z.

In a mapping of the type in equation (3), such as, for example

$$\xi = X + 2Y,$$
$$\eta = 3X + 5Y,$$
$$\zeta = 4X + 7Y,$$

the difficulty of getting back from ξ, η, ζ to X, Y is of the opposite kind. Now we have too many equations to satisfy rather than too few. For example, we cannot find any X, Y to give $\xi = 3$, $\eta = 8$, $\zeta = 2$.

Determinants also are not defined except for square matrices.

Partition of matrices

In §27 dealing with determinants, we sometimes spoke of determinant $\begin{pmatrix} p & q \\ r & s \end{pmatrix}$ and sometimes of $\det(u, v)$ where

$$u = \begin{pmatrix} p \\ r \end{pmatrix}, \qquad v = \begin{pmatrix} q \\ s \end{pmatrix}.$$

Thus as it were, we drew a line down the middle of the matrix, and fused the entries p and r together into the vector u, and the entries q and s into the vector v.

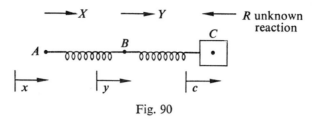

Fig. 90

We can in fact write the matrix $\begin{pmatrix} p & q \\ r & s \end{pmatrix}$ as (u, v) and the procedures of matrix algebra will work just as well. If this matrix represents a transformation mapping x, y to x^*, y^* we can write

$$\begin{pmatrix} x^* \\ y^* \end{pmatrix} = (u\,v)\begin{pmatrix} x \\ y \end{pmatrix} = ux + vy$$

by the usual row and column multiplication procedure. This again can be written

$$\begin{pmatrix} x^* \\ y^* \end{pmatrix} = \begin{pmatrix} p \\ r \end{pmatrix}x + \begin{pmatrix} q \\ s \end{pmatrix}y = \begin{pmatrix} px \\ rx \end{pmatrix} + \begin{pmatrix} qy \\ sy \end{pmatrix} = \begin{pmatrix} px + qy \\ rx + sy \end{pmatrix}$$

by the usual procedures for calculations involving vectors. These are the correct equations, corresponding to the matrix in its original form. Note incidentally how well the expression $ux + vy$ occurring above brings out the meaning of the columns u, v, which we noted in §22.

In fact we have considerable freedom to chop a matrix up and to regard it as a mosaic of smaller matrices. This process is known as *partitioning*.

We may wish to partition a matrix because the quantities on which it acts fall into separate classes. For instance, consider a very much simplified problem in the theory of structures.* We have two springs, which lie in a line, as shown in Fig. 90.

The point C is fastened; it has a prescribed displacement, c. Known forces X and Y act at A and B. They call into existence an unknown reaction R at C. The points A and B experience unknown displacements x and y. Thus both the displacements and the forces fall naturally into two classes, known and unknown. We naturally separate x, y from c; X, Y from R.

The quantities in this problem are linked by the equations

$$\begin{aligned} X &= k_{11}x + k_{12}y \;\vdots\; + k_{13}c \\ Y &= k_{21}x + k_{22}y \;\vdots\; + k_{23}c \\ \hline R &= k_{31}x + k_{32}y \;\vdots\; + k_{33}c \end{aligned}$$

where the coefficients k_{rs} can be calculated from the spring constants.

In view of the separation we have noticed, it would be natural to introduce a vector

---

* See H. C. Martin – *Introduction to Matrix Methods of Structural Analysis* chapter 2 (McGraw-Hill, 1966).

$v = \begin{pmatrix} x \\ y \end{pmatrix}$ to specify unknown displacements, and a vector $F = \begin{pmatrix} X \\ Y \end{pmatrix}$ to specify known forces. In the equations above, the first two columns on the right-hand side are associated with the vector v, the first two rows with the vector F. The dotted lines indicate this situation. We now look at the various compartments in these equations. The contents of each compartment go very easily into matrix notation:

$$\begin{matrix} k_{11}x + k_{12}y = \\ k_{21}x + k_{22}y = \end{matrix} \begin{pmatrix} k_{11} & k_{12} \\ k_{21} & k_{22} \end{pmatrix} \begin{pmatrix} x \\ y \end{pmatrix} = Hv, \quad \text{say.}$$

This deals with the north-west box. In the north-east we see

$$\begin{pmatrix} k_{13}c \\ k_{23}c \end{pmatrix} = \begin{pmatrix} k_{13} \\ k_{23} \end{pmatrix} c = Lc, \quad \text{say.}$$

Similarly we have

$$k_{31}x + k_{32}y = (k_{31} \quad k_{32}) \begin{pmatrix} x \\ y \end{pmatrix} = Gv, \quad \text{say.}$$

We do not need any new notation for $k_{33}c$ which is a single number. Note incidentally that a single number, k_{33}, can be regarded as a 1×1 matrix, mapping 1-dimensional space to 1-dimensional space.

Our equations can now be written

$$F = Hv + Lc, \tag{6}$$

$$R = Gv + k_{33}c. \tag{7}$$

These can be combined into the form of a single matrix

$$\begin{pmatrix} F \\ R \end{pmatrix} = \begin{pmatrix} H & L \\ G & k_{33} \end{pmatrix} \begin{pmatrix} v \\ c \end{pmatrix}. \tag{8}$$

Equations (6) and (7) might be convenient for finding v and R. If we solve equation (6) for v we find $v = H^{-1}(F - Lc)$. All the quantities on the right-hand side are known. Note here that we are making use of the inverse, H^{-1}. Substituting in equation (7) gives us the unknown reaction R as $GH^{-1}(F - Lc) + k_{33}c$.

Of course this problem is so simple that we would in fact never solve it this way. It is intended to provide an illustration and a model for much more complicated problems involving hundreds of variables. The importance of equation (8) is that the whole procedure becomes stereotyped. The computer will be programmed to deal with this single matrix equation, the form of which is such that the machine can recognize which quantities are known and which are unknowns. The computer would be programmed to calculate the matrices H, L, G and the numbers k_{33} from raw data about the structure involved – the number of bars and plates involved, the manner in which they are joined, their elastic constants and so forth.

In this particular example the columns and rows have been split in the same manner. In each case we have separated the first two from the third. It is not in any way necessary that this similarity should exist. For instance, in our first example of partition we had the division

$$\begin{pmatrix} p & \vdots & q \\ \cdots & & \cdots \\ r & \vdots & s \end{pmatrix}.$$

We divided the columns; the rows were not divided at all. So long as we are dealing with a single matrix, we can divide the rows in any way we like and the columns in any way we like.

When we are dealing with the product of matrices, or with a matrix acting as a vector, we must make sure that the partition is such that the necessary multiplications can in fact be carried out. For instance in the product

$$\begin{pmatrix} 1 & 2 & 3 & \vdots & 4 & 5 \\ 6 & 7 & 8 & \vdots & 9 & 10 \\ 11 & 12 & 13 & \vdots & 14 & 15 \\ 16 & 17 & 18 & \vdots & 19 & 20 \end{pmatrix} \begin{pmatrix} 1 & 2 & 3 \\ 4 & 5 & 6 \\ 7 & 8 & 9 \\ \hdashline 10 & 11 & 12 \\ 13 & 14 & 15 \end{pmatrix}$$

if we have decided to split the second matrix into 3 rows and 2 rows, we are forced to split the first matrix into 3 columns and 2 columns. Otherwise we shall not be able to form the necessary products in the scheme

$$(S \vdots T)\begin{pmatrix} U \\ \cdots \\ V \end{pmatrix} = SU + TV.$$

In a product of two matrices, such as the example just given, the restriction does not in any way affect our freedom to chop up the rows of the first matrix and the columns of the second. The partition shown above is perfectly satisfactory: we can, if we wish, leave it as it is. However, if we have some reason for doing so, we can draw further horizontal divisions in the first matrix, and further vertical divisions in the second, in any way we like.

Questions

1   Calculate the products

(a) $\begin{pmatrix} 1 & 1 & 1 \\ 1 & 1 & 0 \end{pmatrix} \begin{pmatrix} 1 & 4 & 6 & 3 \\ 2 & 5 & 5 & 2 \\ 3 & 6 & 4 & 1 \end{pmatrix}$ (b) $\begin{pmatrix} 1 & 1 & 1 \\ 1 & 1 & 0 \end{pmatrix} \begin{pmatrix} 1 & 4 \\ 2 & 5 \\ 3 & 6 \end{pmatrix}$

(c) $\begin{pmatrix} 1 & 1 \\ 1 & 1 \\ 1 & 0 \end{pmatrix} \begin{pmatrix} 1 & 3 & 5 \\ 2 & 4 & 6 \end{pmatrix}$ (d) $\begin{pmatrix} 1 \\ 2 \\ 3 \end{pmatrix} (1 \quad 3 \quad 5)$

(e) $(1 \quad 3 \quad 5) \begin{pmatrix} 1 \\ 2 \\ 3 \end{pmatrix}$
(f) $\begin{pmatrix} a & b & c \\ x & y & z \end{pmatrix} \begin{pmatrix} a & x \\ b & y \\ c & z \end{pmatrix}$

(g) $\begin{pmatrix} a & x \\ b & y \\ c & z \end{pmatrix} \begin{pmatrix} a & b & c \\ x & y & z \end{pmatrix}$.

2   In Question 1, nine different matrices occur as factors of the various products. Discuss which of these matrices are capable of being added together.

3   Calculate

$$\begin{pmatrix} 1 & 1 & 1 \\ a & b & c \end{pmatrix} \begin{pmatrix} 1 & a \\ 1 & b \\ 1 & c \end{pmatrix} \quad \text{and} \quad \begin{pmatrix} 1 & 1 & 1 & 1 \\ a & b & c & d \\ a^2 & b^2 & c^2 & d^2 \end{pmatrix} \begin{pmatrix} 1 & a & a^2 \\ 1 & b & b^2 \\ 1 & c & c^2 \\ 1 & d & d^2 \end{pmatrix};$$

the answers can be written conveniently by using the abbreviation S_n for $a^n + b^n + \cdots$ The products so found belong to a type of matrix that occurs when curves are fitted to observed data by the method of Least Squares.

4   Calculate

$$\begin{pmatrix} 1 & 1 & 2 \\ 1 & -2 & 1 \end{pmatrix} \begin{pmatrix} -1 & -3 \\ 0 & -1 \\ 1 & 2 \end{pmatrix} \quad \text{and} \quad \begin{pmatrix} 1 & 1 & 2 \\ 1 & -2 & 1 \end{pmatrix} \begin{pmatrix} 4 & 2 \\ 1 & 0 \\ -2 & -1 \end{pmatrix}.$$

Does the result throw any light on the question of whether a matrix with 2 rows and 3 columns can have an inverse?

5 P is a matrix with 2 rows and 4 columns. Q is a matrix, and the products PQ and QP are both meaningful. How many rows and columns has Q? What are the numbers of rows and columns in PQ? And in QP?

6   Let

$$O = \begin{pmatrix} 0 & 0 \\ 0 & 0 \end{pmatrix} \quad \text{and} \quad I = \begin{pmatrix} 1 & 0 \\ 0 & 1 \end{pmatrix}.$$

Let

$$P = \begin{pmatrix} O & I \\ I & O \end{pmatrix}, \quad Q = \begin{pmatrix} O & -I \\ I & O \end{pmatrix}, \quad R = \begin{pmatrix} I & -I \\ O & I \end{pmatrix}$$

so that P, Q, R are matrices with 4 rows and 4 columns. Calculate P^2, Q^2, R^2, R^3, PQ, PR and QR by means of the partitioned forms given above. Also write P, Q, R out in full and calculate the results directly, without the use of partitions. Compare the labour involved in the two methods.

The solution for the first of the required products is given here.

Solution: to find P^2.
First method.

$$P^2 = \begin{pmatrix} O & I \\ I & O \end{pmatrix} \begin{pmatrix} O & I \\ I & O \end{pmatrix} = \begin{pmatrix} I & O \\ O & I \end{pmatrix} = \begin{pmatrix} 1 & 0 & 0 & 0 \\ 0 & 1 & 0 & 0 \\ 0 & 0 & 0 & 0 \\ 0 & 0 & 0 & 1 \end{pmatrix}.$$

Second method.

$$P^2 = \begin{pmatrix} 0 & 0 & 1 & 0 \\ 0 & 0 & 0 & 1 \\ 1 & 0 & 0 & 0 \\ 0 & 1 & 0 & 0 \end{pmatrix} \begin{pmatrix} 0 & 0 & 1 & 0 \\ 0 & 0 & 0 & 1 \\ 1 & 0 & 0 & 0 \\ 0 & 1 & 0 & 0 \end{pmatrix} = \begin{pmatrix} 1 & 0 & 0 & 0 \\ 0 & 1 & 0 & 0 \\ 0 & 0 & 1 & 0 \\ 0 & 0 & 0 & 1 \end{pmatrix}.$$

7   Let

$$\mathbf{I} = \begin{pmatrix} 1 & 0 \\ 0 & 1 \end{pmatrix}, \qquad \mathbf{A} = \begin{pmatrix} p & q \\ r & s \end{pmatrix} \text{ and } \mathbf{O} = \begin{pmatrix} 0 & 0 \\ 0 & 0 \end{pmatrix}.$$

Find the product

$$\begin{pmatrix} \mathbf{I} & \mathbf{A} \\ \mathbf{O} & \mathbf{I} \end{pmatrix} \begin{pmatrix} \mathbf{I} & -\mathbf{A} \\ \mathbf{O} & \mathbf{I} \end{pmatrix}.$$

What is the inverse of the matrix

$$\begin{pmatrix} 1 & 0 & 2 & 3 \\ 0 & 1 & 4 & 5 \\ 0 & 0 & 1 & 0 \\ 0 & 0 & 0 & 1 \end{pmatrix}?$$

8   Calculate the product

$$\begin{pmatrix} a_1 & a_2 & a_3 \\ \hline b_1 & b_2 & b_3 \end{pmatrix} \begin{pmatrix} u_1 & v_1 \\ u_2 & v_2 \\ u_3 & v_3 \end{pmatrix}$$

directly, i.e. without any reference to partitioning. By partitioning in the manner indicated by the broken lines, this product may be written as $\begin{pmatrix} a' \\ b' \end{pmatrix}(u \quad v)$ where a', b' represent the rows of the first matrix and u, v the columns of the second. Use this form to obtain the product. Observe that the product, in this second form obtained by partitioning, gives a useful abbreviation for the product as obtained directly.

9   Let

$$M = \begin{pmatrix} a & d \\ b & e \\ c & f \end{pmatrix} \begin{pmatrix} 1 & 3 & 5 \\ 2 & 4 & 6 \end{pmatrix}.$$

We can write M in the abbreviated form

$$M = (u\ v)\begin{pmatrix} 1 & 3 & 5 \\ 2 & 4 & 6 \end{pmatrix}$$

where u and v represent the columns of the first factor in M. Use this last equation to write the product M. What can be said, in geometrical terms, about the three vectors that occur in the columns of M? What is the value of det M? Can any similar conclusion be drawn about det (PQ) where P is any 3×2 matrix and Q any 2×3 matrix?

29   Subscript and summation notation

In §28, equations (3) and (4) came close to using the entire resources of the alphabet. In problems where large numbers of variables are involved, we are faced not only with the difficulty of finding enough symbols; we also need some systematic way of choosing and employing the symbols. For instance, in an expression such as $ax + by + cz$ we can see that a is the coefficient of the first variable x, b of the second variable y and c of the third variable z. However, we cannot proceed in this way if we have 100 variables and 100 coefficients corresponding to them. If we are communicating with a human being we may write $ax + by + \cdots cz$ where the dots indicate 'and so on, until the 100th variable z is reached'. Such explanations are useless if we are dealing with a computer: explicit instructions are needed then.

The natural numbers provide a convenient means for generating arbitrarily large collections of symbols. If we have 100 unknowns, we can write them as x_1, x_2, x_3 up to x_{100}. If we wish to form a linear expression we can use $a_1x_1 + a_2x_2 + \cdots + a_{100}x_{100}$, and it is immediately clear that the coefficient of, say, x_{57} is a_{57}. We still have dots indicating 'and so on', but it is quite easy to programme instructions that make a computer do exactly what is required.

The summation sign, $\sum$ or S, provides a convenient abbreviation. '$\sum$' is the Greek equivalent of 'S', which is the initial letter of 'Sum'. By the expression

$$\sum x_r, \quad 1 \leqslant r \leqslant 3$$

we understand $x_1 + x_2 + x_3$. The instruction $1 \leqslant r \leqslant 3$ tells us that in x_r we are to replace r in turn by 1, 2 and 3, so we get x_1, x_2, x_3. The instruction $\sum$ tells us that these are to be summed to give $x_1 + x_2 + x_3$. The linear expression mentioned in the previous paragraph would be shown as $\sum a_r x_r; 1 \leqslant r \leqslant 100$. Sometimes the numbers, at which r is to begin and end, are put below and above the summation sign. This is convenient enough for handwriting but is very inconvenient for the printer.

When we are dealing with a linear transformation, that is to be expressed by a matrix, a certain scheme of subscripts is almost invariably used. It is seen in the following example:

$$\left.\begin{aligned}
x_1^* &= a_{11}x_1 + a_{12}x_2 + a_{13}x_3 + a_{14}x_4 \\
x_2^* &= a_{21}x_1 + a_{22}x_2 + a_{23}x_3 + a_{24}x_4 \\
x_3^* &= a_{31}x_1 + a_{32}x_2 + a_{33}x_3 + a_{34}x_4
\end{aligned}\right\} \tag{1}$$

Here the first subscript tells us which row we are in; for example, $a_{31}, a_{32}, a_{33}, a_{34}$ all occur in the equation for x_3^* and so are found in the third row. The second subscript tells us in which column the coefficient occurs; thus a_{12}, a_{22}, a_{32} all occur as coefficients of x_2 and so are in the second column of the coefficient arrangement.

The whole of equations (1) can be written in the very compact form

$$x_r^* = \sum_s a_{rs}x_s; \qquad 1 \leqslant r \leqslant 3; 1 \leqslant s \leqslant 4.$$

Here we have to write s under the summation sign to indicate that we are summing only for s, not for r. If we plan to make a further transformation

$$\left. \begin{array}{l} x_1^{**} = b_{11}x_1^* + b_{12}x_2^* + b_{13}x_3^* \\ x_2^{**} = b_{21}x_1^* + b_{22}x_2^* + b_{23}x_3^* \end{array} \right\} \qquad (2)$$

this transformation can be written

$$x_q^{**} = \sum_r b_{qr}x_r^*; \qquad 1 \leqslant q \leqslant 2; 1 \leqslant r \leqslant 3.$$

Let A and B represent transformations (1) and (2), and let C stand for the product transformation BA. It is useful for anyone unfamiliar with this symbolism to write the matrices for B and A, and to calculate sample entries in the matrix for C. There are 8 entries in C and each involves the sum of 3 products. Nothing is gained by writing out the entire matrix for C. Enough should be written for the pattern to become apparent. For instance, we can check that c_{23}, the entry in the second row and third column of C, is given by

$$c_{23} = b_{21}a_{13} + b_{22}a_{23} + b_{23}a_{33}.$$

The pattern here may appear more clearly if we suppress the middle subscripts, and write first

$$c_{23} = b_{2.}a_{.3} + b_{2.}a_{.3} + b_{2.}a_{.3}$$

The pattern in the last equation should be evident. To get to the complete equation, we replace the dots in the first term by 1 and 1, in the second term by 2 and 2, and in the third term by 3 and 3. Our results may be shown in compact form as

$$c_{23} = \sum_r b_{2r}a_{r3}.$$

The numbers 2 and 3 enjoy no special privileges. If we replace them by the symbols q and s we obtain the formula for any entry in the matrix for C;

$$c_{qs} = \sum_r b_{qr}a_{rs}. \qquad (3)$$

In our particular example we would have to indicate that q could stand for 1 or 2, s for 1, 2, 3 or 4, and that we had to sum r over the values 1, 2, 3.

For *any* matrix product, $C = BA$, equation (3) is valid. It needs to be supplemented only by details as to the values through which q, r and s are to run.

Obviously this notation is particularly valuable when matrices with many rows and columns are involved. It can also be useful when certain theoretical results are being obtained. For instance, we shall need it when we come to define and to calculate with the transpose of a matrix.

Notice that, with this symbolism, the subscripts over which we sum are always adjacent. In equation (3) above, no other subscript separates the two subscripts r involved in the summation. Similarly, in the compact form of equations (1),

$$x_r^* = \sum_s a_{rs} x_s,$$

we are summing over s, and we find the subscripts s close together. (This does not apply to the letter s written under the summation sign; this is not a subscript.)

Repeated summations may be met. For instance, if in the equations $x_q^{**} = \sum_s c_{qs} x_s$, which represent the transformation C we substitute the values of c_{qs} given by equation (3), we find

$$x_q^{**} = \sum_r \sum_s b_{qr} a_{rs} x_s. \tag{4}$$

These equations give x_1^{**}, x_2^{**} in terms of x_1, x_2, x_3, x_4 and could be obtained by a primitive method, namely by substituting in equations (2) the values of x_1^*, x_2^*, x_3^* given by equations (1). Anyone who is not clear about the effect of the double summation sign might find it helpful to carry through this primitive method far enough to see what the expressions mean.

Quite generally, if some symbolism involving summation signs does not convey a clear message to the student, the best thing to do is to go back from the compact notation to the more cumbersome, but more familiar, elementary notation, such as that in equations (1) and (2).

Notice that equation (4) still shows the subscripts, over which we are to sum, as neighbours. We are summing over r and s; the sequence of subscripts is $q; r, r; s, s$.

30   *Row and column vectors*

At some stage of his reading a student is likely to meet references to row vectors and column vectors. For instance in 2 dimensions we have already spoken of the 2×1 matrix $\begin{pmatrix} x \\ y \end{pmatrix}$ as a column vector. The 1×2 matrix (a, b) is often spoken of as a row vector.

In matrix formalism it is possible to add two column vectors: we have already done this on many occasions. It is also possible to add two row vectors,

$$(a_1, b_1) + (a_2, b_2) = (a_1 + a_2, b_1 + b_2).$$

However, it is meaningless to speak of the sum of a row vector and a column vector, for the lay-out of the entries differs in the two cases. With multiplication the situation is rather the reverse. Matrix rules give no meaning to the product of (a_1, b_1) and (a_1, b_2). We cannot carry out row and column multiplication. Similarly no meaning attaches to the product of two column vectors. However, it is possible to form the product of a row vector and a column vector $(a, b)\begin{pmatrix} x \\ y \end{pmatrix} = ax + by$.

It may seem very strange that there should exist two different types of vector, and that there should be such restrictions on operations with them. However, there are situations in which such a distinction arises quite naturally. Suppose for instance that a and b represent forces acting on some structure, while x and y represent displacements. It is quite natural that we should add displacements together, or add forces together, but never add displacement to force. Again force multiplied by displacement represents work done, and in fact the expression $ax + by$ does arise in this connection. However, force multiplied by force does not suggest any significant physical property, nor, in this immediate context, does displacement multiplied by displacement. Matrix formalism therefore would seem to fit quite snugly to the idea of identifying forces with row vectors and displacements with column vectors.

The idea comes out even more strongly in a very elementary application. Suppose $\begin{pmatrix} x \\ y \end{pmatrix}$ represents a purchase of x nuts and y bolts, and that (a, b) represents a price table – a dollars for a nut, b dollars for a bolt. Then $ax + by$, which is $(a, b)\begin{pmatrix} x \\ y \end{pmatrix}$, represents the cost of a purchase $\begin{pmatrix} x \\ y \end{pmatrix}$ when the price table is (a, b). We do not expect to multiply purchases by purchases or dollars by dollars. Equally we may add purchases together, or we may add price schemes, but we do not add purchases to price schemes.

This application is helpful when we come to consider the graphical representation of column and row vectors. A purchase, we represent, as usual, by a point on the graph paper. Thus the point shown in Fig. 91 represents a purchase of 4 nuts and 1 bolt.

How are we to represent a price scheme such as $(0.5, 1)$, $0.50 for a nut and $1 for a bolt? We cannot mark the point with $x = 0.5$ and $y = 1$. That would represent a purchase (if such a thing were possible) of half a nut and a bolt. We cannot mark a

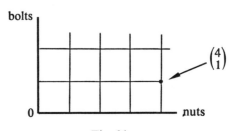

Fig. 91

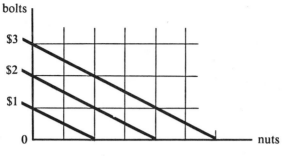

Fig. 92

point on the graph for purchases; knowing the prices at which objects were bought tells us nothing about how many were bought.

We could meet the difficulty by having an entirely separate chart on which prices were marked, and this idea can be developed in a fruitful manner. However, we shall not do this here. We want to bring out the relevance of the price scheme to the purchase diagram, and this is easily stated. A price scheme does not tell us what we have purchased; it does tell us what it would cost us if we decided to make some particular purchase. This we can show on the graph. We could mark on each point the cost corresponding to the purchase that point represents. It might be rather more convenient to follow the procedure of a weather map, in which pressures are indicated by isobars linking points with equal pressures. The diagram would then appear as in Fig. 92.

From this diagram we can read off the cost of each purchase. Thus the point for 4 nuts and 1 bolt, which we saw on the diagram before this, falls on the line marked 3. It would cost $3. Some points of course fall between the lines marked. Thus 3 nuts and 1 bolt cost $2.5. It can be seen that the corresponding point falls halfway between the lines marked 2 and 3. Thus in fact this graph does convey all the information contained in the price scheme.

The equations of the lines are $0.5x + y = 1$, $0.5x + y = 2$, $0.5x + y = 3$. Thus, in reading off the numbers marked on the lines we are simply obtaining the value of $0.5x + y$, the expression with coefficients 0.5 and 1. Thus the row vector $(0.5, 1)$ can be regarded as a coefficient scheme for such an expression.

A price scheme tells us how to work out the cost of any purchase. Thus a price scheme establishes a mapping purchase → cost. In our example above this mapping may be specified as

$$\begin{pmatrix} x \\ y \end{pmatrix} \rightarrow 0.5x + y.$$

Quite generally we can regard the row vector (a, b) as something that acts on a column vector and gives a number as the output; this ties in easily with the equation

$$(a, b)\begin{pmatrix} x \\ y \end{pmatrix} = ax + by.$$

We may interpret this as 'the operation (a, b) acting on the vector $\begin{pmatrix} x \\ y \end{pmatrix}$ gives the number $ax + by$'.

The fact that a column vector is represented by a point while a row vector is seen as a system of lines brings out the fact that they are two very different things. What they have in common is that they both require only two numbers to specify them. In 3 dimensions they would each require 3 numbers. In 3 dimensions, the column vector would still be represented by a point; the row vector would be represented by a system of evenly spaced parallel planes corresponding to the equations $ax + by + cz = n$, with n an integer.

Change of axes

In some piece of work, we may begin with a certain system of axes, and points specified by co-ordinates in those axes. We think of these co-ordinates as written in the form of a column. At some stage we may wish to introduce a new system of axes. We still wish to deal with the same points as before, and so we ask what new co-ordinates will now be needed to represent the same points. In this way we arrive at formulas connecting the co-ordinates in the old and the new system. This was done in §9.

In the same way, there may be some physical object which is specified in the original axes by a row vector. In 2 dimensions, this row vector may be represented by a system of lines, that enable us to attach a number to each point of the plane. These numbers have physical significance, so naturally, if we decide to use co-ordinates in some other system of axes, we must make sure that the row vector used gives the same numbers, the same system of lines. Thus the lines provide the link between the two systems.

An example may be helpful. Suppose our two systems are as shown in Fig. 93.

The dotted lines, labelled with the numbers 1, 2, 3, represent a row vector. It can be checked that in the first system of co-ordinates (based on squares) the dotted lines correspond to the expression $(x/2) + (y/2)$. In the second system (oblique co-ordinates) the line marked 1 contains the points $X = 2$, $Y = 0$ and $X = 0$, $Y = 1$. It is not hard to show that the equation of this line must be $(X/2) + Y = 1$ and in fact the expression

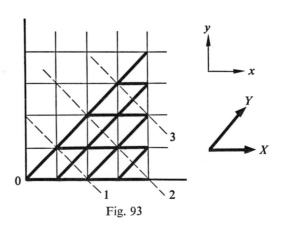

Fig. 93

$(X/2) + Y$ is the one we want for this system of co-ordinates. Reading off the coefficients in the two cases, we see that the row vector $(\frac{1}{2}, \frac{1}{2})$ in the first system represents the same object as the row vector $(\frac{1}{2}, 1)$ in the second system.

Exercise

By the method of §9, find equations giving x, y in terms of X, Y and verify that, for any X, Y, $(x/2) + (y/2) = (X/2) + Y$. This shows that the conclusion reached above could also have been found directly by algebra.

We will now consider the general case. Suppose we have a row vector (a, b) in a system with co-ordinates x and y, and that new co-ordinates, X and Y, are brought in by means of the equations

$$\left.\begin{array}{l} x = pX + qY, \\ y = rX + sY. \end{array}\right\} \tag{1}$$

We wish to find (A, B), the row vector as specified in the new system. This can be found very quickly. Any difficulty that may arise in this connection is not in carrying out the calculations, but in seeing clearly what is being done.

The row vector (a, b) assigns to the point $\begin{pmatrix} x \\ y \end{pmatrix}$ the number $ax + by$. By the equations above, this number is the same as $a(pX + qY) + b(rX + sY)$, which in turn equals $(ap + br)X + (aq + bs)Y$, where $\begin{pmatrix} X \\ Y \end{pmatrix}$ is the specification of the same point in the second system. Now, as seen by users of the second system, this point should be assigned the number $AX + BY$. These two results will agree if $A = ap + br$ and $B = aq + bs$. These equations tell us how to deal with row vectors when we change from one system to the other.

It may be noticed that the equations just obtained can be written in matrix form as

$$(A \quad B) = (a \quad b) \begin{pmatrix} p & q \\ r & s \end{pmatrix}. \tag{2}$$

Now equations (1) can be written as

$$\begin{pmatrix} x \\ y \end{pmatrix} = \begin{pmatrix} p & q \\ r & s \end{pmatrix} \begin{pmatrix} X \\ Y \end{pmatrix}. \tag{3}$$

Thus the same matrix appears in equations (2) and (3); in equations (3) it is written *before* the column vector, in equations (2) *after* the row vector. Note also a difference between equations (2) and (3); in equations (3) we can substitute the *new* co-ordinates (X and Y) to find the old co-ordinates (x and y), but in equations (2) it is the other way round – we can substitute the *old* in order to obtain the new.

All of this can be shown very conveniently by bringing in vector and matrix abbreviations; this incidentally provides a proof of these results for any number of dimensions.

We write

$$v = \begin{pmatrix} x \\ y \end{pmatrix}, \qquad V = \begin{pmatrix} X \\ Y \end{pmatrix}, \qquad M = \begin{pmatrix} p & q \\ r & s \end{pmatrix},$$

$$h = (a, b), \qquad H = (A, B).$$

Then $ax + by$ is written hv; this is the number associated with the vector that is called v in the first system. That vector is called V in the second system and is assigned the number HV. The number marked at any point is independent of the co-ordinates that some observer is using to describe that point, so we must have $hv = HV$.

Equations (3) may be written $v = MV$. If we substitute for v in the equation $hv = HV$, we get $hMV = HV$. Since this is to be so for every V, we must have $hM = H$, and this is equivalent to equations (2).

The equation $hM = H$ may also be written $h = HM^{-1}$. (Since M is a matrix suitable for changing axes, it must have an inverse; it must be possible to go back to our original system if we want to.)

Thus we can write both equations so as to give old in terms of new. The equations are thus, for column vectors $v = MV$, for row vectors $h = HM^{-1}$. It will be seen that our basic requirement, $hv = HV$, is automatically met, for

$$hv = (HM^{-1})(MV) = H(M^{-1}M)V = HV,$$

since $M^{-1}M = I$; the inverse, M^{-1}, wipes out M.

31   Affine and Euclidean geometry

The geometry systematized by Euclid and still taught in schools is essentially mathematical physics; that is to say, it is an attempt to give a logical account of our actual experiences with shapes and sizes. Both in geometry lessons and in everyday life we take it for granted that if there are 2 points there must be a distance between them, and if there are 2 lines that intersect, there must be an angle between them, that we can measure with a protractor. And of course, if we are surveying a region or designing some structure, these are perfectly sound and satisfactory ideas.

However, as we have seen, the idea of vector space is a very general one. The geometry of everyday life gives us an example of a vector space of 3 dimensions but this is far from being the only example. Many of the vector spaces we have met have appeared very remote from geometry as it is usually understood. We have in fact defined a vector space as being any collection of objects, $u, v, \ldots$, with a satisfactory definition of sums, such as $u + v$, and multiplication by a number, leading to expressions such as kv. In such a collection it by no means follows that we are able to define the length of a vector or the angle between two vectors.

For instance we have found it perfectly possible to work with a vector space in which $xA + yB$ signifies 'x copies of an article A and y copies of an article B'. In diagrams related to this space, we may very well use ordinary graph paper simply because it is readily available. In such a diagram the lines QA and QB will be of equal length and perpendicular. It is important to realize that the diagram has brought in aspects which are entirely foreign to the vector space in question, in which such ideas as *length* and *angle* are undefined and are totally without meaning. Suppose, for instance, we mark a dot on the point $(3, 4)$ of ordinary graph paper to represent the vector '3 nuts and 4 bolts'. This dot will be at a distance 5 units from the origin, and we may be tempted to think of 5 as the length or size of the vector. But there is no sensible procedure that would lead us to introduce 5 as a measure of magnitude for the consignment, 3 nuts and 4 bolts. Any meaningful property of the vector space, in which x nuts and y bolts is the typical vector, will appear equally well with graph paper in which the axes are not perpendicular and different units are employed on the x-axis and the y-axis. We shall still find that $u + v$ completes the parallelogram with 3 corners at 0, u, v, and that kv is reached by taking k steps, each equal to v, from the origin. Observe for instance that in §9, dealing with change of axes, in neither of the systems shown are the axes perpendicular.

A mathematical theory deals with situations in which certain features can be recognized. It may demand a long list of features; in this case very many results will be proved, but it may not be possible to apply the theory very often, since it will be rarely that all the necessary features will be found together in an actual situation. At the other extreme, very few features may be required; then relatively few results will follow but many applications may be expected, since only a few tests have to be passed. Of course, the nature of the features demanded is important. Mathematicians, and particularly of course those mathematicians of most interest to engineers and scientists, study the logical consequences of those features which have frequently occurred in real problems in the past or are likely to occur in the future. Great mathematicians sometimes seem to have an uncanny instinct for studying what will prove important in the future, as Riemann in 1854 laid one of the foundations for Einstein's theory of general relativity, and Hilbert, around 1905, provided mathematical procedures needed for quantum theory after 1925.

Most of the work we have done so far has been based on very modest demands; the only features we have required have been reasonable definitions of $u + v$ and kv. We may refer to this as *pure vector theory*. The geometry that develops from pure vector theory is known as Affine geometry. The word 'affine' is derived from 'affinity'; it deals with figures that resemble each other in a rather loose sense, as contrasted with the relations of congruence and similarity studied in Euclid's geometry. In affine geometry there is no mention of length or perpendicularity. The basic ideas are those

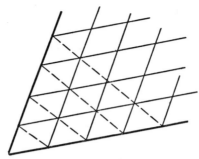

Fig. 94

of *line* and *parallel lines*. A grid, suitable for co-ordinate work, can be specified purely in terms of these two ideas, as Fig. 94 suggests. Notice how the parallel dotted lines here fix the whole grid once the first parallelogram has been drawn; they prevent an uneven spacing of the intervals along the axes.

From time to time we have used concepts that belong to Euclidean rather than affine geometry; we may have used squared graph paper, or spoken of reflections and rotations. This has been done purely in order to draw examples from situations familiar to students. All such references could have been omitted, so far as mathematical development was concerned. A linear mapping was defined by requiring

$$u + v \rightarrow u^* + v^*,$$

$ku \rightarrow ku^*$; the only operations involved here are the two basic operations of pure vector theory. When dealing with determinants, we seemed to use ideas involving perpendicularity, such as area = base times height. But this could have been avoided. The ratio of areas is a perfectly good affine concept. In Fig. 95, we can see that the parallelogram $OA^*C^*B^*$ can be chopped up into four pieces which can be moved by translation (an affine concept) so as to cover the parallelogram OAA^*B twice. Thus we can conclude, without going outside affine geometry, that the transformation $A \rightarrow A^*$, $B \rightarrow B^*$, doubles all areas and hence has determinant 2.

Students will no doubt meet the expressions $u \cdot v$, the scalar or dot product, and $u \times v$, the vector or cross product. It should be understood that these do not belong to pure vector theory and affine geometry. The dot product, $u \cdot v$, can be defined only in

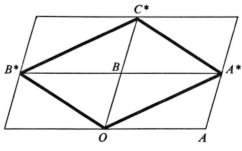

Fig. 95

situations to which Euclidean geometry applies; it requires for its definition the ideas of *length* and *perpendicular*. The cross product $u \times v$ is very much more restricted; it can be defined only in Euclidean space of 3 dimensions.

Until now, apart from occasional references and examples, we have kept within the bounds of affine geometry. From now on, we shall make use of Euclidean concepts whenever they are required.

32 Scalar products

In order to pass from affine geometry to Euclid's geometry a new idea has to be brought in, the *length* of a vector.

If we are using ordinary graph paper we can define the length, L, of a vector by the equation $L^2 = x^2 + y^2$. Similarly in 3 dimensions we would use $L^2 = x^2 + y^2 + z^2$ and in any number, n, of dimensions $L^2 = \sum x_r^2, 1 \leqslant r \leqslant n$.

In affine geometry, we had a great freedom to change axes; we could go from any basis of a space to any other basis without affecting our formulas. In Euclidean geometry this freedom is considerably less. Suppose for instance we decided to introduce new co-ordinates in 2 dimensions by means of the equations $x = X + Y, y = Y$. We would then find $L^2 = x^2 + y^2 = X^2 + 2XY + 2Y^2$. It would therefore be quite incorrect to take $X^2 + Y^2$ as giving L^2; the definition, in the form $L^2 = x^2 + y^2$ holds only in certain privileged co-ordinates systems.

From the definition of length we derive the definition of distance; *the distance between P and Q is the length of the vector Q − P.* If we use s for this distance we find in 2 dimensions $s^2 = (x_2 - x_1)^2 + (y_2 - y_1)^2$ in a well known notation. In 3 dimensions the definition gives $s^2 = (x_2 - x_1)^2 + (y_2 - y_1)^2 + (z_2 - z_1)^2$. In n dimensions there is a similar result; we have to change our notation so as to make use of subscripts; $s^2 = \sum (u_r - v_r)^2, 1 \leqslant r \leqslant n$ for vectors u, v.

The symbol $\|v\|$ indicates the length of the vector v. In the traditional notation of geometry, there is some uncertainty. We may write AB to indicate the line AB, the vector AB that runs from A to B, or the length AB. The length AB is measured by a number, and in our work it is always important to be clear whether a symbol stands for a transformation, a vector or a number. The symbols $\|v\|$ or $\|AB\|$ make it clear that a number is intended; $\|v\|^2$ or $\|AB\|^2$ indicates the square of that number. The symbol v^2 is meaningless; we cannot multiply a vector by a vector. Probably AB^2 would convey something to most readers, owing to their recollections of traditional geometry. However, writing $\|AB\|^2$ we put the matter beyond doubt, if it is understood here that AB means the vector from A to B.

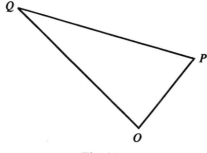

Fig. 96

Our definition of length is obviously suggested by our earlier experiences with Pythagoras' Theorem. We can use Pythagoras' Theorem, so to speak in reverse, to get a definition of perpendicular. We will say that $OP \perp OQ$ if $\|PQ\|^2 = \|OP\|^2 + \|OQ\|^2$, Fig. 96. This defines 'perpendicular' in terms of lengths, and as we have a formula for the length of any vector, we can use this definition to obtain an algebraic condition for OP and OQ to be perpendicular.

Exercises

1 In 3 dimensions, if P has co-ordinates (p_1, p_2, p_3) and Q has (q_1, q_2, q_3) show that $OP \perp OQ$ when $p_1q_1 + p_2q_2 + p_3q_3 = 0$.

2 In n dimensions, find the condition for $u \perp v$, where u has co-ordinates $(u_1, \ldots, u_n)$ and v has $(v_1, \ldots, v_n)$.

Work done by a force

Suppose that, in 3 dimensions a force (F_1, F_2, F_3) moves its point of application through a displacement (x_1, x_2, x_3).

In Fig. 97, suppose the particle on which the force is acting moves from A to D by the

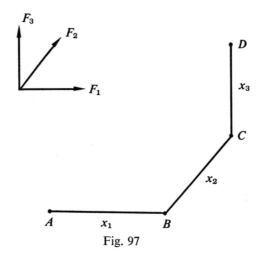

Fig. 97

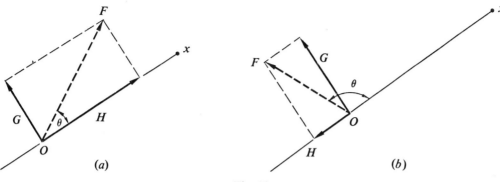

Fig. 98

path $ABCD$, in which AB, BC and CD are parallel to the first, second and third axes respectively. In the displacement AB only the component F_1 does any work; the work done in this part of the displacement is F_1x_1. Similar considerations apply to BC and CD. The total work done, as the particle goes from A to D, is thus $F_1x_1 + F_2x_2 + F_3x_3$.

If the force F were perpendicular to the displacement AD, the work done of course would be zero. This ties in well with the condition proved in Exercise 1 above. It shows further that this kind of expression has a significance even when the vectors are not perpendicular.

The work done by the force F in the displacement x is commonly explained in one of two ways. We may suppose F equivalent to a force G perpendicular to x, and a force H in the same line as x. In the first form of the definition the work done by F in the displacement x is defined as $\|H\| \cdot \|x\|$ in case (a), and $-\|H\| \cdot \|x\|$ in case (b) of Fig. 98. The second definition follows easily from this. If θ denotes the angle between x and F, in both cases we find the work done is given by $\|x\| \cdot \|F\| \cos \theta$, the product of the lengths of the vectors and the cosine of the angle between them. We are thus led to associate the quantity just discussed with the expression found earlier. $F_1x_1 + F_2x_2 + F_3x_3$, and either one of these may be referred to as the 'scalar product' or 'dot product' $F \cdot x$.

The argument given above, for $F_1x_1 + F_2x_2 + F_3x_3$ as being the work done, is useful as a way of seeing why such an expression is to be expected and as a way of remembering it. However, it is not entirely convincing as a proof. How do we know that the work done, if we go from A to D by the route $ABCD$, is the same as if we went from A to D in the most natural way, along the straight line from A to D?

Accordingly, some alternative ways of proving this result will be indicated. These do not make any appeal to mechanics. Apart from leading to the desired result, these proofs are useful exercises in themselves and are presented as such; the general procedure is outlined and the details are to be provided by the student.

Exercises

1 (a) The sides of a triangle are of lengths a, b, c. State, or derive, the formula for $\cos C$ in terms of a, b, c.

(b) Deduce the value of $ab \cos C$ in terms of a, b, c.

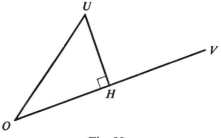

Fig. 99

(c) A triangle is formed by the origin, O, the point $u = (u_1, u_2, u_3)$ and the point $v = (v_1, v_2, v_3)$. If $a = \|u\|$, $b = \|v\|$ and $c = \|u - v\|$, find a^2, b^2 and c^2 in terms of the co-ordinates of u and v.

(d) Use (b) and (c) to calculate $\|u\| \cdot \|v\| \cdot \cos C$ in terms of the co-ordinates of u and v. Here C is the angle between the vectors u and v, in accordance with the usual conventions for the triangle considered in (c); so here we are calculating $u \cdot v$ from the second definition given above.

2   The first definition allows us to find $u \cdot v$ without any appeal to trigonometry. We do, however, have to use the condition for perpendicularity found in the first exercise of this section.

Let U have co-ordinates (u_1, u_2, u_3) and V have co-ordinates (v_1, v_2, v_3). Let H be the foot of the perpendicular from U to OV, Fig. 99. We shall switch freely from geometrical to vector notation. By OV we mean the line joining the points O and V; but we shall also speak of the vector V, the vector H and the vector U.

(a) The fact that H lies on the line OV may be expressed by the vector equation $H = tV$, where t stands for some number that will be found later.

(b) What vector represents the step from H to U? Find its co-ordinates; take account of the equation $H = tV$; that is, we want an answer depending only on t and the co-ordinates of U and V.

(c) Express algebraically $HU \perp OV$, and deduce the value of t in terms of the co-ordinates of U and V.

(d) On the assumption that OH and OV point in the same direction so that $t > 0$, find $\|H\| \cdot \|V\|$, which gives $U \cdot V$ for this case. Fortunately the case $t < 0$, corresponding to case (b) in Fig. 99, leads to the same answer, the algebra automatically compensates for this change.

3   Show that the procedure of Exercise 2 applies equally well in n dimensions, for any natural number n.

Applications of scalar product

When we are dealing with a number of points, specified by their co-ordinates in 3 dimensions, we may wish to visualize the figure they form. This can be done with the aid of the distance formula and the scalar product, as is shown in the following worked examples.

Worked example 1

What figure is formed by $OPQR$ where

$$O = (0, 0, 0), P = (-13, 18, 6), Q = (6, -3, 22), R = (18, 14, -3)?$$

Solution   This question can be answered by using the distance formula alone. We have $\|OP\|^2 = 13^2 + 18^2 + 6^2 = 529$, so $\|OP\| = 23$.

Also $\|PQ\|^2 = (-13 - 6)^2 + (18 + 3)^2 + (6 - 22)^2 = 1058$. Hence $\|PQ\| = 23\sqrt{2}$. In the same way we can find the lengths of the remaining edges of the figure. The result is that PO, OQ and OR are all of length 23, while PQ, QR and PR are all of length $23\sqrt{2}$. It will be seen that PQR is an equilateral triangle; also the lengths of the sides of the triangles OPQ, OPR and OQR indicate that each of these is right-angled with the right angle at O. Thus a cube of side 23 could be placed with one corner at O and adjacent corners at P, Q, R.

We could also have discovered that OP, OQ and OR are mutually perpendicular by using scalar products. For instance

$$P \cdot Q = (-13)(6) + (18)(-3) + (6)(22) = -78 - 54 + 132 = 0$$

which shows $OP \perp OQ$. The scalar products $P \cdot R$ and $Q \cdot R$ will also be found to be zero. Establishing perpendicularity by means of scalar products is here seen to be somewhat quicker and easier than the method first used. Logically the methods are closely related, since the properties of scalar products are proved by appealing to the distance formula.

Worked example 2

Let $A = (10, 5, 5)$, $B = (-11, 2, 5)$, $C = (1, -7, -10)$, $N = (1, -7, 5)$. Compute the scalar products involved and find the lengths of OA, OB, OC and ON. Hence discuss the figure formed by A, B, C, N and O.

Solution

$A \cdot N = 10 - 35 + 25 = 0$. $B \cdot N = -11 - 14 + 25 = 0$. $C \cdot N = 1 + 49 - 50 = 0$.

Thus ON is perpendicular to OA, to OB and to OC, i.e. A, B and C lie in the plane through the origin perpendicular to ON.

We find also $A \cdot B = -75$, $A \cdot C = -75$, $B \cdot C = -75$ and that OA, OB and OC are all of length $\sqrt{150}$.

Let α denote the angle between OA and OB. We know $A \cdot B = \|OA\| \cdot \|OB\| \cos \alpha$. As $A \cdot B = -75$ and $\|OA\| = \|OB\| = \sqrt{150}$, we find $\cos \alpha = -75/150 = -0.5$. Hence the angle between OA and OB is $120°$. Exactly the same calculation shows the angle between OB and OC, and that between OC and OA, to be $120°$.

It is now seen that ABC is an equilateral triangle with O as its centre, lying in the plane perpendicular to ON.

The length of ON is found to be $\sqrt{75}$.

Scalar products provide a simple way of dealing with the equations of planes in 3 dimensions. This topic will be taken up again in §39, but it seems useful to introduce it now, by the Worked examples 3 and 4.

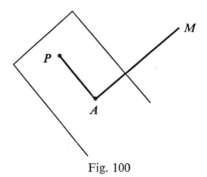

Fig. 100

Worked example 3

Find (a) the equation of the plane through the origin perpendicular to ON, where $N = (1, 2, 3)$, (b) the equation of the plane through $(4, 5, 6)$ perpendicular to $(1, 2, 3)$.

Solution (a) Let $P = (x, y, z)$. P will lie in the plane through O perpendicular to ON if OP is perpendicular to ON. This will be so if the scalar product $P \cdot N$ is zero, that is, if $x \cdot 1 + y \cdot 2 + z \cdot 3 = 0$. Thus $x + 2y + 3z = 0$ is the required equation.
(b) Let $A = (4, 5, 6)$ and let AM be the normal to the required plane, as in Fig. 100. Thus the vector AM is $(1, 2, 3)$.

Students sometimes find difficulty with vectors that start from points other than the origin. In this example it may help to imagine a mass acted upon by a constant force $(1, 2, 3)$; then AM represents the force acting on the mass when the mass is at A. The force does no work if the mass is displaced from A to P, where P lies in the plane through A perpendicular to AM. As $A = (4, 5, 6)$ and $P = (x, y, z)$, the displacement from A to P is $(x - 4, y - 5, z - 6)$. If the force $(1, 2, 3)$ does no work in such a displacement we have $(x - 4) \cdot 1 + (y - 5) \cdot 2 + (z - 6) \cdot 3 = 0$. This then is the condition for P to lie in the plane specified. It simplifies to $x + 2y + 3z - 32 = 0$.

Worked example 4

What is the orientation of the plane $2x - 5y + 7z = 0$?

Solution It is evident from the work in the solution to Worked example 3(a) that we would arrive at this equation if we sought the equation of the plane through the origin perpendicular to $(2, -5, 7)$. So our answer is immediate; this is the plane with the normal $(2, -5, 7)$.

Worked example 5

A crystal has a cubic lattice; there is an atom at every point (x, y, z) for which x, y and z are whole numbers. Draw a diagram showing how atoms occur in the plane $x = y$.

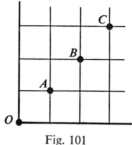

Fig. 101

Solution   Consider first the atoms lying in the plane $z = 0$. If these also lie in the plane $x = y$, they will be at positions such as $(0, 0, 0)$, $(1, 1, 0)$, $(2, 2, 0)$, $(3, 3, 0)$, as shown in Fig. 101. If this book is lying on a table, so that this illustration is in a horizontal plane, the plane $x = y$ will be vertical. It will contain the points O, A, B, C. It will also contain atoms at $(0, 0, 1)$, $(1, 1, 1)$, $(2, 2, 1)$, $(3, 3, 1)$, one unit vertically above O, A, B, C. Rising a further unit we shall find atoms at $(0, 0, 2)$, $(1, 1, 2)$, $(2, 2, 2)$, $(3, 3, 2)$. Thus, as we go up a vertical line, we find atoms spaced at unit intervals.

The points O, A, B, C are spaced at intervals of length $\sqrt{2}$. Thus the section of the crystal by the plane $x = y$ shows atoms arranged as in Fig. 102.

Algebraic properties of scalar products

From the formula for the scalar product it is easy to verify that results such as $u \cdot v = v \cdot u$ and $u \cdot (av + bw) = a(u \cdot v) + b(u \cdot w)$ hold, whatever the vectors u, v, w, and the numbers a, b. Thus scalar products can be handled very much like products of numbers, as is done in the following discussion.

Fig. 103 shows the point P with co-ordinates x, y, z. We usually visualize the numbers x, y, z with the help of a picture such as this, involving the three vectors OM, MN and NP. However, it is possible to visualize the co-ordinate x without bringing in the point N. For we can see that OMP is a right angle. Thus OM is the projection of OP on the first axis; the number x is obtained by taking $\|OM\|$, the length of this projection, and

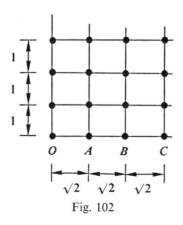

Fig. 102

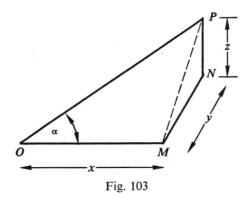

Fig. 103

attaching the appropriate sign, $+$ or $-$, to it. In fact $x = \|OP\| \cos \alpha$ and this equation gives x with the correct sign. The appearance of projection in our considerations suggests that scalar products may be helpful. In fact, let OA be the unit vector along the first axis, so $\|OA\| = 1$. Then $P \cdot A = \|OA\| \cdot \|OP\| \cos \alpha = \|OP\| \cos \alpha = x$. Similarly if OB and OC are unit vectors along the second and third axes we have $P \cdot B = y$ and $P \cdot C = z$.

This argument can be put in purely algebraic form, without any appeal to a geometrical illustration. As was indicated in §6, we can compress the statement, 'P has co-ordinates (x, y, z) in the system based on A, B, C' into the equation

$$P = xA + yB + zC.$$

Take the scalar product of both sides of the equation with A. It follows that

$$P \cdot A = x(A \cdot A) + y(B \cdot A) + z(C \cdot A).$$

As A, B and C are perpendicular axes, $B \cdot A = 0$ and $C \cdot A = 0$. What is $A \cdot A$? The angle between A and itself is zero and $\cos 0 = 1$. Thus $A \cdot A$ reduces simply to $\|OA\|^2$, and as OA is of unit length, this means $A \cdot A = 1$. So our equation for $P \cdot A$ reduces to $P \cdot A = x$, as expected.

The scalar product approach is particularly useful when we wish to change from one set of perpendicular axes to another. If co-ordinates x, y, z are based on perpendicular unit vectors A, B, C and co-ordinates X, Y, Z are based on perpendicular unit vectors D, E, F, we can quickly pass from one system to the other. The argument used above to prove $x = P \cdot A$, $y = P \cdot B$, $z = P \cdot C$ for any point P shows equally well $X = P \cdot D$, $Y = P \cdot E$, $Z = P \cdot F$. Thus the co-ordinates can be found immediately by forming scalar products, as is shown in the following example.

Worked example 6

Let the vectors D, E, F be given (in the A, B, C co-ordinate system) by $D = (-\frac{1}{3}, \frac{2}{3}, \frac{2}{3})$, $E = (\frac{2}{3}, -\frac{1}{3}, \frac{2}{3})$, $F = (\frac{2}{3}, \frac{2}{3}, -\frac{1}{3})$. Verify that D, E, F are perpendicular vectors of unit length and find the X, Y, Z co-ordinates of the point $(1, 4, 10)$.

Solution   Arithmetic only is involved in checking that $D \cdot D = E \cdot E = F \cdot F = 1$ and $D \cdot E = D \cdot F = E \cdot F = 0$, so D, E, F are of unit length and mutually perpendicular. Let $P = (1, 4, 10)$.

We have

$$X = P \cdot D = 1 \cdot (-\tfrac{1}{3}) + 4 \cdot (\tfrac{2}{3}) + 10 \cdot (\tfrac{2}{3}) = 9,$$

$$Y = P \cdot E = 1 \cdot (\tfrac{2}{3}) + 4 \cdot (-\tfrac{1}{3}) + 10 \cdot (\tfrac{2}{3}) = 6,$$

$$Z = P \cdot F = 1 \cdot (\tfrac{2}{3}) + 4 \cdot (\tfrac{2}{3}) + 10 \cdot (-\tfrac{1}{3}) = 0.$$

So P is specified by $X = 9$, $\dot{Y} = 6$, $Z = 0$. We can check this by calculating

$$9D + 6E + 0 \cdot F$$

and seeing that it does give P.

It would have been possible to answer this question by the method of §9 which applies to *any* change of axes. That method, however, gives the 'old' co-ordinates x, y, z in terms of the 'new' co-ordinates X, Y, Z. To find X, Y, Z given $x = 1$, $y = 4$, $z = 10$ we would have to solve 3 simultaneous equations. The present method is very much quicker but it is important to note that it applies *only when we are changing from one system of perpendicular unit vectors to another system of the same kind.*

Exercises

1   In each of the following cases find the scalar product of the two vectors given. Find the magnitude of each vector. Deduce the cosine of the angle between them.

(a) $(1, 1, 0)$ and $(1, 0, 1)$ (b) $(1, -2, 1)$ and $(3, 4, 5)$
(c) $(1, 1, 1)$ and $(1, 0, 0)$ (d) $(3, 4, 5)$ and $(5, 5, 0)$
(e) $(1, 2, -2)$ and $(1, -1, 4)$ (f) $(-2, 3, 6)$ and $(4, 1, 9)$.

2   According to a student, the diagonal of a cube makes an angle of $45°$ with each edge of the cube, Discuss this remark.

3   Find the lengths and the scalar products of the vectors $(1, 1, 0)$, $(-1, 0, -1)$, $(0, -1, 1)$, $(1, -1, -1)$ and discuss the figure they form.

4   Find the lengths of the vectors $P = (1, 5, 1)$, $Q = (-5, -1, 1)$, $R = (1, -1, -5)$, $S = (3, -3, 3)$ and the values of the 6 scalar products that can be formed from them. Describe in geometrical terms the figure formed by the line segments OP, OQ, OR and OS. What figure is formed by the points $PQRS$? How could the answer to this last question be checked directly?

To what chemical situations is this question related?

5   Check that the vectors $D = (-\tfrac{1}{3}, -\tfrac{2}{3}, \tfrac{2}{3})$, $E = (\tfrac{2}{3}, \tfrac{1}{3}, \tfrac{2}{3})$, $F = (-\tfrac{2}{3}, \tfrac{2}{3}, \tfrac{1}{3})$ are perpendicular unit vectors. Let $V = (1, 1, 1)$. Find X, Y, Z, the co-ordinates of V in the system based on D, E, F. Check that $XD + YE + ZF$ does give V.

6   Proceed as in question 5, but with the values given below for D, E, F and V.

(a) $D = (-\tfrac{7}{9}, \tfrac{4}{9}, -\tfrac{4}{9})$, $E = (-\tfrac{4}{9}, \tfrac{1}{9}, \tfrac{8}{9})$, $F = (\tfrac{4}{9}, \tfrac{8}{9}, \tfrac{1}{9})$, $V = (-2, 2, 1)$.
(b) $D = (-\tfrac{6}{7}, -\tfrac{3}{7}, \tfrac{2}{7})$, $E = (\tfrac{3}{7}, -\tfrac{2}{7}, \tfrac{6}{7})$, $F = (-\tfrac{2}{7}, \tfrac{6}{7}, \tfrac{3}{7})$, $V = (-1, 1, 2)$.
(c) $D = (-\tfrac{4}{9}, \tfrac{1}{9}, \tfrac{8}{9})$, $E = (\tfrac{7}{9}, -\tfrac{4}{9}, \tfrac{4}{9})$, $F = (\tfrac{4}{9}, \tfrac{8}{9}, \tfrac{1}{9})$, $V = (3, 7, 4)$.
(d) $D = (-\tfrac{9}{11}, -\tfrac{6}{11}, \tfrac{2}{11})$, $E = (\tfrac{6}{11}, -\tfrac{7}{11}, \tfrac{6}{11})$, $F = (-\tfrac{2}{11}, \tfrac{6}{11}, \tfrac{9}{11})$, $V = (6, 4, -5)$.
(e) $D = (-\tfrac{2}{3}, -\tfrac{2}{3}, \tfrac{1}{3})$, $E = (\tfrac{2}{3}, -\tfrac{1}{3}, \tfrac{2}{3})$, $F = (-\tfrac{1}{3}, \tfrac{2}{3}, \tfrac{2}{3})$, $V = (1, 6, 5)$.

7 In each section of question 6, and in question 5, replace the particular vector V by the general vector $V = (x, y, z)$, and obtain equations giving X, Y, Z in terms of x, y, z.

Also find the equations giving x, y, z in terms of X, Y, Z. This could be done by solving the first set of equations but is much easier by the method of §9.

Compare the matrices associated with the two sets of equations. What do you notice?

8 In a crystal with a cubic lattice there is an atom at every point (x, y, z) for which x, y, z belong to the whole numbers $0, 1, 2, 3, \ldots$ Draw diagrams to show the positions of the atoms in the following planes;

(a) $x = 0$,   (b) $x + y + z = 3$,   (c) $x + y + z = 4$.

9 In a body-centred cubic lattice atoms occupy all the points (x, y, z) with x, y, z whole numbers, and also points such as $(1\frac{1}{2}, 5\frac{1}{2}, 2\frac{1}{2})$ for which x, y, z are *all* of the form 'whole number plus a half'. Draw diagrams showing the atoms in the following plane sections;

(a) $z = 0$ (b) $z = \frac{1}{2}$ (c) $x = y$ (d) $x + y = 3$
(e) $x + y + z = 3$ (f) $x + y + z = 3\frac{1}{2}$ (g) $x + y + 2z = 4$.

10 In a face-centred cubic lattice, atoms are found at all points (x, y, z) with x, y, z whole numbers, and also at points such as $(1\frac{1}{2}, 3, 5\frac{1}{2})$ for which one co-ordinate is a whole number and the other two co-ordinates differ by $\frac{1}{2}$ from whole numbers. Draw diagrams for sections by the planes;

(a) $z = 0$ (b) $z = \frac{1}{2}$ (c) $x = y$ (d) $y = x + \frac{1}{2}$ (e) $x + y + z = 3$.

33   *Transpose; quadratic forms*

In §30 we saw that the matrix notation we have had so far gave us no way of multiplying two column vectors together. If we had a column vector $\begin{pmatrix} x \\ y \end{pmatrix}$ we could get an expression such as $ax + by$ for the product with a row vector (a, b) but there was no way of arriving at quadratic terms such as x^2, xy, y^2, for these would imply that we had formed the product of the vector with itself and this, in pure vector theory (affine geometry), is impossible.

Now there are cases in which we want to consider quadratics. In Euclid's geometry we need $x^2 + y^2$ for the definition of length. If we have two springs of stiffness k_1 and k_2, with extensions x_1 and x_2, the potential energy is given by the quadratic expression $\frac{1}{2}(k_1 x_1^2 + k_2 x_2^2)$. Similarly, if we have currents, i_1 and i_2, in resistances r_1 and r_2, the rate at which heat is generated is given by $r_1 i_1^2 + r_2 i_2^2$. If we are to cope with such situations, we must bring some new machinery into matrix algebra.

The situations mentioned in the last paragraph can be conveniently represented within Euclidean geometry. If the displacements are indicated by plotting the point (x_1, x_2) on ordinary graph paper, the forces acting on the springs can be represented by the point (F_1, F_2) where $F_1 = k_1 x_1, F_2 = k_2 x_2$. The expression $k_1 x_1^2 + k_2 x_2^2$ that occurs in the formula for the potential energy is then the scalar product $F_1 x_1 + F_2 x_2$. Similarly if we represent the currents by the point (i_1, i_2), the voltage drops in the resistances can be represented by the point (V_1, V_2) where $V_1 = r_1i_1$, $V_2 = r_2i_2$; heat generation is then given by the scalar product $V_1i_1 + V_2i_2$. In the purely geometrical question, if we write u for the vector with co-ordinates x, y, then $x^2 + y^2$ is $u \cdot u$, the scalar product of u with itself.

Accordingly, we can cope with situations of this kind if we can find some way of writing scalar products in matrix notation.

Suppose then that we have some point $\binom{a}{b}$, plotted on ordinary graph paper. If $\binom{x}{y}$ denotes a variable point, the scalar product of $\binom{a}{b}$ and $\binom{x}{y}$ is $ax + by$. In this expression, the numbers a and b appear as coefficients. §30 has taught us to associate such coefficients with a row vector, in fact with the row vector (a, b). Thus in Euclidean geometry to each column vector $\binom{a}{b}$ there corresponds a row vector (a, b). This situation is entirely different from that in affine geometry where, fortunately, an order for 200 nuts and 100 bolts (a column vector) does not imply a price scheme of $200 per nut and $100 per bolt (a row vector). But in Euclidean geometry, because scalar product is defined, there is a direct correspondence between column vectors and row vectors. Provided a suitable co-ordinate system (based on squared graph paper) is used, this correspondence is very simple; it is given by $\binom{a}{b} \rightarrow (a, b)$.

If $v = \binom{a}{b}$, then (a, b) is known as the *transpose* of v, and is written v' or v^T.

Transposition can be defined for any matrix (compare Exercise 3 on determinants). It consists in interchanging rows and columns. Thus we may also say that v is the transpose of v'.

Note that transposition is something done to the numbers that represent a vector. If we see $\binom{7}{10}$ we can write down its transpose $(7, 10)$. Whether this transpose will have a significant geometrical or physical meaning will depend on the co-ordinate system in which it occurs. But whether it is meaningful or not, it is still called the transpose.

With the help of the transpose we can express scalar products in matrix notation, as is verified in the following exercises.

Exercises

1 If u and v are two column vectors in 2 dimensions, the co-ordinate system being based on conventional (squared) graph paper, show that $u'v = v'u = $ the scalar product $u \cdot v$.

2 Verify that the same holds in n dimensions for any natural number n.

Quadratic forms

At the beginning of this section we considered quadratic expressions that arose in connection with springs and with electric circuits. The mathematics in the two cases was almost identical. It will be sufficient to study one of these examples, say that of the springs. A student interested in electricity can easily translate the discussion into electrical terms. What processes were involved in forming the expression for the potential energy?

We began with a column vector $\begin{pmatrix} x_1 \\ x_2 \end{pmatrix}$ which gave the extensions of the two springs. We then used Hooke's Law to find F_1, F_2 with $F_1 = k_1 x_1$, $F_2 = k_2 x_2$. We thus arrive at a new vector, F, by means of a linear transformation. As a linear transformation can be specified by a matrix we can write our equations in the form

$$\begin{pmatrix} F_1 \\ F_2 \end{pmatrix} = \begin{pmatrix} k_1 & 0 \\ 0 & k_2 \end{pmatrix} \begin{pmatrix} x_1 \\ x_2 \end{pmatrix}.$$

We may write this more briefly as $F = Kv$, with F the vector specifying forces, v specifying extensions, and K the stiffness matrix. As was pointed out earlier, the expression $k_1 x_1^2 + k_2 x_2^2$, which equals $2E$, is the scalar product $F \cdot v$. Our last exercise showed that $F \cdot v$ could be written $v'F$. Since $F = Kv$, this expression could be written $v'Kv$. This is the kind of symbol we usually employ for a quadratic form, that is to say an expression in which every term is of the second degree in the variables. Thus $ax^2 + bx + c$ is *not* a quadratic form, since bx is only of the first degree, while c is simply a constant. However, $ax^2 + bxy + cy^2$ is a quadratic form, and so is $x^2 - yz$. With the summation notation, a quadratic form can always be expressed as

$$\sum \sum a_{rs} x_r x_s; \qquad 1 \leqslant r \leqslant n, 1 \leqslant s \leqslant n.$$

Exercises

1 $v'Kv$ means $(x_1 \quad x_2) \begin{pmatrix} k_1 & 0 \\ 0 & k_2 \end{pmatrix} \begin{pmatrix} x_1 \\ x_2 \end{pmatrix}.$
Multiply this out by the usual rule for matrix multiplication and check that it does give the result stated above.

2 Multiply out $(x \quad y) \begin{pmatrix} a & h \\ h & b \end{pmatrix} \begin{pmatrix} x \\ y \end{pmatrix}.$

3 Multiply out $(x \quad y) \begin{pmatrix} 0 & 1 \\ -1 & 0 \end{pmatrix} \begin{pmatrix} x \\ y \end{pmatrix}.$

4 Multiply out $(x \quad y \quad z) \begin{pmatrix} 1 & -1 & -1 \\ -1 & 1 & -1 \\ -1 & -1 & 1 \end{pmatrix} \begin{pmatrix} x \\ y \\ z \end{pmatrix}.$

5 Write $x^2 + 6xy + y^2$ in the form $v'Mv$, with the matrix M of the type used in question 2 above.

Quadratic expressions of the form $v'Mv$ are very important. As we shall see in §34, there are situations in which the whole problem can be specified by giving a certain expression, which is either quadratic or partly quadratic and partly linear. Very often at some stage of the work it becomes necessary to bring in some new co-ordinate system. Instead of v we may now use V to specify some physical situation. We shall have a formula for change of axes, $v = SV$, where S is a matrix. In $v'Mv$ we readily see that Mv is to be replaced by MSV, but what are we to do about v'?

Let us look at a particular example. Suppose

$$v = \begin{pmatrix} x \\ y \\ z \end{pmatrix} \quad \text{and} \quad V = \begin{pmatrix} X \\ Y \end{pmatrix}.$$

Here we have not merely a change of axes but even a change of dimension. Such things do happen in engineering practice, as will be seen in an electrical example in §34. We suppose

$$\begin{array}{l} x = aX + bY \\ y = cX + dY \\ z = eX + fY \end{array} \qquad v = \begin{pmatrix} x \\ y \\ z \end{pmatrix} = \begin{pmatrix} a & b \\ c & d \\ e & f \end{pmatrix} \begin{pmatrix} X \\ Y \end{pmatrix} = SV.$$

Now v' is the row vector (x, y, z), which, from the equations above, means

$$(aX + bY, cX + dY, eX + fY).$$

This has 1 row and 3 columns, and 2 products appear in each entry. Such an expression can only arise if the first factor has 1 row and 2 columns, while the second factor has 2 rows and 3 columns. It is pretty clear that the first factor must be (X, Y), and in fact we find we can obtain the result above by multiplying out

$$(X, Y) \begin{pmatrix} a & c & e \\ b & d & f \end{pmatrix}.$$

Both of these we can identify. (X, Y) is V', the transpose of V. The 2×3 matrix is S', the transpose of S, for the columns of this matrix are the same as the rows of S, and the rows of this matrix contain the same elements as the columns of S. Thus $v' = V'S'$. As $v = SV$, this means $(SV)' = V'S'$.

The result we have found here is in fact true for any product of matrices; *the transpose of a product is the product of the transposes, in the reverse order.*

This general statement may be proved in two ways.

In the first way, we begin by satisfying ourselves that the result $(SV)' = V'S'$ above does not depend on the fact that S is a 3×2 matrix, but that it would hold for any $m \times n$ matrix S and any vector V in n dimensions. Provided we are able to do this, we then consider a vector acted upon by any succession of mappings, say, for instance $ABCV$. If we put $S = ABC$, we can see that the transpose of $(ABC)V$ is $V'(ABC)'$. But we can also consider the mappings A, B, C applied one at a time. By using the result $(SV)' = V'S'$ three times we see

$$[A(BCV)]' = (BCV)'A' = (CV)'B'A' = V'C'B'A'.$$

Accordingly $V'(ABC)' = V'C'B'A'$ and since this holds for any vector V we must have $(ABC)' = C'B'A'$. Clearly the same argument could be applied to any number of matrices, and so the general result is established.

The second method of proof uses subscript notation and is found in many books.

Let the matrix A have a_{pq} in the pth row and qth column. Since A' is obtained from A by interchanging rows and columns. A' will have a_{qp} in row p and column q. We express this by writing $a'_{pq} = a_{qp}$. Let $D = ABC$. Then $d_{ps} = \sum_q \sum_r a_{pq} b_{qr} c_{rs}$. Accordingly, for the transpose D' we have $d'_{ps} = d_{sp} = \sum_q \sum_r a_{sq} b_{qr} c_{rp}$. Here we have interchanged p and s in the previous formula. We are working towards a result involving A', B' and C', so naturally our next step is to bring these into the picture by using $a_{sq} = a'_{qs}$, $b_{qr} = b'_{rq}$ and $c_{rp} = c'_{pr}$. Making these substitutions we obtain $d'_{ps} = \sum_q \sum_r a'_{qs} b'_{rq} c'_{pr}$. This result is not at present in a form suitable for a matrix product since (see §29) in a matrix product the subscripts over which we sum are always immediate neighbours. We can remedy this by reversing the order of the factors. We then have

$$d'_{ps} = \sum_q \sum_r c'_{pr} b'_{rq} a'_{qs}.$$

Now we have the correct form for a matrix product; the repeated subscripts, r and q, are not separated and p and s, the subscripts of d', occur on the right-hand side in the first and last places respectively, as they should. In fact the last equation expresses $D' = C'B'A'$, the result we wish to prove.

This result covers the case of a matrix and a vector, $(SV)' = V'S'$, for a vector, V, may be regarded as a matrix with one column, and its transpose, V', as a matrix with one row.

Symmetric and antisymmetric matrices

The matrix $\begin{pmatrix} 1 & 3 \\ 3 & 4 \end{pmatrix}$ is the same as its transpose. A matrix with this property, $M' = M$, is said to be *symmetric*.

A matrix for which $M' = -M$ is said to be *antisymmetric*. An example of such a matrix is

$$\begin{pmatrix} 0 & c & -b \\ -c & 0 & a \\ b & -a & 0 \end{pmatrix}.$$

An antisymmetric matrix is bound to have zeros down the main diagonal, as in the example shown. For, if M has m_{pq} as its entry in row p and column q, $m_{pq} = -m_{qp}$, for each p and each q. If $q = p$, this means $m_{pp} = -m_{pp}$, so $m_{pp} = 0$.

Only square matrices can be symmetric or antisymmetric.

In question 3 of the exercises above, on quadratic forms, an expression $v'Mv$ with antisymmetric M occurred. It gave zero. In fact, this always happens. In questions 2 and 5 the matrix was symmetric. *Any quadratic form can be represented by $v'Mv$ with M*

symmetric, and it is the invariable custom (for which there are good reasons) always to use a symmetric matrix for this purpose.

A single number, a, can be regarded as a matrix of 1 row and 1 column. It is automatically symmetric. We can use this to prove certain results. For instance, the scalar product $u \cdot v$ can be written $u'v$. Its transpose is $v'u$, by our usual rule – form transposes and reverse order. But the scalar product is a single number, hence symmetric, hence equal to its transpose. So $u'v = v'u$, a result that was checked earlier in an exercise.

Again $u'Mv$ is a single number, hence equal to its transpose $v'M'u$. So $u'Mv = v'M'u$. This result holds for any vectors u, v and any matrix M such that the product $u'Mv$ makes sense.

Exercises

1 Prove $(S + T)' = S' + T'$.

2 Prove $(kS)' = kS'$.

3 If M is any square matrix, prove $\frac{1}{2}(M + M')$ is symmetric, and $\frac{1}{2}(M - M')$ is antisymmetric. Hence show that any matrix is the sum of a symmetric and an antisymmetric matrix.

4 If A is an antisymmetric square matrix, prove $v'Av = 0$ whatever the vector v. (*Hint.* Consider the transpose of the number $v'Av$.).

5 If M is any matrix, prove that $M'M$ is symmetric. (Consider its transpose.)

6 If R is a symmetric $m \times m$ matrix, and T is any $m \times n$ matrix, prove that $T'RT$ is symmetric.

7 In the quadratic form $v'Sv$, where S is symmetric, we make the substitution $v = TV$. Find what $v'Sv$ becomes when this substitution is made. Will the matrix that appears in the answer be symmetric or not?

8 Find the result when in the scalar product $u'v$ we substitute $u = TU$, $v = TV$. What condition must T satisfy if for all vectors we are to have $u'v = U'V$?

34 Maximum and minimum principles

Early in a calculus course a student learns how to find the places at which some function takes its maximum or minimum values. He works a number of problems which have some interest – the largest parcel permitted by the postal regulations, the most economical shape for fencing a chicken farm, the shape of a soup tin that uses least metal. It is seen that there may be problems of design in which calculus indicates the most efficient or economical way of dealing with some detail. But very few students appreciate the

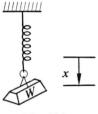

Fig. 104

immense vista that this work in calculus opens up. One of the strange things about this universe is that, in every science that has become sufficiently precise to admit mathematical formulation, there is a principle that some quantity is to be made a maximum or a minimum, and this principle, by itself, sums up the laws of that science. Thus in statics stable equilibrium occurs when potential energy is a minimum; in optics, light travels by the path that takes least time; in dynamics there is the principle of Least Action, which can be adapted so as to apply to electromagnetic phenomena also; in thermodynamics, an isolated system is in equilibrium when the entropy is a maximum, and from this a whole theory applicable to physical and chemical processes can be deduced; in an electrical network the currents distribute themselves in such a way that the generation of heat is a minimum.*

Here we have an outstanding example of the point made in §1, 'that mathematics is concerned not with particular situations but with patterns that occur again and again'. A student who has mastered the basic mathematical tools – differentiation and integration, partial differentiation, calculus of variations – has secured a foothold on an enormous variety of sciences.

We will consider two very simple examples. Suppose a weight W hangs from a spring of stiffness k, and x denotes the extension of the spring (Fig. 104). The potential energy, P, of the system is given by $P = \frac{1}{2}kx^2 - Wx$. This will be a minimum where $0 = dP/dx = kx - W$. Thus the equilibrium condition $W = kx$ follows the principle that potential energy is to be minimized.

Now consider the simple electrical circuit in which a voltage V drives a current i through a resistance r (Fig. 105). Ohm's Law tells us $V = ri$.

If we compare Ohm's Law $V = ri$ and $W = kx$ we see that they have the same

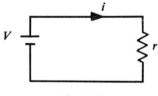

Fig. 105

---

* Strictly speaking, we should perhaps speak of 'stationary' rather than maximum or minimum principles. For example a bead on a wire can be in equilibrium if it is at a point of horizontal inflexion.

mathematical form. If, starting from Hooke's Law, $W = kx$, we replace W by V, k by r, and x by i, we reach Ohm's Law, $V = ri$.

Now we saw that Hooke's Law was a consequence of the requirement that potential energy should be minimized. If we replace the expression $\frac{1}{2}kx^2 - Wx$ by $\frac{1}{2}ri^2 - Vi$, making this latter expression a minimum will ensure that Ohm's Law is obeyed. (This can easily be verified directly.)

Now of course the principle in statics that potential energy is to be minimized does not apply only to a simple system with one spring. It applies equally well to a structure in which there may be hundreds of elastic members. In the same way, for electrical circuits we can find a function that is to be minimized, even though there may be a complicated network of batteries and resistances. This function may be written $\sum \frac{1}{2}ri^2 - \sum Vi$. By this we mean that for each resistance r in the circuit with current i flowing through it, we form the term $\frac{1}{2}ri^2$ and add all these terms together, to give $\sum \frac{1}{2}ri^2$. Similarly $\sum Vi$ means that for each battery of voltage V with current i flowing through it, we form the product Vi, and add all such products together. This is a very natural generalization of the function that was minimized in the case of the simple circuit.

This kind of result allows us to make a certain type of analogue computer. Suppose we have some complex structure. Its equilibrium is determined by minimizing a certain function. If we now design an electrical circuit that depends on essentially the same function (with displacements replaced by currents, stiffness constants by ohmic resistances and applied forces by battery voltages), the behaviour of this circuit will predict the behaviour of the structure. The same idea allows us to establish analogies between other branches of science, and to replace awkward experiments by convenient ones.

The existence of analogies also allows us to make advances in our theories. If some method of calculation has been found useful in the study of electrical circuits, it should be useful in the theory of structures. And in fact precisely this argument was used in a classical paper, 'Analysis of elastic structures by matrix transformation with special regard to semimonocoque structures' by Borje Langefors of the Saab Aircraft Company, in the *Journal of Aeronautical Sciences* for July 1952 (page 451). His article begins by pointing out that identical principles apply to electrical circuits and to elastic structures. Now Kron in 1939 had published a method for solving circuit problems by using matrices. It should therefore be possible to solve problems about structures by matrices. And indeed Kron in 1944 had published the electrical equivalents of various elastic structures. Langefors then proceeded to develop this idea in relation to the design of aircraft.

It seems worthwhile to consider in detail a particular electrical example, since this shows the use both of the quadratic form and the transpose, the two topics discussed in §33.

We will consider the circuit shown in Fig. 106. The resistances r_1, r_2, r_3 and the voltages of the batteries V_1, V_2, V_3 are known; we wish to determine the currents i_1, i_2, i_3. Our purpose of course is not to solve this simple problem, but to find a procedure that can be applied to much more complicated cases.

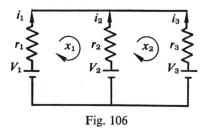

Fig. 106

By considering the potential difference between the wire at the top of the diagram and the wire at the bottom we see

$$V_1 - r_1 i_1 = V_2 - r_2 i_2 = V_3 - r_3 i_3. \tag{1}$$

We have three currents all shown as flowing into the top wire; one of these, at least, must be negative. The currents in fact must satisfy the condition $i_1 + i_2 + i_3 = 0$. This condition can be automatically satisfied by supposing that a current x_1 circulates in the left-hand rectangle, and a current x_2 circulates in the right-hand rectangle. Superposing these we find $i_1 = x_1$, $i_2 = -x_1 + x_2$, $i_3 = -x_2$.

We can express these equations in vector and matrix notation as $i = Tx$, where

$$i = \begin{pmatrix} i_1 \\ i_2 \\ i_3 \end{pmatrix}, \qquad x = \begin{pmatrix} x_1 \\ x_2 \end{pmatrix}, \qquad T = \begin{pmatrix} 1 & 0 \\ -1 & 1 \\ 0 & -1 \end{pmatrix} \tag{2}$$

If we look at the batteries in the left-hand loop, we see V_1 opposed by V_2, so it is natural to write $e_1 = V_1 - V_2$. (The letter e is the initial of 'electromotive force'.) Similarly for the right-hand loop we write $e_2 = V_2 - V_3$. In matrix form these equations are

$$\begin{pmatrix} e_1 \\ e_2 \end{pmatrix} = \begin{pmatrix} 1 & -1 & 0 \\ 0 & 1 & -1 \end{pmatrix} \begin{pmatrix} V_1 \\ V_2 \\ V_3 \end{pmatrix}. \tag{3}$$

The 2×3 matrix that occurs here is T', the transpose of the matrix already introduced. Thus, with a notation that will be understood we may write $e = T'V$.

From equation (1), and the definitions of e_1 and e_2, we find

$$\left. \begin{array}{l} e_1 = V_1 - V_2 = r_1 i_1 - r_2 i_2, \\ e_2 = V_2 - V_3 = \qquad r_2 i_2 - r_3 i_3. \end{array} \right\} \tag{4}$$

We introduce the resistance matrix

$$R = \begin{pmatrix} r_1 & 0 & 0 \\ 0 & r_2 & 0 \\ 0 & 0 & r_3 \end{pmatrix}. \tag{5}$$

Observe that Ri gives the voltage drops across each of the three resistances.

Now the coefficients of i_1, i_2 and i_3 in equations (4) are closely related to the matrices R and T', as may be seen from the equation

$$\begin{pmatrix} r_1 & -r_2 & 0 \\ 0 & r_2 & -r_3 \end{pmatrix} = \begin{pmatrix} 1 & -1 & 0 \\ 0 & 1 & -1 \end{pmatrix} \begin{pmatrix} r_1 & 0 & 0 \\ 0 & r_2 & 0 \\ 0 & 0 & r_3 \end{pmatrix}.$$

The matrix on the left-hand side of this equation is formed from the coefficients of i_1, i_2 and i_3 in equations (4). The right-hand side is $T'R$.

Thus equations (4) may be put in the compact form $e = T'Ri$.

We intend to work with the loop currents, x, rather than the original currents, i, so we make use of our result $i = Tx$ and write $e = T'RTx$.

Our problem is now expressed purely in matrix form. We introduce new variables for currents by means of the equation $i = Tx$. We introduce new specifications of voltage by the equation $e = T'V$. We then have the equation $e = T'RTx$ in which e is known and the vector x has to be found.

The main advantage of this method is its systematic nature. We can programme a computer to form the necessary matrices, even for a very complicated circuit, and to solve the resulting equations.

Clearly some questions arise. Why is the matrix in the voltage equation, $e = T'v$, the transpose of the matrix in the current equation $i = Tx$? Can we be sure that $e = T'RTx$ will be the correct equation for any circuit however complicated?

A well known and good introduction to our present topic is *Matrix Analysis of Electric Networks* by P. le Corbeiller (Harvard University Press, 1950). Kron's work had been presented in rather an abstruse form. Le Corbeiller set out to explain Kron's ideas in a form that the majority of engineers would be able to understand, and this in general he did very well. However, to answer the two questions above required what he himself called 'a rather elaborate train of reasoning'. The demonstration filled three short sections, 19, 20 and 21, and used a strange argument in which a circuit was dissected into many pieces and then peculiar short-circuit arrangements were introduced.

The value of the minimum principle is shown by the ease with which it allows us to answer the two questions. The minimum principle itself we shall assume as having been established in some standard book on electrical theory.

In our explanation, we shall continue to use the particular circuit considered above, and the symbols introduced in connection with it. However, there is nothing special about this circuit; it should be clear how to apply the same procedure to any circuit whatever.

For our circuit we have to minimize the quantity

$$\tfrac{1}{2}r_1 i_1^2 + \tfrac{1}{2}r_2 i_2^2 + \tfrac{1}{2}r_3 i_3^2 - V_1 i_1 - V_2 i_2 - V_3 i_3,$$

which may be abbreviated to $\tfrac{1}{2}i'Ri - i'V$. It is understood that we only consider as possible those currents in which the amount of electricity flowing into any point is balanced by the amount flowing out. In our example, this means $i_1 + i_2 + i_3 = 0$.

Thus we can only choose two of the quantities i_1, i_2 and i_3; the value of the third is then fixed. This condition is automatically met by bringing in the loop currents, x_1 and x_2. They can vary in any way they like; there are no hidden conditions, and this of course simplifies the work considerably.

Accordingly, we get rid of i_1, i_2, i_3 in favour of x_1, x_2 by means of the equation $i = Tx$. Then, by the reversal rule, $i' = x'T'$. Substituting these in $\frac{1}{2}i'Ri - i'V$, we obtain the quantity to be minimized in the form $\frac{1}{2}x'T'RTx - x'T'V$. If we bring in e as an abbreviation for $T'V$, we obtain the expression $\frac{1}{2}x'T'RTx - x'e$.

Our task then is to minimize $\frac{1}{2}x'Mx - x'e$ where $M = T'RT$.

Notice that in this expression we have no terms of more than second degree. We are dealing with a quadratic expression; the problem is a generalization of finding the minimum value of $\frac{1}{2}ax^2 - bx$. It should not be too difficult.

Minimum of a quadratic expression

In calculus we have learnt to find maximum and minimum values in cases where x represented a number; we have not met examples where x was a vector. It is therefore necessary to leave our electrical problem for a moment, and consider how calculus can be adapted to vector problems.

The usual calculus argument amounts more or less to the following. If, in $\frac{1}{2}ax^2 - bx$, we replace x by $x + h$ we get $\frac{1}{2}a(x + h)^2 - b(x + h)$, which can be written as $\frac{1}{2}ax^2 - bx + h(ax - b) + \frac{1}{2}ah^2$. From this we can see the effect of a small change h. When h is small, h^2 is even smaller. For example, if $h = 0.001$, then $h^2 = 0.000001$. If $ax - b$, the coefficient of h, is not zero, for sufficiently small h the term $h(ax - b)$ will outweigh the term $\frac{1}{2}ah^2$, and so will decide whether the change from x to $x + h$ makes the quantity increase or decrease. Now $h(ax - b)$, in these circumstances, changes sign if h changes sign. Accordingly, by choosing the sign of h appropriately, we can make the quantity increase or decrease, whichever we want. In this case, we have neither a maximum nor a minimum. Such things can happen only if $ax - b = 0$. Then our quantity is $(\frac{1}{2}ax^2 - bx) + \frac{1}{2}ah^2$. If $a > 0$, this will be larger than $\frac{1}{2}ax^2 - bx$, whether h is positive or negative. That is to say, we have a minimum corresponding to the value x; any small change from x to $x + h$ produces an increase in the quantity given by our quadratic expression. Of course, if $y = \frac{1}{2}ax^2 - bx$, then $dy/dx = ax - b$. Our discussion, purely in terms of algebra, is establishing that, in this particular case, minimum can occur only if $dy/dx = 0$.

Let us try to apply this procedure to the expression $\frac{1}{2}x'Mx - x'e$, where x now represents a vector. We suppose x replaced by $x + h$, where h is a vector too; all the numbers appearing in h are to be small. We thus get $\frac{1}{2}(x' + h')M(x + h) - (x' + h')e$, which multiplies out to give

$$\frac{1}{2}x'Mx - x'e + (\frac{1}{2}h'Mx + \frac{1}{2}x'Mh - h'e) + \frac{1}{2}h'Mh.$$

Here we see the original expression at the beginning; the bracket in the middle contains the terms which, apart from the exceptional case, determine whether we have an increase; finally we have $\frac{1}{2}h'Mh$, which involves the squares and products of small quantities. We need to do a little work on the middle term. Now $x'Mh$ represents simply a number; it therefore is equal to its transpose $(x'Mh)'$. This is $h'M'x$ and, as we have agreed always to use a symmetric matrix when we are dealing with quadratic forms, we know $M' = M$. Accordingly $\frac{1}{2}x'Mh = \frac{1}{2}h'Mx$. Thus the first two terms in the bracket are equal. Their sum is $h'Mx$. Accordingly the bracket contains the quantity $h'Mx - h'e$ or $h'(Mx - e)$.

We now argue, just as we did in the case where x was a number, that this term will control the size of the change, and can be made positive or negative at will, except when $Mx - e = 0$. (Check for yourself that these assertions remain true in the vector situation.)

We need not go on to investigate whether we have a maximum, minimum or other situation. It will be sufficient for us to have the result corresponding to the calculus result that stationary points occur only when $dy/dx = 0$, namely –

The expression $\frac{1}{2}x'Mx - x'e$, where M is a symmetric matrix, is stationary only when $Mx - e = 0$.

The electrical problem, resumed

In the electrical question, the expression $\frac{1}{2}x'Mx - x'e$ does in fact have a minimum. But we need not go into that. If it is to be a minimum, it certainly must be stationary and so, by our result above, x must satisfy $Mx = e$.

We must tidy up one point. The theorem above requires M to be symmetric. Now, in our electrical question, M is an abbreviation for $T'RT$. Is this in fact symmetrical? It certainly is. In question 6 of the exercises at the end of §33 we saw that, if R is symmetric, then $T'RT$ is automatically symmetric. And our matrix R (see equation (5)) obviously is symmetric.

We have now proved everything we had to do. We have shown that if we put $i = Tx$, where the matrix T is fixed when we dissect our circuit into loops, then it is natural to introduce a vector e defined by $e = T'V$ and a matrix $M = T'RT$, and that this will lead to the equation $Mx = e$.

The power of matrix notation is shown not only by its application to electrical, structural and other problems, but also by the general result we obtained, on the way, about the stationary points for $\frac{1}{2}x'Mx - x'e$. Here, by working on the analogy of a proof in calculus, concerned with a single number x, we obtained a proof that would apply, for instance, to a problem involving a thousand variables, for the notation $Mx = e$ remains the same whether x denotes a vector in one, two or a thousand dimensions.

Exercises

1   For the circuit shown in Fig. 107, write equations expressing the currents i_1, i_2, i_3, i_4, i_5, i_6 in

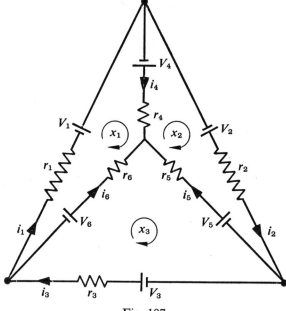

Fig. 107

terms of the loop currents x_1, x_2, x_3. Also write equations giving the loop voltages e_1, e_2, e_3 in terms of V_1, V_2, V_3, V_4, V_5, V_6. Check that the matrices corresponding to these systems of equations are transposes, each of the other.

2   Fig. 108 shows three springs joined together and lying in a straight line. The end points are

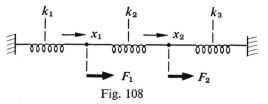

Fig. 108

fixed and the system is not self-stressed. Constant forces F_1 and F_2 act at the junctions of the springs. Write the expression for the potential energy of the system when these junctions have displacements x_1 and x_2, the stiffness constants of the springs being k_1, k_2 and k_3.

3   Fig. 109 shows an electrical circuit. Is it possible to choose the voltages V_1, V_2 and the resistances

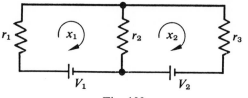

Fig. 109

r_1, r_2, r_3 in such a way that function to be minimized in the electrical situation will correspond exactly to the potential energy which is minimized for equilibrium of the system of springs? (If this can be done, the behaviour of the electrical circuit will predict the behaviour of the structure.)

Check your answer by writing the equations for the two systems, without any appeal to minimum principles, and comparing the equations obtained.

35   *Formal laws of matrix algebra*

So far our work with matrix algebra has been informal. In §20, §21 and §22 exercises were provided by which the student could see that transformations, and the matrices that represented them, when expressed in algebraic symbolism sometimes behaved very much like numbers in elementary algebra, and sometimes behaved very differently. Thus for example in the exercises at the end of §22, question 4 led to

$$(A - I)^2 = A^2 - 2A + I,$$

which we would expect by analogy with elementary algebra, while question 5 showed that of $(A + B)(A - B)$, $(A - B)(A + B)$ and $A^2 - B^2$, no two were equal – an unexpected result. In Exercise 20, we had various examples, intended to give the student a feeling for the way things worked, and on the basis of this experience judgements were invited – does it seem likely that $ST = TS$, that $S + T = T + S$? We have worked with matrix algebra on the basis of these judgements, but we have never set out its laws explicitly and considered their justification. It is now time to turn to this.

If we consider the properties of numbers assumed in elementary algebra, and ask whether similar properties hold for matrices, we find the properties fall into three classes: (1) those that do not apply to matrices, and that we must remember *not* to assume, (2) those that are most easily verified by calculation, (3) those that are most easily verified by a more or less geometrical argument.

Properties that do not apply are easily disposed of. We simply have to produce an example that shows matrices behaving in a way incompatible with those properties. Thus, in the symbolism of §20 and §21, the fact that $MJ_1 \neq J_1M$ immediately indicates that the commutative property of multiplication, $ST = TS$, is *not* universally true for matrices. In fact it is exceptional for matrices to commute when multiplied.

Again, we cannot conclude from $ST = 0$ that $S = 0$ or $T = 0$. Questions 12 and 15 at the end of §21 clearly rule this out.

The commutative and associative properties for the addition of matrices can be checked by routine algebra. These two properties together amount to the 'in-any-order-rule' – if matrices are added in any order the final result is the same. If matrices of specific shape and size are given, the entries being either particular numbers or algebraic symbols such as a, b, c, d, the student will have no difficulty in checking that the order does not affect the sum. Difficulty may be found in establishing the result quite generally, for $m \times n$ matrices. The idea is understood, but it may be hard to clothe it in language. The subscript symbolism may be helpful. If $S = A + B + C$, where A, B, C are matrices with entries a_{rt}, b_{rt}, c_{rt}; $1 \leqslant r \leqslant m$, $1 \leqslant t \leqslant n$, the typical entry in S can be calculated, and it will be seen that no change in the order of A, B, C can lead to a different sum.*

The proof will not be set out in this book, since it uses only routine algebra. A student who cannot work it out for himself will not be able to follow the argument as set out by someone else.

Students who find great difficulty in the linguistic aspects of mathematics – in expressing ideas in words and symbols – may find that, after checking the in-any-order property of addition for say 2×2, 2×3 and 4×2 matrices, with arbitrary entries, a, b, c, $\ldots$, they achieve an insight which enables them to be confident that this property holds for any $m \times n$ matrix. Such students may find it possible to work on, without bothering about the details of the formal general proof. However, it is advantageous if students can master the language and symbolism of formal proofs, since on occasion they may need to consult works using such words and symbols.

Very similar remarks apply to the distributive law, $A(B + C) = AB + AC$. A student can easily verify that this law does in fact hold for, say, all 2×2 matrices. He can soon convince himself that this verification does not depend on the particular number of rows or columns considered, provided of course that the matrices are of types such that the additions and multiplications required are meaningful. Finally, the student should, if possible, write a proof for the general case, where A is an $m \times n$ matrix and B and C are $n \times p$ matrices.

In arithmetic, from $a(b + c) = ab + ac$ we can deduce $(b + c)a = ba + ca$ since multiplication is commutative. With matrices we cannot use this argument. It is therefore necessary to state and prove separately $(B + C)A = BA + CA$. This second distributive law holds for all matrices for which the products involved are meaningful.

In arithmetic, the 'in-any-order' law holds for multiplication. If we have to calculate $2 \times 3 \times 4$ there are six different ways of doing it and they all lead to the same answer 24. For a child learning arithmetic there is no obvious reason why the 'in-any-order' rule should be split into two properties, commutative and associative. But when we come to matrices, this distinction does arise in a very natural way. As we have seen, the 'in-any-order' property ceases to hold; as a rule it is not even true that $ST = TS$ for matrices. So, for a product involving three matrices A, B and C, we might guess that no law at all would hold, that every way of combining them would lead to a distinct result. Actually, this is not so; there is something that can be said. The point in question arises in practice. In an earlier section we were concerned with the quadratic expression $v'Mv$, which in 2 dimensions has the form

$$(x, y) \begin{pmatrix} a & h \\ h & b \end{pmatrix} \begin{pmatrix} x \\ y \end{pmatrix}.$$

---

* It will be necessary to assume the commutative and associative laws for the addition of *numbers*. Students sometimes object that we are assuming what we want to prove and that this is unfair. But we are not 'begging the question'. At this stage, we *know* that the *real numbers* have the properties in question; we are trying to prove that *matrices*, which are built from numbers but are not themselves numbers, have some of these properties.

When confronted with such an expression students frequently ask in what way they are supposed to multiply this out. Should they first work out the product of the row vector and the square matrix? This would give them

$$(ax + hy, hx + by) \ \begin{pmatrix} x \\ y \end{pmatrix}.$$

Or should they first work out the product of the square matrix and the column vector? This would give

$$(x, y) \ \begin{pmatrix} ax + hy \\ hx + by \end{pmatrix}.$$

It turns out not to matter. By either route, the final result is $ax^2 + 2hxy + by^2$.

It should be noted that, in the calculations above, we have nowhere changed the *order* of the matrices. Throughout, the row vector came in front of the square matrix, and the square matrix in front of the column vector; that is, we always kept the order of the product $v'Mv$. The difference in the two calculations above lay, not in changing the order of the matrices, but in the procedure for combining them. In the first calculation, we find $v'M$ and then combine this with v. Symbolically this is recorded as $(v'M)v$. In the second calculation we find Mv and then bring in v'; that is, we calculate $v'(Mv)$. The fact that both ways lead to the same final result is expressed by the equation $(v'M)v = v'(Mv)$.

This is an example of the associative law, and in fact for any three matrices A, B, C it is true that $(AB)C = A(BC)$; here of course we are assuming that the numbers of rows and columns in A, B and C are such as to make these products meaningful.

It is possible to check the assertion just made by direct calculations, but even for moderate-sized matrices the work is very tedious. For instance, if A, B and C are all 2×2 matrices, with arbitrary entries $a, b, c, \ldots$, to express the product $(AB)C$ we have to write 48 letters and 12 plus signs. The general result can be proved by using subscripts and summation signs, and is so proved in many textbooks.

It seems easier and more instructive to note that matrices represent linear mappings, and that all mappings, whether linear or not, automatically obey the associative law. In Fig. 110, p represents any kind of object. Some mapping C sends p to q, then B sends q to r and finally A sends r to s. It seems reasonable that we should be able to write

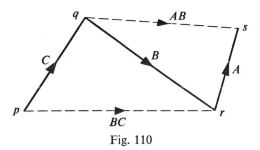

Fig. 110

$s = ABCp$, and that the mapping ABC should have a clear meaning and not require any brackets in its symbol. We can in fact show that $(AB)C$ and $A(BC)$ represent the same mapping. By BC we understand the effect of C followed by B. Either from the diagram, or from the equations $q = Cp$, $r = Bq$, we can see that BC sends p to r. As A sends r to s, it follows that $A(BC)$ sends p to s. Again we can see that AB sends q to s. Thus $(AB)C$, which indicates C followed by AB, sends p to s. Either way, p goes to s. Now p could be any object for which the mappings in question are defined. Thus $(AB)C$ and $A(BC)$ have the same effect on each object; that is, $(AB)C$ and $A(BC)$ represent the same mapping, which was what we wanted to show.

Accordingly, as far as addition and multiplication are concerned, the main thing to remember is that $ST = TS$ is *as a rule, not true* for matrices. We must be careful to preserve the order in which matrix products are written. Apart from this, we can carry out calculations very much as in the elementary algebra of numbers.

In matrix algebra we may meet expressions such as $2A + 3B$. The presence of the numbers, such as 2 and 3, does not lead to any surprises. We have, for example, $4A + 5A = 9A$ and $(2A)(3B) = 6AB$. We can form general laws to cover such results, and these can be checked or given formal proof.

We think of $2A$ as the product of the *number* 2 and the *matrix A*. There is a matrix closely associated with the number 2. For instance, if $A = \begin{pmatrix} a & b \\ c & d \end{pmatrix}$, it is easily verified that

$$\begin{pmatrix} 2 & 0 \\ 0 & 2 \end{pmatrix} \begin{pmatrix} a & b \\ c & d \end{pmatrix} = \begin{pmatrix} a & b \\ c & d \end{pmatrix} \begin{pmatrix} 2 & 0 \\ 0 & 2 \end{pmatrix} = \begin{pmatrix} 2a & 2b \\ 2c & 2d \end{pmatrix} = 2A.$$

Quite generally, if we let $K = kI$, we can verify that, for any square matrix A, $KA = AK = kA$. (If A is $n \times n$, we understand that I also is $n \times n$.)

A matrix of the form kI is known as a *scalar matrix*. Such a matrix commutes with every matrix A.

Subtraction also behaves as expected. We can regard $A - B$ as $A + (-1)B$, so that it is covered by the remarks made above about numerical coefficients.

Accordingly in matrix algebra we can make calculations such as the following:

(1) $(A + B)^2 = (A + B)(A + B) = A(A + B) + B(A + B) = A^2 + AB + BA + B^2$,

(2) $(A + B)(A - B) = A(A - B) + B(A - B) = A^2 - AB + BA - B^2$.

The absence of a commutative law prevents us from simplifying these any further.

Polynomials in one matrix

There is an important case in which matrix algebra is particularly simple; this occurs when we are dealing with expressions involving only one matrix. For example, the matrices A^2 and A^3 *do commute*. Here A^2 indicates that a certain transformation has to be applied twice, A^3 that it has to be applied three times. Both $A^2 \cdot A^3$ and $A^3 \cdot A^2$

indicate that the transformation has to be applied five times. Thus $A^2 \cdot A^3 = A^3 \cdot A^2 = A^5$. The same argument shows that for any natural* numbers m and n we have

$$A^m \cdot A^n = A^n \cdot A^m = A^{m+n}$$

exactly as in elementary algebra. *Thus the powers of a single matrix commute among themselves.* The identity operation I may be regarded as A^0 – applying the transformation 'no times', to put it ungrammatically. (I commutes with every matrix; $IM = MI$; it does not matter whether you 'leave things alone' before applying M or after.)

Thus for polynomials in a single matrix A we have *all the properties on which elementary algebra is based*; the computations are identical in form with those in the algebra of numbers.

Division

Division is essentially different for matrices from what it is for numbers. In arithmetic $12 \div 3$ corresponds to the question '3 times what is 12?' In slightly more sophisticated terminology this question invites us to solve the equation $3x = 12$. In the algebra of real numbers, the equation $ax = b$ always has a solution, except when $a = 0$. Thus b/a, the solution of this equation, is defined unless $a = 0$.

With matrices the situation is far otherwise. If we wish to be sure that the equation $AX = B$ has a solution X, (A, B, X being matrices), it is not by any means sufficient to check $A \neq 0$. We have already seen a particular case of this in §26, when we were considering the inverses of square matrices. The inverse of A, when it exists, satisfies the equation $AX = I$. No inverse exists when $\det A = 0$, that is, when A maps the space into a space of lower dimension. Fig. 111 indicates what happens in such a case. Several vectors u, v, w will be sent by A to the same point p, while no vector is sent to some other point q. Notice in this situation we have $p = Au = Av = Aw$. However, it would be incorrect to 'cancel A' in the equation $Au = Av$ and conclude $u = v$.

Exercise

Construct an example to illustrate the possibility $Au = Av$ with $u \neq v$ and $A \neq 0$.

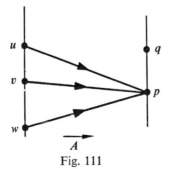

A

Fig. 111

* A natural number is a positive whole number 1, 2, 3, . . .

A vector can be regarded as an $m \times 1$ matrix. If such complications can arise in the equation $AX = B$ even when X and B are vectors, we would naturally expect complications in the general case, and in fact such complications do arise: the equation may have no solution at all, or it may have an infinity. Either way, we are unable to define 'B divided by A'.

The only straightforward case is when A has the inverse A^{-1}. Then $AX = B$ can have only one solution. For if $AX = B$, then $A^{-1}(AX) = A^{-1}B$, and this simplifies to $X = A^{-1}B$. So $AX = B \Rightarrow X = A^{-1}B$. By substitution we verify that

$$X = A^{-1}B \Rightarrow AX = B.$$

With ordinary numbers we could do without fraction symbolism b/a and write ba^{-1} or $a^{-1}b$ for this fraction. With numbers ba^{-1} and $a^{-1}b$ mean the same thing. With matrices, as multiplication is not commutative, BA^{-1} and $A^{-1}B$ are liable to mean different things; $A^{-1}B$ gives the solution of $AX = B$ and BA^{-1} the solution of $XA = B$.

It will be realized that the inverse A^{-1} can exist only for a square matrix A. The inverse exists if $\det A \neq 0$. However, a determinant is usually a very inconvenient thing to calculate; later we will give more suitable ways for seeing whether an inverse exists, and for calculating it when it does exist.

If A^{-1} exists, it is legitimate to 'cancel A' in an equation such as $AX = AY$. For we have $A^{-1}(AX) = A^{-1}(AY)$, which simplifies to $X = Y$. However, it is wisest to use the brief argument just given rather than to think of any 'rule for cancelling'. If we proceed from $AX = AY$ to $A^{-1}(AX) = A^{-1}(AY)$ we are reminded that this step is allowable only when A^{-1} exists, and we are less likely to commit the error of supposing X must equal Y in other circumstances.

In elementary algebra a quadratic equation never has more than two solutions. For example, if $x^2 = 1$, we can deduce $(x - 1)(x + 1) = 0$. The product is zero only when one factor is zero, so $x - 1 = 0$ or $x + 1 = 0$ and the only solutions are 1 and -1. *For matrices this argument fails.* It is easily verified that, for

$$X = \begin{pmatrix} \cos\theta & \sin\theta \\ \sin\theta & -\cos\theta \end{pmatrix},$$

we have $X^2 = I$. Thus, as θ may have any value, this equation has an infinity of solutions. Geometrically, it is evident this must be so. The matrix X given above represents reflection in the line through the origin at angle $\theta/2$; any reflection, done twice, brings us back to our starting point, so $X^2 = I$. It is instructive to see at which point the proof that $x^2 - 1 = 0$ has only two solutions fails if we try to apply it to $X^2 - I = 0$. Now I and $-I$ do satisfy the equation; they correspond to the solutions 1 and -1 in the algebra of numbers. But we also get a solution if we take, say, $X = \begin{pmatrix} 0 & 1 \\ 1 & 0 \end{pmatrix}$, and we can use this to find out where the argument breaks down. The early parts are quite in order.

With this value for X, we do in fact have $(X - I)(X + I) = 0$. For

$$X - I = \begin{pmatrix} -1 & 1 \\ 1 & -1 \end{pmatrix} \quad \text{and} \quad X + I = \begin{pmatrix} 1 & 1 \\ 1 & 1 \end{pmatrix}.$$

It is easily checked that the product of these two matrices is 0. The argument breaks down at the next step; the product is zero but neither factor is zero. (It may be noted that neither $X - I$ nor $X + I$ has an inverse.)

Exercise

Show that if $AB = 0$ for two matrices A, B, and one of the matrices has an inverse, then the other matrix must be 0.

The simplest example in which $PQ = 0$ without either factor being zero is obtained by taking

$$P = \begin{pmatrix} 1 & 0 \\ 0 & 0 \end{pmatrix}, \quad Q = \begin{pmatrix} 0 & 0 \\ 0 & 1 \end{pmatrix}.$$

P represents projection on the x-axis; it sends the point (x, y) to the point $(x, 0)$; that is, it wipes out the y co-ordinate. Similarly Q wipes out the x co-ordinate. Thus the operation PQ wipes out both co-ordinates, and sends every point to the origin; thus $PQ = 0$. But this effect is achieved in two stages; first Q replaces y by 0, then P replaces x by 0. But neither P nor Q is the matrix 0.

Exercise

Interpret geometrically the transformations $(X - I)/2$ and $(X + I)/2$, where $X = \begin{pmatrix} 0 & 1 \\ 1 & 0 \end{pmatrix}$, in terms of standard graph paper. What is their product, in either order?

Functions involving inverses

If two matrices commute, so do their powers. If $AB = BA$, we have

$$A^2B = A(AB) = A(BA) = (AB)A = (BA)A = BA^2$$

and it is not difficult to extend this argument to show $A^m B^n = B^n A^m$ for any natural numbers m, n.

It follows from this, that, if we are working with polynomials in two commuting matrices, A and B, we can carry out additions, subtractions and multiplications exactly as in elementary algebra.

Now we saw in §26 that, if A has an inverse A^{-1}, then $AA^{-1} = A^{-1}A = I$. So $AA^{-1} = A^{-1}A$, that is, A commutes with its inverse. Accordingly, we may take $B = A^{-1}$

in the result of the previous paragraph. A polynomial in A and A^{-1} is an expression involving positive and negative powers of A such as, for instance,

$$5A^2 + 3A - 4I + 7A^{-1} - A^{-3}.$$

Any two such expressions commute, and we can work with them by the familiar rules of elementary algebra, as regards addition, subtraction and multiplication. As we saw earlier, caution is required in any process involving division of matrices.

For reference purposes it may be helpful to list the properties of real numbers that, consciously or unconsciously, we assume when doing algebra, and the situation in matrix algebra in relation to each.

Real numbers	*Matrices*
$a + b$ defined always	$A + B$ defined always
$a - b$ defined always	$A - B$ defined always
'In-any-order rule' holds for $a + b + c$	The same for $A + B + C$
ab defined always	AB defined always
$ab = ba$ always	$AB = BA$ only in special cases
$a(bc) = (ab)c$ always	$A(BC) = (AB)C$ always
$a(b + c) = ab + ac$ always	$A(B + C) = AB + AC$ always
$(b + c)a = ba + ca$ always	$(B + C)A = BA + CA$ always
a/b defined unless $b = 0$	AB^{-1} and $B^{-1}A$ defined when B^{-1} exists
$bx = by$ implies $x = y$, if $b \neq 0$	$BX = BY$ implies $X = Y$ only when B^{-1} exists
$xy = 0$ implies $x = 0$ or $y = 0$	$XY = 0$ can happen with $X \neq 0$ and $Y \neq 0$
—	If $K = kI$, $KA = AK$ for any A
—	Powers, positive or negative, of a single matrix commute $A^m \cdot A^n = A^n \cdot A^m = A^{n+m}$ for any integers m, n

36   *Orthogonal transformations*

In §5 we considered the geometrical effect of a linear transformation of a plane. This was shown by a diagram in which a network of squares was transformed into a network of parallelograms. If we wanted to demonstrate such a transformation by applying it to an actual object, we would have to choose something, like an elastic sheet, that was capable of considerable deformations. Some linear transformations, such as rotations and reflections, can be demonstrated with a rigid body, such as a piece of cardboard. (Cardboard of course is not absolutely rigid – nothing is! – but its behaviour is sufficiently different from that of an elastic sheet to make our distinction meaningful.) Such transformations, and their generalizations to space of any number of dimensions, form a special class, known as orthogonal transformations. These transformations naturally occur when we are dealing with rigid bodies, or indeed with any objects in actual physical space, which we think of as being Euclidean. Orthogonal transformations also have certain mathematical properties that cause them to occur in situations that have no direct connection with rigid bodies or physical space. Statistics is one example of this. Vibration problems are another; it may be objected that vibrations are concerned with actual rigid bodies (though this hardly applies to electrical vibrations), but the relevance of orthogonal transformations does not arise from this aspect of the situation. It is due rather to analogies that exist between vibration problems and the purely mathematical idea of rotations in space of n dimensions (where n may be bigger than 3).

Let us consider the simplest and most familiar case, a rotation in a plane, specified with the help of conventional graph paper.

Suppose we rotate the plane until the point P, which originally had co-ordinates (x, y) lands at the position (X, Y) as shown in Fig. 112. If the angle of rotation is θ, it can be shown (by projections, or by considering the fate of the unit basis vectors) that

$$\begin{pmatrix} X \\ Y \end{pmatrix} = \begin{pmatrix} \cos\theta & -\sin\theta \\ \sin\theta & \cos\theta \end{pmatrix} \begin{pmatrix} x \\ y \end{pmatrix}. \tag{1}$$

If we wished to obtain (x, y) in terms of (X, Y) we might consider that $(X, Y) \rightarrow (x, y)$ is a rotation of $-\theta$, or we might use projection, or we might solve the equations (1). Whichever we did, we would arrive at the result

$$\begin{pmatrix} x \\ y \end{pmatrix} = \begin{pmatrix} \cos\theta & \sin\theta \\ -\sin\theta & \cos\theta \end{pmatrix} \begin{pmatrix} X \\ Y \end{pmatrix}. \tag{2}$$

The matrix occurring in equations (2) is the inverse of the matrix occurring in equations (1), for it represents the transformation that undoes the effect of rotation through θ. Thus if L denotes the matrix in equations (1), L^{-1} denotes the matrix in equations (2). But if we examine these matrices as they appear in the two equations, we notice that the second matrix is simply the transpose of the first. Accordingly we have $L' = L^{-1}$. *This is the characteristic property of an orthogonal matrix; its inverse is the same as its transpose.* This incidentally is an extremely convenient property. In 2 dimensions it is easy to calculate the inverse of a matrix, but with increasing number of dimensions the

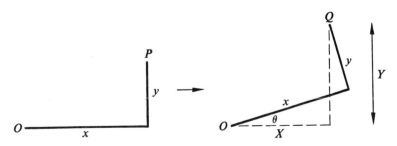

Fig. 112

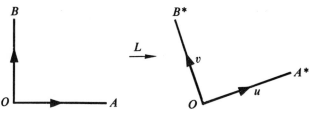

Fig. 113

calculation of an inverse becomes extremely messy. On the other hand, a transpose can be written down immediately. Accordingly, the appearance of an orthogonal matrix in a problem is always welcomed.

Work in 2 dimensions uses very special methods. In 2 dimensions we can specify a rotation by giving a single number, θ; we can specify a direction by giving a single number, the value of m in $y = mx$. In 3 or more dimensions such simple procedures cease to be possible.

It is therefore useful to re-examine the situation in 2 dimensions, and to try to explain $L' = L^{-1}$ by methods that work equally well in 3 or more dimensions.

Accordingly, let us consider a 2×2 matrix

$$L = \begin{pmatrix} u_1 & v_1 \\ u_2 & v_2 \end{pmatrix}$$

and try to interpret what it means if for this matrix $L^{-1} = L'$. This means $L'L = I$, so we have

$$\begin{pmatrix} u_1 & u_2 \\ v_1 & v_2 \end{pmatrix} \begin{pmatrix} u_1 & v_1 \\ u_2 & v_2 \end{pmatrix} = \begin{pmatrix} u_1^2 + u_2^2 & u_1v_1 + u_2v_2 \\ u_1v_1 + u_2v_2 & v_1^2 + v_2^2 \end{pmatrix} = \begin{pmatrix} 1 & 0 \\ 0 & 1 \end{pmatrix}.$$

This means $u_1^2 + u_2^2 = 1$, $u_1v_1 + u_2v_2 = 0$, $v_1^2 + v_2^2 = 1$. These three equations are readily interpreted geometrically. The first and the last indicate that the vectors u and v are of unit length; the middle equation involves a scalar product and indicates that u and v are perpendicular.

Now the vectors u and v occur in the columns of the matrix L. In §22, under the heading 'The columns in a matrix' it was pointed out that these columns represented the vectors to which the unit vectors along the axes were mapped. Thus we may represent the effect of L as in Fig. 113.

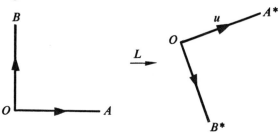

Fig. 114

The condition $L'L = I$ assures us that the lines $OA*$ and $OB*$ will be of unit length and at right angles, which of course is so for a rotation. The condition, however, does *not* tell us that we *must* be dealing with a rotation. It is perfectly consistent with a diagram such as Fig. 114. Here a reflection is involved.

The argument we have used does not in any way depend on the special features of 2 dimensions. It is easily adapted to show that, in any number of dimensions, the condition $L'L = I$ means that L sends the unit vectors along the axes to vectors that are of unit length and perpendicular to each other.

Exercise

Write the 3 × 3 matrix L, whose columns are the vectors u, v, w. Check that $L'L = I$ does mean that u, v, w are all of unit length and perpendicular to each other.

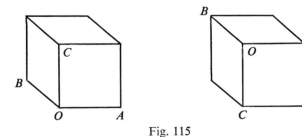

Fig. 115

For each of the two cubes shown in Fig. 115, the lines OA, OB, OC are of equal length and are perpendicular to each other. Yet it is impossible to turn the first cube in such a way that the letters will appear as in the second cube. The second cube is a reflection of the first.

Thus, as was already noted in connection with 2 dimensions, the condition $L'L = I$ does not tell us that L represents a rotation; it may represent a rotation followed by a reflection. But either way we can assert that the transformation represented by L *preserves lengths unchanged*; to use the technical term, it is an *isometry* (iso-, the same; -metry, measure).

It seems reasonable that L should preserve all distances unchanged. If OA, OB, OC are unit vectors along the axes, they can be visualized as the edges of a unit cube. If $L'L = I$, L sends these vectors to $OA*$, $OB*$, $OC*$ which are of unit length and per- pendicular – that is to say, are edges of some unit cube. We can bring the original unit cube to the new position by a rotation, perhaps combined with a reflection. We saw in §10 that the fate of the unit vectors along the axes determined the fate of every vector, in a linear transformation. So, if we think of all the vectors as being embedded in some rigid body, we can show the effect of L by rotating this rigid body, and, if necessary, following this by a reflection. In such a process, all lengths stay unchanged.

It is sometimes necessary to consider orthogonal transformations in, for example, 6 dimensions, and here we may not feel the same confidence in our physical picture of what is meant by a rotation or a reflection. So it is desirable to provide a proof by algebra that, when $L'L = I$, the transformation L preserves the distances between points. The proof is very simple, and illustrates a type of matrix calculation we may often meet.

We first show that, if L sends $p \to p^*$, $q \to q^*$, for any vectors p, q, then the scalar product of p^* and q^* equals the scalar product of p and q. More briefly, L *preserves all scalar products.*

The proof is immediate. $p^* = Lp$, $q^* = Lq$. The scalar product is, by §33, $(Lp)'Lq$. This is, by the reverse-order rule for transposes, $p'L'Lq$. But $L'L = I$. So we have $p'Iq$, which is $p'q$, the scalar product of the original vectors.

Note the convenience of matrix notation. Without any heavy work or special trick, we have proved the result for any number of dimensions. Even in 3 dimensions the notation effects a considerable economy. This proof, written out in full as was done in the older textbooks on 'co-ordinate solid geometry' would need 9 symbols for the entries in the 3×3 matrix L, and 3 symbols each for the vectors p and q. Several equations would be needed to express our single equation $L'L = I$.

Note also that the proof is making free use of the associative property. If we retained brackets, we would obtain $(Lp)'(Lq)$ first as $(p'L')(Lq)$, and would then have to re-arrange brackets to get $(L'L)$ appearing and replaced by I. The associative property allows us simply to dispense with brackets.

Our proof is now nearly complete, for all lengths can be expressed as scalar products. For any vector p, we have $\|p\|^2 = p'p$. If p represents the vector OP, $\|p\|$ equals the length $\|OP\|$. As this length is expressible by means of the scalar product $p'p$, and as L preserves scalar products, it follows that, when L acts, the distance of any point from the origin is unchanged. We wish also to show that the length $\|PQ\|$ is preserved for any points P, Q. Let q denote the vector OQ. Then $q - p$ denotes the vector PQ. The scalar product of this vector with itself gives $\|PQ\|^2$. So

$$\|PQ\|^2 = (q - p)'(q - p) = (q' - p')(q - p) = q'q - p'q - q'p + p'p.$$

It is sufficient to observe that each term here is a scalar product, hence unchanged when p is replaced by p^* and q by q^*. The expression can in fact be simplified, since $q'p = p'q$. (See §33, near end.) It may be written

$$\|PQ\|^2 = q'q - 2p'q + p'p.$$

This last result in fact is essentially the same as a standard formula in trigonometry, $c^2 = a^2 - 2ab \cos C + b^2$.

Determinants

The determinant of L, $\det L$, gives the ratio in which volumes (or the corresponding concept in n dimensions) are changed. A rotation keeps volumes unchanged, so for a rotation $\det L = +1$. A reflection, with the convention established in §27, multiplies volumes by -1. So if L represents the effect of a rotation followed by a reflection,

$\det L = -1$. Thus the determinant enables us to tell to which of these two classes any given orthogonal transformation, L, belongs.

Exercise

Apply this test to the matrices

$$\begin{pmatrix} \cos\theta & -\sin\theta \\ \sin\theta & \cos\theta \end{pmatrix} \quad \text{and} \quad \begin{pmatrix} \cos\theta & \sin\theta \\ \sin\theta & -\cos\theta \end{pmatrix}.$$

By a reflection in n dimensions we understand a situation in which one vector is reversed, and all vectors perpendicular to it remain unchanged. Thus, for example, in 4 dimensions the equations $x^* = -x$, $y^* = y$, $z^* = z$, $t^* = t$ would define a reflection. Any reflection can be expressed in this form by choosing axes suitably related to its eigenvectors.

Change of axes

At the beginning of this section we had equations (1), $X = x \cos\theta - y \sin\theta$, $Y = x \sin\theta + y \cos\theta$. These gave us the co-ordinates (X, Y) to which a point (x, y) was sent by a rotation through θ. Part of the diagram used to illustrate this rotation appeared as in Fig. 116.

If we look at this part of the diagram by itself we see the point Q; we do not see the point P from which it came by rotation. Someone seeing only this part of the diagram might easily imagine that equations (1) were intended to specify, not a rotation, but a change of axes. For the point Q has co-ordinates (X, Y) in horizontal and vertical axes, and co-ordinates (x, y) in axes inclined at an angle θ. In fact equations (1) could be used for this purpose. Thus the same equations arise in describing a rotation and in specifying a change of axes. In practice, orthogonal transformations are probably used more often for introducing more convenient axes than for describing a rotation of some object.

In lectures on matrices it is customary to mention a mild joke – that a matrix can either be an alibi or an alias. An alibi means 'somewhere else', and so might refer to a transformation that sends each point to a new position. An alias means a new name, and

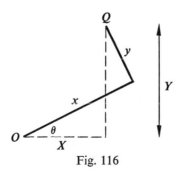

Fig. 116

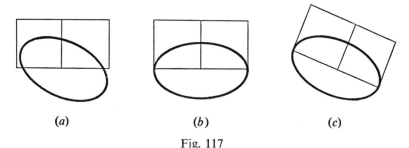

(a)   (b)   (c)

Fig. 117

is appropriate to the situation just discussed where the point Q has two names; Q is called (x, y) in one system of axes and (X, Y) in another.

We can see why the same equations arise in the two cases.

In Fig. 117 (a) we see an ellipse and some squares which we may suppose drawn on a transparent sheet placed over the ellipse. In (b) the ellipse has been rotated until its principal axes coincide with lines on the graph paper. In (c) the same effect has been achieved by bringing the graph paper to the ellipse. Thus from (a) to (b) represents a rotation of the ellipse, with fixed graph paper. From (a) to (c) represents a rotation of the graph paper (a change of axes) with a fixed ellipse. The relative motion of the ellipse and the graph paper is the same in both cases. Note that if in going from (a) to (b) the ellipse rotates through an angle θ, in going to (a) to (c) the graph paper rotates through an angle $-\theta$, equal and opposite. In work of this kind it is always necessary to check carefully that the matrix you have written represents the transformation you want, and not its inverse.

Worked example

Test for orthogonality the matrix

$$M = \begin{pmatrix} -\frac{2}{3} & \frac{1}{3} & \frac{2}{3} \\ \frac{1}{3} & -\frac{2}{3} & \frac{2}{3} \\ \frac{2}{3} & \frac{2}{3} & \frac{1}{3} \end{pmatrix}.$$

Investigate fully the geometrical meaning of the transformation of 3-dimensional space it represents.

Solution   The matrix will be orthogonal if the vectors that occur in its columns are perpendicular and of unit length. This is easily verified. So M is orthogonal and $M'M = I$. But we notice M is symmetric, with $M' = M$. Accordingly $M^2 = I$. So M represents some operation that, done twice, returns each point of space to its original position.

Is the operation a rotation or a transformation involving reflection? The sign of the determinant will tell us which; det $M = (4 + 4 + 4 + 8 - 1 + 8)/27 = 1$. As det M is positive, M represents a rotation.

Any rotation in 3 dimensions must be about some axis. Every point on the axis stays still, so any point v on the axis must satisfy $Mv = v$. If $v = (x, y, z)$ this means

$$x = (-\tfrac{2}{3})x + (\tfrac{1}{3})y + (\tfrac{2}{3})z,$$
$$y = (\tfrac{1}{3})x + (-\tfrac{2}{3})y + (\tfrac{2}{3})z,$$
$$z = (\tfrac{2}{3})x + (\tfrac{2}{3})y + (\tfrac{1}{3})z.$$

These equations are equivalent to $x = y$, $z = 2x$. So, for example, if we choose $x = 1$, we see that $(1, 1, 2)$ is on the axis of rotation.

Earlier we had $M^2 = I$. A rotation which, done twice, brings us back to the starting situation, must be through $0°$ or $180°$. (We need not consider $360°$, $540°$ and so on, since these have the same final effect.) Now $0°$ would indicate $M = I$, which is clearly not so, so M must represent a rotation through $180°$.

We can confirm this conclusion. A rotation of $180°$ reverses every vector in the plane perpendicular to the axis of rotation. Thus every vector in this plane should satisfy the equation $Mv = -v$. If $v = (x, y, z)$ this means

$$-x = (-\tfrac{2}{3})x + (\tfrac{1}{3})y + (\tfrac{2}{3})z,$$
$$-y = (\tfrac{1}{3})x + (-\tfrac{2}{3})y + (\tfrac{2}{3})z,$$
$$-z = (\tfrac{2}{3})x + (\tfrac{2}{3})y + (\tfrac{1}{3})z.$$

Each of these equations reduces to $x + y + 2z = 0$. This is the equation of the plane perpendicular to $(1, 1, 2)$, since it expresses that the scalar product of $(1, 1, 2)$ and (x, y, z) is zero.

It will be noticed that the non-zero vectors lying in the axis of rotation are eigenvectors with $\lambda = 1$, while those perpendicular to the axis are eigenvectors with $\lambda = -1$. This illustrates one of the most important maxims of linear algebra – if you want to know what a matrix does, look for its eigenvectors and eigenvalues.

In this particular case, by observing special features of the problem, we were able to avoid the computation involved in the standard routine of solving $\det (M - \lambda I) = 0$, and then finding the eigenvectors. If this routine had been followed, we would have found $\det (M - \lambda I) = -(\lambda - 1)(\lambda + 1)^2$, and been led to the same results, though with a little more computation.

Questions

1 Let P be the point $(0.8, 0.6)$ and Q the point $(-0.6, 0.8)$ on conventional graph paper. Find the lengths OP and OQ and the angle between OP and OQ. Is the matrix $\begin{pmatrix} 0.8 & -0.6 \\ 0.6 & 0.8 \end{pmatrix}$ orthogonal or not?
  What is its inverse?

2 Which of the following matrices are orthogonal?

$$\begin{pmatrix} 0 & 1 \\ 1 & 0 \end{pmatrix}, \quad \begin{pmatrix} 0 & -1 \\ 1 & 0 \end{pmatrix}, \quad \begin{pmatrix} 0.8 & 0.6 \\ 0.6 & -0.8 \end{pmatrix}, \quad \begin{pmatrix} 1 & 1 \\ 0 & 1 \end{pmatrix}.$$

Which of them represent rotations? Which of them represent reflections?

3 Test for orthogonality the matrix

$$\begin{pmatrix} -\frac{1}{3} & \frac{2}{3} & \frac{2}{3} \\ \frac{2}{3} & -\frac{1}{3} & \frac{2}{3} \\ \frac{2}{3} & \frac{2}{3} & -\frac{1}{3} \end{pmatrix}.$$

What is its inverse? What does this matrix do to the vectors

$$\begin{pmatrix} 1 \\ 1 \\ 1 \end{pmatrix}, \quad \begin{pmatrix} 1 \\ -1 \\ 0 \end{pmatrix}, \quad \begin{pmatrix} 1 \\ 0 \\ -1 \end{pmatrix}.$$

How can its effect be described in simple geometrical terms?

4 A transformation T sends (x, y, z) to (x^*, y^*, z^*) where $x^* = y$, $y^* = x$, $z^* = -z$. Write the matrix for T. Is it orthogonal? What is the geometrical meaning of T? What are its eigenvectors and eigenvalues?

5 The matrices L and M are both orthogonal. Which of the following matrices can we be certain will be orthogonal?
   (a) LM,   (b) L^{-1},   (c) $L + M$.
(Much harder.) Can we be certain that any of the above matrices will *not* be orthogonal?

6 Let

$$S = \begin{pmatrix} 0 & k \\ -k & 0 \end{pmatrix}.$$

Calculate the matrix $L = (I + S)(I - S)^{-1}$ and check that L is orthogonal. (It can in fact be proved that for any antisymmetric matrix S, acting in n dimensions, the matrix L given by this formula will be orthogonal.)

7 Is the matrix

$$\begin{pmatrix} -\frac{3}{7} & \frac{6}{7} & \frac{2}{7} \\ -\frac{2}{7} & -\frac{3}{7} & \frac{6}{7} \\ \frac{6}{7} & \frac{2}{7} & \frac{3}{7} \end{pmatrix}$$

orthogonal or not? Apply the transformation specified by this matrix to the points $(-5, 3, 8)$, $(-7, 0, 14)$, $(-8, 9, 10)$, $(-10, 6, 16)$. What geometrical figure results? What can be said about the figure formed by the original four points?

8 Show that the four points $(0, 0, 0)$, $(0, 1, 1)$, $(1, 0, 1)$, $(1, 1, 0)$ are the corners of a regular tetrahedron. Find where these points are sent by the transformation

$$\begin{pmatrix} -\frac{2}{3} & \frac{2}{3} & \frac{1}{3} \\ -\frac{1}{3} & -\frac{2}{3} & \frac{2}{3} \\ \frac{2}{3} & \frac{1}{3} & \frac{2}{3} \end{pmatrix}.$$

Would you expect the transformed points to be at the corners of a regular tetrahedron? If so, why? Are they in fact so situated?

9 Find the length of, and the angles between, the vectors given by the columns of the matrix

$$\begin{pmatrix} -7 & 4 & -4 \\ -4 & 1 & 8 \\ 4 & 8 & 1 \end{pmatrix}.$$

This matrix is not orthogonal. Is there any combination of simple geometrical operations that would describe its effect? To what would it transform the four points mentioned in Question 8?

10 Do either of the matrices mentioned in Questions 7 and 8 represent a reflection or a rotation, and if so about what axis?

11 What vectors are unaltered when the transformation

$$\begin{pmatrix} \frac{1}{3} & -\frac{2}{3} & -\frac{2}{3} \\ -\frac{2}{3} & \frac{1}{3} & -\frac{2}{3} \\ -\frac{2}{3} & -\frac{2}{3} & \frac{1}{3} \end{pmatrix}$$

acts on them? What vectors are reversed by it? What geometrical operation does this matrix represent? Is it an orthogonal matrix? Would it be possible to say, without the labour of carrying out the calculation, whether its determinant is positive or negative?

12 In question 6 at the end of §32, each part specified 3 vectors D, E, F. In each case, if a matrix were formed having D, E, F as its columns, would this matrix be orthogonal? If a matrix were formed having D, E and F as its rows, would this matrix be orthogonal?

13 A rotation is represented by the matrix

$$M = \begin{pmatrix} -\frac{6}{7} & \frac{3}{7} & -\frac{2}{7} \\ -\frac{3}{7} & -\frac{2}{7} & \frac{6}{7} \\ \frac{2}{7} & \frac{6}{7} & \frac{3}{7} \end{pmatrix}.$$

Find the axis about which this rotation takes place. Let P be the point $(1, 3, -2)$. Find P^*, the point to which P is mapped by M. What is the cosine of the angle between OP and OP^*? If M represents a rotation through the angle α, does α equal the size of the angle between OP and OP^*? Justify what you assert.

14 A rotation is represented by the matrix

$$M = \begin{pmatrix} 0 & 0 & 1 \\ 1 & 0 & 0 \\ 0 & 1 & 0 \end{pmatrix}.$$

Find the axis of rotation. Find the angle of rotation produced by M by considering the powers of M. Let $P = (1, -1, 0)$ and $Q = (0, 1, -1)$. Find P^* and Q^*, the points to which P and Q map under the action of M. Find the angle between OP and OP^* and the angle between OQ and OQ^*. Explain the result.

Show that M has only one real eigenvalue and consequently only one real eigenvector. Explain why this result is to be expected on geometrical grounds.

15 The 8 corners of a cube have the co-ordinates $(\pm 1, \pm 1, \pm 1)$, in which it is understood that every possible combination of plus and minus signs is to be taken. Apply to these 8 points the transformation represented by the matrix M in Question 13. Delete the third co-ordinate in each point so obtained, i.e. change (x, y, z) into (x, y), and plot the 8 resulting points (x, y) on ordinary squared paper. Do these points, when suitably joined, look like a picture of a cube? Explain what you observe.

Note. Further exercises of this kind can be obtained, if desired, by taking for M the matrix mentioned in some other question of this section.

37   Finding the simplest expressions for quadratic forms

In most secondary school courses, students meet problems about ellipses and hyperbolas. The ancient Greeks studied these curves and named them conic sections, since they could be obtained by taking a plane section of a cone. Chopping cones by planes is not an activity that occupies a large part of an engineer's time, and one is inclined to regard conic sections as an entertainment for pure mathematicians. This view, though natural, is entirely mistaken. Conics have a way of turning up in all kinds of practical situations.

Before dealing with their applications today, we may note that the whole development of modern science would probably have been considerably delayed if the Greeks had not established conics as a traditional part of mathematical education. In the seventeenth century, when Newton put the science of mechanics into systematic form, he was very much helped by Kepler's work on the orbits of the planets. Now Kepler had been faced by a difficult problem. In astronomy the motion of, say, Mars is observed from the earth, itself travelling in a curved path with varying speed. The problem is to find assumptions about the orbits of Mars and the earth that will explain the observations of Mars made from the earth. At first Kepler tried to fit the data by assuming the orbits to be circles, perhaps with centres not at the sun, and perhaps described with variable speeds. Long and patient calculations showed that circles would not do. But if not circles, what then? There are other possible curves in great variety. How to decide which kind to try next? Very fortunately, the ellipse, which was already known to Kepler as a curve resembling the circle, turned out to be the correct guess. Without this fortunate coincidence, Kepler might have spent his entire life trying different curves and never landing on the right one.

For an understanding of the modern importance of conics the Greek definition 'section of a cone' is not particularly helpful. For us the important thing is that any equation of the second degree gives a conic. The general equation of the second degree is $ax^2 + 2hxy + by^2 + 2gx + 2fy + c = 0$. Here a, b, c, f, g, h are constants. By 'the second degree' we understand that terms may be present, such as x^2 or xy, in which two variables are multiplied together, but that we never have terms, such as x^4, x^2y or y^3, with three or more variables as factors. Such an equation may represent an ellipse, parabola or hyperbola; there are also certain special cases such as, for example, a pair of straight lines. For engineering work the ellipse is probably the case most often met.

The definition just given is easily generalized. In 3 dimensions we can consider equations of the second degree; they may contain linear terms, squares such as x^2, y^2, z^2, and products such as xy, xz and yz, but no higher terms. Such an equation defines a *quadric surface*; in particular it may define an *ellipsoid*. An ellipsoid may be visualized as a sphere that has been subjected to a particular kind of distortion. Suppose we start with the unit sphere, $X^2 + Y^2 + Z^2 = 1$. Then we enlarge the scale on the first axis a times, on the second axis b times, and on the third axis c times. That is, we send (X, Y, Z) to (x, y, z) where $x = aX$, $y = bY$, $z = cZ$. We obtain the ellipsoid with equation $(x^2/a^2) + (y^2/b^2) + (z^2/c^2) = 1$. This surface in 3 dimensions obviously has analogies with the ellipse in 2 dimensions.

We saw in §2 that many scientific laws can be well approximated by linear functions and that this is why we are likely to meet simple linear expressions for such things as momentum, mv, force in a stretched spring, kx, and voltage drop in a resistance, ri. Now the integrals of these expressions also have physical significance; the momentum, mv, is the rate of change of $\frac{1}{2}mv^2$, the kinetic energy; the force, kx, is the rate of change of potential energy, $\frac{1}{2}kx^2$; the voltage drop, ri, is the derivative of $\frac{1}{2}ri^2$, which is proportional to the rate of generation of heat; the angular momentum, $I\omega$, is the derivative of $\frac{1}{2}I\omega^2$, the kinetic energy of a rotating flywheel. It will be noticed that the expressions related to energy are all of the second degree. This happens also when we are considering systems involving several variables. Thus we are not surprised to find that in much scientific work we are confronted by some formula involving squares and products, a formula that can be shown graphically by means of an ellipsoid or other quadric surface. Thus in the theory of elasticity we meet the stress-ellipsoid; both in connection with the balancing of machines and with objects hurled through space we meet the momental ellipsoid; in the theory of vibrations we have to deal with a pair of ellipses or ellipsoids representing kinetic and potential energy; in statistics the dots of scatter diagrams frequently form egg-shaped clusters; in quantum theory we meet something resembling an ellipsoid, adapted to the needs of that strange subject (the Hermitian form).

In §24 we considered a vibration problem with two masses and three springs. It was shown that the differential equations for the motion of these masses could be brought to their simplest form by means of the substitution $x = X + Y$, $y = X - Y$. We could have been led to this substitution by a different argument, which does not even use calculus, if we had posed the question in another way: find the co-ordinate system in which *the expressions for the kinetic and potential energies take the simplest form*.

Let u and v stand for the velocities of the two unit masses in the problem of §24. (Note that u and v here stand for single numbers, not for vectors in 2 or 3 dimensions as they often have.) The kinetic energy, T, is given by $T = \frac{1}{2}(u^2 + v^2)$. The potential energy is $P = \frac{1}{2}kx^2 + \frac{1}{2}k(y - x)^2 + \frac{1}{2}ky^2$, which simplifies to $P = k(x^2 - xy + y^2)$.

We can show the positions of the two masses by plotting (x, y) on graph paper. The vector (u, v) will then show the velocity with which the plotted point moves (Fig. 118). The quantities u and v are taken parallel to the same axes as those used for x and y. Accordingly, if we introduce new co-ordinates X, Y for position by means of the equations $x = aX + bY$, $y = cX + dY$ we must also introduce new co-ordinates for velocity, U, V, with $u = aU + bV$, $v = cU + dV$. (The same conclusion could be reached by the calculus argument that $u = \dot{x}$ and $v = \dot{y}$, where the dot indicates differentiation with respect to time; if $U = \dot{X}$ and $V = \dot{Y}$, we can obtain the equations

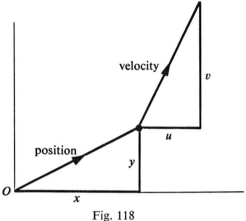

Fig. 118

connecting u, v with U, V by differentiating the equations $x = aX + bY$, $y = cX + dY$.)

Accordingly, we make diagrams to illustrate kinetic and potential energy using the same graph paper to plot (x, y) and (u, v). We will show (u, v) as a vector springing from the origin $(0, 0)$, not as in the last diagram, springing from the point (x, y) and indicating the direction of motion of that point.

The kinetic energy, T, is already in a very simple form. The curves $T =$ constant have equations $u^2 + v^2 =$ constant, that is to say they are circles. The curves $P =$ constant have equations $x^2 - xy + y^2 =$ constant. These curves are in fact ellipses, tilted as shown in Fig. 119. The diagram shows just one ellipse, and the two circles that touch it. The points of contact lie on the lines $y = x$ and $y = -x$ respectively. Thus if we had some machine that automatically drew the curves $T =$ constant and $P =$ constant for us, we would obtain a diagram containing many circles and many ellipses which were enlarged or reduced copies of those shown. At most points the circles would cross the ellipses, but along the lines $y = x$ and $y = -x$ the curves would touch. Thus the machine would draw our attention to the fact that the north-east and south-east direc-

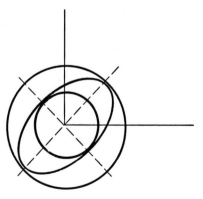

Fig. 119

tions had a special physical significance. The vectors $\begin{pmatrix}1\\1\end{pmatrix}$ and $\begin{pmatrix}1\\-1\end{pmatrix}$ lie in these directions. If we take them as a basis for new co-ordinates (X, Y) we obtain the equations

$$\begin{pmatrix}x\\y\end{pmatrix} = X\begin{pmatrix}1\\1\end{pmatrix} + Y\begin{pmatrix}1\\-1\end{pmatrix}.$$

That is, we have $x = X + Y, y = X - Y$ as in §24. If we substitute these expressions for x and y in our equation for P, and the corresponding expressions $u = U + V$, $v = U - V$ in the equation for T, we find $T = U^2 + V^2, P = k(X^2 + 3Y^2)$.

Thus this substitution achieves two goals: (1) it brings P to a simple form, with no product XY, (2) it does this without spoiling the simple form that T had from the outset.

Now of course in practice we do not usually have a machine that will sketch graphs for us; we may have computers that will calculate such transformations for us, but they do it by a rather different method. The purpose of the discussion so far has been, not to provide a method of computation, but to show the kind of situation that exists when we have two physically significant quantities (such as kinetic and potential energy) represented by quadratic forms. In such a situation it is frequently possible to choose axes in such a way that one quadratic form is of the type we associate with a circle or sphere, while the other is of the type we associate with a conic referred to its principal axes. Thus in 3 dimensions we can often find axes that bring two quadratic expressions to the forms $X^2 + Y^2 + Z^2$ and $aX^2 + bY^2 + cZ^2$ when both involve the same variables, or to $U^2 + V^2 + W^2$ and $aX^2 + bY^2 + cZ^2$ in cases, such as that considered above, where one quadratic form involves velocity and the other position.

We cannot prove this here, but in fact by choosing the axes that put kinetic and potential energy into the simplest possible form, we are automatically choosing the axes that put the differential equations for the motion into the simplest possible form.

We have used vibration problems as an illustration because in §24 we worked such a problem out in detail. The mathematical ideas apply equally well to any problem in which two quadratic forms are involved. Sometimes one of the quadratic forms may be present but camouflaged. For instance, in any question involving the motion of a rigid body (balancing of machinery, projectiles, flight of aircraft) the values of the moments and products of inertia can be shown graphically by a certain ellipsoid embedded in the body. We are interested in finding the principal axes of this ellipsoid, that is to say, the axes in which its equation takes the form

$$(X^2/a^2) + (Y^2/b^2) + (Z^2/c^2) = 1.$$

Here we seem to be dealing with only one quadratic form, that appearing in the equation of the ellipsoid. But in fact a second quadratic is involved. The body is *rigid*; the distances between any two particles in it is fixed. Now the distance of a particle at (x, y, z) from the origin is given by $\sqrt{(x^2 + y^2 + z^2)}$. When we bring in new co-ordinates X, Y, Z we want this distance to be given by $\sqrt{(X^2 + Y^2 + Z^2)}$; otherwise we are liable to slide into all kinds of mistakes, because our usual formulas for distances, scalar products, perpendicularity and so forth will cease to apply. So what in fact we are trying to do when we search for the principal axes of the ellipsoid is to find co-ordinates that simplify the equation of the ellipsoid *without spoiling the simplicity of the distance formula* $s^2 = x^2 + y^2 + z^2$. Compare this with our earlier work on the vibration problem, where we managed to bring the potential energy P to a simple form without disturbing the simple form that T already had. It will be seen that the two problems, so different in physical terms, involve essentially the same mathematical problem.

Students sometimes think that all they need learn in engineering mathematics is procedures for carrying out calculations. But this is in fact not enough. We need first of all an assurance that the problem does have a solution. If we apply some standard procedure to an impossible problem difficulties of one kind or another are bound to arise. Again, even when a problem is possible, it may be that there are certain exceptional cases where the solution is of an unexpected kind, or where the computing procedure has to be modified. It is clearly desirable to be aware of these possibilities.

The first task, then, is not to learn an algorithm, but to survey the scene and classify problems so that we can recognize the general case, the exceptional case and perhaps the impossible case.

When dealing with quadratic forms, we can get quite a good idea of how things will work out in any number of dimensions by considering fairly familiar examples in 2 or 3 dimensions. In 2 dimensions we very readily believe that any ellipse has a major and a minor axis, and that these are perpendicular. Accordingly, if we choose points C and D on these axes, at unit distance from the origin O, OC and OD will be unit vectors at right angles (Fig. 120). If OA and OB are the basis vectors of the original co-ordinate system, we need only rotate axes to arrive at OC and OD. Rotation of axes does not alter the equation of a circle. The circle $x^2 + y^2 = 1$ in the system based on OA and OB will have the equation $X^2 + Y^2 = 1$ in the system based on OC and OD.

Thus our impression is that there will be no impossible problems. Given a quadratic form $ax^2 + 2hxy + by^2$, we expect to be able to rotate axes (thus preserving the simplicity of $x^2 + y^2$) so as to obtain an expression of the form $AX^2 + BY^2$. Of course, this expression may not correspond to ellipses. For example, we might obtain $X^2 - Y^2$ and the curves $X^2 - Y^2 = $ constant would be hyperbolas.

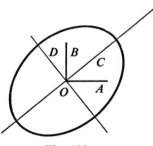

Fig. 120

Exercise

Prove the statement that the problem is never impossible; that is, show that with the substitution $x = X \cos \theta - Y \sin \theta$, $y = X \sin \theta + Y \cos \theta$ we can always choose θ so that $ax^2 + 2hxy + by^2$ becomes an expression in which the XY term is absent. Check also that $x^2 + y^2 = X^2 + Y^2$.

While there is no impossible case, there do exist exceptional cases. A very simple example would be the expression x^2. This is already in its simplest possible form. The equation $x^2 = 1$ represents a pair of vertical lines (Fig. 121). This may be thought of as a limiting case, in which the major axis of the ellipse has become of infinite length. Such a situation arises if in $AX^2 + BY^2$ either $A = 0$ or $B = 0$. Of course it is not necessary that the lines should be vertical as in our example using x^2. By rotating the diagram we can obtain an example with parallel lines pointing in any direction. For example $x^2 - 2xy + y^2 = 1$ is the same as $(x - y)^2 = 1$, which breaks up into the two lines $x - y = 1$ and $x - y = -1$.

In 3 dimensions we have an element of variety in the special cases. We can always rotate axes so as to get the form $AX^2 + BY^2 + CZ^2$. In the general case, A, B, C are all different from zero; if they all happen to be positive, $AX^2 + BY^2 + CZ^2 = 1$ will represent an ellipsoid. If one of them, say C, is zero, the equation reduces to

$$AX^2 + BY^2 = 1.$$

Now Z does not appear in this, so Z is completely arbitrary; by this we mean that if some particular point, $(X_0, Y_0, 0)$ satisfies the equation, then the point (X_0, Y_0, t) will also satisfy it, for every value of t. Thus the surface consists of complete lines parallel to the Z-axis. It is a cylinder – not necessarily a cylinder with circular cross-section (Fig. 122). If A and B are both positive, the section will be an ellipse, as in the illustration here. If A and B have opposite signs, the section will be a hyperbola.

If two of the quantities A, B, C are zero, we have an even more special case. For example, if $A = 1$, $B = 0$, $C = 0$ we have $X^2 = 1$. Now Y and Z can vary at will; the surface breaks up into the two planes $X = 1$ and $X = -1$.

It will be noticed that we have sometimes spoken simply of an *expression* such as

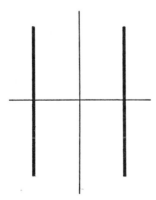

Fig. 121   Graph of $x^2 = 1$.

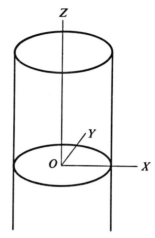

Fig. 122

$ax^2 + 2hxy + by^2$ and sometimes of the *equation* $ax^2 + 2hxy + by^2 = 1$. Let us consider a particularly simple vibration problem, that of a particle sliding around near the bottom of a smooth bowl. This situation is not particularly significant in itself: it is important, however, as giving us a way both of *visualizing and simulating what happens in a large class of vibration problems*. The potential energy of the particle is given by its vertical height above the lowest point of the bowl; suppose that, when it is at the position (x, y), its vertical height is given by $P = 9x^2 + 4y^2$. The shape of the bowl may be represented by drawing the contours $P = 1$, $P = 2$, $P = 3$ and so on. As shown in Fig. 123 these are all ellipses. They have the same axes, and they differ only in the scale on which they are drawn. Accordingly, if we can find the principal axes for any ellipse, say for $P = 1$, we have the axes in which the whole system appears most simply.

It will be noticed that the axes correspond to the points on any ellipse that are

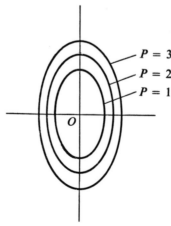

Fig. 123   Contours for P, with $P = 9x^2 + 4y^2$.

closest to and farthest from the origin O. This provides an idea that can be used both to prove that, in any number of dimensions, perpendicular principal axes always exist, and also, on occasion, as a way of calculating the positions of these axes.

Before we develop this idea we had better look at a possible complication. Suppose we are confronted with a situation involving not ellipses but hyperbolas, for example $P = x^2 - y^2$. The contours would now run somewhat as in Fig. 124. We now no longer have a bowl, but something like a saddle or mountain pass. The origin lies in the middle of the mountain pass. If you go east or west from O, you are climbing the sides of the pass. If you go north or south from O you are descending; this is where the road through the mountains would be. It is evident that such a situation is not likely to require detailed analysis in engineering work. It represents an unstable situation; a particle dislodged from O would be most likely to roll down the hillside and never be seen again. As a rule our concern would be simply to avoid a situation in which such a collapse could occur; we would not be interested in knowing exactly how fast the collapse developed.

Even in this case the x-axis and the y-axis can be identified by a minimum principle. The x-axis is given by the points on the contour $P = 1$ nearest to O; the y-axis is given by the points on the contour $P = -1$ nearest to O. Accordingly our guiding idea can be adapted to this case, by considering the two contours $P = 1$ and $P = -1$. This complication would certainly have to be considered in any exhaustive mathematical treatment. As this complication does not occur in the most important practical situations, it seems justified to simplify and shorten our explanation by excluding it. From now on, we will consider only situations in which hyperbolas and their generalizations are excluded; this is equivalent to saying that *we shall consider only quadratic forms that never take negative values*. These are technically known as *non-negative definite quadratic forms*.

We have to establish a connection between the geometrical idea of 'the point nearest to the origin' and the algebraic idea of getting rid of products such as xy in the equation of an ellipse.

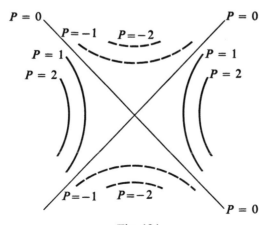

Fig. 124

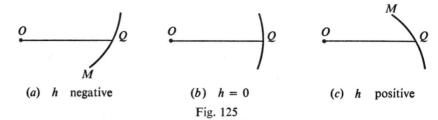

(a) *h* negative   (b) *h* = 0   (c) *h* positive

Fig. 125

This connection depends upon the following fact in plane geometry. The behaviour of the curve $ax^2 + 2hxy + by^2 = 1$ near the point where it crosses the *x*-axis depends on the coefficient *h* as shown in the three diagrams of Fig. 125 (we suppose $a > 0$). Only when $h = 0$ can *Q* be at a minimum distance from *O*. To prove this we change to polar co-ordinates (Fig. 126), by putting $x = r \cos \theta$, $y = r \sin \theta$ in the equation $ax^2 + 2hxy + by^2 = 1$. From this we find

$$\frac{1}{r^2} = a \cos^2 \theta + 2h \cos \theta \sin \theta + b \sin^2 \theta.$$

So

$$\frac{d}{d\theta} \left(\frac{1}{r^2} \right) = 2(b - a) \sin \theta \cos \theta + 2h(\cos^2 \theta - \sin^2 \theta).$$

When $\theta = 0$ this derivative has the value $2h$. When *r* is a minimum, $1/r^2$ is a maximum, and its derivative must be zero. So only when $h = 0$ can this happen for $\theta = 0$, corresponding to the point *Q*. (The same argument would show that only when $h = 0$ can the distance be a maximum at *Q*.)

Thus we have the result: *if* $(q, 0)$ *is the point of the curve* $ax^2 + 2hxy + by^2 = 1$ *nearest to the origin, then* $h = 0$.

In 2 dimensions this is enough to show that the equation must be in its simplest form $ax^2 + by^2 = 1$; the result also has consequences in spaces of higher dimension.

Suppose we are told that no point of the surface

$$ax^2 + by^2 + cz^2 + 2fyz + 2gxz + 2hxy = 1$$

is nearer to the origin than the point $(q, 0, 0)$ of this surface. (The point being in the surface implies that $aq^2 = 1$, but we shall not need to use this fact.) Consider the points of the surface that lie in the plane $z = 0$. They satisfy $ax^2 + by^2 + 2hxy = 1$. If $h \neq 0$, by our result there will be some point of the ellipse $ax^2 + by^2 + 2hxy = 1$ in the plane $z = 0$ that lies nearer to the origin than the point of this ellipse on the *x*-axis. Such a

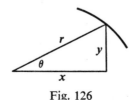

Fig. 126

point would lie on the surface of the ellipsoid, and this would contradict our assumption. Accordingly we must have $h = 0$.

Again consider the intersection of the ellipsoid with the plane $y = 0$. This intersection is the ellipse $ax^2 + cz^2 + 2gxz = 1$. The same argument shows $g = 0$.

Accordingly the ellipsoid must have an equation of the form

$$ax^2 + (by^2 + cz^2 + 2fyz) = 1,$$

if the point $(q, 0, 0)$ is the point closest to the origin. Thus, if the nearest point lies on the x-axis the equation must break into two parts; in one part we have an x^2 term, in the other part we have only the other variables y and z; nowhere do we find the product of x with another variable, such as xy or xz.

This argument generalizes to any number of dimensions. For instance, if we had four variables, x, y, z, t we would consider points with $z = 0$, $t = 0$; then points with $y = 0$, $t = 0$; finally points with $y = 0$, $z = 0$. From these we would deduce in turn that the coefficients of xy, of xz, and of xt were all zero.

Note that all this argument is based on the assumption that we are using a 'conventional' system of co-ordinates, with perpendicular axes and the same units on all axes; otherwise expressions such as $x^2 + y^2$ or $x^2 + y^2 + z^2$ for the square of a distance, and equations such as $x = r \cos \theta$, $y = r \sin \theta$ cease to be true. We assume that we start with such conventional axes and that we try to bring our quadratic expression to its simplest form by rotating the axes, not by distorting them. By a suitable rotation we can always get rid of the product terms xy, xz, yz, and our result above allows us to prove this very easily.

Suppose then we have an ellipsoid $ax^2 + by^2 + cz^2 + 2fyz + 2gxz + 2hxy = 1$. Somewhere on this ellipsoid there will be a point Q at the minimum distance from O. We get a new system of axes by rotating the old ones until the first axis lies along OQ. The other two axes will automatically be in the plane perpendicular to OQ, and for the moment we do not care how they lie in that plane. Thus in this first step we choose any set of perpendicular unit vectors, demanding only that the first vector lies along OQ. In these axes the point Q will be $(q, 0, 0)$; as it is at minimum distance from O, the equation (by our earlier result) must reduce to the form $ax^2 + by^2 + cz^2 + 2fyz = 1$. The quantities a, b, c, f here will of course be different from those in the original equation. We are concerned only with the *type* of equation, not with the actual values of the coefficients.

We are now satisfied with our choice of OQ as the first axis and turn our attention to what is happening in the plane perpendicular to OQ. The equation of that plane is now $x = 0$; by substituting $x = 0$ in the (new) equation for the ellipsoid we see that points in that plane lie on the ellipse $by^2 + cz^2 + 2fyz = 1$. Our next step is to rotate the axes about OQ in such a way as to give this ellipse the simplest possible form. So we are confronted with a problem of the same type, but in one dimension less. Accordingly we proceed in the same way. On the ellipse there will be some point R that is nearest to the origin; that is, it is the point of the *ellipse* nearest to O; it will probably be farther from O than Q, but we are now concerned only with points in the plane $x = 0$ (Fig. 127).

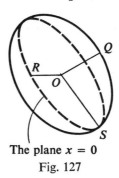

The plane $x = 0$

Fig. 127

We choose OR as our second axis. By our earlier result, this gets rid of the yz term, and the ellipse comes to the form $by^2 + cz^2 = 1$. Rotating the axes about OQ does not change the x co-ordinate in any way, so the equation of the ellipsoid now has the form $ax^2 + by^2 + cz^2 = 1$ that we were hoping for.

There may be some difficulty in seeing that by bringing the *ellipse* in the plane $x = 0$ to its simplest form we are automatically achieving the desired result for the ellipsoid. It may help to look at the work in terms of algebra. The first step brings the equation to the form $ax^2 + by^2 + cz^2 + 2fyz = 1$. When we rotate axes about OQ we are bringing in new co-ordinates, say (X, Y, Z), related to the old by the equations

$$x = X, \qquad y = Y \cos\theta - Z \cos\theta, \qquad z = Y \sin\theta + Z \cos\theta.$$

The first term ax^2 is thus replaced simply by aX^2. The terms $by^2 + cz^2 + 2fyz$ are subjected to exactly the same substitution that we would use in a problem in plane geometry. We know that, by suitable choice of the angle θ, these terms can be made to give a result of the form $BY^2 + CZ^2$. Thus the whole expression on the left-hand side of our equation gives us $aX^2 + BY^2 + CZ^2$, as predicted.

In the most general case, where the principal axes are all of different lengths, the procedure will go through exactly as above. The first step will give us OQ, the shortest semi-axis. The next step will give us OR, the semi-axis of medium length. The longest semi-axis, OS, will remain and will be given automatically by the direction perpendicular both to OQ and OR.

Certain special cases may arise. For instance, it may happen that the ellipsoid has the shape of the football used in Canadian or rugby football. In that case 'a point at minimum distance from O' does not fix a definite point Q (Fig. 128). This does not matter. Anywhere on the circle shown in the illustration will do for Q. The terms xy and xz disappear provided there is no point of the ellipsoid *nearer to O than Q*. It does not matter if there are other points *as near*. Our earlier proof depended on noticing that $2h = \mathrm{d}/\mathrm{d}\theta(1/r^2)$ at $\theta = 0$. If $1/r^2$ is constant we still have $h = 0$.

The extreme case is that of a soccer football, a sphere. In that case *any* three perpendicular directions will serve as principal axes, OQ, OR and OS.

The general ellipsoid, the rugby football (ellipsoid of revolution) and the soccer football (the sphere) are easily visualized. It is important to learn this easy way of remembering the special cases that may arise. In §38 we shall be concerned with the

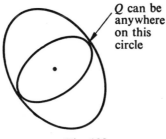

Q can be
anywhere
on this
circle

Fig. 128

actual calculation of the principal axes; naturally these calculations turn out rather differently in the special cases and it is necessary to be prepared for these. Thus, for example, a student who has become accustomed to the general case, in which OQ is a definite line, is likely to be disturbed if he meets the 'rugby football' case in which the equations no longer lead to any one solution for OQ.

Another type of special case has already been mentioned in which the standard form, $AX^2 + BY^2 + CZ^2$, contains one or more zero coefficients. In this case the process of finding OQ, OR, OS unexpectedly terminates. For instance suppose we have the equation $x^2 + y^2 + z^2 + 2xy + 2xz + 2yz = 1$. This in fact is $(x + y + z)^2 = 1$ and represents the planes $x + y + z = 1$ and $x + y + z = -1$. The points in these planes nearest to the origin lie in the direction of the vector $(1, 1, 1)$. If we take the X-axis in this direction we find the equation becomes simply $3X^2 = 1$. The coefficients of Y^2, YZ and Z^2 are all zero. Our work is complete. In the same way, it may happen that, after the second stage of the work, we arrive at an equation of the form $AX^2 + BY^2 = 1$.

We have considered in detail how things work out in 3 dimensions, but the procedure is perfectly general. Given an equation, in any number of dimensions, there will be a point on the corresponding 'surface' nearest to the origin. Take the first axis in the direction of this point, and the remaining axes in any convenient manner (the basis vectors must be perpendicular and of unit length). The equation, in these new axes, will not contain any products x_1x_2, x_1x_3, $\ldots$, x_1x_n. We then put $x_1 = 0$ and find the point $(0, x_2, x_3, \ldots, x_n)$ that is on the surface and nearest to the origin. This point gives the direction for the second axis. We continue in this way. At each stage we are concerned with finding the point nearest to the origin on a certain surface; at each stage, we are working in a dimension one less than in the previous stage.

As mentioned earlier, the object of this section has been to survey the possibilities that may arise when we are seeking the simplest form of a quadratic expression. The next section will be concerned with the actual process of finding the best axes by calculation.

38 Principal axes and eigenvectors

As we saw in §37, a quadratic form $ax^2 + by^2 + cz^2 + 2fyz + 2gxz + 2hyz$ can be brought to the simpler form $AX^2 + BY^2 + CZ^2$ by a suitable change of axes. The quadratic form may be written as $v'Mv$ where

$$v = \begin{pmatrix} x \\ y \\ z \end{pmatrix} \qquad M = \begin{pmatrix} a & h & g \\ h & b & f \\ g & f & c \end{pmatrix},$$

and the simplified form as $V'DV$ where

$$V = \begin{pmatrix} X \\ Y \\ Z \end{pmatrix} \qquad D = \begin{pmatrix} A & 0 & 0 \\ 0 & B & 0 \\ 0 & 0 & C \end{pmatrix}.$$

We have chosen the letter D for this last matrix because this matrix is in *diagonal form*; that is, all its elements are zero except those on the main diagonal.

We met matrices in the first place as ways of specifying mappings, and square matrices in particular were associated with transformations. Thus the matrix M can be used to specify the transformation $v \rightarrow v^*$ where $v^* = Mv$, and it can also be used, as above, to specify the bilinear form $v'Mv$ which we visualize by means of some surface, such as an ellipsoid. We naturally wonder whether there is some relationship between the transformation $v \rightarrow Mv$ and the surface $v'Mv = $ constant.

There is in fact such a connection, and the transformation $v \rightarrow Mv$ in fact gives us a very useful way of finding the simplest expression for the quadratic form $v'Mv$.

The connection between the quadratic form and the transformation was well known long before mathematicians began to use matrix notation,* for a question that arises naturally in many sciences leads to this connection being noticed. It will be easiest to discuss this in 2 dimensions; the argument applies equally well in any number.

In §37, we had a number of ellipses which represented the contours of a bowl. Let us consider the general question of a point on the side of a fairly smooth hill. Through this point there is a line that goes directly up the hill. This is known as the line of greatest slope. If you are very strong and vigorous you may decide to climb in this direction. If you wish for an easier climb, or if you are choosing a route for a road or railway, you may prefer to climb obliquely. The question arises: how does the steepness of the climb vary as you increase the angle between your direction of navigation and the direction of steepest slope?

* Matrix notation was first explained by the English mathematician A. Cayley in 1855, but in looking at mathematical and scientific work during the century before this date one can see that much of it would have gone very nicely in matrix symbolism.

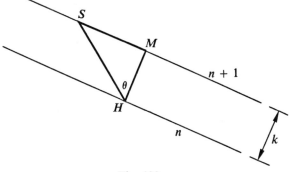

Fig. 129

It is usually sound policy to attack a problem by considering the simplest possible case. Let us simplify drastically by supposing the hillside to be an inclined plane. The contours will then be parallel straight lines (Fig. 129). We suppose we are at H on the contour for height n and that contour $n + 1$ is a parallel line at distance k from contour n. If we take the shortest and steepest path HM, we shall travel a horizontal distance k in order to rise through unit height: $1/k$ thus measures the steepness of this path. If instead we decide to use path HS, this is of length $k/\cos \theta$. The steepness is obtained by dividing 1 by this quantity; thus the steepness for HS is $(1/k) \cos \theta$. We observe that the steepness falls off with the cosine of the angle, exactly as when we are finding the resolved part of a force. Thus, if we draw a vector of length $1/k$ in the direction of HM, the steepness in any direction is given by the component of the vector in the direction in question (Fig. 130). If $z = f(x, y)$ gives the height z at the point (x, y) of the contour map, the vector just described is called 'the gradient of z' or 'grad z' for short. Grad z is in the direction of greatest slope, and its magnitude equals the steepness (the gradient) in that direction.

In applications of gradient we need not be concerned only with heights of hillsides. In an expression grad V, the symbol V may stand for temperature, gravitational, electric or magnetic potential, or a potential function used in hydrodynamics and aerodynamics.

With all these applications existing, the question has naturally been studied – how do we calculate grad z when we need it? Suppose, in our example of the inclined plane, we have $z = px + qy$ as the equation of the plane. Let H be the point (x_0, y_0). We have

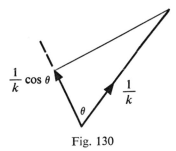

Fig. 130

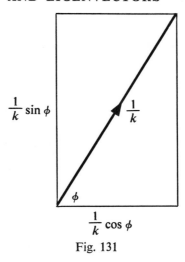

$\frac{1}{k} \sin \phi$

$\frac{1}{k}$

ϕ

$\frac{1}{k} \cos \phi$

Fig. 131

already used $1/k$ to indicate the magnitude of grad z; we suppose that grad z makes the angle ϕ with the x-axis (Fig. 131). Accordingly the components of grad z are $(1/k) \cos \phi$ and $(1/k) \sin \phi$. We know something about these components; they represent the steepness of the hillside in the directions parallel to the x-axis and the y-axis. Now it is not hard to calculate these gradients. If we start at (x_0, y_0) and move parallel to the x-axis, the x co-ordinate will change but the y co-ordinate will stay fixed. Thus we shall reach points given by (x, y_0) with varying x. So, for these points, $z = px + qy_0$. Differentiating, we find $dz/dx = p$, the gradient for motion parallel to the x-axis. In the same way, we find for motion parallel to the y-axis the gradient q. Accordingly the components of grad z are p, q. We may, if we wish, write $p = (1/k) \cos \phi$, $q = (1/k) \sin \phi$ or we may prefer to write simply grad $z = (p, q)$. We notice that grad z is constant; it does not depend on (x_0, y_0); this is reasonable, for in respect of steepness any part of an inclined plane is like any other part.

It was mentioned in §2 that one of the guiding ideas of calculus was that, for a wide class of situations, a small part of a curve could be efficiently approximated by a straight line, and a small part of a surface by a plane. For a curved surface the contours will be as shown in Fig. 132. The arc MS is no longer straight, the angle SMH is no longer

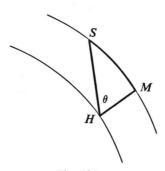

Fig. 132

exactly a right angle. However, it will often happen that, in the limit, as M is taken closer and closer to H, the diagram will become more and more like the one we had earlier. To find out and prove in exactly what conditions that will happen will not be attempted here. We will state, however, that for $z = f(x, y)$, where $f(x, y)$ is a polynomial in x and y, our main conclusions about grad z still hold. In particular, (1) the vector grad z is perpendicular to the contour $z = $ constant, (2) the components of grad z can be found by calculating the rates of change of z, for motion parallel to each axis in turn; that is, by the procedure used above. Of course for a curved surface, these components will not be constants, like p and q, but will depend on the position of the point (x_0, y_0), as will be illustrated in a moment.

We now come to our main question, the behaviour of the contours for

$$z = ax^2 + 2hxy + by^2.$$

If we consider points (x, y_0), with fixed y_0 and varying x, we have

$$z = ax^2 + 2hxy_0 + by_0^2.$$

In this expression a, h, b, y_0 all represent constants, x is the only variable – the problem is only that of differentiating a quadratic (the profusion of symbols sometimes obscures this simple fact). Accordingly $dz/dx = 2ax + 2hy_0$. We are interested in what happens at the point (x_0, y_0), so we put $x = x_0$, and obtain $2ax_0 + 2hy_0$. This gives the x-component of grad z at (x_0, y_0). In the same way, by considering points (x_0, y) and eventually putting $y = y_0$ we find the y-component of grad z to be $2hx_0 + 2by_0$.

Thus at (x_0, y_0) we have grad $z = (2ax_0 + 2hy_0, 2hx_0 + 2by_0)$. (See Fig. 133.) We know that grad z is perpendicular to the contour through (x_0, y_0). Multiplying a vector by a number does not change its direction, we may remove the factor 2, and thus reach the result; *the vector $(ax_0 + hy_0, hx_0 + by_0)$ gives the direction of the perpendicular at (x_0, y_0) to the contour, $ax^2 + 2hxy + by^2 = $ constant, passing through (x_0, y_0).* In the diagram, the vector HJ is perpendicular to the curve. The line HJ is often called 'the normal at H'.

In much of our work we think of a vector as a point, and, if we wish to draw an arrow, we draw it from the origin, O, to that point. In the present case, it is obviously more helpful to draw the vector $(ax_0 + hy_0, hx_0 + by_0)$ with its beginning at the point H. We could, if we liked, draw it from the origin: it would then represent a direction parallel to the normal at H. Pictorially this is less effective.

The following purely graphical exercise may help to fix this result in the memory.

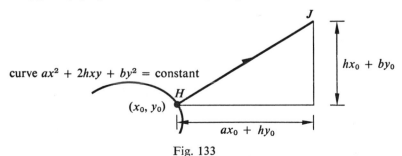

curve $ax^2 + 2hxy + by^2 = $ constant

(x_0, y_0)

H

J

$hx_0 + by_0$

$ax_0 + hy_0$

Fig. 133

Exercise

On graph paper plot the following points (50, 0), (48, 7), (40, 15), (30, 20), (14, 24), (0, 25). These lie on the ellipse $(x^2/4) + y^2 = 625$ and thus correspond to $a = \frac{1}{4}$, $h = 0$, $b = 1$. The vectors along the normals are thus given by the formula $(x_0/4, y_0)$. So, for example, from the point (48, 7) we draw the vector with components (12, 7). The resulting diagram appears as in Fig. 134. Small pieces of the tangents are shown, perpendicular to these vectors. These facilitate the freehand drawing of the ellipse.

We will now express this result in matrix notation. If we let

$$v = \begin{pmatrix} x \\ y \end{pmatrix} \qquad v_0 = \begin{pmatrix} x_0 \\ y_0 \end{pmatrix} \qquad M = \begin{pmatrix} a & h \\ h & b \end{pmatrix}$$

then $ax^2 + 2hxy + by^2 = $ constant may be written $v'Mv = $ constant; the constant is to be chosen so that the curve passes through the particular point v_0. The vector $\begin{pmatrix} ax_0 + hy_0 \\ hx_0 + by_0 \end{pmatrix}$ that gives the normal at this point is Mv_0. The transformation $v_0 \to Mv_0$ thus associates with any point v_0 a vector that gives the direction of the normal at that point (Fig. 135).

The matrix M that appears here must be symmetrical. This diagram gives us a useful way of visualizing the effect of a transformation, represented by a symmetric matrix.

What happens if v_0 should chance to lie along a principal axis of the conic, $v'Mv = $ constant? Evidently the normal, Mv_0, will be in the same direction as v_0; that is $Mv_0 = \lambda v_0$, and so v_0 *is an eigenvector of the matrix M* (Fig. 136).

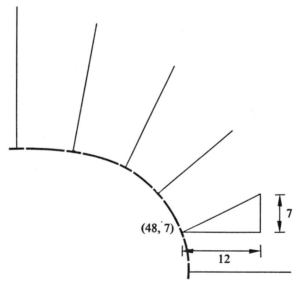

Fig. 134

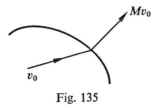

Fig. 135

Accordingly no new method is required for finding the principal axes, which give the simplest form of $v'Mv$. We simply have to look for the eigenvectors of M.

In this section we have considered an expression $ax^2 + 2hxy + by^2$ involving only two variables x, y. The arguments however are perfectly general and apply to $v'Mv$, where v can be a column vector in any number of dimensions.

The previous section showed us that we have to be prepared for various special cases. With only two variables, these special cases can be spotted immediately. By turning the axes, we can bring our expression to the form $AX^2 + BY^2$. If $A \neq B$, $A \neq 0$, $B \neq 0$, we have the general case. If $A = B \neq 0$, we have a system of circles $A(X^2 + Y^2) =$ constant and our original equations must have been $Ax^2 + Ay^2 =$ constant, since turning the axes has no effect on the equation of a circle. We get a pair of lines if $B = 0$, $A \neq 0$. This would happen, for instance, if the original expression were $x^2 + 6xy + 9y^2$; this is the same as $(x + 3y)^2$, and such a situation can always be spotted by the fact that the quadratic form is a perfect square. Finally, if $A = 0$, $B = 0$ in the simplified form, the original expression must have been $0x^2 + 0xy + 0y^2$, which we would immediately recognize (and rarely if ever meet in practice).

Accordingly it is only when we come to quadratic forms in three variables that worthwhile illustrations can be found of the special cases that may arise. These we can visualize as corresponding to the rugby football, the sphere, the cylinder, and the pair of planes. The sphere we need not discuss; it can only arise from a quadratic form of the type $kx^2 + ky^2 + kz^2$, which we recognize at once.

The general case, and the other special cases, we will show by means of worked examples.

Worked example 1: the general case

Find the axes in which $v'Mv = 10x^2 - 12xy + 23y^2 + 12yz + 16z^2$ takes its simplest form

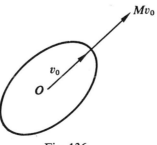

Fig 136

Solution   Here

$$M = \begin{pmatrix} 10 & -6 & 0 \\ -6 & 23 & 6 \\ 0 & 6 & 16 \end{pmatrix}.$$

We are looking for eigenvectors of M, that is, non-zero vectors that satisfy some equation of the form $Mv = \lambda v$. As explained near the end of §26, this equation may be written $(M - \lambda I)v = 0$ and can only have a non-zero solution v if λ is chosen to satisfy the characteristic equation det $(M - \lambda I) = 0$.

If we write this determinant and work it out as described in §27 (no neat method is available) we get

$$\det (M - \lambda I) = \begin{vmatrix} 10 - \lambda & -6 & 0 \\ -6 & 23 - \lambda & 6 \\ 0 & 6 & 16 - \lambda \end{vmatrix}$$
$$= (10 - \lambda)(23 - \lambda)(16 - \lambda) - 36(10 - \lambda) - 36(16 - \lambda)$$
$$= -\lambda^3 + 49\lambda^2 - 686\lambda + 2744.$$

This expression is zero if λ equals 7, 14 or 28. These then are the eigenvalues.

It will be noticed that the eigenvalues are not very small numbers. In fact arithmetical simplicity cannot be expected here. The characteristic equation is a cubic; in a realistic situation, the solutions of it would probably be irrational numbers, and we would use some form of automatic computation to arrive at them. We have avoided irrational numbers artificially by selecting M carefully, and of course the same has been done in the later examples.

Having the eigenvalues, we now proceed to find the eigenvectors. By solving $Mv = 28v$, we find the column vector with components $-2k$, $6k$, $3k$, where k may be any non-zero number, corresponds to $\lambda = 28$; similarly any non-zero column vector with components in the ratios $3: -2: 6$ is an eigenvector for $\lambda = 14$, and one with ratios $6: 3: -2$ is an eigenvector for $\lambda = 7$.

It can be checked that these three vectors are mutually perpendicular, as we expect for the principal axes of an ellipsoid. We could arrive at the simple form

$$AX^2 + BY^2 + CZ^2$$

by changing axes to these principal axes. This is mildly laborious but by no means impossible. When carried out it leads to the result $28X^2 + 14Y^2 + 7Z^2$.

This result suggests very strongly that the rather lengthy calculation for change of axes was quite unnecessary, except perhaps as a check on accuracy, for the coefficients 28, 14 and 7 in $28X^2 + 14Y^2 + 7Z^2$ are exactly the same as the eigenvalues λ that have already been found. This suggests that we might try to prove, this always happens. This is done in the following theorem.

Theorem

Let the perpendicular vectors u_1, u_2, u_3, each of unit length, be eigenvectors of the 3×3 matrix M, so that $Mu_1 = \lambda_1 u_1$, $Mu_2 = \lambda_2 u_2$, $Mu_3 = \lambda_3 u_3$. If (X, Y, Z) are co-ordinates based on u_1, u_2, u_3, the quadratic form $v'Mv = \lambda_1 X^2 + \lambda_2 Y^2 + \lambda_3 Z^2$.

Proof If v has co-ordinates X, Y, Z in the system based on u_1, u_2, u_3, we have $v = Xu_1 + Yu_2 + Zu_3$.
Accordingly

$$v'Mv = (Xu_1' + Yu_2' + Zu_3')M(Xu_1 + Yu_2 + Zu_3).$$

As

$$Mu_1 = \lambda_1 u_1, \qquad Mu_2 = \lambda_2 u_2 \quad \text{and} \quad Mu_3 = \lambda_3 u_3,$$

we have

$$v'Mv = (Xu_1' + Yu_2' + Zu_3')(X\lambda_1 u_1 + Y\lambda_2 u_2 + Z\lambda_3 u_3).$$

When this expression is multiplied out, 9 terms result, but 6 of these are zero; for example, since $u_1 \perp u_2$, the scalar product $u_1'u_2 = 0$. By this and similar considerations we see that the expression reduces to $X^2\lambda_1 u_1'u_1 + Y^2\lambda_2 u_2'u_2 + Z^2\lambda_3 u_3'u_3$ Now $u_1'u_1$ is the scalar product of u_1 with itself. As u_1 is of unit length, this scalar product must be 1. Similarly we find $u_2'u_2 = 1$ and $u_3'u_3 = 1$. Accordingly $v'Mv = \lambda_1 X^2 + \lambda_2 Y^2 + \lambda_3 Z^2$.

This was what we wanted to prove – the coefficients are given by the eigenvalues.

Note that the above proof does not employ any ingenuity. We are asked to show that something happens when a certain system of axes is used. The equation

$$v = Xu_1 + Yu_2 + Zu_3$$

is our regular starting point when we want to work with co-ordinates based on u_1, u_2, u_3. (Compare the equation $S = XP + YQ$ in §9, 'Change of axes', to express that the point S has co-ordinates X, Y in axes based on P and Q.) In the rest of the proof, we simply substitute this expression for v in $v'Mv$ and use the information we have; the equations such as $Mu_1 = \lambda_1 u_1$ tell us where M sends u_1, u_2 and u_3. After using this information, we are left simply with numbers and scalar products. The values of the scalar products follow immediately from our information that u_1, u_2 and u_3 are perpendicular and of unit length, i.e. that they are axes of the type conventional in elementary work.

The best way to remember the above theorem, and its proof, is to observe how simple and straightforward the strategy is, and to think the proof out from time to time.

It should be noted that nothing whatever is assumed about the eigenvalues λ_1, λ_2, λ_3; it does not matter whether they are equal or unequal, zero or non-zero, positive or negative. Thus this theorem applies equally well to general and special cases, to ellipsoids and to hyperboloids.

Worked example 2: the rugby football

Find the axes for which $v'Mv = 2x^2 + 2y^2 + 2z^2 + 2xy + 2yz + 2zx$ takes the simplest form.

Solution   Here

$$M = \begin{pmatrix} 2 & 1 & 1 \\ 1 & 2 & 1 \\ 1 & 1 & 2 \end{pmatrix}.$$

As in the first example, we calculate the characteristic equation, $\det(M - \lambda I) = 0$. We find

$$0 = -\lambda^3 + 6\lambda^2 - 9\lambda + 4 = -(\lambda - 4)(\lambda - 1)^2.$$

Taking $\lambda = 4$, we solve $Mv = 4v$ and find that any non-zero column vector with components k, k, k is an eigenvector for $\lambda = 4$. This part of the procedure works out exactly as in the general case; the equation $Mv = 4v$ is equivalent to 2 equations in the 3 unknowns x, y, z. The new feature appears when we come to the other solution, $\lambda = 1$, for $Mv = v$ gives the same equation $x + y + z = 0$ three times. Thus we shall get an eigenvector by choosing any numbers, x, y, z with sum zero. We have an unusual variety of eigenvectors with $\lambda = 1$; they do not all lie in a line, as in the general case. In fact, we can interpret the equation $x + y + z = 0$, for $x + y + z$ is the scalar product of (x, y, z) and $(1, 1, 1)$; the equation tells us that these vectors are to be perpendicular. *Thus any vector perpendicular to* $(1, 1, 1)$ *is an eigenvector with* $\lambda = 1$. These vectors fill a plane.

What is happening in this plane, that causes *every* line through the origin to be a principal axis? Usually, if we have an ellipse in a plane, there are two quite definite principal axes of this ellipse. The only exception is the circle; for it, any diameter is a principal axis; no direction is favoured above any other. Thus we are led to guess that the intersections of the surfaces $v'Mv =$ constant with the plane $x + y + z = 0$ are circles.

We can establish this quite definitely by using the theorem given above. To apply this theorem, we must consider three perpendicular unit vectors, u_1, u_2, u_3, that are eigenvectors of M. We take u_1 in the direction $(1, 1, 1)$; u_2 and u_3 can be any two unit vectors, that are perpendicular to each other, and lie in the plane perpendicular to u_1. We know, by the work above, that $Mu_1 = 4u_1$, $Mu_2 = u_2$, $Mu_3 = u_3$. So, by our theorem, if we take u_1, u_2, u_3 as a basis for co-ordinates, $v'Mv = 4X^2 + Y^2 + Z^2$. Where a surface $v'Mv =$ constant meets the plane $X = 0$, we have $Y^2 + Z^2 =$ constant, the equation of a circle.

A surface $v'Mv =$ constant could be obtained by spinning an ellipse about the line $x = y = z$; it is an ellipsoid of revolution. This particular ellipsoid is more like a flying saucer than a football.

Worked example 3: the sphere

As already mentioned, there is nothing to say about this. A sphere corresponds to $k(x^2 + y^2 + z^2)$. It is already in its simplest form. If we rotated the axes, it would remain in this form; we would obtain $k(X^2 + Y^2 + Z^2)$. Any three perpendicular directions will serve as principal axes.

Worked example 4: the cylinder

Find the axes for which

$$v'Mv = 2x^2 + 2xy + 2y^2 - 6xz - 6yz + 6z^2$$

takes the simplest form.

Solution   Here we find

$$\det(M - \lambda I) = \begin{vmatrix} 2 - \lambda & 1 & -3 \\ 1 & 2 - \lambda & -3 \\ -3 & -3 & 6 - \lambda \end{vmatrix}$$

$$= (2 - \lambda)^2(6 - \lambda) + 9 + 9 - 9(2 - \lambda) - 9(2 - \lambda) - (6 - \lambda)$$
$$= -9\lambda + 10\lambda^2 - \lambda^3$$
$$= -\lambda(\lambda - 1)(\lambda - 9).$$

The eigenvalues are 9, 1, 0. $Mv = 9v$ leads us to an eigenvector with components in the ratios $1:1:-2$. $Mv = v$ gives the ratios $1:-1:0$, while $Mv = 0$ gives $1:1:1$. If we take unit vectors with these directions as our basis, we find $v'Mv = 9X^2 + Y^2 + 0 \cdot Z^2$, that is, simply $9X^2 + Y^2$.

It can be checked that the eigenvectors just found are perpendicular to each other.

The only special feature in this situation is that we obtain $\lambda = 0$ as one of the solutions of the characteristic equation.

There is a particular case of the cylinder which also has the peculiarity discussed in Worked example 2, the rugby football. The rugby football had circular sections perpendicular to one of its principal axes; we can quite easily have a cylinder with circular sections perpendicular to its axis – in fact, in everyday life this is what we understand by a cylinder, as when we speak of the cylinders in a car. In such a case we would have $X^2 + Y^2 = $ constant, and any vector in the plane $Z = 0$ will be an eigenvector of M. In such a case, the calculations will run very much as they did in Worked example 2. Instead of a characteristic equation such as $0 = -(\lambda - 4)(\lambda - 1)^2$ we shall have perhaps $0 = -\lambda(\lambda - 1)^2$. The fact that $\lambda = 0$ is a solution will not affect our general strategy at all.

Worked example 5: a pair of planes

Find the axes in which $x^2 + y^2 + z^2 + 2xy + 2xz + 2yz = 1$ takes its simplest form.

Solution   Here we have the characteristic equation

$$0 = \det(M - \lambda I) = \begin{vmatrix} 1 - \lambda & 1 & 1 \\ 1 & 1 - \lambda & 1 \\ 1 & 1 & 1 - \lambda \end{vmatrix}$$

$$= (1 - \lambda)^3 + 2 - 3(1 - \lambda)$$
$$= -\lambda^3 + 3\lambda^2 = -\lambda^2(\lambda - 3).$$

For $Mv = 3v$ we find the eigenvector with components in the ratios $1:1:1$. If $\lambda = 0$, we get, just as we did in Worked example 2 for $\lambda = 1$, the same equation $x + y + z = 0$ three times. Thus any vector in this plane is an eigenvector with $\lambda = 0$. Here again, the fact that $\lambda = 0$ does not call for any change in our strategy. We can argue exactly as we did in Worked example 2, and conclude that the simplified form must be

$$3X^2 + 0 \cdot Y^2 + 0 \cdot Z^2.$$

Naturally, we write this as $3X^2$.

Summing up

If we look at these examples, we can see that two considerations are involved.
(1) Do we have a repeated root of the characteristic equation? That is, are we dealing with an equation such as $0 = -(\lambda - a)(\lambda - b)^2$ or even $0 = -(\lambda - a)^3$?
(2) Does zero occur as a root of the characteristic equation?

When all the roots are distinct, we get eigenvectors lying in definite directions, which are fixed by the equations $Mv = \lambda v$. When a repeated root occurs, we no longer have definite eigenvectors: we may be free to choose any pair of perpendicular axes in some plane, or, in the extreme case of the sphere, any set of perpendicular axes whatever. Thus repeated roots affect the kind of calculations we have to make.

Quite the opposite is true for the second consideration, whether $\lambda = 0$ is a solution or not. When zero eigenvalues occur the surfaces $v'Mv = $ constant certainly *look* different from the other cases. An infinite cylinder, for example, would hardly be convenient for use as a rugby football. But the procedure for calculating the axes would be exactly the same. The procedure of Worked example 1 can be applied unaltered to the problem stated at the beginning of Worked example 4; all we have to remember is not to be upset by the appearance of the solution $\lambda = 0$. In the same way, the procedure of Worked example 2 (rugby football) can be applied unaltered to the particular case mentioned at the end of Worked example 4 (circular cylinder). In fact, we can usefully think of the elliptical cylinder as a special or limiting case of the general ellipsoid, and the circular cylinder as a limiting case of the rugby football.

How this comes about can be seen by considering a sequence of ellipses in 2 dimensions. The equation $(x^2/a^2) + (y^2/b^2) = 1$ represents an ellipse with axes of length $2a$ and $2b$. We will fix b at the value 1, but consider in turn $a = 10$, $a = 100$, $a = 1000$, and so on. In this sequence of ellipses, the major axis is becoming longer and longer. The equations of the ellipses are $0.01x^2 + y^2 = 1$, $0.0001x^2 + y^2 = 1$, $0.000001x^2 + y^2 = 1$ and so on. Clearly the coefficient of x^2 is approaching 0. If we consider the limiting case, in which the coefficient of x^2 is made actually equal to zero, we have $y^2 = 1$, which represents the pair of lines $y = 1$ and $y = -1$.

The business of finding eigenvectors in the limiting case is no different from what it was for any ellipse in the sequence. For example, the first ellipse corresponds to the matrix

$$\begin{pmatrix} 0.01 & 0 \\ 0 & 1 \end{pmatrix}.$$

We have an eigenvector in the y-axis with $\lambda = 1$, and one in the x-axis with $\lambda = 0.01$. For the pair of lines $y^2 = 1$ we have the matrix

$$\begin{pmatrix} 0 & 0 \\ 0 & 1 \end{pmatrix},$$

which has an eigenvector in the y-axis with $\lambda = 1$ and one in the x-axis with $\lambda = 0$.

Many paradoxes and incorrect results can come from proceeding rashly to a limit or from talking loosely about infinity. In the example just given we have checked carefully that there is no sudden jump made by the eigenvectors or eigenvalues as we go to the limit. It is legitimate to think of the pair of lines, $y^2 = 1$, as an ellipse for which the major axis has become infinite, or as a limiting case of an ellipse.

Thus in 2 dimensions we could classify the possibilities as follows:

Type 1   Roots for λ unequal; ellipse or hyperbola (limiting case, when $\lambda = 0$ is one solution, pair of lines).

Type 2   Roots for λ equal; circle.

For type 1 we have two definite principal axes; for type 2, any pair of perpendicular lines are principal axes.

In 3 dimensions, we can similarly justify the inclusion of limiting cases. Our classification then runs:

Type 1   Roots for λ distinct: the general case (limiting case, elliptic or hyperbolic cylinder).

Type 2   Equation of form $(\lambda - a)(\lambda - b)^2 = 0$; rugby football (limiting cases, circular cylinder if $a = 0$; pair of planes, if $b = 0$.)

Type 3   Equation of form $(\lambda - a)^3 = 0$; sphere.

For type 1, there are definite principal axes; for type 2, one axis is in a definite direction, about which the other axes are free to rotate; for type 3, any three perpendicular directions will do for principal axes.

The cylinders, type 1 or type 2, occur when one axis has become of infinite length; the pair of planes occurs when two axes have become infinite in length.

Note that the rugby football and the flying saucer both belong to the same type, type 2; which case we get depends on whether $a < b$ or $a > b$.

Questions

1   Find the eigenvalues and eigenvectors of the matrix

$$\begin{pmatrix} 5 & 4 \\ 4 & 5 \end{pmatrix}.$$

Hence find the principal axes of the conic $5x^2 + 8xy + 5y^2 = 9$. Find the equation of the conic referred to its principal axes and sketch the curve. Give the (x, y) co-ordinates of the ends of the principal axes, and check that they do satisfy the original equation.

2   Do the same for the matrix

$$\begin{pmatrix} 5 & -4 \\ -4 & 5 \end{pmatrix}$$

and the equation $5x^2 - 8xy + 5y^2 = 9$. How does this curve differ from the curve in Question 1?

3   Do as in Question 1 for the matrix

$$\begin{pmatrix} 4 & 5 \\ 5 & 4 \end{pmatrix}$$

and the equation $4x^2 + 10xy + 4y^2 = 9$.

4   Find the equation of the conic $3x^2 - 4xy = 4$ referred to its principal axes and sketch the curve.

5   As in Question 4 investigate the conic

$$8x^2 - 4xy + 5y^2 = 36.$$

6   Write down a vector that has the same direction as the normal to the conic

$$10x^2 - 12xy + 5y^2 = 13$$

at the point (x_0, y_0) on the curve. Find this normal vector for the point $(2, 3)$ on the curve. What conclusion can be drawn about the directions of the principal axes of this conic?

7   Do as in Question 6, but with the conic $22x^2 - 6xy + 14y^2 = 130$ and the point $(1, 3)$ on it.

8   Find vectors that give the directions of the normals to the conic $16x^2 + 40xy + 25y^2 = 196$ at the points $(1, 2)$ and $(6, -2)$ on this conic. What are the slopes of these normals? Explain your result.

9   As was mentioned earlier there is difficulty in finding 3-dimensional examples in which attention is not distracted from the principles involved by purely arithmetical complications. In the selection of the following examples an attempt has been made to keep the numbers involved as small as possible. The characteristic equation being a cubic, it will often be necessary to search for some simple solution, such as ± 1, ± 2 or ± 3, and then to remove the corresponding factor; a quadratic equation then remains to be solved.

Each of the equations listed below is of the form $v'Mv = $ constant. In each case, write the matrix M and find its eigenvalues and eigenvectors. Write the equation of the quadric referred to its principal axes, and indicate clearly which vectors (as specified in the x, y, z system) are the new X, Y and Z axes. Say to which type the quadric belongs.

(a) $5x^2 + 4xy + 4y^2 - 4yz + 3z^2 = 28$
(b) $3x^2 + 8xy + 5y^2 - 8xz + z^2 = 1$
(c) $2x^2 + 2y^2 + 2z^2 - 2xy - 2xz - 2yz = 1$
(d) $2xy + 2xz + 2yz = 1$
(e) $2x^2 + 4xy + y^2 - 4yz = 4$
(f) $5x^2 + 8xy + 5y^2 - 4xz - 4yz + 2z^2 = 1$
(g) $4x^2 + 4xy + 5y^2 + 4xz + 3z^2 = 28$
(h) $x^2 + 4y^2 + 9z^2 + 4xy + 6xz + 12yz = 25$
(i) $4xy + y^2 + 4xz - z^2 = 3$

(*j*) $x^2 + 4xy - 4yz - z^2 = 3$

(*k*) $7x^2 + 4xy + 6y^2 - 4yz + 5z^2 = 3$, it being given that $(2, 2, -1)$ lies along a principal axis.

10 Stable equilibrium occurs only when the contours for potential energy are ellipsoids. If the expressions in Question 9 represent potential energy, in which cases is equilibrium stable?

---

39 Lines, planes and subspaces; vector product

When finding the eigenvectors of a matrix, as §38 showed, we may meet various situations. In the general case we find that the eigenvectors, corresponding to some particular value of λ, lie in a line through the origin O. But, for the rugby football and the flying saucer, they may lie in a plane. And for the sphere, they are scattered through the whole space of 3 dimensions. Now in all the cases considered in §38, we were solving 3 equations in 3 unknowns; thus all the cases looked very much alike, yet they led to several different kinds of result. In engineering applications, we may meet m equations in n unknowns, where m and n are any natural numbers whatever. It is evident, from our experience in §38 with 3 equations in 3 unknowns, that we must expect to meet a variety of situations and be prepared to cope with them. Accordingly the problem arises quite naturally – *to classify the possibilities that may arise in solving systems of linear equations.* This problem we shall take up in §40. Before coming to that it will be wise to clear the ground by considering another question. In §38, the solutions of the equations were associated with a line, a plane, or the whole space. It seems desirable to discuss how we specify a line or a plane in 3 dimensions, and the corresponding ideas for space of n dimensions. That will be done in the present section.

Lines and planes can be specified by the methods of affine geometry – that is, by means of the basic vector operations, giving $u + v$ and ku, and without any reference to such Euclidean ideas as length or angle. We shall use this approach first. Its advantage is its flexibility. In 3 dimensions we can use *any* 3 vectors as a basis for co-ordinates, so long as they do not lie in a plane; we need not worry whether they are perpendicular or not. After finding out as much as we can by affine methods, we will turn to Euclidean ideas and look at such questions as, for example, the equation of a plane perpendicular to a given vector. Such results are useful, for example, in problems concerned with actual objects in physical space, which has approximately the properties required by Euclid.

Our main interest will be in lines and planes through the origin.

Suppose then we are dealing with some object that we cannot handle directly, but which we can control by dials on some piece of equipment. To begin with, suppose there is only one dial. The object is mounted on rails. It starts at the origin, and the rails carry it in the direction of some fixed vector u. The rails are very long; they run right through the origin, and by turning our dial suitably, we can get the object anywhere we like on the line containing the vector u. Any point on this line is specified by a vector su, where s is some real number, positive, negative or zero. By turning the dial an amount corresponding to s, we can bring the object to the position su.

Thus s is at our disposal. We can vary it as we like by turning the dial. A quantity that we can vary at will is often called a *parameter*. Thus s is a parameter, and when we say that the line consists of the points su, where s runs through all real values, we are specifying the line *parametrically*.

Since there is *one* number s at our disposal, this situation is described as having *one degree of freedom*.

There is another way of specifying a line. Instead of describing the freedom it has, we may describe the restrictions put on it. Suppose we take any two vectors, v and w, such that u, v, w constitute a basis, and let x, y, z be co-ordinates in this system; that is, (x, y, z) denotes the point $xu + yv + zw$. The points su, that can be reached by the object when we turn the dial, have co-ordinates of the form $(s, 0, 0)$. Thus we have restrictions imposed on us; we can only reach points for which $y = 0$, $z = 0$. Thus the line can be specified by *a pair of equations*. Such specification can be done in many ways; for example, the equations $y - z = 0$, $y + z = 0$ are equivalent to $y = 0$, $z = 0$ and specify the same line.

A convenient way of visualizing and remembering the parametric form su is to imagine an object travelling with constant velocity u; s seconds after passing through the origin it would reach the ·position su. Its path would be a straight line. Negative values of s would give the places on the line reached before it came to the origin.

Now of course we cannot always arrange to have one axis along the line we are studying; there may be other aspects of the problem that compel us to use some other system of axes. Suppose then that we have to use co-ordinates X, Y, Z based on axes U, V, W, where $U = au + bv + cw$, $V = du + ev + fw$, $W = gu + hv + iw$ connect the two systems. We can change axes by the method of §9. We have

$$xu + yv + zw = XU + YV + ZW$$
$$= X(au + bv + cw) + Y(du + ev + fw) + Z(gu + hv + iw).$$

By considering the amount of V on each side we find $y = bX + eY + hZ$, and from the amount of w we get $z = cX + fY + iZ$. Thus the equations $y = 0$, $z = 0$ are replaced by $bX + eY + hZ = 0$, $cX + fY + iZ = 0$.

So, in the X, Y, Z system of co-ordinates we still have 2 equations for the line through the origin.

It is reasonable that there should be 2 equations. If a point is free to roam anywhere in space of 3 dimensions, it has 3 degrees of freedom; we are free to choose each of the 3 co-ordinates. Each restriction, that is, each equation, destroys one degree of freedom, so to cut down the freedoms of the point from 3 to 1, we require 2 equations. And, incidentally, these two equations must be independent. It will not do to have equations like $X + Y + Z = 0, 2X + 2Y + 2Z = 0$ where the second equation is a consequence of the first. In the work above, one could show that there would be a contradiction if anything like this occurred, for $y = 0, z = 0$, the equations from which we started are independent.

The change-of-axes argument used above is a convenient way of showing that a line must have two equations, since it does not involve us in any very detailed algebraic calculations. In any particular example, where we wanted to find 2 equations for a line, we would not go through the trouble of changing axes.

Worked example

Find the equation of the line joining the origin to the point (2, 3, 4).

Solution If a point travels with velocity (2, 3, 4) for s seconds, it will reach the point with $x = 2s, y = 3s, z = 4s$. This point satisfies the equations $3x = 2y, 4y = 3z$.

Conversely it can be seen that any point satisfying these equations does lie on the line in question. For these equations imply that $x/2, y/3$ and $z/4$ are all equal. Take this common value for s. Then $x = 2s, y = 3s, z = 4s$, as required.

It may seem that the algebra used in this example is so simple that we might dispense with the change-of-axes argument altogether and simply do the algebra. However, it should be borne in mind that our study of the straight line, given by 2 equations in 3 unknowns, is only the curtain raiser for the general case of m equations in n unknowns. In such a vague and general situation, it is difficult, perhaps impossible, to perform algebraic calculations in the way we do for simple, particular cases. It is a great advantage if we can suppose axes chosen in such a way that the equations become something like $x_k = 0, x_{k+1} = 0, \ldots, x_{n-1} = 0, x_n = 0$.

Planes in 3 dimensions

We now return to our apparatus for moving an object, and suppose two dials are provided. The rails now form a kind of grid. When the first dial is turned, the object is displaced in a line parallel to a fixed vector u, when the second dial is turned, the object is displaced parallel to a vector v. Thus, if the object starts at the origin, by turning the dials suitably we can bring it to any position $su + tv$, where s and t are real numbers.

In some cities, such as Toronto, the public transport system permits a passenger to travel a desired distance east or west, and then transfer to another vehicle which carries him or her north or south. This has much in common with the object moving apparatus with two dials; setting the first dial corresponds to deciding how far to travel east or west: setting the second dial corresponds to choosing a distance north or south. Of course the correspondence is not exact; a passenger cannot choose to travel through π stops, not even through $\frac{1}{4}$ of a stop. Again, the apparatus would allow us to simulate conditions in a city that lay on an inclined plane, with roads that ran in two directions not necessarily perpendicular.

For our purposes, it will be convenient to *define* a plane through the origin as consisting of all the points $su + tv$, where u and v are two fixed, independent vectors, and s, t are arbitrary real numbers.

Thus a plane appears first of all in terms of the 2 parameters s, t. It has 2 degrees of freedom; we have 2 numbers at our disposal when we are choosing a point of the plane.

What about specifying it by an equation or equations? We already have 2 fixed vectors u and v, pointing in different directions; if we take any vector w not lying in the plane of u and v, we obtain a basis u, v, w for co-ordinates x, y, z. (Engineers will regard this as geometrically obvious; in a treatise more formal than this, the possibility of obtaining a basis in this way would appear as a theorem, stated in a very general way as applying to any number of vectors in any number of dimensions.)

In the co-ordinate system just introduced the point $su + tv$ will appear as $(s, t, 0)$. Thus the x and y co-ordinates are completely at our disposal. The only restriction is $z = 0$, which is necessary and sufficient; that is to say, the statements '(x, y, z) lies in the plane containing the origin and the vectors u and v' and '$z = 0$' are completely equivalent. If (x, y, z) lies in the plane, then z must be zero; if $z = 0$, then (x, y, z) lies in the plane.

Thus a plane has *one* equation only.

If we are working with axes that are not so conveniently related to the position of the plane, we will do as we did earlier with the line and call our co-ordinates X, Y, Z. With the same symbolism as before, the equation $z = 0$ will appear in our system as $cX + fY + iZ = 0$. Thus we shall still have just one equation for the plane.

This interpretation of a single equation helps us to see the meaning of our earlier equations for a line. We had $y = 0$, $z = 0$. Each of these equations represents a plane.

$y = 0$ is obviously satisfied by the points $(1, 0, 0)$ and $(0, 0, 1)$, thus $y = 0$ is the equation of the plane containing the first and third axes. Similarly $z = 0$ is the equation of the plane containing the first and second axes. When we require $y = 0$ and $z = 0$ simultaneously we are insisting that the point (x, y, z) should lie in both these planes, that is, in their intersection which is in fact the first axis.

Fig. 137 uses a picture of a cube with the origin at one corner to illustrate this, because

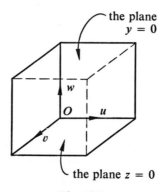

Fig. 137

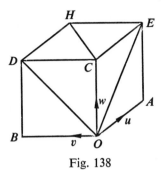

Fig. 138

perpendicular axes are familiar to us and easily visualized. It should be emphasized that perpendicularity is in no way necessary and is indeed totally irrelevant to the result.

The line, with only 1 degree of freedom, in 3 dimensions requires 2 equations to remove 2 freedoms. The plane, having 2 freedoms, is sacrificing only 1 degree of freedom and obeys only 1 equation. As an extreme case, we could consider the whole space of 3 dimensions; then x, y, z are all arbitrary; all 3 freedoms are preserved and no equation restricts our movements. At the other end, a point has sacrificed all its freedom; the origin is specified by $(0, 0, 0)$ in which no parameter occurs, and satisfies the 3 equations $x = 0$, $y = 0$, $z = 0$.

When we are counting equations, it is essential to check that they are independent. For instance, if we have $x - y = 0$, $y - z = 0$, $z - x = 0$ we seem to be dealing with 3 equations, and we might, at first sight, expect these to fix a single point, the origin. But in fact (s, s, s) satisfies all three equations. Any point on the line joining the origin to $(1, 1, 1)$ lies in the planes $x - y = 0$, $y - z = 0$, $z - x = 0$. In fact the first two equations tell us $x = y$ and $y = z$, from which it follows that $x = z$. The third equation does not bring us any new information.

Fig. 138, which uses a cube with the same reservations as before, the equation $x = y$ corresponds to the plane OCH, $y = z$ to ODH and $x = z$ to OEH. These three planes all contain the line OH.

Thus three planes with dependent equations may, as a rule, be pictured as having a line in common, rather like the pages of a pamphlet that have been somewhat spread out. We have to say 'as a rule' because there is an extreme case of dependence in which all 3 equations represent the same plane.

Subspaces

A line is a vector space of 1 dimension, and a plane is a vector space of 2 dimensions.

In this section we have considered both lines and planes as objects in space of 3 dimensions. This is not the only way to study lines and planes. In much of school geometry – for example, in Pythagoras' Theorem – we are simply concerned with the plane itself; we do not think of the diagrams as lying in a horizontal, or a vertical, or an inclined plane. They are just in 'the' plane; we are in a universe of 2 dimensions.

We can work with a line in the same way – for instance, the 'real number line'; no one discusses the compass bearing of this line!

When we consider a vector space as lying inside a vector space of higher dimension (like a line or a plane in a space of 3 dimensions) we call it a *subspace*. Thus, for instance, if we consider the plane, consisting of all the points $xu + yv$, as part of the 3-dimensional space, containing all the points $xu + yv + zw$, then this plane is a subspace of the 3-dimensional space.

Towards the end of §7 we gave the requirements for a vector space; if p and q were two objects in a vector space, we had to have 'sensible definitions' of $p + q$ and kp, for any number k. It was understood that $p + q$ and kp would also belong to the vector space; they would not be something outside it. Those requirements imply that, for any numbers a and b, the 'mixture' $ap + bq$ belongs to the space. For by multiplication of p and q we can obtain ap and bq, and by addition of these we reach $ap + bq$.

A subspace is a collection of some of the points of a vector space, which itself meets the qualifications for a vector space. If we restrict ourselves to vectors in the subspace, we can add, and multiply by a number, without ever needing to go outside the subspace. (Consider, for example, the horizontal plane through the origin in physical space of 3 dimensions.) So, if p and q are in a subspace, $ap + bq$ must also be in that subspace. In particular, we can consider $a = 0$, $b = 0$. This gives us the origin, O. Accordingly, *a subspace is bound to contain the origin O.*

In §12, when we were dealing with eigenvectors and eigenvalues, we made the convention that invariant line was to mean only an invariant line *through the origin*. The reason for this is now beginning to appear. Invariant lines are a particular case of a more general idea, invariant subspace. Since a subspace automatically contains the origin, it was useful to make this stipulation for the invariant lines considered in §12.

We can now generalize our earlier results about the degrees of freedom and equations of lines and planes in 3-dimensional space.

A line appeared earlier as consisting of all the points su, with u a fixed vector and s an arbitrary number; a plane appeared similarly as a collection of points $su + tv$. A line is a subspace of 1 dimension, a plane a subspace of 2 dimensions.

Suppose now we want to define a subspace of r dimensions in a vector space of n dimensions. We suppose r to be smaller than n.

We begin with r independent vectors, $u_1, u_2, \ldots, u_r$, and r arbitrary numbers $s_1, s_2, \ldots, s_r$, and form the mixture $s_1u_1 + s_1u_1 + s_2u_2 + \cdots + s_ru_r$.

All these points form a vector space: it is easy to see that you can add two of these vectors, or multiply one of them by a number, without reaching anything of a different type. So this collection does qualify as a subspace; it is in fact a subspace of r dimensions. Now $r < n$, so, on the analogy with a line or plane in 3-dimensional space, we assume that this subspace, of dimension r, does not in fact fill the whole space of dimension n. We will further assume, on the analogy of what we did with lines and planes, that we can find a basis for *the whole space*, the first r vectors of which are $u_1, \ldots, u_r$, which are the basis of the subspace. The vectors $u_{r+1}, \ldots, u_r$ will lie outside the subspace and can be chosen in many different ways. Thus, in the whole space we have co-ordinates $(x_1, x_2, \ldots, x_n)$ for the point $x_1u_1 + x_2u_2 + \cdots x_nu_n$. The points of the subspace appear as $(s_1, s_2, \ldots, s_r, 0, 0, \ldots, 0)$. The first r co-ordinates are arbitrary numbers, the rest are all zero. Thus the subspace gives us r degrees of freedom, corresponding to the r parameters $s_1, s_2, \ldots, s_r$. There are $n - r$ equations, expressing the fact that for points in the subspace, the remaining co-ordinates are zero. These equations are $x_{r+1} = 0,\ x_{r+2} = 0, \ldots, x_{n-1} = 0,\ x_n = 0$. If we change axes, we shall obtain equations of a less simple type, but there will still be $n - r$ independent equations.

The converse can be proved; if we start with $n - r$ independent linear equations (with no constant terms), each equation kills one degree of freedom, and we are left with r degrees of freedom in a subspace of r dimensions.

To prove, or even to justify in part, the assumptions made above requires machinery that will be developed in §40. For the moment, we are not attempting to prove the majority of our statements; we are surveying the scene and reporting on it. Most of our remarks deal with results which a mathematician, investigating this topic for the first time, would guess to be true on the basis of analogy and a few experiments in calculation.

All these considerations are relevant to our work in §38. When we are looking for eigenvectors, we first find a suitable value for λ and then solve the equations $(M - \lambda I)v = 0$. Here we have (in the general case) n equations in n unknowns. Some of these equations will be redundant, that is, they will be consequences of other equations in this set. Weeding these out, we are left with a certain number of independent equations, say m equations. Accordingly the solutions will form a subspace of $n - m$ dimensions.

Lines and planes not through the origin, in 3 dimensions

To deal with lines not through the origin we need only very small modifications of our earlier work. If a point moving with constant velocity, given by the vector u, starts not at the origin but at a position, given by the vector p, then, after s seconds it will be at the position $p + su$. Thus the parametric specification of the line it moves along can be written down immediately.

The equations of the line can be obtained from the parametric form. These equations express the condition that the line will pass through some given point (x, y, z). Suppose for example, a point starts at $(1, 2, 3)$ and moves with constant velocity $(4, 5, 6)$. After s seconds it will be at $(1 + 4s, 2 + 5s, 3 + 6s)$. Now suppose the question arises; does the line pass through the point $(13, 12, 9)$? This is the same as asking, will there be a time, s seconds, when the point is at $(13, 12, 9)$? This question can be answered in more than one way. We might, for instance, find s by equating the x co-ordinates. This gives us $1 + 4s = 13$, so the moving point has $x = 13$ when $s = 3$. Putting $s = 3$, we find the point is then at $(13, 17, 21)$ which is not the same as $(13, 12, 9)$; choosing the time that makes the x value right makes y and z wrong, so the point is not on the line.

Another way would be to write down the 3 equations that hold if $(1 + 4s, 2 + 5s, 3 + 6s)$ coincides with $(13, 12, 9)$. These equations are $1 + 4s = 13,\ 2 + 5s = 12,\ 3 + 6s = 9$. The first equation requires $s = 3$, the second requires $s = 2$, the third

$s = 1$. Thus the co-ordinates x, y, z take the required values *at different times*; there is no time when they all have the required values; this means the line does *not* go through the point.

We could use either method to find conditions for the line to go through *any* point (x, y, z). The second method leads to a result that is simpler and more easily remembered, and most textbooks accordingly give the equations of a line in the form obtained by this method.

Suppose then we are given a certain point (x, y, z) and asked to find whether the line passes through it. We write the 3 equations $x = 1 + 4s$, $y = 2 + 5s$, $z = 3 + 6s$ and solve each of them. The first gives $s = (x - 1)/4$, the second $(y - 2)/5$, the third $(z - 3)/6$. The point will pass through (x, y, z) if these 3 equations are satisfied *at one and the same time*. So the 3 values of s found above must be the same, for this to happen. That means

$$\frac{x - 1}{4} = \frac{y - 2}{5} = \frac{z - 3}{6}.$$

Note the significance and position of the numbers here; $(1, 2, 3)$ is the place where the point starts, (x, y, z) is a point that it reaches (if the conditions are satisfied); $(4, 5, 6)$ specify the direction of the line, by giving the velocity of the moving point. What is the significance of the vector $(x - 1, y - 2, z - 3)$ whose components appear in the numerators? If the equations are satisfied, in what relation does this vector stand to the velocity? Is the result a reasonable one?

Note also that these equations were obtained by finding a common value for s. If we want to get back from the equations to the parametric form, all we need do is to write s for this common value: let

$$\frac{x - 1}{4} = \frac{y - 2}{5} = \frac{z - 3}{6} = s.$$

Then $x = 1 + 4s$, $y = 2 + 5s$, $z = 3 + 6s$ follows.

Planes

A plane through the origin consists of all the points $su + tv$; these are the points we can get to by going an arbitrary amount in the direction of the vector u, and then an arbitrary amount in the direction of v. If we do the same starting at position p instead of at the origin, we obtain points of the form $p + su + tv$. This then is the parametric specification of a plane that need not contain the origin. (It might contain the origin; we have not started at O, but our wanderings might bring us there.)

There are two types of question that now arise. We may want to know simply, what kind of equation will such a plane have? Or we may be dealing with particular vectors p, u, v and want to know, what exactly is the equation of this particular plane?

We begin with the first question, because the answer to it will help with the second question.

We have already seen that the plane through the origin, consisting of all points of the form $su + tv$, has some equation, $ax + by + cz = 0$. If we add p to $su + tv$ we obtain

$p + su + tv$. Thus the plane we are interested in, consisting of all the points $p + su + tv$, can be obtained by displacing a plane through the origin; the displacement is given by the vector p (Fig. 139).

Questions about displacement should always be done slowly; it is very easy to make a slip and in fact give a displacement opposite to what you intended. For instance, in 2 dimensions, the mistake is often made of supposing that $y = (x + 1)^2$ is the result of displacing the parabola $y = x^2$ one unit to the right; it is not, rather it is the result of a displacement one unit to the *left*, i.e. in the direction of the *negative* x-axis.

Suppose then that the displacement p has components f, g, h. Any point (x, y, z) in the plane we are investigating can be obtained by giving the displacement p to some point, say (x_0, y_0, z_0), in the plane through the origin. Thus $x = x_0 + f$, $y = y_0 + g$, $z = z_0 + h$. As (x_0, y_0, z_0) lies in the plane through the origin, $ax_0 + by_0 + cz_0 = 0$. Our problem is: what restrictions are placed on (x, y, z) if the equations above hold for some (x_0, y_0, z_0)? We are looking for a condition that (x, y, z) lies in a certain plane, so only x, y, z can appear in the equation we want; we have to get rid of all reference to (x_0, y_0, z_0). To do this, we can solve for x_0, y_0, z_0; we find $x_0 = x - f$, $y_0 = y - g$, $z_0 = z - h$. (This makes sense; it shows that we get from (x, y, z) back to the point (x_0, y_0, z_0) by reversing the displacement p.) Substituting in $ax_0 + by_0 + cz_0 = 0$ we obtain the required equation $a(x - f) + b(y - g) + c(z - h) = 0$. If we multiply out, and introduce the abbreviation d for $af + bg + ch$, this equation can be written $ax + by + cz = d$, a formula that covers every possible plane.

We now come to our second question. Suppose for example, we start at $(1, 2, 3)$ and allow our point to move from there for s seconds with velocity $(1, 1, 1)$ and then for t seconds with velocity $(4, 5, 7)$. However s and t are chosen, the point will end up lying in a certain plane. What is the equation of this plane? If (x, y, z) is the point reached, we have

$$x = 1 + s + 4t,$$
$$y = 2 + s + 5t,$$
$$z = 3 + s + 7t.$$

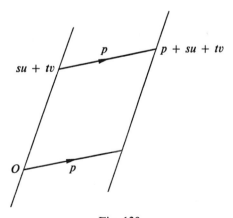

Fig. 139

These points all lie in a fixed plane, so they must satisfy some equation

$$ax + by + cz = d,$$

whatever values s and t may have. Now, from our equations for x, y, and z, we see that

$$ax + by + cz = a + 2b + 3c + (a + b + c)s + (4a + 5b + 7c)t.$$

This is to have a fixed value, d, regardless of the values of s and t. There is only one way this can happen; the coefficients of s and t must be zero. This means $a + b + c = 0$ and $4a + 5b + 7c = 0$. We can make this so by taking $a = 2, b = -3, c = 1$. (Any other solution of the two equations would, in principle, do equally well, and would lead to the same final result.) Thus we are led to $2x - 3y + z = -1$ as the equation satisfied by every point in the plane.

In dealing with lines and planes, certain questions may arise. Some problem has led us to specifications of a line and a plane. It may be important for some purpose to know how they are related. Does the line meet the plane in one point, or does it lie in the plane, or is it parallel to the plane? If we have two lines, do they meet or not? If they do not meet, is it because they are parallel lines or because they are skew? If we have three lines through a point, do they lie in a plane? If we have a point and a line, what is the equation of the plane containing them both?

There are a great variety of such questions. Often there are several ways of answering a question and it is only by practice and by experimenting with different approaches that a student, who needs such skills, can gain facility in translating geometrical problems into algebraic equations and judgement as to the best method to use in any particular case.

There are certain considerations that it is useful to bear in mind; these are discussed in the following paragraphs.

The same line can be traversed at many different speeds. Thus the points (s, s, s) and the points $(2s, 2s, 2s)$, where s takes all real values, are both parametric specifications of the line $x = y = z$. In the second specification the velocity is twice what it is in the first. Thus, for velocity (u_1, u_2, u_3) it is only the *ratios* $u_1 : u_2 : u_3$ that are important for the direction of the line.

Two lines are parallel (or possibly coincident) if they have the same direction. Thus the lines with parametric representations $(11 + t, 6 + 2t, -1 + 3t)$ and $(2 + 10t, 7 + 20t, 4 + 30t)$ are parallel. For if u denotes $(1, 2, 3)$, the velocity of the point tracing the first line, the velocity for the point tracing the second line is $10u$, so it is moving in the same direction but ten times as fast. Before we can assert that the lines are parallel we must check that they are not identical. The point $(2, 7, 4)$ lies on the second line. It is easily checked that we cannot choose t in $(11 + t, 6 + 2t, -1 + 3t)$ to make it coincide with $(2, 7, 4)$. This is sufficient to show that the lines are not identical, so they must be parallel.

We obtained the equation $ax + by + cz = d$ for an arbitrary plane by taking the plane $ax + by + cz = 0$ and displacing it by a translation. Thus the plane

$$ax + by + cz = d$$

is parallel to $ax + by + cz = 0$. Thus the constant d is important when we are talking about the *position* of a plane, but may be neglected in any question about the *orientation* of the plane.

Some questions can be translated immediately into a convenient algebraic form. For instance, is the line specified by $(1 + 4t, 2 + 5t, 3 + 6t)$ parallel or not to the plane $4x - 2y - z = 20$? This simply means – does the line meet the plane? So we try to find the point of intersection and see whether we run into any difficulty. When we substitute the co-ordinates of the moving point into the equation of the plane, we get $-3 = 20$, which can certainly not be satisfied. Thus there is no time t for which the moving point lies in the plane. This means that the line traced by the point must be parallel to the plane.

We could apply this method to the general problem, is the line specified by $(e + pt, f + qt, g + rt)$ parallel to the plane $ax + by + cz = d$? Putting the co-ordinates of the moving point into the equation of the plane we obtain

$$(ap + bq + cr)t + ae + bf + cg = d.$$

We can certainly solve this for t, except when $ap + bq + cr = 0$. So this equation indicates that something unusual is happening. If, as in our numerical example above, $ae + bf + cg \neq d$, there is no value t that satisfies the equation; the line is parallel to the plane. If $ae + bf + cg = d$, every t satisfies the equation: the moving point is always in the plane: the line lies in the plane.

We can check our conclusion by a quite different argument. It will be noticed that our condition $ap + bq + cr = 0$ does not involve d, e, f or g. It depends only on a, b, c, which specify the orientation of the plane, and p, q, r, which fix the direction of the line. In fact, our condition amounts to saying that (p, q, r), the velocity of the point, is a vector lying in $ax + by + cz = 0$, the plane through the origin parallel to the given plane. And this is reasonable on geometrical grounds. If a line is parallel to a plane, it is parallel to some line lying in that plane.

There is a certain trap to avoid when parametric specifications are being used. Suppose, for instance, we wish to find out whether the line

$$(x - 1)/4 = (y - 2)/5 = (z - 3)/6$$

meets the line $(x + 1)/2 = (y + 2)/3 = (z + 3)/4$. The first line can be specified as consisting of all the points $(1 + 4t, 2 + 5t, 3 + 6t)$, and the second one all the points $(-1 + 2t, -2 + 3t, -3 + 4t)$. If we now look for a common point by equating these, we find the x co-ordinates are equal only when $t = -1$, the y co-ordinates only for $t = -2$, and the z co-ordinates only when $t = -3$. There is no time when all three co-ordinates coincide. This seems to indicate that the lines have no common point. But this conclusion is certainly false. If we try $x = 5, y = 7, z = 9$ we find that original equations are all satisfied; the point $(5, 7, 9)$ lies on both lines.

One way to avoid this difficulty would be to avoid the parametric approach altogether. Our equations contain 4 'equals' signs; we have 4 equations for 3 unknowns. We could solve 3 of them for x, y, z and then see whether these values satisfied the remaining equation. This would be quite a good way of doing things, but the question remains – what was the fallacy in our first method?

If we look back at our parametric specifications we will see that these do make the lines have the common point $(5, 7, 9)$. The first line was specified by the moving point $(1 + 4t, 2 + 5t, 3 + 6t)$, and this point is at $(5, 7, 9)$ when $t = 1$. The second moving point was specified as $(-1 + 2t, -2 + 3t, -3 + 4t)$ and this is at $(5, 7, 9)$ when $t = 3$. Here we have the key to our difficulty; there is a place that both points pass through, but they pass *at different times*; we were asking too much when we equated the co-ordinates – that required not merely a common point on the two itineraries, but that it should be reached at the same time. *It is essential not to tie the movements of the two points together by using the same parameter t for both.* We should ask rather – is there a place that the first point reaches after s seconds and the second after t seconds? Our first line would then be specified as all the points $(1 + 4s, 2 + 5s, 3 + 6s)$ and the second as all the points $(-1 + 2t, -2 + 3t, -3 + 4t)$. Our equations would be $1 + 4s = -1 + 2t$, $2 + 5s = -2 + 3t$, $3 + 6s = -3 + 4t$. Solving the first two equations, we find $s = 1, t = 3$, and these values do satisfy the third equation. The discrepancy has been cleared up.

There is a useful device that can be used in a problem where, for instance, we are given two planes and asked to find a plane through their intersection that satisfies some further condition. This device may have been met already in 2-dimensional co-ordinate work, and we will explain it first in that context. Suppose we are dealing with the two lines, $x + 4y - 9 = 0$ and $2x + 3y - 8 = 0$. As may be checked by putting $x = 1$, $y = 2$, both these lines pass through the point $(1, 2)$. Now consider the equation $(x + 4y - 9) + k(2x + 3y - 8) = 0$. Put $x = 1, y = 2$ in this. We obtain $0 + 0k = 0$, which is true, whatever k. Thus, for any k, the equation above gives a line passing through the point $(1, 2)$.

The principle used here is perfectly general. If $f(x, y) = 0$ and $F(x, y) = 0$ are any two curves in the plane, the equation $f(x, y) + kF(x, y) = 0$ is automatically satisfied by any (x, y) for which $f(x, y) = 0$ and $F(x, y) = 0$; that is, $f(x, y) + kF(x, y) = 0$ always represents a curve that passes through all the intersections of $f(x, y) = 0$ and $F(x, y) = 0$.

Worked example

$x^2 + y^2 - 1 = 0$ is the unit circle. $xy = 0$ gives the axes. Any curve
$$x^2 + y^2 + kxy - 1 = 0$$
is bound to pass through the intersections of the unit circle with the axes. In Fig. 140, the dotted graph shows such a curve for one value of k.

The same principle applies in 3 dimensions. If we are given $13x - 11y + 73z - 2 = 0$ and $5x + 2y - 8z - 9 = 0$, the equation
$$(13x - 11y + 73z - 2) + k(5x + 2y - 8z - 9) = 0$$

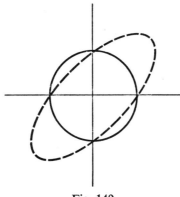

Fig. 140

is bound to represent a surface passing through the intersection of the given planes. It is clear that the equation in fact represents a plane through that intersection. The value of k might then be fixed by some further condition, for instance that the plane pass through the origin. This would lead to $k = -\frac{2}{9}$.

It is perhaps useful to repeat here what was said earlier; all the considerations so far given belong to affine geometry, to 'pure' vector theory. They do not in any way depend on the Euclidean concepts of length, angle, perpendicularity. They can therefore be used in any system of axes based on congruent parallelograms; it is not necessary for rectangles or squares to be involved in the network.

The diamond crystal, illustrating the use of oblique axes

Crystals exist in our everyday physical space, in which 'length' and 'perpendicular' have meanings, so one might expect that conventional perpendicular axes would give the best way of dealing with crystals, and that it would pay to use the full machinery of Euclidean geometry. However, this is not always so; there are some questions for which it pays to forget about right angles and scalar products, to use only the expressions of pure vector theory, $u + v$ and ku, and to work with oblique axes, that is, axes which are not perpendicular.

To illustrate this a partial investigation will be made of the structure of diamond. Naturally, one of the best ways of understanding this structure is to examine a physical model of this crystal; nevertheless, there are some problems for which we need to use co-ordinate methods. In the following work, we compare two possible approaches. We begin by using perpendicular axes; the investigation is left to the reader to complete. For comparison, we then begin again, using oblique axes.

We first describe a procedure for making a model of a diamond crystal. Let the vectors a, b, c, d be given by $a = (-1, 1, 1)$, $b = (1, -1, 1)$, $c = (1, 1, -1)$, $d = (-1, -1, -1)$ in a conventional system with perpendicular axes. The model is made from red and black balls, with rods joining them. The balls represent carbon atoms; the colours have no chemical significance, they simply facilitate the description of the model.

A red ball is placed at the origin. Rods emerge from it, representing the vectors a, b, c, d. At the end of each rod a black ball is placed. Thus we have black balls at $(-1, 1, 1)$, $(1, -1, 1)$, $(1, 1, -1)$ and $(-1, -1, -1)$.

Now from each black ball there emerge rods representing the vectors $-a, -b, -c$, $-d$, and a red ball is placed at the end of each of these. One of these positions requires no attention; the vector $-a$ brings us back to the origin, where there is already a red ball. We have to place red balls at the ends of the vectors $-b$, $-c$ and $-d$, that is at the points $(-2, 2, 0)$, $(-2, 0, 2)$ and $(0, 2, 2)$.

The construction now continues indefinitely. Whenever we have a red ball, we allow vectors a, b, c, d to sprout from it and mark their ends with black balls. From each black ball vectors $-a, -b, -c, -d$ sprout with red balls at their ends. This construction in fact leads to a reasonable framework. Models of this kind (without the distinction of red and black) can often be seen in chemical departments of schools and universities.

It is instructive to work out the positions of a few balls, as given by this construction, and to try to see some order in it. Can you find a rule for determining whether any given point, for example $(21, 13, 7)$, will have a ball at it, and if so, what the colour of the ball will be? It is desirable to try this before reading on.

$$* \qquad * \qquad *$$

We now look at this question with the help of oblique axes. The vectors a, b, c, d which represent the chemical bonds are of course not perpendicular; we cannot have 4 perpendicular vectors in 3 dimensions. We can get a co-ordinate system appropriate to the carbon atom by taking axes in the directions of three of our vectors, say a, b and c. Note that $d = -a - b - c$, so in this system d has the co-ordinates $(-1, -1, -1)$. It is purely by chance that these co-ordinates of d happen to coincide with the co-ordinates of d in the system we used first.

Let us put d on one side for the moment and consider where we can get to by starting at the origin and using only a, b, c when leaving a red ball and $-a, -b, -c$ when leaving a black ball. Positive and negative steps will be taken alternately. Thus if we used the sequence $+a, -b, +c, -b$ we would meet a black ball at position a relative to the origin, a red ball at $a - b$, a black ball at $a - b + c$, and finally a red ball at $a - 2b + c$. It is noticeable that when we leave a red ball a coefficient changes by $+1$, while when we leave a black ball a coefficient changes by -1. After two steps the sum of the coefficients is unchanged. Thus for all the red balls the coefficient sum is the same, 0, while for all the black balls the sum is 1. Thus the red balls that we can reach without using d at all will be at points $xa + yb + zc$ with $x + y + z = 0$, and the black balls that can be reached without using d will be at points with $x + y + z = 1$. (Question for investigation; does *every* point with $x + y + z = 0$ have a red ball on it?)

Now let us gradually bring d into the picture. Suppose that at one step in a sequence such as that just considered we replace a step $-b$ by $-d$. Now $-d = a + b + c$. Thus, instead of a coefficient change of -1 we shall have an increase of $+3$. Thus replacing $-b$ by $-d$ increases the coefficient sum by 4. Accordingly we shall find some red balls at points with $x + y + z = 4$ and some black balls at points with

$$x + y + z = 5.$$

The argument can now be extended. If in a sequence of steps we replace a number of negative steps, $-a$, $-b$ or $-c$, by $-d$, each such change increases the eventual coefficient sum by 4. If, on the other hand, we replace positive steps, $+a$, $+b$ or $+c$, by $+d$, each such change reduces the coefficient sum by 4. Thus we find that red balls cannot occur anywhere except in planes of the form $x + y + z = 4n$ and black balls cannot occur except in planes $x + y + z = 4n + 1$, where n is some integer.

We will not go on to obtain a complete and detailed picture of the structure of diamond, though this can be done by vector methods. Readers interested in this question may like to work out some further details.

Exercises on lines and planes

1   An object moves with constant velocity. At time $t = 0$ it is at $(1, 1, 1)$, at time $t = 1$ at $(2, 3, 4)$. What vector specifies its velocity? Where will it be at time t? The answer to this last question provides a parametric specification of the line joining $(1, 1, 1)$ and $(2, 3, 4)$. Find a specification of this line by a pair of equations (there are many possible correct answers).

2   The equation $x - 2y + z = 0$ specifies a plane. Find at what times (if any) objects moving as specified below will lie in this plane;
   Object A begins at $(0, 1, 0)$ and has velocity $(1, 1, 2)$.
   Object B begins at $(0, 1, 0)$ and has velocity $(1, 1, 1)$.
   Object C begins at $(1, 2, 3)$ and has velocity $(4, 5, 6)$.
In what relation to the plane do the lines described by these three moving objects stand?

3   Express the equations of the lines mentioned in questions 1 and 2 in the form

$$(x - a)/d = (y - b)/e = (z - c)/f.$$

Specify a starting position and velocity that would cause an object to move in the line

$$(x - 10)/2 = (y - 20)/3 = (z - 30)/4.$$

Does this question have one or many correct answers?

4   An object starts at the point $(5, 1, 7)$ and moves with velocity $(1, 2, 3)$. Will it pass through the point $(8, 7, 16)$ or not? What relation exists between the lines

$$(x - 5)/1 = (y - 1)/2 = (z - 7)/3$$

and

$$(x - 8)/1 = (y - 7)/2 = (z - 16)/3?$$

5   An object starts at the point $(0, 4, 6)$ and moves with velocity $(1, 0, 0)$. Another object starts at $(5, 0, 0)$ at the same time and moves with velocity $(0, 2, 3)$. Will the objects collide? Do the lines in which they move meet each other?

Lines and planes in Euclidean space of 3 dimensions

All the considerations so far given remain true in Euclid's geometry, so we start our present topic with a good deal of information and we have not much to add to it.

We will suppose, for the remainder of this section, that we are in Euclidean geometry and are using 'standard' axes, based on squares or cubes. We are thus entitled to use $u_1v_1 + u_2v_2 + u_3v_3$, the scalar product of vectors u and v, and to give it the usual geometrical or mechanical interpretation. (By the mechanical interpretation is meant 'work done'.)

Let us take some fixed vector, (p, q, r) and consider the vectors (x, y, z) perpendicular to it. The condition for perpendicularity is that the scalar product be zero;

$$px + qy + rz = 0.$$

This is the equation of a plane. It verifies, what we would expect on geometrical grounds, that all the vectors perpendicular to a fixed vector do lie in a plane. It also gives us a way of finding the normal to a plane, the equation of which we know; the normal to $px + qy + rz = 0$ is (p, q, r).

As we saw above, a plane $px + qy + rz = d$ is parallel to $px + qy + rz = 0$, so the constant d does not affect the result; when it is present, the normal is still (p, q, r).

If we need to write down the equation of the plane, through (a, b, c), perpendicular to the direction (p, q, r), we can proceed as follows. The plane must be

$$px + qy + rz = d$$

for some value of d; this ensures that (p, q, r) will be the normal. The plane will go through (a, b, c) if $pa + qb + rc = d$, so we choose this value for d. The required plane is $px + qy + rz = pa + qb + rc$.

This result could be written $p(x - a) + q(y - b) + r(z - c) = 0$, which we can interpret as a scalar product. The equation states that $(x - a, y - b, z - c)$ is perpendicular to (p, q, r). Now $(x - a, y - b, z - c)$ is the vector going from the fixed point (a, b, c) to the variable point (x, y, z), and if (x, y, z) wanders about in the plane in question, this vector ought to be perpendicular to (p, q, r). This argument provides an alternative way of arriving at the required equation.

An engineer's calculations lead to practical action, and if the calculations are incorrect, the consequences can be disastrous. It is therefore very wise not to rely on a single method blindly applied, but whenever possible to use two different methods and check whether they lead to the same conclusion. When a result has been obtained it is good to examine it, to see if it seems reasonable, and whether it can be interpreted in some way akin to what we did with the result $px + qy + rz = pa + qb + rc$. We should not be content to work blindly with equations but should try to see their meaning with the help of drawings, models and perhaps additional calculations.

For example, consider the equation $x + y + z = 1$. We know this is a plane, perpendicular to $(1, 1, 1)$. But we can also observe that the equation is satisfied by the points $(1, 0, 0)$, $(0, 1, 0)$, $(0, 0, 1)$, shown in Fig. 141 as A, B and C. So the plane contains the equilateral triangle ABC, which, as the sketch of the cube shows, is indeed perpendicular to the diagonal OG, where G is $(1, 1, 1)$. The vector $(-1, 1, 0)$ corresponds to the journey from A to B, and $(-1, 0, 1)$ to that from A to C. We could use AB and

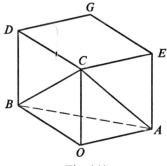

Fig. 141

AC as basis vectors for an oblique co-ordinate system in the plane ABC. Any point of that plane can be shown as $A + sAB + tAC$ for some s, t; that is, as

$$(1, 0, 0) + s(-1, 1, 0) + t(-1, 0, 1)$$

or $(1 - s - t, s, t)$. This checks; if $x = 1 - s - t, y = s, z = t$ we do have

$$x + y + z = 1.$$

We have in fact the parametric form, which can be reached geometrically, as above, or by a purely algebraic argument.

   The lines AB, BC and CA are the intersections of the plane $x + y + z = 1$ with the co-ordinate planes $z = 0$, $x = 0$ and $y = 0$. Considering the intersections of a plane with the co-ordinate planes is often a useful way of seeing how the plane is situated in space.

Worked example

A drawing is to be made showing the section of some object by the plane passing through the points $(5, 0, 0)$, $(0, 5, 0)$, $(0, 0, 5)$. On this drawing it is desired to mark all the points (x, y, z) whose co-ordinates x, y, z are whole numbers. Find a parametric representation of the plane suitable for this purpose and make a diagram of the section with the points in question marked.

Solution   Let A, B, C in Fig. 142 be the points $(5, 0, 0)$, $(0, 5, 0)$ and $(0, 0, 5)$. Fig. 142 shows the plane ABC and the co-ordinate grid in the plane $y = 0$. It is clear that if we go from A to C by the straight route, the first point with whole number co-ordinates we meet after leaving A will be P, the point $(4, 0, 1)$. If we use u to denote the vector AP, the points on the line AC that interest us may be written as A, $A + u$, $A + 2u$, $A + 3u$, $A + 4u$ and $A + 5u$, the last of these being C. Thus u presents itself as a vector lying in the plane ABC that we can profitably use when making our parametric representation. The vector $u = (-1, 0, 1)$.

   Similarly, if we consider the plane OAB, our attention is drawn to the vector $v = (-1, 1, 0)$, which takes us from A in the direction AB.

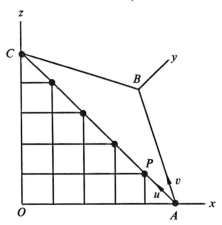

Fig. 142

If we start at A and travel for s seconds with velocity u and then for t seconds with velocity v, we shall end somewhere in the plane ABC; our position will be given by $A + su + tv$. What values of s and t will give us points with whole number co-ordinates x, y, z? To answer this we must express $A + su + tv$ in co-ordinate form as

$$(5, 0, 0) + s(-1, 0, 1) + t(-1, 1, 0);$$

this simplifies to $(5 - s - t, t, s)$. The co-ordinates here will be whole numbers if s and t are whole numbers and $s + t \leqslant 5$.

Now s and t are co-ordinates in the plane ABC in the system with origin at A and axes in the directions AC and AB. Obviously the triangle ABC is equilateral, so the axes AC and AB are at an angle of $60°$. The basis vectors u and v are of equal length. If we draw the grid for the s, t co-ordinate system and use the expression $(5 - s - t, t, s)$ to give us the x, y, z labels for the points of the grid, we obtain the diagram shown in Fig. 143. If we wish to mark any further point, the co-ordinates (x, y, z) of which are known, this is easily done, since we have $x = 5 - s - t, y = t, z = s$. The values of s and t are given directly by those of z and y. The value of x will automatically be correct, since the equations imply $x + y + z = 5$ and it is only points of this plane that appear in the section of the object that we are considering.

Exercises

1 For each of the following planes write a vector normal to it;
(a) $2x + 3y + 4z = 0$ (b) $2x + 3y + 4z = 1$ (c) $x + y + z = 17$
(d) $5x - 2y + 11z = 10$ (e) $x + y = 3$ (f) $z = 0$.

2 Find the equation of the plane through the points $(a, 0, 0)$, $(0, b, 0)$, $(0, 0, c)$. Find the normal to this plane.

3 State the vector normal to the plane $x + 2y + 3z = 56$. Give the parametric representation of the line through the origin perpendicular to this plane. Find the point where this line meets the plane.

Also give a parametric representation for the line through the point $(3, 4, 1)$ perpendicular to the plane, and find where this line meets the plane.

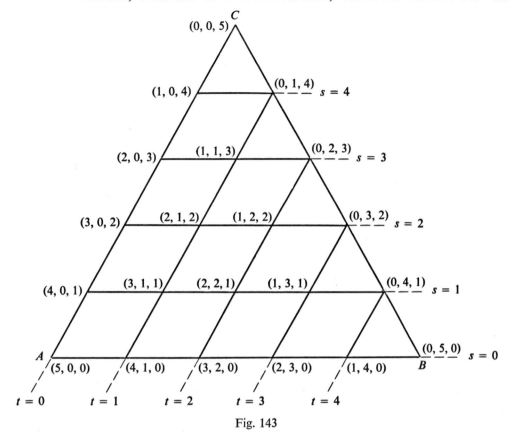

Fig. 143

4   Carry through the work of question 2 but for the plane $ax + by + cz = d$, and with (x_0, y_0, z_0) instead of (3, 4, 1).

(The second part of this question leads to a formula for the perpendicular projection of any point onto any plane.)

5   By means of your result in question 4 find the point obtained by projecting (x_0, y_0, z_0) perpendicularly onto the plane $x + y + z = 0$. Hence obtain the matrix that represents the operation of perpendicular projection onto this plane, i.e. the matrix M such that if v is any point, Mv is its projection.

Check the accuracy of your work by verifying that the determinant, eigenvectors and eigenvalues of M are what they ought to be.

6   (*a*) Let $D = (4, 0, 0)$, $E = (0, 4, 0)$, $F = (0, 0, 4)$. Draw a diagram for the points with whole number co-ordinates in the plane DEF similar to that drawn for the plane ABC in the worked example above.

(*b*) Find the co-ordinates for the perpendicular projection of (x_0, y_0, z_0) onto the plane $x + y + z = 5$.

(*c*) Copy the diagram for triangle ABC found in the Worked example, and then mark on it the perpendicular projections onto the plane ABC of all the points in the plane DEF that were shown in your answer to (*a*).

7   Let $P = (x_0, y_0, z_0)$ and $K = (1, 1, 1)$. We can find the perpendicular projection of P onto the line OK by constructing the plane through P that is perpendicular to OK and taking its intersection,

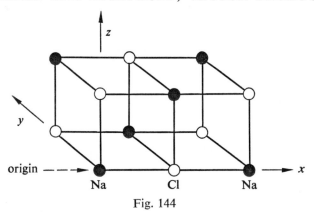

Fig. 144

N, with OK. Find the equation of that plane, the parametric representation of the line OK, and hence the co-ordinates of N.

Write the matrix M that represents the transformation $P \rightarrow N$, and check its accuracy in the way suggested in question 5.

8 A cube has faces lying in the planes $x = 1$, $y = 1$, $z = 1$, $x = -1$, $y = -1$, and $z = -1$. The plane $x + y + z = 0$ divides the cube into two congruent pieces. Find the co-ordinates of the points where this plane meets the edges of the cube. What plane figure is the section of the cube by this plane? Make a sketch showing clearly the cube and the intersection of this plane with its faces.

9 Fig. 144 shows the arrangement of sodium and chlorine atoms in a crystal of common salt, with black balls for sodium and white for chlorine. If the cubes have unit side, atoms occur at points with whole number co-ordinates. Given any point, for example, (87, 1230, 29), how can one tell whether this point is occupied by a sodium or a chlorine atom? Certain planes are occupied by sodium atoms alone, others by chlorine atoms alone. Identify some such planes and give their equations; a variety of possible orientations should be sought. How can one prove that such planes, however far they may be continued, will contain only one kind of atom?

10 If P is any point and M is its perpendicular projection onto a plane, the reflection of P in the plane is reached from P by going twice the distance $\|PM\|$ in the direction of PM. Find the co-ordinates of the reflection of (x_0, y_0, z_0) in the plane $ax + by + cz = d$.

Find the matrix that represents reflection in the plane $x + 2y + 3z = 0$. Do you expect this matrix to be (a) orthogonal, (b) symmetric?

Vector product

As was mentioned earlier, vector product is a very restricted concept; it has meaning only in Euclidean geometry of 3 dimensions. It arises very naturally in statics and dynamics and also in electromagnetic theory.

In Fig. 145 a force F acts in the plane of the paper, at a point whose position, relative to the origin, is specified by a vector r. The force exerts a twisting effect about the origin. It has a tendency to make things turn about a line through the origin perpendicular to the plane of the paper. The magnitude of this torque is pF where p is the perpendicular from O to the line of action of F. Thus the custom has arisen of representing the turning effect of F about the origin by a vector of length pF perpendicular to the plane of the paper. This vector is known as the vector product of r and F, and is written $r \wedge F$ or $r \times F$.

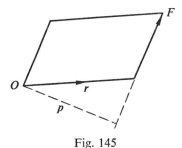

Fig. 145

Note that pF represents the area of the parallelogram contained between the vectors r and F, and so the magnitude of $r \times F$ can be defined as equal to the area of this parallelogram. Note incidentally that 'a length = an area' does not conform to common practices in regard to 'dimensions'. We have to suppose our units fixed; the number of (say) metres in the length is to equal the number of square metres in the area of the parallelogram.

Suppose we are given two vectors u, v, and we wish to find a formula for their vector product, $w = u \times v$.

The first requirement on w is that it should be perpendicular to both u and v, just as, in our discussion above, the torque is perpendicular to the plane of r and F. Let u be a (column) vector with components a, b, c, and let v have components d, e, f. As w is perpendicular to both, we must have $u'w = 0$ and $v'w = 0$. If w has components x, y, z, this means

$$ax + by + cz = 0, \qquad dx + ey + fz = 0.$$

At the end of §27, we saw that the solution of these equations was $x = k(bf - ce)$, $y = -k(af - cd)$, $z = k(ae - bd)$, where k could be any number.

Let A stand for the area of the parallelogram between u and v, and L for the length of w. The definition of w instructs us to make $L = A$. Now the area of a parallelogram between two arbitrary vectors u, v can be calculated, but the calculation is moderately long. The area of the parallelogram involves lengths and the sine of the angle between u and v. From the dot product $u \cdot v$ we can arrive at the cosine of the angle, and from the cosine we can get the sine – a somewhat roundabout procedure.

We can bypass this by considering the volume of the cell contained by u, v, w. Since w is perpendicular to the plane of u and v, taking A as the area of the base, L will give the perpendicular height, so the volume is LA. As we want $L = A$, this means the volume must be simply L^2. We must choose k to make this so.

Now

$$L^2 = x^2 + y^2 + z^2 = k^2\{(bf - ce)^2 + (af - cd)^2 + (ae - bd)^2\}.$$

Fortunately for the simplicity of our final formula, the expression above, involving the sum of 3 squares, also appears in the expression for the volume of the cell, and cancels. For the volume of the cell is

$$\det (u, v, w) = -\det (u, w, v) = \det (w, u, v)$$

$$= \begin{vmatrix} x & a & d \\ y & b & e \\ z & c & f \end{vmatrix}$$

$$= x(bf - ce) - y(af - cd) + z(ae + bd).$$

Substituting the values found for x, y, z we obtain

$$k\{(bf - ce)^2 + (af - cd)^2 + (ae - bd)^2\}.$$

Comparing this with our expression for L^2 above, which the volume of the cell has to equal, we find $k^2 = k$. Obviously we do not want the solution $k = 0$, so we are left with the very simple conclusion, $k = 1$. Thus the components of w are simply $bf - ce$, $-(af - cd)$, $ae - bd$. There is a simple rule for obtaining these. We write first of all

$$\begin{matrix} a & b & c \\ d & e & f. \end{matrix}$$

To obtain the first component of $u \times v$, we cover the first column, a and d, and write the determinant of the matrix we then see. For the second component, we cover b and e, and write the determinant of what we then see, *but with a minus sign*. For the third component, we cover c and f, and write the determinant of what remains visible. The alternation of signs, $+, -, +$, is exactly the same as in the rules for multiplying out a determinant.

One last point remains to be settled. Suppose for example u represents a unit due east, and v a unit due north in the horizontal plane. The parallelogram is then a square of unit area, and we know $u \times v$ should be perpendicular to the horizontal plane, that is, vertical. But should it be a unit vector vertically upwards or vertically downwards? If we write $1, 0, 0$ for a, b, c and $0, 1, 0$ for d, e, f in the scheme explained above, we obtain $0, 0, 1$ for the components of $u \times v$; thus $u \times v$ is vertically upwards. But how are we to interpret our formula when u and v do not lie so conveniently, but are any two vectors? Our earlier work gives us a way to answer this question. We found k by using the equation $\det (u, v, w) = L^2$. *This means that u, v, w must always lie in such a way that $\det (u, v, w)$ is positive.*

To interpret this result geometrically, we use two very simple properties of determinants. A determinant is given by a polynomial function of the numbers $a, b, c, \dots$ that appear in it. Accordingly, it is given by a continuous function; little changes in a, b, c and the other numbers produce only little changes in the determinant. Now the value of a continuous function can change from $+$ to $-$ only by passing through the value 0 (Fig. 146). So if we have three vectors whose determinant is positive and we change them gradually, with care never to let the volume of the cell become 0, we shall be sure of ending up with a positive determinant.

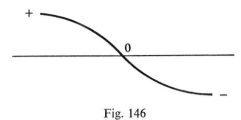

Fig. 146

Now we have seen that u to the east, v to the north and w upwards gives a positive value to the determinant. Suppose we keep the directions of u and w the same, but allow v to swing round in the horizontal plane. The determinant will become zero only if v swings so far that its direction becomes either east or west. If we forbid this, if we insist that v points somewhere between just north of east and just north of west, we have removed all possibility of a change of sign; with this condition, det (u, v, w) will certainly be positive.

Now consider a different kind of variation. Suppose we have our vectors u, v, w in any position, and that we represent them by three rods, welded together to form a rigid body. We then gradually rotate this rigid system in any manner whatever. All the vectors then change gradually, and so the value of the determinant can never jump suddenly. But, since the system is rigid, the volume of the cell is fixed in size; its magnitude cannot change. If, at the start, det (u, v, w) was K, at any later time it must be either K or $-K$. But one can only get from K to $-K$ by a sudden jump; there is no way of going gradually from K to $-K$ without passing through other numbers. But we showed earlier that gradual rotation can never make the determinant's value jump; there is only one way to meet both requirements: during rotation, the determinant's value must stay unaltered at K. Rotation does not change the magnitude, or the sign, of $\det(u, v, w)$.

Accordingly let us take our three welded rods and turn them until the rod u points due east. Then turn them, keeping u due east, until v lies in the horizontal plane through u, and in the half of that plane described above, from just north of east to just north of west. This operation can always be carried out. The third vector w equals $u \times v$, so $\det(u, v, w)$ must be positive. This means w must point vertically upwards, not vertically downwards.

The relationship of u, v, w, in the situation just described, is often embodied in the Right-Hand Rule; if we turn a screw head in such a way that a lever, originally pointing in the direction of u, comes to the direction v *by the shortest route*, then the screw will advance in the direction of w, where $w = u \times v$, as in Fig. 147.

If we now suppose our welded rods returned to their original position, this statement will remain correct; it is not affected by a rotation of the rigid object.

In all the discussion above we have assumed that u and v point in different directions. If u and v point in the same direction, the area of the 'parallelogram' between them is 0, and $u \times v = 0$. No further discussion is called for.

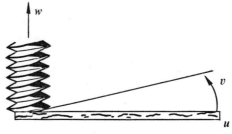

Fig. 147

Exercises

1 Show, both algebraically and from the geometrical definition, that $v \times u = -(u \times v)$.

2 If u, v, r are any three vectors, we can form the scalar product of r with the vector $u \times v$. The resulting number is written $r \cdot (u \times v)$. Show that, if r lies in the plane of u and v, then $r \cdot (u \times v) = 0$. (This comes without any calculation by a simple geometrical argument.)

3 Show, by algebra, that $r \cdot (u \times v) = \det(r, u, v)$. Is this result reasonable on geometrical grounds?

4 How do we express, by algebraic symbols, that r lies in the plane of u and v? What theorem about determinants comes by combining the statements in questions 3 and 4?

Note The arguments used in this section justify the statement made in §36, that a rotation has determinant $+1$ while a reflection has determinant -1. The essential reason for this is that any rotation can be carried out gradually. We can bring a body to any required position by a series of tiny changes. A 'tiny change' has a matrix that is very nearly the identity matrix, I. Since $\det I = +1$, and the determinant is a continuous function, the determinant of a tiny rotation must be $+1$; it cannot jump to -1. Accordingly a rotation, which can be brought about by a succession of small rotations, must also have determinant $+1$, for determinant of a product equals product of determinants.

A reflection on the other hand is a sudden change. There is, for instance, no way in which we can *gradually* change a car with right-hand drive into one with left-hand drive.

Exercises on vector products

1 Find the following vector products, and check that they do lie in the directions to be expected
(*a*) $(1, 1, 0) \times (-1, 1, 0)$ (*b*) $(1, 1, 1) \times (1, 1, 2)$ (*c*) $(-1, -2, 3) \times (4, 5, -9)$
(*d*) $(1, 1, 3) \times (5, 5, 16)$ (*e*) $(1, 2, 3) \times (5, 10, 15)$.

2 By means of vector product theory find a vector normal to the plane containing the vecto $(2, 6, 9)$ and $(-6, -7, 6)$. What is the area of the parallelogram that has these vectors as two of sides?
What figure in fact is that parallelogram, and how can you confirm your answer by determin its area without any use of vector products?

3 In each of the following cases find a vector normal to the plane of the two given vectors, also the area of the parallelogram with the two vectors as sides;
(*a*) $(1, -1, 0)$ and $(0, 1, -1)$ (*b*) $(1, -1, 0)$ and $(1, 0, -1)$,   (*c*) $(1, 0, -1)$ and $(0, 1,$

In each case, deduce the sine of the angle between the two given vectors. Also, by means of scalar products, find the cosine of the angle between them. Check the consistency of these results.

Discuss the geometrical figure formed by the 3 vectors that are mentioned in (a), (b) and (c) above.

4 From any point O of the paper you are using draw 2 arrows to represent vectors, u and v, in the plane of the paper. Let $p = u \times v$. Indicate on your drawing where you would expect $u \times p$ and $v \times p$ to lie. Test your conclusion by the example $u = (1, 0, 0)$, $v = (1, 1, 0)$, it being understood that $z = 0$ is the plane of the paper.

5 Let $u = (1, 2, 3)$, $v = (2, 3, 4)$, $w = (3, 4, 5)$. Find $p = u \times v$, and the scalar product $w \cdot p$. Does w lie in the plane of u and v?

6 If a force F acts at the point R its moment about an axis through the origin is given by $U \cdot (R \times F)$, where U is a vector of unit length along the axis.

(a) Find the moment of the force $(6, 3, 7)$ acting at $(4, 2, 5)$ about the axis joining the origin to the point $(1, 2, 2)$.

(b) Show that if the force in (a) were changed to $(-4, 10, 1)$, the other data remaining unaltered, there would be no moment about the axis.

7 A set of three perpendicular vectors can be constructed in the following way. Choose any non-zero vector u. Choose v perpendicular to it. Let $w = u \times v$. Then u, v, w constitute a system of mutually perpendicular vectors. Construct a system of this kind corresponding to each of the vectors u listed here;

(a) $u = (1, -1, 0)$,   (b) $u = (1, 1, 1)$,   (c) $u = (1, 2, 2)$,   (d) $u = (1, 2, 3)$.

Note. This procedure can be used to construct examples of orthogonal matrices, if in addition we require the vectors u and v to be of unit length; we then form the matrix with columns u, v, w.

40 *Systems of linear equations*

Suppose we have the equations

$$x + 3y = 25, \tag{1}$$

$$2x + 7y = 55. \tag{2}$$

If we subtract twice equation (1) from equation (2), and leave equation (1) unaltered, we obtain

$$x + 3y = 25, \tag{3}$$

$$y = 5. \tag{4}$$

If we now subtract 3 times equation (4) from equation (3), we bring our pair of equations to the form

$$x \quad = 10, \tag{5}$$

$$y = 5. \tag{6}$$

If we write

$$M = \begin{pmatrix} 1 & 3 \\ 2 & 7 \end{pmatrix}, \qquad v = \begin{pmatrix} x \\ y \end{pmatrix},$$

it will be seen that equations (1) and (2) have, on the left, the vector Mv, while equations (5) and (6) have simply Iv, where I is, as usual, the identity matrix.

A remarkable verbiage has grown up around this calculation, and its generalization to m equations in n unknowns. It is called the Gauss–Jordan elimination procedure, and the matrix I, appearing in equations (5) and (6) is known as the Row–Echelon normal form of the original matrix M.

Many textbooks, both for school and university courses, begin with this procedure and use it to introduce the idea of matrix. We have done the opposite; we have kept linear equations until near the end of the course, and this has been done for a very definite reason. The use of linear equations to introduce the idea of matrix has led to confusion in the minds of some students which the authors of the textbooks certainly did not foresee or intend. That approach, and the use of the term 'row–echelon normal form of the matrix M', led some students to believe that somehow the final matrix, I, was *the same as* the original matrix M. Now this is clearly false; whatever else it may be, the matrix $\begin{pmatrix} 1 & 3 \\ 2 & 7 \end{pmatrix}$ is most certainly *not* the identity matrix I.

This confusion is very likely if a student identifies a matrix with a system of equations. The equations (5) and (6) are *equivalent* to the equations (1) and (2). The pair of statements, (5) and (6), contains exactly the same information as the pair of statements (1) and (2).

Let us examine in more detail the actual relationship between the matrices. We will begin with the equations

$$x + 3y = a_1, \tag{7}$$

$$2x + 7y = a_2. \tag{8}$$

Applying the same procedure as before, we obtain first

$$x + 3y = a_1 \qquad\quad = b_1, \quad \text{say.} \tag{9}$$

$$y = -2a_1 + a_2 = b_2, \quad \text{say.} \tag{10}$$

Next we have

$$x \qquad = b_1 - 3b_2 = c_1, \quad \text{say} \tag{11}$$

$$y = \qquad\quad b_2 = c_2, \quad \text{say.} \tag{12}$$

Equations (7) and (8) can be combined in a single vector equation, $Mv = a$, where the meaning of a should be self-explanatory. The vector b, involved in equations (9) and (10), can be written $b = Qa$, where $Q = \begin{pmatrix} 1 & 0 \\ -2 & 1 \end{pmatrix}$. Since $a = Mv$, we must have $b = QMv$, and equations (9) and (10) may be put in the form $QMv = b$. (If desired, this can be verified by actual calculation of QM.) Similarly, we have $c = Pb$, where

$P = \begin{pmatrix} 1 & -3 \\ 0 & 1 \end{pmatrix}$. Accordingly, on substituting for b, we obtain $PQMv = c$ as a concise form of equations (11), (12). Now equations (11) and (12), as we see by looking at the actual equations, are in fact $Iv = c$. Comparing the two equations we see that I can be identified, not with M, but with PQM. We have $PQM = I$. Thus $PQ = M^{-1}$.

The calculations indeed pass from $Mv = a$ to $QMv = Qa$ and finally to

$$PQM = PQa.$$

As $PQ = M^{-1}$, the overall effect is that we go from $Mv = a$ to $v = M^{-1}a$. The procedure in fact is one that can conveniently be used for calculating the inverse matrix, M^{-1}.

Many books give a procedure for calculating the inverse matrix, in which matrices only are written. The logic of this procedure may be seen by considering the process of solution described above. The only change is that we write Ia instead of a. The successive pairs of equations can then be shown as follows:

$$Mv = \quad Ia, \qquad\qquad (7), (8)$$

$$QMv = \quad QIa, \qquad\qquad (9), (10)$$

$$PQMv = PQIa. \qquad\qquad (11), (12)$$

That is

$$Iv = PQIa.$$

Suppose we now agree to omit the vectors v and a, since they remain unchanged throughout. We can then record the calculation as

$$M \ , \quad I \ .$$
$$QM \ , \quad QI \ .$$
$$PQM \ , PQI \ .$$
$$I \ , PQI \ (= M^{-1}).$$

In carrying out such a calculation, we do not usually think of multiplying by the matrices Q and P, but rather of operations done to the rows. For if we consider any matrix

$$K = \begin{pmatrix} d & e \\ f & g \end{pmatrix},$$

we have

$$QK = \begin{pmatrix} 1 & 0 \\ -2 & 1 \end{pmatrix} \begin{pmatrix} d & e \\ f & g \end{pmatrix} = \begin{pmatrix} d & e \\ -2d + f & -2e + g \end{pmatrix}.$$

Thus going from K to QK is achieved by *subtracting twice the first row from the second*. This in fact was how the matrix Q first came into the story. (Look back at our original solution of the equations, and see how it led us to consider Q.) The multiplication by P also indicates an operation on the rows.

The procedure for calculating M^{-1} may be summarized as follows; choose a sequence of operations on rows that will change M into I; this sequence of operations will change I into M^{-1}.

We assume here that M does have an inverse; if M^{-1} does not exist, we shall not be able to find a sequence of operations that takes us from M to I.

All the work of this section is best understood by thinking about the solution of equations. All the operations we shall consider will be calculations that any student would naturally perform if confronted with a number of linear equations.

The matrix form of this work adds no essential idea, but simply puts the material in a form suitable for an electronic computer. A computer does not record equations such as $x + 3y = 25$, $2x + 7y = 55$ in the way we do. It simply records the numbers involved at appropriate addresses, and so deals with a scheme of the type

$$\begin{matrix} 1 & 3 & 25 \\ 2 & 7 & 55. \end{matrix}$$

Such a scheme to us looks like a matrix and can indeed be handled by rules of matrix algebra. This change of notation is a very minor affair, as it was also earlier when we agreed to drop v and a from our equations and simply understand that they were supposed to be there. At any time when we are uncertain why some particular step is being taken, it will usually be sufficient to go back from the matrix symbols to the equations they represent, and the reason will become clear.

Equivalence of equations

When we operate on a system of equations, we want to end up with another system of equations that says neither more nor less than our original set: we do not want to bring in any condition that was not in the original set, for this might lead us to overlook some perfectly good solutions; on the other hand, we do not want to lose any condition, for that might lead us to accept something which in fact was not a solution of the original equations. For example, suppose we have to attack the following system of equations:

$$3x + 4y = 20, \tag{13}$$

$$4x + 5y = 27, \tag{14}$$

$$4x + 3y = 23. \tag{15}$$

Students sometimes hand in the following calculation. Subtract equation (13) from (14), and also subtract (14) from (15). This gives

$$x + y = 7, \tag{16}$$

$$x - y = 3. \tag{17}$$

These equations lead to $x = 5$, $y = 2$, and these values are handed in as a solution. But if we try these values in equations (13), (14), (15), we find that not one of the equations is satisfied. In fact, equations (13), (14), (15) are inconsistent; they have no solution. The trouble is that our argument is not reversible; equations (16) and (17) do follow from equations (13), (14), (15), but equations (13), (14), (15) cannot be deduced from equations (16), (17). So (16), (17) is *not* a system equivalent to (13), (14), (15).

We would have obtained an equivalent system if we had not merely written (16) and (17), but had copied (13) down again. Equations (13), (16), (17) are equivalent to (13), (14), (15); for we can get back to (14) by adding (13) and (16), and back to (15) by adding (13) and (17).

The situation can be shown in the language of symbolic logic as follows. In the first argument we have

$$(13) \ \& \ (14) \ \& \ (15) \Rightarrow (16) \ \& \ (17).$$

That is, (16) and (17) follow from (13), (14), (15), but we do not have any authority for going in the opposite direction. The statement shows that any solution (if it existed) of (13), (14), (15) would satisfy (16), (17). This gives us a kind of elimination procedure; it tells us that it is no use considering values x, y *unless* they satisfy (16) and (17). So $x = 5$, $y = 2$ is all we need try: either it is a solution, or there is no solution.
In the second way of doing things we have

$$(13) \ \& \ (14) \ \& \ (15) \Leftrightarrow (13) \ \& \ (16) \ \& \ (17).$$

That is, each set of statements can be deduced from the other set. The information contained in (13) & (16) & (17) is exactly the same as that in (13) & (14) & (15). Now this of course is the situation we want to achieve – to bring our equations to a simpler form, and to be sure that we have neither brought in things that appear to be solutions and in fact are not (like $x = 5$, $y = 2$ in our example) nor ruled out perfectly good solutions of the original equations; *we want our arguments to be reversible*; we want $\Leftrightarrow$, not the one-way street of $\Rightarrow$.

Given any system of equations, there are three quite simple and obvious reversible operations that we can apply to them.

(1) We may alter the order of the equations. To reverse this, we simply go back to the original order.

(2) We may replace any equation by that equation multiplied by a non-zero number. To reverse this is always possible; for instance, if the equation had been replaced by 3 times itself, we would get back by using $\frac{1}{3}$ times the new equation.

(3) We can replace any equation by that equation together with any multiple of another equation. Here a shorthand arrangement is convenient; if for example, we add 10 times equation (73) to equation (52), we may denote this by $(52) + 10(73) = (52^*)$, where (52^*) is the new equation that replaced equation (52). If we want to get back to our original equation we can do so, for $(52) = (52^*) - 10(73)$. We simply have to subtract equation (73), which is still there in the new system, from the new equation (52^*).

Our concern in the remainder of this section will be to see how we can use these simple operations to bring a system of equations to the most convenient form. The general objective can be seen from the following example. Suppose we are given some equations in 5 unknowns, and we find they are equivalent to the following 3 equations:

$$\begin{aligned} x_1 \quad - 5x_3 \quad + 2x_5 &= 0, \\ x_2 + \quad x_3 \quad + 7x_5 &= 0, \\ x_4 - 4x_5 &= 0. \end{aligned}$$

Here x_1, x_2 and x_4 are in a special position; x_1 occurs in the first equation only, x_2 in the second only, x_4 in the third only. Suppose we allow someone to choose any values he likes for x_3 and x_5. We can meet his requirements and still find a solution. For we can make the first equation correct by putting $x_1 = 5x_3 - 2x_5$. As x_1 occurs only in the first equation, this will have no repercussions on the other equations. Similarly, by taking $x_2 = -x_3 - 7x_5$ and $x_4 = 4x_5$ we can satisfy the other two equations. Clearly, then, there are a lot of solutions; we have 2 degrees of freedom – free choice of x_3 and x_5. If we put $x_3 = s$ and $x_5 = t$, we have the 2 parameters s and t, and the solution $x_1 = 5s - 2t$, $x_2 = -s - 7t$, $x_3 = s$, $x_4 = 4t$, $x_5 = t$. In fact these solutions fill a plane; if we put $s = 1$, $t = 0$ we get a particular solution corresponding to the point $(5, -1, 1, 0, 0)$, while for $s = 0$, $t = 1$, we get $(-2, -7, 0, 4, 1)$. If we denote these points by the (column) vectors u, v, every solution is of the form $su + tv$. The solutions fill a subspace of 2 dimensions.

Our work falls into three parts, (1) bringing the left-hand sides of the equations to the simplest form, (2) interpreting the results when the right-hand sides are all 0, (3) interpreting the results when non-zero numbers occur on the right-hand side.

Obtaining a simple form for the left-hand sides

The procedure for simplification will usually be carried out by a computer. We therefore have to be careful to make sure that the procedure can always be carried through, and that we do not overlook any possibilities that might arise.

We begin by considering the first variable, x_1, in the first equation. The situation is straightforward if we see a term ax_1, where $a \neq 0$. In this case, we divide by a, so our equation begins $x_1 + \cdots$ By subtracting suitable multiples of this equation from the other equations, we can get rid of x_1 in all equations except the first.

Worked example

Given equations $10x_1 + 4x_2 - x_3 = 0$, $3x_1 - 2x_2 + 11x_3 = 4$, $7x_1 + 22x_2 + 5x_3 = 9$. Multiply the first equation by 0.1 to obtain $x_1 + 0.4x_2 - 0.1x_3 = 0$. Subtracting 3 times this equation from the second equation, and 7 times this equation from the third equation leads to a system where x_1 has coefficient 1 in the first equation, and coefficients 0 in the other equations. (For an understanding of this point it is not necessary actually to calculate the coefficients of x_2 and x_3 in the resulting equations.)

First obstacle; suppose x_1 does not appear in the first equation. In this case, we look down the equations until we find one that does contain a term ax_1. We bring this equation up to first place, and then proceed as before.

Second obstacle; suppose we cannot do this – that is, suppose x_1 does not appear in any of the equations. In this case, we ignore x_1 and try to get x_2 in the first equation with a coefficient 1, and no x_2 in any other equation. If x_2 occurs nowhere, we go on to x_3 and so on.

Let us assume, for simplicity of description that x_1 actually does occur, so that we have our first equation $x_1 + \cdots$, and no x_1 anywhere else. The remaining equations contain the $(n-1)$ unknowns $x_2, x_3, \ldots, x_n$, and in effect we are faced with the same problem again, but with one variable less. So we try to arrange for a second equation beginning with $x_2 + \cdots$, and for x_2 to be absent from the remaining equations, *including the first equation of all*. If the second equation starts $x_2 + \cdots$, we certainly can get rid of x_2 everywhere else by appropriate subtractions of multiples of the second equation. However, it may happen that x_2 does not occur at all in any equation except the first; in that case there is nothing more we can do, we leave x_2 in the first equation and go on to consider x_3.

Thus we may reach a set of equations that begin with, for example, something like

$$x_1 \qquad + 2x_3 + 3x_4 = 4,$$
$$x_2 + 5x_3 + 6x_4 = 7,$$

or we may have something like

$$x_1 + 8x_2 \qquad + 2x_4 = 4,$$
$$x_3 + 3x_4 = 7.$$

In both cases, x_1 occurs only in the first equation. In the first case x_2 occurs only in the second equation; in the second case it is x_3 that occurs there only.

It could of course happen that x_2 never occurs at all. By erasing $8x_2$ in the example above, we can see how the first two equations might appear in such a case.

The process continues, always using very much the same ideas. Suppose, at a certain stage, we have dealt with a number of equations; the remaining equations do not contain $x_1, x_2, \ldots, x_{p-1}$. They do, however, contain x_p. We choose an equation that contains x_p, bring it to the top, if it is not already there, and multiply it by the number that makes the coefficient of x_p equal to 1. By using this equation, we get rid of x_p in all the other equations, whether they come before or after the equation in question. This equation has now been 'dealt with'. The equations that come after it are now free of $x_1, x_2, \ldots, x_{p-1}$ and x_p. If x_q is the first variable that occurs in them, we repeat the procedure, with x_q playing the role that x_p has just played. The process terminates when either there are no more equations, or when every variable in the remaining equations has coefficient 0. These possibilities are illustrated in the following examples.

In order not to distract attention by arithmetical complications, these equations have been 'faked' to work out with whole numbers. In real life, of course, we would expect to work to several places of decimals.

Worked example 1

Put the following system of equations into standard form:

$$x_1 + 2x_2 + 2x_3 + 5x_4 + 36x_5 = 0, \tag{18}$$

$$3x_1 + 6x_2 + 8x_3 + 19x_4 + 136x_5 = 0, \tag{19}$$

$$4x_1 + 8x_2 + 7x_3 + 19x_4 + 135x_5 = 0. \tag{20}$$

Solution   The first equation already contains x_1 with coefficient 1. So we repeat equation (18) as equation (21). To get rid of x_1 in the other equations, we take $(22) = (19) - 3(18)$ and $(23) = (20) - 4(18)$. This gives

$$x_1 + 2x_2 + 2x_3 + 5x_4 + 36x_5 = 0, \tag{21}$$

$$2x_3 + 4x_4 + 28x_5 = 0, \tag{22}$$

$$- x_3 - x_4 - 9x_5 = 0. \tag{23}$$

Now x_3 is the first variable that remains in the last pair of equations. Halving equation (22) gives us equation (25) with coefficient 1 for x_3. To get rid of x_3 in the other equations we form $(24) = (21) - 2(25)$ and $(26) = (23) + (25)$. This leads to

$$x_1 + 2x_2 + x_4 + 8x_5 = 0, \tag{24}$$

$$x_3 + 2x_4 + 14x_5 = 0, \tag{25}$$

$$x_4 + 5x_5 = 0. \tag{26}$$

As equation (24) contains the next variable, x_4, already with coefficient 1, we retain this equation unaltered as equation (26a). To get rid of x_4 elsewhere, we take

$$(24a) = (24) - (26)$$

and $(25a) = (25) - 2(26)$. We thus finish with

$$x_1 + 2x_2 + 3x_5 = 0, \tag{24a}$$

$$x_3 + 4x_5 = 0, \tag{25a}$$

$$x_4 + 5x_5 = 0. \tag{26a}$$

The process is now complete, for there are no more equations to deal with.

Worked example 2

Simplify the system

$$x_1 + 2x_2 + 3x_3 = 0, \tag{27}$$

$$3x_1 + 6x_2 + 7x_3 = 0, \tag{28}$$

$$2x_1 + 4x_2 + 10x_3 = 0. \tag{29}$$

Solution In equation (27) we already have x_1 with coefficient 1, so we form simply (28) $-$ 3(27), which is $-2x_3 = 0$, and (29) $-$ 2(27), which is $4x_3 = 0$. We divide the first of these by -2 and so obtain

$$x_1 + 2x_2 + 3x_3 = 0, \tag{30}$$

$$x_3 = 0, \tag{31}$$

$$4x_3 = 0. \tag{32}$$

Forming (30) $-$ 3(31) gets rid of x_3 in the first of these equations. Also (32) $-$ 4(31) gets rid of x_3 in the third equation, but then nothing at all remains in that equation. The original 3 equations are thus equivalent to the 2 equations

$$x_1 + 2x_2 \qquad = 0, \tag{33}$$

$$x_3 = 0. \tag{34}$$

The first example we had of equations in standard form was

$$\begin{aligned}
x_1 \quad - 5x_3 \quad + 2x_5 &= 0, \\
x_2 + \ x_3 \quad + 7x_5 &= 0, \\
x_4 - 4x_5 &= 0.
\end{aligned}$$

The special variables here were x_1, x_2 and x_4. The form of the equations is easier to see if x_1, x_2, x_4 are written first, then x_3 and x_5. We can bring this about by introducing new symbols. Let $y_1 = x_1$, $y_2 = x_2$, $y_3 = x_4$, $y_4 = x_3$, $y_5 = x_5$; that is, we give priority to the special variables. The equations then appear like this;

$$\begin{aligned}
y_1 \quad - 5y_4 + 2y_5 &= 0, \\
y_2 \quad + \ y_4 + 7y_5 &= 0, \\
y_3 \quad - 4y_5 &= 0.
\end{aligned}$$

It will simplify our later discussions to suppose that, if it is needed, such a substitution has been made. We shall not bother to use the letter y; we shall simply assume that the variable at the beginning of the first equation is called x_1, that at the beginning of the second is called x_2, and so on. Then we shall always see a pattern like that of the equations last written. If there are k equations, $x_1, x_2, \ldots, x_k$ will stand alone, then $x_{k+1}, \ldots, x_n$ will occur with arbitrary coefficients. Some of these coefficients may happen to be 0, as for y_4 in the third equation above, but this is of no significance. The extreme cases, $k = n$ and $k = 0$ may occur.

The important thing to notice is that the reduction to standard form *can always be carried through*. It may be that all the coefficients in all the given equations are 0; in that case, we have no equations at all; this is a standard case, with $k = 0$. If this extreme case does not occur, then some equation contains some variable with non-zero coefficient. We write that equation first, call the variable x_1, and get the equation into the form $x_1 + \cdots$ By subtracting suitable multiples of this from the other equations we can make sure that x_1 is never mentioned again. The remaining equations now contain only the other variables, and we have the same alternative again. Are all the coefficients 0? Then our job is finished. If not, there is some variable that actually occurs in some equation; we call it x_2, get the equation into the form $x_2 + \cdots$ and so proceed.

Solutions when all right-hand sides are 0

The situation is particularly simple when only 0 occurs after the equals signs. For example, suppose we have some equations in 5 unknowns, and that this system, when reduced to standard form gives the 3 equations

$$
\begin{aligned}
x_1 \qquad\qquad - ax_4 - dx_5 &= 0, \\
x_2 \qquad - bx_4 - ex_5 &= 0, \\
x_3 - cx_4 - fx_5 &= 0.
\end{aligned}
$$

As we saw earlier, we can allow someone to prescribe arbitrary values for x_4 and x_5, and still have a solution of the equations. In fact, if he chooses $x_4 = s$, $x_5 = t$ we take $x_1 = as + dt$, $x_2 = bs + et$, $x_3 = cs + ft$, as the equations force us to do. Now every solution is covered by this procedure. Suppose, for instance, we are told that g_1, g_2, g_3, g_4, g_5 is a solution of the system of equations. We will take $s = g_4$, $t = g_5$. This gives the correct values for x_4 and x_5. But once x_4 and x_5 have been chosen, the equations fix x_1, x_2 and x_3. So, if g_1, g_2, g_3, g_4, g_5 really is a solution, our formula is bound to give the values g_1, g_2, g_3 for x_1, x_2, x_3.

We may write our solution in the form

$$
\begin{pmatrix} x_1 \\ x_2 \\ x_3 \\ x_4 \\ x_5 \end{pmatrix} = s \begin{pmatrix} a \\ b \\ c \\ 1 \\ 0 \end{pmatrix} + t \begin{pmatrix} d \\ e \\ f \\ 0 \\ 1 \end{pmatrix}.
$$

If we call the three vectors occurring here v, u_1, u_2, the above equation may be written $v = su_1 + tu_2$. So a solution, v, results if we start at the origin, and go an arbitrary distance in the direction of the vector u_1, and then an arbitrary distance in the direction of u_2. That is to say, the solutions fill a plane, a subspace of 2 dimensions.

We can see where this number, 2, comes from. We have 5 unknowns. There are 3 equations, and in these x_1, x_2 and x_3 occur, each only once. We are left with $5 - 3 = 2$ unknowns, x_4 and x_5, that can be chosen at will.

This argument can be generalized. If we have n unknowns, and the standard form contains k equations, then $x_1, x_2, \ldots, x_k$ will occur each once only, and we shall be able to choose $x_{k+1}, \ldots, x_n$ at will; once chosen, these fix the values of $x_1, \ldots, x_k$. You should satisfy yourself that, if we choose numbers $t_1, t_2, \ldots, t_{n-k}$ for $x_{k+1}, x_{k+2}, \ldots, x_n$, then $x_1, \ldots, x_k$ are given by expressions linear in $t_1, \ldots, t_{n-k}$ and the general solution may be written in vector form as $v = t_1 u_1 + t_2 u_2 + \cdots + t_{n-k} u_{n-k}$, where $u_1, u_2, \ldots, u_{n-k}$ are *fixed* vectors and $t_1, t_2, \ldots, t_{n-k}$ are arbitrary parameters.

Are we now in a position to assert the theorem: *if there are k equations in the standard form for a system of equations in n variables* (with 0 only on right-hand sides), *then the solutions fill a subspace of n − k dimensions?* We are very close to having proved this, but there is one point still to settle. In the equation $v = t_1 u_1 + \cdots + t_{n-k} u_{n-k}$ we have something that looks very much like the parametric specification of a subspace of $n - k$ dimensions. But what would happen if the vectors $u_1, \ldots, u_{n-k}$ were not independent? As an extreme case, if these vectors were all 0, v would be confined to a single point, the origin. If $u_1, \ldots, u_{n-k}$ were all multiples of the single vector u_1 we would have a subspace of 1 dimension only. If any one of them were a mixture of the others, the dimension would fall below $n - k$. However, none of these things can happen. If you look back at the particular example we had at the beginning of this discussion, you will see that u_1 had the components $a, b, c, 1, 0$ and u_2 had $d, e, f, 0, 1$. So we can certainly dismiss the possibility of u_1 and u_2 both being 0. Admittedly a, b, c, d, e, f might all be 0 (this is a perfectly possible case) but then we would be left with 0, 0, 0, 1, 0 and 0, 0, 0, 0, 1. There is no way of making either vector 0; we cannot get rid of these components with value 1. Nor is there any way of choosing a, b, c, d, e, f to make these vectors linearly dependent. If we try to make $u_2 = h u_1$, for some number h, the equations corresponding to the first three equations do not present any obstacle; however, when we come to the fourth and fifth components we get the equations $0 = h \cdot 1$, $1 = h \cdot 0$, which it is impossible to satisfy; things turn out much the same if we try to make $u_1 = h u_2$.

In the same way, if $n - k = 3$, we find that the last three components of u_1 are $\cdots 1, 0, 0$; those of u_2 are $\cdots 0, 1, 0$; and those of u_3 are $\cdots 0, 0, 1$. By considering these components alone we can see that it is impossible for any one of u_1, u_2, u_3 to be a mixture of the other two.

The same argument applies in every case and so the theorem suggested above is in fact true.

Notice that the number k in the expression $n - k$ is the number of equations *in the standard form*, not the number of equations as originally given.

Non-zero solutions

Equations of the type we are considering always have a solution in which every variable is 0. We often want to know (for instance, if the equations arise in a search for eigenvectors) whether any non-zero vector v provides a solution. Here again, we have a very definite, simple answer: *if k < n, a non-zero solution certainly exists.* This looks very reasonable, for when $k < n$, the number $n - k$ must be either 1 or larger than 1; accordingly the solutions fill a line, a plane, or subspace of higher dimension, so must include vectors other than 0. We can put the matter beyond doubt by appealing to the algebra.

The general solution is $v = t_1 u_1 + \cdots + t_{n-k} u_{n-k}$. If we put $t_1 = 1$ and the other parameters 0, we get $v = u_1$ and, as we saw, u_1 is never 0. Indeed we can see this even more simply by looking at the standard form of the equations. When we had in the standard form 3 equations for 5 variables, we saw that x_4 and x_5 could be chosen at will and a solution obtained by solving for x_1, x_2 and x_3. So we can assign x_4 and x_5 any non-zero values we like, say $x_4 = 2$ and $x_5 = 3$, and obtain a solution. The last two components being 2 and 3, we clearly are not getting 0 as our vector.

Now suppose we are given m equations in n variables, but not in standard form. Without making calculations we cannot be sure exactly what value k has, but there is one thing of which we can be certain; k is not going to be larger than m. Our process of reducing to standard form may decrease the number of equations, but it never increases them. So, if at the start $m < n$, we can be sure $k < n$, and so, by our theorem above, there must be a non-zero solution. Accordingly we have – *if the number of equations is less than the number of variables, a non-zero solution is bound to exist.*

It is important to remember that, at the moment, we are considering only the case in which all equations have 0 on the right-hand side. We can use matrix notation to save repeating this phrase. Our equations can be put in the concise form $Mv = 0$. If we have m equations in n variables, M will be an $m \times n$ matrix (i.e. m rows, n columns). Obviously, we may often need to refer to k, the number of equations in the standard form; k is called the *rank* of the matrix M. Our theorems may then be stated thus:

Theorem 1 The solutions of $Mv = 0$ fill a subspace of dimension $n - k$. (If $n = k$, this number is 0. By a subspace of dimension 0 we understand a single point, the origin.)

Theorem 2 If $k < n$, a non-zero solution exists. (In fact, an infinity of non-zero solutions will exist.)

Theorem 3 If $m < n$, a non-zero solution exists.

In Theorem 1 we met the number $n - k$, and this may make us wonder; what happens if $k > n$? What do we mean by a subspace of negative dimension? This question need not occupy us, for $k > n$ can never happen. Suppose we apply the procedure for getting the standard form and at some stage of it we have written as many as n equations. These equations must be simply $x_1 = 0, x_2 = 0, \ldots, x_n = 0$. This must represent the end of the process. There may still be equations to deal with, but they cannot produce anything new. For suppose such an equation is

$$a_1x_1 + a_2x_2 + \cdots + a_nx_n = 0.$$

Subtracting from this equation a_1 times our first equation ($x_1 = 0$), a_2 times our second equation ($x_2 = 0$), and so on, up to a_n times our last equation ($x_n = 0$), nothing at all remains.

Note. The results above allow us to prove formally some statements that were presented as 'reasonable' in §7.

The first statement is that you cannot have $n + 1$ linearly independent vectors in space of n dimensions. A set of vectors is linearly dependent if one of the vectors can be expressed as a mixture of the others. Another way of expressing this, which is often useful, is to say that vectors $u_1, u_2, \ldots, u_p$ are linearly dependent if there is an equation $c_1u_1 + c_2u_2 + \cdots + c_pu_p = 0$ where the numbers $c_1, c_2, \ldots, c_p$ are *not all* 0. For suppose one of these numbers, say c_q, is not 0. Then we can solve for u_q by dividing by c_q, and get u_q expressed as a mixture of the other vectors.

Now we come to our statement. Suppose we are given any $n + 1$ vectors in n dimensions, $u_1, u_2, \ldots, u_{n+1}$. Each vector is a column vector with n components. Can we find numbers $c_1, c_2, \ldots, c_{n+1}$, not all zero, for which $c_1u_1 + c_2u_2 + \cdots + c_{n+1}u_{n+1} = 0$? If we write this equation out in full we shall have n equations, since each vector has n components. But the unknowns, $c_1, c_2, \ldots, c_{n+1}$ are $n + 1$ in number. There are more unknowns than equations; this, as we saw earlier, means that a non-zero solution exists. So the $n + 1$ vectors are bound to be linearly dependent.

It would be very strange if this statement were not true. For $n + 1$ independent vectors form a basis for a space of $n + 1$ dimensions; if we could find such a basis in space of n dimensions it would mean that a space of $n + 1$ dimensions could be found within a space of n dimensions, which, as mentioned in §7, we certainly do not expect to be possible.

The second statement we can now prove is that any n independent vectors in space of n dimensions form a basis for that space, that is to say, every vector in the space can be expressed as a mixture of them, and this in only one way. Suppose then we have n independent vectors, $u_1, u_2, \ldots, u_n$, in n dimensions, and let v be any vector in that space. We want to prove that v must be a mixture of $u_1, u_2, \ldots, u_n$. The proof is not long. First of all, $u_1, u_2, \ldots, u_n, v$ are $n + 1$ vectors in the space of n dimensions. By our first result, these vectors must be linearly dependent, that is, we must have an equation $c_1u_1 + c_2u_2 + \cdots + c_nu_n + cv = 0$, where the coefficients are not all 0. If $c \neq 0$, we can solve for v, and get v as a mixture of $u_1, \ldots, u_n$, and all will be well. So our result will be proved if we show that it is impossible for c to be 0. Suppose it were; then we would have $c_1u_1 + c_2u_2 + \cdots + c_nu_n = 0$, and this would mean that $u_1, \ldots, u_n$ were not linearly independent, in contradiction to the information given.

Questions

1 The equation $c_1u_1 + c_2u_2 + \cdots + c_nu_n = 0$ would not prove $u_1, \ldots, u_n$ linearly dependent, if $c_1 = c_2 = \cdots = c_n = 0$. Why can we rule out this situation?

2 How can we complete the proof by showing that v can only be expressed in one way as a mixture of $u_1, \ldots, u_n$?

If this second statement were not true, it would mean that we could have one space of n dimensions inside another space of n dimensions, and not filling it.

Exercises

1 In each system given below reduce the equations to standard form, determine the general solution, and describe in geometrical terms the subspace filled by the solutions (e.g. a plane in 7 dimensions; a subspace of 3 dimensions in 5-dimensional space.)

(a) $x - y = 0$, $y - z = 0$, $z - x = 0$.

(b) $2x - 3y = 0$, $5x - 3z = 0$, $5y - 2z = 0$.

(c) $2x + 7y - 4z = 0$, $16x - 9y - 6z = 0$, $7x + 2y - 5z = 0$.

(d) $x + y + z = 0$, $x + 2y + 3z = 0$, $x + 4y + 5z = 0$.

(e)
$$x + \ y + \ z + \quad u + \ v = 0,$$
$$2x + \ 3y + \ z + \ 5u \qquad = 0,$$
$$3x + \ 7y - \ z + 15u - 5v = 0,$$
$$8x + 11y + 5z + 17u + 2v = 0.$$

(f)
$$x + \quad 7y - \ \cdot13z = 0,$$
$$13x + \ 91y - 169z = 0,$$
$$17x + 119y - 221z = 0.$$

(g)
$$x_1 + \ 2x_2 + \ 2x_3 + \quad x_4 = 0,$$
$$2x_1 + \ 7x_2 + 11x_3 + \ 3x_4 = 0,$$
$$x_1 + \ 3x_2 + \ 6x_3 + \ 8x_4 = 0,$$
$$4x_1 + 11x_2 + 19x_3 + 21x_4 = 0.$$

(h)
$$2x_1 + \ 3x_2 + 4x_3 + \ 8x_4 + \ 6x_5 + \ 7x_6 = 0,$$
$$4x_1 + \ 6x_2 + 3x_3 + 11x_4 + \ 7x_5 + \ 9x_6 = 0,$$
$$8x_1 + 12x_2 + 7x_3 + 23x_4 + 15x_5 + 19x_6 = 0,$$
$$2x_1 + 3x_2 + \ x_3 + \ 5x_4 + \ 3x_5 + \ 5x_6 = 0.$$

2 (a) What equations would we have to solve to determine whether the vectors $(1, 1, 1)$, $(2, 3, 2)$, $(4, 9, 4)$ are linearly dependent or not? Are they in fact linearly dependent?

(b) Are $(1, 1, 1)$, $(2, 3, 5)$, $(4, 9, 25)$ linearly dependent or not?

(c) What are the conditions for $(1, 1, 1)$, (a, b, c), (a^2, b^2, c^2) to be linearly dependent? What are the conditions for these vectors to form a basis in 3 dimensions?

(d) What are the conditions for $(1, a, a^2)$, $(1, b, b^2)$, $(1, c, c^2)$ to form a basis in 3 dimensions?

(e) In how many ways can a basis for space of 3 dimensions be selected from the 4 vectors $(1, 2, 4)$, $(1, 3, 9)$, $(1, 4, 16)$, $(1, 5, 25)$?

3 Let $u_1 = (1, 1, 1)$, $u_2 = (2, 3, 5)$, $u_3 = (4, 9, 25)$, $u_4 = (8, 27, 125)$. Any 4 vectors in 3 dimensions must be linearly dependent. Find an equation of the form $c_1u_1 + c_2u_2 + c_3u_3 + c_4u_4 = 0$, where c_1, c_2, c_3 and c_4 are constants, that expresses the linear dependence of u_1, u_2, u_3 and u_4.

4 In questions 2(a) and 2(b) we investigated the possible linear dependence of three vectors. It would also be possible to settle this question by finding the value of the determinant of the three vectors. Compare the work involved in answering this question (a) by the method of the present section, (b) by multiplying the determinant right out, (c) by using the properties of determinants. Is there any resemblance between any of these methods?

Equations with non-zero numbers on right-hand side

The equations $Mv = 0$ considered so far show two possibilities; there may be just the one solution, $v = 0$, or there may be an infinity of solutions. We are now going to consider equations of the form $Mv = h$, where h is some given vector. Here again there may be just one solution, or there may be an infinity of solutions; there is also a third possibility – there may be no solution at all.

It is obvious that this possibility exists. We have such simple examples as the pair of equations $x + y = 1$, $x + y = 2$, where the second statement contradicts the first. Clearly no solution can exist. Sometimes we need to be aware of this possibility only in order to avoid it; for instance in designing a structure we may want it to carry a large weight while itself weighing as little as possible. There is clearly a potential conflict between these aims; if we ask for too much we may be demanding the impossible – that is; producing equations with no solution. But sometimes we have a positive interest in there being no solution. In our example above, $x + y = 1$ and $x + y = 2$ are the equations of parallel lines. If in the course of some work we needed the equation of the line through $(2, 0)$ parallel to $x + y = 1$, we might check our answer, $x + y = 2$, by trying to solve the pair of equations. We now hope there will be no solution. If there were one, it would mean we had made a mistake; the lines would then have an intersection and not be parallel.

Fortunately, there is very little to add to our earlier remarks. If we apply to equations of the form $Mv = h$ the procedure already described for bringing the *left-hand side* to the standard form, this will automatically tell us whether we have 0, 1 or ∞ solutions.

The very first example in this section illustrates the case where there is just one solution. In equations (1) and (2), we had $x + 3y = 25$, $2x + 7y = 55$. Some simple operations brought the left-hand side to the standard form. By reading through the calculations there made, it will be seen that at no stage was reference made to the numbers 25 and 55 on the right of the equals signs. The successive multiplications and subtractions were determined purely by the coefficients of x and y. This is always so. The sequence of calculations made to bring $Mv = h$ to standard form is exactly what would be done to bring $Mv = 0$ to standard form. If, as in this first example, $Mv = h$ is changed by these operations to an equivalent system of equations of the form $Iv = g$, then there is clearly one solution only, namely $v = g$.

We can obtain an illustration of equations with an infinity of solutions by amending equations (27), (28), (29) in worked example 2, considered earlier in this section. We keep the coefficients of x_1, x_2 and x_3 the same, but replace the zeros on the right-hand side. Accordingly, let us consider the equations

$$x_1 + 2x_2 + 3x_3 = 13, \tag{27a}$$

$$3x_1 + 6x_2 + 7x_3 = 37, \tag{28a}$$

$$2x_1 + 4x_2 + 10x_3 = 30. \tag{29a}$$

If we now carry out the same series of operations as those used in Exercise 2, the same simplification of the left-hand sides will result, but certain numbers will automatically appear on the right. It is left to you to check that we eventually get the equations

$$x_1 + 2x_2 \qquad = 10, \tag{33a}$$

$$x_3 = 1. \tag{34a}$$

Here x_1 and x_3 are the 'special' variables, and we may put $x_2 = t$, an arbitrary parameter. We thus find $x_1 = 10 - 2t$, $x_2 = t$, $x_3 = 1$ as the general solution. These solutions fill a line; it is the line that would be described by a point starting at $(10, 0, 1)$ and travelling with velocity $(-2, 1, 0)$. In vector form it could be represented as $v = u_0 + tu_1$.

This line is not a subspace, for it does not go through the origin. It represents rather a displacement of a subspace; the points tu_1 represent a subspace, a line through the origin. If we displace every point of this subspace by the same vector u_0, we obtain the line filled by the solutions.

In other problems we will also find that the solutions fill a region obtained by displacing a subspace.

Equations with no solutions Now consider the effect on equations (27a), (28a), (29a) of making a slight change in one of them. Suppose that in the last equation we replace 30 by 31, and so consider the equations

$$x_1 + 2x_2 + 3x_3 = 13, \tag{27b}$$

$$3x_1 + 6x_2 + 7x_3 = 37, \tag{28b}$$

$$2x_1 + 4x_2 + 10x_3 = 31. \tag{29b}$$

As before, let us carry out the sequence of operations prescribed in Worked example 2. You will find that we then reach the equations

$$x_1 + 2x_2 + 3x_3 = 13, \tag{30b}$$

$$x_3 = 1, \tag{31b}$$

$$4x_3 = 5. \tag{32b}$$

Our next instruction is to get rid of x_3 in the last equation by subtracting 4 times (31b). And here something essentially new happens; we get $0 = 1$, which is impossible. There is no solution.

The standard form gives a system of equations equivalent to the original system. If the original system makes impossible demands, the standard form must do the same. The only way in which the standard form can require an impossibility is by containing one or more equations of the type $0x_1 + 0x_2 + 0x_3 = a$, where the number $a \neq 0$. Thus the fact that there are no solutions will come to light in the course of the usual procedure.

Topic for discussion

The equations (27a), (28a), (29a) have an infinity of solutions. Each equation represents a plane. What figure do these planes form?

By changing the constant in the last equation, we obtain (27b), (28b), (29b), a system with no solution. What figure do these planes form?

Questions

1 Reduce each of the following systems of equations to standard form. Give the general solution or state that no solution exists.

(a) $2x - y - z = 1,$
  $-x + 2y - z = 2,$
  $-x - y + 2z = 3.$

(b) $2x + 5y + 3z = 0,$
  $4x + 11y + z = 6,$
  $10x + 26y + 10z = 6.$

(c) $x + 2y + z + 2u + v = 7,$
  $3x + y + 4z + 3u + 5v = 16,$
  $15x + 10y + 19z + 18u + 23v = 80.$

(d) $x + 2y - 3z = 1,$
  $2x + 3y + z = 8,$
  $5x + 9y - 8z = 11,$
  $4x + 5y + 9z = 22.$

2 Three planes are given by the equations

$$2x + 3y - 5z = 23, \qquad x - 2y + z = 1, \qquad 10x - 6y - 4z = 10.$$

Do these planes have any point in common? Each pair of planes intersects in a line; find a parametric representation for each such line. What can be said about the three lines in question?

3 Make up equations for three planes that are related to each other like the planes in Question 2, but involve different numbers.

4 In some experimental work it is thought probable that a parabola $y = a + bx + cx^2$ can be found which passes through the points $(1, 10), (2, 23), (4, 79), (-1, 14), (-3, 58)$. Can this in fact be done?

5 (a) What solutions (if any) have the equations $3z - 5y = 8, 5x - 2z = 3, 2y - 3x = -5$?
  (b) What solutions (if any) have the equations $3z - 5y = 4, 5x - 2z = 1, 2y - 3x = 3$?
  (c) What conditions must a, b, c satisfy if the equations $3z - 5y = a, 5x - 2z = b, 2y - 3x = c$ are to have a solution or solutions? Is it possible for these equations to have exactly one solution?
  (d) Interpret (c) in terms of vector products and your answer in terms of scalar products. Show that your answer to (c) could have been predicted by purely geometrical reasoning.

6 Consider a rigid piece of material subjected to forces which lie in a plane. The material is in equilibrium (that is, it does not translate or rotate) when the resultant force acting on it is zero. The condition that the resultant force is zero can be represented by three scalar equations. For example the equation $M_A = 0$, stating that the moment of the forces about A is zero, indicates that the resultant force must pass through A. Adding the condition $R_H = 0$, (the horizontal component of the resultant is zero) would only permit a vertical force through A. Finally the equation $R_V = 0$, stating that the vertical component of the resultant is zero, would ensure no resultant at all and thus that the body was in equilibrium under the given force system. Some sets of three equations ensure equilibrium and others do not. The following example will illustrate dependent and independent sets of equations.

By isolating the frame shown in Fig. 148 as a free body the six following equations are readily obtained.

$M_A = 0$	$3X_2 + 8X_3 - 72 = 0,$	(1)
$M_B =$	$4X_1 + 4X_3 - 36 = 0,$	(2)
$M_C =$	$8X_1 - 3X_2 = 0,$	(3)
$M_D = 0$	$8X_1 - 36 = 0,$	(4)
$R_H = 0$	$X_2 - 12 = 0,$	(5)
$R_V = 0$	$X_1 - X_3 = 0.$	(6)

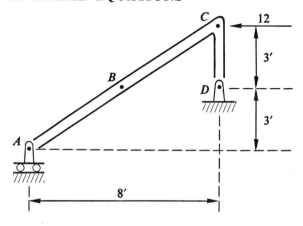

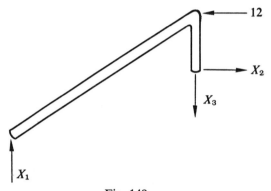

Fig. 148

Determine by the row–echelon method, and also by physical considerations, which of the following sets of equations are independent.

(*a*) equations (1), (2) and (4).   (*b*) equations (1), (2) and (3)
(*c*) equations (1), (5) and (6)   (*d*) equations (3), (4) and (5)
(*e*) equations (1), (3) and (5)   (*f*) equations (2), (4) and (5)
(*g*) equations (2), (4) and (6)

41 Null space, column space, row space of a matrix

In §40 we were concerned, almost exclusively, with certain routines for calculation. In this section we try to show pictorially certain aspects of the topics considered in §40. This also provides an opportunity for explaining the meanings of certain technical terms that may be met in the literature on matrices.

In §40, Exercise 2, with minor variations, was discussed on three separate occasions. We will therefore take the matrix occurring in this exercise as our theme. It may appear a somewhat special case, since it is a 3×3 matrix, and in general we are concerned with $m \times n$ matrices. A 3×3 matrix has the advantage that it is easily visualized in terms of physical space. Most of the features we shall observe can be generalized to the case of an $m \times n$ matrix.

Accordingly we consider the matrix M and vector v, where, as in Exercise 2,

$$M = \begin{pmatrix} 1 & 2 & 3 \\ 3 & 6 & 7 \\ 2 & 4 & 10 \end{pmatrix}, \qquad v = \begin{pmatrix} x_1 \\ x_2 \\ x_3 \end{pmatrix}.$$

We saw in Exercise 2, in its original form, that the equations $Mv = 0$ were equivalent to the pair of equations $x_1 + 2x_2 = 0$, $x_3 = 0$. Thus $x_1 = -2$, $x_2 = 1$, $x_3 = 0$ is a solution, and the subspace filled by the solutions consists of the line through the origin and the point $(-2, 1, 0)$. Any point of this line is mapped to the origin under the mapping M.

Now when a line is sent to the origin in this way, the mapping always gives a space of lower dimension. This may be proved as follows. Let u_1 be a vector for which $Mu_1 = 0$. Choose any two vectors u_2, u_3 so that u_1, u_2, u_3 is a basis for the space. Then every point can be represented as $a_1u_1 + a_2u_2 + a_3u_3$ for some numbers, a_1, a_2, a_3. M sends this point to $a_1u_1^* + a_2u_2^* + a_3u_3^*$. However, $u_1^* = 0$, so every point in the output is of the form $a_2u_2^* + a_3u_3^*$; that is, it lies in the plane containing the origin, u_2^* and u_3^*. (This proof is for 3 dimensions, but it generalizes, in an obvious way, to n dimensions.)

This consideration applies to the particular mapping M, that we are using as an example. Let $Mv = w$, where w has components y_1, y_2, y_3. The equation $Mv = w$, written in full, is the system of equations

$$x_1 + 2x_2 + 3x_3 = y_1, \tag{27c}$$

$$3x_1 + 6x_2 + 7x_3 = y_2, \tag{28c}$$

$$2x_1 + 4x_2 + 10x_3 = y_3. \tag{29c}$$

Note that these equations cover the three situations considered in §40 in relation to Worked example 2. In each situation the output w was specified, and we tried to find what inputs would give it. In the original example, we required an output with $y_1 = 0$, $y_2 = 0$, $y_3 = 0$. Then we considered the output $y_1 = 13$, $y_2 = 37$, $y_3 = 30$, and finally (the system with no solution) $y_1 = 13$, $y_2 = 37$, $y_3 = 31$.

We have seen from the argument above that the output, Mv, must lie in a certain plane, whatever the input v. This suggests the reason why we cannot find any solution in the third problem; it seems that $y_1 = 13$, $y_2 = 37$, $y_3 = 30$ specifies a point not lying in the plane containing all possible outputs; that is why we cannot find any input v to produce it.

The procedure for finding the standard form of the equations gives us a way of finding the actual equation of the plane in question. If we apply the steps given in Worked example 2 of §40 to the general equations (27c), (28c), (29c) we find,

$$x_1 + 2x_2 + 3x_3 = y_1, \tag{30c}$$

$$x_3 = -(y_2 - 3y_1)/2, \tag{31c}$$

$$4x_3 = y_3 - 2y_1. \tag{32c}$$

Now, forming $(32c) - 4(31c)$ as instructed, we get

$$0 = (y_3 - 2y_1) + 2(y_2 - 3y_1) = y_3 + 2y_2 - 8y_1.$$

So every possible output satisfies the equation $y_3 + 2y_2 - 8y_1 = 0$. We can check this by going back to our original equations, (27c), (28c) and (29c). If we calculate $y_3 + 2y_2 - 8y_1$ we get $0x_1 + 0x_2 + 0x_3$, so whatever the input x_1, x_2, x_3 may be, the

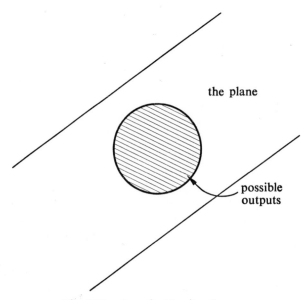

the plane

possible outputs

Fig. 149   Conceivable situation.

output must satisfy $y_3 + 2y_2 - 8y_1 = 0$. It can be checked that $y_1 = 13$, $y_2 = 37$, $y_3 = 30$ does satisfy this equation, and so seems to be a possible output, while $y_1 = 13$, $y_2 = 37$, $y_3 = 31$ does not satisfy it and so is certainly not a possible output.

The rather cautious phrase 'seems to be a possible output' was used because a question of logic is involved. We have proved that every output lies in the plane

$$y_3 + 2y_2 - 8y_1 = 0,$$

but by itself this does not prove that every point of that plane is a possible output. For instance, it is imaginable that the possible outputs could all lie within a certain circle in that plane, as shown in Fig. 149. Then all the outputs would lie in the plane, but it would be quite false to assume that every point in the plane was automatically a possible output, for such a point might happen to lie outside the shaded circular region. *Now in fact this is not how things are*; actually, every point of the plane is an output for some input, but we must prove that this is so.

If we take equations (30c), (31c), (32c) and carry out the remaining instructions given in Exercise 2, we find the standard form to be the following:

$$x_1 + 2x_2 \qquad = -\tfrac{7}{2}y_1 + \tfrac{3}{2}y_2, \tag{33c}$$

$$x_3 = \quad \tfrac{3}{2}y_1 - \tfrac{1}{2}y_2, \tag{34c}$$

$$0 = -8y_1 + 2y_2 + \; y_3. \tag{35c}$$

If the output point lies in the plane in question, equation (35c) is satisfied. In equations (33c) and (34c), whatever values y_1 and y_2 may have, we can find x_1 and x_3 to make the equations hold, and indeed we can do this for an arbitrary value of x_2. Thus if we put $x_2 = t$, we can find not merely *a* solution but an infinity of solutions depending on the parameter t. This is in agreement with our experience in §40, when we found that the output $y_1 = 13$, $y_2 = 37$, $y_3 = 30$ gave the solution $x_1 = 10 - 2t$, $x_2 = t$, $x_3 = 1$. It was pointed out in §40 that this solution was of the form $v = u_0 + tu_1$.

We can see why, if there is one solution of $Mv = w$ there must be an infinity of solutions. At the beginning of this section we noted that the input $x_1 = -2$, $x_2 = 1$, $x_3 = 0$ gave zero output. Call this input u_1. Then $Mu_1 = 0$. Now suppose $Mv = w$ has a solution u_0; this means that input u_0 gives output w; that is $Mu_0 = w$. Now consider what output comes from the input $u_0 + tu_1$. We have

$$M(u_0 + tu_1) = Mu_0 + t \cdot Mu_1.$$

This equals Mu_0 since $Mu_1 = 0$; also $Mu_0 = w$. So $M(u_0 + tu_1) = w$; the input $u_0 + tu_1$ gives the output w, whatever number may be chosen for t. The points $u_0 + tu_1$ fill a line, parallel to the vector u_1. Thus we may think of the input space as being full of lines, all pointing in the same direction; if two inputs lie on the same line, their outputs will be identical.

In the output space we have a plane; it is impossible to produce an output that does not lie in this plane, but in compensation for this, every output that does lie in the plane can be produced by infinitely many different inputs.

Thus, when we are trying to visualize the effect of this particular mapping M, two geometrical objects come naturally to our attention. In the input space there is a line through the origin, consisting of all the points tu_1. Each of these points maps to the origin in the output space; if $v = tu_1$, then $Mv = 0$. In the output space there is a plane; whatever the input may be, the output lies in this plane.

For *any* matrix M corresponding objects exist, and have special names. All the vectors v for which $Mv = 0$ constitute the *Null Space of M*. By Theorem 1 of §40, if k is the rank of the matrix M (that is, the number of equations in the standard form of the equation system $Mv = 0$) then the null space of M is a subspace of dimension $n - k$, the matrix M being of type $m \times n$. The null space of M lies in the input space, which is of n dimensions. If $k = 0$, which happens only for $M = 0$, the null space is the entire input space. If $k = n$, the null space contains only the origin, O.

In the output space, corresponding to the plane $y_3 + 2y_2 - 8y_1 = 0$ in our particular example, there is an object formed by all possible outputs, that is, all the points Mv that can be obtained from some v in the input space. This is known as the *Column Space of M*; we can at the same time show why this name is used and demonstrate that it is in fact a subspace of the output space. The essential ideas can be seen by considering our particular example, M. This matrix expresses in the form $Mv = w$, the equations $(27c)$, $(28c)$, $(29c)$ considered near the beginning of this section. Now these equations can be written like this:

$$x_1 \begin{pmatrix} 1 \\ 3 \\ 2 \end{pmatrix} + x_2 \begin{pmatrix} 2 \\ 6 \\ 4 \end{pmatrix} + x_3 \begin{pmatrix} 3 \\ 7 \\ 10 \end{pmatrix} = \begin{pmatrix} y_1 \\ y_2 \\ y_3 \end{pmatrix}.$$

If we write C_1, C_2, C_3 for the 3 columns of the matrix M, this equation can be written $x_1C_1 + x_2C_2 + x_3C_3 = w$. Thus every possible output w, is a mixture of the three vectors given by the columns of M. That is why the name 'column space' is given to the geometrical object consisting of all possible outputs. We see too that this object is a subspace; taking arbitrary mixtures of any collection of vectors is bound to produce a subspace. (The formal proof would verify that, if w is in such a collection, so is tw for any number t, and that, if w_1 and w_2 are in it, so is their sum $w_1 + w_2$. Thus the collection is a vector space, hence a subspace.)

Notice, however, that C_1, C_2, C_3 is *not* a basis for the column space. If it were, the column space would be of 3 dimensions. However, C_1, C_2, C_3 are not independent; in fact $C_2 = 2C_1$, so that a mixture of C_1, C_2, C_3 is simply a mixture of C_1 and C_3; C_1 and C_3 are in fact a basis of the column space.

The column space is said to be *spanned* by C_1, C_2, C_3; this means that every vector in the column space is a mixture of C_1, C_2, C_3. When we say that a space is spanned by a certain collection of vectors we imply that the space contains every mixture of these vectors, and nothing else. As will be clear from the example of C_1, C_2 and C_3, we are *not* implying that the collection of vectors is a basis of the space.

Returning to the question of the general $m \times n$ matrix M, we know that it acts on an

n-dimensional space and maps a subspace, the null space, of $n - k$ dimensions, to zero. Seeing M wipes out $n - k$ dimensions, we rather expect its outputs to form a space of k dimensions, since $n - (n - k) = k$. We can prove that this does indeed happen. The proof depends on selecting a convenient basis for the input space. We need n independent vectors, $v_1, \ldots, v_n$, for a basis. We know the null space has $n - k$ dimensions, so we begin by selecting $n - k$ vectors that form a basis for the null space. We call these $v_{k+1}, \ldots, v_n$. We then choose k more vectors, not lying in the null space, and call these $v_1, \ldots, v_k$; these are to be chosen in such a way that the whole collection, $v_1, \ldots, v_n$, forms a basis of the whole space. (In a fully formal treatment we would have to prove that such a selection is always possible.) Any point of the input space can now be represented as $x_1 v_1 + \cdots + x_n v_n$; M will send this point to

$$x_1 v_1^* + \cdots + x_n v_n^*.$$

Now $v_{k+1}, \ldots, v_n$ all lie in the null space, so $v_{k+1}^*, \ldots, v_n^*$ are all 0. Thus the output point is simply $x_1 v_1^* + \cdots + x_k v_k^*$. We seem to have proved what we want – that the outputs fill a space of k dimensions. There is still, however, a possible objection to meet; if $v_1^*, \ldots, v_k^*$ are linearly dependent, they will span the output space, but some of them will be superfluous, the space will be of less than k dimensions. So we have to investigate whether $v_1^*, \ldots, v_k^*$ could be linearly dependent. Suppose they were. Then these would be numbers, not all 0, such that $c_1 v_1^* + \cdots + c_k v_k^* = 0$. Now the left-hand side of this equation is the output corresponding to $c_1 v_1 + \cdots + c_k v_k$, and according to the equation this output is 0. That means the input must lie in the null space, and so be a mixture of $v_{k+1}, \ldots, v_n$. But this means that $v_1, \ldots, v_n$ cannot be a basis, for the point in question could then be represented in two quite distinct ways, one as

$$c_1 v_1 + \cdots + c_k v_k$$

and one as a mixture of $v_{k+1}, \ldots, v_n$. So supposing $v_1^*, \ldots, v_k^*$ linearly dependent leads to a contradiction. Accordingly the objection has been overcome, and we know that the column space, the space formed by the outputs, is of k dimensions exactly.

Column space and row space

The number k, the rank of M, arose first of all as the number of equations in the standard form of the system $Mv = 0$. Each equation corresponds to a *row* in M; the standard form is obtained by calculating suitable mixtures of the rows.

We have just found that k is the dimension of the column space, consisting of all possible mixtures of the columns. There is evidently some connection between the rows of a matrix and the columns, since the same number k occurs in relation to both.

The column space was obtained by partitioning the matrix into column vectors, and considering the space spanned by these. In the same way, we can define the row space; we partition the matrix into row vectors, and consider the space spanned by them; this is called the row space.

Now k is in fact the dimension of the row space. An example will show why this is so. In §40, under the heading 'Solutions when all right-hand sides are 0' we considered the equations of the standard form

$$
\begin{aligned}
x_1 \qquad\qquad - ax_4 - dx_5 &= 0, \\
x_2 \qquad - bx_4 - ex_5 &= 0, \\
x_3 - cx_4 - fx_5 &= 0.
\end{aligned}
$$

If we form the matrix of the coefficients and partition it into rows, we get the 3 row vectors r_1, r_2, r_3 where

$$
\begin{aligned}
r_1 &= (1, 0, 0, -a, -d), \\
r_2 &= (0, 1, 0, -b, -e), \\
r_3 &= (0, 0, 1, -c, -f).
\end{aligned}
$$

Now the standard form is obtained from the original system of equations, $Mv = 0$, by means of mixtures of those equations, and each equation corresponds to a row in M; that is to say, r_1, r_2, r_3 are mixtures of the rows of the original matrix M. This means that r_1, r_2, r_3 lie in the row space of M. Now r_1, r_2, r_3 are clearly independent. For any mixture of r_2 and r_3 must have 0 in the first place; as r_1 begins with 1, it is impossible that r_1 should be a mixture of r_2 and r_3. Similarly, by looking at the numbers in the second place we see that r_2 cannot be obtained by mixing r_1 and r_3, while from the third place we conclude that r_3 is not a mixture of r_1 and r_2. So r_1, r_2, r_3 are independent vectors. Thus the row space of the original matrix M contains three independent vectors, so it must be of 3 dimensions or more.

But we were careful to make the process of forming the standard equations reversible. We can get back to the rows of M by taking suitable mixtures of r_1, r_2, r_3. Any mixture of the rows of M must therefore be a mixture of r_1, r_2, r_3. This means that every vector in the row space is a mixture of the 3 vectors r_1, r_2, r_3: the dimension of the row space is therefore 3 or less.

The two arguments together show that the dimension of the row space is exactly 3.

Nothing in the arguments used depends on any special property of the numbers 3 and 5. If we have any $m \times n$ matrix M, with k equations in the standard system equivalent to $Mv = 0$, we can reason in exactly the same way, and conclude that the row space will be of dimension k.

Thus we have a theorem for any matrix; the column space and the row space have the same number of dimensions.

In some books the term 'column rank' is introduced for the number of dimensions of the column space, and 'row rank' for the number of dimensions of the row space. It is then proved that these two numbers are equal, and their common value is called simply 'the rank of the matrix'.

We have already seen how to find the dimension of the row space. It is simply k, the number of equations in the standard form obtained from the equation system $Mv = 0$. When we are speaking of the 'row space' we tend to think of the rows as containing the components of vectors, rather than as giving the coefficients in equations. This therefore seems an appropriate place to mention that the row–echelon procedure, used to bring an equation system to its standard form, can equally well be used to determine the dimension of the space spanned by a set of vectors and to give a basis of that space.

Suppose, for example, we wish to study the space spanned by the three row vectors

$$r_1 = (1, 1, 1, 1, 1),$$
$$r_2 = (1, 2, 3, 4, 5),$$
$$r_3 = (3, 5, 7, 9, 11).$$

We write

$$r_1 = (1, 1, 1, 1, 1),$$
$$r_2 - r_1 = (0, 1, 2, 3, 4) = u_1, \quad \text{say,}$$
$$r_3 - 3r_1 = (0, 2, 4, 6, 8) = u_2, \quad \text{say.}$$

We have here treated the first components of the vectors in the same way that we treated the coefficients of x_1 in the equation systems. We would next replace u_2 by $u_2 - 2u_1$ in order to get a vector beginning $0, 0, \ldots$ In this example, we get a vector that is zero throughout. Thus we reach a standard form in which r_1 and u_1 alone appear. These two vectors form a basis for the space spanned by r_1, r_2 and r_3, so the space must be of 2 dimensions.

It would be possible to give as a basis the vectors $r_1 - u_1$ and u_1, that is to say

$$(1, 0, -1, -2, -3) = v_1, \quad \text{say,}$$
$$(0, 1, 2, 3, 4) = v_2, \quad \text{say.}$$

This would correspond most closely to what we did with the equations. It has the advantage that we can tell immediately what mixture of v_1 and v_2 gives any vector in the space simply by looking at the first two components. For instance r_3 begins with 3, 5; it can be no other than $3v_1 + 5v_2$. However, enough useful information can often be obtained without going to this stage.

A similar procedure is of course possible with column vectors.

Worked example

Find a basis for the column space of the matrix whose rows are r_1, r_2, r_3 as given just above.

Solution   We have the vectors

$$\begin{pmatrix} 1 \\ 1 \\ 3 \end{pmatrix}, \begin{pmatrix} 1 \\ 2 \\ 5 \end{pmatrix}, \begin{pmatrix} 1 \\ 3 \\ 7 \end{pmatrix}, \begin{pmatrix} 1 \\ 4 \\ 9 \end{pmatrix}, \begin{pmatrix} 1 \\ 5 \\ 11 \end{pmatrix}.$$

Subtract the first vector from each of the others. This gives

$$\begin{pmatrix} 1 \\ 1 \\ 3 \end{pmatrix}, \begin{pmatrix} 0 \\ 1 \\ 2 \end{pmatrix}, \begin{pmatrix} 0 \\ 2 \\ 4 \end{pmatrix}, \begin{pmatrix} 0 \\ 3 \\ 6 \end{pmatrix}, \begin{pmatrix} 0 \\ 4 \\ 8 \end{pmatrix}.$$

It is now noticeable that the last 3 vectors are all multiples of the second. Thus the first 2 vectors provide a basis for the column space.

Questions

1 Let

$$M = \begin{pmatrix} 1 & 1 & 3 & 3 & 5 \\ 1 & 0 & 2 & 1 & 3 \\ 0 & 1 & 1 & 2 & 2 \end{pmatrix}.$$

Of how many dimensions is the row space of M? What number gives the rank of M? Of how many dimensions do you expect the column space of M to be? Find a basis for the column space. Find the condition or conditions for

$$\begin{pmatrix} x \\ y \\ z \end{pmatrix}$$

to be in the column space.

Let

$$u_1 = \begin{pmatrix} 1 \\ 1 \\ 1 \end{pmatrix}, \quad u_2 = \begin{pmatrix} 4 \\ 3 \\ 2 \end{pmatrix}, \quad u_3 = \begin{pmatrix} 7 \\ 5 \\ 2 \end{pmatrix}, \quad u_4 = \begin{pmatrix} 0 \\ 1 \\ -1 \end{pmatrix}.$$

For which of these does the equation $Mv = u_s$ have a solution v? ($s = 1, 2, 3$ or 4.)

Of how many dimensions would you expect the null space of M to be? (Give your reason.) Find the equations that specify the null space, and find a basis for the null space.

Discuss the solution set of the equation $Mv = u_s$ for each of the vectors u_s given above.

2 For

$$M = \begin{pmatrix} 1 & 1 & 2 \\ 2 & 3 & 3 \\ 1 & 2 & 1 \\ 1 & 0 & 3 \end{pmatrix}$$

find the dimension and a basis for (*a*) the row space, (*b*) the column space, (*c*) the null space.

Find the condition or conditions a vector u must satisfy if $Mv = u$ is to have a solution or solutions, v.

Show that $Mv = u$ has solutions if

$$u = \begin{pmatrix} 3 \\ 8 \\ 5 \\ 1 \end{pmatrix}.$$

Find the general solution for v in this case.

If v_0 is a particular column vector in 3 dimensions and $u_0 = Mv_0$, what is the general solution of $Mv = u_0$?

3 Find a basis for (a) the row space and (b) the null space of the matrix

$$M = \begin{pmatrix} 1 & 1 & 1 & 0 & 0 \\ 1 & 0 & 1 & 1 & 1 \\ 2 & 1 & 2 & 1 & 1 \end{pmatrix}.$$

For each row vector, R, included in your answer to (a) and for each column vector, N, in your answer to (b) discuss the product RN. What is its value? Is it even meaningful? Is any general conclusion suggested?

42 Illustrating the Importance of orthogonal matrices

A trap to avoid

In §38, we saw that, for a symmetric matrix M, the transformation $v \rightarrow Mv$ could be pictured with the help of the curves $v'Mv = $ constant. Here the matrix M plays a double role; it specifies a transformation, and it specifies a system of curves. This double role, if not properly understood, can lead us into errors. The present section has the aim of showing how such errors arise and how to avoid them.

As an illustration we will use the matrix $M = \begin{pmatrix} 1 & -1 \\ -1 & 2 \end{pmatrix}$; the transformation associated with this has $x^* = x - y, y^* = -x + 2y$; the curves associated with it are $x^2 - 2xy + 2y^2 = $ constant. In our illustrations we shall use the procedure of §38; that is, the vector Mv will be shown sprouting out of the end of the vector v. In Fig. 150(a) we see a number of points lying on the ellipse $x^2 - 2xy + 2y^2 = 25$. Attached to each point, v, we see the vector Mv, and in each case, as we expect from §38, the vector Mv is normal to the ellipse. The table here shows (x, y), the co-ordinates of v, and (x^*, y^*), the components of Mv.

(x, y)	$(5, \ 0)$	$(7, \ 3)$	$(7, 4)$	$(5, 5)$	$(1, \ 4)$	$(-1, 3)$
(x^*, y^*)	$(5, -5)$	$(4, -1)$	$(3, 1)$	$(0, 5)$	$(-3, 7)$	$(-4, 7)$

Some person, other than ourselves, is also interested in this transformation, but for his purposes the important quantities are not x, y but X, Y where $X = x - y$ and $Y = y$. Accordingly he takes the table above and uses it to construct a table, showing what happens to his co-ordinates X, Y. Simply by substitution he arrives at the following specification of the transformation $(X, Y) \rightarrow (X^*, Y^*)$.

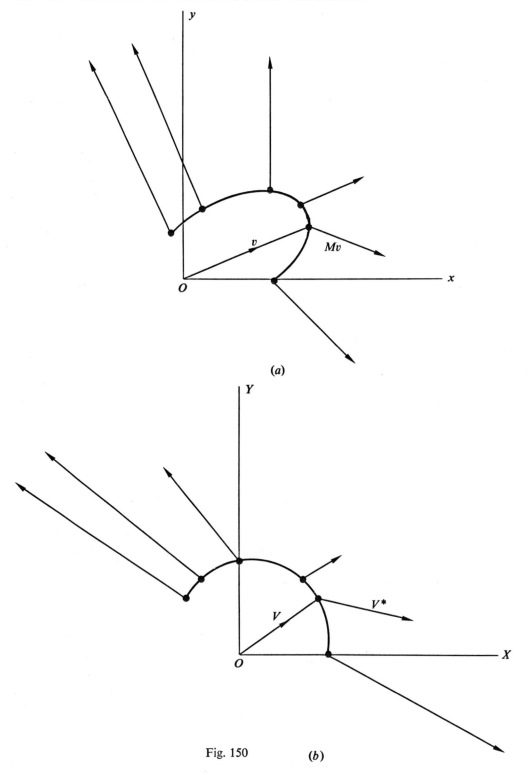

(a)

Fig. 150   (b)

(X, Y)	$(5, 0)$	$(4, 3)$	$(3, 4)$	$(0, 5)$	$(-3, 4)$	$(-4, 3)$
(X^*, Y^*)	$(10, -5)$	$(5, -1)$	$(2, 1)$	$(-5, 5)$	$(-10, 7)$	$(-11, 7)$

It may be checked that this table fits the formula $X^* = 2X - Y$, $Y^* = -X + Y$. These equations specify the transformation as it appears in his co-ordinates: these equations can of course be found directly by algebra: how to do this will be considered later in this section. For the moment it is not necessary to enquire how these equations were obtained, but only to observe that they do fit the data.

Fig. 150(b) shows the six points as they might be drawn by the person using (X, Y) co-ordinates. Attached to each vector V, with co-ordinates (X, Y), is a vector V^* with components (X^*, Y^*). It will be seen that in diagram (b) the six points lie on a circle. And this may be checked by calculation. The equations $X = x - y$, $Y = y$ imply $x = X + Y$, $y = Y$. If we substitute these values in the equation $x^2 - 2xy + 2y^2 = 25$ we obtain $(X + Y)^2 - 2(X + Y)Y + 2Y^2 = 25$ which simplifies to $X^2 + Y^2 = 25$.

Now the quadratric form $X^2 + Y^2$ corresponds to the matrix I, while the transformation, expressed in terms of X and Y, corresponds to the matrix $\begin{pmatrix} 2 & -1 \\ -1 & 1 \end{pmatrix}$.

In the original (x, y) system the transformation $v \to Mv$ and the curves $v'Mv = $ constant were specified by the same matrix M. When we go over to the (X, Y) system, we find the curves are specified by the identity matrix I while the transformation is specified by a matrix which most certainly is not the identity.

One could easily imagine someone believing that, in any system of co-ordinates, the same matrix would specify both the transformation and the curves. Such a person might calculate, quite correctly, the matrix I for the curves, and expect to find that the transformation $V \to V^*$ was also specified by I. In this particular instance, the mistake is so glaring that it would immediately be recognized, for clearly the transformation is not the identity transformation. But with some other change of co-ordinate system, the curves might be specified by some matrix that had no obvious interpretation. In such a situation the error might well escape detection, and the person might continue his calculations in the belief that the matrix he had calculated for the curves would also specify the transformation in the (X, Y) system.

The questions thus arise; if in the (x, y) system, we have the transformation $v \to Mv$ and the curves $v'Mv = $ constant, what matrices will specify the transformation and the curves in some other co-ordinate system? Why are we liable to get two matrices, one for the transformation and the other for the curves? Does this always happen, or are there special co-ordinate systems for which the two matrices remain equal?

We already have part of the answer. Exercise 7, at the end of §33, raised the question of what happens to a quadratic form when axes are changed. This question is answered as follows. We will suppose that new co-ordinates X, Y, are related to the old co-ordinates x, y, by equations of the form $x = pX + qY$, $y = rX + sY$. We can express these concisely as $v = TV$ where

$$v = \begin{pmatrix} x \\ y \end{pmatrix}, \qquad V = \begin{pmatrix} X \\ Y \end{pmatrix}, \qquad T = \begin{pmatrix} p & q \\ r & s \end{pmatrix}.$$

Note that T here establishes an alias, not an alibi. The numbers, x and y, that occur in v, are used in one system, to specify some object. The numbers, X and Y, that occur in V are used in another system to specify *the same object*. The matrix T enables us to act as interpreter between the systems; if X and Y occur in the label in the new system, by applying the matrix T we can find the label that will be understood by users of the old system.

To find how the quadratic form $v'Mv$ is represented in the new system, we need only substitute $v = TV$. Then, by the 'reversal rule for transposes', $v' = V'T'$. Accordingly, $v'Mv = V'T'MTV$. The matrix at the heart of this is $T'MT$; this is the matrix that will specify the curves when the X, Y co-ordinate system is used.

Exercise

At the beginning of this section we used the example $v'Mv = x^2 - 2xy + 2y^2$ and the equations $x = X + Y, y = Y$ for change of axes. Write the matrices M and T and verify that $T'MT$ does give the identity matrix I.

Change of axes for transformations

The principle involved, when a transformation $v \to Mv$ has to be expressed in terms of new axes, is a very general one that occurs in many branches of mathematics. It can be explained by a very simple illustration without any references to axes, co-ordinates or numerical calculations.

We will suppose we have a number of objects, labelled by Greek letters, and a mapping M involving these objects. Thus our mapping might send object α to object β, object π to object δ, object ν to object ϕ. A specification of M is available; it is a list, $\alpha \to \beta, \pi \to \delta, \nu \to \phi$ and so forth.

Now it becomes necessary for these objects to be handled by someone who uses our everyday letters as labels. He sees a package labelled N; to what object should it be mapped? He is provided with a little dictionary which tells him the Greek equivalent of every letter. He finds that ν is Greek for N. The specification of M tells him $\nu \to \phi$. But which object is ϕ? He uses his little dictionary in reverse, and finds that ϕ is the Greek for F. So, in his language, the mapping sends object N to object F; $N \to F$.

The process involved is as shown.

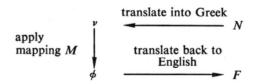

We will now express this process in symbols. Let T indicate translation from English to Greek. Thus $TA = \alpha$, $TB = \beta$, $TP = \pi$, $TD = \delta$, $TN = \nu$ and $TF = \phi$. We need, at one stage of the work to translate in the opposite direction, from Greek to English.

This is the inverse operation, T^{-1}. We have $A = T^{-1}\alpha$, $B = T^{-1}\beta$ and so on, to $F = T^{-1}\phi$.

Thus the steps in our process are as follows; $v = TN$; $Mv = \phi$; $T^{-1}\phi = F$. Notice that only in the middle stage do we actually go from one object to another. In the first step we establish that the object called N in one language is called v in the other; in the last step we are again concerned with two different names, ϕ and F, for the same object.

If we combine the results of our equations, we find

$$F = T^{-1}\phi = T^{-1}Mv = T^{-1}MT \cdot N.$$

Thus the mapping $N \to F$, which expresses in English exactly the same idea as the Greek $v \to \phi$, is given by the symbol $T^{-1}MT$. Thus $T^{-1}MT$ is the form taken in the new system by the mapping that appeared as M in the old.

Now it is clear that this account is a very general one. It does not describe at all the objects on which the mapping M acts. Nor does it depend at all on the grammar of the Greek and English languages. All that it assumes is a 1–1 correspondence between the symbols, there must be just one Greek letter corresponding to the letter N; there must be just one English letter corresponding to ϕ.

Accordingly a symbol such as $T^{-1}MT$ is likely to arise whenever we have a mapping, and two different systems in which the objects involved can be specified.

In our problem, as described earlier, we have a mapping that appears in one system as $v \to v^*$, in the other as $V \to V^*$. We have a matrix T that connects the two systems; $v = TV$, $v^* = TV^*$. And we have a mapping specified by $v^* = Mv$.

Here again we have a scheme similar to the earlier one.

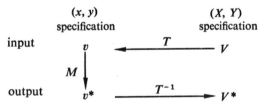

Either from this diagram, or from the equations immediately preceding it, we see that $V^* = T^{-1}MT \cdot V$. The transformation, represented by M in the (x, y) system, appears as $T^{-1}MT$ in the (X, Y) system.

Exercise

With M and T appropriate to the situation at the beginning of the section (under the heading 'A trap to avoid') check that $T^{-1}MT$ does give the matrix relating (X, Y) to (X^*, Y^*).

The role of orthogonal matrices

We now see why it is that we get led to two different matrices when we consider the effect of change of axes on the transformation $v \to Mv$ and the quadratic form $v'Mv$.

In the new axes, the transformation is specified by the matrix $T^{-1}MT$ while the quadratic form is specified by $T'MT$. As a rule, these matrices will be unequal, since usually T^{-1} is different from T'.

However, it can happen that $T' = T^{-1}$. This in fact is the condition for T to be an orthogonal matrix that we had in §36. Accordingly, if we are going to make an *arbitrary* change of axes, we must distinguish between a matrix specifying a transformation and a matrix specifying a quadratic form; however, if we restrict ourselves to using orthogonal matrices only, we can forget this complication.

It is not surprising that orthogonal matrices should appear in this connection. In §38 we used the contours $v'Mv =$ constant to give us a way of visualizing the equation $v^* = Mv$, in which M plays the role of a transformation. We showed, in effect, that if $z = \frac{1}{2}v'Mv$, then $Mv = \operatorname{grad} z$. Now grad z was a vector giving the direction of the steepest slope. This direction is *perpendicular* to the contour through v, the point in question. On a map, one might find the direction of steepest slope by going from a point on one contour *to the nearest point* on the next contour. The words *perpendicular* and *the nearest point* involve ideas of angle and distance; that is, they belong to Euclidean geometry; they cannot be defined in terms of affine geometry, that is, of pure vector theory involving only the definitions of $u + v$ and ku. In affine geometry, we can make any change of axes – that is, we may go from any basis to any other basis. In Euclidean geometry, it is possible to use oblique axes, but this involves reconsidering all the formulas we use. If we are to go ahead without such complications, we have to stick to orthogonal matrices for change of axes; then lengths and angles are found by exactly the same formulas as in the original 'standard' system of axes.

43 Linear programming

Linear programming is a procedure for use in situations where something can be done in very many ways and we wish to find which way is the best, as judged by some agreed standard. Suppose for example a group of students have to provide food for themselves for a certain period of time. Even after ruling out articles of diet which they do not like, a very considerable choice remains. They are very scientifically minded, and want to consume an adequate amount of protein, carbohydrates, vitamins and other ingredients of a satisfactory diet. They wish to do this at the lowest possible cost. There may be certain restrictions on what they can do; some articles need to be stored in a refrigerator, and they have limited refrigerator space. They are busy, and so insist that the total time spent preparing the meals be kept below a certain amount. How do their choices appear in symbolic form? Suppose the local stores offer them 100 articles. They have to choose the amounts of each to buy, say $x_1, x_2, \ldots, x_{100}$. Each one has a certain cost, given by $c_1, c_2, \ldots, c_{100}$. Their total expenditure is

$$c_1 x_1 + c_2 x_2 + \cdots + c_{100} x_{100},$$

and this they want to keep as low as possible. The amount of protein contained in the various articles is specified by the numbers $p_1, p_2, \ldots, p_{100}$. They require

$$p_1 x_1 + p_2 x_2 + \cdots + p_{100} x_{100} \geqslant P,$$

where P is the total amount of protein needed by the student commune. Some articles might contain no protein: for them the protein coefficient would be zero. Similar inequalities would hold for the other nutritional ingredients. Each article would require a certain amount of refrigerator space, so there would be numbers $r_1, r_2, \ldots, r_{100}$ for this aspect. If some article did not require refrigeration, the number r corresponding to it would be 0. It would be necessary to have $r_1 x_1 + r_2 x_2 + \cdots + r_{100} x_{100} \leqslant R$, where R was the total refrigerator space. (If the refrigerator were divided into compartments with specialized purposes, this condition might have to be replaced by several different inequalities.) Each article would involve a certain time of preparation; there would be a condition $t_1 x_1 + \cdots + t_{100} x_{100} \leqslant T$, where T was the total time the students felt it was reasonable to spend in the kitchen.

Such a problem then appears as that of making the linear expression

$$c_1 x_1 + \cdots + c_{100} x_{100}$$

as small as possible, subject to a number of linear inequalities.

Similar problems arise in industry and in government. In a chemical plant, various raw materials may be available, containing desired substances in various proportions. The desired substances correspond to the proteins, vitamins and so forth of the dietary problem. There may be restrictions of various kinds in relation to plant available, the number of competent workers in different departments, and so forth. There may be a restriction on how much of a particular by-product the market can absorb. Similarly, in any kind of enterprise, the problem may arise of using most effectively the resources available. Very often, of course, some simplification and distortion of the actual situation may be needed to get the problem into the form of linear expressions. The engineer must use his judgement to see whether such simplification is a useful approximation to reality, or whether it will lead to entirely misleading conclusions.

Questions of linear programming may appear as minimum or maximum problems. In the diet question we wished to achieve adequate nutrition for the *least possible cost* – a minimum problem. However, the same problem could be posed in terms of a maximum – how to save as much as possible on food expenditure. Other problems may arise naturally as maximum problems – how to do something and obtain the maximum benefit.

It will be convenient to suppose that every problem we consider has been put in the form of making something a maximum. This will save us continually having to give two explanations, one for a minimum, and one for a maximum. The conversion is always possible. If some quantity f is to be made as small as possible, then $k - f$ is to be as large as possible; here k is any convenient constant.

So, from now on, we suppose our task is to make some quantity, M, as large as possible.

Fig. 151 represents a simple problem in linear programming. We have two numbers, x_1 and x_2, at our disposal, and our aim is to make $M = x_1 + 2x_2$ as large as possible, subject to the restrictions imposed by some situation in which we find ourselves. First of all, x_1 and x_2 represent quantities of some actual substances, so they cannot be negative. They are available in limited quantities; x_1 cannot be larger than 4 and x_2 cannot exceed 3. Further, there is some restriction on the total amount of these substances; we know $x_1 + x_2 \leqslant 6$. The diagram shows these restrictions graphically. The point (x_1, x_2) must lie in the first quadrant since neither x_1 nor x_2 can be negative; $x_1 \leqslant 4$ means we must keep to the left of the line ABF with equation $x_1 = 4$, or be on that line. The restriction $x_2 \leqslant 3$ similarly means that we must not go above the line DCF, while $x_1 + x_2 \leqslant 6$ means that we must keep to the left of CBE, or perhaps be on that line. Thus we are restricted to the region $OABCD$; points on the boundary of this region are permissible.

As our aim is to make $M = x_1 + 2x_2$ as large as possible, the contours $M = $ constant are shown. To avoid confusion in the figure, only a small part of each contour has been shown; the broken lines should be imagined as extended to go right across the diagram. Thus the line $M = 6$ would pass through the point E; E in fact is not a possible situation, since for it $x_1 = 6$, which is too large.

It will be clear that, for points within the permitted region, the largest value of M is taken at C. In such a simple problem, the solution can easily be obtained by this graphical method. But, as a rule, such a direct attack is not possible. Suppose, for instance, that we have to choose, not two, but three numbers. We shall have to make a

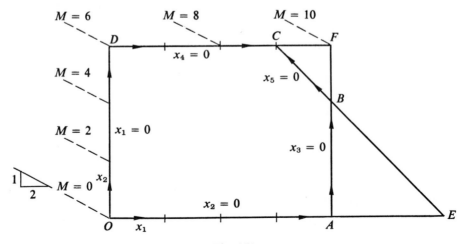

Fig. 151

model in three dimensions. The region will be bounded by plane faces; if we have many inequalities to satisfy, it will be quite hard to tell which intersections of the planes give the edges and corners of the region. If we have more than three numbers to choose, we shall not be able to make a physical model at all. And such a possibility is very real; for instance, in the manufacture of some substance, there might well be ten raw materials available, and we might wish to find the most effective and economical proportions in which to blend them.

Evidently, then, a diagram such as the one above, can serve only as a guide to suggest and illustrate methods, which in practice would involve many variables and many restrictions. This is in line with all our work in linear algebra; we can realize geometrically vector spaces of 2 or 3 dimensions; these suggest *analogies* that may be helpful when we are dealing with problems involving more than three variables. Starting from our diagram, we have to find a procedure that can be carried through purely by calculations with algebraic expressions.

In the diagram, it will be noticed that arrows have been marked on the sides of the region $OABCD$. These arrows denote the direction that leads to an increase in M. For instance, if we go from O to D, we pass in turn the contours $M = 0$, $M = 2$, $M = 4$, $M = 6$. Clearly, M is increasing, so the arrow points from O to D. If we think of the contours as being like the contours on a map, we can say that from O to D is 'uphill'.

Now, if you take a map of a piece of country, choose some point on it, and go uphill from there as far as you can, you will end at the top of some hill. However, it may be a modest mound; by descending from it, and climbing another hill, you might reach a greater height. However, in linear programming, this complication cannot arise. It can be shown that, if you start anywhere on the boundary, and proceed along the edges, always rising, you will arrive at the absolute maximum. In our diagram C is the highest point; from O you may go from O to D to C, or from O to A to B to C, but either way you arrive at C by a path that never descends. It can be proved, for problems in which linear expressions alone occur, that the maximum will always occur on the boundary of the region, and that it can be reached from any point of the boundary by following an uphill path.

In the diagram we can see that A, for example, is not the highest point, because the arrow on AB indicates that we can climb higher by going in the direction AB. On the other hand it is evident that C is the highest point, for both the arrows at C point inwards towards C. If we left C, either for D or B, we would be going against the arrows, that is to say, downhill.

Our first problem then is this; in a complicated problem, where we cannot make an effective diagram, how are we going to tell whether a point gives a maximum, like C, or is capable of improvement, like B? How are we going to translate the arrows into algebra?

This is done by a simple device known as the introduction of *slack* variables. In our example, we have the condition $x_1 \leqslant 4$. If we choose any suitable value of x_1, this falls short of 4 by $4 - x_1$; so to speak, we have $4 - x_1$ in hand, so far as this condition is concerned. We write $x_3 = 4 - x_1$, and replace the inequality $x_1 \leqslant 4$ by the equation

$x_1 + x_3 = 4$, with the understanding that no variable is ever to become negative. We deal with the other two inequalities in the same way, and so the restrictions are expressed by the equations

$$x_1 \qquad + x_3 \qquad\qquad = 4, \tag{1}$$

$$x_2 \qquad + x_4 \qquad = 3, \tag{2}$$

$$x_1 + x_2 \qquad\qquad + x_5 = 6. \tag{3}$$

In the diagram, the sides AB, CD and BC have been labelled $x_3 = 0$, $x_4 = 0$ and $x_5 = 0$ respectively. It is natural that such labels should occur on the boundary, for being on the boundary means that we are pressing hard against some restriction. On AB for instance, we have $x_1 = 4$; the condition $x_1 \leqslant 4$ does not allow us to go any further to the right; we have nothing left in reserve, and so $x_3 = 0$. Similarly on BC, we are making the fullest use of the condition $x_1 + x_2 \leqslant 6$ by taking $x_1 + x_2 = 6$, and so $x_5 = 0$ expresses that we have no slack to take in; this is verified by equation (3).

We now have five symbols x_1, x_2, x_3, x_4, x_5 and three equations, (1), (2), (3), restricting them. Accordingly, we can choose any two as variables, and the values of the others will be fixed automatically by the equations. It will appear in a moment that we can find out what is happening near any corner of the region by choosing a suitable pair of variables.

Our aim is to make $M = x_1 + 2x_2$ as large as possible. This equation tells us how the arrows run near the origin, O. For $x_1 = 0$, $x_2 = 0$ means we are at the origin. The equation $M = x_1 + 2x_2$ shows that if either x_1 or x_2 gets larger, M will increase. In other words, if we move either in the direction OA or OD, we are going uphill. So we can improve the value of M by going either from O to A or from O to D. Suppose then we go to A; how will the arrows run there? A lies on the sides of the region labelled $x_2 = 0$ and $x_3 = 0$, so we want x_2 and x_3 as variables. That means, we want to get rid of x_1 and bring in x_3, in the expression for M. Equation (1) tells us $x_1 = 4 - x_3$, so substituting this in $M = x_1 + 2x_2$ we obtain $M = 4 + 2x_2 - x_3$. Now x_3 has a negative sign; if we let x_3 grow, M will decrease, which we do not want. Our best plan is to keep $x_3 = 0$, that is, to stay on AB. But increasing x_2 will make M larger. Increasing x_2 takes us to the north, along AB. We cannot go beyond B without leaving the region, so we stop at B. Now B lies on the lines $x_3 = 0$, $x_5 = 0$, so we express M in terms of x_3 and x_5. We have already seen $x_1 = 4 - x_3$, and equation (3) gives $x_2 = 6 - x_5 - x_1$, from which we get $x_2 = 6 - x_5 - (4 - x_3) = 2 + x_3 - x_5$. Accordingly,

$$M = x_1 + 2x_2 = 8 + x_3 - 2x_5.$$

The negative coefficient of x_5 shows that it would be a mistake to let x_5 grow, but the term $+x_3$ shows that it will pay to let x_3 grow. Keeping $x_5 = 0$ and letting x_3 grow takes us along BC.

At this point it may be useful to bear in mind that (x_3, x_5) are oblique co-ordinates with origin at B, and to consider what their co-ordinate grid looks like. This is shown in Fig. 152. The lines are obtained by drawing the contours for $x_3 = 4 - x_1$ and

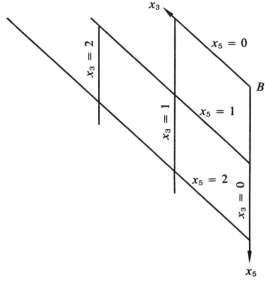

Fig. 152

$x_5 = 6 - x_1 - x_2$. It may be a useful exercise to draw the corresponding grids for (x_4, x_5) at C and (x_1, x_4) at D.

When we come to C, we find it convenient to use x_4, x_5 as variables. We find, with the help of equations (2) and (3), that $M = 9 - x_4 - x_5$. Then $x_4 = 0$, $x_5 = 0$ makes $M = 9$ and it would clearly be a mistake to go away from this situation; any increase in either x_4 or x_5 would drag M downwards. The negative coefficients in x_4 and x_5 thus express the same facts as did the arrows pointing in towards C.

Our origin problem was – what values of x_1 and x_2 make M a maximum? It would therefore be necessary to calculate x_1 and x_2 corresponding to $x_4 = 0$, $x_5 = 0$. Equations (2) and (3) easily give us $x_1 = 3$, $x_2 = 3$.

The equation $M = 9 - x_4 - x_5$ by itself proves that C, where $x_4 = x_5 = 0$, is the best place to be. It would be possible to search for this point by computing the expressions for M corresponding to each of the corners O, A, B, C, D and seeing in which of them only negative coefficients appeared. However, in an actual problem, the number of corners would be extremely large, and this procedure would be very inefficient. It is better to work along the lines already suggested – to choose some corner and work uphill, stage by stage, as we did from O to A to B to C.

In doing this, another problem arises. When, starting at O, we found it would pay to increase x_1, we went along the line OAE until we came to A. As the diagram showed, if we had gone any further we would have left the region of permissible choices. But in terms of algebraic calculation, the situation is not so clear; A is the point with $x_2 = 0$, $x_3 = 0$; E is the point with $x_2 = 0$, $x_5 = 0$; how are we to know that we should stop when we reach $x_3 = 0$ rather than when we reach $x_5 = 0$?

It is not too difficult to answer this question in terms of algebra alone, without any appeal to the diagram. Suppose we start at O and move along the line OAE at unit speed. After t seconds we shall have $x_1 = t$, $x_2 = 0$. We calculate the other variables x_3, x_4, x_5 from equations (1), (2), (3). We find $x_3 = 4 - t$, $x_4 = 3$, $x_5 = 6 - t$. The essential thing about the variables is that they must not be negative. Now x_3 becomes 0 when $t = 4$, and is negative thereafter; x_5 becomes negative when $t > 6$; x_4 never gives any trouble. Accordingly, x_3 threatens to become negative before x_5 does, and we must stop moving along the line when $x_3 = 0$.

These considerations can be embodied in a computational rule, to illustrate which we shall need a slightly more complicated example. Suppose that, after introducing slack variables, we have the equations

$$4x_1 + x_2 + x_3 = 44, \tag{4}$$

$$3x_1 + 2x_2 + x_4 = 38, \tag{5}$$

$$2x_1 + 3x_2 + x_5 = 37, \tag{6}$$

$$ x_2 + x_6 = 9, \tag{7}$$

$$-x_1 + x_2 + x_7 = 6. \tag{8}$$

The point (x_1, x_2) has to lie in the region $OABCDEF$ (Fig. 153). M could still be $x_1 + 2x_2$, so that, as before, it would pay us to leave the origin and move in the direction OA. After time t, as before, we have $x_1 = t$, $x_2 = 0$. From the equations we find $x_3 = 44 - 4t$, $x_4 = 38 - 3t$, $x_5 = 37 - 2t$, $x_6 = 9$, $x_7 = 6 + t$. Thus x_3 will become negative if we continue to move after $\frac{44}{4} = 11$ seconds; x_4 if we go past $\frac{38}{3} = 12.67$ seconds; x_5 after $\frac{37}{2} = 18.5$ seconds; x_6 and x_7 never become negative. The smallest time, 11 seconds, occurs when $x_3 = 0$, so x_3 is the first variable to become negative. By pure calculation we have found a result that is evident in the figure. Thus we decide to move from O, with $x_1 = 0$, $x_2 = 0$ to A, with $x_2 = 0$, $x_3 = 0$.

The numbers we calculated were $\frac{44}{4}$, $\frac{38}{3}$ and $\frac{37}{2}$. It will be seen that the numerators

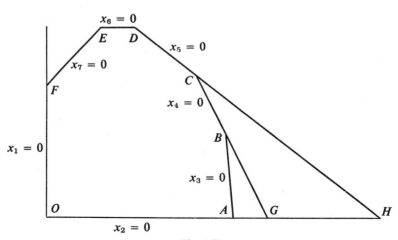

Fig. 153

44, 38 and 37 all come from the constant terms on the right; the denominators 4, 3, 2 are all coefficients of x_1. In going away from O, we decided to let x_1 grow; so the denominators are the coefficients of the variable that is allowed to grow.

In calculating these numbers, we ignore equation (7) in which the coefficient of x_1 is 0, and equation (8) in which the coefficient is negative. These correspond to $x_6 = 9 + 0t$ and $x_7 = 6 + t$, neither of which can ever become negative. This agrees with the diagram, for $x_6 = 0$ corresponds to ED which is parallel to OA; however far we go along $OAGH$, we can never cross ED to make x_6 negative, while FE is sloping away from $OAGH$, so the further we go in the direction OA, the larger and more positive x_7 gets.

We may sum up the procedure as follows: in each equation we expect to find x_1, x_2, which are zero at O, and *one other variable*. We ignore any equation in which x_1, the variable that is going to grow to make M larger, has zero or negative coefficient. In the remaining equations, we divide the constant term by the coefficient of x_1, and see for which equation the resulting quotient is least. That equation tells us which variable should become zero at the next stage. For instance, in our example, the first equation gives $\frac{44}{4}$, which is 11. If $x_1 = 11$, $x_2 = 0$, the first equation shows $x_3 = 0$. So $x_2 = 0$, $x_3 = 0$ gives us the point, A, to which we should go next.

When we get to A we find there that M can be increased if we move in the direction AB, and the same kind of question arises again, – how to tell (without using the diagram) how far we can go along AB without leaving the permissible region. We do not need any new routine to answer this question; previously we began at the origin, with $x_1 = 0$, $x_2 = 0$ and investigated how much x_1 could grow without violating the conditions; when we consider A, with $x_2 = 0$, $x_3 = 0$ the question is – how much can x_2 be increased without violating the conditions? Notice the symmetry of the diagram; there is a variable corresponding to every side of the polygon $OABCDEF$; the point O is in no way superior or significantly different from any other corner of the polygon. Admittedly, we have made the angle at O a right angle, but this is quite irrelevant; the diagram would be equally helpful if we had used oblique axes for plotting (x_1, x_2). This is a good example of the distinction we have made several times between affine and Euclidean geometry. Linear programming belongs to affine geometry; our equations can all be expressed by means of pure vector theory; scalar products, perpendiculars, distances do not enter into the question at all. Accordingly, the question of what we shall do at A, where $x_2 = 0$, $x_3 = 0$ is on exactly the same footing as the problem already considered, of what to do at O, with $x_1 = 0$, $x_2 = 0$.

This is particularly convenient for purposes of computer programming. We simply have to devise a subroutine that can be applied at any corner of the region.

As has been mentioned already, the equations (4)–(8) have a special form. In each equation we may have x_1 and x_2 (the coefficient of one of these may happen to be 0, as in equation (7)); apart from that, we have only one other variable in each equation, and that has coefficient $+1$.

Accordingly, we shall be able to apply our subroutine at corner A provided we first get the equations into the appropriate form, with x_2 and x_3 now playing the role that

x_1 and x_2 did before. That is, in each equation, x_2 and x_3 may occur, but only one other variable may appear, and that with coefficient $+1$.

If we divide equation (4) by 4, we obtain

$$x_1 + \tfrac{1}{4}x_2 + \tfrac{1}{4}x_3 = 11. \tag{4a}$$

This is a suitable form; we have x_2 and x_3, and the 'other variable' is x_1 with coefficient $+1$. In the remaining equations the 'other variables' are x_4, x_5, x_6, x_7. We must get rid of x_1 in these equations, or else we shall have two 'other variables'. This is easily done. If we subtract 3 times equation (4a) from equation (5) we eliminate x_1. Similarly we can get rid of x_1 in equations (6) and (8) by subtracting (or adding, if you prefer) suitable multiples of equation (4a). Equation (7) we leave as it is, since it is already free of x_1. Thus we find

$$\tfrac{5}{4}x_2 - \tfrac{3}{4}x_3 + x_4 \qquad\qquad = 5, \tag{5a}$$

$$\tfrac{5}{2}x_2 - \tfrac{1}{2}x_3 \qquad + x_5 \qquad = 15, \tag{6a}$$

$$x_2 \qquad\qquad\quad + x_6 \quad = 9, \tag{7a}$$

$$\tfrac{5}{4}x_2 + \tfrac{1}{4}x_3 \qquad\qquad + x_7 = 17. \tag{8a}$$

We also have to express M in a form suitable for work near A. We can write our original equation for M in the form $M - x_1 - 2x_2 = 0$. We want, in this equation also, to get rid of x_1 and permit x_3 to enter in its place. If we simply add equation (4a) we obtain

$$M - \tfrac{7}{4}x_2 + \tfrac{1}{4}x_3 = 11. \tag{9a}$$

We could write this as $M = 11 + \tfrac{7}{4}x_2 - \tfrac{1}{4}x_3$, from which we see (as expected from the diagram) that M can be further increased by letting x_2 grow, but that x_3 should be held to its value 0. However, there is an established convention in linear programming to suppose the variables written on the left of the equals sign, as in equation (9a). This of course reverses the signs, and so it is important to remember that, when things are written this way, the *negative* coefficient of x_2 in equation (9a) means that it will pay to increase x_2, while the *positive* coefficient of x_3 in equation (9a) means that x_3 should be held down to 0 for the next step.

So x_2 is the variable that is to grow. Accordingly, by the rule given earlier, we divide the constant terms by the coefficients of x_2. Applying this, in equations (4a) to (8a), we get the quotients $11/\tfrac{1}{4} = 44$; $5/\tfrac{5}{4} = 4$; $15/\tfrac{5}{2} = 6$; $9/1 = 9$; $17/\tfrac{5}{4} = 13.6$. The smallest value here is 4, arising from equation (5a), which contains the 'other variable' x_4. So we should travel until x_4 becomes 0. Our next corner is that with $x_3 = 0$, $x_4 = 0$.

Such calculations are often presented in an abbreviated form; the main idea is simply that of omitting the variables $x_1, \ldots, x_7$, and recording only the coefficients and the constants in appropriate positions. The calculation then takes a routine shape and appears as below, except for the comments and explanations. We indicate, for instance, the variables to which the coefficients relate, and the equations in which the coefficients occurred when we first explained the procedure. Such things, of course, would not appear in routine work. In practice, of course, we might not see the steps at all, as they would be operations done inside a computer. Here, then, in more concise form, is the calculation arising from equations (4)–(8).

Equation	x_1	x_2	x_3	x_4	x_5	x_6	x_7	Constant	Quotients
(4)	4	1	1	0	0	0	0	44	← 11
(5)	3	2	0	1	0	0	0	38	12.67
(6)	2	3	0	0	1	0	0	37	18.5
(7)	0	1	0	0	0	1	0	9	–
(8)	−1	1	0	0	0	0	1	6	–
(9)	−1	−2	0	0	0	0	0	0	
	*								

Equation (9) indicates that M is given by $M - x_1 - 2x_2 = 0$. The equation always begins with M, so we do not show this term, M, in the scheme at all. Looking along row (9) we notice the negative coefficient, -1. This means that M will increase if x_1 is allowed to grow. The asterisk indicates that we are going to operate on the first column, that is, the coefficients of x_1. How much can we let x_1 grow? We saw that this can be answered by forming quotients; in each row the last entry (the constant) is to be divided by the entry in the * column. Note that these quotients can only be entered *after* we have made a suitable choice for the * column. In this case, we were not forced to choose the first column. In row (9) there is a negative number, -2, in the second column. We could, if we like, put the asterisk against the second column. Some authors suggest that we should choose the most negative number in this row. Others, probably rightly, say that the extra computer time spent in looking for the most negative number is not justified, for it does not guarantee that the calculation will be shorter. They recommend that as soon as the computer has located *any* negative coefficient in the M equation, the corresponding column should be selected and dealt with, without further delay.

No quotients are entered in the rows (7) and (8), since the numbers in the * column in these rows are zero and negative respectively.

The smallest quotient, 11, occurs in row (4). This indicates that x_1 should increase until x_3 becomes zero, for x_3 is the variable other than x_1, x_2 with a non-zero coefficient in row (4). This corresponds to the move we discussed earlier, from O with $x_1 = x_2 = 0$ to A with $x_2 = x_3 = 0$. We are going to consider whether M increases if x_2 or x_3 grows, so we want now to have M in terms of x_2 and x_3, rather than x_1 and x_2. Thus the asterisk shows the variable, x_1, that we are getting rid of in the equation for M; the arrow, opposite the smallest quotient, 11, leads us to x_3, the variable we are bringing into the the equation for M. Now x_1 has to drop back to the role of 'another variable', that is, in its column, we want to find a single 1 and the rest 0. The 1, naturally enough, must be in the row singled out by the arrow.

Our next step, then, is to multiply the 'arrow' row by $\frac{1}{4}$, so as to get 1 in the x_1 column. We then subtract suitable multiples of this row (or equation) from the other rows, so as to make all the entries in the * column 0, – apart of course from the 1 we already have. This leads us to the following table, which is simply a concise way of recording equations (4a) to (9a). This time we will omit the column headings.

									Quotients
(4a)	1	$\frac{1}{4}$	$\frac{1}{4}$	0	0	0	0	11	44
(5a)	0	$\frac{5}{4}$	$-\frac{3}{4}$	1	0	0	0	5	← 4
(6a)	0	$\frac{5}{2}$	$-\frac{1}{2}$	0	1	0	0	15	6
(7a)	0	1	0	0	0	1	0	9	9
(8a)	0	$\frac{5}{4}$	$\frac{1}{4}$	0	0	0	1	17	13.6
(9a)	0	$-\frac{7}{4}$	$\frac{1}{4}$	0	0	0	0	11	
		*							

This time we have no choice. We must put the asterisk against the second column, which is the only one with a negative number in the bottom row. The quotients can then be worked out, and the least quotient, 4, indicates the second row. We multiply this row by $\frac{4}{5}$, to get 1 in the * column, and enter this as row (5b) in our next scheme. We combine suitable multiples of the new row, (5b), with the remaining rows of our present scheme, (4a), (6a), (7a), (8a), (9a), to get 0 in the second column of the next table.

Before going to this next stage, it is worthwhile to examine the appearance of this table. The table is concerned with the action we should take after reaching position A. The point A has $x_2 = 0$, $x_3 = 0$. We notice the appearance of the columns corresponding to x_2 and x_3; they contain an assortment of numbers, most of which are different from 0. The third column does contain zero in the row (7a); it could happen – so to speak by chance that quite a number of zeros occurred in these columns. It does not affect the work if this should happen. Now suppose we remove these two columns, and ignore the last column. We then see a definite pattern, which is in no way due to chance; in each row except the last there is a single 1, in each column there is a single 1, everywhere else, 0. This is the hallmark for the columns not associated with the point where we are. If you are studying a record of a calculation, and enough zeros occur in the columns corresponding to our x_2, x_3 to make it hard to identify these, you should mark the columns that contain a single 1 and zeros. In this way you can discover the columns *not* corresponding to x_2, x_3 in our example.

The numbers in the last column have a meaning that should be understood. The top row, (4a), in our last table corresponds to the equation we had earlier,

$$x_1 + \tfrac{1}{4}x_2 + \tfrac{1}{4}x_3 = 11. \tag{4a}$$

At A, $x_2 = x_3 = 0$. It follows that, at A, $x_1 = 11$. In the same way, the numbers in rows (5a), (6a), (7a), (8a) in the last column tell us that, at A, we have $x_4 = 5$, $x_5 = 15$, $x_6 = 9$, $x_7 = 17$, should we need this information. The last row corresponds to our earlier equation

$$M - \tfrac{7}{4}x_2 + \tfrac{1}{4}x_3 = 11. \tag{9a}$$

Once again using the fact that $x_2 = x_3 = 0$ at A, we see that, in the situation represented by A, the quantity M that we are trying to maximize takes the value 11. So this final entry in the table tells us what progress we have made in our search for the largest possible M. Accordingly we can say, by examining this table in the way indicated, that the value $M = 11$ can be achieved by choosing $x_1 = 11$, $x_2 = 0$. The point A in fact is $(11, 0)$. It will be remembered that x_1 and x_2 are the quantities that appear in the problem as originally stated. The other variables, x_3, x_4, x_5, x_6, x_7, we have introduced to help us solve the problem.

We now return to our calculation. Carrying out the operations described, just after the last table was given, we arrive at the following table.

									Quotients
(4b)	1	0	$\frac{2}{5}$	$-\frac{1}{5}$	0	0	0	10	25
(5b)	0	1	$-\frac{3}{5}$	$\frac{4}{5}$	0	0	0	4	–
(6b)	0	0	1	-2	1	0	0	5	← 5
(7b)	0	0	$\frac{3}{5}$	$-\frac{4}{5}$	0	1	0	5	8.33
(8b)	0	0	1	-1	0	0	1	12	12
(9b)	0	0	$-\frac{4}{5}$	$\frac{7}{5}$	0	0	0	18	
			*						

This table is concerned with the situation and the action required when we reach the point B. At B, $x_3 = x_4 = 0$, and we observe the difference between the third and fourth columns and the others. We see from the final entry, 18, that by coming to B we have attained $M = 18$. From the first two numbers in the last column we see that this happens for $x_1 = 10$, $x_2 = 4$.

The only negative number in the last row is $-\frac{4}{5}$; this occurs in the third column, and indicates that x_3 should be allowed to grow away from the value 0. The arrow indicates the row (6b). Here, very conveniently, the coefficient of x_3 is already 1, so our new row, (6c), in the next table will be identical with (6b). Suitable multiples of row (6c), combined with the rows of the present table, will allow us to get zeros in the third column, except of course in row (6c). Owing to the 1 that occurs in the fifth column of row (6b), and so also in (6c), this mixing of rows will get rid of the zeros that at present occur in the fifth column. Thus in the new table, the fourth and fifth columns will stand out as differing from the others. We shall in fact have moved to C, for which $x_4 = x_5 = 0$.

In the list of quotients above, it will be noticed that no entry is made opposite row (5b). This of course comes from the rule that no entry is made when the number in the * column is negative, as $-\frac{3}{5}$ is here. It is useful to look back at the diagram and see what this means. The table relates to the point B. The asterisk indicates that x_3 is to grow; but $x_4 = 0$ is to remain. Thus we are going to leave B in the direction BC. The quotient 25 in row (4b) indicates that x_3 can grow to the value 25 without x_1 becoming negative. The quotient in row (5b), if there were one, would tell us how far we could go in the direction BC without x_2 becoming negative. But x_2 is the co-ordinate to the north, and BC is taking us somewhat west of north; the further we go along BC, the larger x_2 will be; there is no danger of x_2 ever becoming negative. In fact G, the intersection of BC with the line $x_2 = 0$ is already behind us when we set off from B with our faces towards C.

In the same way, when we get to C, and find that it pays to move on in the direction CD, there will be no quotient entries in rows (5c) and (6c), for the intersections of CD with $x_2 = 0$ and $x_3 = 0$ are behind us, and we are moving away from them.

The routine of linear programming can best be learned by working a fairly large number of problems, all the time having the graph before you and thinking what the meanings of the steps are. It is a defect of the method that a mistake made at any stage will spoil all the later work and make it incorrect. It is therefore wise, when learning the routine, to work with a problem where you know what should happen, and will recognize if anything goes wrong.

For instance, the problem we have been considering was composed by first of all choosing points on squared paper that would make the sides of the polygon $OABCDEF$ have convenient slopes. These points were in fact $A = (11, 0)$, $B = (10, 4)$, $C = (8, 7)$, $D = (5, 9)$, $E = (3, 9)$; $F = (0, 6)$. The equations of the sides AB, BC and so on were worked out; these gave the inequalities and hence the equations for the slack variables x_3, x_4, x_5, x_6, x_7, namely equations (4) to (8) above. From these equations, together with the equation $M = x_1 + 2x_2$, we can complete the following table, giving the values of all the variables at the points A, B, C, D, E, F.

	x_1	x_2	x_3	x_4	x_5	x_6	x_7	M
A	11	0	0	5	15	9	17	11
B	10	4	0	0	5	5	12	18
C	8	7	5	0	0	2	7	22
D	5	9	15	5	0	0	2	23
E	3	9	23	11	4	0	0	21
F	0	6	38	26	19	3	0	12

It is clear from the last column that the maximum for M will be found at D. Looking back at the table with rows (4a) to (9a), we see that the numbers in the final column of that table agree with the numbers in the row for A above. The column contains the numbers 11, 5, 15, 9, 17, 11, and so gives the values of $x_1, x_4, x_5, x_6, x_7, M$ at A; it skips over $x_2 = 0$, $x_3 = 0$. In the same way, from the row for B above we can check the accuracy of the final column in the table with rows (4b) to (9b).

It should be understood that this way of checking is purely a learning device; it is intended as a way of avoiding errors in exercises worked to gain familiarity with the routine. It helps also to bring out the significance of the numbers in the final column of the table. In any actual problem, the labour involved in such a procedure would be much too great. In practice, the computations would be carried out by a computer programme, into which devices for recognizing the occurrence of any error would have to be incorporated.

Exercises

1 Complete the calculations for the problem discussed above; that is, obtain the tables corresponding to the points C and D. Check the accuracy of your work by means of the table above, that gives the values of the variables at the corners of the polygon.

2 Do the same problem, by going back to the beginning of the work, and placing the asterisk under the second column rather than the first. Check accuracy in the same way. By what route do you now reach the optimal point, D?

3 The first problem we considered in this section had

$$x_1 \leqslant 4, \qquad x_2 \leqslant 3, \qquad x_1 + x_2 \leqslant 6, \qquad M - x_1 - 2x_2 = 0.$$

Using the slack variables and the diagram given, solve this problem by the routine method with tables (*a*) using the route O, A, B, C (*b*) using the route O, D, C. If accuracy is checked by the method suggested, observe that the final columns give the values of x_1, x_2, x_3, x_4, x_5 but not necessarily in that order. Explain this, by writing the equations corresponding to the rows of the table.

4 On a ship a volume V is available for cargo and the weight of the cargo must not exceed W. Two commodities are available for loading. The units of money, volume and weight are so chosen that unit value of commodity A has unit volume and unit weight. In this system of measurement, unit value of commodity B has volume v and weight w. It is desired to load the ship in such a way that the cargo carried will have maximum value.

For each of the situations listed below solve this problem by means of a diagram, and also by the method of calculation explained in §43. Check that the results agree. Although no heavy calculation is needed here it is useful to draw the diagram first, as this will indicate the most economical route to the desired solution.

Situation (*a*): $v = 2$, $\quad w = \frac{1}{2}$, $\quad V = 20$, $\quad W = 11$.
Situation (*b*): $v = 3$, $\quad w = 2$, $\quad V = 33$, $\quad W = 26$.
Situation (*c*): $v = \frac{1}{2}$, $\quad w = \frac{1}{3}$, $\quad V = 14$, $\quad W = 11$.
Situation (*d*): $v = 3$, $\quad w = 2$, $\quad V = 12$, $\quad W = 20$.

5 600 tons of a certain material are available at warehouse A and 800 tons of the same material at warehouse B; 300 tons are to be delivered at a place P, 500 tons at Q and 600 tons at R. The distances in miles are as shown in the table.

	P	Q	R
A	10	20	100
B	70	90	30

It is required to arrange the delivery in such a way that the total transport required is a minimum. (A journey in which t tons are carried m miles counts as mt units of transport.)

(*Suggested method.* If x_1 tons are taken from A to P and x_2 tons from A to Q, the other quantities involved can be expressed in terms of x_1 and x_2. Certain inequalities must be satisfied. Taking $x_1 = 0$, $x_2 = 0$ provides a starting point for the calculation, but is obviously far from the best arrangement. The problem should be treated both graphically and by calculation.)

6 What difference would it make to the arrangements in question 5 if the distance from B to P were 85 miles, all the other data being as before?

7 On a ship 25 units of volume are available for cargo, the weight of which must not exceed 23 units. Unit value of commodity A has unit volume and weight 4 units; unit value of commodity B has volume 2 units and weight 2 units; unit value of commodity C has volume 3 units and weight 1 unit. Calculate the cargo that gives the greatest value.

If we have commodity A to the value x_1, commodity B to the value x_2 and commodity C to the value x_3, the restrictions on space and weight require the point (x_1, x_2, x_3) to lie in a region bounded by 5 planes. Sketch this region and find the co-ordinates of its vertices. Find the value of the cargo that corresponds to each vertex, and check that none of these values exceeds that found by linear programming.

44   Linear programming, continued

The work of §43 was based on a simple idea; start at a corner of the permitted region, and then keep moving along the edges of the region in such a way that the situation continually improves. The corner we started at was the origin. Now the origin had no claim to represent a particularly efficient arrangement. For instance, in a problem of assigning work to men in a factory, the origin would correspond to complete idleness – put no men on each machine, so nothing is produced. The only reason for choosing the origin is *to get the calculation started*; it gives us a point from which we can proceed to other, more satisfactory points. Idleness automatically satisfies the conditions; it can always be achieved; whatever restrictions may exist on manpower, supply of materials, working hours, available finance and so forth. It constitutes what is called a *feasible solution*, an arrangement that meets the restrictions under which we have to work. It is a starting point in our search for the *optimal feasible solution*, that is, the best arrangement possible in the circumstances.

However, there are problems in which the origin does not provide us with a feasible solution. Consider the problem we had earlier of feeding a student commune. The origin would correspond to leaving the supermarket empty-handed, to spending nothing on each article on the list. This would not satisfy the conditions that adequate amounts of essential nutrients were to be provided. In such a problem we do not have a feasible solution ready-made to hand. Indeed, it is possible that no feasible solution exists. Perhaps the budget committee were insufficiently aware of the rising cost of living. They may have allocated to the food buyer a sum too small to provide an adequate diet however much wisdom he may use in his spending. So the problem is – to locate a feasible solution, if such a thing exists, as a starting point for the procedure of §43, or to demonstrate that the demands made are impossible.

Fig. 154 illustrates a simple problem where impossible requirements are imposed. The restrictions are

$$x_1 + 2x_2 \leqslant 8, \qquad 3x_1 + x_2 \leqslant 9, \qquad x_1 + x_2 \geqslant 6.$$

The first two inequalities require us to be inside the quadrilateral $OABC$. The last condition requires us to be above the line DE. Evidently no point meets these requirements.

How can we reach this conclusion without the aid of a diagram? The first two conditions offer no difficulty; they are satisfied by the origin, $x_1 = 0, x_2 = 0$. The question now is – can we move around in such a way that, while still satisfying the first two conditions, we meet the third also? That is, can we keep in $OABC$, and still make $x_1 + x_2 \geqslant 6$? We decide we will do our best to meet it; 'doing our best' is itself a problem of maximizing, and we can formulate it as follows. Suppose some generous person, seeing we are in difficulties with the third condition, offers to make up whatever

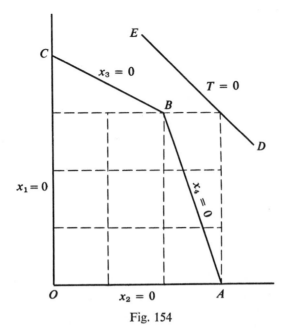

Fig. 154

amount we are short, but on the understanding that we will not ask for more assistance than is absolutely necessary. Let T stand for the deficit this person has to make good. Our problem now reads

$$x_1 + 2x_2 \leqslant 8, \qquad 3x_1 + x_2 \leqslant 9, \qquad x_1 + x_2 + T \geqslant 6,$$

where T is to be as small as possible; that is, we have to maximize $-T$.

This problem is of the kind considered in §43. We bring in slack variables x_3, x_4, x_5 and write

$$x_1 + 2x_2 + x_3 \qquad\qquad = 8, \tag{1}$$

$$3x_1 + x_2 \qquad + x_4 \qquad = 9, \tag{2}$$

$$x_1 + x_2 \qquad\qquad - x_5 + T = 6. \tag{3}$$

Note here the minus sign with x_5; this is because we have $\geqslant$ in our third condition; x_5 measures the amount by which $x_1 + x_2 + T$ *exceeds* 6.

Let $N = -T$, the amount to be maximized. We have $N + T = 0$, and this may cause some dismay; in §43, the work was finished when all the signs in the bottom row (corresponding to this equation) were positive. But that was based on the fact that the variables corresponding to the corner in question were all zero, so could not be made any smaller. But in our case we are starting from the situation $x_1 = 0$, $x_2 = 0$, $T = 6$. This is feasible, since the equations show $x_3 = 8$, $x_4 = 9$, $x_5 = 0$; no variable is negative. We have not made N a maximum, as T may well be capable of decreasing from its present value of 6. We do have three zero quantities, x_1, x_2 and x_5. So the sensible thing is to express everything in terms of these. Equation (3) gives us

$$T = 6 - x_1 - x_2 + x_5,$$

accordingly $N + T = 0$ becomes

$$N - x_1 - x_2 + x_5 = -6. \tag{4}$$

The table now appears as shown.

x_1	x_2	x_3	x_4	x_5	T		Quotients
1	2	1	0	0	0	8	8
3	1	0	1	0	0	9	← 3
1	1	0	0	−1	1	6	6
−1	−1	0	0	1	0	−6	
*							

Observe here that the columns x_3, x_4, T each contain a single 1 and the other entries in them are all zero. Thus x_3, x_4, T are easily found in terms of the remaining variables x_1, x_2, x_5. The quantities above the line are all positive, which agrees with the observation already made that $x_1 = x_2 = x_5 = 0$ gives a feasible solution, for this makes $x_3 = 8$, $x_4 = 9$, $T = 6$. The table is now completely in the form used in §43.

If we put the asterisk under x_1, the smallest quotient occurs in the second row, and our next table is as shown below.

0	$\frac{5}{3}$	1	$-\frac{1}{3}$	0	0	5	← 3
1	$\frac{1}{3}$	0	$\frac{1}{3}$	0	0	3	9
0	$\frac{2}{3}$	0	$-\frac{1}{3}$	−1	1	3	4.5
0	$-\frac{2}{3}$	0	$\frac{1}{3}$	1	0	−3	
	*						

Continuing we obtain the next table.

0	1	$\frac{3}{5}$	$-\frac{1}{5}$	0	0	3	
1	0	$-\frac{1}{5}$	$\frac{2}{5}$	0	0	2	
0	0	$-\frac{2}{5}$	$-\frac{1}{5}$	−1	1	1	
0	0	$\frac{2}{5}$	$\frac{1}{5}$	1	0	−1	

Now the coefficients in the bottom row are all positive, and this time we really have finished. The bottom line indicates $N + \frac{2}{5}x_3 + \frac{1}{5}x_4 + x_5 = -1$. The best we can do is take $x_3 = x_4 = x_5 = 0$, but even so we are only able to get $N = -1$. We still fall short, by 1, of meeting the third condition.

The first two rows of the table indicate that, when $x_3 = x_4 = x_5 = 0$, we have $x_1 = 2$, $x_2 = 3$, that is, we are at the point B. This is the nearest we can get to the line DE. The rather small coefficients, $\frac{2}{5}$ and $\frac{1}{5}$, of x_3 and x_4, correspond to the fact that, if we leave B along BA or BC, we are indeed getting further from the line DE, but only rather slowly. At B, $x_1 + x_2 = 5$, which, as expected, falls short by 1 of the demanded value, 6.

If the third condition is replaced by $x_1 + x_2 \leqslant 4.5$, the problem becomes possible. There is then a triangle within which (x_1, x_2) must lie. In this case, by maximizing N we would arrive at a corner of this triangle. Since no 'subsidy' T would be necessary, the maximum value of N would be 0, shown by zero appearing as the last entry in the bottom row of the table.

It will be seen from the tables that T plays exactly the same role as a sixth variable, x_6. In fact this is the usual notation. A variable representing a subsidy that may be needed to rectify an unsatisfied demand is known as an *artificial variable*. Like all the other variables, it is supposed never to be negative. Our generous person may not need to subsidize us, but he does not demand that any surplus be handed to him. Accordingly, the maximum value of N can never exceed 0. When N attains the maximum value of 0, this indicates that the conditions can be fulfilled and a feasible solution has been located.

Notice that M, the quantity we eventually wish to maximize, has not been mentioned at all in this work. The procedure just explained has the purpose of locating a point within the realm of possibilities. There is no suggestion that this point will be the best point for maximizing M. It hardly could be, seeing the function defining M is nowhere mentioned in the procedure.

Exercises

1 Replacing the condition $x_1 + x_2 \geqslant 6$ by $x_1 + x_2 \geqslant 4.5$ carry through the procedure for maximizing N, and check that it does lead to a feasible solution. The work will, of course, be almost identical with that done above, so the table given there can be used as a check of numerical accuracy.

2 Find a feasible solution for $x_1 \leqslant 2$, $x_1 + x_2 \leqslant 4$, $2x_1 + x_2 \geqslant 5$ by means of an artificial variable. After doing this, draw a diagram, and trace the path followed by the point (x_1, x_2) at the various stages of the calculation.

Beginning of solution. We write the equations

$$x_1 + x_3 = 2, \qquad x_1 + x_2 + x_4 = 4, \qquad 2x_1 + x_2 - x_5 + x_6 = 5.$$

Here x_3, x_4, x_5 are slack variables and x_6 is an artificial variable, brought in because $x_1 = 0$, $x_2 = 0$, which satisfies the first two conditions, leaves a deficit in the third condition. $N + x_6 = 0$, so $N - 2x_1 - x_2 + x_5 = -5$. Maximize N.

3 Investigate, first by calculation, then by a diagram, what happens if the third condition in question 2 is replaced by $2x_1 + x_2 \geqslant 7$.

4 What happens if, in question 2, we replace the last condition by $2x_1 + x_2 \geqslant 6$?

5 By common sense, or by a diagram, consider whether a feasible solution can be found for
(a) $x_1 \leqslant 3$, $x_2 \leqslant 2$, $x_1 + x_2 \geqslant 9$,
(b) $x_1 \leqslant 3$, $x_2 \leqslant 2$, $x_1 + x_2 \geqslant 4$.
Apply artificial variable method to these, and trace the path on the diagram followed by (x_1, x_2) at the successive improvements of N.

Problems with two artificial variables

It may happen that we have more than one inequality that is not satisfied at the origin, or at some other point where we are starting our search. Fig. 155 illustrates the problem $x_1 \leqslant 4$, $x_2 \leqslant 2$, $2x_1 - x_2 \geqslant 1$, $2x_2 - x_1 \geqslant 1$. The last two conditions are not satisfied by the origin. So we bring in slack variables x_3, x_4, x_5, x_6 and artificial variables x_7, x_8.

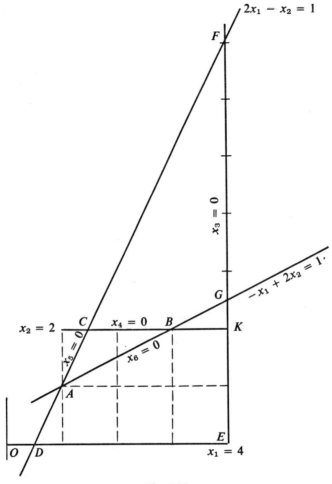

Fig. 155

We then have

$$
\begin{aligned}
x_1 \quad\quad + x_3 \quad\quad\quad\quad\quad\quad\quad\quad &= 4, \\
x_2 \quad\quad + x_4 \quad\quad\quad\quad\quad &= 2, \\
2x_1 - \; x_2 \quad\quad\quad - x_5 \quad + x_7 \quad\;\; &= 1, \\
-x_1 + 2x_2 \quad\quad\quad\quad - x_6 \quad\;\; + x_8 &= 1.
\end{aligned}
$$

Our total deficit is $x_7 + x_8$, so we take $N = -x_7 - x_8$. Maximizing N minimizes the total deficit. Suppose we reach a situation that gives $N = 0$. As x_7 and x_8 cannot be negative, $x_7 + x_8 = 0$ implies $x_7 = 0$ and $x_8 = 0$. Accordingly we have a feasible solution, since there is no deficit in any condition.

Suppose we find the maximum value of N turns out to be negative. This would mean the conditions were impossible. For if a feasible solution exists, it makes the deficits zero, so $x_7 = 0$, $x_8 = 0$ can be achieved, and so $N = 0$ can be reached. A negative maximum for N clearly rules out this possibility.

So the maximum value for N will tell us whether the conditions are possible or not.

In this particular problem, a feasible solution does exist and is reached in two stages. The artificial variable calculation takes us from the origin to D, with $x_1 = \frac{1}{2}$, $x_2 = 0$, and then to A, with $x_1 = 1$, $x_2 = 1$, which meets all the conditions. In making the calculation, we know this because we find $N = 0$ from the bottom row.

Exercises

1 Carry out the calculation just described.

2 Find a feasible solution, or prove that none exists, for the conditions

$$x_1 \leqslant 3, \qquad x_2 \leqslant 4, \qquad 2x_1 + x_2 \geqslant 6, \qquad x_1 + x_2 \geqslant 4.$$

Is the method wasteful?

There is a point in the above procedure that may strike you as strange. The first problem in this section considered the conditions $x_1 + 2x_2 \leqslant 8$, $3x_1 + x_2 \leqslant 9$, $x_1 + x_2 \geqslant 6$. Now in fact, as may be seen from the diagram, the first two conditions made it impossible for the third one to be fulfilled. We had to bring in an artificial variable, which we then called T and would now call x_6, as a sort of subsidy to meet our deficit on this condition. Accordingly, as x_6 has to be provided to bring $x_1 + x_2$ up to 6, we might expect to write the equation $x_1 + x_2 + x_6 = 6$. But this is not the equation that our procedure tells us to write. We consider the condition $x_1 + x_2 + x_6 \geqslant 6$ and bring in the slack variable x_5 to express this inequality; so we write $x_1 + x_2 - x_5 + x_6 = 6$. In this equation, x_5 represents the *excess* of $x_1 + x_2 + x_6$ over 6. So it seems that at one and the same time we are asking for a subsidy x_6, and anticipating a surplus x_5. If x_6 were an actual subsidy, this might be regarded as a piece of adroit dishonesty. But x_6 does not represent some piece of wealth that we wish to acquire by fair means or foul. It is simply a variable in a calculation, and if we are indeed using two variables, x_5 and x_6, where one would do, we are involving ourselves in unnecessary arithmetic and extra computer time.

Now, in fact, in this particular problem it is not necessary to bring x_5 in. We could write $x_1 + x_2 + x_6 = 6$ and proceed to maximize $N = -x_6$ and be led to the point B that makes the deficit least.

Exercise

Check this statement.

There are in fact a number of situations in which we could dispense with the slack variables that represent surpluses. For instance, in our problem with two artificial variables, $x_1 \leqslant 4$, $x_2 \leqslant 2$, $2x_1 - x_2 \geqslant 1$, $2x_2 - x_1 \geqslant 1$, we could do this. We could drop x_5 and x_6, replace our earlier equations by $x_1 + x_3 = 4$, $x_2 + x_4 = 2$, $2x_1 - x_2 + x_7 = 1$, $-x_1 + 2x_2 + x_8 = 1$ and maximize $N = -x_7 - x_8$. This would lead us to the feasible solution represented by the point A in Fig. 155.

Exercise

Check this.

The reason why this simplified procedure works in this case is that the point A *just* meets the two conditions; at A we have $x_1 = 1$, $x_2 = 1$ and $2x_1 - x_2 = 1$, $2x_2 - x_1 = 1$.

In the diagram we notice that we can reach A without crossing either of the lines $2x_1 - x_2 = 1$, $2x_2 - x_1 = 1$. At A there is no surplus, and so the variables x_5 and x_6, included to meet the contingency of a surplus, can be dropped without any damage being done.

Now one might think that this would always happen. If we have difficulty in meeting a set of conditions why should we anticipate meeting them and having something to spare?

The reason can be seen from Fig. 156, which illustrates the set of conditions

$$x_1 + x_2 \leqslant 5, \qquad 3x_1 + x_2 \geqslant 3, \qquad 2x_1 + x_2 \geqslant 4.$$

Feasible solutions exist for these conditions; the point (x_1, x_2) must lie in the

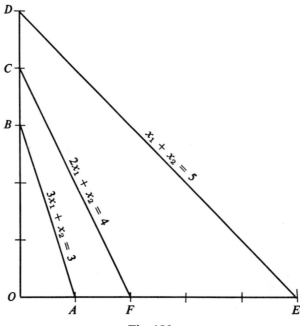

Fig. 156

quadrilateral $CDEF$, or on its boundary. The condition $3x_1 + x_2 \geqslant 3$ requires the point to lie above or to the right of the line AB. The condition $2x_1 + x_2 \geqslant 4$ requires it to lie above or to the right of CF. But any point on CF is already well past AB; this means that we can only satisfy the condition $2x_1 + x_2 \geqslant 4$ if we are prepared to satisfy $3x_1 + x_2 \geqslant 3$ with a surplus.

It is instructive to work through this problem, first by the standard method, using the equations $x_1 + x_2 + x_3 = 5$, $3x_1 + x_2 - x_4 + x_6 = 3$, $2x_1 + x_2 - x_5 + x_7 = 4$, and by the method just proposed, with x_4 and x_5 dropped. The standard method can lead you from O to A to B to C or, by another choice, from O to B to C. Either way, we reach the feasible solution C, with $N = 0$. The other method may take us from O to A to B, or directly from O to B, depending whether we put the asterisk under the first or second column. But, with this modified method we find we cannot get any result better than $N = -1$. The reason is that, by writing $3x_1 + x_2 + x_6 = 3$, we have committed ourselves to the region $3x_1 + x_2 \leqslant 3$; we cannot get past the line AB, and so we cannot get onto the line CF, or beyond it. Thus we have shut ourselves out from the feasible region $CDEF$.

The values of the seven variables at C are $x_1 = 0$, $x_2 = 4$, $x_3 = 1$, $x_4 = 1$, $x_5 = 0$, $x_6 = 0$, $x_7 = 0$. The value 1 for x_3 indicates that we are not pressing against the boundary DE; we have some slack in the condition $x_1 + x_2 \leqslant 5$, since, at C,

$$x_1 + x_2 = 4.$$

From $x_5 = x_7 = 0$ we see that we have neither a surplus nor a deficit in regard to the condition $2x_1 + x_2 \geqslant 4$; we are right on the line CF. From $x_6 = 0$ we see that we have no deficit in $3x_1 + x_2 \geqslant 3$, rather, as $x_4 = 1$ indicates, we have a surplus of 1, since at C the value of $3x_1 + x_2$ is 4, and so in excess of 3. If this excess were forbidden, it would be impossible to produce a feasible solution.

Accordingly the presence of *two* extra variables in an equation such as

$$3x_1 + x_2 - x_4 + x_6 = 3$$

is justified in certain situations. In a complicated problem it is extremely likely that we shall have to permit a surplus in some condition, so that the variables provided to meet this contingency are fulfilling a useful purpose. In any case, if it did happen in a particular problem that no surplus arose, we would not be able to recognize this simply by examining the equations. Before we could use the method without the surplus variables, we would have to apply a subroutine to determine that the situation was suitable for such a procedure, and this would almost certainly be more time-consuming than applying the standard method.

Accordingly, it does not seem that any economy of calculation could be achieved by trying to trim down the number of variables. This is satisfactory from the viewpoint of the student; we have the standard method, and do not have to bother about variations to meet special situations.

The solution as a whole

We now wish to put the pieces together. At the beginning of §43, we considered a problem where feasible solutions lay within or on a particular polygon. We showed how to work from any corner of that polygon until we arrived at the corner most suited to our purposes. In the present section we have seen how to locate a corner that can be used as a starting point, when there is no obvious corner of the permitted region. We now need to see how these two procedures fit together in detail. This is shown by means of two worked examples.

Worked example 1

It is required to maximize $M = x_1 + 4x_2$, where x_1 and x_2 are not negative and are subject to the restrictions $x_1 \leqslant 4$, $x_2 \leqslant 2$, $2x_1 - x_2 \geqslant 1$, $-x_1 + 2x_2 \geqslant 1$.

Solution   With so few and so simple conditions it is of course easy to sketch the region and spot a suitable corner. However, we are not concerned to find the solution so much as to use this simple problem to illustrate the general procedure.

We write the conditions with the help of 4 slack variables, and also 2 artificial variables needed for the last 2 inequalities. We are thus led to the table shown below.

x_1	x_2	x_3	x_4	x_5	x_6	x_7	x_8	
1	0	1	0	0	0	0	0	4
0	1	0	1	0	0	0	0	2
2	-1	0	0	-1	0	1	0	1 ←
-1	2	0	0	0	-1	0	1	1
-1	-1	0	0	1	1	0	0	-2
*								

The final row of the table is not related in any way to the quantity M that is to be maximized. We are simply trying to locate any corner of the permitted region, one that makes our deficits, x_7 and x_8, both zero. So we set out to maximize

$$N = -x_7 - x_8 = x_1 + x_2 - x_5 - x_6 - 2,$$

and the last row of the table expresses this.

If we put the asterisk under the x_1 column, the arrow will have to point to the third row. At the next stage our table becomes as shown.

0	$\frac{1}{2}$	1	0	$\frac{1}{2}$	0	$-\frac{1}{2}$	0	$3\frac{1}{2}$
0	1	0	1	0	0	0	0	2
1	$-\frac{1}{2}$	0	0	$-\frac{1}{2}$	0	$\frac{1}{2}$	0	$\frac{1}{2}$
0	$1\frac{1}{2}$	0	0	$-\frac{1}{2}$	-1	$\frac{1}{2}$	1	$1\frac{1}{2}$ ←
0	$-1\frac{1}{2}$	0	0	$\frac{1}{2}$	1	$\frac{1}{2}$	0	$-1\frac{1}{2}$
	*							

The next step gives the table below. The zero at the end of the last row makes it clear

0	0	1	0	$\frac{2}{3}$	$\frac{1}{3}$	$-\frac{2}{3}$	$-\frac{1}{3}$	3
0	0	0	1	$\frac{1}{3}$	$\frac{2}{3}$	$-\frac{1}{3}$	$-\frac{2}{3}$	1
1	0	0	0	$-\frac{2}{3}$	$-\frac{1}{3}$	$\frac{2}{3}$	$\frac{1}{3}$	1
0	1	0	0	$-\frac{1}{3}$	$-\frac{2}{3}$	$\frac{1}{3}$	$\frac{2}{3}$	1
0	0	0	0	0	0	1	1	0

that we have succeeded in making our deficit zero. The columns corresponding to x_5, x_6, x_7 and x_8 have the characteristic solid appearance, which indicates that we should take these variables as zero. We thus see that our original equations are satisfied by $x_1 = 1$, $x_2 = 1$, $x_3 = 3$, $x_4 = 1$, $x_5 = 0$, $x_6 = 0$, $x_7 = 0$, $x_8 = 0$. The fact that

$$x_7 = x_8 = 0$$

indicates that we are not working on a deficit; we have managed to meet the conditions originally specified without any need for a subsidy. We now have no further need of x_7 and x_8. Having got into the permitted region we intend to stay there; we have no intention of going into debt again. Accordingly, we now drop x_7 and x_8. In our last table, we can delete the columns corresponding to x_7 and x_8. That this step is justified can be seen by writing the equations that the table represents. Since $x_7 = 0$ and $x_8 = 0$, and since we intend to keep these values permanently, the terms containing x_7 and x_8 will simply lead to blank spaces in the equations.

Now that we are on the permitted region we can begin to think about improving our position; for the first time, we can turn our attention to M. Now $M = x_1 + 4x_2$, and we have discovered that $x_1 = 1$, $x_2 = 1$ is a feasible solution. However, x_1 and x_2 are not convenient variables in which to work from the position where we are. In §43, when we were at a corner of the permitted region, we found it was most convenient to work with the variables that were zero at that point. Now at the corner we have located we know the values of the variables. We have decided to drop x_7 and x_8; the values of the variables that still interest us are $x_1 = 1$, $x_2 = 1$, $x_3 = 3$, $x_4 = 1$, $x_5 = 0$, $x_6 = 0$, as we saw earlier. Thus the variables we want to work with are x_5 and x_6. These correspond to the 'solid looking' columns in the table – the only such columns that remain after the x_7 and x_8 columns have been deleted. This is very convenient, since our last table gives the values of x_1, x_2, x_3 and x_4 in terms of x_5 and x_6. In fact we can read off, from the third and fourth rows of that table, the equations

$$x_1 = 1 + \tfrac{2}{3}x_5 + \tfrac{1}{3}x_6$$
$$x_2 = 1 + \tfrac{1}{3}x_5 + \tfrac{2}{3}x_6,$$

from which it follows that

$$M = x_1 + 4x_2 = 5 + 2x_5 + 3x_6.$$

If we now write below the table a line corresponding to this result, we obtain the new table below.

0	0	1	0	$\frac{2}{3}$	$\frac{1}{3}$	3
0	0	0	1	$\frac{1}{3}$	$\frac{2}{3}$	1
1	0	0	0	$-\frac{2}{3}$	$-\frac{1}{3}$	1
0	1	0	0	$-\frac{1}{3}$	$-\frac{2}{3}$	1
0	0	0	0	-2	-3	5

If you wish to work entirely in terms of the tables, and not to write equations at all this last table can be reached in the following way. Write the bottom line in the form corresponding to our original equation, $M = x_1 + 4x_2$. This line will contain the numbers

$$-1 \;\overset{\bullet}{} -4 \quad 0 \quad 0 \quad 0 \quad 0 \quad 0.$$

Then operate according to the usual rules, with the asterisk first under the -1 and then under the -4. This will lead to the same result as our work above.

We now have only to apply the procedure of §43. If the asterisk is placed under the -2 in the bottom row, the arrow will have to be placed against the second row. The table is then brought to the form shown below.

0	0	1	-2	0	-1	1
0	0	0	3	1	2	3
1	0	0	2	0	1	3
0	1	0	1	0	0	2
0	0	0	6	0	1	11

All the signs in the bottom row are now positive, apart from the zeros, so we have reached the optimal solution, $x_1 = 3$, $x_2 = 2$, for which M takes its largest possible value, 11.

Worked example 2

Maximize $M = 2x_1 + x_2$ subject to the conditions $x_1 \leqslant 1$, $x_2 \leqslant 1$, $x_1 + x_2 \geqslant 1$.

Solution The permitted region is shown in Fig. 157. Here again the solution is graphically obvious, and our only purpose in using this problem is to discuss features of the procedure that arise here and might cause difficulty in a less simple situation.

In the routine treatment we begin by searching for a feasible solution. We need a slack variable for each of the conditions and an artificial variable, x_6, for the last condition. We proceed to minimize $N = -x_6$. We thus obtain the following table.

x_1	x_2	x_3	x_4	x_5	x_6		
1	0	1	0	0	0	1	←
0	1	0	1	0	0	1	
1	1	0	0	-1	1	1	
-1	-1	0	0	1	0	-1	
*							

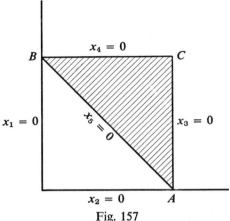

Fig. 157

The bottom line expresses the equation

$$N - x_1 - x_2 + x_5 = -1,$$

which follows from the equation corresponding to the third row of the table and the fact that $N = -x_6$.

If we put the asterisk under the x_1 column, the quotient $1/1 = 1$ occurs both in the first and the third row, so we have a choice. We have (for no special reason) chosen the first row. The next table is as shown.

1	0	1	0	0	0	1
0	1	0	1	0	0	1
0	1	−1	0	−1	1	0
0	−1	1	0	1	0	0

The last entry in the bottom row is 0, which indicates that our first objective has been attained; we have reduced our deficit, x_6, to zero. But a paradox appears to be arising; we still have -1 for the second entry in the bottom line, and this indicates that we should be able to increase N still further by putting an asterisk under this -1 and continuing by our usual procedure. But it is impossible that N should increase any further. For already $N = 0$; if N increases it will become positive. As $N = -x_6$, this means x_6 will become negative, and this is impossible, for the basic assumption is that no x ever becomes negative, and the whole routine is arranged to ensure that a negative value never arises.

Question for discussion

What does happen if we put an asterisk under the -1 and seek to continue the improvement of the value of N?

For the moment we turn our thoughts from this apparent paradox and concentrate on the fact that we have found a feasible solution, which the table shows to be $x_1 = 1$, $x_4 = 1$, the other variables x_2, x_3 and x_5 being zero. (We are no longer interested in x_6, which has done its work.) Here we have an unusual feature. Usually at the corner of the polygon which surrounds the permitted region we have *two* lines meeting and the two variables corresponding to these zero. However, in Fig. 157 it will be seen that the three lines $x_2 = 0$, $x_3 = 0$, $x_5 = 0$ all pass through A. When this happens we are said to have a *degenerate case*. It is because of this situation that we have the special features already observed, a choice in the placing of the arrow right at the outset, and a minus sign surviving in the bottom row even after $N = 0$ had been achieved.

It is now time to turn our attention to $M = 2x_1 + x_2$, the quantity we wish to maximize. We take our last table and remove the x_6 column and the bottom line, both of which were relevant only to our first objective, locating a point on the boundary of the permitted region. We write a new bottom line, corresponding to $M - 2x_1 - x_2 = 0$. (Here we are following the procedure mentioned in our first example as a possible alternative.) This gives us the table below. Now we have 3 equations (above the line)

1	0	1	0	0	1
0	1	0	1	0	1
0	1	−1	0	−1	0
−2	−1	0	0	0	0

in 5 variables. We should be able to choose the values for 2 variables at will and solve for the other three. Examining the top three rows, we see that x_1, x_4 and x_5 each occur in one equation only, while the columns for x_2 and x_3 look reasonably 'solid'. Accordingly it is indicated that we try to get M expressed in terms of x_2 and x_3 alone. This is easily done. We need only add twice the top row to the bottom row, to bring the table to the form shown.

1	0	1	0	0	1
0	1	0	1	0	1
0	1	−1	0	−1	0 ←
0	−1	2	0	0	2
	*				

This table now looks very much like the tables we had in §43 when we were part way through the calculation. We have the solid columns for x_2 and x_3, while in the other columns we have only one non-zero entry under x_1, x_4 and x_5. The only difference is that while the first column contains the entries 1, 0, 0, 0 and the fourth column has 0, 1, 0, 0 (both as usual) the fifth column has 0, 0, −1, 0. In the contexts of §43 this would have caused trouble; on putting $x_2 = x_3 = 0$ we would have obtained a negative value for x_5. The situation is saved by the zero in the sixth column. When $x_2 = x_3 = 0$, the equations represented by the first three rows reduce to $x_1 = 1$, $x_4 = 1$ and $-x_5 = 0$, so we are not confronted with any negative value.

We now proceed in the usual way. With the asterisk and arrow as indicated we next bring the table to the form shown below. This result is a little disturbing. We are still

1	0	1	0	0	1
0	0	1	1	1	1	←
0	1	−1	0	−1	0
0	0	1	0	−1	2
		*			

at the point A with $x_1 = 1$, $x_2 = 0$. The only difference is that before we considered it as $x_2 = x_3 = 0$ while now we call it $x_3 = x_5 = 0$.

Our method will have failed if at the next step we find we are still at A, but calling it $x_2 = x_3 = 0$ once again. Fortunately this does not happen. The next step gives us the table below. We have completed our search, there are no negative entries left in the

1	0	1	0	0	1
0	0	1	1	1	1
0	1	0	1	0	1
0	0	2	1	0	3

bottom line. We have escaped from the point A and reached C with $x_1 = 1$, $x_2 = 1$, to yield the maximum value $M = 3$.

Usually things seem to work out as they have done here; in spite of degeneracy, the routine usually brings us to the solution of the problem. Theoretically the possibility does exist of the computation going round and round and never emerging from a closed cycle of situations. An example illustrating this possibility will be found on page 84 of *Mathematical Programming* by S. Vajda (Addison–Wesley, 1961). This book gives a very full bibliography of books and papers on linear programming. S. Vajda has also written a smaller book, that students will find easier to read, called *Planning by Mathematics* (Sir Isaac Pitman and Sons, 1969). This book, in about a hundred pages presents twenty-one short articles dealing with practical problems that can be handled by linear programming or related techniques. The topics range from blending aviation gasolines to betting at race-courses.

Exercises

It should be borne in mind throughout that not all linear programming objectives are attainable. An answer to a problem therefore consists either in giving a solution, or in demonstrating that the demands made are impossible.

1 Maximize $3x_1 + x_2$, subject to $x_1 + x_2 \leqslant 3$, $x_1 \geqslant 1$, $x_2 \geqslant 1$.

2 Maximize $2x_1 + x_2$, subject to $x_2 \leqslant 2$, $x_1 + x_2 \leqslant 4$, $x_1 + 2x_2 \geqslant 5$.

3 Maximize $x_1 + x_2$ subject to $3x_1 + 2x_2 \leqslant 15$, $2x_1 + 3x_2 \leqslant 15$, $3x_1 + 4x_2 \geqslant 22$.

4 Subject to $x_1 + 3x_2 \leqslant 19$, $3x_1 + x_2 \leqslant 17$, $3x_1 + 2x_2 \geqslant 15$, $x_1 + 2x_2 \geqslant 9$, minimize $x_1 + x_2$. (Maximize $-x_1 - x_2$.)

5 With the same conditions as in question 4, maximize $x_1 + x_2$.

6 Maximize $11x_1 + 17x_2$ subject to $x_1 + x_2 \leqslant 3$.

7 Minimize $11x_1 + 17x_2$ subject to $x_1 + x_2 \geqslant 3$.

8 Maximize $5x_1 + 3x_2$ subject to $x_1 \leqslant 4, -x_1 + x_2 \leqslant 1, x_1 + 2x_2 \leqslant 5, 4x_1 + 9x_2 \geqslant 23$.

9 Maximize $x_1 + 4x_2 + 3x_3$ subject to

$$x_1 + x_2 + x_3 \leqslant 11, \qquad x_1 + 2x_2 + 3x_3 \leqslant 26, \qquad x_1 - x_2 + x_3 \geqslant 5.$$

10 Maximize $x_1 + x_2 + 3x_3 + 5x_4$ subject to

$$x_1 \leqslant x_2, \qquad x_2 \leqslant x_3, \qquad x_3 \leqslant x_4, \qquad x_1 + x_2 + x_3 + x_4 \leqslant 24, \qquad x_1 + x_2 \geqslant 9.$$

11 100 tons of a certain material are available in a warehouse P and 80 tons in a warehouse Q. It is required to deliver 50 tons to a place A, 60 tons to a place B and 70 tons to a place C with the minimum use of transport. The distances in miles are given by the table below.

	A	B	C
P	100	50	20
Q	110	50	10

Find how much is delivered to each place from each warehouse in the optimal arrangement. (*Possible method.* Suppose x_1 tons go from P to A and x_2 tons from P to B. The amounts of the remaining deliveries can be calculated in terms of x_1 and x_2.)

Answers

Page 9
13 With the usual definitions of addition and multiplication of functions, (*a*), (*b*) and (*c*) but not (*d*) give linear mappings.

Page 14
1, 2, 3, 4, 7, 11, 12, 13 are true for mappings 1, 2, 3; 1 and 4 hold for mappings 4, 5. (It is possible to regard some of the others as true for 4 and 5, if limiting cases are accepted.)

Page 21
1 Three vectors OP, OQ, OR fail to provide a basis if they lie in a plane through the origin. This covers the more extreme cases where the vectors lie in a line through O, or where some or all of P, Q, R coincide with O.
2 Yes. Let v_1 = a nut, v_2 = a bolt, v_3 = a washer and v_4 = a nail. The definitions in the text then apply with v_1, v_2, v_3, v_4 as a basis. The space is of 4 dimensions.
3 (*a*) (i) fails. (*b*) is basis. (*c*) (ii) fails. (*d*) (i) and (ii) fail. (*e*) (i) fails. (*f*) (ii) fails. (*g*) is basis. (*h*) (i) and (ii) fail.

Page 24
1 (*a*) $2P + t(P + Q)$; $y = x - 2$. (*b*) $Q + t(P + Q)$; $y = x + 1$. (*c*) $3Q + t(P - Q)$; $x + y = 3$. (*d*) $P + 3Q + t(P - Q)$; $x + y = 4$. (*e*) tQ; $x = 0$. (*f*) $P + tQ$; $x = 1$. (*g*) $2Q + tP$; $y = 2$.
2 Q; P; $QFGHJ$; $(t, 1)$; $y = 1$.
3 (*a*) $SLGD$. (*b*) QLU. (*c*) QMW. (*d*) SMJ. (*e*) KGE.

Page 27
4 (*a*) $X^2 + Y^2 = 1$; note that this is *not* a circle, as the (X, Y) system uses axes that are not perpendicular. (*b*) $X^2 + 3Y^2 = 1$. (*c*) $X^2 + 2Y^2 = 1$.
6 Since every quadratic is a mixture of x^2, x and 1, the most obvious basis consists of x^2, x, 1, and the space is of 3 dimensions. $(x - 1)^2$, x^2 and $(x + 1)^2$, since every quadratic can be expressed, in one way only, in the form $p(x - 1)^2 + q(x^2) + r(x + 1)^2$, does constitute a basis.
7 $x = Y + 2Z$, $y = 3X + 4Z$, $z = 5X + 6Y$. Plane, $X = 0$.
8 $x = X + Y + Z$, $y = 10X + 20Y - 10Z$, $z = 5X - 5Y + 15Z$. (*a*) $Z = 0$. (*b*) $Y = 0$. (*c*) $X + Y = 0$.
9 $x = X + Y + Z$, $y = X + 2Y + 4Z$, $z = X + 3Y + 9Z$. (*a*) $Y = 0$. (*b*) $X = 0$. (*c*) $x - 2y + z = 0$.

Page 30
3 (*a*) $A^* = A$, $B^* = A + B$. (*b*) $A^* = B$, $B^* = A$. (*c*) $A^* = B$, $B^* = -A$. (*d*) $A^* = A + B$, $B^* = -A$.

Pages 32–3
3 $p + qx + \frac{1}{2}rx(x - 1) + \frac{1}{6}sx(x - 1)(x - 2)$.
4 Yes; $y = p(x - t)/(s - t) + q(x - s)/(t - s)$.
5 Hint: the equation corresponding to $p = 1, q = 0, r = 0$ is $y = (x - t)(x - u)/[(s - t)(s - u)]$.

Pages 36–7
1 (*a*) 2. (*b*) 6. (*c*) 2.
2 (*a*) $x \rightarrow 4x$, $x \rightarrow 8x$. (*b*) $x \rightarrow 20$, $x \rightarrow 30$. (*c*) $x \rightarrow x^4$, $x \rightarrow x^8$.
3 (*a*), (*c*), (*d*), (*f*).
4 $T^2 = I$.
5 (*a*) Yes. (*b*) Yes. (*c*) No.

7 (a) T^n: $A \to A + nB$, $B \to B$. (b) On squared paper, T represents a rotation of 45° combined with enlargement of scale $\sqrt{2}$ times. Thus T^n is a rotation of $n \cdot 45°$ and enlargement $(\sqrt{2})^n$. (c) $T^n = I$ when n is even, $T^n = T$ for n odd.

8 T^n: $1 \to 1 + x + x^2 + \cdots + x^n$.

9 T^n, acting on 1, gives the terms of the series for e^z as far as the term $x^n/(n!)$.

Pages 43–4

1 (a) = (i) = (m) = (n). (b) = (j) = (k) = (l). (d) = (h) = (o). (e) = (p) = (s). (f) = (q) = (r). (g) = (t). (c). (u). (v).

2 None for (a), (c), (i), (m), (n), (u), (v). Every vector is an eigenvector with $\lambda = -1$ for (b), (j), (k), (l); rotation through 180° has an infinity of invariant lines. (d), (h), (o); $A, \lambda = 1$; $B, \lambda = -1$. (e), (p), (s); $A, \lambda = -1$; $B, \lambda = 1$. (f), (q), (r); $C, \lambda = 1$; $J, \lambda = -1$. (g), (t); $C, \lambda = -1$; $J, \lambda = 1$.

3 All lines through the origin are invariant.

4 Horizontal and vertical axes, $A, \lambda = 1$. $B, \lambda = 0$.

5 Horizontal axis and line through origin at 45°; $A, \lambda = 1$. $C, \lambda = 0$.

6 Horizontal axis is the only invariant line.

7 $C, \lambda = 3$. $J, \lambda = 1$..

8 $B \to -A$. $T^3 = -I$. $T^6 = I$.

9 Reflection in OJ.

10 Reflection in OC.

11 (a) Invariant line, $y = 0$; $(t, 0)$ is eigenvector with $\lambda = 0$, for any number $t \neq 0$. (b) $x + y = 0$; $(t, -t)$. (c) $x = 0$; $(0, t)$. (d) $y = 0$; $(t, 0)$. (e) $x + y = 0$; $(t, -t)$. (f) $x = y$; (t, t).

Page 51

4 c^{-1} is at distance $1/r$ and angle $-\theta$.

5 $(r \cos \theta) + j(r \sin \theta)$.

6 No; see §14.

Page 54 (first set)

(a) $\pm(1 + j)/\sqrt{2}$. (b) $\pm j$. (c) $\pm(-1 + j)/\sqrt{2}$. (d) $-1, \frac{1}{2} \pm j(\frac{1}{2}\sqrt{3})$. (e) $1, -\frac{1}{2} \pm j(\frac{1}{2}\sqrt{3})$. (f) $1, j, -1, -j$. (g) $(\pm 1 + j)/\sqrt{2}, (\pm 1 - j)/\sqrt{2}$. (h) Combine answers to (d) and (e). (i) $\pm\{(\sqrt{3}) + j\}$. (j) $1 + j$, $\sqrt{2} \cdot (-\cos 15° + j \sin 15°)$, $\sqrt{2} \cdot (\cos 75° - j \sin 75°)$. (k) $-2, 1 \pm j\sqrt{3}$. (l) $3 \pm 3j, -3 \pm 3j$.

Page 54 (second set)

(a) $(\frac{5}{13}) - j(\frac{1}{13})$. (b) $0.6 - j(0.8)$. (c) $-j$. (d) $\frac{1}{2} - j(\frac{1}{2})$. (e) $\cos \theta + j \sin \theta$. (f) $\cos^2 \theta - j \sin \theta \cos \theta$.

Page 62

1
(a) $\frac{1}{2}[ab + 1/(ab)]$
(b) $\frac{1}{2}[(a/b) + (b/a)]$
(c) $[ab - 1/(ab)]/(2j)$
(d) $[(a/b) - (b/a)]/(2j)$
(e) $\frac{1}{2}[a^2 + 1/(a^2)]$
(f) $[a^2 - 1/(a^2)]/(2j)$
(g) $\frac{1}{2}[a^3 + 1/(a^3)]$
(h) $[a^3 - 1/(a^3)]/(2j)$
(i) $\frac{1}{2}[a^4 + 1/(a^4)]$
(j) $[a^4 - 1/(a^4)]/(2j)$
(k) $\frac{1}{2}[ab^2 + 1/(ab^2)]$
(l) $[ab^2 - 1/(ab^2)]/(2j)$
(m) $\frac{1}{2}[(a/b^2) + (b^2/a)]$
(n) $[(a/b^2) - (b^2/a)]/(2j)$
(o) $\frac{1}{2}(ab^{-2}c^3 + a^{-1}b^2c^{-3})$
(p) $(ab^{-2}c^3 - a^{-1}b^2c^{-3})/(2j)$
(q) $\frac{1}{2}(ab^{-2}c^3d^{-4} + a^{-1}b^2c^{-3}d^4)$
(r) $(ab^{-2}c^3d^{-4} - a^{-1}b^2c^{-3}d^4)/(2j)$

2 (c) and (e) are incorrect

Page 66

1 (a) 1. (b) 1. (c) 5. (d) 5. (e) $\sqrt{2}$. (f) 3. (g) $\sqrt{(a^2 + b^2)}$. (h) 1. (i) 1.

2 (a) $\frac{1}{2}\pi$. (b) π. (c) $-\frac{1}{2}\pi$. (d) $\frac{1}{4}\pi$. (e) $\frac{3}{4}\pi$. (f) $\frac{1}{6}\pi$. (g) $-\frac{1}{3}\pi$ or $\frac{5}{3}\pi$. (h) θ. (i) θ.

3 (a) 2. (b) 3. (c) a. (d) b. (e) $\frac{1}{2}$. (f) $-\frac{1}{2}$. (g) $\cos \theta$. (h) $\sin \theta$. (i) $\cos \theta$. (j) $-\sin \theta$.

Page 68

3 Division is possible except when $a = b = 0$

Page 72

1 (a) Reflection in $y = x$; $(t, t), \lambda = 1$; $(t, -t), \lambda = -1$. (b) Reflection in $y = -x$; $(t, t), \lambda = -1$;

$(t, -t))$, $\lambda = 1$. (c) Same as (a). (d) Rotation of $180°$ about O; every non-zero vector, $\lambda = -1$. (e) Same as (d). (f) Identity transformation, I; every non-zero vector, $\lambda = 1$. (g) Same as (f). (h) Same as (d). (i) Same as (f). (j) Project on y-axis, then double; $(0, t)$, $\lambda = 2$; $(t, 0)$, $\lambda = 0$. (k) Rotate $45°$ about O and enlarge $\sqrt{2}$ times. (l) Same as (f). (m) All points map to O; every non-zero vector, $\lambda = 0$. (n) Same as (m).

2 (a) False. (b) False. (c) True. (d) True. (e) True.

3 $ST = TS$ sometimes. $S + T = T + S$ always.

4 Yes.

5 (a) Perpendicular projection onto x-axis. (b) Perpendicular projection onto y-axis. (c) Rotation of $45°$ about O; compare multiplication by $(1 + j)/(\sqrt{2})$ on Argand diagram.

Pages 77–8

1 $(2x, -2y)$.

2 (x, y); I.

3 $(4x, 0)$.

4 $(4x, 0)$.

5 They are the same.

6 $(-x, -y)$.

7 $(-x, y)$.

8 Both equal $-I$.

9 $(0, 0)$; 0.

10 $(0, 2y)$.

11 $(0, 0)$; 0.

12 Both equal 0.

13 $(x - y, x - y)$.

14 $(4, 4)$; $(0, 0)$; $(0, 0)$.

15 Every point maps to $(0, 0)$.

16 (y, x); no.

17 $(x, y) \rightarrow (2y, 2x)$; no.

18 Same as (b).

19 Infinitely many solutions, including reflection in $y = mx$ for any m.

20 See 11 and 15.

Pages 84–6

1 $x^* = 2x + 5y$, $y^* = 8x + 3y$.

2 $\begin{pmatrix} x^* \\ y^* \end{pmatrix} = \begin{pmatrix} 4 & 9 \\ 16 & 25 \end{pmatrix} \begin{pmatrix} x \\ y \end{pmatrix}$.

3 (a) $\begin{pmatrix} 21 \\ 13 \end{pmatrix}$. (b) $\begin{pmatrix} 2 \\ 6 \end{pmatrix}$. (c) $\begin{pmatrix} 60 \\ 30 \end{pmatrix}$. (d) $\begin{pmatrix} 62 \\ 36 \end{pmatrix}$. (e) $\begin{pmatrix} k \\ 3k \end{pmatrix}$. (f) $\begin{pmatrix} 20m \\ 10m \end{pmatrix}$. (g) $\begin{pmatrix} k + 20m \\ 3k + 10m \end{pmatrix}$. (h) $\begin{pmatrix} -19 \\ -7 \end{pmatrix}$.

4 (a) $\begin{pmatrix} 1 & 0 \\ 0 & 1 \end{pmatrix}$. (b) $\begin{pmatrix} 0 & 0 \\ 0 & 0 \end{pmatrix}$. (c) $\begin{pmatrix} 1 & 0 \\ 0 & 1 \end{pmatrix}$. (d) $\begin{pmatrix} a + c & b + d \\ c & d \end{pmatrix}$. (e) $\begin{pmatrix} a & a + b \\ c & c + d \end{pmatrix}$.

(f) $\begin{pmatrix} ad - bc & 0 \\ 0 & ad - bc \end{pmatrix}$. (g) $\begin{pmatrix} a + 2c & b + 2d \\ 3a + 4c & 3b + 4d \end{pmatrix}$. (h) $\begin{pmatrix} a + 3b & 2a + 4b \\ c + 3d & 2c + 4d \end{pmatrix}$.

5 (a) $\begin{pmatrix} 1 & 0 \\ 0 & 1 \end{pmatrix}$. (b) $\begin{pmatrix} 4 & 0 \\ 0 & 9 \end{pmatrix}$. (c) $\begin{pmatrix} -1 & 0 \\ 0 & -1 \end{pmatrix}$. (d) $\begin{pmatrix} 1 & 0 \\ 0 & 1 \end{pmatrix}$. (e) $\begin{pmatrix} 1 & 0 \\ 0 & 1 \end{pmatrix}$.

(f) $\begin{pmatrix} 1 + bc & 0 \\ 0 & 1 + bc \end{pmatrix}$. (g) $\begin{pmatrix} 0 & 0 \\ 0 & 0 \end{pmatrix}$. (h) $\begin{pmatrix} 0 & 0 \\ 0 & 0 \end{pmatrix}$. (i) $\begin{pmatrix} 0 & 0 \\ 0 & 0 \end{pmatrix}$.

6 $\begin{pmatrix} 1 & 0 \\ 0 & 1 \end{pmatrix}$, $\begin{pmatrix} 0 & 0 \\ 0 & 0 \end{pmatrix}$.

7 $M_1 = \begin{pmatrix} 1 & 0 \\ 0 & -1 \end{pmatrix}$, $M_2 = \begin{pmatrix} -1 & 0 \\ 0 & 1 \end{pmatrix}$, $J = \begin{pmatrix} 0 & -1 \\ 1 & 0 \end{pmatrix}$.

9 (a) $\begin{pmatrix} 1 & 0 \\ 0 & -1 \end{pmatrix}$. (b) $\begin{pmatrix} -1 & 0 \\ 0 & 1 \end{pmatrix}$. (c) $\begin{pmatrix} 0 & 1 \\ 1 & 0 \end{pmatrix}$. (d) $\begin{pmatrix} -1 & 0 \\ 0 & -1 \end{pmatrix}$. (e) $\begin{pmatrix} 0 & -1 \\ 1 & 0 \end{pmatrix}$. (f) $\begin{pmatrix} 0 & 1 \\ -1 & 0 \end{pmatrix}$.

(g) $\begin{pmatrix} 1/\sqrt{2} & 1/\sqrt{2} \\ -1/\sqrt{2} & 1/\sqrt{2} \end{pmatrix}$. (h) $\begin{pmatrix} \frac{1}{2}\sqrt{3} & -\frac{1}{2} \\ \frac{1}{2} & \frac{1}{2}\sqrt{3} \end{pmatrix}$. (i) $\begin{pmatrix} \cos \alpha & -\sin \alpha \\ \sin \alpha & \cos \alpha \end{pmatrix}$. (j) $\begin{pmatrix} \cos 2\alpha & \sin 2\alpha \\ \sin 2\alpha & -\cos 2\alpha \end{pmatrix}$.

Pages 88–9

1 $\begin{pmatrix} 81 & 62 \\ 73 & 54 \end{pmatrix}$.

2 $\begin{pmatrix} 5 & 14 \\ 23 & 32 \end{pmatrix}$.

3 $\begin{pmatrix} 2 & 0 \\ 0 & 2 \end{pmatrix}$.

4 $0, 0$; yes.

5 $\begin{pmatrix} -1 & 2 \\ -2 & 1 \end{pmatrix}$, $\begin{pmatrix} 1 & 2 \\ -2 & -1 \end{pmatrix}$, $\begin{pmatrix} 0 & 2 \\ -2 & 0 \end{pmatrix}$; no.

6 0, 0; yes.   **7** No; see 4 and 6 above.   **8** Yes.

9 0.   **10** $\begin{pmatrix} 0 & 1 \\ 0 & -1 \end{pmatrix}, \begin{pmatrix} 1 & 0 \\ 0 & 1 \end{pmatrix}.$ **11** Correct.

12 $\begin{pmatrix} 4 & 2 \\ 6 & 4 \end{pmatrix}, \begin{pmatrix} 6 & 3 \\ 9 & 6 \end{pmatrix}, \begin{pmatrix} 8 & 4 \\ 12 & 8 \end{pmatrix}, \begin{pmatrix} 10 & 5 \\ 15 & 10 \end{pmatrix}.$ **13** Yes; $k = 4$.

14 0.   **15** $p = 3, q = -2$.   **16** Yes; yes.

Pages 91–2

1 $\begin{pmatrix} 13 & -21 \\ -8 & 13 \end{pmatrix}.$

2 For a symmetrical circuit, if an input v, i gives an output V, I, then an input V, $-I$ gives an output v, $-i$. The negative signs are due to the fact that input current is regarded as entering by the upper wire, while the opposite is true for output current. If $\begin{pmatrix} a & b \\ c & d \end{pmatrix}$ represents a symmetric circuit, $ad - bc = 1, a = d$.

3 $BAB = \begin{pmatrix} 2 & -1 \\ -3 & 2 \end{pmatrix}.$

4 No.

Page 101

1 (a) $\begin{pmatrix} 3t \\ t \end{pmatrix}, \begin{pmatrix} t \\ t \end{pmatrix}, \begin{pmatrix} 1 & 0 \\ 0 & -1 \end{pmatrix}.$ (b) $\begin{pmatrix} 2t \\ 5t \end{pmatrix}, \begin{pmatrix} t \\ 3t \end{pmatrix}, \begin{pmatrix} 1 & 0 \\ 0 & 2 \end{pmatrix}.$ (c) $\begin{pmatrix} t \\ t \end{pmatrix}, \begin{pmatrix} t \\ 0 \end{pmatrix}, \begin{pmatrix} 2 & 0 \\ 0 & 3 \end{pmatrix}.$

(d) $\begin{pmatrix} t \\ -t \end{pmatrix}, \begin{pmatrix} t \\ t \end{pmatrix}, \begin{pmatrix} 1 & 0 \\ 0 & 3 \end{pmatrix}.$ (e) $\begin{pmatrix} 2t \\ t \end{pmatrix}, \begin{pmatrix} t \\ -2t \end{pmatrix}, \begin{pmatrix} 3 & 0 \\ 0 & -2 \end{pmatrix}.$ (f) $\begin{pmatrix} t \\ t \end{pmatrix}, \begin{pmatrix} t \\ -t \end{pmatrix}, \begin{pmatrix} 2 & 0 \\ 0 & 0 \end{pmatrix}.$

(g) $\begin{pmatrix} t \\ t \end{pmatrix}, \begin{pmatrix} t \\ 0 \end{pmatrix}, \begin{pmatrix} 0 & 0 \\ 0 & 1 \end{pmatrix}.$

Page 104

(a) $\begin{pmatrix} t \\ -t \end{pmatrix}, \lambda = 2; \begin{pmatrix} t \\ t \end{pmatrix}, \lambda = 4.$ (b) $\begin{pmatrix} t \\ -t \end{pmatrix}, \lambda = 3; \begin{pmatrix} t \\ t \end{pmatrix}, \lambda = 5.$ (c) $\begin{pmatrix} 2t \\ t \end{pmatrix}, \lambda = 1; \begin{pmatrix} t \\ 2t \end{pmatrix}, \lambda = 4.$

(d) $\begin{pmatrix} 2t \\ t \end{pmatrix}, \lambda = -1; \begin{pmatrix} t \\ 3t \end{pmatrix}, \lambda = 4.$ (e) $\begin{pmatrix} t \\ -t \end{pmatrix}, \lambda = 0; \begin{pmatrix} t \\ 2t \end{pmatrix}, \lambda = 3.$ (f) $\begin{pmatrix} t \\ 0 \end{pmatrix}, \lambda = 2; \begin{pmatrix} t \\ t \end{pmatrix}, \lambda = 3.$

(g) $\begin{pmatrix} t \\ -t \end{pmatrix}, \lambda = a - b; \begin{pmatrix} t \\ t \end{pmatrix}, \lambda = a + b.$ (h) Real eigenvectors only when $b = 0$.

Pages 113–4

4 (a), (c), (f) and (g) have no inverses.

7 (a) The plane maps to $y = 0$, which is an invariant line with $\lambda = 1$. The line $x = 0$ maps to the origin; it is an invariant line with $\lambda = 0$. (b) $y = x$, $\lambda = 1$; $x = 0$, $\lambda = 0$. (c) $y = x$, $\lambda = 1$; $x + y = 0$, $\lambda = 0$. (d) $y = x$, $\lambda = 2$; $x + y = 0$, (e) $y = 2x$, $\lambda = 1$; $x + 2y = 0$, $\lambda = 0$.

Pages 125–7

1 (a) -1. (b) 1. (c) 0. (d) 0. (e) 0.

4 (a) 0. (b) 0.

5 (a) 240. (b) -10. (c) 10. (d) 0. (e) 10.

6 (a) 0. (b) -2.

7 (a) -30. (b) 0. (c) 0.

8 (a) $\begin{pmatrix} 1 \\ 1 \\ 1 \end{pmatrix}.$ (b) $\begin{pmatrix} 1 \\ 2 \\ 3 \end{pmatrix}.$ (c) $\begin{pmatrix} 1 \\ -2 \\ 1 \end{pmatrix}.$ (d) $\begin{pmatrix} 1 \\ -2 \\ 1 \end{pmatrix}.$ (e) $\begin{pmatrix} 1 \\ -2 \\ 1 \end{pmatrix}.$

9 Many answers.

10 $\begin{pmatrix} 1 \\ -2 \\ -2 \end{pmatrix}, \begin{pmatrix} -2 \\ 1 \\ -2 \end{pmatrix}, \begin{pmatrix} -2 \\ -2 \\ 1 \end{pmatrix}.$

11 $\lambda = 2 \quad \lambda = 0 \quad \lambda = -2$

$\begin{pmatrix} 1 \\ \sqrt{2} \\ 1 \end{pmatrix} \quad \begin{pmatrix} 1 \\ 0 \\ -1 \end{pmatrix} \quad \begin{pmatrix} 1 \\ -\sqrt{2} \\ 1 \end{pmatrix}.$

12 $\lambda = 3 \quad \lambda = 0 \quad \lambda = -3$

$\begin{pmatrix} 1 \\ -2 \\ -2 \end{pmatrix} \quad \begin{pmatrix} -2 \\ 1 \\ -2 \end{pmatrix} \quad \begin{pmatrix} -2 \\ -2 \\ 1 \end{pmatrix}.$

13 $\lambda = -3 \quad \lambda = 4 \quad \lambda = 6$

$\begin{pmatrix} 1 \\ -1 \\ 1 \end{pmatrix} \quad \begin{pmatrix} 1 \\ 1 \\ 0 \end{pmatrix} \quad \begin{pmatrix} 1 \\ -1 \\ -2 \end{pmatrix}.$

14 $\lambda = 3 \quad \lambda = -3 \quad \lambda = 9$

$\begin{pmatrix} 1 \\ -2 \\ -2 \end{pmatrix} \quad \begin{pmatrix} -2 \\ 1 \\ -2 \end{pmatrix} \quad \begin{pmatrix} -2 \\ -2 \\ 1 \end{pmatrix}.$

15 $\lambda = 1$

$\begin{pmatrix} 1 \\ 1 \\ 1 \end{pmatrix};$ for $\lambda = 0$, every non-zero vector in the plane $x + y + z = 0$.

16 $\lambda = 1$

$\begin{pmatrix} 1 \\ 1 \\ 2 \end{pmatrix};$ for $\lambda = 0$, every non-zero vector in the plane $x + y + 2z = 0$.

Pages 134–6

1 (a) $\begin{pmatrix} 6 & 15 & 15 & 6 \\ 3 & 9 & 11 & 5 \end{pmatrix}.$ (b) $\begin{pmatrix} 6 & 15 \\ 3 & 9 \end{pmatrix}.$ (c) $\begin{pmatrix} 3 & 7 & 11 \\ 3 & 7 & 11 \\ 1 & 3 & 5 \end{pmatrix}$ (d) $\begin{pmatrix} 1 & 3 & 5 \\ 2 & 6 & 10 \\ 3 & 9 & 15 \end{pmatrix}.$ (e) (22).

(f) $\begin{pmatrix} a^2 + b^2 + c^2 & ax + by + cz \\ ax + by + cz & x^2 + y^2 + z^2 \end{pmatrix}$ (g) $\begin{pmatrix} a^2 + x^2 & ab + xy & ac + xz \\ ab + xy & b^2 + y^2 & bc + yz \\ ac + xz & bc + yz & c^2 + z^2 \end{pmatrix}.$

3 $\begin{pmatrix} 3 & S_1 \\ S_1 & S_2 \end{pmatrix}, \begin{pmatrix} 4 & S_1 & S_2 \\ S_1 & S_2 & S_3 \\ S_2 & S_3 & S_4 \end{pmatrix}.$

4 Both products are $\begin{pmatrix} 1 & 0 \\ 0 & 1 \end{pmatrix}.$

7 $\begin{pmatrix} 1 & 0 & -2 & -3 \\ 0 & 1 & -4 & -5 \\ 0 & 0 & 1 & 0 \\ 0 & 0 & 0 & 1 \end{pmatrix}.$

8 $\begin{pmatrix} a_1u_1 + a_2u_2 + a_3u_3 & a_1v_1 + a_2v_2 + a_3v_3 \\ b_1u_1 + b_2u_2 + b_3u_3 & b_1v_1 + b_2v_2 + b_3v_3 \end{pmatrix}, \begin{pmatrix} a'u & a'v \\ b'u & b'v \end{pmatrix}.$

9 $M = (u + 2v, 3u + 4v, 5u + 6v)$. Det $M = 0$, since each column represents a vector lying in the plane of u and v. Same argument for $\det(PQ)$.

Page 148

2 $u_1v_1 + u_2v_2 + \cdots + u_nv_n = 0.$

Page 155

1 (a) 1, $\sqrt{2}$, $\sqrt{2}$, $\frac{1}{2}$. (b) 0, $\sqrt{6}$, $\sqrt{50}$, 0. (c) 1, $\sqrt{3}$, 1, $1/\sqrt{3}$. (d) 35, $\sqrt{50}$, $\sqrt{50}$, 0.7. (e) -9, 3, $3\sqrt{2}$, $-1/\sqrt{2}$. (f) 49, 7, $7\sqrt{2}$, $1/\sqrt{2}$.

5 $(-\frac{1}{3}, \frac{5}{3}, \frac{1}{3}).$

6 (a) (2,2,1). (b) (1,1,2). (c) (3,1,8). (d) $(-8, -2, -3)$. (e) $(-3, 2, 7)$.

Page 158

2 $ax^2 + 2hxy + by^2.$

3 0.

4 $x^2 + y^2 + z^2 - 2xy - 2xz - 2yz.$

5 $M = \begin{pmatrix} 1 & 3 \\ 3 & 1 \end{pmatrix}$.

Page 161

7 $V'T'STV$; symmetric (see question 6). **8** $U'T'TV$; $T'T = I$.

Pages 183–5

1 1, 1, 90°; yes; its transpose.

2 The first three; the second; the first and third.

5 The first two must be; the third is not usually, but might be.

6 $L = \begin{pmatrix} (1 - k^2)/(1 + k^2) & 2k/(1 + k^2) \\ -2k/(1 + k^2) & (1 - k^2)/(1 + k^2) \end{pmatrix}$.

8 $(0, 0, 0), (1, 0, 1), (-\frac{1}{3}, \frac{1}{3}, \frac{4}{3}), (0, -1, 1)$.

9 The matrix corresponds to a rotation combined with an enlargement of scale by a factor 9; $(0, 0, 0), (0, 9, 9), (-11, 4, 5), (-3, -3, 12)$.

10 Both are rotations; axes $(1, 1, 2), (1, 1, 3)$.

11 Reflection in the plane $x + y + z = 0$, so determinant negative.

12 Yes; yes.

13 Axis $(0, 2, 3)$; $P^* = (1, -3, 2)$; $\cos POP^* = -\frac{6}{7}$; yes.

Pages 210–11

1 Unit eigenvectors are $(1/\sqrt{2}, 1/\sqrt{2})$, $\lambda = 9$; $(-1/\sqrt{2}, 1/\sqrt{2})$, $\lambda = 1$; $X^2 + (Y^2/9) = 1$; $(1/\sqrt{2}, 1/\sqrt{2}), (-3/\sqrt{2}, 3/\sqrt{2})$.

2 $(1/\sqrt{2}, 1/\sqrt{2})$, $\lambda = 1$; $(-1/\sqrt{2}, 1/\sqrt{2})$, $\lambda = 9$; $(X^2/9) + Y^2 = 1$. $(3/\sqrt{2}, 3/\sqrt{2}), (-1/\sqrt{2}, 1/\sqrt{2})$; by a rotation of 90°.

3 $(1/\sqrt{2}, 1/\sqrt{2})$, $\lambda = 9$; $(-1/\sqrt{2}, 1/\sqrt{2})$, $\lambda = -1$; $X^2 - (Y^2/9) = 1$; $(1/\sqrt{2}, 1/\sqrt{2})$; hyperbola.

4 Unit vectors along new axes are $(-2/\sqrt{5}, 1/\sqrt{5}), (1/\sqrt{5}, 2/\sqrt{5})$; $4X^2 - Y^2 = 4$; hyperbola.

5 $(X^2/4) + (Y^2/9) = 1$; axes as in 4.

6 $(10x_0 - 6y_0, -6x_0 + 5y_0)$; $(2, 3)$.

7 $(22x_0 - 3y_0, -3x_0 + 14y_0)$; $(13, 39)$.

8 $(56, 70), (56, 70)$.

9 (a) $X^2 + 4Y^2 + 7Z^2 = 28$; unit vectors along axes $(-\frac{1}{3}, \frac{2}{3}, \frac{2}{3}), (\frac{2}{3}, -\frac{1}{3}, \frac{2}{3}), (\frac{2}{3}, \frac{2}{3}, -\frac{1}{3})$; ellipsoid. (b) $3X^2 - 3Y^2 + 9Z^2 = 1$; axes as in (a); hyperboloid. (c) $Y^2 + Z^2 = \frac{1}{3}$; circular cylinder; axis, $(1, 1, 1)$. (d) $2X^2 - Y^2 - Z^2 = 1$; hyperboloid of revolution; axis $(1, 1, 1)$. (e) $-2X^2 + Y^2 + 4Z^2 = 4$; hyperboloid; axes as in (a). (f) $10X^2 + Y^2 + Z^2 = 1$; flying saucer; axis $(2, 2, -1)$. (g) $X^2 + 4Y^2 + 7Z^2 = 28$; axes $(-\frac{2}{3}, \frac{1}{3}, \frac{2}{3}), (\frac{1}{3}, -\frac{2}{3}, \frac{2}{3}), (\frac{2}{3}, \frac{2}{3}, \frac{1}{3})$; ellipsoid. (h) $14Z^2 = 25$; pair of planes, perpendicular to $(1, 2, 3)$. (i) $-X^2 + Z^2 = 1$; hyperbolic cylinder; axes as in (g). (j) $-X^2 + Z^2 = 1$; hyperbolic cylinder; axes as in (a). (k) $X^2 + 2Y^2 + 3Z^2 = 1$; ellipsoid; axes as in (a).

10 (a), (f), (g), (k).

Page 225

1 $(1, 2, 3)$; $(1 + t, 1 + 2t, 1 + 3t)$; $y = 2x - 1, z = 3x - 2$.

2 $t = 2$; never; always; intersects; parallel; in plane.

3 $x - 1 = (y - 1)/2 = (z - 1)/3$; $x = y - 1 = z/2$; $x = y - 1 = z$; $(x - 1)/4 = (y - 2)/5 = (z - 3)/6$; $(10, 20, 30), (2, 3, 4)$; many.

4 Yes; they are the same line. **5** No; yes.

Page 228

1 (a) $(2, 3, 4)$. (b) $(2, 3, 4)$. (c) $(1, 1, 1)$. (d) $(5, -2, 11)$. (e) $(1, 1, 0)$. (f) $(0, 0, 1)$.

2 $(x/a) + (y/b) + (z/c) = 1$; $(1/a, 1/b, 1/c)$.

3 $(1, 2, 3)$; $(t, 2t, 3t)$; $(4, 8, 12)$; $(3 + t, 4 + 2t, 1 + 3t)$; $(6, 10, 10)$.

Page 234

4 $r = au + bv$ for some a, b; $\det (au + bv, u, v) = 0$.

Pages 234–5
1 (a) (0, 0, 2). (b) (1, −1, 0). (c) (3, 3, 3). (d) (1, −1, 0). (e) (0, 0, 0).
2 (99, −66, 22); 121.
3 (a) (1, 1, 1), $\sqrt{3}$; $\frac{1}{2}\sqrt{3}$, −$\frac{1}{2}$. (b) (1, 1, 1), $\sqrt{3}$; $\frac{1}{2}\sqrt{3}$, $\frac{1}{2}$. (c) (1, 1, 1), $\sqrt{3}$; $\frac{1}{2}\sqrt{3}$, $\frac{1}{2}$.
5 (−1, 2, −1), 0, yes.
6 (a) 1.

Page 248
1 (a) (t, t, t); line in 3 dimensions. (b) $(3t, 2t, 5t)$; line in 3 dimensions. (c) Same as (b). (d) (0, 0, 0);
point. (e) $x + 2z − 2u + 3v = 0, y − z + 3u − 2v = 0$; subspace of 3 dimensions in 5 dimensions.
(f) $x + 7y − 13z = 0$; plane in 3 dimensions. (g) $(−11t, 9t, −4t, t)$; line in 4 dimensions. (h)
$x_1 + \frac{3}{2}x_2 + 2x_4 + x_5 = 0, x_3 + x_4 + x_5 = 0, x_6 = 0$; subspace of 3 dimensions in 6 dimensions.
2 (a) Yes. (b) Independent. (c) $a = b$ or $a = c$ or $b = c$; a, b, c all distinct. (d) a, b, c all distinct.
(e) 4.
3 $30u_1 − 31u_2 + 10u_3 − u_4 = 0$.
4 (a) resembles (c).

Pages 250–2
1 (a) No solution. (b) $(−14t − 15, 5t + 6, t)$. (c) No solution. (d) $(13 − 11t, 7t − 6, t)$.
2 No; $(7 + t, 3 + t, t), (1 + t, t, t), (4 + t, 5 + t, t)$; parallel lines.
4 Yes; $y = 7 − 2x + 5x^2$.
5 (a) $(1 + 2t, −1 + 3t, 1 + 5t)$. (b) None. (c) $2a + 3b + 5c = 0$; no.
6 (a), (c), (e), (f).

Pages 260–1
1 2, 2, 2; $\begin{pmatrix} 1 \\ 1 \\ 0 \end{pmatrix}, \begin{pmatrix} 0 \\ −1 \\ 1 \end{pmatrix}$; $x = y + z$; u_3, u_4; $5 − 2 = 3$;

$x_1 + 2x_3 + x_4 + 3x_5 = 0, x_2 + x_3 + 2x_4 + 2x_5 = 0$; $\begin{pmatrix} −2 \\ −1 \\ 1 \\ 0 \\ 0 \end{pmatrix}, \begin{pmatrix} −1 \\ −2 \\ 0 \\ 1 \\ 0 \end{pmatrix}, \begin{pmatrix} −3 \\ −2 \\ 0 \\ 0 \\ 1 \end{pmatrix}$;

for $Mv = u_3, x_1 = 5 − 2x_3 − x_4 − 3x_5, x_2 = 2 − x_3 − 2x_4 − 2x_5$, with x_3, x_4, x_5 arbitrary; for
$Mv = u_4, x_1 = 1 − 2x_3 − x_4 − 3x_5, x_2 = −1 − x_3 − 2x_4 − 2x_5$.
2 (a) 2; (1, 0, 3), (0, 1, −1). (b) 2; $\begin{pmatrix} 1 \\ 0 \\ −1 \\ 3 \end{pmatrix}, \begin{pmatrix} 0 \\ 1 \\ 1 \\ −1 \end{pmatrix}$. (c) 1; $\begin{pmatrix} −3 \\ 1 \\ 1 \end{pmatrix}$; $u_3 = u_2 − u_1$,

$u_4 = 3u_1 − u_2$; $\begin{pmatrix} 1 − 3t \\ 2 + t \\ t \end{pmatrix}$; $v_0 + t\begin{pmatrix} −3 \\ 1 \\ 1 \end{pmatrix}$.
3 (a) (1, 1, 1, 0, 0), (1, 0, 1, 1, 1). (b) $\begin{pmatrix} −1 \\ 0 \\ 1 \\ 0 \\ 0 \end{pmatrix}, \begin{pmatrix} −1 \\ 1 \\ 0 \\ 1 \\ 0 \end{pmatrix}, \begin{pmatrix} −1 \\ 1 \\ 0 \\ 0 \\ 1 \end{pmatrix}$.

Page 279
4 Take A of value x_1, B of value x_2. Then (a) $x_1 = 8, x_2 = 6, M = 14$. (b) $x_1 = 26, x_2 = 0$,
$M = 26$. (c) $x_1 = 0, x_2 = 28, M = 28$. (d) $x_1 = 12, x_2 = 0, M = 12$.

5 $x_1 = 100, x_2 = 500$.
6 $x_1 = 300, x_2 = 300$.
7 For $(0, 11, 1)$ $M = 12$. Other vertices are $(0, 0, 0)$, $(5\frac{3}{4}, 0, 0)$, $(0, 11\frac{1}{2}, 0)$. $(0, 0, 8\frac{1}{3})$, $(4, 0, 7)$.

Page 283
1 $x_1 = 2.25, x_2 = 2.25$.
2 $x_1 = 2, x_2 = 1$.
3 No solution; $x_1 = 2, x_2 = 2$ minimizes deficit.
4 $x_1 = 2, x_2 = 2$.
5 (*a*) No solution. (*b*) $x_1 = 3; x_2 = 1$.

Page 285
2 $x_1 = 2, x_2 = 2$.

Pages 293–4
1 $x_1 = 2, x_2 = 1$.
2 $x_1 = 3, x_2 = 1$.
3 No solution.
4 $x_1 = 3, x_2 = 3$.
5 $x_1 = 4, x_2 = 5$.
6 $x_1 = 0, x_2 = 3$.
7 $x_1 = 3, x_2 = 0$.
8 No solution.
9 $x_1 = 2, x_2 = 3, x_3 = 6$.
10 $x_1 = 4\frac{1}{2}, x_2 = 4\frac{1}{2}, x_3 = 4\frac{1}{2}, x_4 = 10\frac{1}{2}$.
11 $x_1 = 50, x_2 = 50$.

Index